Principles of Laser Spectroscopy and Quantum Optics

Principles of Laser Spectroscopy and Quantum Optics

Paul R. Berman
Vladimir S. Malinovsky

PRINCETON UNIVERSITY PRESS • PRINCETON AND OXFORD

Published by Princeton University Press, 41 William Street,
Princeton, New Jersey 08540

In the United Kingdom: Princeton University Press, 6 Oxford Street,
Woodstock, Oxfordshire OX20 1TW
press.princeton.edu

Library of Congress Cataloging-in-Publication Data
Berman, Paul R., 1945–
Principles of laser spectroscopy and quantum optics / Paul R. Berman,
Vladimir S. Malinovsky.
p. cm.
Includes bibliographical references and index.
ISBN 978-0-691-14056-8 (hardback : alk. paper) 1. Quantum optics.
2. Laser spectroscopy. I. Malinovsky, Vladimir S., 1962– II. Title.
QC446.2.B45 2011
535′.15—dc22 2010014005

British Library Cataloging-in-Publication Data is available

This book has been composed in Sabon
Printed on acid-free paper. ∞

Typeset by S R Nova Pvt Ltd, Bangalore, India
Printed in the United States of America

10 9 8 7 6 5 4 3 2 1

To the memory of
Belle and Solomon Berman

To the memory of
Antonida Malinovskaya and Sergey Bortkevich

Contents

Preface

This book is based on a course that has been given by one of us (PRB) for more years than he would like to admit. The basic subject matter of the book is the interaction of optical fields with atoms. This book is divided roughly into two parts. In the first half of the book, fields are treated classically, while atoms are described using quantum mechanics. In the context of this *semiclassical* theory of matter–field interactions, we establish the basic formalism of the theory and go on to discuss several applications. Most of the applications can be grouped under the general heading of *laser spectroscopy*, although both atom optics and atom interferometry are discussed as well. An emphasis is placed on introducing the physical concepts one encounters in considering the interaction of radiation with matter. In the second half of this book, the electromagnetic field is quantized, and problems are discussed in which it is necessary to use a fully quantized picture of matter–field interactions. Spontaneous emission is a prototypical problem in which a quantized field approach is needed. We examine in detail the radiation pattern and atomic dynamics that accompany spontaneous emission. An extension of this work to optical pumping, sub-Doppler laser cooling, and light scattering is also included.

This book is intended to serve a dual purpose. First and foremost, it can be used as a text in a course that follows an introductory graduate-level quantum mechanics course. There is undoubtedly too much material in the book for a one-semester course, but the core of a one-semester course could include chapters 1 to 8, 10, 12 to 16, and 19. The heart of this book is chapters 2 and 3, where the basic formalism is introduced for both atomic state amplitudes and density matrix elements. Chapters on slow light, atom optics and interferometry, optical pumping, sub-Doppler laser cooling, light scattering, and entanglement can be added as time permits. The second purpose of this book is to provide a reference for graduate students and others working in atomic, molecular, and optical physics.

There are many excellent texts available that cover the fields of laser spectroscopy and quantum optics. While presenting topics that are covered in many of these texts, we try to complement the approaches that have been given by other authors. In particular, we give a detailed description of different representations that can be used to analyze problems involving matter–field interactions. A semiclassical dressed-state basis is also defined that allows us to effectively solve problems involving strong fields. The chapters on atom optics and interferometry, optical pumping, light scattering, and sub-Doppler laser cooling offer material that may not be readily available in other introductory texts. The advantages and use of irreducible tensor formalism are explained and encouraged. On the other hand, we discuss only briefly, or not at all, such topics as superradiance, laser theory, bistability, nonlinear optics, Bose condensates, and pulse propagation. To keep this book to a manageable size, we chose to concentrate on a limited number of fundamental applications. Moreover, although references to experimental results are given, there is no reproduction of experimental data.

Each chapter contains a problems section. The problems are an integral part of any course based on this book. They extend and illustrate the material presented in the text. Many of the problems are far from trivial, requiring an intensive effort. Students who work through these problems will be rewarded with an improved understanding of matter–field interactions. Many problems require the students to use computational techniques. We plan to post Mathematica notebooks that contain algorithms for some of the calculations needed in the problems on the website associated with this book (http://press.princeton.edu/titles/9376.html). Moreover, we will use the website to post errata, offer additional problems, and discuss any topics that have been brought up by readers.

We would like to thank Yvan Castin, Bill Ford, Galina Khitrova, Jean-Louis Le Gouët, Rodney Louden, Hal Metcalf, Peter Milonni, Ignacio Sola, and Kelly Younge for their helpful comments. PRB would especially like to acknowledge the many discussions he had with Duncan Steel on topics contained in this book, as well as his encouragement in the endeavor of writing this text. We would also like to thank Boris Dubetsky for a careful reading and his critique of chapters 10 and 11 and Michael Martin and Jun Ye for their comments on section 10.3.

Last, but not least, we benefited from the continual support of our wives (Debra and Svetlana) and families.

Ann Arbor, MI; Hoboken, NJ
June 2010

Paul Berman
Vladimir Malinovsky

Principles of Laser Spectroscopy and Quantum Optics

1

Preliminaries

1.1 Atoms and Fields

As any worker knows, when you come to a job, you have to have the proper tools to get the job done right. More than that, you must come to the job with the proper attitude and a high set of standards. The idea is not simply to get the job done but to achieve an end result of which you can be proud. You must be content with knowing that you are putting out your best possible effort. Physics is an extraordinarily difficult "job." To understand the underlying physical origin of many seemingly simple processes is sometimes all but impossible. Yet the satisfaction that one gets in arriving at that understanding can be exhilarating. In this book, we hope to provide a foundation on which you can build a working knowledge of atom–field interactions, with specific applications to linear and nonlinear spectroscopy. Among the topics to be discussed are absorption, emission and scattering of light, the mechanical effects of light, and quantum properties of the radiation field.

This book is divided roughly into two parts. In the first part, we examine the interaction of *classical* electromagnetic fields with *quantum-mechanical* atoms. The external fields, such as laser fields, can be monochromatic, quasi-monochromatic, or pulsed in nature, and can even contain noise, but any *quantum* noise effects associated with the fields are neglected. Theories in which the fields are treated classically and the atoms quantum-mechanically are often referred to as *semiclassical* theories. For virtually all problems in laser spectroscopy, the semiclassical approach is all that is needed. Processes such as the photoelectric effect and Compton scattering, which are often offered as evidence for *photons* and the quantum nature of the radiation field can, in fact, be explained rather simply with the use of classical external fields. The price one pays in the semiclassical approach is the use of a time-dependent Hamiltonian for which the energy is no longer a constant of the motion.

Although the semiclassical approach is sufficient for a wide range of problems, it is not always possible to consider optical fields as classical in nature. One might ask when such *quantum optics* effects begin to play a role. Atoms are remarkable devices. If you place an atom in an excited state, it radiates a uniquely quantum-type field, the *one-photon state*. One of the authors (PRB) is a former student of Willis Lamb, who claimed that it should be necessary for people to apply for a license before they can use the word *photon*. Lamb was not opposed to the idea of a quantized field mode, but he felt that the word *photon* was misused on a regular basis. We will try to explain the distinction between a one-photon field and a photon when we begin our discussion of the quantized radiation field.

The field radiated by an atom in an excited state has a uniquely quantum character. In fact, any field in which the average value of the number operator for the field (average number of photons in the field) is less than or on the order of the number of atoms with which the field interacts must usually be treated using a quantized field approach. Thus, the second, or quantum optics, part of this book incorporates a fully quantized approach, one in which both the atoms and the fields are treated as quantum-mechanical entities. The advantage of using quantized fields is that one recovers a Hamiltonian that is perfectly Hermitian and independent of time. The most common quantum optics effects are those associated with spontaneous emission, scattering of external fields by atoms, quantum noise, and cavity quantum electrodynamics. There is another class of problems related to quantized field effects involving van der Waals forces and Casimir effects, but we do not discuss these in any detail [1].

1.2 Important Parameters

Why did the invention of the laser cause such a revolution in physics? Laser fields differ from conventional optical sources in their coherence properties and intensity. In this book, we look at applications that exploit the coherence properties of lasers, although complementary textbooks could be written in which the emphasis is on strong field–matter interactions. Moreover, we touch only briefly on the current advances in atto-second science that have been enabled using nonlinear atom–field interactions. Even if we deal mainly with the coherence properties of the fields, our plate is quite full. Historically, the coherence properties of optical fields have been one of the limiting factors in determining the ultimate *resolution* one can achieve in characterizing the transition frequencies of atomic, molecular, and condensed phase systems. It will prove useful to list some of the relevant frequencies that one encounters in considering such problems.

First and foremost are the transition frequencies themselves. We focus mainly on *optical* transitions in this text, for which the transition frequencies are of order $\omega_0/2\pi \simeq 5 \times 10^{14}$ Hz. The laser fields needed to probe such transitions must have comparable frequencies. The first gas and solid-state lasers had a very limited range of tunability, but the invention of the dye laser allowed for an expanded range of tunability in the visible part of the spectrum. One might even go so far as to say that it was the dye laser that really launched the field of laser spectroscopy. Since that time, the development of tunable semiconductor-based and titanium-sapphire lasers operating at infrared frequencies, combined with frequency doublers (nonlinear optical crystals) and frequency dividers (optical parametric amplifiers and oscillators), has enabled the creation of tunable coherent sources over a wide range of frequencies from the ultraviolet to the far-infrared.

Assuming for the moment that such sources are nearly monochromatic (typical line widths range from kHz to GHz), there are still underlying processes that limit the resolution one can achieve using laser sources to probe atoms. In other words, suppose that two transition frequencies in an atom differ by an amount Δf. What is the minimum value of Δf for which the transitions can be resolved? The ultimate limiting factor for any transition is the *natural width* associated with that transition. The natural width arises from interactions of atoms with the vacuum radiation field, leading to spontaneous emission. Typical natural widths for allowed optical transitions are in the range $\gamma_2/2\pi \simeq 10^7 - 10^8$ Hz, where γ_2 is a spontaneous emission decay rate. For "forbidden" transitions, such as those envisioned as the basis for optical frequency standards, natural line widths can be as small as a Hz or so. The fact that an allowed transition has a natural width equal to 10^8 Hz does *not* imply that the transition frequency can be determined only to this accuracy. By fitting experimental line shapes to theory, one can hope to reduce this resolution by a factor of 100 or more.

The natural width is referred to as a *homogeneous width* since it is the same for all atoms in a sample and cannot be circumvented. Another example of a homogeneous width in a vapor is the collision line width that arises as a result of energy shifts of atomic levels that occur during collisions. If the collision duration (typically of order 5 ps) is much less than all relevant timescales in the problem, except the optical period, then collisions add a homogeneous width of order 10 MHz per Torr of perturber gas pressure [1 Torr = (1/760) atm $\approx$ 133 Pa $\approx$ 1 mmHg]. This width is often referred to as a *pressure broadening width*.

Even if there are no collisions in a vapor, linear absorption or emission line shapes can be broadened by an *inhomogeneous* line broadening mechanism, as was first appreciated by Maxwell [2]. In a vapor, the moving atoms are characterized by a velocity distribution. As viewed in the laboratory frame, any radiation emitted by an atom is Doppler shifted by an amount $(\omega_0/2\pi)(v/c)$ (Hz), where v is the atom's speed and c is the speed of light. For a typical vapor at room temperature, the velocity width is of order 5×10^2 m/s, leading to a Doppler width of order 1.0 GHz or so. In a solid, crystal strain and fluctuating fields can give rise to inhomogeneous widths that can be factors of 10 to 100 times larger than Doppler widths in vapors. As you will see, it is possible to eliminate inhomogeneous contributions to line widths using methods of nonlinear laser spectroscopy.

Another contribution to absorption or stimulated emission line widths is so-called power broadening. The atom–field interaction strength in frequency units is $\Omega_0/2\pi \simeq \mu_{12}E/h$, where μ_{12} is a dipole moment matrix element, E is the amplitude of the applied field that is driving the transition, $h = 2\pi\hbar = 6.63 \times 10^{-34}\,\text{J}\cdot\text{s}$ is Planck's constant, and Ω_0 is referred to as the *Rabi frequency*.[1] For a 1-mW laser focused to a 1-mm^2 spot size, $\Omega_0/2\pi$ is of the order of several MHz and grows as the square root of the intensity. Of course, power broadening can be reduced by using weaker fields.

For vapors, there is an additional cause of line broadening. Owing to their motion, atoms may stay in the atom–field interaction region for a finite time τ, which

[1] We refer to quantities such as the transition frequency ω_0, the optical field frequency ω, the Rabi frequency Ω_0, and the detuning δ as "frequencies," even though they are actually angular frequencies, having units of s^{-1}. To obtain frequencies in Hz, one must divide these quantities by 2π.

TABLE 1.1
Typical Values for Line Widths and Shifts.

Width or shift	Typical value
Natural width	5 to 100 MHz
Doppler width	1 GHz
Collision broadening	10 MHz/Torr
Power broadening	0 to 10 MHz
Transit-time broadening	1 to 10^5 Hz
Light shifts	1 to 10^6 Hz
Zeeman shifts	14 GHz/T
Recoil shift	10 to 100 kHz

gives rise to a broadening in Hz of order $1/(2\pi\tau)$. For laser-cooled atoms, such *transit-time broadening* is usually negligible (on the order of a Hz or so), but in a thermal vapor it can be as large as a hundred KHz for laser beam diameters equal to 1 mm.

The broadening limits the resolution that one can achieve in probing atomic transitions with optical fields. One must also contend with *shifts* of the optical transition frequency resulting from atom–field interactions. If the optical fields are sufficiently strong, they can give rise to *light shifts* of the transition frequency that are of order $\Omega_0^2/(2\pi\delta)$ (Hz), where $\delta/2\pi$ is the frequency mismatch between the the atomic transition and the applied field frequencies in Hz (assumed here to be larger than the natural or Doppler widths). Light shifts range from 1 Hz to 1 MHz for typical powers of continuous-wave laser fields.

Magnetic fields also result in a shift and splitting of energy levels, commonly referred to as a *Zeeman splitting*. The magnetic interaction strength in frequency units is of order $\mu_B/h \simeq 14$ GHz/T, where $\mu_B = 9.27 \times 10^{-24}\,\mathrm{JT}^{-1}$ is the Bohr magneton. As a consequence, typical level splittings in the Earth's magnetic field are on the order of a MHz.

Last, there is a small shift associated with the recoil that an atom undergoes when it absorbs, emits, or scatters radiation. This *recoil shift* in Hz is of order $(\hbar k)^2/(2hM)$, where $\hbar k$ is the momentum associated with a photon in the radiation field, and M is the atomic mass. Typical recoil shifts are in the 10 to 100 kHz range.

These frequency widths and shifts are summarized in table 1.1. The resolution achievable in a given experiment depends on the manner in which these shifts or widths affect the overall absorption, emission, or scattering line shapes.

As we go through applications, the approximations that we can use are dictated by the values of these parameters. If you keep these values stored in *your* memory, you will be well on your way to understanding the relative contributions of these terms and the validity of the approximations that will be employed.

1.3 Maxwell's Equations

Throughout this text, we are interested in situations where there are no free currents or free charges in the volume of interest. That is, we often look at situations where an external field is applied to an ensemble of atoms that induces a polarization in the ensemble. We set $\mathbf{B} = \mu_0\mathbf{H}$ (neglecting any effects arising from magnetization), but do not take $\mathbf{D} = \varepsilon_0\mathbf{E}$. Rather, we set $\mathbf{D} = \varepsilon_0\mathbf{E} + \mathbf{P}$, where the polarization $\mathbf{P}$ is the electric dipole moment per unit volume. We adopt this approach since the polarization is calculated using a theory in which the atomic medium is treated quantum-mechanically.

With no free currents or charges and with $\mathbf{B} = \mu_0\mathbf{H}$, Maxwell's equations can be written as

$$\nabla\cdot(\varepsilon_0\mathbf{E} + \mathbf{P}) = 0, \tag{1.1a}$$

$$\nabla\times\mathbf{E} = -\frac{\partial\mathbf{B}}{\partial t}, \tag{1.1b}$$

$$\nabla\times\mathbf{B} = \mu_0\frac{\partial(\varepsilon_0\mathbf{E} + \mathbf{P})}{\partial t}, \tag{1.1c}$$

$$\nabla\cdot\mathbf{B} = 0. \tag{1.1d}$$

The quantity

$$\mu_0 = 4\pi\times10^{-7}\ \mathrm{T\cdot m/A} \tag{1.2}$$

is the permeability of free space, while

$$\varepsilon_0 \approx 8.85\times10^{-7}\ \mathrm{C^2/N\cdot m^2} \tag{1.3}$$

is the permittivity of free space. All field variables are assumed to be functions of position $\mathbf{R}$ and time t.

From equation (1.1), we find

$$\begin{aligned}\nabla\times(\nabla\times\mathbf{E}) &= \nabla(\nabla\cdot\mathbf{E}) - \nabla^2\mathbf{E}\\ &= -\frac{\partial}{\partial t}(\nabla\times\mathbf{B}) = -\mu_0\varepsilon_0\frac{\partial^2\mathbf{E}}{\partial t^2} - \mu_0\frac{\partial^2\mathbf{P}}{\partial t^2}\end{aligned} \tag{1.4}$$

or

$$-\nabla(\nabla\cdot\mathbf{E}) + \nabla^2\mathbf{E} - \mu_0\varepsilon_0\frac{\partial^2\mathbf{E}}{\partial t^2} = \mu_0\frac{\partial^2\mathbf{P}}{\partial t^2}. \tag{1.5}$$

In free space, $\nabla\cdot\mathbf{E} = 0$ and $\mathbf{P} = 0$, leading to the *wave equation*

$$\nabla^2\mathbf{E} - \frac{1}{v_0^2}\frac{\partial^2\mathbf{E}}{\partial t^2} = 0, \tag{1.6}$$

where the wave propagation speed in free space is equal to

$$v_0 = \frac{1}{\sqrt{\mu_0\varepsilon_0}}. \tag{1.7}$$

Historically, by comparing the electromagnetic (i.e., that based on the force between electrical circuits) and the electrostatic units of electrical charge, Wilhelm Weber had shown by 1855 that the value of $1/(\mu_0\varepsilon_0)^{1/2}$ was equal to the speed of light within experimental error. This led Maxwell to conjecture that light is an electromagnetic phenomenon [3]. One can only imagine the excitement Maxwell felt at this discovery.

We return to Maxwell's equations later in this text, but for now, let us consider *plane-wave solutions* of equations (1.5) and (1.1) for which we can take $\nabla \cdot \mathbf{E} = 0$. We still do not have enough information to solve equation (1.5) since we do not know the relationship between $\mathbf{P}(\mathbf{R}, t)$ and $\mathbf{E}(\mathbf{R}, t)$. In general, one can write $\mathbf{P}(\mathbf{R}, t) = \varepsilon_0 \boldsymbol{\chi}_e \cdot \mathbf{E}(\mathbf{R}, t)$, where $\boldsymbol{\chi}_e$ is the *electric susceptibility tensor*, but this does not resolve our problem, since $\boldsymbol{\chi}_e$ is not yet specified. To obtain an expression for $\boldsymbol{\chi}_e$, one must model the medium–field interaction in some manner. Ultimately, we calculate $\boldsymbol{\chi}_e$ using a quantum-mechanical theory to describe the atomic medium.

For the time being, however, let us the assume that the medium is linear, homogeneous, and isotropic, implying that $\boldsymbol{\chi}_e$ is a constant times the unit tensor and independent of the electric field intensity. Moreover, if we neglect dispersion and assume that χ_e is independent of frequency over the range of incident field frequencies, then it is convenient to rewrite χ_e as

$$\chi_e = n^2 - 1, \tag{1.8}$$

where n is the index of refraction of the medium. In these limits, equation (1.5) reduces to

$$\nabla^2 \mathbf{E} - \frac{n^2}{c^2} \frac{\partial^2 \mathbf{E}}{\partial t^2} = 0, \tag{1.9}$$

where c is the speed of light in vacuum. Neglecting dispersion, the fields propagate in the medium with speed $v = c/n$, as expected.

For a monochromatic or nearly monochromatic field having angular frequency centered at ω, the magnetic field (or, more precisely, the magnetic induction) $\mathbf{B}$ is related to the electric field via

$$\mathbf{B} = \frac{\mathbf{k} \times \mathbf{E}}{\omega}, \tag{1.10}$$

where $\mathbf{k}$ is the propagation vector having magnitude $k = n\omega/c$. It then follows that the time average of the Poynting vector, $\mathbf{S} = \mathbf{E} \times \mathbf{H} = \mathbf{E} \times \mathbf{B}/\mu_0$, is equal to

$$\langle \mathbf{S} \rangle = \frac{|E|^2 n}{2c\mu_0} \hat{\mathbf{k}} = \frac{1}{2} n \varepsilon_0 c\, |E|^2\, \hat{\mathbf{k}} \tag{1.11}$$

for optical fields having electric field amplitude $|E|$ and propagation direction $\hat{\mathbf{k}} = \mathbf{k}/k$.

By using the Poynting vector, one can calculate the electric field amplitude from the field intensity using

$$E = \sqrt{2c\mu_0\, |\langle \mathbf{S} \rangle|} \simeq 27.5\sqrt{S}\ \text{V/m}, \tag{1.12}$$

where $S \equiv |\langle \mathbf{S} \rangle|$ is expressed in W/m^2, and we have taken $n = 1$. At the surface of the sun, $S \approx 6.4 \times 10^7\ \text{W/m}^2$, giving a value $E \approx 2.2 \times 10^5\ \text{V/m}$. This is to be compared with the value $E \approx 5 \times 10^{11}\ \text{V/m}$ at the Bohr radius of the hydrogen atom and a value $E \approx 1 \times 10^6\ \text{V/m}$, which is the breakdown voltage of air. For a He-Ne laser having 1 mW of *continuous-wave* (cw) power focused in $1\ \text{mm}^2$, $S \approx 10^3\ \text{W/m}^2$, and for an Ar ion laser having 10 W of cw power focused in $1\ \text{mm}^2$, $S \approx 10^7\ \text{W/m}^2$. Semiconductor diode lasers produce tunable cw output in the mW to W range, having central frequencies that can range from the near-ultraviolet to the infrared. Ti:sapphire lasers produce several watts of tunable cw radiation centered in the infrared. Pulsed lasers provide much higher powers (but for short intervals of time so that the average energy in the pulse rarely exceeds a Joule or so). In 1965, Nd:YAG lasers produced 1 mJ in 1 μs—if focused to 1 mm^2, $S \approx 10^9\,\mathrm{W/m^2}$, which produces an E field on the order of the breakdown voltage of air. Currently, Ti:sapphire lasers produce pulsed output with average powers as large as a few watts and pulse lengths as short as a few fs. In 2007, the power output of the Hercules laser at the University of Michigan was on the order of $100\,\mathrm{TW} = 10^{14}\,\mathrm{W}$, with power densities greater than $10^{22}\,\mathrm{W/cm^2}$.

From the field amplitude E and the dipole moment matrix element μ_{12} associated with the atomic transition that is being driven by the field, one can calculate the Rabi frequency $\Omega_0 = \mu_{12}E/\hbar$. Typically, μ_{12} is of order ea_0, where $e = 1.60 \times 10^{-19}\,\mathrm{C}$ is the *magnitude* of the charge of the electron, and $a_0 = 5.29 \times 10^{-11}\,\mathrm{m}$ is the Bohr radius. A power of $1\,\mathrm{W/cm^2}$ corresponds to $E \approx 3 \times 10^3\,\mathrm{V/m}$, which in turn corresponds to a Rabi frequency on the order of $\Omega_0 \approx 10^8\,\mathrm{s^{-1}}$ or $\Omega_0/2\pi \approx 10^7\,\mathrm{Hz} = 10\,\mathrm{MHz}$.

1.4 Atom–Field Hamiltonian

In dealing with problems involving the interaction of optical fields with atoms, one often makes the *dipole approximation*, based on the fact that the wavelength of the optical field is much larger than the size of an atom. You may recall that the leading term in the interaction between a neutral charge distribution and an electric field that varies slowly on the length scale of the charge distribution is the dipole coupling, $-\boldsymbol{\mu} \cdot \mathbf{E}$, where $\boldsymbol{\mu}$ is the dipole moment of the charge distribution, and $\mathbf{E}$ is the electric field evaluated at the center of the charge distribution.

Thus, it is not unreasonable to take as the Hamiltonian for an N-electron atom interacting with an optical field having electric field $\mathbf{E}(\mathbf{R}, t)$ a Hamiltonian of the form

$$\hat{H} = \frac{\hat{\mathbf{P}}_{CM}^2}{2M} + \hat{H}_{atom} + \hat{V}_{AF}, \tag{1.13}$$

where

$$\hat{H}_{atom} = \sum_{j=1}^{N} \frac{\hat{\mathbf{p}}_j^2}{2m} + \hat{V}_C \tag{1.14}$$

is the atomic Hamiltonian, and

$$\hat{V}_{AF} = -\hat{\boldsymbol{\mu}} \cdot \mathbf{E}(\mathbf{R}_{CM}, t) \tag{1.15}$$

is the atom–field interaction Hamiltonian. In these equations, $\mathbf{R}_{CM}$ is the position and $\hat{\mathbf{P}}_{CM}$ the momentum operator associated with the center of mass of an atom having mass M, $\hat{\mathbf{p}}_j$ is the momentum operator of the jth electron in the atom, m is the electron mass,

$$\hat{\boldsymbol{\mu}} = -e\sum_{j=1}^{N} \mathbf{r}_j \tag{1.16}$$

is the *electric dipole moment operator* of the atom, $\mathbf{r}_j$ is the coordinate of the jth electron relative to the nucleus, $\hat{V}_C$ is the Coulomb interaction between the charges in the atom, and $\mathbf{A}^2 \equiv \mathbf{A} \cdot \mathbf{A}$ for any vector $\mathbf{A}$. To a good approximation, $\mathbf{R}_{CM}$ coincides with the position of the nucleus. The Hamiltonian (1.13) provides the starting point for semiclassical calculations of atom–field interactions in the dipole approximation. You are urged to study the appendix in this chapter, where further justification for the choice of this Hamiltonian is given.

1.5 Dirac Notation

It is assumed that anyone reading this text has been exposed to Dirac notation. Dirac developed a powerful formalism for representing state vectors in quantum mechanics. Students leaving an introductory course in quantum mechanics often can *use* Dirac notation but may not appreciate its significance. It is not our intent to go into a detailed discussion of Dirac notation. Instead, we would like to remind you of some of the features that are especially relevant to this text.

It is probably easiest to think of Dirac notation in analogy with a three-dimensional vector space. Any three-dimensional vector can be written as

$$\mathbf{A} = A_x\mathbf{i} + A_y\mathbf{j} + A_z\mathbf{k}, \tag{1.17}$$

where A_x, A_y, A_z are the components of the vector in this x, y, z basis. We can represent the unit vectors as column vectors,

$$\mathbf{i} = \begin{pmatrix} 1 \\ 0 \\ 0 \end{pmatrix}, \quad \mathbf{j} = \begin{pmatrix} 0 \\ 1 \\ 0 \end{pmatrix}, \quad \mathbf{k} = \begin{pmatrix} 0 \\ 0 \\ 1 \end{pmatrix}, \tag{1.18}$$

such that the vector $\mathbf{A}$ can be written as

$$\mathbf{A} = \begin{pmatrix} A_x \\ A_y \\ A_z \end{pmatrix}. \tag{1.19}$$

Of course, the basis vectors $\mathbf{i}, \mathbf{j}, \mathbf{k}$ are not unique; any set of three noncollinear unit vectors would do as well. Let us call one such set $\mathbf{u}_1, \mathbf{u}_2, \mathbf{u}_3$, such that

$$\mathbf{A} = A_1\mathbf{u}_1 + A_2\mathbf{u}_2 + A_3\mathbf{u}_3. \tag{1.20}$$

The vector $\mathbf{A}$ is absolute in the sense that it is basis-independent. For a given basis, the components of $\mathbf{A}$ change in precisely the correct manner to ensure that $\mathbf{A}$ remains unchanged. We are at liberty to represent the basis vectors as

$$\begin{pmatrix} 1 \\ 0 \\ 0 \end{pmatrix}, \begin{pmatrix} 0 \\ 1 \\ 0 \end{pmatrix}, \begin{pmatrix} 0 \\ 0 \\ 1 \end{pmatrix} \tag{1.21}$$

in any *one* basis, but once we choose this basis, we must express all other unit vectors in terms of this specific basis. The example in the problems should make this clear.

In quantum mechanics, we express a state vector in a specific basis as

$$|\psi\rangle = \sum_{n=1}^{N} A_n|n\rangle, \tag{1.22}$$

where the sum is over all possible states of the system. In contrast to the case of three-dimensional vectors, this expansion rarely has a simple geometrical interpretation. Rather, the abstract state vector or *ket* $|\psi\rangle$ is expanded in terms of a basis set of eigenkets. In analogy with the case of vectors, one can take $|n\rangle$ as a column matrix in which there is a 1 in the nth row and a zero everywhere else. We are free to choose only *one* set of basis functions with this representation.

In Dirac notation, state vectors are represented by column vectors, and operators are represented by matrices. Thus, an operator **B** can be written as

$$\mathbf{B} = \sum_{n,m=1}^{N} B_{nm}\,|n\rangle\,\langle m|, \tag{1.23}$$

where the *bra* $\langle m|$ can be represented as a row matrix with a 1 in the mth location and a zero everywhere else. The *basis operator* $|n\rangle\,\langle m|$ is then an $N \times N$ matrix with a 1 in the nmth location and zeros everywhere else. Typically, one writes a matrix element of **B** as $B_{nm} = \langle n|\,B\,|m\rangle$. This tells you nothing about how to calculate these matrix elements; moreover, the matrix elements depend on the basis that is chosen.

In general, we know only that any Hermitian operator has an associated set of eigenkets, such that the operator is diagonal in the basis of these eigenkets. For example, the states $|E\rangle$ are eigenkets of the energy operator $\hat{H}$; the fact that $\hat{H}$ is diagonal in the $|E\rangle$ basis does not provide any prescription for calculating the diagonal elements (eigenvalues). In essence, one must often revert to the Schrödinger equation in coordinate space to obtain the eigenvalues, although it is sometimes possible to use operator techniques (as in the case of a harmonic oscillator) to deduce the energy spectrum.

1.6 Where Do We Go from Here?

Now that we have reviewed some of the concepts that are needed in the following chapters, it might prove useful to formulate a strategy for optimizing the benefits that you can derive from this text. There are many excellent texts on quantum mechanics, laser spectroscopy, lasers, nonlinear optics, and quantum optics on the market. Several of these are listed in the bibliography at the end of this chapter. Some of the material that we present overlaps with that in other texts, so you may prefer one treatment to another. You are urged to consult other texts to complement the material presented herein. In fact, there are many topics that we barely touch on at all, such as collective effects, laser theory, optical bistability, and quantum information.

The problems form an integral part of this text. Some of the problems are far from trivial and require considerable effort, but the more problems you are able to solve, the better will be your understanding. Hopefully, the text will provide a useful reference to which you can return as needed. Some of the calculations that would disrupt the development are included as appendices in the chapters.

The first few chapters are devoted to a study of a classical electromagnetic field interacting with a "two-level" atom. These chapters are really the heart of the material. They provide the fundamental underlying formalism and must be mastered if the various applications are to be appreciated. Let's get started!

1.7 Appendix: Atom–Field Hamiltonian

The atom–field Hamiltonian can be written using different degrees of sophistication. In Coulomb gauge, one can choose the "minimal-coupling" Hamiltonian for a neutral atom containing N electrons interacting with an external, classical optical field as

$$\hat{H} = \frac{1}{2M}[\hat{\mathbf{P}} - Ne\mathbf{A}(\mathbf{R}, t)]^2 + \frac{1}{2m}\sum_{j=1}^{N}[\hat{\mathbf{P}}_j + e\mathbf{A}(\mathbf{R}_j, t)]^2 + \hat{V}_C, \tag{1.24}$$

where $\mathbf{A}(\mathbf{R}, t)$ is the vector potential of the external field, $\hat{\mathbf{P}}$ is the momentum operator of the nucleus having mass M and coordinate $\mathbf{R}$, $\hat{\mathbf{P}}_j$ is the momentum operator of the jth electron having mass m and coordinate $\mathbf{R}_j$, e is the magnitude of the electron charge, and

$$\hat{V}_C = -\frac{1}{4\pi\epsilon_0}\sum_{j=1}^{N}\frac{Ne^2}{|\mathbf{R}_j - \mathbf{R}|} + \frac{1}{8\pi\epsilon_0}\sum_{i,j=1;i\neq j}^{N}\frac{e^2}{R_{ij}} \tag{1.25}$$

is the Coulomb potential energy of the charges in the atom ($\mathbf{R}_{ij} = \mathbf{R}_i - \mathbf{R}_j$). You are probably familiar with Hamiltonians of the form (1.24) from your quantum mechanics course. The external electric field is transverse and is related to the vector potential via $\mathbf{E}_\perp(\mathbf{R}, \mathbf{t}) = -\partial\mathbf{A}(\mathbf{R}, t)/\partial t$. The magnetic field is given by $\mathbf{B}(\mathbf{R}, t) = \nabla \times \mathbf{A}(\mathbf{R}, t)$.

The Hamiltonian (1.24) leads to the correct force law for the time rate of change of the average momentum of the charges. To show this, we recall that the expectation value of any quantum-mechanical operator $\hat{\mathrm{O}}$ evolves as

$$d\langle\hat{\mathrm{O}}\rangle/dt = \frac{1}{i\hbar}\langle[\hat{\mathrm{O}}, \hat{H}]\rangle, \tag{1.26}$$

where $[\hat{\mathrm{O}}, \hat{H}]$ is the commutator of $\hat{\mathrm{O}}$ and $\hat{H}$. As such, one finds

$$d\langle\hat{\mathbf{R}}\rangle/dt = \frac{1}{i\hbar}\langle[\hat{\mathbf{R}}, \hat{H}]\rangle = \frac{\langle\hat{\mathbf{P}}\rangle - Ne\mathbf{A}(\mathbf{R}, t)}{M} \equiv \langle\hat{\mathbf{v}}\rangle, \tag{1.27a}$$

$$d\langle\hat{\mathbf{R}}_j\rangle/dt = \frac{1}{i\hbar}\langle[\hat{\mathbf{R}}_j, \hat{H}]\rangle = \frac{\langle\hat{\mathbf{P}}_j\rangle + e\mathbf{A}(\mathbf{R}_j, t)}{m} \equiv \langle\hat{\mathbf{v}}_j\rangle, \tag{1.27b}$$

$$\begin{aligned} d\langle\hat{\mathbf{P}}\rangle/dt &= \frac{1}{i\hbar}\langle[\hat{\mathbf{P}}, \hat{H}]\rangle = -\frac{1}{2M}\langle\nabla_{\mathbf{R}}[\hat{\mathbf{P}} - Ne\mathbf{A}(\mathbf{R}, t)]^2\rangle - \langle\nabla_{\mathbf{R}}\hat{V}_C\rangle \\ &= Ne\langle(\hat{\mathbf{v}}\cdot\nabla_{\mathbf{R}})\mathbf{A}(\mathbf{R}, t)\rangle + Ne\langle\hat{\mathbf{v}}\rangle \times \mathbf{B}(\mathbf{R}, t) - \langle\nabla_{\mathbf{R}}\hat{V}_C\rangle, \end{aligned} \tag{1.27c}$$

where $\hat{\mathbf{v}} = [\hat{\mathbf{P}} - Ne\mathbf{A}(\mathbf{R},t)]/M$ is the operator associated with the nuclear velocity, $\hat{\mathbf{v}}_j = [\hat{\mathbf{P}}_j + e\mathbf{A}(\mathbf{R}_j,t)]/m$ is the operator associated with the velocity of electron j, and the vector identity $\nabla(\mathbf{F}\cdot\mathbf{F}) = 2[(\mathbf{F}\cdot\nabla)\mathbf{F} + \mathbf{F}\times(\nabla\times\mathbf{F})]$ has been used in equation (1.27c). It then follows that

$$\begin{aligned} M\frac{d\langle\hat{\mathbf{v}}\rangle}{dt} &= \frac{d\langle\hat{\mathbf{P}}\rangle}{dt} - Ne\frac{\partial\mathbf{A}(\mathbf{R},t)}{\partial t} \\ &= \frac{d\langle\hat{\mathbf{P}}\rangle}{dt} - Ne\left[\frac{\partial\mathbf{A}(\mathbf{R},t)}{\partial t} + \langle(\hat{\mathbf{v}}\cdot\nabla_{\mathbf{R}})\mathbf{A}(\mathbf{R},t)\rangle\right] \\ &= Ne[\mathbf{E}_\perp(\mathbf{R},\mathbf{t}) + \langle\hat{\mathbf{v}}\rangle\times\mathbf{B}(\mathbf{R},\mathbf{t})] - \langle\nabla_{\mathbf{R}}\hat{V}_C\rangle. \end{aligned} \tag{1.28}$$

Similarly, one can find that the force on the jth electron is

$$m\frac{d\langle\hat{\mathbf{v}}_j\rangle}{dt} = -e[\mathbf{E}_\perp(\mathbf{R}_j,t) + \langle\hat{\mathbf{v}}_j\rangle\times\mathbf{B}(\mathbf{R}_j,t)] - \langle\nabla_{\mathbf{R}_j}\hat{V}_C\rangle. \tag{1.29}$$

Equations (1.28) and (1.29) constitute the Lorentz force law. This provides some justification for use of the Hamiltonian (1.24).

In most cases to be considered in this text, the wavelength of the optical field is much larger than the size of the atom. In this limit, one can make the *dipole approximation* and set $\mathbf{A}(\mathbf{R}_j,t) \approx \mathbf{A}(\mathbf{R},t)$ such that the Hamiltonian (1.24) becomes

$$\hat{H} = \frac{1}{2M}[\hat{\mathbf{P}} - Ne\mathbf{A}(\mathbf{R},t)]^2 + \frac{1}{2m}\sum_{j=1}^{N}[\hat{\mathbf{P}}_j + e\mathbf{A}(\mathbf{R},t)]^2 + \hat{V}_C, \tag{1.30}$$

along with the corresponding Schrödinger equation

$$i\hbar\frac{\partial\psi(\mathbf{r}_j,\mathbf{R},t)}{\partial t} = \hat{H}\psi(\mathbf{r}_j,\mathbf{R},t), \tag{1.31}$$

where $\mathbf{r}_j = \mathbf{R}_j - \mathbf{R}$ is the relative coordinate of electron j. The Coulomb potential is a function of all the r_j's and r_{ij}'s, where $r_{ij} = |\mathbf{r}_i - \mathbf{r}_j|$.

If we are dealing with a more than one electron atom, there is no way in which the problem can be separated into motion of the center of mass and motion of each electron relative to the center of mass. Since such a separation is possible only for a two-body system, we consider that case first and then generalize the results to an N-electron system. For a one-electron atom,

$$\begin{aligned} \hat{H} &= \frac{1}{2M}[\hat{\mathbf{P}} - e\mathbf{A}(\mathbf{R},t)]^2 + \frac{1}{2m}[\hat{\mathbf{P}}_1 + e\mathbf{A}(\mathbf{R},t)]^2 + \hat{V}_C(r_1) \\ &= \frac{\hat{\mathbf{P}}^2}{2M} + \frac{\hat{\mathbf{P}}_1^2}{2m} + e\mathbf{A}(\mathbf{R},t)\cdot\left(\frac{\hat{\mathbf{P}}_1}{m} - \frac{\hat{\mathbf{P}}}{M}\right) \\ &\quad + \frac{e^2|\mathbf{A}(\mathbf{R},t)|^2}{2}\left(\frac{1}{m} + \frac{1}{M}\right) + \hat{V}_C(r_1). \end{aligned} \tag{1.32}$$

Defining conjugate coordinates and momenta for the center-of-mass motion via

$$\mathbf{R}_{CM} = \frac{M\mathbf{R} + m\mathbf{R}_1}{(m+M)}, \qquad \hat{\mathbf{P}}_{CM} = \hat{\mathbf{P}} + \hat{\mathbf{P}}_1, \tag{1.33}$$

and that for the relative coordinates via

$$\mathbf{r} = \mathbf{r}_1 = \mathbf{R}_1 - \mathbf{R}, \quad \hat{\mathbf{p}} = \frac{M\hat{\mathbf{P}}_1 - m\hat{\mathbf{P}}}{(m+M)}, \tag{1.34}$$

one finds that the Hamiltonian (1.32) is transformed into

$$\begin{aligned} \hat{H} &= \frac{\hat{\mathbf{P}}_{CM}^2}{2(m+M)} + \frac{1}{2}\left(\frac{1}{m} + \frac{1}{M}\right)[\hat{\mathbf{p}} + e\mathbf{A}(\mathbf{R}, t)]^2 + \hat{V}_C(r) \\ &\approx \frac{\hat{\mathbf{P}}_{CM}^2}{2M} + \frac{1}{2m}[\hat{\mathbf{p}} + e\mathbf{A}(\mathbf{R}_{CM}, t)]^2 + \hat{V}_C(r). \end{aligned} \tag{1.35}$$

This result suggests that, for an N-electron atom interacting with an external field, we take as our Hamiltonian

$$\hat{H} = \frac{\hat{\mathbf{P}}_{CM}^2}{2M} + \frac{1}{2m}\sum_{j=1}^{N}\left[\hat{\mathbf{p}}_j + e\mathbf{A}(\mathbf{R}_{CM}, t)\right]^2 + \hat{V}_C, \tag{1.36}$$

where the $\hat{\mathbf{p}}_j$ are momenta conjugate to the relative coordinates $\mathbf{r}_j$, and the Coulomb potential is

$$\hat{V}_C = -\frac{1}{4\pi\epsilon_0}\sum_{j=1}^{N}\frac{Ne^2}{r_j} + \frac{1}{8\pi\epsilon_0}\sum_{i,j=1;i\neq j}^{N}\frac{e^2}{r_{ij}}. \tag{1.37}$$

This Hamiltonian can be put in a somewhat simpler form if we carry out a unitary transformation of the wave function given by

$$\Psi(\mathbf{r}_j, \mathbf{R}_{CM}, t) = \hat{U}(t)\psi(\mathbf{r}_j, \mathbf{R}_{CM}, t), \tag{1.38}$$

where the unitary operator $\hat{U}(t)$ is defined by

$$\hat{U}(t) = \exp\left[-\frac{i}{\hbar}\hat{\boldsymbol{\mu}} \cdot \mathbf{A}(\mathbf{R}_{CM}, t)\right], \tag{1.39}$$

with

$$\hat{\boldsymbol{\mu}} = -e\sum_{j=1}^{N}\mathbf{r}_j \tag{1.40}$$

the *electric dipole moment operator* for the atom. The dependence of $\hat{U}$ on the time has been noted explicitly.

Under this transformation,

$$i\hbar\frac{\partial\Psi(\mathbf{r}_j, \mathbf{R}_{CM}, t)}{\partial t} = \hat{U}(t)\hat{H}\hat{U}^\dagger(t)\Psi(\mathbf{r}_j, \mathbf{R}_{CM}, t) + i\hbar\frac{\partial\hat{U}(t)}{\partial t}\hat{U}^\dagger(t)\Psi(\mathbf{r}_j, \mathbf{R}_{CM}, t) \tag{1.41}$$

and using the fact that

$$e^A B e^{-A} = B + [A, B] + \frac{1}{2!}[A, [A, B]] + \cdots \tag{1.42}$$

and

$$i\hbar\frac{\partial\hat{U}(t)}{\partial t}\hat{U}^{\dagger}(t)=\hat{\boldsymbol{\mu}}\cdot\frac{\partial\mathbf{A}(\mathbf{R}_{CM},t)}{\partial t}\hat{U}(t)\hat{U}^{\dagger}(t)=-\hat{\boldsymbol{\mu}}\cdot\mathbf{E}_{\perp}(\mathbf{R}_{CM},t),\tag{1.43}$$

one finds

$$i\hbar\frac{\partial\Psi(\mathbf{r}_j,\mathbf{R}_{CM},t)}{\partial t}=\left[\frac{\hat{\mathbf{P}}_{CM}^2}{2M}+\sum_{j=1}^{N}\frac{\hat{\mathbf{p}}_j^2}{2m}+\hat{V}_C-\hat{\boldsymbol{\mu}}\cdot\mathbf{E}_{\perp}(\mathbf{R}_{CM},t)\right]\Psi(\mathbf{r}_j,\mathbf{R}_{CM},t),\tag{1.44}$$

neglecting terms that are v_{CM}/c smaller than the interaction term that arise from the commutator $[-\frac{i}{\hbar}\hat{\boldsymbol{\mu}}\cdot\mathbf{A}(\mathbf{R}_{CM},t),\hat{\mathbf{P}}_{CM}^2/2M]$.

As a consequence, the effective Hamiltonian for this system can be written as

$$\hat{H}=\frac{\hat{\mathbf{P}}_{CM}^2}{2M}+\hat{H}_{atom}+\hat{V}_{AF},\tag{1.45}$$

where

$$\hat{H}_{atom}=\sum_{j=1}^{N}\frac{\hat{\mathbf{p}}_j^2}{2m}+\hat{V}_C\tag{1.46}$$

is the atomic Hamiltonian, and

$$\hat{V}_{AF}=-\hat{\boldsymbol{\mu}}\cdot\mathbf{E}_{\perp}(\mathbf{R}_{CM},t)\tag{1.47}$$

is the atom–field interaction Hamiltonian. Note that the Hamiltonians (1.36) and (1.45) lead to the same values for expectation values of operators, even if the wave function transformation given by equation (1.38) can be quite complicated.

Problems

1. Go online or to other sources to determine the fine and hyperfine separations in the 3S and 3P levels (as well as the $3S_{1/2}$-$3P_{1/2,3/2}$ separations) in ^{23}Na and the fine and hyperfine separations in the 5S and 5P levels (as well as the $5S_{1/2}$-$5P_{1/2,3/2}$ separations) in ^{85}Rb.
2. Estimate the Doppler width and collision width on the 3S-3P transition in ^{23}Na. To estimate the Doppler width, assume a temperature of 300 K. To estimate the collision width per Torr of perturber gas, assume that the perturbers undergoing collisions with the sodium atoms are much more massive than sodium and that the collision cross section is 10 Å^2. Compare your answer with data on broadening of the sodium resonance line by rare gas perturbers.
3. Estimate the recoil frequency in Na and Rb.
4. Look up the transition matrix elements for ^{85}Rb to estimate the Rabi frequency for a laser field having 10 mW of power focused to a spot size of 50 μm^2.
5. Consider the two-dimensional vector $\mathbf{A}=\mathbf{i}+2\mathbf{j}$. Take as your basis states

$$\mathbf{u}_1=\begin{pmatrix}1\\0\end{pmatrix},\quad \mathbf{u}_2=\begin{pmatrix}0\\1\end{pmatrix},$$

where

$$\mathbf{u}_{1,2} = \frac{\mathbf{i} \pm \mathbf{j}}{\sqrt{2}}.$$

Express the unit vectors **i** and **j** in this basis, and find the coordinates. Show explicitly that $A_1\mathbf{u}_1 + A_2\mathbf{u}_2 = A_x\mathbf{i} + A_y\mathbf{j}$.

6. Derive equation (1.27c).
7. Prove that

$$\frac{d\mathbf{A}(\mathbf{R}, t)}{dt} = \frac{\partial \mathbf{A}(\mathbf{R}, t)}{\partial t} + (\mathbf{v} \cdot \nabla)\, \mathbf{A}(\mathbf{R}, t)$$

for a vector function $\mathbf{A}(\mathbf{R}, t)$, with $\mathbf{v} = \dot{\mathbf{R}}$.

8. Derive equation (1.44) from equation (1.41).
9. Show that the analogue of the wave equation (1.5) for the displacement vector $\mathbf{D}(\mathbf{R}, t)$ is

$$\nabla^2\mathbf{D} - \mu_0\varepsilon_0\frac{\partial^2\mathbf{D}}{\partial t^2} = -\nabla\times(\nabla\times\mathbf{P}).$$

10. For an infinite square well potential, show that an arbitrary initial wave packet will return to its initial state at integral multiples of a *revival time* $\tau = (4ma^2)/\pi\hbar$, where m is the mass of the particle in the well, and a is the width of the well.
11. The radiative reaction rate γ_2 for a classical oscillator having charge e, mass m, and frequency ω is given by

$$\gamma_2 = \frac{1}{4\pi\epsilon_0}\frac{2}{3}\frac{e^2\omega^2}{c^3}.$$

Show that

$$\frac{\gamma_2}{\omega} = \alpha_{FS}\frac{\hbar\omega}{mc^2}, \quad \alpha_{FS} = \frac{1}{4\pi\epsilon_0}\frac{e^2}{\hbar c} \approx \frac{1}{137},$$

and estimate this ratio for an electron oscillator having a frequency corresponding to an optical frequency.

References

[1] For an introduction to this topic, see, for example, P. W. Milonni and M.-L. Shih, *Casimir Forces*, Contemporary Physics **33**, 313–322 (1992); K. A. Milton, *The Casimir Effect* (World Scientific, Singapore, 2001). A bibliography can be found at the following website: http://www.cfa.harvard.edu/~babb/casimir-bib.html.

[2] J. C. Maxwell, *Note on a natural limit to the sharpness of spectral lines*, Nature **VIII**, 474–475 (1873).

[3] J. C. Maxwell, *A Treatise on Electricity and Magnetism*, vol. 2 (Dover, New York, 1954), chap. XX, article 781.

Bibliography

Listed here is a representative bibliography of some general reference texts, books on laser spectroscopy, and books on quantum optics.

General Reference Texts

L. Allen and J. H. Eberly, *Optical Resonance and Two-Level Atoms* (Wiley, New York, 1985).

R. Balian, S. Haroche, and S. Liberman, Eds., *Frontiers in Laser Spectroscopy*, Les Houches Session XXVII, vols. 1 and 2 (North-Holland, Amsterdam, 1975).

N. Bloembergen, *Nonlinear Optics* (W. A. Benjamin, New York, 1965).

M. Born and E. Wolf, *Principles of Optics*, 7th ed. (Cambridge University Press, Cambridge, 1999).

R. W. Boyd, *Nonlinear Optics*, 3rd ed. (Academic Press, Burlington, MA, 2008).

D. Budker, D. F. Kimball, and D. P. DeMille, *Atomic Physics: An Exploration through Problems and Solutions* (Oxford University Press, Oxford, UK, 2006).

C. Cohen-Tannoudji, J. Dupont-Roc, and G. Grynberg, *Atom-Photon Interactions* (Wiley-Interscience, New York, 1992).

———, *Photons and Atoms—Introduction to Quantum Electrodynamics* (Wiley-Interscience, New York, 1989).

C. DeWitt, A. Blandin, and C. Cohen-Tannoudji, Eds., *Quantum Optics and Electronics* (Gordon and Breach, New York, 1964).

C. J. Foot, *Atomic Physics* (Oxford University Press, Oxford, UK, 2005).

H. Haken, *Laser Theory* (Springer-Verlag, Berlin, 1984).

S. Haroche and J.-M. Raimond, *Exploring the Quantum: Atoms, Cavities, and Photons* (Oxford University Press, Oxford, UK, 2006).

W. Heitler, *The Quantum Theory of Radiation*, 3rd ed. (Oxford University Press, London, 1954).

L. Mandel and E. Wolf, *Optical Coherence and Quantum Optics* (Cambridge University Press, Cambridge, UK, 1995).

H. J. Metcalf and P. van der Straten, *Laser Cooling and Trapping* (Springer-Verlag, New York, 1999).

P. Meystre and M. Sargent III, *Elements of Quantum Optics*, 4th ed. (Springer-Verlag, Berlin, 2007).

P. W. Milonni, *The Quantum Vacuum: An Introduction to Quantum Electrodynamics* (Academic Press, San Diego, CA, 1993).

P. W. Milonni and J. H. Eberly, *Lasers (Wiley Series in Pure and Applied Optics): Laser Physics* (Wiley, Hoboken, 2010).

S. Mukamel, *Principles of Nonlinear Laser Spectroscopy* (Oxford University Press, Oxford, UK, 1955).

E. A. Power, *Introductory Quantum Electrodynamics* (American Elsevier Publishing, New York, 1965).

M. Sargent III, M. O. Scully, and W. E. Lamb Jr., *Laser Physics* (Addison-Wesley, Reading, MA, 1974).

Y. R. Shen, *The Principles of Nonlinear Optics* (Wiley-Interscience, New York, 1984).

B. W. Shore, *The Theory of Coherent Excitation*, vols. 1 and 2 (Wiley-Interscience, New York, 1990). This encyclopedic work contains a wealth of references.

A. E. Siegman, *Lasers* (University Science Books, Mill Valley, CA, 1986).

Laser Spectroscopy Texts

A. Corney, *Atomic and Laser Spectroscopy* (Oxford Classics Series, Oxford, UK, 2006).

W. Demtröder, *Laser Spectroscopy*, 4th ed., vol. 1, *Basic Principles*, and vol. 2, *Experimental Techniques* (Springer-Verlag, Berlin, 2008).

V. S. Letokhov and V. P. Chebotaev, *Non-Linear Laser Spectroscopy* (Springer-Verlag, Berlin, 1977).

M. D. Levenson, *Introduction to Nonlinear Laser Spectroscopy* (Academic Press, New York, 1982).

S. Stenholm, *Foundations of Laser Spectroscopy* (Wiley, New York, 1984; Dover, Mineola, NY, 2005).

Quantum Optics Texts

S. M. Barnett and P. M. Radmore, *Methods in Theoretical Quantum Optics* (Clarendon Press, Oxford, UK, 1997).

M. Fox, *Quantum Optics: An Introduction* (Oxford University Press, Oxford, UK, 2006).

J. C. Garrison and R. Y. Chiao, *Quantum Optics* (Oxford University Press, Oxford, UK, 2006).

C. C. Gerry and P. L. Knight, *Introductory Quantum Optics* (Cambridge University Press, Cambridge, UK, 2005).

J. R. Klauder and E. C. G. Sudarshan, *Fundamentals of Quantum Optics* (W. A. Benjamin, New York, 1968; Dover, Mineola, NY, 2006).

P. L. Knight and L. Allen, *Concepts of Quantum Optics* (Pergamon, New York, 1985).

R. Loudon, *The Quantum Theory of Light*, 3rd ed. (Oxford University Press, Oxford, UK, 2003).

W. Louisell, *Quantum Statistical Properties of Radiation* (Wiley, New York, 1973).

G. J. Milburn and D. F. Walls, *Quantum Optics* (Springer-Verlag, Berlin, 1994).

H. M. Nussenzweig, *Introduction to Quantum Optics* (Gordon and Breach, London, 1973).

W. P. Schleich, *Quantum Optics in Phase Space* (Wiley-VCH, Berlin, 2001).

M. O. Scully and M. S. Zubairy, *Quantum Optics* (Cambridge University Press, Cambridge, UK, 1997).

2

Two-Level Quantum Systems

The general subject matter of this text is the interaction of radiation with matter. A "two-level" atom driven by an optical field is considered to be a prototypical system. We examine this problem from several different points of view and use different analytical tools to solve the relevant equations. It may seem like a bit of overkill, but this is a building block problem that must be understood if further progress is to be achieved. Moreover, the problem of a two-level atom interacting with a radiation field has many more surprises than you might expect. As a result, you will learn some interesting physics as we go along. Many different mathematical representations are used, and you might ask if this is really necessary. It turns out that each of these representations is well-suited to specific classes of problems involving the interaction of radiation with matter. At first, we consider generic quantum systems, but focus eventually on the two-level atom. Appendix A contains a summary of the various representations that are introduced.

2.1 Review of Quantum Mechanics

2.1.1 Time-Independent Problems

We are interested in problems that can be termed *semiclassical* in nature. In such problems, the atoms are treated quantum mechanically, but the external fields with which they interact are treated classically. Before discussing time-dependent Hamiltonians, let us review time-independent Hamiltonians, $\hat{H} = \hat{H}(\mathbf{r})$, for an effective one-electron atom with the electron's position denoted by $\mathbf{r}$.

For such Hamiltonians, an arbitrary wave function $\psi(\mathbf{r}, t)$ can be expanded as

$$\psi(\mathbf{r}, t) = \sum_n a_n(t)\psi_n(\mathbf{r}), \tag{2.1}$$

and an arbitrary state vector $|\psi(t)\rangle$ as

$$|\psi(t)\rangle = \sum_n a_n(t)|n\rangle, \tag{2.2}$$

where $\psi_n(\mathbf{r})$ is the eigenfunction, $|n\rangle$ is the eigenket, and $a_n(t)$ is the probability amplitude associated with state n. The eigenfunctions and eigenkets are solutions of the time-independent Schrödinger equation,

$$\hat{H}\psi_n(\mathbf{r}) = E_n\psi_n(\mathbf{r}), \tag{2.3}$$

where $\hat{H}$ is an operator, or

$$\mathbf{H}|n\rangle = E_n|n\rangle\,, \tag{2.4}$$

where $\mathbf{H}$ is a matrix. Recall that the eigenfunctions are related to the eigenkets via

$$\psi_n(\mathbf{r}) = \langle \mathbf{r}|n\rangle, \tag{2.5}$$

where $|\mathbf{r}\rangle$ is the eigenket of the position operator.

It follows from the Schrödinger equation

$$i\hbar\frac{\partial\psi(\mathbf{r},t)}{\partial t} = \hat{H}\psi(\mathbf{r},t) \tag{2.6}$$

and equation (2.1) that the probability amplitudes obey the differential equation

$$i\hbar\dot{a}_n(t) = E_n a_n(t), \tag{2.7}$$

where a dot above a symbol indicates differentiation with respect to time. The solution of this equation is

$$a_n(t) = \exp(-iE_n t/\hbar)\, a_n(0). \tag{2.8}$$

In Dirac notation, $\mathbf{H}$ is a matrix whose elements depend on the representation chosen (it is diagonal in the energy representation), and the Schrödinger equation can be written

$$i\hbar\frac{\partial|\psi(t)\rangle}{\partial t} = \mathbf{H}|\psi(t)\rangle. \tag{2.9}$$

If we try a solution of the type (2.2) in Eq. (2.9), then the $a_n(t)$, arranged as a column vector $\mathbf{a}(t)$, obey the differential equation

$$i\hbar\dot{\mathbf{a}}(t) = \mathbf{H}\mathbf{a}(t), \tag{2.10}$$

which has as its solution

$$\mathbf{a}(t) = \exp(-i\mathbf{H}t/\hbar)\,\mathbf{a}(0), \tag{2.11}$$

where $\exp(-i\mathbf{H}t/\hbar)$ is defined by its series expansion,

$$\exp(-i\mathbf{H}t/\hbar) = 1 - i\frac{\mathbf{H}t}{\hbar} + \frac{1}{2!}\left(-i\frac{\mathbf{H}t}{\hbar}\right)^2 + \cdots\,. \tag{2.12}$$

Of course, the solutions (2.8) and (2.11) are equivalent, since

$$\langle n| \exp(-i\mathbf{H}t/\hbar)\, |n'\rangle = \exp(-i E_n t/\hbar)\, \delta_{n,n'}, \tag{2.13}$$

where $\delta_{n,n'}$ is the Kronecker delta, equal to 1 if $n = n'$ and zero otherwise.

These results imply that the state populations $|a_n(t)|^2$ are constant. In other words, the populations of the eigenstates of a time-independent Hamiltonian do not change in time. Even though the populations remain constant, this does not imply that the quantum system is just sitting around doing nothing. You already know that a free-particle wave packet spreads in time, for example. The *dynamics* of a quantum system is determined by *both* the absolute value of the state amplitudes (which are constant) *and* the relative phases of these amplitudes (which vary linearly with time). Moreover, the expectation values of Hermitian operators depend on bilinear products of the probability amplitudes and their conjugates, as is discussed in chapter 3. Since physical observables are associated with Hermitian operators, the average values of these quantities can be functions of time. For example, the average dipole moment or average momentum of an harmonic oscillator is periodic with the oscillator period if the oscillator is prepared initially in a superposition of eigenstates.

If we can solve the Schrödinger equation for the eigenstates and expand the initial state in terms of the eigenstates, the expansion coefficients totally specify the time evolution of the state. Unfortunately, it is often difficult to obtain analytic solutions, and one must rely on approximate or numerical solutions. Fortunately, with the availability of high-speed computers and assorted software, numerical solutions that were once a challenge can be obtained with a few keystrokes.

2.1.2 Time-Dependent Problems

Often, we are confronted with problems where we can solve for the eigenstates of *part* of the Hamiltonian. Suppose that we can write

$$\hat{H}(\mathbf{r}, t) = \hat{H}_0(\mathbf{r}) + \hat{V}(\mathbf{r}, t), \tag{2.14}$$

$$\mathbf{H}(t) = \mathbf{H}_0 + \mathbf{V}(t), \tag{2.15}$$

where $\hat{V}(\mathbf{r}, t)$ represents the interaction of a classical, time-dependent field with the quantum system, and $\hat{H}_0(\mathbf{r})$ is the Hamiltonian for the quantum system in the absence of the interaction. For example, $\hat{H}_0(\mathbf{r})$ can be the Hamiltonian of an isolated atom, and $\hat{V}(\mathbf{r}, t)$ can be the interaction energy associated with an atom driven by a classical optical field. If $\hat{H}(\mathbf{r}, t)$ depends on time, the energy is no longer a constant of the motion. Let the eigenstates of $\hat{H}_0(\mathbf{r})$ be noted by $\psi_n(\mathbf{r})$ and the eigenkets of $\mathbf{H}_0$ by $|n\rangle$, such that

$$\hat{H}_0(\mathbf{r})\psi_n(\mathbf{r}) = E_n\psi_n(\mathbf{r}), \tag{2.16}$$

or, in Dirac notation,

$$\mathbf{H}_0|n\rangle = E_n|n\rangle. \tag{2.17}$$

Again, we expand

$$\psi(\mathbf{r}, t) = \sum_n a_n(t)\psi_n(\mathbf{r}). \tag{2.18}$$

From the Schrödinger equation

$$i\hbar \frac{\partial \psi(\mathbf{r}, t)}{\partial t} = [\hat{H}_0(\mathbf{r}) + \hat{V}(\mathbf{r}, t)]\psi(\mathbf{r}, t), \tag{2.19}$$

it then follows that

$$i\hbar \sum_n \dot{a}_n(t)\psi_n(\mathbf{r}) = \hat{H}_0(\mathbf{r}) \sum_n a_n(t)\psi_n(\mathbf{r}) + \hat{V}(\mathbf{r}, t) \sum_n a_n(t)\psi_n(\mathbf{r}). \tag{2.20}$$

In Dirac notation, the analogous equations are

$$|\psi(\mathbf{r}, t)\rangle = \sum_n a_n(t)|n\rangle, \tag{2.21}$$

$$i\hbar \frac{\partial |\psi(\mathbf{r}, t)\rangle}{\partial t} = [\mathbf{H}_0 + \mathbf{V}(t)]\,|\psi(\mathbf{r}, t)\rangle, \tag{2.22}$$

$$i\hbar \sum_n \dot{a}_n(t)|n\rangle = \mathbf{H}_0 \sum_n a_n(t)|n\rangle + \mathbf{V}(t) \sum_n a_n(t)|n\rangle. \tag{2.23}$$

Using the orthogonality of the eigenfunctions or eigenkets [that is, multiplying equation (2.20) by $\psi^*_{n'}(\mathbf{r})$ and integrating over $\mathbf{r}$, or equation (2.23) by $\langle n'|$], we find that the state amplitudes evolve as

$$i\hbar \dot{a}_n(t) = E_n a_n(t) + \sum_m V_{nm}(t) a_m(t), \tag{2.24}$$

where the *matrix element* $V_{nm}(t)$ is defined as

$$\begin{aligned} V_{nm}(t) &= \langle n|\mathbf{V}(t)|m\rangle \\ &= \int \langle n|\mathbf{r}\rangle\langle \mathbf{r}|\mathbf{V}(t)|\mathbf{r}'\rangle\langle \mathbf{r}'|m\rangle d\mathbf{r}d\mathbf{r}' \\ &= \int d\mathbf{r}d\mathbf{r}'\,\psi_n(\mathbf{r})\,V(\mathbf{r}, t)\delta(\mathbf{r} - \mathbf{r}')\psi_m(\mathbf{r}') \\ &= \int \psi^*_n(\mathbf{r})\,V(\mathbf{r}, t)\psi_m(\mathbf{r})\,d\mathbf{r}. \end{aligned} \tag{2.25}$$

We can write equation (2.24) as the matrix equation

$$i\hbar \dot{\mathbf{a}}(t) = \mathbf{E}\mathbf{a}(t) + \mathbf{V}(t)\mathbf{a}(t), \tag{2.26}$$

where $\mathbf{E}$ is a diagonal matrix whose elements are the eigenvalues of $\hat{H}_0(\mathbf{r})$ ($\mathbf{E}$ is simply equal to $\mathbf{H}_0$ written in the energy representation), and $\mathbf{V}(t)$ is a matrix having elements $V_{nm}(t)$. The fact that $\mathbf{V}(t)$ is not diagonal, in general, implies that there are *transitions* between the eigenstates of $\mathbf{H}_0$.

It makes sense to talk about transitions between eigenstates of $\mathbf{H}_0$ only if the interaction $\mathbf{V}(t)$ has not distorted the original quantum system to a state where it is unrecognizable. Otherwise, it might not be possible to measure a physical quantity that corresponds to properties of $\mathbf{H}_0$. Consider, for example, the hydrogen atom in a *static* electric field. There are no stationary states for this system; however, for sufficiently small values of the field, it makes perfect sense to talk about this field inducing transitions between states of the hydrogen atom. If the field interaction strength becomes comparable with the energy separation of populated states of the atom, however, a proper description would involve the use of the eigenstates of the composite system of atom plus field.

To obtain the dynamics, one must solve equation (2.26) for the state amplitudes. If $\mathbf{V}(t)$ is a finite matrix, the coupled equations can be solved by computer. As with any differential equation, you can obtain an *analytic* solution only if you already *know* the solution. You can guess a solution based on what others have learned in the past. For example, based on the solution of the *scalar* equation

$$i\hbar\dot{x}(t) = f(t)x(t), \tag{2.27}$$

which is

$$x(t) = \exp\left[-\frac{i}{\hbar}\int_0^t f(t')\,dt'\right]x(0), \tag{2.28}$$

you might think (and you would have some company) that a solution to equation (2.26) is

$$\mathbf{a}(t) = \exp\left\{-\frac{i}{\hbar}\left[\mathbf{E}t + \int_0^t \mathbf{V}(t')\,dt'\right]\right\}\mathbf{a}(0), \tag{2.29}$$

but you would be wrong, as is discussed in section 2.7.3. Only in the limiting case that $\mathbf{V}$ is independent of time is it possible to write the solution as

$$\mathbf{a}(t) = \exp\left[-\frac{i}{\hbar}(\mathbf{E}+\mathbf{V})t\right]\mathbf{a}(0). \tag{2.30}$$

In general, however, it is impossible to obtain analytic solutions to equation (2.26) when $\mathbf{V}$ is a function of time.

2.2 Interaction Representation

In some sense, we have finished. Either we can solve equation (2.26) or we cannot. That does not prevent us from modifying these equations into what may be more convenient forms. Remember, however, that modifying the equations does not make them solvable, but it may reveal a structure where the solution is more apparent. The first such modification that we use, applicable to any time-dependent quantum problem, involves an *interaction representation*. The idea behind the interaction representation is to have the state amplitudes be *constant* in the absence of the interaction $\mathbf{V}(t)$. To accomplish this, one must remove the rapidly varying phase factor $\exp(-iE_nt/\hbar)$ from the state amplitudes by writing

$$|\psi(t)\rangle = \sum_n c_n(t)e^{-iE_nt/\hbar} \tag{2.31}$$

or

$$\psi(\mathbf{r},t) = \sum_n c_n(t)e^{-iE_nt/\hbar}\psi_n(\mathbf{r}). \tag{2.32}$$

It then follows from the Schrödinger equation that the amplitudes $c_n(t)$ of the interaction representation obey the differential equation

$$i\hbar\dot{c}_n(t) = \sum_m V_{nm}(t)c_m(t)e^{i\omega_{nm}t} \equiv \sum_m \left[V^I(t)\right]_{nm} c_m(t), \tag{2.33}$$

where

$$\omega_{nm} = (E_n - E_m)/\hbar \tag{2.34}$$

is a transition frequency, and

$$\left[V^I(t)\right]_{nm} = e^{i\omega_{nm}t}V_{nm}(t) \tag{2.35}$$

is a matrix element in the interaction representation.

From equations (2.21) and (2.31), one sees that the amplitudes $a_n(t)$ and $c_n(t)$ are related by

$$a_n(t) = c_n(t)e^{-iE_nt/\hbar}. \tag{2.36}$$

In matrix form, equation (2.36) can be written as

$$\mathbf{a}(t) = \mathbf{U}_0(t)\mathbf{c}(t), \tag{2.37}$$

where

$$\mathbf{U}_0(t) = e^{-i\mathbf{H}_0t/\hbar} = e^{-i\mathbf{E}t/\hbar} \tag{2.38}$$

is an evolution operator associated with $\mathbf{H}_0$ that satisfies the differential equation

$$i\hbar\dot{\mathbf{U}}_0(t) = \mathbf{H}_0\mathbf{U}_0(t) = \mathbf{E}\mathbf{U}_0(t), \tag{2.39}$$

subject to the initial condition $\mathbf{U}_0(0) = \mathbf{1}$. Combining equations (2.26) and (2.37), we find

$$\begin{aligned} i\hbar\dot{\mathbf{a}}(t) &= i\hbar[\mathbf{U}_0(t)\dot{\mathbf{c}}(t) + \dot{\mathbf{U}}_0(t)\mathbf{c}(t)] \\ &= i\hbar\mathbf{U}_0(t)\dot{\mathbf{c}}(\mathrm{t}) + \mathbf{E}\mathbf{a}(t) = \mathbf{E}\mathbf{a}(t) + \mathbf{V}(t)\mathbf{a}(t), \end{aligned} \tag{2.40}$$

from which it follows that

$$i\hbar\dot{\mathbf{c}}(t) = \mathbf{U}_0^\dagger(t)\mathbf{V}(t)\mathbf{U}_0(t)\mathbf{c}(t) = \mathbf{V}^I(t)\mathbf{c}(t), \tag{2.41}$$

where

$$\mathbf{V}^I(t) = \mathbf{U}_0^\dagger(t)\mathbf{V}(t)\mathbf{U}_0(t) = e^{i\mathbf{H}_0t/\hbar}\mathbf{V}(t)e^{-i\mathbf{H}_0t/\hbar}, \tag{2.42}$$

and we have used the fact that $\mathbf{U}_0^\dagger(t)\mathbf{U}_0(t) = \mathbf{1}$. Equation (2.41) is equivalent to equation (2.26). The interaction representation does not make the problem easier to solve—it only simplifies the notation. Transitions from state n to m are driven effectively when $V_{nm}(t)$ has Fourier components at the frequency separation ω_{mn}.

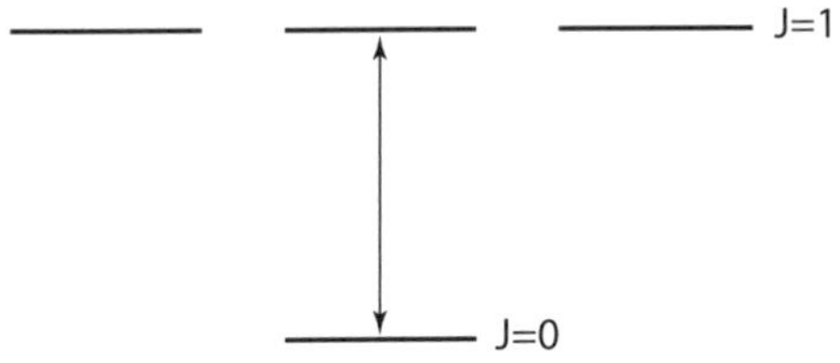

Figure 2.1. For $\hat{\mathbf{z}}$-polarization of an incident optical field nearly resonant with an atomic transition from a $J = 0$ ground state to a $J = 1$ excited state, the atom can be approximated as a two-level quantum system.

A final point to note:

$$|\psi(t)\rangle = \sum_n c_n(t)e^{-iE_nt/\hbar}|n\rangle \equiv \sum_n c_n(t)|n^I(t)\rangle, \tag{2.43}$$

where $|n^I(t)\rangle = \exp(-iE_nt/\hbar)|n\rangle$. In the interaction picture, the eigenkets $|n^I(t)\rangle$ have time dependence. It is important not to forget this time dependence when calculating expectation values of operators. In general, the interaction representation is used often in numerical solutions rather than the Schrödinger representation; in this manner, one need not start the integration until the interaction is turned on. In the Schrödinger representation, the phases of the state amplitudes evolve even in the absence of the interaction.

2.3 Two-Level Atom

To make some of these concepts more concrete, we consider now a prototypical system, a "two-level" atom interacting with a radiation field. It is not difficult to imagine a situation where such a two-level approximation is valid. For example, if an optical field is nearly resonant with the ground to first excited state transition frequency of an atom whose ground and excited states have angular momentum quantum numbers $J = 0$ and $J = 1$, respectively, and if the field is z-polarized, then the field interacts effectively with only two levels of the atom (see figure 2.1), the ground state and the $m = 0$ sublevel of the excited state. To make matters simple, we can think of the atom as a one-electron atom whose nucleus is located at position $\mathbf{R}$. The position of the electron relative to the nucleus is denoted by $\mathbf{r}$.

In dipole approximation, the interaction Hamiltonian is given by equation (1.15),

$$\hat{V}(\mathbf{R}, t) \approx \hat{V}_{AF}(\mathbf{R}, t) \approx -\hat{\boldsymbol{\mu}} \cdot \mathbf{E}(\mathbf{R}, t) = e\hat{\mathbf{r}} \cdot \mathbf{E}(\mathbf{R}, t), \tag{2.44}$$

where $\hat{\boldsymbol{\mu}} = -e\hat{\mathbf{r}}$ is the atomic dipole moment operator (a matrix in the Dirac picture), and $\mathbf{E}(\mathbf{R}, t)$ is the electric field of the applied field, evaluated at the nuclear position. Recall that the charge of the electron is $-e$ in our notation. If atomic motion is neglected, as we assume in this chapter, we can set $\mathbf{R} = 0$.

The applied electric field at the nucleus of the atom is assumed to vary as

$$\mathbf{E}(t) = \hat{\mathbf{z}}\,|E_0(t)| \cos\left[\omega t - \varphi(t)\right] = \frac{1}{2}\hat{\mathbf{z}}|E_0(t)|\left[e^{i\varphi(t)}e^{-i\omega t} + e^{-i\varphi(t)}e^{i\omega t}\right], \tag{2.45}$$

where

$$\frac{1}{2}E_0(t)e^{-i\omega t} = \frac{1}{2}|E_0(t)|\, e^{i\varphi(t)}e^{-i\omega t} \tag{2.46}$$

is the *positive frequency component* of the field, $E_0(t) = |E_0(t)|\, e^{i\varphi(t)}$ is the complex amplitude of the field, ω is the carrier frequency of the field, and $\varphi(t)$ is the phase of the field. Both the amplitude and the phase can be functions of time. A time-varying amplitude could correspond to a pulse envelope, while a time-varying phase gives rise to a frequency "chirp" (a frequency that varies in time). With this choice of field, the interaction Hamiltonian becomes

$$\hat{V}(\mathbf{r}, t) = e\hat{z}|E_0(t)|\cos[\omega t - \varphi(t)], \tag{2.47}$$

where $\hat{z}$ is the z-component of the position operator.

For our two-level atom, the energy of the ground state is taken as $-\hbar\omega_0/2$ and that of the excited state as $\hbar\omega_0/2$. Denoting the ground-state eigenket as $|1\rangle$ and the excited-state eigenket as $|2\rangle$, we can write the probability amplitudes and matrix elements of the interaction Hamiltonian as

$$\mathbf{a} = \begin{pmatrix} a_1 \\ a_2 \end{pmatrix} \tag{2.48}$$

and

$$V_{12} = ez_{12}\,|E_0(t)|\cos[\omega t - \varphi(t)]\,, \tag{2.49a}$$

$$V_{21} = ez_{21}\,|E_0(t)|\cos[\omega t - \varphi(t)]\,, \tag{2.49b}$$

$$V_{11} = V_{22} = 0, \tag{2.49c}$$

where

$$z_{12} = \langle 1|\hat{z}|2\rangle = \langle 2|\hat{z}|1\rangle^* = z_{21}^*\,. \tag{2.50}$$

The diagonal elements of the interaction Hamiltonian vanish since the operator $\hat{z}$ has odd parity. In general, the matrix element z_{12} is complex, but any *single* transition matrix element can be taken as real by an appropriate choice of phase in the wave function. (However, if z_{12} is taken to be real, then we are not at liberty to take x_{12} as real, since the phase of the electronic part of the wave function has been fixed—the matrix element of *one* component only of $\mathbf{r}_{12}$ can be taken as real, and this choice determines whether the other components are real or complex.) Therefore, we can set

$$ez_{12} = ez_{21} = -(\mu_z)_{12} \;\; \text{(real)}, \tag{2.51}$$

$$V_{12} = V_{21} = -(\mu_z)_{12}|E_0(t)|\cos[\omega t - \phi(t)], \tag{2.52}$$

and write the Hamiltonian as

$$\mathbf{H}(t) = \mathbf{H}_0 + \mathbf{V}(t) = \frac{\hbar}{2}\begin{pmatrix} -\omega_0 & 0 \\ 0 & \omega_0 \end{pmatrix} + \hbar\begin{pmatrix} 0 & |\Omega_0(t)|\cos[\omega t - \phi(t)] \\ |\Omega_0(t)|\cos[\omega t - \phi(t)] & 0 \end{pmatrix}, \tag{2.53}$$

TABLE 2.1
Commonly Used Symbols.

$\mathbf{E}(t) = \frac{1}{2}\hat{\boldsymbol{\epsilon}}\left[E_0(t)e^{-i\omega t} + E_0^*(t)e^{i\omega t}\right]$	Electric field
$E_0(t) = \lvert E_0(t)\rvert\, e^{i\phi(t)}$	Complex electric field amplitude
$\hat{\boldsymbol{\mu}}$	Atomic dipole moment operator
$\Omega_0(t) = -(\hat{\boldsymbol{\mu}})_{21}\cdot\hat{\boldsymbol{\epsilon}}\, E_0(t)/\hbar$	Rabi frequency
$\chi(t) = -(\hat{\boldsymbol{\mu}})_{21}\cdot\hat{\boldsymbol{\epsilon}}\, E_0(t)/2\hbar$	Rabi frequency/2
$\omega_0 = \omega_{21}$	Atomic transition frequency
$\delta = \omega_0 - \omega$	Atom-field detuning

where

$$\Omega_0(t) = -(\mu_z)_{21}E_0(t)/\hbar = \lvert\Omega_0(t)\rvert\, e^{i\varphi(t)} \tag{2.54}$$

is known as the Rabi frequency and is a measure of the atom–field interaction strength in frequency units. The Rabi frequency is defined such that it is positive for positive $E_0(t)$ and z_{12}. Equation (2.26) for the probability amplitude $\mathbf{a}(t)$ can be written as

$$i\hbar\dot{\mathbf{a}}(t) = \frac{\hbar}{2}\begin{pmatrix} -\omega_0 & 2\lvert\Omega_0(t)\rvert\cos\left[\omega t - \phi(t)\right] \\ 2\lvert\Omega_0(t)\rvert\cos\left[\omega t - \phi(t)\right] & \omega_0 \end{pmatrix}\mathbf{a}(t)\,. \tag{2.55}$$

This equation can be solved numerically.

Note that the Hamiltonian can be recast as

$$\mathbf{H}(t) = -\frac{\hbar\omega_0}{2}\boldsymbol{\sigma}_z + \hbar\lvert\Omega_0(t)\rvert\cos\left[\omega t - \phi(t)\right]\boldsymbol{\sigma}_x\,, \tag{2.56}$$

where the Pauli spin matrices are defined as

$$\boldsymbol{\sigma}_x = \begin{pmatrix} 0 & 1 \\ 1 & 0 \end{pmatrix},\quad \boldsymbol{\sigma}_y = \begin{pmatrix} 0 & -i \\ i & 0 \end{pmatrix},\quad \boldsymbol{\sigma}_z = \begin{pmatrix} 1 & 0 \\ 0 & -1 \end{pmatrix}. \tag{2.57}$$

This is the same type of Hamiltonian that one encounters for the interaction of the spin of the electron with a magnetic field, a problem that is considered in appendix B.

Before we move on, it might be useful to list some of the symbols we introduce in this chapter and in chapter 3. You can refer to table 2.1 to remind yourself of the definitions of these commonly used symbols.

2.4 Rotating-Wave or Resonance Approximation

Although equation (2.55) can be solved numerically, it is best to gain some physical insight into this equation before launching into any solutions. You should not be deceived by the apparent simplicity of these coupled equations. There are books devoted to their solution [1, 2], and even numerical solutions can be difficult to obtain in certain limits [2]. Equation (2.55) characterizes the interaction of an optical field with an atom.

Without solving the problem, one can ask under what conditions the field is effective in driving transitions between levels 1 and 2. Let us assume that the amplitude $|\Omega_0(t)|$ and phase $\phi(t)$ of the field are slowly varying on a timescale of order ω^{-1} and that $|(\omega_0 - \omega)/(\omega_0 + \omega)| \ll 1$ and $|\Omega_0(t)/(\omega_0 + \omega)| \ll 1$. In that case, the field can be considered to be quasi-monochromatic and is effective in driving the 1–2 transition, provided that $(\omega_0 - \omega)$ is small compared with $|\Omega_0(t)|$.

The equation for $\dot{\mathrm{a}}(t)$ can be written as

$$i\hbar\dot{\mathrm{a}}(t) = \frac{\hbar}{2}\begin{pmatrix} -\omega_0 & \Omega_0(t)e^{-i\omega t} + \Omega_0^*(t)e^{i\omega t} \\ \Omega_0(t)e^{-i\omega t} + \Omega_0^*(t)e^{i\omega t} & \omega_0 \end{pmatrix}\mathrm{a}(t). \tag{2.58}$$

In the interaction representation, the corresponding equation for $\dot{\mathrm{c}}(t)$ is

$$i\hbar\dot{\mathrm{c}}(t) = \frac{\hbar}{2}\begin{pmatrix} 0 & \Omega_0(t)e^{-i(\omega_0+\omega)t} + \Omega_0^*(t)e^{-i\delta t} \\ \Omega_0(t)e^{i\delta t} + \Omega_0^*(t)e^{i(\omega_0+\omega)t} & 0 \end{pmatrix}\mathrm{c}(t), \tag{2.59}$$

where

$$\mathrm{c}(t) = \begin{pmatrix} c_1 \\ c_2 \end{pmatrix}, \tag{2.60}$$

and

$$\delta = \omega_0 - \omega \tag{2.61}$$

is the atom–field detuning. In the interaction representation, we see that there are terms that oscillate with frequency $\omega_0 + \omega$ and those that oscillate at frequency δ. Moreover, we expect that there can also be oscillation at frequency $\Omega_0(t)$. As long as $|\Omega_0(t)/(\omega_0 + \omega)| \ll 1$, $|\delta/(\omega_0 + \omega)| \ll 1$, the rapidly oscillating terms do not contribute much since they average to zero in a very short time. In other words, if we take a *coarse-grain time average* over a time interval much greater than $1/(\omega_0 + \omega)$, the contribution from these rapidly varying terms would be negligibly small compared with the slowly varying terms. The neglect of such terms is called the rotating-wave approximation (RWA) or resonance approximation. The reason for the nomenclature rotating wave will soon become apparent. In the RWA, equations (2.58) and (2.59) become

$$i\hbar\dot{\mathrm{a}}(t) = \frac{\hbar}{2}\begin{pmatrix} -\omega_0 & \Omega_0^*(t)e^{i\omega t} \\ \Omega_0(t)e^{-i\omega t} & \omega_0 \end{pmatrix}\mathrm{a}(t), \tag{2.62}$$

$$i\hbar\dot{\mathrm{c}}(t) = \frac{\hbar}{2}\begin{pmatrix} 0 & \Omega_0^*(t)e^{-i\delta t} \\ \Omega_0(t)e^{i\delta t} & 0 \end{pmatrix}\mathrm{c}(t). \tag{2.63}$$

At this point, it is useful to remind oneself that small is a relative term and, even more importantly, that just being relatively small does not mean that a term can automatically be neglected. (A "small" dog can still give you a serious bite.) As a simple example, consider $\exp[i(1000 + 1)]$. Even though $1 \ll 1000$, neglecting the second term in the exponential produces totally erroneous results. When parameters appear in *exponents*, their absolute value must be much less than unity before they can be neglected.

With this reminder, let us estimate the corrections to equations (2.62) and (2.63) produced by the rapidly varying (or counterrotating) terms as follows. The amplitude equations in the interaction representation are

$$\dot{c}_1(t) = \left[-i\chi^*(t)e^{-i\delta t} - i\chi(t)e^{-i(\omega_0+\omega)t}\right] c_2(t), \tag{2.64a}$$

$$\dot{c}_2(t) = \left[-i\chi(t)e^{i\delta t} - i\chi^*(t)e^{i(\omega_0+\omega)t}\right] c_1(t), \tag{2.64b}$$

where

$$\chi(t) = \Omega_0(t)/2. \tag{2.65}$$

Formally integrating equation (2.64b) for $c_2(t)$,

$$c_2(t) = c_2(0) - i\int_0^t \left[\chi(t')e^{i\delta t'} + \chi^*(t')e^{i(\omega_0+\omega)t'}\right] c_1(t')\,dt', \tag{2.66}$$

and substituting this result into equation (2.64a), we obtain

$$\begin{aligned}\dot{c}_1(t) &= -i\chi^* e^{-i\delta t}c_2(t) - \chi(t)\int_0^t e^{-i(\omega_0+\omega)(t-t')}\chi^*(t')c_1(t')\,dt' \\ &\quad -\chi(t)e^{-i(\omega_0+\omega)t}\int_0^t e^{i\delta t'}\chi(t')c_1(t')\,dt' - i\chi(t)e^{-i(\omega_0+\omega)t}c_2(0).\end{aligned} \tag{2.67}$$

If $c_{1,2}(t)$ and $\chi(t)$ are slowly varying with respect to $e^{-i(\omega_0+\omega)t}$, then the third and fourth terms are rapidly varying and can be neglected in this order of approximation.[1] Integrating the second term in equation (2.67) by parts, we find

$$\begin{aligned}\int_0^t e^{-i(\omega_0+\omega)(t-t')}\chi^*(t')c_1(t')\,dt' &\approx \frac{\chi^*(t)c_1(t) - e^{-i(\omega_0+\omega)t}\chi^*(0)c_1(0)}{i(\omega_0+\omega)} \\ &\quad - \frac{1}{i(\omega_0+\omega)}\int_0^t e^{-i(\omega_0+\omega)(t-t')}\frac{d}{dt}\left[\chi^*(t')c_1(t')\right]dt'.\end{aligned} \tag{2.68}$$

Neglecting the last term in equation (2.68) as well as the rapidly varying term proportional to $\chi^*(0)c_1(0)$, we obtain

$$\int_0^t e^{-i(\omega_0+\omega)(t-t')}\chi^*(t')c_1(t')\,dt' \approx \frac{\chi^*(t)c_1(t)}{i(\omega_0+\omega)}. \tag{2.69}$$

Equation (2.67) for $\dot{c}_1(t)$ takes the form

$$i\hbar\dot{c}_1(t) = -\hbar\frac{|\chi(t)|^2}{\omega_0+\omega}c_1(t) + \hbar\chi^*(t)e^{-i\delta t}c_2(t)\,. \tag{2.70}$$

Similarly, for $\dot{c}_2(t)$ we find

$$i\hbar\dot{c}_2(t) = \hbar\frac{|\chi(t)|^2}{\omega_0+\omega}c_2(t) + \hbar\chi(t)e^{i\delta t}c_1(t). \tag{2.71}$$

From equations (2.70) and (2.71), we see that energy of level 1 is shifted down by $\hbar|\chi(t)|^2/(\omega_0+\omega)$ and energy of level 2 is shifted up by $\hbar|\chi(t)|^2/(\omega_0+\omega)$. These level shifts are known as *Bloch-Siegert shifts* [4], which tend to be more important

[1] Actually, the third term can contribute a correction of order $|\chi(t)|/\omega$; however, this correction would depend on the phase of the field and would vanish on averaging over a random distribution of field phases—for a good discussion of this problem, see the article by Shirley [3].

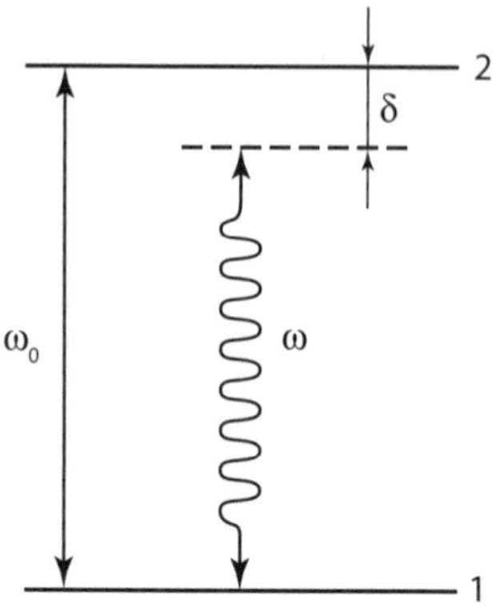

Figure 2.2. Two-level quantum system with applied field frequency ω, transition frequency ω_0, and detuning $\delta = \omega_0 - \omega$.

in magnetic field interactions since the level spacings are much smaller. The shift $\hbar|\chi(t)|^2/(\omega_0 + \omega)$ is the lead term in a power series expansion. There are other shifts (light shifts) that arise from virtual transitions to states outside the two-level subspace that are comparable in magnitude and must be included if the correct value for the shift is to be obtained. If $\chi(t) \approx 10^8\ \mathrm{s}^{-1}$, the shift is of order $1.0\ \mathrm{s}^{-1}$, or about 10^{-15} of an optical frequency. The stability and precision of lasers has reached the point where such resolution is achievable.

If the Bloch-Siegert shifts are neglected, the equations in the RWA become

$$\dot{c}_1(t) = -i\chi^*(t)e^{-i\delta t}c_2(t)\,, \tag{2.72a}$$

$$\dot{c}_2(t) = -i\chi(t)e^{i\delta t}c_1(t)\,. \tag{2.72b}$$

These equations look deceptively simple. For a wide range of parameters, they are easy to solve numerically; however, if the envelope $\chi(t)$ corresponds to a pulse having duration T and if $|\delta|T \gg 1$, the numerical solutions can become extremely challenging. The reason for this is that the transition amplitudes are exponentially small in $|\delta|T$, requiring very small round-off errors, while the step size required for the calculations varies inversely with $|\delta|T$. The effective two-level system is depicted in figure 2.2.

2.4.1 Analytic Solutions

When $\chi(t) = |\chi(t)|e^{i\phi(t)}$ is a function of time, there are very few analytic solutions of equation (2.72), although there are certain combinations of $|\chi(t)|$ and $\phi(t)$ for which the equations can be solved analytically [5–7]. If $\phi(t)$ is constant and $\delta \neq 0$, the only smooth symmetric pulse shape for which an analytic solution is possible is the hyperbolic secant pulse shape [8]. In that case, the amplitudes can be expressed as hypergeometric functions. Analytic solutions are also possible for $\phi(t) = 0$ [$\chi(t)$ real] and $\delta = 0$, or for $\chi(t) =$ constant.

2.4.1.1 $\phi = 0,\ \delta = 0$

In this case, the amplitude equations (2.72) become

$$\dot{c}_1(t) = -i\chi(t)c_2(t)\,, \tag{2.73a}$$

$$\dot{c}_2(t) = -i\chi(t)c_1(t)\,. \tag{2.73b}$$

Making the transformation

$$c_\pm(t) = c_1(t) \pm c_2(t), \tag{2.74}$$

we obtain differential equations of the form

$$\dot{c}_\pm(t) = \mp i\chi(t)c_\pm(t), \tag{2.75}$$

which have solutions

$$c_\pm(t) = e^{\mp i\int_0^t \chi(t')\,dt'} c_\pm(0), \tag{2.76}$$

where $c_\pm(0) = c_1(0) \pm c_2(0)$ are the initial conditions for the probability amplitudes. In terms of $c_1(t)$ and $c_2(t)$, the solution is

$$c_1(t) = \cos[\theta(t)]\,c_1(0) - i\sin[\theta(t)]\,c_2(0)\,, \tag{2.77a}$$

$$c_2(t) = -i\sin[\theta(t)]\,c_1(0) + \cos[\theta(t)]\,c_2(0)\,, \tag{2.77b}$$

where

$$\theta(t) = \int_0^t \chi(t')\,dt'. \tag{2.78}$$

The interpretation of these equations is straightforward. The two levels are degenerate in the interaction representation, and the driving field couples the states. The dynamics does not depend on the details of the pulse at any time, only on the integrated value $\theta(t)$.

In the case of an applied pulse such as a Gaussian which "turns on" at $t = -\infty$, the initial conditions should be taken at $t = -\infty$, in which case the solution becomes

$$c_1(t) = \cos[\theta(t)]\,c_1(-\infty) - i\sin[\theta(t)]\,c_2(-\infty)\,, \tag{2.79a}$$

$$c_2(t) = -i\sin[\theta(t)]\,c_1(-\infty) + \cos[\theta(t)]\,c_2(-\infty)\,, \tag{2.79b}$$

where $\theta(t) = \int_{-\infty}^t \chi(t')\,dt'$. Note that, if at $t = -\infty$, $c_1(-\infty) = 1$ and $c_2(-\infty) = 0$, then

$$|c_2(\infty)|^2 = \sin^2[\theta(\infty)]\,, \tag{2.80}$$

where

$$\theta(\infty) \equiv A/2 = \int_{-\infty}^{\infty} \chi(t')\,dt', \tag{2.81}$$

and A is the referred to as the *pulse area*. The pulse area determines how much population is transferred from the initial to final state. For reasons to be discussed in connection with the Bloch vector in chapter 3, A is defined such that a pulse area of π corresponds to a complete inversion, $|c_1(\infty)| = 0$, $|c_2(\infty)| = 1$; a pulse area of $\pi/2$ results in an equal superposition of ground and excited states, $|c_1(\infty)| = |c_2(\infty)| = 1/\sqrt{2}$.

We have arrived at a fairly interesting result. By controlling the pulse area, one can determine the degree of excitation that is achieved. We will explore applications involving a series of pulses in the context of optical coherent transients. At this point, however, it might be useful to point out that the use of a π pulse for level inversion is not a "robust" method. One must ensure that the pulse intensity is uniform over the entire sample and that the pulse area is exactly equal to π to ensure that all the atoms are inverted.

2.4.1.2 $\chi(t) = \Omega_0/2$ is constant

In this case, equations (2.72) reduce to

$$\dot{c}_1(t) = -\frac{i}{2}\Omega_0^* e^{-i\delta t} c_2(t), \tag{2.82a}$$

$$\dot{c}_2(t) = -\frac{i}{2}\Omega_0 e^{i\delta t} c_1(t). \tag{2.82b}$$

Taking the derivative of equation (2.82b) and using equation (2.82a), we obtain

$$\ddot{c}_2(t) - i\delta\dot{c}_2(t) + \frac{|\Omega_0|^2}{4} c_2(t) = 0. \tag{2.83}$$

The characteristic equation of Eq. (2.83) has roots

$$r_{1,2} = i\frac{\delta \pm \Omega}{2}, \tag{2.84}$$

where

$$\Omega = \sqrt{\delta^2 + |\Omega_0|^2} \tag{2.85}$$

is known as the *generalized Rabi frequency*. It then follows that the solution of equation (2.83) is

$$c_2(t) = e^{i\delta t/2}\left[A\cos\left(\frac{\Omega t}{2}\right) + B\sin\left(\frac{\Omega t}{2}\right)\right]. \tag{2.86}$$

Similarly, we obtain a solution for $c_1(t)$ as

$$c_1(t) = e^{-i\delta t/2}\left[D\cos\left(\frac{\Omega t}{2}\right) + E\sin\left(\frac{\Omega t}{2}\right)\right]. \tag{2.87}$$

Only two of the coefficients A, B, D, and E can be independent since we started with two, first-order coupled differential equations—the constants are related through the differential equations. It is convenient to take A and D as independent since, clearly, $A = c_2(0)$ and $D = c_1(0)$. Using equations (2.82), (2.83), and (2.86), we can write

$$\dot{c}_1(0) = -i\frac{\Omega_0^*}{2}c_2(0) = -i\frac{\Omega_0^*}{2}A = -i\frac{\delta}{2}D + \frac{\Omega}{2}E, \tag{2.88}$$

which gives

$$E = i\frac{\delta}{\Omega}c_1(0) - i\frac{\Omega_0^*}{\Omega}c_2(0). \tag{2.89}$$

Similarly,

$$B = -i\frac{\Omega_0}{\Omega}c_1(0) - i\frac{\delta}{\Omega}c_2(0). \tag{2.90}$$

Last, the solution of equations (2.82) is

$$c_1(t) = e^{-i\delta t/2}\left\{\left[\cos\left(\frac{\Omega t}{2}\right) + i\frac{\delta}{\Omega}\sin\left(\frac{\Omega t}{2}\right)\right]c_1(0) - i\frac{\Omega_0^*}{\Omega}\sin\left(\frac{\Omega t}{2}\right)c_2(0)\right\}, \tag{2.91a}$$

$$c_2(t) = e^{i\delta t/2}\left\{-i\frac{\Omega_0}{\Omega}\sin\left(\frac{\Omega t}{2}\right)c_1(0) + \left[\cos\left(\frac{\Omega t}{2}\right) - i\frac{\delta}{\Omega}\sin\left(\frac{\Omega t}{2}\right)\right]c_2(0)\right\}. \tag{2.91b}$$

This solution is of fundamental importance since it gives the response of a two-level atom to a monochromatic field. Note that the amplitudes depend *nonlinearly* on the applied amplitude—the atom acts as a nonlinear device, in contrast to a harmonic oscillator.

If $c_1(0) = 1, c_2(0) = 0$, then

$$|c_2(t)|^2 = \frac{|\Omega_0|^2}{\Omega^2}\sin^2\left(\frac{\Omega}{2}t\right). \tag{2.92}$$

The population oscillates as a function of time. This is known as *Rabi flopping*. On time average,

$$\overline{|c_2(t)|^2} = \frac{1}{2}\frac{|\Omega_0|^2}{\Omega^2} = \frac{1}{2}\frac{|\Omega_0|^2}{\delta^2 + |\Omega_0|^2}. \tag{2.93}$$

Graphs of $|c_2(t)|^2$ as a function of time and the time-averaged population $\overline{|c_2(t)|^2}$ versus δ are shown in figures 2.3 and 2.4, respectively.

For $|\delta| \gg |\Omega_0|$, $\overline{|c_2(t)|^2} \approx |\Omega_0|^2/(2\delta^2)$. This is a somewhat surprising result. Even though the field is off resonance, the transition probability falls off inversely only as $1/\delta^2$. In fact, we can turn off the field at a time $t = T = \pi/\Omega$ and find that $|c_2(T)|^2 \approx |\Omega_0|^2/\delta^2$. From the energy-time uncertainty principle (which is not a rigorous result, as are uncertainty relations involving noncommuting operators in quantum mechanics), we might expect that the transition probability would vanish at least *exponentially* as $|\delta|T$. (Recall that increasing the light intensity in the photoelectric experiment does not increase the number of photoelectrons if the field frequency is not sufficiently high.) What is going on?

Had the field been turned on and off *smoothly*, one would indeed find that $|c_2(t)|^2 \sim e^{-f(|\delta|T)}$, where f is some positive function. However, in the calculation that was carried out, the field is turned on *instantaneously* at $t = 0$ and turned off *instantaneously* at $t = T$. Owing to the discontinuities in the derivative of the step function, the Fourier components of a step function vary inversely with δT, rather than as an exponentially decaying function of $|\delta|T$.

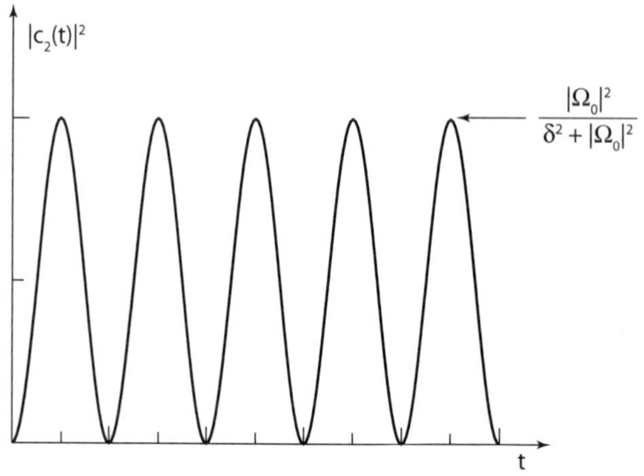

Figure 2.3. The upper state population oscillates as a function of time (Rabi flopping) for a constant field amplitude.

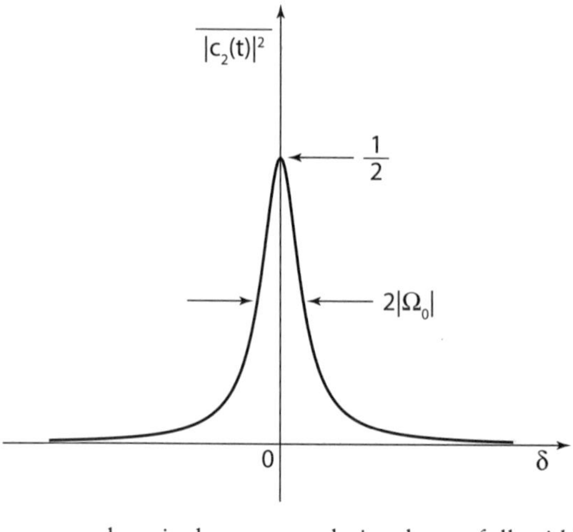

Figure 2.4. The time-averaged excited state population has a full width at half maximum equal to $2\,|\Omega_0|$.

2.5 Field Interaction Representation

There is another representation that is especially useful when a single quasi-monochromatic field drives transitions between two levels or two manifolds of levels. Instead of extracting the atomic frequency, we go into a frame rotating at the laser frequency and write

$$
\begin{aligned}
|\psi(t)\rangle &= \tilde{c}_1(t)e^{i\omega t/2}|1\rangle + \tilde{c}_2(t)e^{-i\omega t/2}|2\rangle \\
&\equiv \tilde{c}_1(t)|\tilde{1}(t)\rangle + \tilde{c}_2(t)|\tilde{2}(t)\rangle, \qquad (2.94)
\end{aligned}
$$

where

$$|\tilde{1}(t)\rangle = e^{i\omega t/2}|1\rangle, \tag{2.95a}$$

$$|\tilde{2}(t)\rangle = e^{-i\omega t/2}|2\rangle \tag{2.95b}$$

are time-dependent kets, as were the standard interaction representation kets. The transformation from the Schrödinger and interaction representation to the field interaction representation can be written using a Pauli matrix as

$$\mathbf{a}(t) = e^{i\omega\sigma_z t/2}\mathbf{c}(t) = e^{i\omega_0\sigma_z t/2}\mathbf{c}(t) \tag{2.96}$$

or

$$\tilde{\mathbf{c}}(t) = e^{i\delta\sigma_z t/2}\mathbf{c}(t)\,. \tag{2.97}$$

It follows from equations (2.97) and (2.64) that

$$\dot{\tilde{c}}_1(t) = i\frac{\delta}{2}\tilde{c}_1(t) - i\chi^*(t)\tilde{c}_2(t) - i\chi(t)e^{-2i\omega t}\tilde{c}_2(t)\,, \tag{2.98a}$$

$$\dot{\tilde{c}}_2(t) = -i\frac{\delta}{2}\tilde{c}_2(t) - i\chi(t)\tilde{c}_1(t) - i\chi^*(t)e^{-2i\omega t}\tilde{c}_1(t)\,. \tag{2.98b}$$

In a frame rotating at the field frequency, there are rapidly varying terms oscillating at twice the field frequency. The neglect of such terms in this rotating frame is the origin of the nomenclature RWA. In the RWA, the equations reduce to

$$\dot{\tilde{c}}_1(t) = i\frac{\delta}{2}\tilde{c}_1(t) - i\chi^*(t)\tilde{c}_2(t)\,, \tag{2.99a}$$

$$\dot{\tilde{c}}_2(t) = -i\frac{\delta}{2}\tilde{c}_2(t) - i\chi(t)\tilde{c}_1(t)\,. \tag{2.99b}$$

We can arrive at these equations in another fashion. The equation for $\dot{\mathbf{a}}(t)$ is

$$i\hbar\dot{\mathbf{a}}(t) = \mathbf{H}(t)\mathbf{a}(t), \tag{2.100}$$

where

$$\mathbf{H}(t) = -\frac{\hbar\omega_0}{2}\boldsymbol{\sigma}_z + \hbar\left[\chi(t)e^{-i\omega t} + \chi(t)^* e^{i\omega t}\right]\boldsymbol{\sigma}_x. \tag{2.101}$$

Taking into account equations (2.96) and (2.97), we obtain

$$i\hbar\left[i\frac{\omega}{2}\boldsymbol{\sigma}_z e^{i\omega\sigma_z t/2}\tilde{\mathbf{c}}(t) + e^{i\omega\sigma_z t/2}\dot{\tilde{\mathbf{c}}}\right] = \mathbf{H}(t)e^{i\omega\sigma_z t/2}\tilde{\mathbf{c}}(t), \tag{2.102}$$

or

$$i\hbar\dot{\tilde{\mathbf{c}}} = \tilde{\mathbf{H}}(t)\tilde{\mathbf{c}}(t)\,, \tag{2.103}$$

where

$$\tilde{\mathbf{H}}(t) = \frac{\hbar\omega}{2}\boldsymbol{\sigma}_z + e^{-i\omega\sigma_z t/2}\mathbf{H}(t)e^{i\omega\sigma_z t/2}\,. \tag{2.104}$$

Using the fact that [9]

$$e^{-i\mathbf{n}\cdot\boldsymbol{\sigma}\theta/2}\left(\boldsymbol{\beta}\cdot\boldsymbol{\sigma}\right)e^{i\mathbf{n}\cdot\boldsymbol{\sigma}\theta/2} = \left(\boldsymbol{\beta}\cdot\mathbf{n}\right)\left(\mathbf{n}\cdot\boldsymbol{\sigma}\right) - \boldsymbol{\beta}\cdot\left[\mathbf{n}\times\left(\mathbf{n}\times\boldsymbol{\sigma}\right)\right]\cos\theta - \boldsymbol{\beta}\cdot\left(\mathbf{n}\times\boldsymbol{\sigma}\right)\sin\theta, \tag{2.105}$$

where $\mathbf{n}$ is a unit vector, $\boldsymbol{\beta}$ is an arbitrary vector, and $\boldsymbol{\sigma} = \boldsymbol{\sigma}_x\hat{\mathbf{x}} + \boldsymbol{\sigma}_y\hat{\mathbf{y}} + \boldsymbol{\sigma}_z\hat{\mathbf{z}}$, one can evaluate the second term in equation (2.104) with $\mathbf{n} = \hat{\mathbf{z}}$, $\theta = \omega t$, and

$$\boldsymbol{\beta} = -\frac{\hbar\omega_0}{2}\hat{\mathbf{z}} + \hbar\left[\chi(t)e^{-i\omega t} + \chi(t)^* e^{i\omega t}\right]\hat{\mathbf{x}}, \tag{2.106}$$

to obtain

$$\begin{aligned} e^{-i\omega\boldsymbol{\sigma}_z t/2}\mathbf{H}(t)e^{i\omega\boldsymbol{\sigma}_z t/2} &= -\frac{\hbar\omega_0}{2}\boldsymbol{\sigma}_z + \hbar\left[\chi(t)e^{-i\omega t} + \chi(t)^* e^{i\omega t}\right] \\ &\quad \times \left(\boldsymbol{\sigma}_x \cos\omega t + \boldsymbol{\sigma}_y \sin\omega t\right). \end{aligned} \tag{2.107}$$

As a result, one finds

$$\tilde{\mathbf{H}}(t) = \frac{\hbar}{2}\begin{pmatrix} -\delta & 2\chi(t)e^{-2i\omega t} + 2\chi(t)^* \\ 2\chi(t) + 2\chi(t)^* e^{2i\omega t} & \delta \end{pmatrix}, \tag{2.108}$$

or, in the RWA,

$$\tilde{\mathbf{H}}(t) = \hbar\begin{pmatrix} -\frac{\delta}{2} & \chi(t)^* \\ \chi(t) & \frac{\delta}{2} \end{pmatrix} = \frac{\hbar}{2}\begin{pmatrix} -\delta & \Omega_0^*(t) \\ \Omega_0(t) & \delta \end{pmatrix}. \tag{2.109}$$

In terms of the Pauli matrices, the RWA Hamiltonian is

$$\tilde{\mathbf{H}}(t) = \frac{\hbar}{2}\left\{-\delta\boldsymbol{\sigma}_z + \mathrm{Re}\left[\Omega_0(t)\right]\boldsymbol{\sigma}_x + \mathrm{Im}\left[\Omega_0(t)\right]\boldsymbol{\sigma}_y\right\}. \tag{2.110}$$

The usefulness of the field interaction representation is apparent when $\Omega_0(t)$ is constant. In that limit, $\tilde{\mathbf{H}}$ is time-independent, and the solution of equation (2.103) is simply

$$\tilde{\mathbf{c}}(t) = e^{-\frac{i}{\hbar}\tilde{\mathbf{H}}t}\tilde{\mathbf{c}}(0), \tag{2.111}$$

where

$$e^{-\frac{i}{\hbar}\tilde{\mathbf{H}}t} = e^{-\frac{i}{2}\left[-\delta\boldsymbol{\sigma}_z + \mathrm{Re}(\Omega_0)\boldsymbol{\sigma}_x + \mathrm{Im}(\Omega_0)\boldsymbol{\sigma}_y\right]t} = e^{-i\frac{\Omega t}{2}\mathbf{n}\cdot\boldsymbol{\sigma}}, \tag{2.112}$$

and

$$\mathbf{n} = \frac{\mathrm{Re}\left(\Omega_0\right)\hat{\mathbf{x}} + \mathrm{Im}\left(\Omega_0\right)\hat{\mathbf{y}} - \delta\hat{\mathbf{z}}}{\Omega}. \tag{2.113}$$

Since [9]

$$e^{-i\frac{\Omega t}{2}\mathbf{n}\cdot\boldsymbol{\sigma}} = \mathbf{I}\cos\frac{\Omega t}{2} - i\mathbf{n}\cdot\boldsymbol{\sigma}\sin\frac{\Omega t}{2}, \tag{2.114}$$

where $\mathbf{I}$ is the identity matrix, one can combine equations (2.111) and (2.112) to obtain

$$\tilde{\mathbf{c}}(t) = \left[\mathbf{I}\cos\left(\frac{\Omega t}{2}\right) - i\frac{\mathrm{Re}\left(\Omega_0\right)\boldsymbol{\sigma}_x + \mathrm{Im}\left(\Omega_0\right)\boldsymbol{\sigma}_y - \delta\boldsymbol{\sigma}_z}{\Omega}\sin\left(\frac{\Omega t}{2}\right)\right]\tilde{\mathbf{c}}(0), \tag{2.115}$$

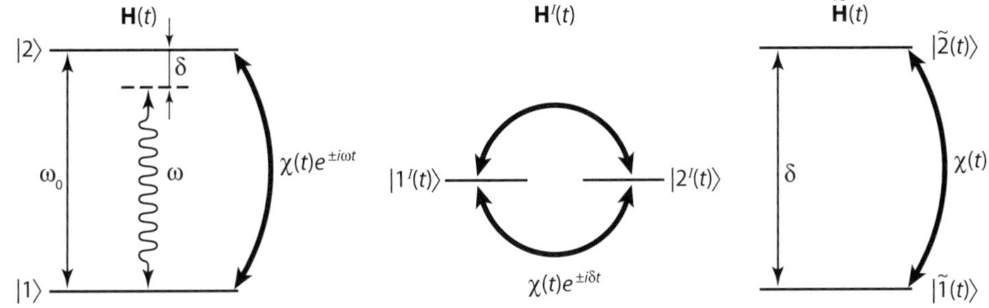

Figure 2.5. A schematic picture of two-level atoms interacting with a radiation field in different representations. The left panel is the Schrödinger representation, the center panel is the interaction representation, and the right panel is the field interaction representation.

or, in matrix form,

$$\tilde{c}(t) = \begin{pmatrix} \cos\left(\frac{\Omega t}{2}\right) + \frac{i\delta}{\Omega}\sin\left(\frac{\Omega t}{2}\right) & -\frac{i\Omega_0^*}{\Omega}\sin\left(\frac{\Omega t}{2}\right) \\ -\frac{i\Omega_0}{\Omega}\sin\left(\frac{\Omega t}{2}\right) & \cos\left(\frac{\Omega t}{2}\right) - \frac{i\delta}{\Omega}\sin\left(\frac{\Omega t}{2}\right) \end{pmatrix} \tilde{c}(0), \tag{2.116}$$

which is consistent with equations (2.91). Recall that $\Omega = [\delta^2 + |\Omega_0|^2]^{1/2}$.

At this juncture, it is useful to take a breath and to compare the Hamiltonians in the various representations within the RWA approximation, (figure 2.5). The Hamiltonian in the Schrödinger representation is

$$\mathbf{H}(t) = \hbar \begin{pmatrix} -\frac{\omega_0}{2} & \chi(t)^* e^{i\omega t} \\ \chi(t)e^{-i\omega t} & \frac{\omega_0}{2} \end{pmatrix}. \tag{2.117}$$

In this representation, the two levels are separated in frequency by ω_0 and are coupled by a field having carrier frequency ω. Resonance occurs when $\omega = \omega_0$. The eigenkets in this case are proper, time-independent eigenkets.

When we move to the interaction representation, the effective Hamiltonian is

$$\mathbf{H}^I(t) = \hbar \begin{pmatrix} 0 & \chi(t)^* e^{-i\delta t} \\ \chi(t)e^{i\delta t} & 0 \end{pmatrix}. \tag{2.118}$$

In the interaction representation, the states are degenerate, and the levels are coupled by an effective field having frequency δ. Resonance occurs for $\delta = 0$. In this case, the eigenkets for the effective Hamiltonian are time-dependent.

Last, in the field interaction representation, one has

$$\tilde{\mathbf{H}}(t) = \hbar \begin{pmatrix} -\frac{\delta}{2} & \chi(t)^* \\ \chi(t) & \frac{\delta}{2} \end{pmatrix}. \tag{2.119}$$

In the field interaction representation, the levels are separated in frequency by δ and are coupled only by the (complex) field envelope. Resonance occurs for $\delta = 0$, and the eigenkets for the effective Hamiltonian are time-dependent.

You might ask which representation is the best to use. The answer is, "It depends." Rarely does one use the Schrödinger representation, except in formal manipulations of the equations. For the two-level problem and arbitrary $\chi(t)$, there is not much difference between the interaction and field interaction representations. Differences between the two representations arise in problems involving more than two levels and more than a single field. Generally speaking, the field interaction representation is most useful when a single field drives transitions between two manifolds of levels, while the interaction representation should be used when fields having two or more frequencies drive transitions between two levels or two manifolds of levels. The examples discussed in the remainder of this text help to amplify these ideas.

2.6 Semiclassical Dressed States

The form of the Hamiltonian in the field interaction representation,

$$\tilde{\mathrm{H}}(t) = \hbar \begin{pmatrix} -\frac{\delta}{2} & \chi(t)^* \\ \chi(t) & \frac{\delta}{2} \end{pmatrix}, \tag{2.120}$$

is the springboard for yet *another* representation. If $\chi(t)$ were constant, one would normally solve the equation $i\hbar\dot{\tilde{\mathbf{c}}} = \tilde{\mathrm{H}}\tilde{\mathbf{c}}(t)$ by diagonalizing $\tilde{\mathrm{H}}$ and obtaining the eigenvalues and eigenstates. Some caution is needed since the original eigenkets in the field interaction basis are time-dependent—as a result, the new eigenkets are time-dependent as well. The new basis is called the *semiclassical dressed-state basis* since the atomic states are said to be "dressed" by the field [10–12]. A true dressed-state basis with time-independent eigenkets is discussed in chapter 15 after we quantize the radiation field. The semiclassical dressed-state representation is particularly useful when the field amplitude and phase vary slowly compared with detunings or Rabi frequencies.

The semiclassical dressed states can be given the simplest interpretation if we redefine the field interaction representation slightly using

$$|\psi(t)\rangle = \tilde{c}_1(t)e^{i[\omega t-\phi(t)]/2}|1\rangle + \tilde{c}_2(t)e^{-i[\omega t-\phi(t)]/2}|2\rangle \tag{2.121}$$

[recall that $\Omega_0(t) = |\Omega_0(t)|e^{i\phi(t)}$], in which the phase of the complex field amplitude is included explicitly. With this definition, everything proceeds as before, except that δ is replaced by

$$\delta(t) = \delta + \dot{\phi}(t), \tag{2.122}$$

and $\Omega_0(t)$ by $|\Omega_0(t)|$. In other words, one obtains

$$i\hbar\dot{\tilde{\mathbf{c}}}(t) = \tilde{\mathrm{H}}(t)\,\tilde{\mathbf{c}}(t)\,, \tag{2.123}$$

where

$$\tilde{\mathrm{H}}(t) = \frac{\hbar}{2}\begin{pmatrix} -\delta(t) & |\Omega_0(t)| \\ |\Omega_0(t)| & \delta(t) \end{pmatrix}. \tag{2.124}$$

In this picture, the state amplitudes are coupled by the field amplitude, and the detuning is a function of time that depends on the field phase. If $\phi(t) = 0$, both field interaction representations are equivalent.

For the most part, we will use the original definition (2.94) of the field interaction representation, perhaps with a *spatial phase* extracted when we consider moving atoms. The second definition (2.121) will be reserved for problems involving adiabatic states and semiclassical dressed states. To distinguish between the two field interaction representations, we add primes to the amplitudes and state vectors appearing in equation (2.94). Admittedly, this can be a bit confusing, but we retain the distinction between the two field interaction representations in this chapter and in chapter 3, since the representation (2.121) is particularly well-suited for a discussion of adiabatic states and the Bloch vector. From chapter 4 onward, we revert to equation (2.94) and drop all the primes.

With this modified definition, equations (2.95) are replaced by

$$|\tilde{1}(t)\rangle = e^{i[\omega t-\phi(t)]/2}|1\rangle\,, \tag{2.125a}$$

$$|\tilde{2}(t)\rangle = e^{-i[\omega t-\phi(t)]/2}|2\rangle\,. \tag{2.125b}$$

We introduce a new basis via

$$\begin{pmatrix} |I(t)\rangle \\ |II(t)\rangle \end{pmatrix} = \mathbf{T}(t)\begin{pmatrix} |\tilde{1}(t)\rangle \\ |\tilde{2}(t)\rangle \end{pmatrix}, \tag{2.126}$$

where $\mathbf{T}(t)$ is an orthogonal, real matrix of the form

$$\mathbf{T}(t) = \begin{pmatrix} \cos\theta(t) & -\sin\theta(t) \\ \sin\theta(t) & \cos\theta(t) \end{pmatrix}. \tag{2.127}$$

The angle $\theta(t)$ is chosen such that $\mathbf{T}(t)$ diagonalizes $\tilde{\mathbf{H}}(t)$ at time t.

Before calculating $\theta(t)$, we first expand the state vector as

$$|\psi(t)\rangle = \sum_{n=1}^{2}\tilde{c}_n(t)|\tilde{n}(t)\rangle = \sum_{i=I,II} c_{di}(t)|i(t)\rangle = \tilde{\mathbf{c}}^T(t)|\tilde{\mathbf{n}}(t)\rangle = \mathbf{c}_d^T(t)|\mathbf{D}(t)\rangle, \tag{2.128}$$

where the superscript ^{T} means transpose,

$$|\tilde{\mathbf{n}}(t)\rangle = \begin{pmatrix} |\tilde{1}(t)\rangle \\ |\tilde{2}(t)\rangle \end{pmatrix}, \tag{2.129}$$

and

$$|\mathbf{D}(t)\rangle = \begin{pmatrix} |I(t)\rangle \\ |II(t)\rangle \end{pmatrix} = \mathbf{T}(t)|\tilde{\mathbf{n}}(t)\rangle\,. \tag{2.130}$$

It follows from equations (2.128) and (2.130) that the field interaction and dressed-state amplitudes are related by

$$\tilde{\mathbf{c}}^T(t) = \mathbf{c}_d^T\mathbf{T}(t)\,, \tag{2.131}$$

or, since $\mathbf{T}^T(t) = \mathbf{T}^\dagger(t)$,

$$\tilde{\mathbf{c}}(t) = \mathbf{T}^\dagger(t)\mathbf{c}_d(t)\,, \qquad \mathbf{c}_d(t) = \mathbf{T}(t)\tilde{\mathbf{c}}(t)\,. \tag{2.132}$$

It then follows from equation (2.123) that

$$i\hbar[\dot{\mathbf{T}}^\dagger(t)\mathbf{c}_d(t) + \mathbf{T}^\dagger(t)\dot{\mathbf{c}}_d] = \tilde{\mathbf{H}}(\mathrm{t})\mathbf{T}^\dagger(t)\mathbf{c}_d(t)\,, \tag{2.133}$$

or

$$i\hbar\dot{\mathbf{c}}_d(t) = \mathbf{T}(t)\tilde{\mathbf{H}}(t)\mathbf{T}^\dagger(t)\mathbf{c}_d(t) - i\hbar\mathbf{T}(t)\dot{\mathbf{T}}^\dagger(t)\mathbf{c}_d(t)\,. \tag{2.134}$$

The matrix $\mathbf{T}(t)$ is chosen to diagonalize $\tilde{\mathbf{H}}(t)$. By demanding that $\mathbf{T}(t)\tilde{\mathbf{H}}(t)\mathbf{T}^\dagger(t)$ is diagonal, one finds that the choice

$$\tan[2\theta(t)] = \frac{|\Omega_0(t)|}{\delta(t)}\,, \tag{2.135a}$$

$$\sin\theta(t) = \sqrt{\frac{1}{2}\left[1 - \frac{\delta(t)}{\Omega(t)}\right]}\,, \tag{2.135b}$$

$$\cos\theta(t) = \sqrt{\frac{1}{2}\left[1 + \frac{\delta(t)}{\Omega(t)}\right]}\,, \tag{2.135c}$$

with

$$\Omega(t) = \sqrt{\delta(t)^2 + |\Omega_0(t)|^2}\,, \tag{2.136}$$

leads to

$$\mathbf{T}(t)\tilde{\mathbf{H}}(t)\mathbf{T}^\dagger(t) = \frac{\hbar}{2}\begin{pmatrix} -\Omega(t) & 0 \\ 0 & \Omega(t) \end{pmatrix}. \tag{2.137}$$

Moreover, since

$$\mathbf{T}(t)\dot{\mathbf{T}}^\dagger(t) = \begin{pmatrix} 0 & 1 \\ -1 & 0 \end{pmatrix}\dot{\theta}(t) = i\boldsymbol{\sigma}_y\dot{\theta}(t)\,, \tag{2.138}$$

equation (2.134) becomes

$$i\hbar\dot{\mathbf{c}}_d(t) = -\frac{\hbar\Omega(t)}{2}\boldsymbol{\sigma}_z\mathbf{c}_d(t) + \hbar\boldsymbol{\sigma}_y\dot{\theta}(t)\mathbf{c}_d(t) = \mathbf{H}_d(t)\mathbf{c}_d(t)\,, \tag{2.139}$$

where

$$\mathbf{H}_d(t) = \hbar\begin{pmatrix} -\frac{\Omega(t)}{2} & -i\dot{\theta}(t) \\ i\dot{\theta}(t) & \frac{\Omega(t)}{2} \end{pmatrix}. \tag{2.140}$$

This Hamiltonian is represented schematically in figure 2.6.

The two states (which are time-dependent) are separated in frequency by $\Omega(t)$ and coupled by a term proportional to $\dot{\theta}(t)$. If $\dot{\theta}(t) = 0$ [that is, if $\Omega(t)$ is constant], there is no coupling, and the dressed states become eigenstates of the effective Hamiltonian. Note that equation (2.139) is rigorously equivalent to equation (2.123); often it is faster numerically to use equation (2.139).

Recall that in the field interaction representation, state $|\tilde{2}(t)\rangle$ lies above state $|\tilde{1}(t)\rangle$ by $\delta(t)$. If $\delta(t)$ is negative, state $|\tilde{2}(t)\rangle$ has a *lower* energy than state $|\tilde{1}(t)\rangle$. By convention, we define the semiclassical dressed states such that state $|II(t)\rangle$ *always* has a greater energy than state $|I(t)\rangle$—the energy separation between the two levels

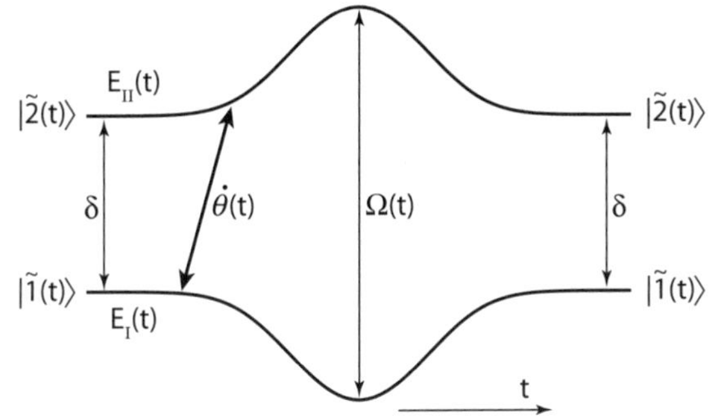

Figure 2.6. Semiclassical dressed energy levels as a function of time for a pulsed field. The frequency separation of the levels is $\Omega(t)$ and the coupling of the levels is $\dot{\theta}(t)$, in frequency units.

is $\hbar\Omega(t) = \hbar\left[\delta^2(t) + |\Omega_0(t)|^2\right]^{1/2} > 0$. From equation (2.126), it follows that

$$|I(t)\rangle = \cos\theta(t)|\tilde{1}(t)\rangle - \sin\theta(t)|\tilde{2}(t)\rangle\,, \tag{2.141a}$$

$$|II(t)\rangle = \sin\theta(t)|\tilde{1}(t)\rangle + \cos\theta(t)|\tilde{2}(t)\rangle\,. \tag{2.141b}$$

As a consequence, we must restrict $\theta(t)$ to the range with $0 \leqslant \theta(t) \leqslant \pi/2$. In this way, the lower energy state $|I(t)\rangle \sim |\tilde{1}(t)\rangle$ as $\Omega_0 \to 0$ for $\delta(t) > 0$ and $|I(t)\rangle \sim |\tilde{2}(t)\rangle$ as $\Omega_0 \to 0$ for $\delta(t) < 0$, in agreement with the field interaction representation.

If $\dot{\theta}(t) = 0$, we should recover the results for a constant field. In this limit, the solution of equation (2.139) reduces to

$$c_{d_1}(t) = e^{i\Omega t/2}c_{d_1}(0)\,, \tag{2.142a}$$

$$c_{d_2}(t) = e^{-i\Omega t/2}c_{d_2}(0)\,, \tag{2.142b}$$

or

$$\mathbf{c}_d(t) = e^{i\frac{\Omega}{2}\sigma_z t}\mathbf{c}_d(0)\,. \tag{2.143}$$

Using equation (2.132), one can show that this result is consistent with equations (2.91).

2.6.1 Adiabatic Following

There are situations in which the (complex) field envelope varies slowly compared with other timescales in the problem. In this limit, it is possible to use the semiclassical dressed-state representation to get an approximate solution that agrees very well with the exact solution. It is almost magical. To understand the conditions under which this *adiabatic approximation* holds, one can make reference to figure 2.6. Since the levels are separated in frequency by $\Omega(t)$, and since the coupling between the levels varies as $\dot{\theta}(t)$, the coupling is not effective unless $\dot{\theta}(t)$ has Fourier components near $\Omega(t)$. Note that $\dot{\theta}(t)$ may be obtained from equation (2.135a) as

$$\dot{\theta}(t) = \frac{\delta(t)|\dot{\Omega}_0| - |\Omega_0(t)|\dot{\delta}(t)}{2\Omega^2(t)} . \tag{2.144}$$

If both $|\dot{\Omega}_0(t)|/\Omega^2(t) \ll 1$ and $|\dot{\delta}(t)|/\Omega^2(t) \ll 1$, then

$$\left|\dot{\theta}(t)/\Omega(t)\right| \ll 1. \tag{2.145}$$

When $|\Omega_0(t)|$ and $\delta(t)$ vary smoothly over some interval T, the $\dot{\theta}(t)$ term is small, and adiabaticity holds provided that $\Omega_{\text{max}}(t)T > 1$. If the $\dot{\theta}(t)$ term in equation (2.139) can be neglected, then the solution of that equation is

$$c_{d_1}(t) = e^{i\int_{t_0}^{t} \Omega(t')\,dt'/2} c_{d_1}(t_0) , \tag{2.146a}$$

$$c_{d_2}(t) = e^{-i\int_{t_0}^{t} \Omega(t')\,dt'/2} c_{d_2}(t_0) , \tag{2.146b}$$

and the dressed-state *populations* are unchanged. In this limit, the dressed states are approximate eigenstates of $\mathbf{H}_d(t)$.

Some examples can help to illustrate this idea. If the atom starts in state $|1\rangle$ and a pulse having a smooth envelope function $|\Omega_0(t)|$ is applied, with $\delta(t) = \delta =$ constant, then the atom stays in dressed state $|I(t)\rangle$ provided that condition (2.145) holds. Since

$$|I(t)\rangle = \cos\theta(t)|\tilde{1}(t)\rangle - \sin\theta(t)|\tilde{2}(t)\rangle , \tag{2.147}$$

it follows that

$$|\tilde{c}_2(t)|^2 = \sin^2\theta(t) = [1 - \delta/\Omega(t)]/2. \tag{2.148}$$

As $t \sim \infty$, $\sin[\theta(t = \infty)] \sim 0$, which means that the atom is back in state $|1\rangle$. A comparison of this approximate solution (2.148) with the exact numerical solution for a Gaussian pulse $\Omega_0(t) = \Omega_0 e^{-t^2/T^2}$ is shown in figure 2.7. For $\delta T = \Omega_0 T = 10$, the adiabatic and exact solutions very nearly coincide.

On other hand, imagine that, for the same pulse envelope, we sweep $\delta(t)$ from $+\delta_0$ to $-\delta_0$ ($\delta_0 > 0$). Since $\delta(t = -\infty) = \delta_0 > 0$, the atom starts in state $|I(t)\rangle$. Owing to adiabaticity, the atom remains in state $|I(t)\rangle$ given by equation (2.147), and the approximate solution for $|\tilde{c}_2(t)|^2$ is still given by equation (2.148), with $\delta(t)$ replacing δ; however, the fact that $\delta(t = \infty) = -\delta_0 < 0$ implies that $\sin[\theta(t = \infty)] \sim 1$ and $\cos[\theta(t = \infty)] \sim 0$. At $t = \infty$, the atom is now in state $|2\rangle$! This is *adiabatic switching* and is illustrated in figure 2.8. Adiabatic switching is a good method for totally transferring population from one state to another. It is more "robust" than using a π-pulse, since it relies solely on adiabaticity; the exact strength of the field amplitude is not critical. A comparison of this approximate solution with the exact solution for a chirped (frequency-swept) Gaussian pulse is shown in figure 2.9.

A word of caution is in order when using adiabatic states. Consider the first case when the atom ended up back in state $|1\rangle$ at $t = +\infty$. This cannot be totally correct. From first-order perturbation theory, we know that

$$\tilde{c}_2(\infty) \sim -i\int_{-\infty}^{\infty} \chi(t')e^{i\delta t'}dt' , \tag{2.149}$$

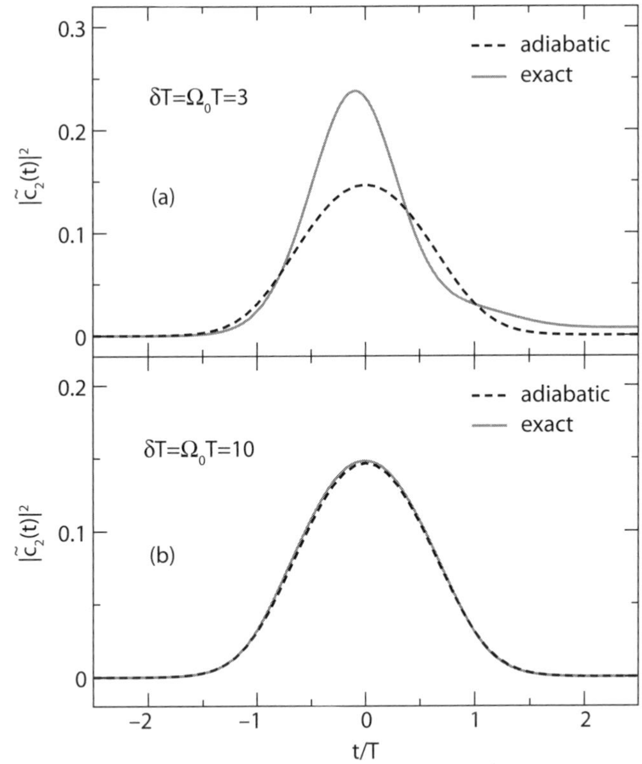

Figure 2.7. Comparison between exact and adiabatic solutions for a Gaussian pulse envelope $\Omega(t) = \Omega_0 e^{-t^2/T^2}$ and constant detuning δ. (a) $\delta T = \Omega_0 T = 3$; (b) $\delta T = \Omega_0 T = 10$.

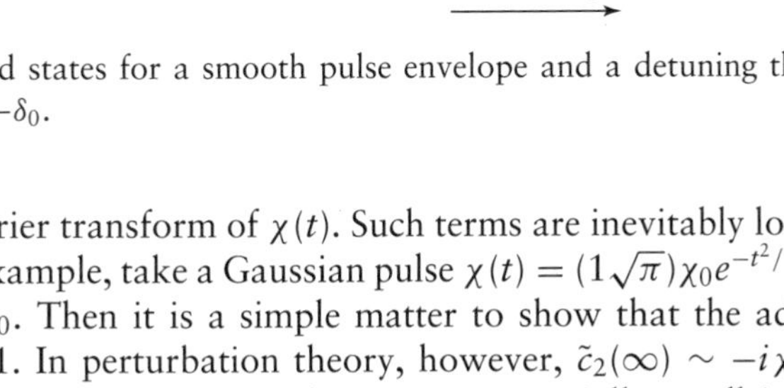

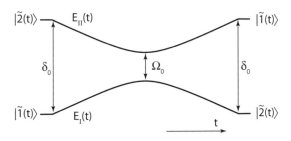

Figure 2.8. Dressed states for a smooth pulse envelope and a detuning that is adiabatically varied from δ_0 to $-\delta_0$.

which is the Fourier transform of $\chi(t)$. Such terms are inevitably lost in an adiabatic treatment. For example, take a Gaussian pulse $\chi(t) = (1\sqrt{\pi})\chi_0 e^{-t^2/T^2}$ and a constant detuning $\delta \gg \chi_0$. Then it is a simple matter to show that the adiabatic theory is valid for $\delta T > 1$. In perturbation theory, however, $\tilde{c}_2(\infty) \sim -i\chi_0 T e^{-\delta^2 T^2/4} \ll 1$, but is not identically zero. Terms that are exponentially small in the adiabaticity parameter are lost when using the adiabatic approximation, as in any asymptotic series.

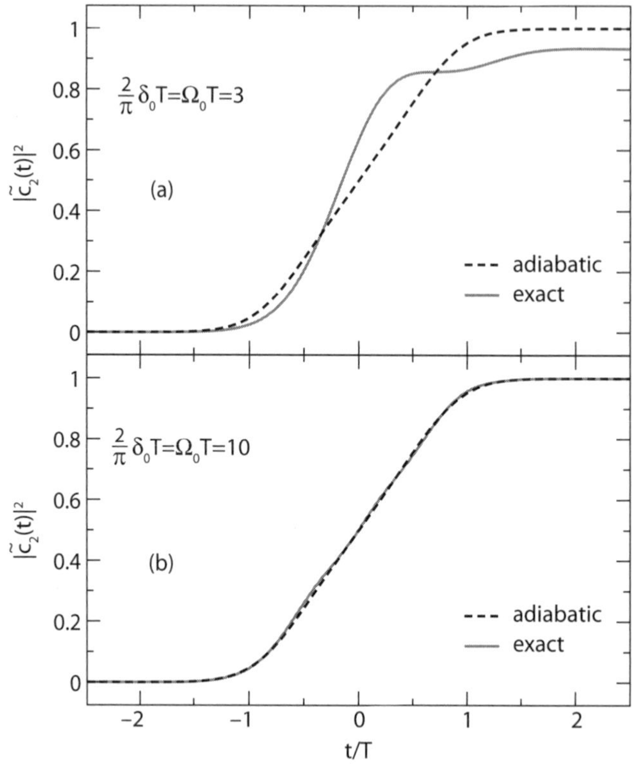

Figure 2.9. Comparison between exact and adiabatic solutions for a chirped Gaussian pulse having $\Omega(t) = \Omega_0 e^{-t^2/T^2}$ and $\delta(t) = -\frac{2}{\pi}\delta_0 \tan^{-1}(t/T)$. (a) $\frac{2}{\pi}\delta_0 T = \Omega_0 T = 3$; (b) $\frac{2}{\pi}\delta_0 T = \Omega_0 T = 10$.

In the case of adiabatic switching using a chirped pulse, one can estimate the exponentially small, nonadiabatic corrections to the excited state population using the Landau-Zener formula [13, 14],

$$|c_2(\infty)|^2 = 1 - \exp\left[-\pi|\Omega_0|^2/2|\dot{\delta}(0)|\right]. \tag{2.150}$$

Equation (2.150) is exact for a constant Rabi frequency and a linearly chirped pulse frequency extending over an infinite time interval. For pulsed fields, it is approximately correct provided that the pulse duration T is larger than the transition time $|\Omega_0|/|\dot{\delta}(0)|$ [15, 16]. As such, equation (2.150) correctly illustrates the importance of the adiabaticity parameter $|\Omega_0|^2/|\dot{\delta}(0)|$ [see equations (2.144) and (2.145)] in determining the final state population.

2.7 General Remarks on Solution of the Matrix Equation $\dot{\mathbf{y}}(t) = \mathbf{A}(t)\mathbf{y}(t)$

We have now looked at a number of ways of solving the two-state problem. It may be of some use to put the solutions in context by considering the general matrix equation

$$\dot{\mathbf{y}}(t) = -i\mathbf{A}(t)\mathbf{y}(t)\,. \tag{2.151}$$

For any finite matrix, this equation can be solved numerically. If $\mathbf{A}(t)$ is Hermitian, $|\mathbf{y}(t)|^2$ is constant. What approximation techniques can be used to solve equation (2.151)?

2.7.1 Perturbation Theory

To first order, a perturbative solution of equation (2.151) is

$$\mathbf{y}(t) \approx \mathbf{y}(0) - i\int_0^t \mathbf{A}(t')\,dt'\mathbf{y}(0), \tag{2.152}$$

valid only if $|\int_0^t \mathbf{A}(t')\,dt'| \ll 1$.

2.7.2 Adiabatic Approximation

We set

$$\mathbf{y}(t) = \mathbf{T}(t)\mathbf{x}(t), \tag{2.153}$$

where $\mathbf{T}^\dagger(t)\mathbf{A}(t)\mathbf{T}(t) = \boldsymbol{\Lambda}(t)$ is diagonal. Then

$$\dot{\mathbf{x}}(t) = -i\boldsymbol{\Lambda}(t)\mathbf{x}(t) - \mathbf{T}^\dagger(t)\dot{\mathbf{T}}(\mathbf{t})\mathbf{x}(t)\,. \tag{2.154}$$

If matrix elements of $\mathbf{T}^\dagger(t)\dot{\mathbf{T}}(t)$ are much less than the eigenvalue spacing of $\Lambda(t)$, then one can neglect the $\mathbf{T}^\dagger(t)\dot{\mathbf{T}}(t)$ term or treat it in perturbation theory. Note that

$$\Lambda_{1,2}(t) = \pm\sqrt{\delta(t)^2 + |\Omega_0(t)|^2}/2 = \pm\Omega(t)/2$$

in the two-level problem. This adiabatic approach is useful only if the $\mathbf{T}^\dagger(t)\dot{\mathbf{T}}(t)$ term can be neglected. Taking first-order perturbation theory in $\mathbf{T}^\dagger(t)\dot{\mathbf{T}}(t)$ can be more time consuming (numerically) than directly integrating the differential equations.

2.7.3 Magnus Approximation

For a single variable, the differential equation

$$\dot{x} = -if(t)x \tag{2.155}$$

has solution

$$x(t) = e^{-i\int_0^t f(t')dt'}x(0). \tag{2.156}$$

One might think that the analogous vector solution of equation (2.151) is

$$\mathbf{y}^{(1)}(t) = e^{-i\int_0^t \mathbf{A}(t')dt'}\mathbf{y}(0). \tag{2.157}$$

This expression is called the first Magnus approximation [17]. We can check whether it works. Solving equation (2.151) iteratively gives

$$\mathbf{y}(t) = \left[1 - i\int_0^t \mathbf{A}(t')dt' - \int_0^t dt' \int_0^{t'} dt''\mathbf{A}(t')\mathbf{A}(t'') + \cdots\right]\mathbf{y}(0), \qquad (2.158)$$

while the first Magnus approximation is

$$\mathbf{y}^{(1)}(t) = e^{-i\int_0^t \mathbf{A}(t')dt'}\mathbf{y}(0) = \left\{1 - i\int_0^t \mathbf{A}(t')dt' - \frac{1}{2}\left[\int_0^t \mathbf{A}(t')dt'\right]^2 + \cdots\right\}\mathbf{y}(0). \qquad (2.159)$$

The two solutions agree to first order in **A**. Let us compare the second-order terms by evaluating the third term in equation (2.159) as

$$\begin{aligned}
\frac{1}{2}\left[\int_0^t \mathbf{A}(t')dt'\right]^2 &= \frac{1}{2}\int_0^t dt' \int_0^t dt''\mathbf{A}(t')\mathbf{A}(t'') \\
&= \frac{1}{2}\int_0^t dt'\mathbf{A}(t')\left[\int_0^{t'} dt''\mathbf{A}(t'') + \int_{t'}^t dt''\mathbf{A}(t'')\right] \\
&= \frac{1}{2}\int_0^t dt' \int_0^{t'} dt''\mathbf{A}(t')\mathbf{A}(t'') + \frac{1}{2}\int_0^t dt'' \int_0^{t''} dt'\mathbf{A}(t')\mathbf{A}(t'') \\
&= \frac{1}{2}\int_0^t dt' \int_0^{t'} dt''\mathbf{A}(t')\mathbf{A}(t'') + \frac{1}{2}\int_0^t dt' \int_0^{t'} dt''\mathbf{A}(t'')\mathbf{A}(t'), \qquad (2.160)
\end{aligned}$$

where the order of integration was interchanged in going from line 2 to line 3. Equation (2.160) would agree with the third term in equation (2.158) *if* $\mathbf{A}(t'')\mathbf{A}(t') = \mathbf{A}(t')\mathbf{A}(t'')$, but $\mathbf{A}(t'')\mathbf{A}(t') \neq \mathbf{A}(t')\mathbf{A}(t'')$, in general. The difference between the two terms can be obtained by writing $\mathbf{A}(t'')\mathbf{A}(t') = \mathbf{A}(t')\mathbf{A}(t'') + [\mathbf{A}(t''), \mathbf{A}(t')]$. When this is substituted into equation (2.160), we find

$$\begin{aligned}
\frac{1}{2}\left[\int_0^t \mathbf{A}(t')dt'\right]^2 &= \int_0^t dt' \int_0^{t'} dt''\mathbf{A}(t')\mathbf{A}(t'') \\
&\quad + \frac{1}{2}\int_0^t dt' \int_0^{t'} dt''[\mathbf{A}(t''), \mathbf{A}(t')]. \qquad (2.161)
\end{aligned}$$

Thus, there is a correction related to the unequal time commutator, and the solution (2.159) is *not* correct. Instead, we can try a solution of the form

$$\mathbf{y}^{(2)}(t) = e^{-i\int_0^t \mathbf{A}(t')dt' - \frac{1}{2!}\int_0^t dt' \int_0^{t'} dt''[\mathbf{A}(t''),\mathbf{A}(t')]}\mathbf{y}(0), \qquad (2.162)$$

which is guaranteed to be correct to second order in **A**. The solution (2.162) is the known as the second Magnus approximation.

We can continue the process. The next correction is

$$\frac{(-i)^3}{3!}\int_0^t dt' \int_0^{t'} dt'' \int_0^{t''} dt''' \left\{\left[\mathbf{A}(t'''), \left[\mathbf{A}(t''), \mathbf{A}(t')\right]\right] + \left[\mathbf{A}(t'''), \mathbf{A}(t'')\right], \mathbf{A}(t')\right]\right\},$$

and this can be added to the exponential to produce a result that is correct to third order in **A**. In practice, many people use the first Magnus approximation with no

TABLE 2.2
Different Representations

$\lvert\psi(t)\rangle$	Representation
$a_1(t)\lvert 1\rangle + a_2(t)\lvert 2\rangle$	Schrödinger
$c_1(t)e^{i\omega_0 t/2}\lvert 1\rangle + c_2(t)e^{-i\omega_0 t/2}\lvert 2\rangle$	Interaction
$\tilde{c}_1'(t)e^{i\omega t/2}\lvert 1\rangle + \tilde{c}_2'(t)e^{-i\omega t/2}\lvert 2\rangle$	Field interaction 1
$\tilde{c}_1(t)e^{i\omega t/2-i\phi/2}\lvert 1\rangle + \tilde{c}_2(t)e^{-i\omega t/2+i\phi/2}\lvert 2\rangle$	Field interaction 2
$c_{d_1}(t)\lvert I(t)\rangle + c_{d_2}(t)\lvert II(t)\rangle$	Semiclassical dressed state

idea of the errors introduced. It has the advantage that if $\mathbf{A}(t)$ is Hermitian, then the solution maintains unitarity. As mentioned earlier, the disadvantage is that it is difficult to estimate the errors.

2.8 Summary

We have introduced several representations for solving the problem of the interaction of a classical optical field with a "two-level" atom. The various representations and their interconnections are summarized in appendix A for easy reference. As you become more familiar with the material, you will be able to choose the most convenient representation with little difficulty. In very rough terms, however, the following prescription might prove helpful: (1) a single quasi-monochromatic field drives transitions between two levels or two manifolds of levels—use the field interaction representation; (2) *two or more* quasi-monochromatic fields having different frequencies drive transitions between two levels or two manifolds of levels—use the interaction representation; and (3) pulsed fields whose envelopes and phases vary slowly compared to their average Rabi frequencies or atom–field detunings drive transitions between two levels—use the semiclassical dressed-state representation.

Armed with these representations and your understanding of atom–field dynamics, you are now ready to attack the important problem of extracting the physically relevant quantities from the solutions we have obtained so far. Moreover, by defining the *density matrix* associated with the two-level atom, you will see how one can introduce irreversible decay processes into the atom–field dynamics.

2.9 Appendix A: Representations

The representations are summarized in table 2.2. Here we use primes on the $\tilde{c}$'s to distinguish the two field interaction representations.

All of the following equations are written in the rotating-wave approximation (RWA), with

$$\mathbf{E}(t) = \frac{\hat{\mathbf{z}}}{2}[E_0(t)e^{-i\omega t} + E_0^*(t)e^{i\omega t}] = \frac{\hat{\mathbf{z}}}{2}[\lvert E_0(t)\rvert\, e^{-i\omega t+i\phi(t)} + \lvert E_0(t)\rvert\, e^{i\omega t-i\phi(t)}], \quad (2.163)$$

and

$$\hat{V}(t) = \frac{-\hat{\mu}_z}{2}[|E_0(t)|\, e^{-i\omega t+i\phi(t)} + |E_0(t)|\, e^{i\omega t-i\phi(t)}]. \tag{2.164}$$

In the Schrödinger representation,

$$i\hbar\dot{\mathbf{a}}(t) = \mathbf{H}(t)\mathbf{a}(t),$$
$$\mathbf{H}(t) = (\hbar/2)\begin{pmatrix} -\omega_0 & \Omega_0(t)^* e^{i\omega t} \\ \Omega_0(t)e^{-i\omega t} & \omega_0 \end{pmatrix}, \tag{2.165}$$

where

$$\Omega_0(t) = -\langle 2|\mu_z|1\rangle E_0(t)/\hbar = 2\chi(t).$$

In the interaction representation,

$$i\hbar d\mathbf{c}(t)/dt = \mathbf{V}^I(t)\mathbf{c}(t), \tag{2.166}$$

$$\mathbf{V}^I(t) = (\hbar/2)\begin{pmatrix} 0 & \Omega_0(t)^* e^{-i\delta t} \\ \Omega_0(t)e^{i\delta t} & 0 \end{pmatrix}, \tag{2.167}$$

where $\delta = \omega_0 - \omega$.

In the field interaction representation,

$$i\hbar d\tilde{\mathbf{c}}'(t)/dt = \tilde{\mathbf{H}}'(t)\,\tilde{\mathbf{c}}'(t), \qquad \tilde{\mathbf{H}}'(t) = (\hbar/2)\begin{pmatrix} -\delta & \Omega_0(t)^* \\ \Omega_0(t) & \delta \end{pmatrix}, \tag{2.168}$$

or

$$i\hbar d\tilde{\mathbf{c}}(t)/dt = \tilde{\mathbf{H}}(t)\tilde{\mathbf{c}}(t), \qquad \tilde{\mathbf{H}}(t) = (\hbar/2)\begin{pmatrix} -\delta(t) & |\Omega_0(t)| \\ |\Omega_0(t)| & \delta(t) \end{pmatrix}, \tag{2.169}$$

where

$$\delta(t) = \omega_0 - \omega + d\phi/dt.$$

In the semiclassical dressed-state representation,

$$i\hbar d\mathbf{c}_d(t)/dt = \mathbf{H}_d(t)\mathbf{c}_d(t), \qquad \mathbf{H}_d(t) = \hbar\begin{pmatrix} -\frac{\Omega(t)}{2} & -i\dot{\theta}(t) \\ i\dot{\theta}(t) & \frac{\Omega(t)}{2} \end{pmatrix}, \tag{2.170}$$

where $\Omega(t) = \left[|\Omega_0(t)|^2 + \delta(t)^2\right]^{1/2}$ and $\tan[2\theta(t)] = |\Omega_0(t)|\,/\delta(t)$.

2.9.1 Relationships between the Representations

Let $\mathrm{U}_0(t) = \exp(-i\mathrm{H}_0 t/\hbar)$, where $\mathrm{H}(t) = \mathrm{H}_0 + \mathrm{V}(t)$. Then

$$\mathbf{c}(t) = \mathbf{U}_0^\dagger(t)\mathbf{a}(t) = e^{i\omega_0 t\sigma_z/2}\mathbf{a}(t), \qquad \mathbf{V}^I(t) = \mathbf{U}_0^\dagger(t)\mathbf{V}(t)\mathbf{U}_0(t), \tag{2.171}$$

and

$$\tilde{\mathrm{c}}'(t) = e^{-i\omega t\sigma_z/2}\,\mathrm{a}(t) = e^{i\delta t\sigma_z/2}\mathrm{c}(t)\,, \tag{2.172a}$$

$$\tilde{\mathrm{H}}'(t) = (\hbar\omega/2)\boldsymbol{\sigma}_z + e^{-i\omega t\sigma_z/2}\mathrm{H}(t)e^{i\omega t\sigma_z/2}\,, \tag{2.172b}$$

$$\tilde{\mathrm{c}}(t) = e^{(-i\omega t/2+i\phi/2)\sigma_z}\mathrm{a}(t) = e^{(i\delta t/2+i\phi/2)\sigma_z}\mathrm{c}(t)\,, \tag{2.172c}$$

$$\tilde{\mathrm{H}}(t) = \hbar(\omega/2 - \dot{\phi}/2)\boldsymbol{\sigma}_z + e^{(-i\omega t/2+i\phi/2)\sigma_z}\mathrm{H}(t)e^{(i\omega t/2-i\phi/2)\sigma_z}\,, \tag{2.172d}$$

$$\mathrm{c}_d(t) = \mathrm{T}(t)\tilde{\mathrm{c}}(t), \quad \mathrm{T}(t) = \begin{pmatrix} \cos\theta(t) & -\sin\theta(t) \\ \sin\theta(t) & \cos\theta(t) \end{pmatrix}, \tag{2.172e}$$

$$\begin{pmatrix} |I(t)\rangle \\ |II(t)\rangle \end{pmatrix} = \mathrm{T}(t)\begin{pmatrix} e^{i[\omega t-\phi(t)]/2}|1\rangle \\ e^{-i[\omega t-\phi(t)]/2}|2\rangle \end{pmatrix}, \tag{2.172f}$$

$$\mathrm{H}_d(t) = \mathrm{T}(t)\tilde{\mathrm{H}}(t)\mathrm{T}^\dagger(t) - i\hbar\mathrm{T}(t)\dot{\mathrm{T}}^\dagger(t). \tag{2.172g}$$

2.10 Appendix B: Spin Half Quantum System in a Magnetic Field

The Hamiltonian that characterizes the interaction of a magnetic field,

$$\mathbf{B}(t) = B_0\hat{\mathbf{z}} + |B_x(t)|\cos[\omega t - \phi(t)]\,\hat{\mathbf{x}}, \tag{2.173}$$

with the spin magnetic moment of an electron is

$$\mathrm{H}_B = -\boldsymbol{\mu}_{\mathrm{mag}}\cdot\mathbf{B}\,, \tag{2.174}$$

where

$$\boldsymbol{\mu}_{\mathrm{mag}} = -\frac{e\mathrm{S}}{m} = -\frac{e\hbar}{2m}\boldsymbol{\sigma} = -\mu_B\boldsymbol{\sigma} \tag{2.175}$$

is the magnetic moment operator of the electron, m is the electron mass,

$$\mu_B = \frac{e\hbar}{2m} = 9.2740154\times 10^{-24} J\,T^{-1} \tag{2.176}$$

is the Bohr magneton, $\mathrm{S} = \hbar\boldsymbol{\sigma}/2$ is the spin operator, and $\boldsymbol{\sigma} = \boldsymbol{\sigma}_x\hat{\mathbf{x}} + \boldsymbol{\sigma}_y\hat{\mathbf{y}} + \boldsymbol{\sigma}_z\hat{\mathbf{z}}$.

In terms of the Pauli matrices, the Hamiltonian can be written

$$\mathrm{H}_B(t) = \mu_B B_0\boldsymbol{\sigma}_z + \mu_B|B_x(t)|\cos[\omega t - \phi(t)]\,\boldsymbol{\sigma}_x\,. \tag{2.177}$$

The connection with equation (2.56) is now apparent if we identify the level spacing as

$$\omega_0 = \frac{2\mu_B B_0}{\hbar} = \frac{e}{m}B_0 = 1.76\times 10^{11}\,B_0(\text{Tesla})\,s^{-1} = 1.76\times 10^{7}\,B_0(\text{Gauss})\,s^{-1}\,. \tag{2.178}$$

Note that the sign of the lead terms in the Hamiltonians (2.56) and (2.177) differ, since for the optical case, we have chosen the basis $\mathbf{a} = (a_1, a_2)$, while for the magnetic case, the standard convention for the Pauli matrices requires that we take

$\mathbf{a}_B = (a_\uparrow, a_\downarrow)$, where the up (down) arrow refers to the state having spin projection number 1/2 (−1/2). In frequency units, $\omega_0/2\pi = 28$ GHz/T.

Although the equations for the two-level atom interacting with an electric field and the spin system interacting with a magnetic field look the same, the physical values of the parameters for the two systems differ markedly. Thus, the frequency separation of the spin up and spin down states in a constant magnetic field can range between 0 Hz and 10 GHz, while radio-frequency (rf) coupling strengths $[\mu_B |B_x(t)| / h]$ are typically less than 1.0 MHz. In the optical case, electronic transition frequencies are of order 10^{14} to 10^{16} Hz, and coupling strengths vary, but are typically much less than the frequency separation of the levels. Only for intense pulses ($>10^{17}$ W/cm^2) can the coupling strength be comparable to the optical frequency separations. For a typical cw laser having a few mW of power, coupling strengths are typically in the MHz to GHz range. Given this qualitative difference in the magnetic and electric cases, it is not surprising that different approximation schemes are used in the two cases. We have seen that the RWA approximation is usually a good approximation for atom–optical field interactions, but this is not necessarily so for the magnetic case.

In the interaction representation, the equations for the state amplitudes are

$$i\hbar\dot{\mathbf{c}}_B(t) = \hbar \begin{pmatrix} 0 & \omega_x(t)\cos[\omega t - \phi(t)]e^{i\omega_0 t} \\ \omega_x(t)\cos[\omega t - \phi(t)]e^{-i\omega_0 t} & 0 \end{pmatrix} \mathbf{c}_B(t), \quad (2.179)$$

with $\mathbf{c}_B = (c_\uparrow, c_\downarrow)$, $\omega_0 = \omega_{\uparrow\downarrow} = \omega_\uparrow - \omega_\downarrow$, and

$$\omega_x(t) = \frac{\mu_B |B_x(t)|}{\hbar}. \quad (2.180)$$

All the solutions that were obtained in the RWA can be taken over to the magnetic case, so we concentrate on situations where the RWA is *not* applicable.

2.10.1 Analytic Solutions—Magnetic Case

2.10.1.1 $\omega_0 = 0$

If there is no longitudinal field [$B_z = 0$], the energy levels are degenerate, and the states are coupled by the field oscillating in the x direction. Equations for probability amplitudes have the form

$$\dot{c}_\uparrow(t) = -if(t)c_\downarrow(t), \quad (2.181a)$$

$$\dot{c}_\downarrow(t) = -if(t)c_\uparrow(t), \quad (2.181b)$$

where

$$f(t) = \omega_x(t)\cos[\omega t - \phi(t)]. \quad (2.182)$$

In analogy with equation (2.79), we find a solution

$$c_\uparrow(t) = \cos[\theta(t)]\, c_\uparrow(0) - i\sin[\theta(t)]\, c_\downarrow(0), \quad (2.183a)$$

$$c_\downarrow(t) = -i\sin[\theta(t)]\, c_\uparrow(0) + \cos[\theta(t)]\, c_\downarrow(0), \quad (2.183b)$$

where

$$\theta(t) = \int_0^t f(t')\, dt. \tag{2.184}$$

The interpretation of these equations is straightforward. The two levels are degenerate, and the driving field couples the states.

Although this is a simple solution, it can be used to illustrate some interesting physical concepts. If we take $\phi(t) = 0$ and $\omega_x(t) = \omega_x =$ constant, then

$$\int_0^t f(t')\, dt' = \frac{\omega_x}{\omega} \sin \omega t, \tag{2.185}$$

which means that the probability amplitudes contain *all* harmonics of the field. To see this more clearly, we can expand $\cos[(\omega_x/\omega) \sin \omega t]$ in terms of a series of Bessel functions J_n using

$$\cos(z \sin\alpha) = J_0(z) + 2\sum_{n=1}^{\infty} J_{2n}(z) \cos(2n\alpha)\,. \tag{2.186}$$

For example, if the initial state has $c_\downarrow(0) = 1$, $c_\uparrow(0) = 0$, then

$$\begin{aligned} |c_\uparrow(t)|^2 &= \sin^2\left(\frac{\omega_x}{\omega} \sin\omega t\right) = \frac{1 - \cos\left(\frac{2\omega_x}{\omega} \sin\omega t\right)}{2} \\ &= \frac{1 - J_0\left(\frac{2\omega_x}{\omega}\right)}{2} + \sum_{n=1}^{\infty} J_{2n}\left(\frac{2\omega_x}{\omega}\right) \cos(2n\omega t). \end{aligned} \tag{2.187}$$

In contrast to a harmonic oscillator, which is intrinsically a *linear* device, a two-level quantum system acts as a nonlinear device—the response does not depend linearly on the applied field and contains all harmonics of the driving field frequency. For $\omega_x/\omega \geqslant \pi/2$, there are still times when $|c_\uparrow(t)|^2 = 1$.

The time-averaged, spin-up population is

$$\overline{|c_\uparrow(t)|^2} = \frac{1 - J_0\left(\frac{2\omega_x}{\omega}\right)}{2}. \tag{2.188}$$

The larger the applied frequency, the smaller the time-averaged value of the state up population, provided that $\omega > 0.522\omega_x$. This is not surprising since the degenerate levels are resonant with a *static* field. On the other hand, there are values of ω_x/ω for which $\overline{|c_\uparrow(t)|^2} > 1/2$. The maximum value of $\overline{|c_\uparrow(t)|^2} = 0.7$ occurs for $2\omega_x/\omega = 3.85$.

2.10.1.2 $\omega_x(t) = \omega_x =$ *constant*; $\omega = 0$; $\phi(t) = \phi =$ *constant*

This corresponds in the magnetic case to a constant field in the x direction and a constant field in the z direction,

$$\mathbf{B} = B_0\hat{\mathbf{z}} + B_x\hat{\mathbf{x}}\,, \tag{2.189}$$

such that

$$\mathrm{H}_B = \hbar\left(\frac{\omega_0}{2}\boldsymbol{\sigma}_z + \omega_x \cos\phi \boldsymbol{\sigma}_x\right) = \frac{\hbar}{2}\begin{pmatrix} \omega_0 & 2\omega_x\cos\phi \\ 2\omega_x\cos\phi & -\omega_0 \end{pmatrix}. \tag{2.190}$$

Clearly, this Hamiltonian is identical to that in equation (2.109) if the substitutions $\delta \Leftrightarrow \omega_0$ and $\chi = \Omega_0/2 \Leftrightarrow \omega_x \cos\phi$ are made and the states are interchanged. It then follows that the solution to this problem is [see equation (2.116)]

$$a_\downarrow(t) = \left[\cos\left(\frac{Xt}{2}\right) + i\frac{\omega_0}{X}\sin\left(\frac{Xt}{2}\right)\right]a_\downarrow(0) - i\frac{y}{X}\sin\left(\frac{Xt}{2}\right)a_\uparrow(0), \tag{2.191a}$$

$$a_\uparrow(t) = -i\frac{y}{X}\sin\left(\frac{Xt}{2}\right)a_\downarrow(0) + \left[\cos\left(\frac{Xt}{2}\right) - i\frac{\omega_0}{X}\sin\left(\frac{Xt}{2}\right)\right]a_\uparrow(0), \tag{2.191b}$$

where $y = 2\omega_x\cos\phi$, and $X = (\omega_0^2 + y^2)^{1/2}$.

Problems

Unless noted otherwise (as in problems 4 to 6), the rotating-wave approximation (RWA) can be used.

1. Express the Rabi frequency Ω_0 in terms of the time-averaged power density of the light field. Assuming some reasonable values for the matrix elements, estimate Ω_0 for typical cw and pulsed laser fields.
2. The appropriate Hamiltonian in the field interaction representation is

$$\tilde{\mathrm{H}}(t) = \frac{\hbar}{2}\begin{pmatrix} -\delta(t) & |\Omega_0(t)| \\ |\Omega_0(t)| & \delta(t) \end{pmatrix},$$

where $\Omega_0(t) = |\Omega_0(t)|\,e^{i\varphi(t)}$, and $\delta(t) = \delta + \dot{\phi}(t)$. For $\Omega_0(t) =$ real and constant, explicitly diagonalize $\tilde{\mathrm{H}}$ to obtain its eigenvalues and eigenfunctions. Show that the amplitudes in this field interaction representation evolve as

$$\tilde{\mathbf{c}}(t) = \mathbf{T}^\dagger e^{i\Omega t\sigma_z/2}\mathbf{T}\tilde{\mathbf{c}}(0),$$

where $\Omega^2 = |\Omega_0|^2 + \delta^2$ and $\mathbf{T}\tilde{\mathbf{H}}\mathbf{T}^\dagger$ is diagonal. Using the matrix $\mathbf{T}$ you find, prove that this result agrees with equation (2.116).
3. With $\tilde{\mathrm{H}}$ given as in problem 2, calculate $\tilde{\mathbf{c}}(t/T)$ for $\Omega_0 T = 5$, $\delta T = 3$. Compare your result with a direct evaluation of $e^{-i\tilde{\mathrm{H}}t/\hbar}$ obtained using *MatrixExponential* of Mathematica or some equivalent program and plot $|\tilde{c}_2(t/T)|^2$ as a function of (t/T) given $\tilde{c}_2(0) = 0$.
4. For the two-level problem, including the counterrotating terms [equation (2.55) or (2.179)], solve for $a_1(t)$ and $a_2(t)$ given that $a_1(0) = 1$ and $a_2(0) = 0$ in the following cases:
   a. $\omega = 0$; $\Omega_0 =$ constant
   b. $\omega_0 = 0$; $\Omega_0 =$ real constant.

   For each case, calculate the upper state population and determine the Fourier components present in the population. (The Fourier components could be monitored by using a probe laser beam on the same or a coupled transition.)

5. Numerically integrate the equations in problem 4, and show that the solutions as a function of (t/T) are consistent with the analytic ones for
   a. $\omega T = 0$; $\Omega_0 T = 0.5, 1, 2, 10$; $\omega_0 T = 5$
   b. $\omega_0 T = 0$; $\Omega_0 T = 0.5, 1, 2, 10$; $\omega T = 2$.
6. Numerically integrate equations (2.55) for $\Omega_0 T = 10$, $\omega_0 T = 50$, and $\omega T = 49, 49.5, 50, 50.5, 51$. Show that the maximum value of $|a_2(t/T)|^2$ occurs for $\omega T = 49.5$, and show that this result is *not* consistent with the Bloch-Siegert shift. The reason for the inconsistency is that no average over the phase of the field has been taken. To see evidence for this, try a different phase for the field such as $\pi/2$ and show that the maximum occurs at a different value of ωT. Also note that the amplitudes of the rapid oscillations in the solution are of order Ω_0/ω.

7–8. In the two-level problem, assume that $\Omega_0(t)$ has the form $\Omega_0(t) = (2A/T)\,\text{sech}(\pi t/T)$. Sketch this pulse envelope function, and prove that the pulse area defined by $\mathcal{A} = \int_{-\infty}^{\infty} \Omega_0(t)dt$ is equal to $2A$. Obtain a second-order differential equation for $c_2(t)$ in terms of $\Omega_0(t)$ and $d\Omega_0/dt$. Using the substitution $z = (1/2)[\tanh(\pi t/T) + 1]$, show that the equation reduces to the standard form for the hypergeometric equation and obtain an explicit solution for c_1 and c_2 as functions of z in terms of c_1 and c_2 at $z = 0$ $(t = -\infty)$. For specific initial conditions $c_2(-\infty) = 0$, $c_1(-\infty) = 1$, show that $P_2 \equiv |c_2(\infty)|^2 = \sin^2(\mathcal{A}/2)\,\text{sech}^2(\delta T/2)$. This solution is attributed to Rosen and Zener [8]. Plot P_2 as a function of $\mathcal{A}$ for $|\delta| T \gg 1$ to prove that P_2 saturates even though $P_2 \ll 1$.

9. In the two-level problem, assume that $\Omega_0(t)$ has the form $\Omega_0(t) = \Omega_0 \exp[-(t/T)^2]$. With initial conditions $\tilde{c}_1(-\infty) = 1$, $\tilde{c}_2(-\infty) = 0$, find approximate solutions for $|\tilde{c}_1(t)|$ and $|\tilde{c}_2(t)|$ assuming that $|\Omega(t)T| \gg 1$.

10–11. For the two-level problem, calculate the transition probability for excitation of an atom for Gaussian and Lorentzian pulses. That is, assume that the pulse envelopes are of the form

   a. $\Omega_0(t/T) = \Omega_0 e^{-(t/T)^2}$

   b. $\Omega_0(t/T) = \Omega_0[1 + (t/T)^2]^{-1}$

   and find $P_2 = |c_2(\infty)|^2$ given that $c_1(-\infty) = 1$. For each of the pulses, plot P_2 as a function of $\Omega_0 T$ (go out to values of $\Omega_0 T$ that give at least three minima in P_2) for $\delta T = 0, 0.5, 1, 2$. Compare the qualitative form of the results with that for the hyperbolic secant pulse of problem 7. Note that this problem must be solved numerically for all but $\delta T = 0$.

12. For the two-level problem, numerically calculate the transition probability as a function of time for excitation of an atom with the Gaussian pulse envelope

$$|\Omega_0(t)| = |\Omega_0|\; e^{-(t/T)^2}$$

   and phase

   a. $\phi(t) = 0$

   b. $\phi(t) = -\delta_0 t - \delta_0 t(2/\pi)\tan^{-1}(t/T)$

where δ_0 is the detuning. Take

$$(i)\ \delta_0 T = 3, \qquad |\Omega_0|\, T = 3$$
$$(ii)\ \delta_0 T = 8, \qquad |\Omega_0|\, T = 8$$

to determine whether you are in the adiabatic following limit. Calculate $\delta(t) = \delta_0 + \dot{\phi}$, and plot $\pm\Omega(t)T/2 = \{|\Omega_0(t)|^2 + [\delta(t)]^2\}^{1/2}T/2$. Work in the second field interaction representation and introduce a dimensionless time $\tau = t/T$. Assume that $c_1(-\infty) = 1$, and calculate $|c_2|^2$ as a function of τ for cases a.(i), a.(ii), b.(i), and b.(ii). Show that in case a.(ii), the population returns to the first state at the end of the pulse, while in case b.(ii) (adiabatic switching), the population is transferred to the second state by the pulse. Compare your numerical solutions with the prediction of the dressed-state theory $|c_2(t)|^2 = \sin^2[\theta(t)] = [1 - \delta(t)/\Omega(t)]/2$.

13. Instead of applying a linearly polarized field, imagine that a circularly polarized field drives a $J = 0$ to $J = 1$ transition. At $Z = 0$, take the field to be of the form

$$\mathbf{E}(t) = E_0 \left[\hat{\mathbf{x}}\cos(\omega t) + \hat{\mathbf{y}}\sin(\omega t)\right],$$

where E_0 is constant. Show that one can write

$$\mathbf{E}(t) = E_0 \left(-\hat{\boldsymbol{\epsilon}}_+ e^{-i\omega t} + \hat{\boldsymbol{\epsilon}}_- e^{i\omega t}\right)/\sqrt{2},$$

where

$$\hat{\boldsymbol{\epsilon}}_\pm = \mp\frac{\hat{\mathbf{x}}\pm i\hat{\mathbf{y}}}{\sqrt{2}}.$$

Given that the matrix elements of $\hat{\boldsymbol{\mu}} \cdot \hat{\boldsymbol{\epsilon}}$ satisfy

$$\begin{aligned}
&\langle J = 0; m_J = 0|\, \hat{\boldsymbol{\mu}} \cdot \hat{\boldsymbol{\epsilon}}_-\, |J = 1; m_J\rangle \\
&= -\langle J = 1; m_J|\, \hat{\boldsymbol{\mu}} \cdot \hat{\boldsymbol{\epsilon}}_+\, |J = 0; m_J = 0\rangle^* \propto \delta_{m_J,1}, \\
&\langle J = 0; m_J = 0|\, \hat{\boldsymbol{\mu}} \cdot \hat{\boldsymbol{\epsilon}}_+\, |J = 1; m_J\rangle \\
&= -\langle J = 1; m_J|\, \hat{\boldsymbol{\mu}} \cdot \hat{\boldsymbol{\epsilon}}_-\, |J = 0; m_J = 0\rangle^* \propto \delta_{m_J,-1},
\end{aligned}$$

prove that, if one considers transitions between the $J = 0$ and $J = 1$, $m_J = 1$ levels only, it is *not* necessary to make any RWA to arrive at equations of the form in equation (2.62). On the other hand, show that the *counterrotating* terms drive transitions between the $J = 0$ and $J = 1, m_J = -1$ levels.

14. Consider the differential equation $\dot{\mathbf{y}}(t) = \mathbf{A}\mathbf{y}(t)$, where $\mathbf{y}(t)$ is a column vector and $\mathbf{A}$ is a **constant** matrix.

(a) Show, by direct substitution, that a solution to this equation is $\mathbf{y}(t)=e^{\mathbf{A}t}\mathbf{y}(0)$.

In the following parts, take

$$\mathbf{A} = -i\begin{pmatrix} -a & b \\ b & a \end{pmatrix}, \qquad \mathbf{y}(0) = \begin{pmatrix} y_1(0) \\ y_2(0) \end{pmatrix}, \qquad \mathbf{y}(t) = \begin{pmatrix} y_1(t) \\ y_2(t) \end{pmatrix}.$$

(b) Solve the differential equation directly by assuming a solution of the form $\mathbf{y}(t) = \mathbf{B}e^{\lambda t}$.
(c) Solve the equations using the identity

$$e^{-i\theta\hat{\mathbf{n}}\cdot\boldsymbol{\sigma}} = \mathbf{1}\cos\theta - i\,\hat{\mathbf{n}}\cdot\boldsymbol{\sigma}\sin\theta\,,$$

where $\hat{\mathbf{n}}$ is a unit vector, and $\boldsymbol{\sigma}$ is a vector having matrix components

$$\boldsymbol{\sigma}_x = \begin{pmatrix} 0 & 1 \\ 1 & 0 \end{pmatrix}, \qquad \boldsymbol{\sigma}_y = \begin{pmatrix} 0 & -i \\ i & 0 \end{pmatrix}, \qquad \boldsymbol{\sigma}_z = \begin{pmatrix} 1 & 0 \\ 0 & -1 \end{pmatrix}.$$

(d) Solve the equation using the $MatrixExp[\{\{Ia, -Ib\}, \{-Ib, -Ia\}\}]$ function of Mathematica or some equivalent program.
(e) Find a matrix $\mathbf{T}$ such that $\mathbf{TAT}^{\dagger} = \boldsymbol{\Lambda}$, where

$$\boldsymbol{\Lambda} = \begin{pmatrix} \Lambda_1 & 0 \\ 0 & \Lambda_2 \end{pmatrix}$$

is a diagonal matrix. Prove that

$$\mathbf{y}(t) = \mathbf{T}^{\dagger}e^{\boldsymbol{\Lambda}t}\mathbf{T}\mathbf{y}(0) = \mathbf{T}^{\dagger}\begin{pmatrix} e^{\Lambda_1 t} & 0 \\ 0 & e^{\Lambda_2 t} \end{pmatrix}\mathbf{T}\mathbf{y}(0),$$

and evaluate this explicitly.
(f) Show that all your results give the same solution. Note that the last method can be used for matrices of any dimension.

15. Suppose that we use the first definition of the field interaction representation. Show that equation (2.134) is unchanged provided that we set

$$|\mathbf{D}(t)\rangle = \mathbf{T}^{*}(t)|\tilde{\mathbf{n}}'(t)\rangle, \text{ where}$$

$$|\tilde{\mathbf{n}}'(t)\rangle = \begin{pmatrix} |\tilde{1}'(t)\rangle \\ |\tilde{2}'(t)\rangle \end{pmatrix} = \begin{pmatrix} e^{i\omega t/2}|1\rangle \\ e^{-i\omega t/2}|2\rangle \end{pmatrix}.$$

In this case, $\mathbf{T}(t)$ is a unitary matrix, but not real. Find the explicit form for the matrix $\mathbf{T}(t)$ that diagonalizes $\tilde{\mathbf{H}}'(t)$ in terms of the angles $\theta(t)$ and $\phi(t)$, where $\theta(t)$ is still defined by equation (2.135a) and $\phi(t)$ is the argument of $\Omega_0(t)$—that is, $\Omega_0(t) = |\Omega_0(t)|\,e^{i\phi(t)}$.

16. Return to problem 7 with equations of the form

$$\dot{c}_1(\tau) = -iA\,\mathrm{sech}(\pi\tau)e^{-id\tau}c_2(\tau),$$
$$\dot{c}_2(\tau) = -iA\,\mathrm{sech}(\pi\tau)e^{id\tau}c_1(\tau).$$

Assuming that $c_1(-\infty) = 1$, calculate $|c_2(\infty)|^2$ using the first Magnus approximation, and compare with the exact result, $|c_2(\infty)|^2 = \sin^2(A)\mathrm{sech}^2(d/2)$. Under what conditions do the two results agree? What is the ratio of the results when $d \gg 1$?

17. For the two-level problem, numerically calculate the transition probability as a function of time for excitation of an atom with the Gaussian pulse envelope

$$|\Omega_0(t)| = |\Omega_0|\;e^{-(t/T)^2}$$

and detuning

$$\delta(t) = \begin{cases} \delta_0 \left[1 - e^{(t/T)}\right]^3 & t < 0 \\ 0 & t > 0 \end{cases}.$$

Take $\delta_0 T = 30$ and $|\Omega_0| T = 30$, and calculate $|\varrho_{12}(\tau)| = \left|c_1(\tau)c_2^*(\tau)\right|$ as a function of $\tau = t/T$, assuming that $c_1(-\infty) = 1$. Show that in this case, $|\varrho_{12}(\infty)| \simeq 1/2$, the maximum value it can have. This combined detuning and pulse shape maximizes the "coherence" between states 1 and 2. Calculate the result in the adiabatic approximation, and show that it agrees with the numerical result.

References

[1] See E. E. Nikitin and S. Ya. Umanskii, *Theory of Slow Atomic Collisions* (Springer-Verlag, Berlin, 1984), for a detailed discussion of this problem, with many references.

[2] For a discussion of some of the problems involved in solving the two-level problem for an off-resonant field, see P. R. Berman, L. Yan, K.-H. Chiam, and R. Sung, *Nonadiabatic transitions in a two-level quantum system: pulse shape dependence of the transition probability for a two-level atom driven by a pulsed radiation field,* Physical Review A **57**, 79–92 (1998).

[3] J. H. Shirley, *Solution of the Schrödinger equation with a Hamiltonian periodic in time,* Physical Review **138**, B 979–B 987 (1965).

[4] F. Bloch and A. Siegert, *Magnetic Resonance for Nonrotating Fields*, Physical Review **57**, 522–527 (1940).

[5] B. W. Shore, *The Theory of Coherent Excitation* (Wiley-Interscience, New York, 1990), chap. 5.

[6] A. Bambini and P. R. Berman, *Analytic solutions to the two-state problem for a class of coupling potentials*, Physical Review A **23**, 2496–2501 (1981).

[7] F. T. Hioe and C. E. Carroll, *Analytic solutions to the two-state problem for chirped pulses*, Journal of the Optical Society B **2**, 497–502 (1985).

[8] N. Rosen and C. Zener, *Double Stern-Gerlach experiment and related collision phenomena*, Physical Review **40**, 502–507 (1932).

[9] See, for example, E. Merzbacher, *Quantum Mechanics*, 2nd ed. (Wiley, New York, 1970), sec. 12.6.

[10] E. Courtens and A. Szoke, *Time and spectral resolution in resonance scattering and resonance fluorescence*, Physical Review A **15**, 1588–1603 (1977).

[11] P. R. Berman and R. Salomaa, *Comparison between dressed-atom and bare-atom pictures in laser spectroscopy*, Physical Review A **25**, 2667–2692 (1982).

[12] S. Reynaud and C. Cohen-Tannoudji, *Dressed atom approach to collisional redistribution*, Journal de Physique (Paris) **43**, 1021–1035 (1982).

[13] L. D. Landau, *Theory of energy transfer II*, Physikalische Zeitschrift der Sowjetunion **2**, 46–51 (1932).

[14] C. Zener, *Non-adiabatic crossing of energy levels*, Proceedings of the Royal Society of London, Series A **137**, 696–702 (1932).

[15] N. V. Vitanov and B. M. Garraway, *Landau-Zener model: effects of finite coupling duration*, Physical Review. A **53**, 4288–4304 (1996).

[16] B. M. Garraway and N. V. Vitanov, *Population dynamics and phase effects in periodic level crossings*, Physical Review A 55, 4418–4432 (1997).

[17] W. Magnus, *On the exponential solution of differential equations of a linear operator*, Communications in Pure and Applied Mathematics 7, 649–673 (1954).

Bibliography

B. W. Shore, *The Theory of Coherent Excitation* chapters 3 and 5 and section 14.4, (Wiley-Interscience, New York, 1990). In addition, the two-level problem is discussed in most of the texts listed in the bibliography in chapter 1.

3

Density Matrix for a Single Atom

3.1 Density Matrix

In the previous chapter, we showed in detail how to obtain solutions of the time-dependent Schrödinger equation for the state amplitudes of a two-level atom interacting with an optical field. To complete the story, we need to understand how these state amplitudes are related to the possible outcomes of experimental measurements. In quantum mechanics, it is assumed that one can associate a Hermitian operator to each physical observable. One can then, in principle, measure the eigenvalues of the Hermitian operator and the probabilities for a physical system to be in one of the eigenstates of the operator. Of course, any physical measurement yields real results, while the state amplitudes are complex. Therefore, it is to be expected that measured observables depend on bilinear products of the state amplitudes. As you know already from elementary quantum mechanics, the absolute value of the state amplitude squared gives the probability for measuring the eigenvalue associated with that given state. Moreover, if you were to calculate the expectation value of the electric dipole moment of a two-level atom, you would find that it depends on the bilinear products $a_1^*(t)a_2(t)$ and $a_2^*(t)a_1(t)$, since

$$\begin{aligned}\langle \hat{\mathbf{r}} \rangle &= \langle \Psi(t)|\hat{\mathbf{r}}|\Psi(t)\rangle \\ &= \langle a_1(t)\psi_1(\mathbf{r}) + a_2(t)\psi_2(\mathbf{r})|\hat{\mathbf{r}}|a_1(t)\psi_1(\mathbf{r}) + a_2(t)\psi_2(\mathbf{r})\rangle \\ &= a_1^*(t)a_2(t)\langle 1|\hat{\mathbf{r}}|2\rangle + \text{c.c.},\end{aligned} \tag{3.1}$$

where c.c. stands for complex conjugate, and the fact that $\hat{\mathbf{r}}$ is an odd operator has been used.

Thus, a knowledge of the state amplitudes allows one to calculate the expectation values of any operators. Although all the information is contained in the state amplitudes, we are not necessarily interested in *all* the information. If we measure only *part* of the information content of a system, an amplitude approach is often no longer satisfactory. This concept can be illustrated with several examples.

First, let us look at spontaneous decay in a two-level atom (figure 3.1). As a result of spontaneous decay, the upper-state population $n_2(t) = |a_2(t)|^2$ decreases and the

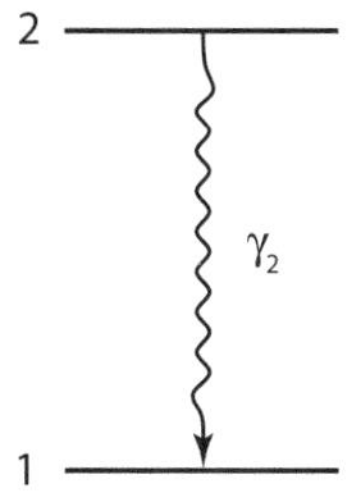

Figure 3.1. Spontaneous emission in a two-level atom.

lower-state population $n_1(t) = |a_1(t)|^2$ increases according to

$$\dot{n}_2(t) = -\gamma_2 n_2(t), \tag{3.2a}$$

$$\dot{n}_1(t) = \gamma_2 n_2(t), \tag{3.2b}$$

where the excited-state decay rate γ_2 is real. Is it possible to account for this decay in an amplitude picture? One can try the equation

$$\dot{a}_2(t) = -\frac{\gamma_2}{2} a_2(t), \tag{3.3}$$

which implies that

$$\frac{d}{dt}|a_2(t)|^2 = \frac{d}{dt}[a_2(t)a_2^*(t)] = \dot{a}_2(t)a_2^*(t) + a_2(t)\dot{a}_2^*(t) = -\gamma_2|a_2(t)|^2. \tag{3.4}$$

It works! But try to reproduce equation (3.2b) in a simple amplitude picture—you will find it impossible to do so.[1]

Second, consider atoms subjected to collisions that change the relative phase of the state amplitudes in a random fashion but do not induce transitions between the states. On average, the collisional contributions to the time rate of change of bilinear products of state amplitudes are

$$\dot{n}_1(t) = \frac{d}{dt}[a_1(t)a_1^*(t)] = 0, \tag{3.5a}$$

$$\dot{n}_2(t) = \frac{d}{dt}[a_2(t)a_2^*(t)] = 0, \tag{3.5b}$$

$$\frac{d}{dt}[a_1(t)a_2^*(t)] = -\Gamma^{(c)}a_1(t)a_2^*(t), \tag{3.5c}$$

where the (complex) collision rate $\Gamma^{(c)} = \Gamma + iS$ is proportional to the collision-averaged value of $(1 - e^{i\phi})$, and ϕ is the relative phase shift of the two state amplitudes in a collision (see appendix B for a derivation of this result). Again, there is no simple way to arrive at these expressions in an amplitude picture, since *each* atom has well-defined state amplitudes—it is only on averaging over many possible collision histories that one arrives at a decay of $a_1(t)a_2^*(t)$. Since $a_1(t)a_2^*(t)$ is related to the expectation value of the electric dipole of the atom, elastic collisions destroy the electric dipole moment but do not affect state probabilities.

[1] It *is* possible to describe spontaneous emission in an amplitude picture using the so-called Monte Carlo or quantum trajectory methods discussed in appendix C in chapter 16.

In problems of this nature, where atoms interact with a bath, such as the vacuum field for spontaneous emission or perturbers for collisions, it is useful to introduce the density matrix to describe the system's evolution. The density matrix arises naturally even when discussing the quantum mechanics of a single atom—so let us start there.

In making measurements in quantum mechanics, one always measures quantities that are related to the expectation value of a Hermitian operator. An arbitrary Hermitian operator, represented by the matrix **A** in Dirac notation, has its expectation value given by

$$\langle\Psi(t)|\mathbf{A}|\Psi(t)\rangle = \sum_{n,m} \langle m|a_m^*(t)\mathbf{A}a_n(t)|n\rangle = \sum_{n,m} a_m^*(t)a_n(t)\langle m|\mathbf{A}|n\rangle\,. \tag{3.6}$$

Let us define a matrix whose elements in the energy basis are equal to $a_n(t)a_m^*(t)$. A matrix satisfying this criterion is the *density matrix*

$$\varrho(t) = |\Psi(t)\rangle\langle\Psi(t)|, \tag{3.7}$$

since

$$\langle n|\varrho(t)|m\rangle = \langle n|\Psi(t)\rangle\langle\Psi(t)|m\rangle = \sum_{p,p'} \langle n|a_p(t)|p\rangle\langle p'|a_{p'}^*(t)|m\rangle = a_n(t)a_m^*(t). \tag{3.8}$$

Note that $\varrho(t)$ is *not* a bona fide Schrödinger operator since it is time-dependent, whereas all operators in the Schrödinger picture are taken as time-independent. With this definition,

$$\langle\Psi(t)|\mathbf{A}|\Psi(t)\rangle = \sum_{nm} \varrho_{nm}(t)A_{mn} = \mathrm{Tr}\,[\varrho(t)\mathbf{A}]\,, \tag{3.9}$$

where Tr stands for "trace."

In the energy basis,

$$\varrho(t) = \varrho_{nm}(t)|n\rangle\langle m|, \tag{3.10}$$

where $|n\rangle\langle m|$ is a matrix with a 1 in the nm location and zeros elsewhere. It is sometimes more convenient to expand $\varrho(t)$ in other bases, such as an irreducible tensor basis, as we do in chapter 16. It is easy to establish some properties for $\varrho(t)$ for our single-atom case. First, we note that

$$\varrho(t) = \mathbf{a}(t)\mathbf{a}^\dagger(t), \tag{3.11}$$

where **a** is interpreted as a column vector and $\mathbf{a}^\dagger$ as a row vector,

$$\mathbf{a} = \begin{pmatrix} a_1 \\ a_2 \\ \vdots \end{pmatrix}, \tag{3.12}$$

$$\mathbf{a}^\dagger = \left(a_1^* a_2^* \cdots\right), \tag{3.13}$$

so that we can write

$$\varrho = \mathbf{a}\mathbf{a}^{\dagger} = \begin{pmatrix} a_1 \\ a_2 \\ \vdots \end{pmatrix} \begin{pmatrix} a_1^* \, a_2^* \cdots \end{pmatrix} = \begin{pmatrix} |a_1|^2 & a_1 a_2^* & \cdots \\ a_2 a_1^* & |a_2|^2 & \cdots \\ \vdots & \vdots & \ddots \end{pmatrix}. \tag{3.14}$$

The density matrix is an idempotent operator, since

$$\varrho^2 = |\Psi\rangle\langle\Psi|\Psi\rangle\langle\Psi| = |\Psi\rangle\langle\Psi| = \varrho. \tag{3.15}$$

Critical to our development is the equation for the time evolution of ϱ,

$$i\hbar\frac{d\varrho}{dt} = i\hbar\frac{d}{dt}(\mathbf{a}\mathbf{a}^{\dagger}) = [\mathbf{H}\mathbf{a}\mathbf{a}^{\dagger} - \mathbf{a}(\mathbf{H}\mathbf{a})^{\dagger}] = \mathbf{H}\varrho - \varrho\mathbf{H} = [\mathbf{H}, \varrho]. \tag{3.16}$$

Note that the sign is *different* from that in the evolution equation for the expectation values of operators,

$$i\hbar\frac{\langle\Psi(t)|\mathbf{A}|\Psi(t)\rangle}{dt} = \langle\Psi(t)|[\mathbf{A}, \mathbf{H}]|\Psi(t)\rangle. \tag{3.17}$$

We have suppressed the explicit time dependence in the amplitudes and operators in equations (3.12) to (3.17). For the most part, we do not indicate such time dependence explicitly from this point onward, although we retain it in some of the equations as a reminder.

As an example, consider a two-level atom interacting with an optical field in the rotating-wave approximation (RWA). In this case,

$$\mathbf{H}(t) = \frac{\hbar}{2}\begin{pmatrix} -\omega_0 & \Omega_0^*(t)e^{i\omega t} \\ \Omega_0(t)e^{-i\omega t} & \omega_0 \end{pmatrix}, \tag{3.18}$$

such that

$$i\hbar\begin{pmatrix} \dot\varrho_{11}(t) & \dot\varrho_{12}(t) \\ \dot\varrho_{21}(t) & \dot\varrho_{22}(t) \end{pmatrix} = \frac{\hbar}{2}\left[\begin{pmatrix} -\omega_0 & \Omega_0^*(t)e^{i\omega t} \\ \Omega_0(t)e^{-i\omega t} & \omega_0 \end{pmatrix}\begin{pmatrix} \varrho_{11}(t) & \varrho_{12}(t) \\ \varrho_{21}(t) & \varrho_{22}(t) \end{pmatrix} - \begin{pmatrix} \varrho_{11}(t) & \varrho_{12}(t) \\ \varrho_{21}(t) & \varrho_{22}(t) \end{pmatrix}\begin{pmatrix} -\omega_0 & \Omega_0^*(t)e^{i\omega t} \\ \Omega_0(t)e^{-i\omega t} & \omega_0 \end{pmatrix}\right], \tag{3.19}$$

or, since $\chi(t) = \Omega_0(t)/2$,

$$\dot\varrho_{11}(t) = -i\chi^*(t)e^{i\omega t}\varrho_{21}(t) + i\chi(t)e^{-i\omega t}\varrho_{12}(t), \tag{3.20a}$$

$$\dot\varrho_{22}(t) = i\chi^*(t)e^{i\omega t}\varrho_{21}(t) - i\chi(t)e^{-i\omega t}\varrho_{12}(t), \tag{3.20b}$$

$$\dot\varrho_{12}(t) = i\omega_0\varrho_{12}(t) - i\chi^*(t)e^{i\omega t}\left[\varrho_{22}(t) - \varrho_{11}(t)\right], \tag{3.20c}$$

$$\dot\varrho_{21}(t) = -i\omega_0\varrho_{21}(t) + i\chi(t)e^{-i\omega t}\left[\varrho_{22}(t) - \varrho_{11}(t)\right]. \tag{3.20d}$$

One can solve these equations for a given $\chi(t)$, but it is easier to solve in the amplitude picture and then simply construct $\varrho_{11}(t) = |a_1(t)|^2$, $\varrho_{22}(t) = |a_2(t)|^2$,

$\varrho_{12}(t) = a_1(t)a_2^*(t)$, and $\varrho_{21}(t) = a_2(t)a_1^*(t)$. It looks like we have not gained *anything*, except making the equations more difficult! In a sense, that is correct. The density matrix becomes useful and essential when dealing with ensembles of particles or particles interacting with a bath.

We defer a discussion of the density matrix for an ensemble of particles until we analyze the polarization of a medium. However, it is not difficult to imagine that a quantity defined by

$$\varrho = \frac{1}{N}\sum_{n=1}^{N} \varrho^{(n)}, \tag{3.21}$$

where N is the total number of atoms, and $\varrho^{(n)}$ is the density matrix of atom n, is of some use.

For the present, let us consider what happens when a single atom interacts with a second system such as a thermal bath. The total Hamiltonian is

$$\mathbf{H} = \mathbf{H}_1 + \mathbf{H}_2 + \mathbf{V}, \tag{3.22}$$

where $\mathbf{H}_1$ is the Hamiltonian of the atomic system (including interactions with any external fields), $\mathbf{H}_2$ is the Hamiltonian of the bath, and $\mathbf{V}$ is the atom–bath interaction energy. The eigenkets of the time-independent part of $\mathbf{H}_1$ are denoted by $|n_1\rangle$, and the eigenkets of $\mathbf{H}_2$ are denoted by $|n_2\rangle$. Suppose that we have an operator $\mathbf{A}_1$ that acts *only* in the space of $\mathbf{H}_1$—that is, $\mathbf{A}_1$ acts only on atomic state variables. For a wave function

$$|\Psi(t)\rangle = \sum_{n_1 n_2} a_{n_1 n_2}(t)|n_1\rangle|n_2\rangle, \tag{3.23}$$

we calculate

$$\begin{aligned}\langle \mathbf{A}_1\rangle = \langle\Psi|\mathbf{A}_1|\Psi\rangle &= \sum_{n_1 n_2}\sum_{n_1' n_2'} a^*_{n_1 n_2}(t)a_{n_1' n_2'}(t)\langle n_1|\langle n_2|\mathbf{A}_1|n_2'\rangle|n_1'\rangle \\ &= \sum_{n_1 n_2}\sum_{n_1' n_2'} a^*_{n_1 n_2}(t)a_{n_1' n_2'}(t)\delta_{n_2,n_2'}\langle n_1|\mathbf{A}_1|n_1'\rangle,\end{aligned} \tag{3.24}$$

or

$$\langle\Psi|\mathbf{A}_1|\Psi\rangle = \sum_{n_1 n_1'}(\mathbf{A}_1)_{n_1 n_1'}\left[\sum_{n_2} a^*_{n_1 n_2}(t)a_{n_1' n_2}(t)\right]. \tag{3.25}$$

We define the *reduced density matrix* for particle 1 as the density matrix traced over the states of the bath,

$$\varrho^{(1)}(t) = \mathrm{Tr}_2\varrho(t), \tag{3.26}$$

or

$$\varrho^{(1)}_{nn'}(t) = \sum_{n_2}\varrho_{nn_2;n'n_2}(t). \tag{3.27}$$

It is clear from equation (3.24) that

$$\langle \mathbf{A}_1\rangle = \mathrm{Tr}\left[\mathbf{A}_1\varrho^{(1)}(t)\right]. \tag{3.28}$$

The key point is that it is often possible to get a simple equation for $\varrho^{(1)}$ that incorporates the effects of the bath. For example, we show explicitly in chapter 19 that the result of spontaneous emission is to introduce terms in the equations of motion for the reduced density matrix elements for the atom given by

$$\dot{\varrho}_{11}(t)_{sp} = \gamma_2\varrho_{22}(t), \tag{3.29a}$$

$$\dot{\varrho}_{22}(t)_{sp} = -\gamma_2\varrho_{22}(t), \tag{3.29b}$$

$$\dot{\varrho}_{12}(t)_{sp} = -\frac{\gamma_2}{2}\varrho_{12}(t), \tag{3.29c}$$

$$\dot{\varrho}_{21}(t)_{sp} = -\frac{\gamma_2}{2}\varrho_{21}(t). \tag{3.29d}$$

These contributions can be added to the terms given in equations (3.20a) to (3.20d). Equation (3.11) loses its meaning for atomic state amplitudes once relaxation is introduced.

Similarly, one can return to the simple collision model in which collisions cause sudden changes in energy of the various levels but are not energetically able to cause transitions between states. In this model, which is discussed in appendix B,

$$\dot{\varrho}_{11}(t)_{coll} = 0, \tag{3.30a}$$

$$\dot{\varrho}_{22}(t)_{coll} = 0, \tag{3.30b}$$

$$\dot{\varrho}_{12}(t)_{coll} = -\Gamma\varrho_{12}(t), \tag{3.30c}$$

$$\dot{\varrho}_{21}(t)_{coll} = -\Gamma\varrho_{21}(t). \tag{3.30d}$$

For simplicity, we take Γ to be real and neglect the collisional shifts that are associated with the imaginary part of Γ. (Often, these can be incorporated by redefining the atom–field detunings to include the collisional shifts.)

Including spontaneous decay and collisions, we have

$$\dot{\varrho}_{11}(t) = -i\chi^*(t)e^{i\omega t}\varrho_{21}(t) + i\chi(t)e^{-i\omega t}\varrho_{12}(t) + \gamma_2\varrho_{22}(t), \tag{3.31a}$$

$$\dot{\varrho}_{22}(t) = i\chi^*(t)e^{i\omega t}\varrho_{21}(t) - i\chi(t)e^{-i\omega t}\varrho_{12}(t) - \gamma_2\varrho_{22}(t), \tag{3.31b}$$

$$\dot{\varrho}_{12}(t) = i\omega_0\varrho_{12}(t) - i\chi^*(t)e^{i\omega t}\left[\varrho_{22}(t) - \varrho_{11}(t)\right] - \gamma\varrho_{12}(t), \tag{3.31c}$$

$$\dot{\varrho}_{21}(t) = -i\omega_0\varrho_{21}(t) + i\chi(t)e^{-i\omega t}\left[\varrho_{22}(t) - \varrho_{11}(t)\right] - \gamma\varrho_{21}(t), \tag{3.31d}$$

where

$$\gamma = \gamma_2/2 + \Gamma. \tag{3.32}$$

Note that $\dot{\varrho}_{11}(t) + \dot{\varrho}_{22}(t) = 0$, consistent with conservation of population. Now we *have* gotten somewhere. It is impossible to write analogous equations using state amplitudes, since equations (3.31) are already averaged over a thermal bath. Equations (3.31) are the starting point for many applications in the interaction of radiation with matter.

We can write equations (3.31) in matrix form as

$$i\hbar\dot{\varrho} = [\mathrm{H}, \varrho] - i\hbar\gamma\,[\sigma_0\varrho + \varrho\sigma_0] + i\hbar\gamma_2\sigma_-\varrho\sigma_+ + 2i\hbar\Gamma\sigma_0\varrho\sigma_0, \tag{3.33}$$

where

$$\sigma_- = \begin{pmatrix} 0 & 1 \\ 0 & 0 \end{pmatrix}, \tag{3.34}$$

$$\sigma_+ = \begin{pmatrix} 0 & 0 \\ 1 & 0 \end{pmatrix}, \tag{3.35}$$

$$\sigma_0 = \sigma_+\sigma_- = \begin{pmatrix} 0 & 0 \\ 0 & 1 \end{pmatrix}, \tag{3.36}$$

and, in the RWA,

$$\mathrm{H}(t) = \frac{\hbar}{2}\begin{pmatrix} -\omega_0 & \Omega_0^*(t)e^{i\omega t} \\ \Omega_0(t)e^{-i\omega t} & \omega_0 \end{pmatrix}. \tag{3.37}$$

An equation of the type (3.33) is often referred to as a *master equation.*

3.2 Interaction Representation

As in the amplitude case, it is useful to introduce several representations. The interaction representation is not restricted to the two-level problem. For a Hamiltonian of the form $\mathrm{H}(t) = \mathrm{H}_0 + \mathrm{V}(t)$, the interaction representation can be defined quite generally by

$$\varrho^{(I)} = e^{i\mathrm{H}_0 t/\hbar}\varrho e^{-i\mathrm{H}_0 t/\hbar}, \tag{3.38}$$

with matrix elements given by

$$\varrho_{mn}^{(I)} = e^{i\omega_{mn}t}\varrho_{mn}. \tag{3.39}$$

Note that $\varrho_{mm}^{(I)} = \varrho_{mm}$. It then follows that

$$i\hbar\dot{\varrho}^{(I)} = -\mathrm{H}_0\varrho^{(I)} + i\hbar e^{i\mathrm{H}_0 t/\hbar}\dot{\varrho}e^{-i\mathrm{H}_0 t/\hbar} + \varrho^{(I)}\mathrm{H}_0, \tag{3.40}$$

or

$$\begin{aligned} i\hbar\dot{\varrho}^{(I)} &= -\left[\mathrm{H}_0, \varrho^{(I)}\right] + \left[e^{i\mathrm{H}_0 t/\hbar}\left\{\mathrm{H}_0 + \mathrm{V}(\mathbf{t})\right\}e^{-i\mathrm{H}_0 t/\hbar}, \varrho^{(I)}\right] + \text{relaxation terms} \\ &= \left[\mathrm{H}^{(I)}(t), \varrho^{(I)}\right] + \text{ relaxation terms}, \end{aligned} \tag{3.41}$$

where

$$\mathrm{H}^{(I)}(t) = \mathrm{V}^{(I)}(t) = e^{i\mathrm{H}_0 t/\hbar}\mathrm{V}(t)e^{-i\mathrm{H}_0 t/\hbar}. \tag{3.42}$$

For a two-level atom interacting with an optical field in the RWA [see equation (2.167)], we have

$$\mathrm{V}^{(I)}(t) = \frac{\hbar}{2}\begin{pmatrix} 0 & \Omega_0^*(t)e^{-i\delta t} \\ \Omega_0(t)e^{i\delta t} & 0 \end{pmatrix}, \tag{3.43}$$

with $\delta = \omega_0 - \omega$. For the relaxation terms, we use equations (3.33) and (3.38) and the fact that $\mathrm{H}_0 = -\hbar\omega_0\boldsymbol{\sigma}_z/2$ to write these terms as

$$e^{-i\frac{\omega_0 t}{2}\sigma_z}\left[-\gamma\left(\boldsymbol{\sigma}_0\boldsymbol{\varrho}+\boldsymbol{\varrho}\boldsymbol{\sigma}_0\right)+\gamma_2\boldsymbol{\sigma}_-\boldsymbol{\varrho}\boldsymbol{\sigma}_+ + 2\Gamma\boldsymbol{\sigma}_0\boldsymbol{\varrho}\boldsymbol{\sigma}_0\right]e^{i\frac{\omega_0 t}{2}\sigma_z}.$$

Since $[\boldsymbol{\sigma}_z, \boldsymbol{\sigma}_0] = 0$ and

$$\boldsymbol{\sigma}_-\boldsymbol{\varrho}\boldsymbol{\sigma}_+ = \begin{pmatrix}\varrho_{22} & 0\\ 0 & 0\end{pmatrix},$$

it follows that

$$e^{-i\frac{\omega_0 t}{2}\sigma_z}\boldsymbol{\sigma}_-\boldsymbol{\varrho}\boldsymbol{\sigma}_+ e^{i\frac{\omega_0 t}{2}\sigma_z} = \boldsymbol{\sigma}_-\boldsymbol{\varrho}^{(I)}\boldsymbol{\sigma}_+. \tag{3.44}$$

As a consequence, the relaxation terms have the same form as in equations (3.31) with $\boldsymbol{\varrho} \to \boldsymbol{\varrho}^{(I)}$, and we can write

$$\dot{\varrho}_{11}(t) = -i\chi^*(t)e^{-i\delta t}\varrho_{21}^{(I)}(t) + i\chi(t)e^{i\delta t}\varrho_{12}^{(I)}(t) + \gamma_2\varrho_{22}(t), \tag{3.45a}$$

$$\dot{\varrho}_{22}(t) = i\chi^*(t)e^{-i\delta t}\varrho_{21}^{(I)}(t) - i\chi(t)e^{i\delta t}\varrho_{12}^{(I)}(t) - \gamma_2\varrho_{22}(t), \tag{3.45b}$$

$$\dot{\varrho}_{12}^{(I)}(t) = -i\chi^*(t)e^{-i\delta t}\left[\varrho_{22}(t) - \varrho_{11}(t)\right] - \gamma\varrho_{12}^{(I)}(t), \tag{3.45c}$$

$$\dot{\varrho}_{21}^{(I)}(t) = \chi(t)e^{i\delta t}\left[\varrho_{22}(t) - \varrho_{11}(t)\right] - \gamma\varrho_{21}^{(I)}(t), \tag{3.45d}$$

where we have used the fact that $\varrho_{mm}^{(I)} = \varrho_{mm}$.

3.3 Field Interaction Representation

In the RWA, the field interaction representation for the two-level problem is defined by

$$\begin{aligned}\tilde{\boldsymbol{\varrho}}' &= e^{-i\frac{\omega t}{2}\sigma_z}\boldsymbol{\varrho}e^{i\frac{\omega t}{2}\sigma_z},\\ \tilde{\varrho}'_{mm} &= \varrho_{mm},\\ \tilde{\varrho}'_{12} &= e^{-i\omega t}\varrho_{12},\\ \tilde{\varrho}'_{21} &= e^{i\omega t}\varrho_{21},\end{aligned} \tag{3.46}$$

or

$$\begin{aligned}\tilde{\boldsymbol{\varrho}} &= e^{-i\frac{\omega t-\phi(t)}{2}\sigma_z}\boldsymbol{\varrho}e^{i\frac{\omega t-\phi(t)}{2}\sigma_z},\\ \tilde{\varrho}_{mm} &= \varrho_{mm},\\ \tilde{\varrho}_{12} &= e^{-i[\omega t-\phi(t)]}\varrho_{12},\\ \tilde{\varrho}_{21} &= e^{i[\omega t-\phi(t)]}\varrho_{21},\end{aligned} \tag{3.47}$$

depending on whether we include the phase in the definition. Note that in contrast to the interaction representation, the field interaction representation is defined for the specific problem of a two-level atom interacting with a nearly resonant field.

With these definitions, it follows from equations (3.33) and (2.107) that

$$i\hbar d\tilde{\varrho}'/dt = [\tilde{\mathbf{H}}', \tilde{\varrho}'] + \text{relaxation terms}, \tag{3.48}$$

or

$$i\hbar d\tilde{\varrho}/dt = [\tilde{\mathbf{H}}, \tilde{\varrho}] + \text{relaxation terms}, \tag{3.49}$$

where

$$\tilde{\mathbf{H}}'(t) = \frac{\hbar}{2}\begin{pmatrix} -\delta & \Omega_0^*(t) \\ \Omega_0(t) & \delta \end{pmatrix}, \tag{3.50}$$

and

$$\tilde{\mathbf{H}}(t) = \frac{\hbar}{2}\begin{pmatrix} -\delta(t) & |\Omega_0(t)| \\ |\Omega_0(t)| & \delta(t) \end{pmatrix}, \tag{3.51}$$

with

$$\delta(t) = \delta + \dot{\phi}(t). \tag{3.52}$$

The relaxation terms take the same form as in equations (3.31), with $\varrho \to \tilde{\varrho}$. Including relaxation, we find

$$\dot{\varrho}_{11}(t) = -i\chi^*(t)\tilde{\varrho}'_{21}(t) + i\chi(t)\tilde{\varrho}'_{12}(t) + \gamma_2\varrho_{22}(t), \tag{3.53a}$$

$$\dot{\varrho}_{22}(t) = i\chi^*(t)\tilde{\varrho}'_{21}(t) - i\chi(t)\tilde{\varrho}'_{12}(t) - \gamma_2\varrho_{22}(t), \tag{3.53b}$$

$$\dot{\tilde{\varrho}}'_{12}(t) = i\delta\tilde{\varrho}'_{12}(t) - i\chi^*(t)\left[\varrho_{22}(t) - \varrho_{11}(t)\right] - \gamma\tilde{\varrho}'_{12}(t), \tag{3.53c}$$

$$\dot{\tilde{\varrho}}'_{21}(t) = -i\delta\tilde{\varrho}'_{21}(t) + i\chi(t)\left[\varrho_{22}(t) - \varrho_{11}(t)\right] - \gamma\tilde{\varrho}'_{21}(t), \tag{3.53d}$$

or, with the second definition of the field interaction representation,

$$\dot{\varrho}_{11}(t) = -i|\chi(t)|\tilde{\varrho}_{21}(t) + i|\chi(t)|\tilde{\varrho}_{12}(t) + \gamma_2\varrho_{22}(t), \tag{3.54a}$$

$$\dot{\varrho}_{22}(t) = i|\chi(t)|\tilde{\varrho}_{21}(t) - i|\chi(t)|\tilde{\varrho}_{12}(t) - \gamma_2\varrho_{22}(t), \tag{3.54b}$$

$$\dot{\tilde{\varrho}}_{12}(t) = i\delta(t)\tilde{\varrho}_{12}(t) - i|\chi(t)|\left[\varrho_{22}(t) - \varrho_{11}(t)\right] - \gamma\tilde{\varrho}_{12}(t), \tag{3.54c}$$

$$\dot{\tilde{\varrho}}_{21}(t) = -i\delta(t)\tilde{\varrho}_{21}(t) + i|\chi(t)|\left[\varrho_{22}(t) - \varrho_{11}(t)\right] - \gamma\tilde{\varrho}_{21}(t), \tag{3.54d}$$

These are equations that we will use often.

Even for constant $\chi = \Omega_0/2$, it is not easy to obtain analytic solutions of equations (3.53). We can eliminate $\varrho_{11}(t)$ from the equations using $\varrho_{11}(t) = 1 - \varrho_{22}(t)$, but we are still faced with solving an auxiliary equation for the roots of the trial solution, $\tilde{\varrho}'_{ij}(t) = \tilde{\varrho}'_{ij}(0)e^{rt}$, that is cubic. On the other hand, if we neglect all relaxation and assume that Ω_0 is constant, the solution of the density matrix equations (3.53) is

$$\varrho_{11}(t) = \left[1 + \frac{\delta^2 + |\Omega_0|^2 \cos \Omega t}{\Omega^2}\right] \frac{\tilde{\varrho}_{11}(0)}{2} + \left[\frac{|\Omega_0|^2 (1 - \cos \Omega t)}{\Omega^2}\right] \frac{\varrho_{22}(0)}{2}$$
$$+ \frac{\Omega_0}{\Omega}\left[\frac{i\Omega \sin \Omega t - \delta (1 - \cos \Omega t)}{\Omega}\right] \frac{\tilde{\varrho}'_{12}(0)}{2}$$
$$- \frac{\Omega_0^*}{\Omega}\left[\frac{i\Omega \sin \Omega t + \delta (1 - \cos \Omega t)}{\Omega}\right] \frac{\tilde{\varrho}'_{21}(0)}{2}, \tag{3.55a}$$

$$\varrho_{22}(t) = \left[\frac{|\Omega_0|^2 (1 - \cos \Omega t)}{\Omega^2}\right] \frac{\varrho_{11}(0)}{2} + \left[1 + \frac{\delta^2 + |\Omega_0|^2 \cos \Omega t}{\Omega^2}\right] \frac{\varrho_{22}(0)}{2}$$
$$- \frac{\Omega_0}{\Omega}\left[\frac{i\Omega \sin \Omega t - \delta (1 - \cos \Omega t)}{\Omega}\right] \frac{\tilde{\varrho}'_{12}(0)}{2}$$
$$+ \frac{\Omega_0^*}{\Omega}\left[\frac{i\Omega \sin \Omega t + \delta (1 - \cos \Omega t)}{\Omega}\right] \frac{\tilde{\varrho}'_{21}(0)}{2}, \tag{3.55b}$$

$$\tilde{\varrho}'_{12}(t) = \frac{\Omega_0^*}{\Omega}\left[\frac{i\Omega \sin \Omega t - \delta (1 - \cos \Omega t)}{\Omega}\right] \frac{\varrho_{11}(0)}{2}$$
$$- \frac{\Omega_0^*}{\Omega}\left[\frac{i\Omega \sin \Omega t - \delta (1 - \cos \Omega t)}{\Omega}\right] \frac{\varrho_{22}(0)}{2}$$
$$+ \left[\frac{|\Omega_0|^2 + \left(\Omega^2 + \delta^2\right) \cos \Omega t}{\Omega^2} + \frac{2i\delta \sin \Omega t}{\Omega}\right] \frac{\tilde{\varrho}'_{12}(0)}{2}$$
$$+ \left[\frac{\Omega_0^{*2} (1 - \cos \Omega t)}{\Omega^2}\right] \frac{\tilde{\varrho}'_{21}(0)}{2}, \tag{3.55c}$$

$$\tilde{\varrho}'_{21}(t) = -\frac{\Omega_0}{\Omega}\left[\frac{i\Omega \sin \Omega t + \delta (1 - \cos \Omega t)}{\Omega}\right] \frac{\varrho_{11}(0)}{2}$$
$$+ \frac{\Omega_0}{\Omega}\left[\frac{i\Omega \sin \Omega t + \delta (1 - \cos \Omega t)}{\Omega}\right] \frac{\varrho_{22}(0)}{2} + \left[\frac{\Omega_0{}^2 (1 - \cos \Omega t)}{\Omega^2}\right] \frac{\tilde{\varrho}'_{12}(0)}{2}$$
$$+ \left[\frac{|\Omega_0|^2 + \left(\Omega^2 + \delta^2\right) \cos \Omega t}{\Omega^2} - \frac{2i\delta \sin \Omega t}{\Omega}\right] \frac{\tilde{\varrho}'_{21}(0)}{2}, \tag{3.55d}$$

which is obtained most easily from the solutions of the amplitude equations. The solution of the density matrix equations (3.54) can be obtained from these equations with the replacements $\tilde{\boldsymbol{\varrho}}' \Longrightarrow \tilde{\boldsymbol{\varrho}}$, $\Omega_0 \Longrightarrow |\Omega_0|$.

3.4 Semiclassical Dressed States

The transformation to the dressed states is given by

$$\varrho_d = \mathrm{T}\tilde{\varrho}\mathrm{T}^\dagger, \tag{3.56}$$

where

$$\mathrm{T} = \begin{pmatrix} \cos\theta & -\sin\theta \\ \sin\theta & \cos\theta \end{pmatrix}. \tag{3.57}$$

We use the second definition of the field interaction representation, with $\delta(t) = \delta + \dot{\phi}(t)$ and $\tan(2\theta) = |\Omega_0(t)|/\delta(t) = 2|\chi(t)|/\delta(t)$. In general, both θ and T are functions of t, although this dependence is not indicated explicitly. Carrying out the matrix multiplication, we find

$$\varrho_d = \begin{pmatrix} c^2\varrho_{11} - sc\,(\tilde{\varrho}_{12} + \tilde{\varrho}_{21}) + s^2\varrho_{22} & c^2\tilde{\varrho}_{12} - sc\,(\varrho_{22} - \varrho_{11}) - s^2\tilde{\varrho}_{21} \\ c^2\tilde{\varrho}_{21} - sc\,(\varrho_{22} - \varrho_{11}) - s^2\tilde{\varrho}_{12} & c^2\varrho_{22} + sc\,(\tilde{\varrho}_{12} + \tilde{\varrho}_{21}) + s^2\varrho_{11} \end{pmatrix}, \tag{3.58}$$

where $c = \cos\theta$, $s = \sin\theta$. The inverse transformation is obtained by letting $\theta \Rightarrow -\theta$.

To get the differential equation for $\dot{\varrho}_d$, we use equations (3.56), (2.139), and (2.140) to write

$$\begin{aligned} i\hbar\dot{\varrho}_d &= [\mathrm{H}_d, \varrho_d] + \text{relaxation terms} \\ &= \left[\left\{\frac{\hbar}{2}\begin{pmatrix} -\Omega(t) & 0 \\ 0 & \Omega(t) \end{pmatrix} + i\hbar\dot{\theta}\begin{pmatrix} 0 & -1 \\ 1 & 0 \end{pmatrix}\right\}, \varrho_d\right] + \text{relaxation terms}, \end{aligned} \tag{3.59}$$

where

$$\Omega(t) = \sqrt{\delta^2(t) + |\Omega_0(t)|^2} = \sqrt{\delta^2(t) + 4|\chi(t)|^2}. \tag{3.60}$$

Since the relaxation terms in the field interaction representation can be written as

$$d\tilde{\varrho}/dt|_{relaxation} = -\gamma\left[\sigma_0\tilde{\varrho} + \tilde{\varrho}\sigma_0\right] + \gamma_2\sigma_-\tilde{\varrho}\sigma_+ + 2\Gamma\sigma_0\tilde{\varrho}\sigma_0, \tag{3.61}$$

the corresponding terms in the dressed representation are obtained by taking

$$d\varrho_d/dt|_{relaxation} = \mathrm{T}\begin{bmatrix} -\gamma\left(\sigma_0\mathrm{T}^\dagger\varrho_d\mathrm{T} + \mathrm{T}^\dagger\varrho_d\mathrm{T}\sigma_0\right) \\ +\gamma_2\sigma_-\mathrm{T}^\dagger\varrho_d\mathrm{T}\sigma_+ + 2\Gamma\sigma_0\mathrm{T}^\dagger\varrho_d\mathrm{T}\sigma_0 \end{bmatrix}\mathrm{T}^\dagger, \tag{3.62}$$

where the inverse transformation $\tilde{\varrho} = \mathrm{T}^\dagger\varrho_d\mathrm{T}$ has been used to express the field interaction density matrix in terms of the semiclassical dressed density matrix. Explicit expressions for the relaxation terms are listed in appendix A; they are more complicated than in the other representations.

The dressed states are useful mainly in the limit that $\gamma/\Omega(t) \ll 1$ and when the $\dot{\theta}$ terms can be neglected. In this *secular approximation*, the off-diagonal, dressed-state density matrix elements are negligibly small. They oscillate at frequency $\Omega(t)$ and decay at rate γ [see equation (3.103) in appendix A], so that their average value is of order $\gamma/\Omega(t)$. As a consequence, one can neglect $\varrho_{I,II}, \varrho_{II,I}$ to lowest order in $\gamma/\Omega(t)$. In general, the $\dot{\theta}$ terms can be neglected if $|\dot{\theta}/\Omega(t)| \ll 1$, as in the discussion of adiabatic following in section 2.6.1. In these limits, the evolution equations for

TABLE 3.1
Commonly Used Symbols

$\mathbf{E}(t) = \frac{1}{2}\hat{\epsilon}\left[E_0(t)e^{-i\omega t} + E_0^*(t)e^{i\omega t}\right]$	Electric field
$E_0(t) = \lvert E_0(t)\rvert\, e^{i\phi(t)}$	Complex electric field amplitude
$\hat{\mu}$	Atomic dipole moment operator
$\Omega_0(t) = -(\hat{\mu})_{21} \cdot \hat{\epsilon} E_0(t)/\hbar$	Rabi frequency
$\chi(t) = -(\hat{\mu})_{21} \cdot \hat{\epsilon} E_0(t)/2\hbar$	Rabi frequency/2
$\omega_0 = \omega_{21}$	Atomic transition frequency
$\delta = \omega_0 - \omega$	Atom-field detuning
$\delta(t) = \delta + \dot{\phi}(t)$	Modified atom-field detuning
$\Omega(t) = \sqrt{\lvert\Omega_0(t)\rvert^2 + \delta^2(t)}$	Generalized Rabi frequency
$\theta(t) = \frac{1}{2}\tan^{-1}\left[\lvert\Omega_0(t)\rvert/\delta(t)\right]$	Dressing angle
γ_2	Excited state decay rate
Γ	Dephasing decay rate
$\gamma_{12} = \gamma = \gamma_2/2 + \Gamma$	Decay rate for coherence ϱ_{12}

the dressed-state populations can be approximated as

$$\begin{aligned}\dot{\varrho}_{I,I} &= -\left[\gamma_2 \sin^2\theta + \frac{1}{4}(2\Gamma - \gamma_2)\sin^2(2\theta)\right]\varrho_{I,I} \\ &\quad + \left[\frac{1}{2}\Gamma\sin^2(2\theta) + \gamma_2\cos^4\theta\right]\varrho_{II,II}, \qquad (3.63a)\end{aligned}$$

$$\begin{aligned}\dot{\varrho}_{II,II} &= -\left[\gamma_2 \cos^2\theta + \frac{1}{4}(2\Gamma - \gamma_2)\sin^2(2\theta)\right]\varrho_{II,II} \\ &\quad + \left[\frac{1}{2}\Gamma\sin^2(2\theta) + \gamma_2\sin^4\theta\right]\varrho_{I,I}. \qquad (3.63b)\end{aligned}$$

These equations can be solved analytically for constant fields or numerically for time-varying fields. Note that, in the absence of relaxation (and with the neglect of the $\dot{\theta}$ terms), the dressed-state populations are constant. Of course, the semiclassical dressed-state representation is *defined* in a way to ensure that this is the case. Dressed states are particularly useful when there is one strong field (or one detuned field) and a second, probe field acting on the atoms. We consider several examples in subsequent chapters.

For convenience, the parameters of table 2.1 are relisted in table 3.1, along with the various decay rates and some dressed-state variables.

3.5 Bloch Vector

For density matrix elements, there is yet another representation that is very useful since it allows for a geometric interpretation of the evolution of the density matrix. Real variables are introduced in terms of density matrix elements in the field interaction representation as

$$\begin{aligned} u' &= \tilde{\varrho}'_{12} + \tilde{\varrho}'_{21}, \\ v' &= i\left(\tilde{\varrho}'_{21} - \tilde{\varrho}'_{12}\right), \\ w' &= \varrho_{22} - \varrho_{11}, \\ m &= \varrho_{22} + \varrho_{11}, \end{aligned} \tag{3.64}$$

with the inverse transform given by

$$\begin{aligned} \tilde{\varrho}'_{12} &= \frac{u' + iv'}{2}, \\ \tilde{\varrho}'_{21} &= \frac{u' - iv'}{2}, \\ \varrho_{11} &= \frac{m - w'}{2}, \\ \varrho_{22} &= \frac{m + w'}{2}, \end{aligned} \tag{3.65}$$

or, with the second definition of the field interaction representation,

$$\begin{aligned} u &= \tilde{\varrho}_{12} + \tilde{\varrho}_{21}, \\ v &= i\left(\tilde{\varrho}_{21} - \tilde{\varrho}_{12}\right), \\ w &= \varrho_{22} - \varrho_{11}, \\ m &= \varrho_{22} + \varrho_{11}, \end{aligned} \tag{3.66}$$

$$\begin{aligned} \tilde{\varrho}_{12} &= \frac{u + iv}{2}, \\ \tilde{\varrho}_{21} &= \frac{u - iv}{2}, \\ \varrho_{11} &= \frac{m - w}{2}, \\ \varrho_{22} &= \frac{m + w}{2}. \end{aligned} \tag{3.67}$$

Note that a matrix element such as $\varrho_{12} = \tilde{\varrho}'_{12}e^{i\omega t} = \tilde{\varrho}_{12}e^{i\omega t - i\phi(t)}$ can be written in terms of these variables as

$$\varrho_{12} = (u' + iv')e^{i\omega t}/2 = (u + iv)e^{i\omega t - i\phi(t)}/2. \tag{3.68}$$

It is especially important not to forget that the *Bloch vector*, having components $[(u, v, w, m)$ or $(u', v', w', m)]$, is related to density matrix elements in the field interaction representation. In calculating expectation values of operators, it is necessary to convert back to the Schrödinger representation.

The elements of the Bloch vector have a simple interpretation. The quantity m is the total population of the levels, and w is the population difference. For the electric dipole transitions under consideration, u and v correspond to components of the atomic dipole moment that are in phase and out of phase with the applied field. One often refers to u and v (as well as $\tilde{\varrho}_{12}$ and $\tilde{\varrho}_{21}$) as *coherence*.

In the presence of relaxation, it follows from these definitions and equations (3.54) that

$$\begin{aligned}
\dot{u}' &= -\delta v' - \mathrm{Im}\,[\Omega_0(t)]\, w' - \gamma u', \\
\dot{v}' &= \delta u' - \mathrm{Re}\,[\Omega_0(t)]\, w' - \gamma v', \\
\dot{w}' &= \mathrm{Re}\,[\Omega_0(t)]\, v' + \mathrm{Im}\,[\Omega_0(t)]\, u' - \gamma_2(w' + 1), \\
\dot{m} &= 0,
\end{aligned} \tag{3.69}$$

or

$$\begin{aligned}
\dot{u} &= -\delta(t)v - \gamma u, \\
\dot{v} &= \delta(t)u - |\Omega_0(t)|w - \gamma v, \\
\dot{w} &= |\Omega_0(t)|v - \gamma_2(w + 1), \\
\dot{m} &= 0.
\end{aligned} \tag{3.70}$$

Conservation of probability is expressed by the relation $m = 1$.

If we construct column vectors

$$\mathbf{\Omega}'(t) = \begin{pmatrix} \mathrm{Re}\,[\Omega_0(t)] \\ -\mathrm{Im}\,[\Omega_0(t)] \\ \delta \end{pmatrix}, \tag{3.71}$$

or

$$\mathbf{\Omega}(t) = \begin{pmatrix} |\Omega_0(t)| \\ 0 \\ \delta(t) \end{pmatrix} \tag{3.72}$$

and

$$\mathbf{B}' = \begin{pmatrix} u' \\ v' \\ w' \end{pmatrix}, \quad \mathbf{B} = \begin{pmatrix} u \\ v \\ w \end{pmatrix}, \tag{3.73}$$

then equations (3.69) and (3.70) take the vectorial form (neglecting relaxation)

$$d\mathbf{B}'/dt = \mathbf{\Omega}'(t) \times \mathbf{B}', \quad d\mathbf{B}/dt = \mathbf{\Omega}(t) \times \mathbf{B}. \tag{3.74}$$

Including relaxation, we have

$$d\mathbf{B}'/dt = \mathbf{\Omega}'(t) \times \mathbf{B}' - \begin{pmatrix} \gamma u' \\ \gamma v' \\ \gamma_2\,(w' + 1) \end{pmatrix}, \tag{3.75}$$

or

$$d\mathbf{B}/dt = \mathbf{\Omega}(t) \times \mathbf{B} - \begin{pmatrix} \gamma u \\ \gamma v \\ \gamma_2\,(w + 1) \end{pmatrix}. \tag{3.76}$$

There is a simple geometric interpretation of equation (3.74). In the absence of relaxation, the Bloch vector $\mathbf{B}$ (or $\mathbf{B}'$) precesses about the *pseudofield vector* $\mathbf{\Omega}(t)$ [or $\mathbf{\Omega}'(t)$] with angular frequency $\Omega(t)$ [or $\Omega'(t)$]. Often, it is easy to picture the interaction (especially in time-dependent problems) using the Bloch vector. To get

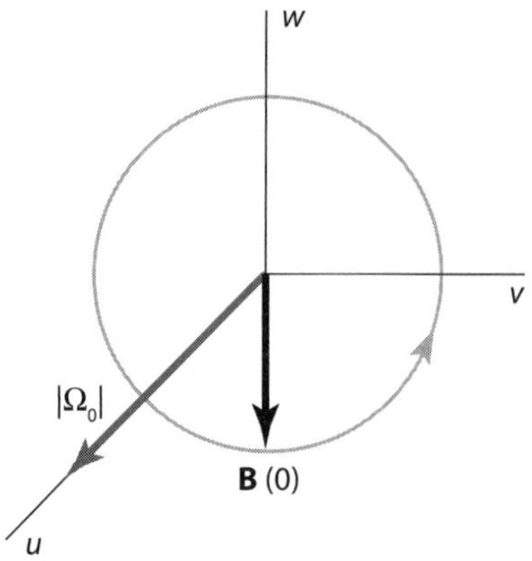

Figure 3.2. When $\delta(t) = 0$ and $|\Omega_0| =$ constant, the Bloch vector $\mathbf{B}(t)$ rotates in the (w, v) plane with angular velocity $|\Omega_0|$.

some idea of the dynamics, let us first consider problems in which relaxation can be neglected. We use equation (3.76) in all that follows. In general, equation (3.76) must be solved numerically.

3.5.1 No Relaxation

Since $\mathbf{B}$ precesses about the pseudofield vector, its magnitude must remain constant. We can show this explicitly by using equation (3.76) to write

$$\frac{d}{dt}|\mathbf{B}|^2 = \frac{d}{dt}(\mathbf{B}\cdot\mathbf{B}) = 2\mathbf{B}\cdot\frac{d\mathbf{B}}{dt} = 2\mathbf{B}\cdot[\boldsymbol{\Omega}(t)\times\mathbf{B}] = 0. \tag{3.77}$$

Therefore,

$$|\mathbf{B}|^2 = u^2 + v^2 + w^2 = \text{constant}. \tag{3.78}$$

With the definitions given in equations (3.66), we find

$$\begin{aligned} |\mathbf{B}|^2 &= u^2 + v^2 + w^2 \\ &= \tilde{\varrho}_{12}^2 + 2\tilde{\varrho}_{12}\tilde{\varrho}_{21} + \tilde{\varrho}_{21}^2 - \tilde{\varrho}_{12}^2 + 2\tilde{\varrho}_{12}\tilde{\varrho}_{21} - \tilde{\varrho}_{21}^2 + \tilde{\varrho}_{22}^2 - 2\tilde{\varrho}_{22}\tilde{\varrho}_{11} + \tilde{\varrho}_{11}^2 \\ &= \tilde{\varrho}_{22}^2 + 2\tilde{\varrho}_{22}\tilde{\varrho}_{11} + \tilde{\varrho}_{11}^2 = (\tilde{\varrho}_{22} + \tilde{\varrho}_{11})^2 = 1, \end{aligned} \tag{3.79}$$

provided that the relationship $\tilde{\varrho}_{12}\tilde{\varrho}_{21} = |a_1|^2|a_2|^2 = \tilde{\varrho}_{11}\tilde{\varrho}_{22}$ is used. Note that this relationship is valid only in the *absence* of any relaxation, allowing us to set $\varrho_{ij} = a_i a_j^*$. Since the magnitude of $\mathbf{B}$ is unity, the Bloch vector traces out a curve on the surface of the *Bloch sphere*, a sphere having radius unity in u, v, w space.

As an example, consider the case when $\delta(t) = 0$ and $|\Omega_0| =$ constant, with initial conditions $\varrho_{11}(0) = 1$ $[\tilde{\varrho}_{12}(0) = \tilde{\varrho}_{21}(0) = \varrho_{22}(0) = 0]$. This implies that $u(0) = v(0) = 0$ and $w(0) = -1$; see figure 3.2. Since $\mathbf{B}$ precesses about $\boldsymbol{\Omega} = |\Omega_0|\hat{\mathbf{u}}$,

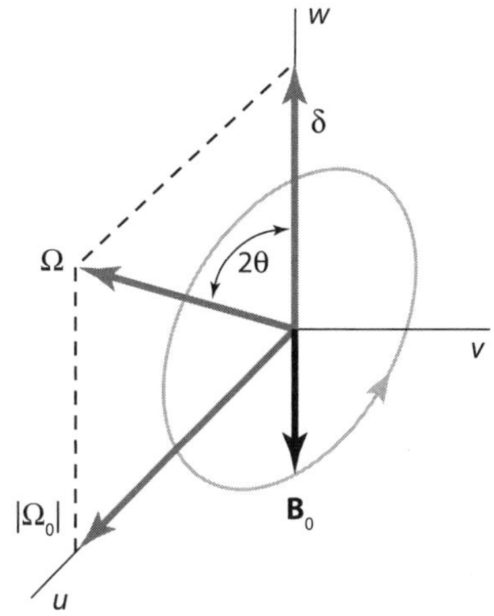

Figure 3.3. When δ = constant and $|\Omega_0|$ = constant, the Bloch vector $\mathbf{B}(t)$ precesses about the pseudofield vector $\boldsymbol{\Omega}$ with angular velocity Ω.

it follows that

$$\begin{aligned} u &= 0, \\ v &= \sin(|\Omega_0|t), \\ w &= -\cos(|\Omega_0|t). \end{aligned} \tag{3.80}$$

If, instead, $|\Omega_0(t)|$ corresponds to a time-varying pulse envelope that starts from 0 at $t = 0$, the precession phase angle at any time is given by

$$A(t) = \int_0^t |\Omega_0(t')|\, dt', \tag{3.81}$$

and equations (3.80) are replaced by

$$\begin{aligned} u &= 0, \\ v &= \sin A(t), \\ w &= -\cos A(t). \end{aligned} \tag{3.82}$$

Note that for times when $A(t) = \pi$, the population is completely inverted ($w = 1$), while for times when $A(t) = \pm\pi/2$, the coherence is at a maximum ($|v| = 1$).

Now let us assume that both the field amplitude and atom-field detuning are constant, $|\Omega_0(t)|$ = constant, δ = constant. The initial condition for the Bloch vector is taken as $\mathbf{B}(t = 0) = \mathbf{B}_0$. The Bloch vector $\mathbf{B}$ precesses about $\boldsymbol{\Omega}$ with rate $\Omega = (|\Omega_0|^2 + \delta^2)^{1/2}$; see figure 3.3. To solve this problem, it is convenient to rotate the w axis so that it is along $\boldsymbol{\Omega}$. This is a *passive transformation*. Alternatively, one can rotate the vector $\boldsymbol{\Omega}$ so that it is along the w axis, which is an *active transformation*. Let us try an active transformation with new vectors denoted by tildes ($\sim$).

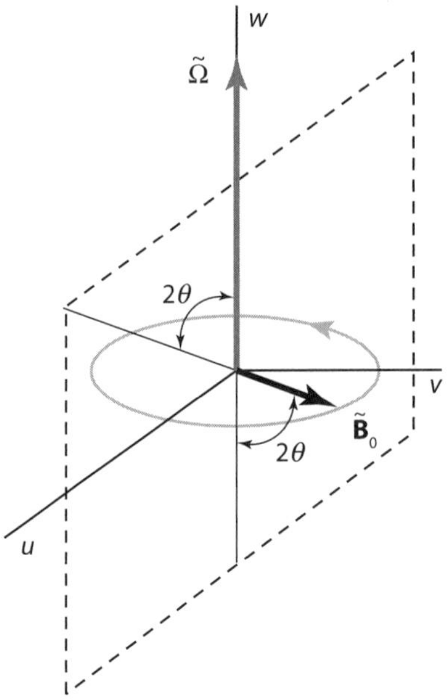

Figure 3.4. Pseudofield and Bloch vectors under the active rotation that aligns the pseudofield vector with the w axis. The Bloch vector $\tilde{\mathbf{B}}_0$ rotates in the plane $w = -\cos(2\theta)$.

The active rotation matrix that accomplishes this transformation is

$$\begin{aligned} \mathbf{D}(\alpha, \beta, \gamma) &= \mathbf{D}(0, -2\theta, 0) \\ &\equiv \mathbf{D}_v(-2\theta) = \begin{pmatrix} \cos 2\theta & 0 & -\sin 2\theta \\ 0 & 1 & 0 \\ \sin 2\theta & 0 & \cos 2\theta \end{pmatrix}, \end{aligned} \tag{3.83}$$

where (α, β, γ) are the Euler angles, and $\mathbf{D}_v(-2\theta)$ corresponds to an active rotation of (-2θ) about the v axis. With this rotation and recalling that $\tan(2\theta) = |\Omega_0|/\delta$, we find

$$\begin{aligned} \tilde{\Omega} &= \mathbf{D}_v(-2\theta)\boldsymbol{\Omega} \\ &= \begin{pmatrix} \cos 2\theta & 0 & -\sin 2\theta \\ 0 & 1 & 0 \\ \sin 2\theta & 0 & \cos 2\theta \end{pmatrix} \begin{pmatrix} |\Omega_0| \\ 0 \\ \delta \end{pmatrix} \\ &= \begin{pmatrix} |\Omega_0| \cos 2\theta - \delta \sin 2\theta \\ 0 \\ |\Omega_0| \sin 2\theta + \delta \cos 2\theta \end{pmatrix} = \begin{pmatrix} 0 \\ 0 \\ \Omega \end{pmatrix}, \end{aligned} \tag{3.84}$$

which brings the pseudofield vector along the w axis (figure 3.4).

Under this transformation,

$$\tilde{\mathbf{B}}_0 = \mathbf{D}_v(-2\theta)\mathbf{B}_0. \tag{3.85}$$

But now $\tilde{\mathbf{B}}_0$ simply precesses about the w axis such that

$$\tilde{\mathbf{B}}(t) = \mathbf{G}(t)\tilde{\mathbf{B}}_0, \tag{3.86}$$

where

$$\mathbf{G}(t) = \begin{pmatrix} \cos\Omega t & -\sin\Omega t & 0 \\ \sin\Omega t & \cos\Omega t & 0 \\ 0 & 0 & 1 \end{pmatrix}. \tag{3.87}$$

Therefore,

$$\tilde{\mathbf{B}} = \mathbf{G}\tilde{\mathbf{B}}_0 = \mathbf{G}\mathbf{D}_v(-2\theta)\mathbf{B}_0, \tag{3.88}$$

such that

$$\mathbf{B} = \mathbf{D}_v^{\dagger}(-2\theta)\tilde{\mathbf{B}} = \mathbf{D}_v(2\theta)\tilde{\mathbf{B}} = \mathbf{U}(\theta, t)\mathbf{B}_0, \tag{3.89}$$

with

$$\begin{aligned} \mathbf{U}(\theta, t) &= \mathbf{D}_v(2\theta)\mathbf{G}(t)\mathbf{D}_v(-2\theta) \\ &= \begin{pmatrix} \dfrac{|\Omega_0|^2 + \delta^2 \cos\Omega t}{\Omega^2} & -\dfrac{\delta}{\Omega}\sin\Omega t & \dfrac{|\Omega_0|\delta}{\Omega^2}(1-\cos\Omega t) \\ \dfrac{\delta}{\Omega}\sin\Omega t & \cos\Omega t & -\dfrac{|\Omega_0|}{\Omega}\sin\Omega t \\ \dfrac{|\Omega_0|\delta}{\Omega^2}(1-\cos\Omega t) & \dfrac{|\Omega_0|}{\Omega}\sin\Omega t & \dfrac{|\Omega_0|^2\cos\Omega t + \delta^2}{\Omega^2} \end{pmatrix}, \end{aligned} \tag{3.90}$$

which implies that

$$\begin{pmatrix} u(t) \\ v(t) \\ w(t) \end{pmatrix} = \begin{pmatrix} \dfrac{|\Omega_0|^2 + \delta^2 \cos\Omega t}{\Omega^2} & -\dfrac{\delta}{\Omega}\sin\Omega t & \dfrac{|\Omega_0|\delta}{\Omega^2}(1-\cos\Omega t) \\ \dfrac{\delta}{\Omega}\sin\Omega t & \cos\Omega t & -\dfrac{|\Omega_0|}{\Omega}\sin\Omega t \\ \dfrac{|\Omega_0|\delta}{\Omega^2}(1-\cos\Omega t) & \dfrac{|\Omega_0|}{\Omega}\sin\Omega t & \dfrac{|\Omega_0|^2\cos\Omega t + \delta^2}{\Omega^2} \end{pmatrix} \begin{pmatrix} u(0) \\ v(0) \\ w(0) \end{pmatrix}. \tag{3.91}$$

This result is of fundamental importance, since it gives the time evolution of density matrix elements for constant field and detuning. The result could have been obtained from the amplitude solutions, equations (2.91), by simply forming the required density matrix elements.

3.5.1.1 Bloch vector with time-dependent $\delta(t)$ and $|\boldsymbol{\Omega}_0(t)|$

It is also possible to use the Bloch vector with time-dependent

$$\boldsymbol{\Omega}(t) = |\Omega_0(t)|\hat{\mathbf{u}} + \delta(t)\hat{\mathbf{w}}. \tag{3.92}$$

Consider first the limit in which $\delta(t) = \delta = \text{constant} > 0$, $|\Omega_0(t = -\infty)| = 0$, and $|\Omega_0(t)|$ is a pulse envelope that varies slowly in time compared with $1/\Omega(t)$. This implies that the Rabi frequency $\Omega(t) = [|\Omega_0(t)|^2 + \delta^2]^{1/2}$ is sufficiently large

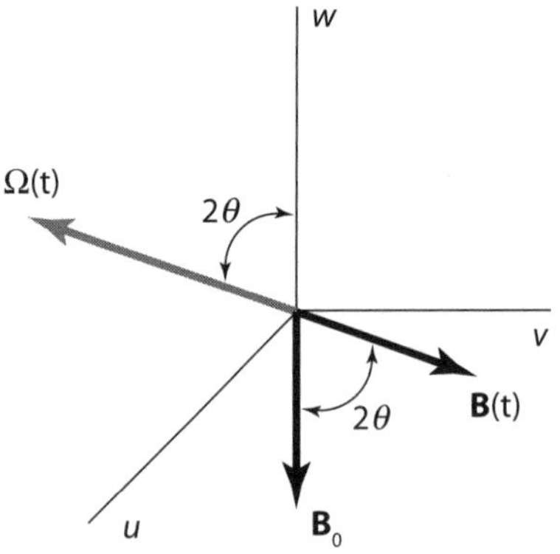

Figure 3.5. Under adiabatic following conditions, the Bloch vector remains aligned with the (negative of the) pseudofield vector, provided that $|\Omega_0(-\infty)| = 0$ and $\delta(-\infty) > 0$ (assuming that the atom is in state 1 at $t = -\infty$).

to ensure that the Bloch vector precesses many times around the pseudofield vector before the pseudofield vector changes as a result of changes in $|\Omega_0(t)|$. Consider what happens when $w(-\infty) = -1$ [$\varrho_{11}(-\infty) = 1$]. At $t = -\infty$, the pseudofield vector is aligned along the positive w axis and the Bloch vector is aligned along the negative w axis, in a direction opposite to the pseudofield vector. As the field $|\Omega_0(t)|$ begins to build, the pseudofield vector rotates sufficiently slowly in the wu plane to ensure that the amplitude of the precession remains very small. In other words, the Bloch vector adiabatically follows the (negative of the) pseudofield vector (figure 3.5).

As a result, the Bloch vector **B** stays in the wu plane, with

$$w(t) = -\cos[2\theta(t)], \quad v(t) = 0, \quad u(t) = -\sin[2\theta(t)], \tag{3.93a}$$

$$\varrho_{22}(t) = \sin^2[\theta(t)], \quad \varrho_{12}(t) = \varrho_{21}(t) = -\sin[\theta(t)]\cos[\theta(t)], \tag{3.93b}$$

where $\tan[2\theta(t)] = |\Omega_0(t)|/\delta$, as in the case of adiabatic following. These are the same results obtained using dressed states, provided that the angle $\theta(t)$ changes at a rate much less than $\Omega(t)$. The Bloch vector gives a geometrical perspective to this result.

Note that if $\delta(t)$ also changes slowly in time and changes sign from δ_0 to $-\delta_0$, then the Bloch vector adiabatically follows the pseudofield vector and rotates into the upper half plane (provided the minimum Rabi frequency is sufficiently large to ensure adiabaticity). At the end of the pulse, $|\Omega_0(\infty)| \sim 0$, the Bloch vector is aligned along the positive w axis—this implies that $\varrho_{22}(\infty) = 1$ and corresponds to adiabatic switching.

3.5.2 Relaxation Included

When relaxation is included, as in equations (3.70), the Bloch vector no longer has constant length, and there is no simple way to find $\mathbf{B}(t)$ for a given $\mathbf{B}_0$. With relaxation, the Bloch vector decreases in length and eventually reaches a steady-state value if the field amplitude and detuning are constant, corresponding to a *single* point *within* the Bloch sphere. Still, one can gain some insight into the dynamics of the atom–field interaction using the Bloch vector.

Consider the case $|\Omega_0| = \text{constant}$, $\delta = 0$, $\varrho_{11}(0) = 1$. If $|\Omega_0| \gg \gamma, \gamma_2$, the Bloch vector rotates many times in the w, v plane before it decays to its steady-state value. In doing so, it decays at a rate that is the average of the decay rates of w and v. Since w decays at rate γ_2 and v at rate γ, one would expect an average decay rate of $(\gamma_2 + \gamma)/2$. For $\gamma = \gamma_2/2$ (no collisions), this rate equals $\frac{3}{4}\gamma_2$.

If, on the other hand, the Bloch vector precesses at a rate that is much less than γ, γ_2, then the Bloch vector approaches its steady-state value in a monotonic fashion. This type of dynamics corresponds to a *rate equation approximation*, as you will see shortly. We will return to the Bloch vector when we consider atoms subjected to a sequence of optical pulses.

3.6 Summary

The single-particle density matrix has been introduced. You have seen that a density matrix approach is needed when we want to include irreversible processes (relaxation) into the atom–field dynamics. In particular, the role played by spontaneous emission and collisions has been discussed. The Bloch vector was also defined, in which complex density matrix elements are replaced by real components that can be given a simple physical interpretation. The Bloch vector also allows one to visualize the atom–field dynamics using a geometric picture. With the introduction of the density matrix, we are now prepared to look at some important applications of atom–field interactions.

3.7 Appendix A: Density Matrix Equations in the Rotating-Wave Approximation

3.7.1 Schrödinger Representation

$$\dot{\varrho}_{11} = -i\chi^*(t)e^{i\omega t}\varrho_{21} + i\chi(t)e^{-i\omega t}\varrho_{12} + \gamma_2\varrho_{22}, \tag{3.94a}$$

$$\dot{\varrho}_{22} = i\chi^*(t)e^{i\omega t}\varrho_{21} - i\chi(t)e^{-i\omega t}\varrho_{12} - \gamma_2\varrho_{22}, \tag{3.94b}$$

$$\dot{\varrho}_{12} = i\omega_0\varrho_{12} - i\chi^*(t)e^{i\omega t}(\varrho_{22} - \varrho_{11}) - \gamma\varrho_{12}, \tag{3.94c}$$

$$\dot{\varrho}_{21} = -i\omega_0\varrho_{21} + i\chi(t)e^{-i\omega t}(\varrho_{22} - \varrho_{11}) - \gamma\varrho_{21}, \tag{3.94d}$$

with

$$\chi(t) = \Omega_0(t)/2, \tag{3.95a}$$

$$\Omega_0(t) = -\mu_{21}E_0(t)/\hbar, \tag{3.95b}$$

$$\delta = \omega_0 - \omega, \tag{3.95c}$$

$$\delta(t) = \omega_0 - \omega + \dot{\phi}(t). \tag{3.95d}$$

3.7.2 Interaction Representation

$$\dot{\varrho}_{11} = -i\chi^*(t)e^{-i\delta t}\varrho^I_{21} + i\chi(t)e^{i\delta t}\varrho^I_{12} + \gamma_2\varrho_{22}, \tag{3.96a}$$

$$\dot{\varrho}_{22} = i\chi^*(t)e^{-i\delta t}\varrho^I_{21} - i\chi(t)e^{i\delta t}\varrho^I_{12} - \gamma_2\varrho_{22}, \tag{3.96b}$$

$$\dot{\varrho}^I_{12} = -i\chi^*(t)e^{-i\delta t}(\varrho_{22} - \varrho_{11}) - \gamma\varrho^I_{12}, \tag{3.96c}$$

$$\dot{\varrho}^I_{21} = i\chi(t)e^{i\delta t}(\varrho_{22} - \varrho_{11}) - \gamma\varrho^I_{21}. \tag{3.96d}$$

3.7.3 Field Interaction Representation

$$\dot{\varrho}_{11} = -i\chi^*(t)\tilde{\varrho}'_{21} + i\chi(t)\tilde{\varrho}'_{12} + \gamma_2\varrho_{22}, \tag{3.97a}$$

$$\dot{\varrho}_{22} = i\chi^*(t)\tilde{\varrho}'_{21} - i\chi(t)\tilde{\varrho}'_{12} - \gamma_2\varrho_{22}, \tag{3.97b}$$

$$d\tilde{\varrho}'_{12}/dt = -i\chi^*(t)(\varrho_{22} - \varrho_{11}) - (\gamma - i\delta)\,\tilde{\varrho}'_{12}, \tag{3.97c}$$

$$d\tilde{\varrho}'_{21}/dt = i\chi(t)(\varrho_{22} - \varrho_{11}) - (\gamma + i\delta)\,\tilde{\varrho}'_{21}, \tag{3.97d}$$

or

$$\dot{\varrho}_{11} = -i\,|\chi(t)|\,(\tilde{\varrho}_{21} - \tilde{\varrho}_{12}) + \gamma_2\varrho_{22}, \tag{3.98a}$$

$$\dot{\varrho}_{22} = i\,|\chi(t)|\,(\tilde{\varrho}_{21} - \tilde{\varrho}_{12}) - \gamma_2\varrho_{22}, \tag{3.98b}$$

$$d\tilde{\varrho}_{12}/dt = -i\,|\chi(t)|\,(\varrho_{22} - \varrho_{11}) - [\gamma - i\delta(t)]\,\tilde{\varrho}_{12}, \tag{3.98c}$$

$$d\tilde{\varrho}_{21}/dt = i\,|\chi(t)|\,(\varrho_{22} - \varrho_{11}) - [\gamma + i\delta(t)]\,\tilde{\varrho}_{21}. \tag{3.98d}$$

3.7.4 Bloch Vector

$$u' = \tilde{\varrho}'_{12} + \tilde{\varrho}'_{21}, \quad \tilde{\varrho}'_{12} = \frac{u' + iv'}{2}, \tag{3.99a}$$

$$v' = i\left(\tilde{\varrho}'_{21} - \tilde{\varrho}'_{12}\right), \quad \tilde{\varrho}'_{12} = \frac{u' - iv'}{2}, \tag{3.99b}$$

$$w' = \varrho_{22} - \varrho_{11}, \quad \varrho_{11} = \frac{m - w'}{2}, \tag{3.99c}$$

$$m = \varrho_{22} + \varrho_{11}, \quad \varrho_{22} = \frac{m + w'}{2}, \tag{3.99d}$$

$$\dot{u}' = -\delta v' - \mathrm{Im}\,[\Omega_0(t)]\,w' - \gamma u', \tag{3.100a}$$

$$\dot{v}' = \delta u' - \mathrm{Re}\,[\Omega_0(t)]\,w' - \gamma v', \tag{3.100b}$$

$$\dot{w}' = \mathrm{Re}\,[\Omega_0(t)]\,v' + \mathrm{Im}\,[\Omega_0(t)]\,u' - \gamma_2(w' + 1), \tag{3.100c}$$

$$\dot{m} = 0, \tag{3.100d}$$

or

$$u = \tilde{\varrho}_{12} + \tilde{\varrho}_{21}, \quad \tilde{\varrho}_{12} = (u + iv)/2, \tag{3.101a}$$

$$v = i(\tilde{\varrho}_{21} - \tilde{\varrho}_{12}), \quad \tilde{\varrho}_{21} = (u - iv)/2, \tag{3.101b}$$

$$w = \varrho_{22} - \varrho_{11}, \quad \varrho_{22} = (m + w)/2, \tag{3.101c}$$

$$m = \varrho_{11} + \varrho_{22}, \quad \varrho_{11} = (m - w)/2, \tag{3.101d}$$

$$\dot{u} = -\delta(t)v - \gamma u, \tag{3.102a}$$

$$\dot{v} = \delta(t)u - |\Omega_0(t)|\, w - \gamma v, \tag{3.102b}$$

$$\dot{w} = |\Omega_0(t)|\, v - \gamma_2(w + 1), \tag{3.102c}$$

$$\dot{m} = 0. \tag{3.102d}$$

3.7.5 Semiclassical Dressed-State Representation

Including decay in the dressed picture leads to rather complicated expressions. Explicitly, using equation (3.62), we find

$$\begin{aligned}\dot{\varrho}_{I,I} = &-\{\gamma_2 \sin^4\theta(t) + (\Gamma/2)\sin^2[2\theta(t)]\}\varrho_{I,I}\\ &-(1/8)\sin[4\theta(t)]\,(\gamma_2 - 2\Gamma)\,(\varrho_{I,II} + \varrho_{II,I})\\ &+\{(\Gamma/2)\sin^2[2\theta(t)] + \gamma_2\cos^4\theta(t)\}\varrho_{II,II}\\ &-\dot{\theta}(t)(\varrho_{I,II} + \varrho_{II,I}),\end{aligned} \tag{3.103a}$$

$$\begin{aligned}\dot{\varrho}_{II,II} = &-\{\gamma_2 \cos^4\theta(t) + (\Gamma/2)\sin^2[2\theta(t)]\}\varrho_{II,II}\\ &+(1/8)\sin[4\theta(t)]\,(\gamma_2 - 2\Gamma)\,(\varrho_{I,II} + \varrho_{II,I})\\ &+\{(\Gamma/2)\sin^2[2\theta(t)] + \gamma_2\sin^4\theta(t)\}\varrho_{I,I}\\ &+\dot{\theta}(t)(\varrho_{I,II} + \varrho_{II,I}),\end{aligned} \tag{3.103b}$$

$$\begin{aligned}\dot{\varrho}_{I,II} = &-\{\gamma - i\Omega(t) - (1/4)(2\Gamma - \gamma_2)\sin^2[2\theta(t)]\}\varrho_{I,II}\\ &+(1/4)\sin[2\theta(t)]\,\{\gamma_2 + 2\Gamma\cos[2\theta(t)] + 2\gamma_2\sin^2\theta(t)\}\varrho_{I,I}\\ &+(1/4)\sin[2\theta(t)]\,\{\gamma_2 - 2\Gamma\cos[2\theta(t)] + 2\gamma_2\cos^2\theta(t)\}\varrho_{II,II}\\ &+\{(1/4)(2\Gamma - \gamma_2)\sin^2[2\theta(t)]\}\varrho_{II,I}\\ &-\dot{\theta}(t)(\varrho_{II,II} - \varrho_{I,I}),\end{aligned} \tag{3.103c}$$

$$\begin{aligned}\dot{\varrho}_{II,I} = &-\{\gamma + i\Omega(t) - (1/4)(2\Gamma - \gamma_2)\sin^2[2\theta(t)]\}\varrho_{II,I}\\ &+(1/4)\sin[2\theta(t)]\,\{\gamma_2 + 2\Gamma\cos[2\theta(t)] + 2\gamma_2\sin^2\theta(t)\}\varrho_{I,I}\\ &+(1/4)\sin[2\theta(t)]\,\{\gamma_2 - 2\Gamma\cos[2\theta(t)] + 2\gamma_2\cos^2\theta(t)\}\varrho_{II,II}\\ &+\{(1/4)(2\Gamma - \gamma_2)\sin^2[2\theta(t)]\}\varrho_{I,II}\\ &-\dot{\theta}(t)(\varrho_{II,II} - \varrho_{I,I}),\end{aligned} \tag{3.103d}$$

where

$$\Omega(t) = \{[\delta(t)]^2 + |\Omega_0(t)|^2\}^{1/2}, \tag{3.104a}$$

$$\tan[2\theta(t)] = |\Omega_0(t)|/\delta(t), \tag{3.104b}$$

$$\gamma = \frac{\gamma_2}{2} + \Gamma, \tag{3.104c}$$

and Γ is the collision rate.

3.8 Appendix B: Collision Model

In many experiments involving atomic vapors, one attempts to work at sufficiently low pressures to eliminate the effects of atomic collisions. There are cases, however, where collisions play an important role in modifying the manner in which atoms interact with external radiation fields. Generally speaking, collisions affect atoms in two, inseparable fashions. During a collision, the energy levels of the atom are shifted, but they return to their unperturbed values following the collision. In addition, the velocity of an atom can change as a result of the collision. Both processes modify the absorption or emission profiles associated with atom–field interactions. In this appendix, we restrict our discussion to changes in atomic energy levels during a collision, but it should be noted that the velocity changes can also modify spectral profiles, especially in the case of nonlinear spectroscopy [1].

Our purpose here is to offer a very simple collision model. It is assumed that the atomic vapor consists of *active atoms* that interact with external fields and a *perturber bath* whose atoms undergo collisions with the active atoms but interact negligibly with the external fields. As a typical system, one can envision alkali metal atoms as the active atoms and rare gas atoms as the perturber bath. For radiation in the visible or near infrared, ground-state rare gas atoms are virtually unaffected by such radiation fields, while alkali metal atoms can absorb readily at specific frequencies in this range.

A number of simplifying assumptions can be made if one considers the various timescales in the problem. For thermal vapors, a typical collision duration is $\tau_c = 1.0$ ps, assuming a collision radius of 5 Å and a relative velocity of 5×10^2 m/s between collision partners. The collision duration τ_c sets the timescale. Since optical transition frequencies ω_0 are much larger than τ_c^{-1}, collisions do not possess the Fourier components needed to effectively induce optical transitions. Thus, one can assume that collisions do not change state populations, as we did in writing equations (3.30a) and (3.30b). Moreover, *during* a collision, any evolution of the atomic state vector in the interaction representation (with the rapidly varying optical frequency removed) is totally negligible provided that $\Omega(t)\tau_c \ll 1$ and $\dot{\Omega}(t)\tau_c/\Omega(t)^2 \ll 1$, where $\Omega(t)$ is the generalized Rabi frequency. Of course, one's ability to talk about individual collisions implicitly relies on the condition that the duration of a collision is much less than the time between collisions, Γ^{-1}, where Γ is the collision rate. Typically, $\Gamma/2\pi$ is of order 10 MHz/Torr of perturber pressure, allowing us to use this *binary collision approximation* for perturber pressures less than a few hundred Torr. Together, the binary collision approximation, $\Gamma\tau_c \ll 1$, and the conditions $\Omega(t)\tau_c \ll 1$ and $\dot{\Omega}(t)\tau_c/\Omega(t)^2 \ll 1$ comprise the *impact approximation*. In the impact approximation, the atomic evolution is frozen during a collision, aside from the rapidly varying oscillation associated with the atomic frequency. In what follows, we assume that the impact approximation holds.

If one neglects the velocity-changing nature of the collisional interaction, it is relatively simple to understand how collisions modify off-diagonal density matrix elements. (As noted earlier, diagonal density matrix elements are unaffected by collisions that do not induce transitions.) In the interaction representation, we consider the change in an off-diagonal density matrix element in a time interval Δt that is much larger than the collision duration τ_c, but much smaller than the time between collisions Γ^{-1}. In other words, it is highly unlikely that two collisions can occur in Δt, but Δt is sufficiently large to contain an entire collision.

During a collision characterized by impact parameter b and relative speed v_r, the state amplitude c_j in the interaction representation evolves according to

$$i\hbar\dot{c}_j = E_j(b, v_r, t)c_j, \tag{3.105}$$

where $E_j(b, v_r, t)$ is the collisional energy shift of level j, resulting, for example, from van der Waals interactions. All other contributions to $\dot{c}_j$ can be neglected during a collision owing to the impact approximation. Thus, as a result of a collision, one finds

$$c_j(t + \Delta t) = e^{i\phi_j(b,v_r)}c_j(t), \tag{3.106}$$

where the phase shift acquired in the collision is

$$\phi_j(b, v_r) = -\frac{1}{\hbar}\int_t^{t+\Delta t} E_j(b, v_r, t')\,dt' \approx -\frac{1}{\hbar}\int_{-\infty}^{\infty} E_j(b, v_r, t)\,dt, \tag{3.107}$$

and one can replace $\int_t^{t+\Delta t}$ by $\int_{-\infty}^{\infty}$, since the interval contains an entire collision.

The change in the density matrix element ϱ_{ij}^I in a time interval Δt averaged over collision parameters is given by

$$\begin{aligned}\left\langle \Delta\varrho_{ij}^I(t)\right\rangle &= \left\langle P(b, v_r)\left[c_i(t+\Delta t)c_j^*(t+\Delta t) - c_i(t)c_j^*(t)\right]\right\rangle \Delta t \\ &= \left\langle P(b, v_r)\left[e^{i\phi_{ij}(b,v_r)} - 1\right] c_i(t)c_j^*(t)\right\rangle \Delta t \\ &= \left\langle P(b, v_r)\left[e^{i\phi_{ij}(b,v_r)} - 1\right] \varrho_{ij}^I(t)\right\rangle \Delta t,\end{aligned} \tag{3.108}$$

where $P(b, v_r) = 2\pi b N v_r$ is the probability density (in impact parameter and relative speed) per unit time for a collision to occur with impact parameter b and relative speed v_r, N is the perturber density, and

$$\phi_{ij}(b, v_r) = \phi_i(b, v_r) - \phi_j(b, v_r). \tag{3.109}$$

If it is assumed that each collision is independent of the past, one can factorize

$$\left\langle P(b, v_r)\left[e^{i\phi_{ij}(b,v_r)} - 1\right] \varrho_{ij}^I(t)\right\rangle = \left\langle P(b, v_r)\left[e^{i\phi_{ij}(b,v_r)} - 1\right]\right\rangle \left\langle \varrho_{ij}^I(t)\right\rangle \tag{3.110}$$

and write

$$\frac{\left\langle \Delta \varrho_{ij}^{I}(t) \right\rangle}{\Delta t} = -\Gamma_{ij}^{(c)}(\mathbf{v}) \left\langle \varrho_{ij}^{I}(t) \right\rangle, \tag{3.111}$$

where

$$\Gamma_{ij}^{(c)}(\mathbf{v}) = \Gamma(\mathbf{v}) + iS(\mathbf{v}) = N \int 2\pi b db \int d\mathbf{v}_p \left| \mathbf{v} - \mathbf{v}_p \right| W_p(\mathbf{v}_p)[1 - e^{i\phi_{ij}(b, |\mathbf{v} - \mathbf{v}_p|)}] \tag{3.112}$$

is a complex collision rate, and $W_p(\mathbf{v}_p)$ is the perturber velocity distribution. Note that the relative velocity $\mathbf{v}_r$ has been replaced by $(\mathbf{v} - \mathbf{v}_p)$, where $\mathbf{v}$ is the active atom velocity, and $\mathbf{v}_p$ is the perturber velocity, allowing us to average over the perturber velocity distribution. As a result, the collision width $\Gamma(\mathbf{v})$ and shift $S(\mathbf{v})$ parameters are actually functions of the active atom velocity. In most treatments of collisions, the velocity dependence is often ignored, and the average in equation (3.112) is carried out using the relative velocity distribution rather than the perturber velocity distribution [2].

Since Δt is, in effect, an infinitesimal time in the interaction representation, we can take the limit $\Delta t \sim 0$ in equation (3.111) to obtain

$$\dot{\varrho}_{ij}^{I}(t)_{coll} = -\Gamma_{ij}^{(c)}(\mathbf{v})\varrho_{ij}^{I}, \tag{3.113}$$

where ϱ_{ij}^{I} is now a reduced density matrix element for the active atoms. In this simple *phase-interrupting* collision model, collisions result in a complex decay rate for off-diagonal, electronic state density matrix elements. Equation (3.113) agrees with equation (3.30) if one neglects the shift $S(\mathbf{v})$ and the velocity dependence in $\Gamma(\mathbf{v})$, effectively replacing $\Gamma_{ij}^{(c)}(\mathbf{v})$ by Γ.

Problems

1. Calculate the expectation value of the position operator $\mathbf{r}(t) = \langle \psi(t)|\hat{\mathbf{r}}|\psi(t)\rangle$ for a two-level atom whose levels have opposite parity in terms of density matrix elements in the Schrödinger, interaction, and field interaction representations.
2. Repeat the calculation of problem 1 in the dressed basis, and show that the expectation value of $\mathbf{r}(t)$ depends on both diagonal and off-diagonal density matrix elements in the dressed basis. Be careful to remember that $\langle \hat{\mathbf{r}} \rangle = \langle \psi(t)|\hat{\mathbf{r}}|\psi(t)\rangle$, but that the transformation to the dressed basis is given in terms of density matrix elements in the field interaction representation. In the secular approximation ($\varrho_{I,II} \sim 0$), show that the expectation value depends only on the difference of dressed-state populations.
3. Write explicit expressions for $\tilde{\varrho}(t)$ in terms of $\tilde{\varrho}(0)$ for the two-level problem without decay, assuming a constant field amplitude and phase.
4. In the two-level problem without decay, assume that $|\chi/\delta| \ll 1$ and $|\dot{\chi}/\chi\delta| \ll 1$. Show that the upper-state population is given approximately by $\varrho_{22}(t) = |\chi(t)/\delta|^2$ if the atom starts in state 1 at $t = -\infty$. The population adiabatically follows the field.
5. Show explicitly that the relaxation terms in equations (3.31a) to (3.31d) can be written as $-\gamma\left[\sigma_0\varrho(t) + \varrho(t)\sigma_0\right] + \gamma_2\sigma_-\varrho(t)\sigma_+ + 2\Gamma\sigma_0\varrho(t)\sigma_0$.

6. Solve the Bloch equations numerically for $\gamma = \gamma_2/2 = 1/2$ and ($\delta = 0.1$, $|\Omega_0(t)| = 0.2$) and ($\delta = 0.1$, $|\Omega_0(t)| = 3$), and plot w as a function of time, assuming that the atom is initially in its ground state. In which case does the Bloch vector approach its steady-state value monotonically?
7. From the Bloch equations without decay, prove that for constant $\boldsymbol{\Omega}$, the angle between the Bloch vector and the pseudofield vector remains constant. What is the geometric interpretation of this result? Now, imagine that $\boldsymbol{\Omega}(t)$ is a slowly varying function of t compared to $1/\Omega(t)$. Prove that if initially the Bloch vector is parallel to the pseudofield vector, it remains so (approximately).
8. Consider the two-level problem with $\gamma = \gamma_2/2$. Show that, if $\delta = 0$ and $\varrho_{11}(0) = 1$, then for a constant field amplitude, the upper-state population is given by

$$\varrho_{22}(t) = \frac{|\Omega_0|^2/2}{2\gamma^2 + |\Omega_0|^2}\left\{1 - [\cos(\lambda t) + \frac{3\gamma}{2\lambda}\sin(\lambda t)]e^{-3\gamma t/2}\right\},$$

where $\lambda = (|\Omega_0|^2 - \gamma^2/4)^{1/2}$. Evaluate ϱ_{22} for $|\Omega_0| \gg \gamma$, and give an interpretation in terms of the Bloch vector.
9. Look at equations (3.103) for the density matrix in the semiclassical dressed-state basis, neglecting the $\dot{\theta}$ terms. You need not derive these equations, which follow from the definition of $\boldsymbol{\varrho}_d$ and some straightforward, but cumbersome algebra. Why are diagonal and off-diagonal density matrix elements coupled by terms proportional to the decay rates? Obtain an approximate, steady-state solution to the dressed-state density matrix equations that is valid when $\Omega(t) \gg \gamma, \gamma_2$—that is, obtain steady-state solutions that are correct to zeroth order in $\gamma/\Omega(t)$, $\gamma_2/\Omega(t)$, and $\Gamma/\Omega(t)$. This limit is known as the *secular limit* and, for all practical purposes, is the only limit where the dressed-state basis is particularly useful in problems involving decay.
10. In the field interaction representation, neglecting relaxation, the state vector can be written quite generally as

$$|\psi(t)\rangle = \cos(\theta/2)|\tilde{1}(t)\rangle + \sin(\theta/2)e^{i\phi}|\tilde{2}(t)\rangle,$$

where θ and ϕ are arbitrary real functions of time with $0 \leqslant \theta \leqslant \pi$ and $0 \leqslant \phi \leqslant 2\pi$. Show that the angles θ and ϕ correspond to the spherical angles of the Bloch vector on the Bloch sphere.
11. Simplify equations (3.103) in the limit that there are no collisions, $\Gamma = 0$, and $\gamma = \gamma_2/2$.

References

[1] For a review, see P. R. Berman, *Collisions in atomic vapors*, in *New Trends in Atomic Physics*, Les Houches, Session XXXVIII, edited by G. Grynberg and R. Stora (North-Holland, Amsterdam, 1984), pp. 451–514.

[2] P. R. Berman, *Speed-dependent collisional width and shift parameters in spectral profiles*, Journal of Quantitative Spectroscopy and Radiative Transfer **12**, 1331–1342 (1972).

Bibliography

The density matrix equations are discussed in most of the books cited in chapter 1.

J. Ostmeyer and J. Gea-Banacloche, http://comp.uark.edu/~jgeabana/blochapps/blocheqs2.html *Bloch spheres and Q-functions* (2006). This website contains an applet that allows you to visualize the evolution of the Bloch vector on the Bloch sphere for arbitrary input parameters.

4

Applications of the Density Matrix Formalism

4.1 Density Matrix for an Ensemble

We now have the basic tools in place to carry out a number of calculations that are associated with problems of fundamental importance in matter–field interactions. The topics to be discussed include the absorption coefficient and spectral width, a simple treatment of atomic motion valid in plane-wave fields, and the rate equation approximation (chapter 4); atoms in standing-wave fields—laser cooling (chapter 5); coupling to the radiation field—Maxwell-Bloch equations (chapter 6); atoms in two fields—saturation spectroscopy (chapter 7); three-level atoms in two fields—Autler-Townes splitting (chapter 8); dark states and slow light (chapter 9); and coherent transients—free induction decay, photon echo, Ramsey fringes, and frequency combs (chapter 10).

Up to this point, we have considered the density matrix of a single atom. At this stage, we need to introduce an *ensemble* of atoms that interact with an external field. As long as the atoms interact *independently* with the external fields and do not interact directly with each other, it is an easy matter to generalize the single-particle density matrix to a density matrix for the ensemble of atoms. One simply defines

$$\mathcal{R}(\mathbf{R}, \mathbf{v}, t) = \frac{1}{N}\sum_j \int d\mathbf{R}_j d\mathbf{v}_j \mathcal{R}^{(j)}(\mathbf{R}_j, \mathbf{v}_j, t)\delta(\mathbf{R} - \mathbf{R}_j)\delta(\mathbf{v} - \mathbf{v}_j), \tag{4.1}$$

where $\delta(\mathbf{R} - \mathbf{R}_j)$ is a Dirac delta function, $\mathcal{R}^{(j)}(\mathbf{R}_j, \mathbf{v}_j, t)$ is the single-particle density matrix of atom j, and N is the total number of atoms. We have replaced the symbol ϱ with $\mathcal{R}$ to indicate that $\mathcal{R}^{(j)}(\mathbf{R}_j, \mathbf{v}_j, t)$ is a density in both coordinate and velocity space, whereas ϱ is a dimensionless, single-particle density matrix.

Equation (4.1) may seem innocuous, but it hides some problems in electrodynamics that have yet to receive a satisfactory solution. For the moment, we consider that $\mathbf{R}$ and $\mathbf{v}$ are classical variables—that is, the atomic center-of-mass motion is treated classically. If this were not the case, it would be impossible to define a single-particle density matrix that is a function of both position and velocity (momentum), since the position and momentum operators do not commute. As is shown in chapter 5, however, this is often not a fundamental problem if one interprets $\mathbf{R}_j$ and $\mathbf{v}_j$ as the average position and velocity of a wave packet associated with the atomic center-of-mass motion. What remains as a fundamental problem, however, is one related to *macroscopic* rather than *microscopic* descriptions of the atomic medium.

For classical center-of-mass motion, the single-particle density matrix can be written as

$$\mathcal{R}^{(j)}(\mathbf{R}_j, \mathbf{v}_j, t) = \varrho^{(j)}(\mathbf{R}_j, \mathbf{v}_j, t)\delta\left[\mathbf{R}_j - \mathbf{R}_j(t)\right]\delta[\mathbf{v}_j - \mathbf{v}_j(t)], \tag{4.2}$$

where $\mathbf{R}_j(t)$ and $\mathbf{v}_j(t)$ are the classical position and velocity of atom j, and $\varrho^{(j)}(\mathbf{R}_j, \mathbf{v}_j, t)$, the single-particle density matrix of atom j, is dimensionless. The density matrix $\mathcal{R}(\mathbf{R}, \mathbf{v}, t)$ consists of a sum of delta functions. Often, one tries to obtain a continuous distribution for quantities of this nature by averaging over a volume that contains many atoms, assuming that macroscopic quantities (such as phase space density) vary only slightly in such a volume. In electrodynamics, the volume is typically taken to be smaller than an optical wavelength, but in our case, such a procedure is not valid since we usually consider densities where there are very few atoms in a sphere whose radius corresponds to an optical wavelength. Instead, to go over to continuous variables, we choose a volume whose dimensions are much larger than a wavelength, but one for which the density $\mathcal{N}(\mathbf{R})$ is approximately constant *in the absence of the applied fields*. In velocity space, the macroscopic "volume" $(\Delta v)^3$ is taken to be much smaller than u^3, where u is the most probable speed associated with the velocity distribution $W_0(\mathbf{v})$ that characterizes the atomic ensemble in the absence of any applied fields.

In other words, we take as our density matrix

$$\mathcal{R}(\mathbf{R}, \mathbf{v}, t) = \mathcal{N}(\mathbf{R})\, W_0(\mathbf{v})\, \varrho(\mathbf{R}, \mathbf{v}, t), \tag{4.3}$$

where $\varrho(\mathbf{R}, \mathbf{v}, t)$ is the single-particle density matrix of an atom located at position $\mathbf{R}$ having velocity $\mathbf{v}$ at time t. It is important to note that, in the *presence* of the applied fields, the single-particle density matrix can vary significantly over distances comparable to a wavelength, as well as in velocity space over velocities whose magnitudes are small compared with u, whereas $\mathcal{N}(\mathbf{R})$ and $W_0(\mathbf{v})$ are approximately constant in these ranges. Conservation of probability in a single atom implies that[1]

$$\sum_{\alpha} \varrho_{\alpha\alpha}(\mathbf{R}, \mathbf{v}, t) = 1. \tag{4.4}$$

Unless noted otherwise, we will assume the density $\mathcal{N}$ is constant in the volume of interest.

Since each atom interacts independently with the external fields, the results we obtained previously for the single-particle density matrix can be taken over directly for the multiple-atom case. The only difference is that we can no longer specify the position as that of a single atom (previously taken as $\mathbf{R} = 0$) but must allow it to be a general variable $\mathbf{R}$. Moreover, we must keep track of the manner in which the atomic response depends on the center-of-mass velocity of the atom.

[1] If collisions are present that redistribute the velocities, equation (4.4) is replaced by $\sum_\alpha \int d\mathbf{v}\, W_0(\mathbf{v})\, \varrho_{\alpha\alpha}(\mathbf{R}, \mathbf{v}, t) = 1$.

For the time being, we neglect atomic motion. In that case, for a plane-wave optical field having electric field vector

$$\begin{aligned}\mathbf{E}(\mathbf{R},t) &= \hat{\boldsymbol{\epsilon}}\,|E_0(t)|\cos\left[\mathbf{k}\cdot\mathbf{R}-\omega t-\varphi(t)\right]\\ &= \frac{1}{2}\hat{\boldsymbol{\epsilon}}|E_0(t)|\left[e^{i\varphi(t)}e^{i(\mathbf{k}\cdot\mathbf{R}-\omega t)}+e^{-i\varphi(t)}e^{-i(\mathbf{k}\cdot\mathbf{R}-\omega t)}\right],\end{aligned} \tag{4.5}$$

where $\mathbf{k}$ is the field propagation vector, and $\hat{\boldsymbol{\epsilon}}$ is the field polarization, with $\mathbf{k}\cdot\hat{\boldsymbol{\epsilon}}=0$ and $k=\omega/c$, all the previous equations for the time evolution of the single-particle density matrix elements remain valid, provided one replaces $\chi(t)$ with $\chi(\mathbf{R},t)=\chi(t)e^{i\mathbf{k}\cdot\mathbf{R}}$ or $\Omega_0(t)$ with $\Omega_0(\mathbf{R},t)=\Omega_0(t)e^{i\mathbf{k}\cdot\mathbf{R}}$ in those equations.

4.2 Absorption Coefficient—Stationary Atoms

We first calculate the absorption coefficient for a monochromatic optical field incident on an ensemble of stationary two-level atoms. This is as basic as an atom–field interaction gets. The approach we follow in this chapter is a simple one—we calculate the steady-state density matrix of the atoms in the presence of the field and infer the loss rate for the field. In chapter 6, we use the density matrix equations, combined with Maxwell's equations, to arrive at the same result.

The term *absorption* can be a bit misleading. What does it mean for optical radiation to be absorbed by a medium? A perfectly reasonable definition would be that the incident energy is transformed into internal energy of the medium, as can happen in solids and liquids. In this case, the medium would heat as a result of the absorption. On the other hand, when light is incident on a low-density atomic vapor, a different type of process occurs. The radiation results in some steady-state excited-state population of the atoms that can be viewed as heating, but this heating is a transient process. Once the steady-state population is established, there is no additional heating; nevertheless, the incident field continues to lose energy. The energy loss mechanism involves *scattering* of the incident field by the atoms into unoccupied modes of the vacuum field. Neglecting atomic motion, the average frequency of this scattered radiation is always the frequency of the incident field and *not* the atomic transition frequency, consistent with a scattering interpretation. We discuss this interpretation in greater detail in Chapter 20. Thus, when we talk of absorption in this section and throughout the text, we actually refer to the loss of energy by the incident field and not to an increase in energy of the medium.

You might ask why it is important to study absorption profiles—that is, the absorption of a field by an ensemble of atoms as a function of the frequency of the applied fields. Historically, such profiles served as a blueprint for atomic structure. The frequencies giving rise to maximum absorption are related to atomic transition frequencies, while the widths of the resonances originate from processes such as spontaneous emission and collisional relaxation. The ability to resolve different atomic transition frequencies and to fully characterize all the energy levels in an atom depends on the line widths of the transitions.

Let us assume that the atoms are distributed uniformly in a cylindrical volume of cross-sectional area A and length L (figure 4.1). The axis of the cylinder is taken as

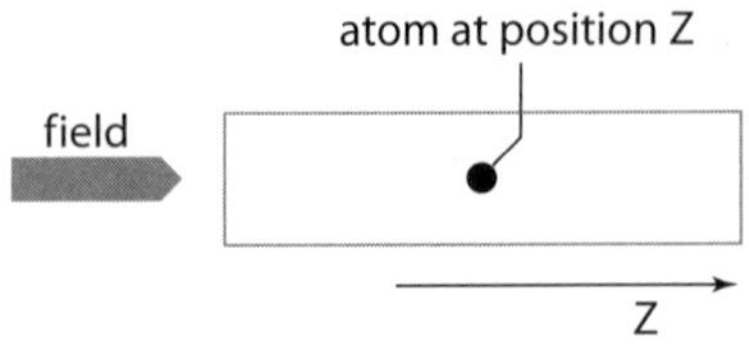

Figure 4.1. A monochromatic field is incident in the $\hat{\mathbf{z}}$ direction on a cell containing atoms that are distributed uniformly in the cell.

the Z axis, and the optical field is incident in this direction such that

$$\mathbf{E}(Z,t) = \frac{1}{2}\hat{\mathbf{x}}\left[E_0(Z,t)e^{i(kZ-\omega t)} + E_0^*(Z,t)e^{-i(kZ-\omega t)}\right], \tag{4.6}$$

where $E_0(Z,t)$ is slowly varying in space over a wavelength and in time over an optical period. Diffraction effects have been neglected on the assumption that $A \gg \lambda^2$.

Although we have gone from z to x polarization for the field, the two-level approximation remains valid for a $J = 0$ to $J = 1$ transition, provided that we take state $|2\rangle$ as

$$|2\rangle = |x\rangle = -\left(|1;1\rangle - |1;-1\rangle\right)/\sqrt{2}. \tag{4.7}$$

An x-polarized field drives transitions between a $J = 0$ ground state and this linear combination of $|J = 1; m_J\rangle$ states.

It is convenient to redefine the field interaction representation as

$$\varrho_{12}(Z,t) = \tilde{\varrho}_{12}(Z,t)e^{-i(kZ-\omega t)}, \tag{4.8}$$

where the rapidly varying spatial and temporal phase has been factored out. Note that this corresponds to the "first" definition of the field interaction representation even though the "primes" have been dropped. If it proves convenient, one can transform to the "second" definition of the field interaction representation by taking $E_0(Z,t)$ as real and positive, replacing δ by $\delta(t) = \delta + \dot{\phi}(t)$, and setting $\varrho_{12}(Z,t) = \tilde{\varrho}_{12}(Z,t)e^{-i[kZ-\omega t+\phi(t)]}$.

The equations of motion [see equations (3.53a) to (3.53d)] in this field interaction representation are

$$\dot{\varrho}_{11}(Z,t) = -i\chi^*(Z,t)\tilde{\varrho}_{21}(Z,t) + i\chi(Z,t)\tilde{\varrho}_{12}(Z,t) + \gamma_2\varrho_{22}(Z,t) \tag{4.9a}$$

$$\dot{\varrho}_{22}(Z,t) = -\gamma_2\varrho_{22}(Z,t) + i\chi^*(Z,t)\tilde{\varrho}_{21}(Z,t) - i\chi(Z,t)\tilde{\varrho}_{12}(Z,t) \tag{4.9b}$$

$$\dot{\tilde{\varrho}}_{12}(Z,t) = -\left(\gamma - i\delta\right)\tilde{\varrho}_{12}(Z,t) - i\chi^*(Z,t)\left[\varrho_{22}(Z,t) - \varrho_{11}(Z,t)\right] \tag{4.9c}$$

$$\dot{\tilde{\varrho}}_{21}(Z,t) = -\left(\gamma + i\delta\right)\tilde{\varrho}_{21}(Z,t) + i\chi(Z,t)\left[\varrho_{22}(Z,t) - \varrho_{11}(Z,t)\right], \tag{4.9d}$$

where

$$\chi(Z,t) = -\left(\mu_x\right)_{21}E_0(Z,t)/2\hbar, \tag{4.10}$$

and $(\mu_x)_{21}$ is a (real) dipole moment matrix element. Recall that $\gamma = \gamma_2/2 + \Gamma$. If we know $E_0(Z,t)$, we can solve this equation at each Z to obtain the density matrix at that point. In this section, we assume that a steady-state regime has been reached such that $E_0(Z,t) = E_0(Z)$ and $\tilde{\boldsymbol{\varrho}}(Z,t) = \tilde{\boldsymbol{\varrho}}(Z)$.

In steady state, all time derivatives are set equal to zero, and we find

$$\tilde{\varrho}_{12}(Z) = \frac{-i\chi^*(Z)\left[\varrho_{22}(Z) - \varrho_{11}(Z)\right]}{\gamma - i\delta}$$

$$= \frac{i\chi^*(Z)}{\gamma - i\delta}\frac{\gamma^2 + \delta^2}{\gamma^2 + \delta^2 + \frac{4\gamma}{\gamma_2}|\chi(Z)|^2}, \tag{4.11a}$$

$$\tilde{\varrho}_{21}(Z) = \frac{i\chi(Z)\left[\varrho_{22}(Z) - \varrho_{11}(Z)\right]}{\gamma + i\delta}$$

$$= \frac{-i\chi(Z)}{\gamma + i\delta}\frac{\gamma^2 + \delta^2}{\gamma^2 + \delta^2 + \frac{4\gamma}{\gamma_2}|\chi(Z)|^2}, \tag{4.11b}$$

$$\varrho_{22}(Z) = \frac{\frac{2\gamma}{\gamma_2}|\chi(Z)|^2}{\gamma^2 + \delta^2 + \frac{4\gamma}{\gamma_2}|\chi(Z)|^2}, \tag{4.11c}$$

$$\varrho_{11}(Z) = 1 - \varrho_{22}(Z) = \frac{\gamma^2 + \delta^2 + \frac{2\gamma}{\gamma_2}|\chi(Z)|^2}{\gamma^2 + \delta^2 + \frac{4\gamma}{\gamma_2}|\chi(Z)|^2}. \tag{4.11d}$$

For

$$\frac{4\gamma}{\gamma_2}|\chi(Z)|^2 \gg \gamma^2 + \delta^2, \tag{4.12}$$

we find that $\varrho_{22}(Z) \approx \varrho_{11}(Z) \approx \frac{1}{2}$, as you might expect, since there is *saturation*, resulting in equal ground- and excited-state populations in strong fields. We can rewrite equation (4.11c) as

$$\varrho_{22}(Z) = \frac{\frac{2\gamma}{\gamma_2}|\chi(Z)|^2}{\gamma_B^2 + \delta^2}, \tag{4.13}$$

where

$$\gamma_B = \gamma\sqrt{1 + \frac{4|\chi(Z)|^2}{\gamma\gamma_2}} \tag{4.14}$$

is referred to as a *power-broadened decay rate*, for reasons to be discussed in this section.

To establish a relationship between $\varrho_{22}(Z)$ and the absorption rate, consider the slab of atoms between Z and $Z+dZ$ shown in figure 4.2. The rate at which energy is lost from the field equals that at which it is scattered by the atoms. In other words, each time a scattered photon is produced, one photon of energy is lost from the field. As a consequence, the rate at which energy $\mathcal{E}$ is lost from the field as a result of scattering by atoms in the slab is given by

$$\frac{d\langle\mathcal{E}\rangle}{dt} = -\mathcal{N}\,\hbar\omega\,\varrho_{22}(Z)\,\gamma_2\,AdZ, \tag{4.15}$$

where $\mathcal{N}$ is the atom density, A is the cross-sectional area of the slab, and the brackets indicate a time average over an optical period. We want to convert this time derivative to a spatial derivative and can do this using the Poynting vector.

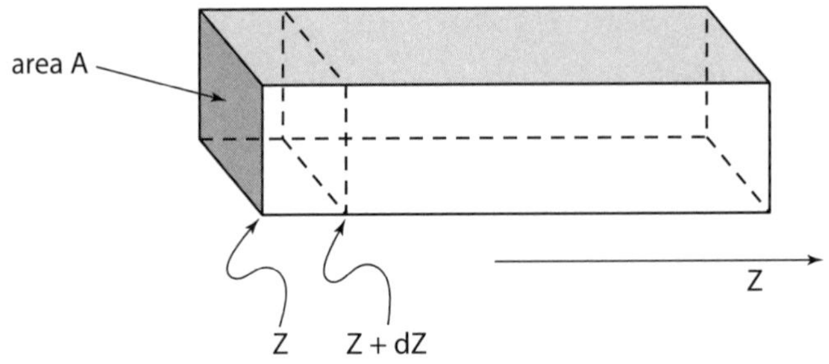

Figure 4.2. A slab of the medium located between positions Z and $Z + dZ$.

In terms of the Poynting vector, $\mathbf{S}(Z) = S(Z)\hat{\mathbf{z}}$,

$$[S(Z + dZ) - S(Z)]\,A = \frac{d\mathcal{E}}{dt}. \tag{4.16}$$

Therefore, using equation (4.13), we have

$$\frac{d\langle S\rangle}{dZ} = \frac{1}{A(dZ)}\frac{d\langle\mathcal{E}\rangle}{dt} = -\mathcal{N}\,\hbar\omega\,\gamma_2\varrho_{22}(Z) = -\mathcal{N}\,\hbar\omega\,\frac{2\gamma|\chi(Z)|^2}{\gamma_B^2 + \delta^2}. \tag{4.17}$$

Since

$$|\chi(Z)|^2 = \frac{|(\mu_x)_{12}E_0(Z)|^2}{4\hbar^2} = \frac{1}{2}\epsilon_0|E_0(Z)|^2\frac{(\mu_x)_{12}^2}{2\epsilon_0\hbar^2}, \tag{4.18}$$

[recall that $(\mu_x)_{21} = (\mu_x)_{12}$ is taken to be real] and since the time-averaged Poynting vector for the plane-wave field (4.6) is equal to

$$\langle S\,(\mathbf{R}, t)\rangle = \frac{1}{2}c\epsilon_0|\,E_0\,(\mathbf{R}, t)\,|^2, \tag{4.19}$$

it follows that

$$|\chi(Z)|^2 = \frac{\langle S(Z)\rangle}{c}\,\frac{(\mu_x)_{12}^2}{2\epsilon_0\hbar^2}. \tag{4.20}$$

Inserting equation (4.20) in equation (4.17), we obtain

$$\frac{d\langle S\rangle}{dZ} = -\alpha(\delta, \langle S\rangle)\langle S\rangle, \tag{4.21}$$

where the (power-dependent) absorption coefficient is defined as

$$\alpha(\delta, \langle S\rangle) = \frac{\mathcal{N}\omega\gamma(\mu_x)_{12}^2}{c\hbar\epsilon_0(\gamma_B^2 + \delta^2)} \equiv \alpha_0\frac{\gamma_2}{2\gamma}\,\frac{\omega}{\omega_0}\,\frac{\gamma^2}{\gamma_B^2 + \delta^2}, \tag{4.22}$$

with

$$\alpha_0 = \frac{2\mathcal{N}\omega_0(\mu_x)_{12}^2}{c\hbar\epsilon_0\gamma_2} \tag{4.23}$$

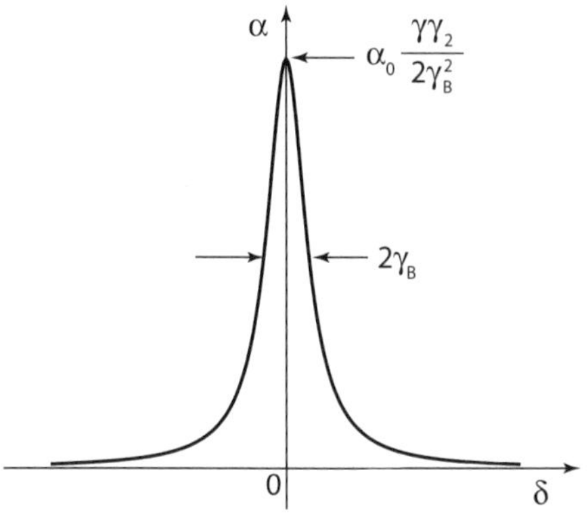

Figure 4.3. Absorption coefficient as a function of detuning.

and

$$\gamma_B^2 = \gamma^2 \left(1 + \frac{4|\chi(Z)|^2}{\gamma\gamma_2}\right) = \gamma^2 \left[1 + \frac{2\langle S(Z)\rangle (\mu_x)_{12}^2}{c\hbar^2 \epsilon_0 \gamma\gamma_2}\right]. \tag{4.24}$$

(From this point onward, we set $\omega/\omega_0 = 1 - \delta/\omega_0 \approx 1$). Note that α and γ_B are both functions of $\langle S(Z)\rangle$ and that α_0 is the absorption coefficient on line center ($\delta = 0$) in the limit that the field strength goes to zero ($\gamma_B = \gamma$) and there are no collisions ($\gamma = \gamma_2/2$). The absorption coefficient associated with the field *amplitude*, rather than the field power, would be $\alpha(\delta, \langle S\rangle)/2$.

Figure 4.3 shows α as a function of δ [see equation (4.22)]. The curve is a Lorentzian with full width at half maximum (FWHM) equal to $2\gamma_B$. With increasing field strength, γ_B increases, and absorption occurs for a larger range of δ. This phenomenon is referred to as *power broadening*, and γ_B is the power-broadened decay rate. The maximum absorption coefficient at line center, $\alpha = \alpha_0 \left(\gamma_2\gamma/2\gamma_B^2\right)$, decreases with increasing field power. Qualitatively similar behavior results from *pressure broadening*. The line profile broadens, and the maximum absorption coefficient decreases with increasing Γ/γ_2. In both cases, the increased line width and decreased maximum absorption coefficient result from the increased relaxation of off-diagonal matrix elements produced either by the strong fields or collisions.

The maximum absorption coefficient is realized in weak fields ($\gamma_B \approx \gamma$), on resonance ($\delta = 0$), and with no collisions ($\gamma = \gamma_2/2$). In that case,

$$\begin{aligned} \alpha(0,0) = \alpha_0 &= \frac{2\mathcal{N}\omega_0 (\mu_x)_{12}^2}{c\hbar\epsilon_0\gamma_2} = 4\pi \frac{2\mathcal{N}\omega_0 e^2 |x_{12}|^2}{4\pi c\hbar\epsilon_0\gamma_2} \\ &= 8\pi\alpha_{FS} \frac{\mathcal{N}\omega_0 |x_{12}|^2}{\gamma_2}, \end{aligned} \tag{4.25}$$

where

$$\alpha_{FS} = \frac{e^2}{4\pi\epsilon_0\hbar c} \approx \frac{1}{137} \tag{4.26}$$

is the fine structure constant.

We show in chapter 16 that

$$\gamma_2 = \frac{4}{3}\alpha_{FS}|x_{12}|^2\frac{\omega_0^3}{c^2}, \tag{4.27}$$

and therefore,

$$\alpha_0 = \frac{3}{4}\frac{8\pi}{k_0^2}\mathcal{N} = \frac{3}{2\pi}\mathcal{N}\lambda_0^2 = 6\pi\bar{\lambda}_0^2\mathcal{N}, \tag{4.28}$$

where

$$k_0 = \omega_0/c = 2\pi/\lambda_0 = 1/\bar{\lambda}_0. \tag{4.29}$$

The resonance absorption cross section $\sigma = \alpha_0/\mathcal{N}$ is of order $\bar{\lambda}_0^2$. For $\lambda_0 = 628$ nm, $\bar{\lambda}_0 = 100$ nm, giving rise to an absorption coefficient

$$\alpha_0 = [6\pi \times 10^{-14}\mathcal{N}\,(\mathrm{m}^{-3})]\,\mathrm{m}^{-1} = [6\pi \times 10^{-10}\mathcal{N}\,(\mathrm{cm}^{-3})]\,\mathrm{cm}^{-1}. \tag{4.30}$$

For $\mathcal{N} = 10^{10}$ atoms/cm^3, $\alpha_0 \approx 20\,\mathrm{cm}^{-1}$. Thus, under these conditions of maximum absorption, a substantial part of the field's energy is lost in a distance of 0.5 mm.

We now return to equation (4.21) to see how the Poynting vector diminishes in the medium. If $\gamma_B \approx \gamma$ (weak field–linear absorption), the solution of equation (4.21) is

$$\langle S(L)\rangle = \langle S(0)\rangle e^{-\alpha(\delta,0)L}, \tag{4.31}$$

where

$$\alpha(\delta, 0) = \alpha_0\frac{\gamma_2}{2\gamma}\frac{\gamma^2}{\gamma^2+\delta^2}. \tag{4.32}$$

In strong fields, we must express γ_B in terms of $\langle S\rangle$ [equation (4.24)]. We can still integrate equation (4.21), which we write in the form

$$\frac{\gamma_B^2+\delta^2}{\gamma^2}\frac{d\langle S(Z)\rangle}{\langle S(Z)\rangle} = -\alpha_0\frac{\gamma_2}{2\gamma}dZ. \tag{4.33}$$

Using equation (4.24), we find

$$\frac{d\langle S(Z)\rangle}{\langle S(Z)\rangle}\frac{\gamma^2\left[1+\frac{2\langle S(Z)\rangle(\mu_x)_{12}^2}{c\hbar^2\epsilon_0\gamma\gamma_2}\right]+\delta^2}{\gamma^2} = -\alpha_0\frac{\gamma_2}{2\gamma}dZ, \tag{4.34}$$

or

$$\frac{d\langle S(Z)\rangle}{\langle S(Z)\rangle}\left[1+\frac{2\langle S(Z)\rangle(\mu_x)_{12}^2\gamma}{c\hbar^2\epsilon_0\gamma_2\left(\gamma^2+\delta^2\right)}\right] = -\alpha_0\frac{\gamma_2}{2\gamma}\frac{\gamma^2}{\gamma^2+\delta^2}dZ. \tag{4.35}$$

After integration, we obtain

$$\ln\frac{\langle S(L)\rangle}{\langle S(0)\rangle}+\left[\frac{\gamma_B^2(0)}{\gamma^2}-1\right]\left[\frac{\langle S(L)\rangle}{\langle S(0)\rangle}-1\right]\frac{\gamma^2}{\gamma^2+\delta^2} = -\alpha_0\frac{\gamma_2}{2\gamma}\frac{\gamma^2}{\gamma^2+\delta^2}L, \tag{4.36}$$

where

$$\gamma_B^2(0) = \gamma^2\left[1+\frac{4|\chi(0)|^2}{\gamma\gamma_2}\right] = \gamma^2\left[1+\frac{2\langle S(0)\rangle(\mu_x)_{12}^2}{c\hbar^2\epsilon_0\gamma\gamma_2}\right]. \tag{4.37}$$

This equation can be solved numerically for $\langle S(L)\rangle/\langle S(0)\rangle$.

Recall that the absorption rate is proportional to $\varrho_{22}(Z)$. As long as the strong field condition (4.12) is valid, the excited-state population $\varrho_{22}(Z)$ is approximately equal to 1/2—the absorption saturates, and the scattering rate is at its maximum. On the other hand, from equation (4.22), it is clear that the absorption coefficient $\alpha(\delta, \langle S(Z)\rangle)$ *decreases* with increasing field strength. Thus, although the scattering rate and total field absorption are at a maximum, the *relative* decrease in field intensity in the medium actually diminishes in strong fields at line center, owing to saturation and power broadening. The medium is said to be *bleached*.

Speaking of power broadening, it is not so obvious why the application of a classical monochromatic field should broaden the absorption profile. In fact, you will see shortly that if we probe a coupled transition, there is a *splitting* rather than a broadening of the spectral profiles. Power broadening is linked to the decay rate of the dipole coherence $\tilde{\varrho}_{12}(Z)$. The strong driving field results in an increase in the decay rate of $\tilde{\varrho}_{12}(Z)$ and a corresponding broadening of the absorption profile. Note that

$$\begin{aligned}\tilde{\varrho}_{12}(Z) &= \frac{-i\chi^*(Z)[\varrho_{22}(Z) - \varrho_{11}(Z)]}{\gamma - i\delta} \\ &= \frac{i\chi^*(Z)[1 - 2\varrho_{22}(Z)]}{\gamma - i\delta} = \frac{i\chi^*(Z)}{\gamma - i\delta}\frac{\gamma^2 + \delta^2}{\gamma_B^2 + \delta^2}.\end{aligned} \tag{4.38}$$

As long as inequality (4.12) holds, the coherence $\tilde{\varrho}_{12}(Z) \sim 0$. In strong fields, the population is equilibrated between the levels, but the coherence goes to zero, since scattering (and collisions) destroy the phase relation between the ground- and excited-state probability amplitudes created by the external field.

4.3 Simple Inclusion of Atomic Motion

There is no escaping the fact that atoms in a vapor are in motion. How does this motion modify the way in which the atoms interact with incident fields? One simple modification is that atoms see a Doppler-shifted frequency for each field. Moreover, if the field *amplitudes* are a function of position (as in a standing-wave optical field), atoms can see different field amplitudes in their lifetime as they move about. Last, the recoil momentum an atom acquires in absorbing, emitting, or scattering light might be comparable to an atom's average momentum—in this case, the center-of-mass motion must be treated quantum-mechanically. All this makes calculations involving atomic motion difficult, the degree of difficulty depending on the specific problem. For the present, we consider only a classical treatment of the center-of-mass motion, assuming that the momentum of the atom is much greater that the photon momentum. This is usually a very good approximation for all but the very coldest atoms (atoms that are laser-cooled to or below microdegree Kelvin temperatures).

If there is only a *single*, *constant-amplitude* field present, it is possible to take into account atomic motion in a simple manner. In the atomic rest frame, the atom sees a Doppler-shifted frequency

$$\omega \to \omega\left(1 - \frac{\mathbf{n}\cdot\mathbf{v}}{c}\right) = \omega - \mathbf{k}\cdot\mathbf{v}, \tag{4.39}$$

where $\mathbf{k} = k\mathbf{n}$ is the field propagation vector. As a result, atomic motion can be accounted for by the replacement

$$\delta = \omega_0 - \omega \to \delta(\mathbf{v}) = \delta + \mathbf{k} \cdot \mathbf{v} . \tag{4.40}$$

In this approximation, equation (4.22) for the absorption coefficient is replaced by

$$\alpha = \alpha_0 \left(\frac{\gamma_2}{2\gamma} \right) \int d\mathbf{v} \frac{W_0(\mathbf{v})\gamma^2}{\gamma_B^2 + (\delta + \mathbf{k} \cdot \mathbf{v})^2}, \tag{4.41}$$

where $W_0(\mathbf{v})$ is the velocity distribution of the atoms. In other words, the absorption coefficient is a weighted average of the absorption coefficients associated with each velocity subclass of atoms in the medium. Let us take $W_0(\mathbf{v})$ to be the Gaussian

$$W_0(\mathbf{v}) = \frac{1}{(\pi u^2)^{3/2}} e^{-v^2/u^2}, \tag{4.42}$$

where u is the most probable speed of the atoms. Then the absorption coefficient takes the form

$$\alpha = \alpha_0 \left(\frac{\gamma_2}{2\gamma} \right) \frac{\gamma^2}{(\pi u^2)^{3/2}} \int d\mathbf{v} \frac{e^{-v^2/u^2}}{\gamma_B^2 + (\delta + \mathbf{k} \cdot \mathbf{v})^2} . \tag{4.43}$$

Taking v_z along $\mathbf{k}$, we obtain

$$\alpha = \alpha_0 \frac{\gamma_2 \gamma}{2\gamma_B^2} I \left(\frac{\gamma_B}{ku}, \frac{\delta}{ku} \right), \tag{4.44}$$

where the absorption profile is given by

$$\begin{aligned} I \left(\frac{\gamma_B}{ku}, \frac{\delta}{ku} \right) &= \frac{\gamma_B^2}{\sqrt{\pi} u} \int_{-\infty}^{\infty} dv_z \frac{e^{-v_z^2/u^2}}{\gamma_B^2 + (\delta + kv_z)^2} \\ &= \frac{\gamma_B^2}{\sqrt{\pi} k^2 u^2} \int_{-\infty}^{\infty} d\xi \frac{e^{-\xi^2}}{\left(\frac{\gamma_B}{ku}\right)^2 + \left(\frac{\delta}{ku} + \xi\right)^2}, \end{aligned} \tag{4.45}$$

with $\xi = v_z/u$. In what follows, we neglect power broadening and set $\gamma_B = \gamma$.

The integral is the convolution of a Gaussian with a Lorentzian (referred to as a *Voigt profile* [1]). This reflects the fact that the absorption coefficient for a single velocity subclass v_z of atoms is a Lorentzian having FWHM equal to 2γ that is centered at the Doppler-shifted frequency $\delta = -kv_z = -ku\xi$, and the total absorption profile is a sum of these individual Lorentzians, weighted with the Doppler distribution of atomic velocities. The width γ is referred to as a *homogeneous* width since it is the same for all atoms and gives a fundamental limitation to the frequency resolution that can be obtained—without collisions and in weak fields, the minimum width in equation (4.45) is $2\gamma = \gamma_2$.

In the limit that $ku \gg \gamma, |\delta|$, the Lorentzian in equation (4.45) is a much narrower function of ξ than the Gaussian and is nonzero only near $\xi = -\delta/ku$ (see figure 4.4); therefore, we can evaluate the Gaussian at $\xi = -\delta/ku$ and obtain

$$I \left(\frac{\gamma}{ku}, \frac{\delta}{ku} \right) \approx \frac{\gamma^2}{\sqrt{\pi} k^2 u^2} e^{-\frac{\delta^2}{k^2 u^2}} \int_{-\infty}^{\infty} d\xi \frac{1}{\left(\frac{\gamma}{ku}\right)^2 + \left(\frac{\delta}{ku} + \xi\right)^2} = \frac{\sqrt{\pi}\gamma}{ku} e^{-\frac{\delta^2}{k^2 u^2}} . \tag{4.46}$$

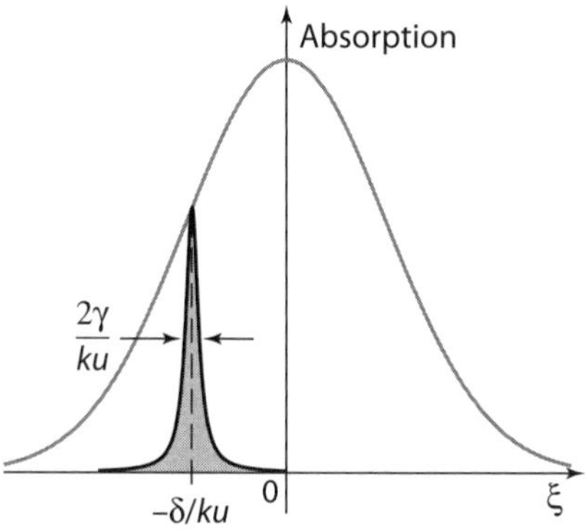

Figure 4.4. For atoms having $\xi = kuv_z = -\delta/ku$, the absorption profile is the Lorentzian shown on the graph. The absorption profile for the vapor is obtained by summing these absorption profiles using the Gaussian weighting function, also shown on the graph.

One can view this result as consequence of taking the lead term in an expansion of the Gaussian $e^{-\xi^2}$ in the integrand about $\xi = -\delta/ku$. Unfortunately, if one tries to take the next term to get a correction, that term diverges, since the integrand varies as $1/\xi$ for large ξ.

The factor γ/ku in equation (4.46) is a reflection of the fact that only a fraction of the velocity distribution is excited for a given δ—namely, those atoms having $|kv + \delta| \lesssim \gamma$ that are Doppler-shifted into resonance with the field. The line profile (4.46), considered as a function of δ/ku, is a Gaussian having FWHM of order $1.67 \gg \gamma/ku$, (figure 4.5). Owing to the different velocity groups, there is an *inhomogeneous* broadening of the line—that is, different velocity atoms are brought into resonance with the applied field as δ is varied. This contrasts with homogeneous broadening, where *all* the atoms interact identically with the applied field. In a thermal vapor for an optical transition, $k \approx 10^5\text{cm}^{-1}$, $u \approx 5 \times 10^4\text{cm/s}$, and $ku \approx 5 \times 10^9\ \text{s}^{-1}$, leading to $ku/2\pi \approx 1\,\text{GHz}$, whereas for a typical transition (assuming negligible collision and power broadening), $\gamma/2\pi \approx 5$ to $50\,\text{MHz}$, implying that $ku/\gamma = 2ku/\gamma_2 \gg 1$. In other words, Doppler broadening dominates the spectral width in the conventional spectroscopy of a thermal vapor if collision broadening is unimportant. As a result, the resolution in conventional linear spectroscopy of thermal atomic vapors is limited by the Doppler width.

The Doppler width is not a fundamental limiting width in resolving the transition frequency—only a practical one in the conventional spectroscopy of vapors (or solids with a distribution of frequencies). The Doppler width can be eliminated by using an atomic beam with $\mathbf{k} \perp \mathbf{v}$ (at a cost of reduced atomic density), by methods of nonlinear spectroscopy, or by cooling the atoms so that the Doppler width is reduced.

Collision broadening rates, $\Gamma/2\pi$, are typically 10 to 50 MHz/Torr of perturber gas pressure. At several hundred Torr, $\gamma \approx \Gamma \gg ku$. In that case, the Lorentzian is much broader than the Gaussian [equation (4.45)], and one can evaluate the Lorentzian at the peak of the Gaussian, $\xi = 0$, to obtain

$$I\left(\frac{\gamma}{ku}, \frac{\delta}{ku}\right) \approx \frac{\Gamma^2}{\sqrt{\pi}k^2u^2} \frac{1}{\left(\frac{\Gamma}{ku}\right)^2 + \left(\frac{\delta}{ku}\right)^2} \int_{-\infty}^{\infty} d\xi e^{-\xi^2} = \frac{\Gamma^2}{\Gamma^2 + \delta^2}. \tag{4.47}$$

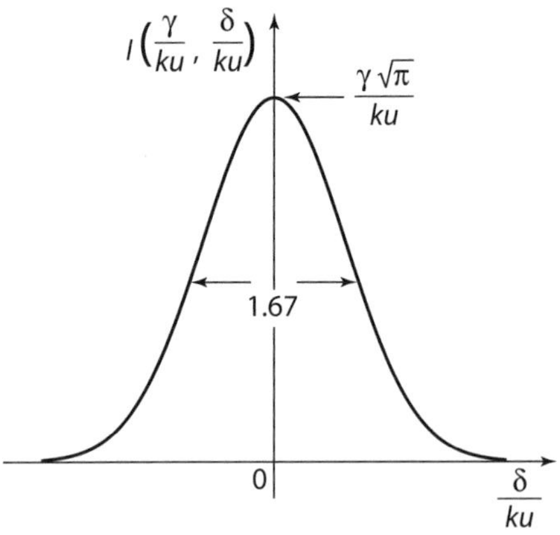

Figure 4.5. Absorption line shape in the limit that $ku \gg \gamma$.

Now all the atoms are used, since the homogeneous width Γ is larger than the Doppler width ku. We recover the Lorentzian line shape associated with homogeneous broadening. The limiting resolution is now determined by the collisional relaxation rate, since $\Gamma \gg ku$.

For arbitrary values of γ_B/ku and δ/ku, the integral (4.45) can be evaluated in terms of the complex error function. If

$$z = \delta/ku + i\gamma_B/ku = x + iy, \tag{4.48}$$

then

$$\begin{aligned} I(z) = I(y, x) &= \frac{y^2}{\sqrt{\pi}} \int_{-\infty}^{\infty} d\xi \frac{e^{-\xi^2}}{y^2 + (\xi + x)^2} \\ &= \frac{iy}{2\sqrt{\pi}} \int_{-\infty}^{\infty} d\xi e^{-\xi^2} \left(\frac{1}{\xi + x + iy} - \frac{1}{\xi + x - iy} \right) \\ &= \frac{y\sqrt{\pi}}{2} w(z) + \text{c.c.}, \end{aligned} \tag{4.49}$$

where

$$w(z) = \frac{i}{\pi} \int_{-\infty}^{\infty} d\xi \frac{e^{-\xi^2}}{z \pm \xi}, \quad \text{with} \quad \text{Im}(z) > 0. \tag{4.50}$$

The function $w(z)$ is sometimes known as the *plasma dispersion function* and can be expressed in terms of the error function as [2]

$$w(z) = e^{-z^2} [1 - \Phi(-iz)], \tag{4.51}$$

where the error function is defined by

$$\Phi(z) = \frac{2}{\sqrt{\pi}} \int_0^z d\xi e^{-\xi^2}. \tag{4.52}$$

Properties of $w(z)$ are given in [2]; the function can usually be evaluated easily using standard computer programs such as Mathematica, Maple, or Matlab.

4.4 Rate Equations

In condensed matter and chemical physics, one often encounters *rate equations* that involve populations of quantum states only. Coherence (off-diagonal density matrix elements) does not appear, and populations approach their steady-state values monotonically. You might ask how this is possible with all the Rabi oscillations we have been talking about.

To arrive at rate equations for state populations, one formally integrates the equation for $\tilde{\varrho}_{12}(t)$, equation (4.9c), to obtain

$$\tilde{\varrho}_{12}(t) = \tilde{\varrho}_{12}(0)e^{-(\gamma - i\delta)t} - i\int_0^t dt' e^{-(\gamma - i\delta)(t-t')}\chi^*(t')\left[\varrho_{22}(t') - \varrho_{11}(t')\right]. \tag{4.53}$$

If $\chi^*(t)\left[\varrho_{22}(t) - \varrho_{11}(t)\right]$ is slowly varying on a timescale $|\gamma - i\delta|^{-1}$ we can integrate the second term by parts, as in the Bloch-Siegert problem. Keeping only the leading term, we find

$$\tilde{\varrho}_{12}(t) = \tilde{\varrho}_{12}(0)e^{-(\gamma - i\delta)t} - i\frac{\chi^*(t)}{\gamma - i\delta}\left[\varrho_{22}(t) - \varrho_{11}(t)\right]. \tag{4.54}$$

The first term is either small or rapidly varying compared with the second if the approximation we used is valid. Neglecting the first term, we obtain

$$\tilde{\varrho}_{12}(t) \approx -i\frac{\chi^*(t)}{\gamma - i\delta}\left[\varrho_{22}(t) - \varrho_{11}(t)\right]. \tag{4.55}$$

That is, the coherence *adiabatically follows* the product of the field amplitude and population difference. The coherence goes to zero in strong fields since the population difference goes to zero as $|\chi(t)|^{-2}$ in this limit. Using equation (4.55) and its complex conjugate in equation (4.9a), (4.9b), we find

$$\begin{aligned}\dot{\varrho}_{11} &= -i\chi^*(t)\tilde{\varrho}_{21}(t) + i\chi(t)\tilde{\varrho}_{12}(t) + \gamma_2\varrho_{22}(t)\\ &= \frac{2\gamma|\chi(t)|^2}{\gamma^2 + \delta^2}\left[\varrho_{22}(t) - \varrho_{11}(t)\right] + \gamma_2\varrho_{22}(t),\end{aligned} \tag{4.56a}$$

$$\begin{aligned}\dot{\varrho}_{22} &= i\chi^*(t)\tilde{\varrho}_{21}(t) - i\chi(t)\tilde{\varrho}_{12}(t) - \gamma_2\varrho_{22}(t)\\ &= -\frac{2\gamma|\chi(t)|^2}{\gamma^2 + \delta^2}\left[\varrho_{22}(t) - \varrho_{11}(t)\right] - \gamma_2\varrho_{22}(t),\end{aligned} \tag{4.56b}$$

or

$$\dot{\varrho}_{11} = \Gamma_f(t)\left[\varrho_{22}(t) - \varrho_{11}(t)\right] + \gamma_2\varrho_{22}(t), \tag{4.57a}$$

$$\dot{\varrho}_{22} = -\Gamma_f(t)\left[\varrho_{22}(t) - \varrho_{11}(t)\right] - \gamma_2\varrho_{22}(t), \tag{4.57b}$$

where

$$\Gamma_f(t) = \frac{2\gamma|\chi(t)|^2}{\gamma^2 + \delta^2} \tag{4.58}$$

is the rate at which the field drives the transitions. Equations (4.57a) and (4.57b) are the rate equations for the populations. Often such equations are used even if they are not fully justified.

What are the validity conditions for the rate equations? Note that the populations approach their steady-state values monotonically. In the Bloch equations, this occurs if the Bloch vector does not undergo a full revolution before it decays to its steady-state value. In other words, the rate equations are valid if $\gamma \gg \Omega$. This is a sufficient, but not necessary, condition. Even if $\Omega \gg \gamma$, the rate equations yield approximate solutions, provided the condition for adiabatic following $\left[\dot{\theta}(t) \ll \Omega(t)\right]$ is satisfied. In this limit, there are many oscillations, but they have small amplitude, and the Bloch vector adiabatically follows the pseudofield vector as it approaches its steady-state values.

As the system approaches equilibrium, all density matrix elements vary slowly in time, and the rate equations become exact. This fact can be used to gain some insight into power broadening. We solve for $\varrho_{22}(t)$ formally using equation (4.9b) to obtain

$$\varrho_{22}(t) = \varrho_{22}(0)e^{-\gamma_2 t} + \int_0^t dt' e^{-\gamma_2(t-t')} i\chi(t') \left[\varrho_{21}(t') - \varrho_{12}(t')\right], \tag{4.59}$$

where, for simplicity, we assume that $\chi(t)$ is real. Near equilibrium, and with $\varrho_{22}(0) = 0$,

$$\varrho_{22}(t) \approx \frac{i\chi(t)}{\gamma_2}\left[\varrho_{21}(t) - \varrho_{12}(t)\right]. \tag{4.60}$$

Therefore,

$$\begin{aligned}\dot{\tilde{\varrho}}_{12}(t) &= -(\gamma - i\delta)\tilde{\varrho}_{12}(t) - i\chi(t)[2\varrho_{22}(t) - 1] \\ &= -(\gamma - i\delta)\tilde{\varrho}_{12}(t) - \frac{2\chi^2(t)}{\gamma_2}\tilde{\varrho}_{12}(t) + \frac{2\chi^2(t)}{\gamma_2}\tilde{\varrho}_{21}(t) + i\chi(t).\end{aligned} \tag{4.61}$$

As such, we find that, near equilibrium, the first and second components of the Bloch vector evolve as

$$\dot{u}(t) = -\gamma u(t) - \delta v(t), \tag{4.62a}$$

$$\begin{aligned}\dot{v}(t) &= -\left[\gamma + \frac{4|\chi(t)|^2}{\gamma_2}\right] v(t) + \delta u(t) + 2\chi(t) \\ &= -\frac{\gamma_B^2}{\gamma} v(t) + \delta u(t) + 2\chi(t),\end{aligned} \tag{4.62b}$$

while

$$\dot{\varrho}_{22}(t) = -\gamma_2\varrho_{22}(t) + \chi v(t). \tag{4.63}$$

In steady state, these equations reproduce equations (4.11). However, when written in this form, the equations support the idea that power broadening is associated with a power-dependent increase in the decay rate of the out-of-phase component of the atomic response—that is, an increase in the decay rate of $v(t)$.

4.5 Summary

We have started to look at applications involving the density matrix. The macroscopic density matrix for an ensemble of atoms was defined in terms of the single particle density matrix, allowing us to draw on the results of chapter 3. Using relatively simple arguments, we were able to derive an expression for the absorption coefficient for an ensemble of stationary atoms. Moreover, we were able to extend the results to allow for atomic motion by a simple inclusion of the Doppler shift of the field frequency, as seen in the rest frame of an atom. Last, we discussed the validity conditions of rate approximations. In the next chapter, we refine our treatment of atomic motion, allowing us to consider cases where there is a spatial variation of the field amplitude. Such spatial variation of the field is of critical importance in atom optics, where the applied fields influence the atomic motion.

Problems

1. Imagine a vapor of two-level atoms having density 3×10^9 atoms/cm^3 contained in a cylindrical volume with cross-sectional area $A = 1\,\text{cm}^2$ and length 10 cm. Neglect collisions, and assume that the excited-state decay rate is $\gamma_2/2\pi = 10\,\text{MHz}$. A monochromatic field is incident on the vapor with propagation vector along the axis of the cylinder. The field is resonant with the atomic transition, and the transition wavelength is 600 nm. Calculate and plot the maximum rate at which energy can be lost from the field at the entry plane as a function of Rabi frequency as $\chi/2\pi$ is varied from 1 to 100 MHz.
2. For the parameters of problem 1, use equation (4.36) to calculate and plot $\langle S(L)\rangle/\langle S(0)\rangle$ as a function of L for $\chi/2\pi = 0.1, 1, 10$, and 100 MHz. Show that when $\gamma_B^2(Z = 0)/\gamma^2 \approx 1$, the dependence is exponential, but when $\gamma_B^2(0)/\gamma^2 \gg 1$, the dependence is *linear* at first and then exponential. Calculate the slope of the linear part. Explain why the dependence is linear at first when $\gamma_B^2(0)/\gamma^2 \gg 1$.
3. Numerically solve the density matrix equations and plot $\varrho_{22}(t)$ and $|\tilde{\varrho}_{12}(t)|$ for $0 \le t \le 10$ for

   a. $\gamma = \gamma_2/2 = 1, \quad \chi = 0.1, \quad \delta = 0.2$
   b. $\gamma = \gamma_2/2 = 1, \quad \chi = 4, \quad \delta = 0.2$
   c. $\gamma = \gamma_2/2 = 1, \quad \chi = 4, \quad \delta = 50.$

   Take $\varrho_{22}(0) = 0$ as the initial condition. In which cases is the rate equation approximation valid? Explain.
4. Numerically solve the density matrix equations and plot $\varrho_{22}(t)$ and $|\tilde{\varrho}_{12}(t)|$ for $0 \le t \le 30$ for $\gamma = \gamma_2/2 = 1, \quad \chi = 4(1 - e^{-t/4})^4$, and $\delta = 50$. Take $\varrho_{22}(0) = 0$ as the initial condition. Compare the result with problem 3c, and explain why the rate equation approximation is approximately valid here, whereas it was not valid in problem 3c.
5. Consider the density matrix equations for the two-level problem as a function of time for $\gamma_2 = 0$, $\delta = 0$, χ = real constant, and $\varrho_{22}(0) = 0$. Show that the steady-state values are $\varrho_{22} = \varrho_{11} = 1/2$, $\tilde{\varrho}_{12} = 0$, but that these values are reached very slowly ($t \gg 1/\gamma$) when $\chi \ll \gamma$. The fact that there is no absorption in steady state is consistent with the scattering interpretation of "absorption," since no scattering occurs in this model.

Problems 6 to 10 refer to the function

$$w(z = x + iy) = e^{-z^2}\left(1 + \frac{2i}{\sqrt{\pi}}\int_0^z e^{t^2}\,dt\right)$$
$$= \frac{i}{\pi}\int_{-\infty}^{\infty}\frac{e^{-t^2}dt}{z \pm t}, \quad y > 0.$$

6. Prove that the two forms for $w(z)$ are equivalent.
7. Prove that $w(z)$ satisfies the differential equation

$$w'(z) = -2zw(z) + 2i/\sqrt{\pi}\,.$$

8. Prove that for $y \ll 1$,

$$w(z) \sim e^{-x^2}(1 - 2ixy)\left(1 + \frac{2i}{\sqrt{\pi}}\int_0^x e^{t^2}\,dt\right) - \frac{2y}{\sqrt{\pi}}\,.$$

9. Prove that for $|z| \gg 1$,

$$w(z) \sim \frac{i}{\sqrt{\pi}z}\left(1 + \frac{1}{2z^2}\right).$$

10. Plot $I(z = x + iy) = y\sqrt{\pi}\mathrm{Re}w(z)$ as a function of x for $y = 0.1$, 1, and 10 to see the transition from a Gaussian to Lorentzian profile.
11. The force on an atom in a radiation field is equal to $\mathbf{F} = \langle\nabla(\hat{\mu}\cdot\mathbf{E})\rangle$. Express this result in terms of density matrix elements for two-level atoms interacting with a linearly polarized monochromatic field. For a traveling-wave field, calculate the time-averaged force $\mathbf{F}$ as a function of detuning for fields of arbitrary intensity. Interpret your result. How can this effect be used to slow down an atomic beam?

References

[1] W. Voigt, *Uber die Intensitatsverteilung innhalb einer Spektrallinie* (*The distribution of intensity within spectral lines*), Physikalische Zeitschrift **14**, 377–381 (1913).

[2] M. Abramowitz and I. A. Stegun, *Handbook of Mathematical Functions with Formulas, Graphs, and Mathematical Tables* (U. S. Department of Commerce, Washington, DC, 1965), chap. 7.

Bibliography

The topics in this chapter are covered in most of the texts listed in chapter 1.

5

Density Matrix Equations:

Atomic Center-of-Mass Motion, Elementary Atom Optics, and Laser Cooling

5.1 Introduction

In the previous chapter, we included the effects of atomic motion for the specific case of atoms interacting with a single plane-wave field. Moreover, it was assumed that effects related to quantization of the atomic center-of-mass motion could be neglected. In this chapter, we retain, for the most part, a classical description of the center-of-mass motion but provide a general framework that will allow us to deal with more complicated atom–field geometries. In particular, we calculate the force on an atom in a standing-wave optical field. To do so, we need a better treatment of atomic motion.

The appendix in this chapter contains a rigorous derivation of the density matrix equations, including a quantized description of the center-of-mass motion, as well as a method for taking a classical limit of these equations in which the atomic position and momentum (or velocity) operators are replaced by classical variables. One can arrive at the same equations by generalizing the results of the previous chapter using a simple heuristic argument. For a moving atom, one can consider the time derivative in the Schrödinger equation as a *total* time derivative. Thus, in taking $d\varrho_{\alpha\alpha'}(\mathbf{R}_i, t)/dt$ for an atom having classical position $\mathbf{R}_i$ and velocity $\mathbf{v}_i = d\mathbf{R}_i/dt$, one sets

$$\begin{aligned} d\varrho_{\alpha\alpha'}(\mathbf{R}_i, t)/dt &= \partial\varrho_{\alpha\alpha'}(\mathbf{R}_i, t)/\partial t + \nabla_{R_i}\varrho_{\alpha\alpha'}(\mathbf{R}_i, t)\cdot d\mathbf{R}_i/dt \\ &= \partial\varrho_{\alpha\alpha'}(\mathbf{R}_i, t)/\partial t + \mathbf{v}_i\cdot\nabla_{R_i}\varrho_{\alpha\alpha'}(\mathbf{R}_i, t). \end{aligned} \tag{5.1}$$

For a single optical field interacting with an ensemble of two-level atoms, equations (4.1) become

$$\frac{\partial \varrho_{11}(\mathbf{R}, \mathbf{v}, t)}{\partial t} + \mathbf{v} \cdot \nabla \varrho_{11}(\mathbf{R}, \mathbf{v}, t) = -i\chi^*(\mathbf{R}, t)\tilde{\varrho}_{21}(\mathbf{R}, \mathbf{v}, t) + i\chi(\mathbf{R}, t)\tilde{\varrho}_{12}(\mathbf{R}, \mathbf{v}, t) + \gamma_2 \varrho_{22}(\mathbf{R}, \mathbf{v}, t), \tag{5.2a}$$

$$\frac{\partial \varrho_{22}(\mathbf{R}, \mathbf{v}, t)}{\partial t} + \mathbf{v} \cdot \nabla \varrho_{22}(\mathbf{R}, \mathbf{v}, t) = i\chi^*(\mathbf{R}, t)\tilde{\varrho}_{21}(\mathbf{R}, \mathbf{v}, t) - i\chi(\mathbf{R}, t)\tilde{\varrho}_{12}(\mathbf{R}, \mathbf{v}, t) - \gamma_2 \varrho_{22}(\mathbf{R}, \mathbf{v}, t), \tag{5.2b}$$

$$\frac{\partial \tilde{\varrho}_{12}(\mathbf{R}, \mathbf{v}, t)}{\partial t} + \mathbf{v} \cdot \nabla \tilde{\varrho}_{12}(\mathbf{R}, \mathbf{v}, t) = -i\chi^*(\mathbf{R}, t)\left[\varrho_{22}(\mathbf{R}, \mathbf{v}, t) - \varrho_{11}(\mathbf{R}, \mathbf{v}, t)\right] - (\gamma - i\delta)\tilde{\varrho}_{12}(\mathbf{R}, \mathbf{v}, t), \tag{5.2c}$$

$$\frac{\partial \tilde{\varrho}_{21}(\mathbf{R}, \mathbf{v}, t)}{\partial t} + \mathbf{v} \cdot \nabla \tilde{\varrho}_{21}(\mathbf{R}, \mathbf{v}, t) = i\chi(\mathbf{R}, t)\left[\varrho_{22}(\mathbf{R}, \mathbf{v}, t) - \varrho_{11}(\mathbf{R}, \mathbf{v}, t)\right] - (\gamma + i\delta)\tilde{\varrho}_{21}(\mathbf{R}, \mathbf{v}, t), \tag{5.2d}$$

where

$$\chi(\mathbf{R}, t) = -\frac{(\hat{\boldsymbol{\epsilon}} \cdot \boldsymbol{\mu}_{21}) E_0(\mathbf{R}, t)}{2\hbar}, \tag{5.3}$$

and

$$\mathbf{E}(\mathbf{R}, t) = \frac{1}{2}\hat{\boldsymbol{\epsilon}} E_0(\mathbf{R}, t) e^{-i\omega t} + \text{c.c.} \tag{5.4}$$

Recall that $\varrho_{ij}(\mathbf{R}, \mathbf{v}, t)$ is a dimensionless, *single-atom* density matrix element. Equations (5.2) are the basic equations for atom–field interactions involving two-level atoms interacting with a quasi-monochromatic optical field. The total population

$$\varrho_{11}(\mathbf{R}, \mathbf{v}, t) + \varrho_{22}(\mathbf{R}, \mathbf{v}, t) = 1 \tag{5.5}$$

is conserved, since the effects of velocity-changing collisions are neglected.

5.2 Atom in a Single Plane-Wave Field

For an external field of the form

$$\mathbf{E}(\mathbf{R}, t) = \frac{1}{2}\hat{\boldsymbol{\epsilon}} E_0 e^{i(\mathbf{k}\cdot\mathbf{R} - \omega t)} + \text{c.c.}, \tag{5.6}$$

with $\hat{\boldsymbol{\epsilon}} \cdot \mathbf{k} = 0$, we set $\chi(\mathbf{R}, t) = \chi e^{i\mathbf{k}\cdot\mathbf{R}}$ in equations (5.2), with

$$\chi = -\frac{(\hat{\boldsymbol{\epsilon}} \cdot \boldsymbol{\mu}_{21}) E_0}{2\hbar}. \tag{5.7}$$

If we assume a solution to equations (5.2) of the form

$$\tilde{\varrho}_{12}(\mathbf{R},\mathbf{v},t)=\tilde{\varrho}_{12}(\mathbf{v},t)e^{-i\mathbf{k}\cdot\mathbf{R}}, \tag{5.8a}$$
$$\tilde{\varrho}_{21}(\mathbf{R},\mathbf{v},t)=\tilde{\varrho}_{21}(\mathbf{v},t)e^{i\mathbf{k}\cdot\mathbf{R}}, \tag{5.8b}$$
$$\varrho_{11}(\mathbf{R},\mathbf{v},t)=\varrho_{11}(\mathbf{v},t), \tag{5.8c}$$
$$\varrho_{22}(\mathbf{R},\mathbf{v},t)=\varrho_{22}(\mathbf{v},t), \tag{5.8d}$$

we find evolution equations

$$\frac{\partial\varrho_{11}(\mathbf{v},t)}{\partial t}=-i\chi^{*}\tilde{\varrho}_{21}(\mathbf{v},t)+i\chi\tilde{\varrho}_{12}(\mathbf{v},t)+\gamma_2\varrho_{11}(\mathbf{v},t), \tag{5.9a}$$
$$\frac{\partial\varrho_{22}(\mathbf{v},t)}{\partial t}=i\chi^{*}\tilde{\varrho}_{21}(\mathbf{v},t)-i\chi\tilde{\varrho}_{12}(\mathbf{v},t)-\gamma_2\varrho_{22}(\mathbf{v},t), \tag{5.9b}$$
$$\frac{\partial\tilde{\varrho}_{12}(\mathbf{v},t)}{\partial t}=i\chi^{*}\left[\varrho_{11}(\mathbf{v},t)-\varrho_{22}(\mathbf{v},t)\right]-\left[\gamma-i(\delta+\mathbf{k}\cdot\mathbf{v})\right]\tilde{\varrho}_{12}(\mathbf{v},t), \tag{5.9c}$$
$$\frac{\partial\tilde{\varrho}_{21}(\mathbf{v},t)}{\partial t}=i\chi\left[\varrho_{22}(\mathbf{v},t)-\varrho_{11}(\mathbf{v},t)\right]-\left[\gamma+i(\delta+\mathbf{k}\cdot\mathbf{v})\right]\tilde{\varrho}_{21}(\mathbf{v},t). \tag{5.9d}$$

As stated in chapter 4, the net effect is to replace the detuning δ by $\delta(\mathbf{v})=\delta+\mathbf{k}\cdot\mathbf{v}$.

5.3 Force on an Atom

When a radiation field interacts with atoms, it can exert forces on the atoms. Forces of this nature are responsible for laser cooling, confinement in optical lattices, and the deflection of atoms. There are times when the atomic center-of-mass motion can be treated classically (as in this chapter) and times when the motion must be quantized. In chapter 11, we explore these regimes in more detail. The manipulation of atomic motion by optical fields or microfabricated gratings belongs to the general field of *atom optics*. As an elementary application of atom optics, we consider now the force on an atom produced by an optical field.

The atom–field interaction energy is $-\hat{\boldsymbol{\mu}}\cdot\mathbf{E}$, so the force on an atom located at position $\mathbf{R}$ having velocity $\mathbf{v}$ at time t is

$$\mathbf{F}(\mathbf{R},\mathbf{v},t)=\langle\nabla(\hat{\boldsymbol{\mu}}\cdot\mathbf{E})\rangle=\mathrm{Tr}\left[\boldsymbol{\varrho}(\mathbf{R},\mathbf{v},t)\nabla(\hat{\boldsymbol{\mu}}\cdot\mathbf{E})\right]. \tag{5.10}$$

Forces can arise because there is a gradient in the field amplitude or as a result of scattering of the incident field by the atoms into unoccupied vacuum field modes. Let us consider three cases: a plane wave, a focused plane wave, and a standing-wave field. In all cases, we take

$$\mathbf{E}(\mathbf{R},t)=\frac{1}{2}\mathbf{E}_0(\mathbf{R})e^{-i\omega t}+\text{c.c.}, \tag{5.11}$$

and

$$\varrho_{12}(\mathbf{R},\mathbf{v},t)=\tilde{\varrho}_{12}(\mathbf{R},\mathbf{v})e^{i\omega t}. \tag{5.12}$$

As a consequence, the force may be written as

$$\begin{aligned}\mathbf{F}(\mathbf{R},\mathbf{v}) &= \mathrm{Tr}\,\{\boldsymbol{\varrho}(\mathbf{R},\mathbf{v},t)\nabla\,[\hat{\boldsymbol{\mu}}\cdot\mathbf{E}(\mathbf{R},\,t)]\}\\ &= \varrho_{12}(\mathbf{R},\mathbf{v},t)\nabla\,[\boldsymbol{\mu}_{21}\cdot\mathbf{E}(\mathbf{R},\,t)] + \text{c.c.}\\ &\approx \frac{1}{2}\tilde{\varrho}_{12}(\mathbf{R},\mathbf{v})\nabla\,[\boldsymbol{\mu}_{21}\cdot\mathbf{E}_0(\mathbf{R})] + \text{c.c.},\end{aligned} \tag{5.13}$$

where rapidly varying terms—that is, terms varying as $\exp[\pm 2i\omega t]$—have been neglected.

5.3.1 Plane Wave

For a plane wave

$$\mathbf{E}_0(\mathbf{R}) = \hat{\boldsymbol{\epsilon}}E_0e^{i\mathbf{k}\cdot\mathbf{R}}, \tag{5.14}$$

with $\hat{\boldsymbol{\epsilon}}\cdot\mathbf{k}=0$, the force is given by

$$\begin{aligned}\mathbf{F}(\mathbf{v}) &= \frac{1}{2}\tilde{\varrho}_{12}(\mathbf{R},\mathbf{v})(\boldsymbol{\mu}_{21}\cdot\hat{\boldsymbol{\epsilon}}E_0)i\mathbf{k}e^{i\mathbf{k}\cdot\mathbf{R}} + \text{c.c.}\\ &= -i\hbar\mathbf{k}\chi\tilde{\varrho}_{12}(\mathbf{v}) + \text{c.c.} = \hbar\mathbf{k}\gamma_2\left[\frac{i\chi^*\tilde{\varrho}_{21}(\mathbf{v}) - i\chi\tilde{\varrho}_{12}(\mathbf{v})}{\gamma_2}\right],\end{aligned} \tag{5.15}$$

where $\chi = -\boldsymbol{\mu}_{21}\cdot\hat{\boldsymbol{\epsilon}}E_0/(2\hbar)$ and $\tilde{\varrho}_{12}(\mathbf{R},\mathbf{v}) = \tilde{\varrho}_{12}(\mathbf{v})e^{-i\mathbf{k}\cdot\mathbf{R}}$. In steady state, the term in brackets is just equal to $\varrho_{22}(\mathbf{v})$ [see equation (5.9b)]; therefore,

$$\mathbf{F}(\mathbf{v}) = \hbar\mathbf{k}\gamma_2\varrho_{22}(\mathbf{v})\,. \tag{5.16}$$

This is just what you might have expected. Every time a photon is scattered by an atom, on average, the atom picks up $\hbar\mathbf{k}$ of momentum from the field, since the scattered dipole radiation has zero average momentum [1]. The scattering rate is just $\gamma_2\varrho_{22}(\mathbf{v})$. Note that it is incorrect to say that the atom absorbs radiation at one frequency and then re-emits via spontaneous emission at another frequency. The correct interpretation is in terms of a scattering process in which the incident field is scattered into modes of the vacuum field that were unoccupied.

The steady-state excited-state population is

$$\varrho_{22}(\mathbf{v}) = \frac{2|\chi|^2\gamma}{\gamma_2}\frac{1}{\gamma_B^2 + (\delta + \mathbf{k}\cdot\mathbf{v})^2}\,. \tag{5.17}$$

As a result, the maximum force occurs for atoms having velocity $\mathbf{v}$ satisfying $\mathbf{k}\cdot\mathbf{v} = -\delta$. For $\delta = \omega_0 - \omega < 0$ (blue detuning), the maximum force occurs for $\mathbf{v}$ and $\mathbf{k}$ parallel and is in the direction of motion (acceleration), while for $\delta = \omega_0 - \omega > 0$ (red detuning), the maximum force occurs for $\mathbf{v}$ and $\mathbf{k}$ antiparallel and opposes the motion (deceleration) (see figure 5.1). This technique for slowing and cooling atoms was proposed originally by Hänsch and Schawlow in 1975 [2].

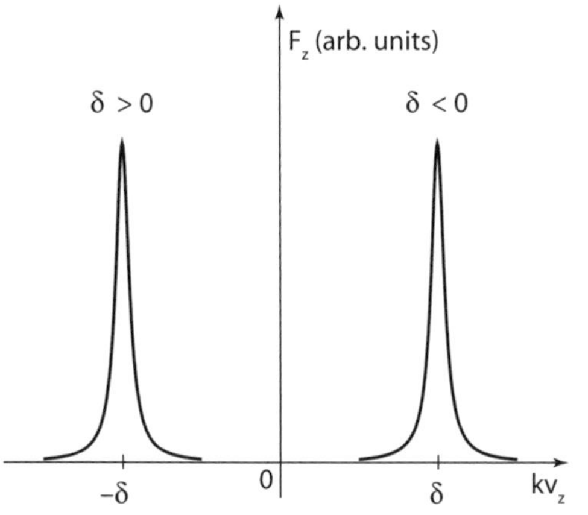

Figure 5.1. Dissipative force experienced by an atom in a plane-wave field. For positive (negative) detuning, the field decelerates (accelerates) the atom.

5.3.2 Focused Plane Wave: Atom Trapping

In the previous example, the net force arose from a *dissipative* process—radiation was scattered irreversibly into modes of the vacuum field. If instead we let the atoms interact with a field having a spatial intensity gradient, both *conservative* and *nonconservative* forces can be experienced by the atoms. To illustrate this point, we assume that the laser field has a Gaussian spatial profile and take $\mathbf{E}_0(\mathbf{R}) = \hat{\boldsymbol{\epsilon}} E_0(R_T) e^{ikZ}$, where

$$\mathbf{R}_T = X\mathbf{i} + Y\mathbf{j} \tag{5.18}$$

is the cylindrical position vector, $\mathbf{k} \cdot \hat{\boldsymbol{\epsilon}} \approx 0$, and

$$E_0(R_T) = E_0 e^{-R_T^2/a^2}, \tag{5.19}$$

with E_0 a real constant. This field does not satisfy Maxwell's equations, but if a is much greater than $\lambda = 2\pi/k$ (allowing one to neglect diffraction of the field), it represents an approximate solution.

Since

$$\nabla \left[\boldsymbol{\mu}_{21} \cdot \mathbf{E}_0(\mathbf{R})\right] = \boldsymbol{\mu}_{21} \cdot \hat{\boldsymbol{\epsilon}} e^{ikZ} \left[ik\hat{\mathbf{z}} E_0(R_T) + \frac{\partial E_0(R_T)}{\partial R_T} \hat{\mathbf{r}}_T \right], \tag{5.20}$$

where $\hat{\mathbf{r}}_T = \mathbf{R}_T / |\mathbf{R}_T|$ is a unit vector, it follows from equation (5.13) that the force is given by

$$\mathbf{F}(\mathbf{R}, \mathbf{v}) = -i\hbar k\hat{\mathbf{z}} \chi(R_T) \tilde{\varrho}_{12}(\mathbf{R}, \mathbf{v}) e^{ikZ} + \frac{1}{2} \boldsymbol{\mu}_{21} \cdot \hat{\boldsymbol{\epsilon}} \tilde{\varrho}_{12}(\mathbf{R}, \mathbf{v}) e^{ikZ} \frac{\partial E_0(R_T)}{\partial R_T} \hat{\mathbf{r}}_T + \text{c.c.}, \tag{5.21}$$

where

$$\chi(R_T) = -(\hat{\boldsymbol{\epsilon}} \cdot \boldsymbol{\mu}_{21}) E_0(R_T)/(2\hbar). \tag{5.22}$$

It is not so easy to solve equations (5.2) for $\tilde{\varrho}_{12}(\mathbf{R}, \mathbf{v})$, since the field is a function of the cylindrical coordinate R_T. Still, the terms in equation (5.21) have an obvious meaning. The first term is associated with scattering of radiation from the incident field into unoccupied modes of the vacuum field, as in section 5.3.1. The second term is a *gradient force* term that arises from spatial variations in the envelope of the field. The origin of the gradient force can be associated with an exchange of momentum between the different modes (directions) of the incident field—it is a conservative force, in contrast to the dissipative force arising from scattering into new vacuum field modes.

In the large detuning limit, $|\delta| \gg ku$, where u is the most probable atomic speed, the Doppler shift is unimportant, and we can solve equation (5.2c) for $\tilde{\varrho}_{12}$ in steady state, neglecting the $\mathbf{v} \cdot \nabla$ term that gives rise to the Doppler shift. If, in addition, $|\delta| \gg \gamma_B$, we can neglect the ϱ_{22} term in equation (5.2c) as well, and obtain

$$\tilde{\varrho}_{12}(Z, R_T) \approx i\chi^*(R_T)\, e^{-ikZ}(\gamma - i\delta)/\delta^2, \tag{5.23}$$

where we have used the fact that $\chi(\mathbf{R}, t) = \chi(R_T)e^{ikZ}$. It then follows from equations (5.21) and (5.23) that

$$\begin{aligned}\mathbf{F}(R_T) &= \frac{2\hbar k\gamma|\chi(R_T)|^2\hat{\mathbf{z}}}{\delta^2} + \frac{1}{\delta}\frac{2\hbar|\chi(R_T)|^2}{E_0(R_T)}\frac{\partial E_0(R_T)}{\partial R_T}\hat{\mathbf{r}}_T, \\ &= \mathbf{F}_{dis}(R_T) + \mathbf{F}_{grad}(R_T),\end{aligned} \tag{5.24}$$

where the first term $\sim\delta^{-2}$ is the dissipative term, and the second one is the gradient term. For large detunings, the gradient force dominates the dissipative force.

From equation (5.19), we calculate

$$\frac{1}{E_0(R_T)}\frac{\partial E_0(R_T)}{\partial R_T} = -\frac{2R_T}{a^2}, \tag{5.25}$$

and

$$\mathbf{F}_{grad}(R_T, \mathbf{v}) = -4\frac{\hbar|\chi(R_T)|^2}{\delta}\frac{R_T}{a^2}\hat{\mathbf{r}}_T. \tag{5.26}$$

For $\delta > 0$ (red detuning), atoms are drawn into the focus. For $\delta < 0$ (blue detuning), atoms are pushed out from the focus. This gradient force is the basis of a far off-resonance optical trap (FORT) [3].

The gradient force (5.26) can be understood as arising from the Stark-shifted energy of the ground state resulting from the field. The ground-state probability amplitudes evolve as

$$\dot{c}_1(R_T, t) = -i\chi^*(R_T, t)e^{-i\delta t}c_2(R_T, t), \tag{5.27a}$$

$$\dot{c}_2(R_T, t) = -i\chi(R_T, t)e^{i\delta t}c_1(R_T, t), \tag{5.27b}$$

where we have allowed for a time dependence in the Rabi frequencies as the field is turned on. If we imagine that the field is turned on slowly in a time compared with δ^{-1}, then

$$c_2(R_T, t) = -i\int_{-\infty}^{t} dt'\chi(R_T, t')e^{i\delta t'}c_1(R_T, t') \approx -\frac{\chi(R_T, t)e^{i\delta t}}{\delta}c_1(R_T, t), \tag{5.28}$$

which, when substituted into (5.27a), yields

$$i\hbar\dot{c}_1(R_T,t) \approx -\frac{\hbar\,|\chi(R_T,t)|^2}{\delta}c_1(R_T,t). \tag{5.29}$$

Once the Rabi frequency has been ramped up to its final value $\chi(R_T)$, the ground-state energy is shifted by an amount $-\hbar|\chi(R_T)|^2/\delta$. The force, evaluated as the negative gradient of this energy shift, agrees with equation (5.26).

5.3.3 Standing-Wave Field: Laser Cooling

We have seen that atoms can be accelerated or slowed by a plane-wave field, depending on their direction relative to the field propagation vector and the atom–field detuning. By combining *two* plane-wave fields in opposite directions (standing-wave field), it is possible to produce slowing along the axis of the propagation vector, regardless of the atom's direction. For a standing-wave optical field, we take[1]

$$\mathbf{E}_0(\mathbf{R}) = 2E_0\hat{\mathbf{x}}\cos kZ, \tag{5.30}$$

such that

$$\nabla\left[\boldsymbol{\mu}_{21}\cdot\mathbf{E}_0(\mathbf{R})\right] = -2k\hat{\mathbf{z}}\left(\boldsymbol{\mu}_{21}\cdot\hat{\mathbf{x}}\right)E_0\sin kZ = 2\hbar k\hat{\mathbf{z}}\chi_s\sin kZ, \tag{5.31}$$

where

$$\chi_s = -\left(\boldsymbol{\mu}_{21}\cdot\hat{\mathbf{x}}\right)E_0/\hbar. \tag{5.32}$$

As a consequence, the force given by equation (5.13) is

$$\mathbf{F}(Z,v_z) = \hbar k\hat{\mathbf{z}}\sin kZ\left[\chi_s\tilde{\varrho}_{12}(Z,v_z) + \chi_s^*\tilde{\varrho}_{21}(Z,v_z)\right]. \tag{5.33}$$

Again, it is not a simple matter to calculate $\tilde{\varrho}_{12}(Z,v_z)$ for moving atoms. In effect, an atom moving with velocity v_z sees two fields in its rest frame, having frequencies $\omega \pm kv_z$. Since the atom acts as a nonlinear device, all harmonics of the beat frequency $2kv_z$ between these fields appear in the atomic response.

For large detunings $[\delta \gg \gamma_B, ku]$, however, the field frequency for all the atoms in their rest frames is approximately the same, regardless of their velocity, and the excited-state population is much less than unity. In the same way we arrived at equation (5.23), we substitute $\chi(\mathbf{R},t) = \chi_s\cos kZ$ into equation (5.2c) and find the approximate solution

$$\tilde{\varrho}_{12}(Z,v_z) \approx -\left(\chi_s^*/\delta\right)\cos kZ. \tag{5.34}$$

From equation (5.33), we calculate the force as

$$\mathbf{F}(Z) = -2\hbar k\hat{\mathbf{z}}\left(|\chi_s|^2/\delta\right)\sin kZ\cos kZ = -\hbar k\hat{\mathbf{z}}\left(|\chi_s|^2/\delta\right)\sin(2kZ). \tag{5.35}$$

This is a gradient force that arises from the standing-wave nature of the field. Its origin can be traced to scattering from one traveling wave component of the

[1] The factor of 2 is introduced for consistency with the notation of subsequent chapters, where a standing-wave field is composed of two traveling-wave fields, *each* having amplitude E_0.

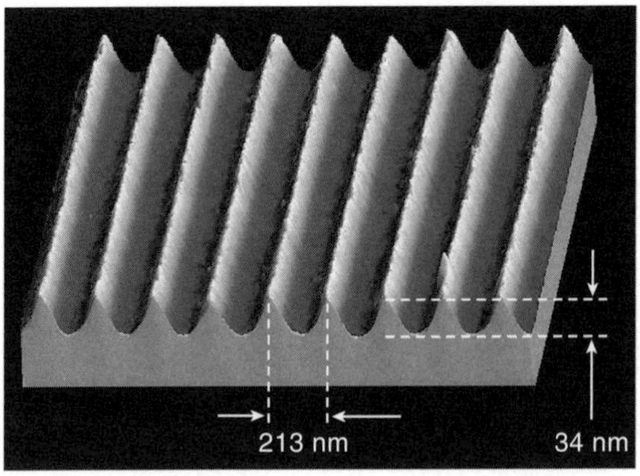

Figure 5.2. Focusing of Cr atoms using a standing-wave laser field. (We would like to thank Jabez McClelland for providing us with a copy of this figure.)

standing-wave field into the other. The dissipative contribution to the force has not been included in equation (5.35).

Alternatively, the force can be interpreted in terms of a spatially modulated energy shift. Equation (5.29) remains valid if $|\chi(R_T, t)|^2$ is replaced by $|\chi_s|^2 \cos^2(kZ)$; as a consequence, the ground-state light shift becomes

$$\Delta E = -\frac{\hbar |\chi_s|^2}{\delta} \cos^2(kZ), \tag{5.36}$$

which implies that the force is

$$\mathbf{F}(Z) = -\nabla(\Delta E) = -2\hbar k \hat{\mathbf{z}} \left(|\chi_s|^2/\delta\right) \sin kZ \cos kZ = -\hbar k \hat{\mathbf{z}} \left(|\chi_s|^2/\delta\right) \sin(2kZ), \tag{5.37}$$

in agreement with equation (5.35). Standing-wave fields can be used to focus atoms, since each lobe of the field acts as a lens (see figure 5.2). We look at this process in detail in chapter 11.

As $|\delta|$ is reduced, the dissipative force begins to play a role. In steady state, one must solve the density matrix equations

$$v_z \frac{\partial \varrho_{11}(Z,v_z)}{\partial Z} = -i \cos kZ \left[\chi_s^* \tilde{\varrho}_{21}(Z,v_z) - \chi_s \tilde{\varrho}_{12}(Z,v_z)\right] + \gamma_2 \varrho_{11}(Z,v_z), \tag{5.38a}$$

$$v_z \frac{\partial \varrho_{22}(Z,v_z)}{\partial Z} = i \cos kZ \left[\chi_s^* \tilde{\varrho}_{21}(Z,v_z) - \chi_s \tilde{\varrho}_{12}(Z,v_z)\right] - \gamma_2 \varrho_{22}(Z,v_z), \tag{5.38b}$$

$$v_z \frac{\partial \tilde{\varrho}_{12}(Z,v_z)}{\partial Z} = -i\chi_s^* \cos kZ [\varrho_{22}(Z,v_z) - \varrho_{11}(Z,v_z)] - (\gamma - i\delta)\, \tilde{\varrho}_{12}(Z,v_z), \tag{5.38c}$$

$$v_z \frac{\partial \tilde{\varrho}_{21}(Z,v_z)}{\partial Z} = i\chi_s \cos kZ [\varrho_{22}(Z,v_z) - \varrho_{11}(Z,v_z)] - (\gamma + i\delta)\, \tilde{\varrho}_{21}(Z,v_z). \tag{5.38d}$$

These equations cannot be solved analytically, but they can be solved in terms of continued fractions. For the present, let us assume that a lowest order perturbation solution (to order χ_s) is sufficient. To zeroth order in χ_s, $\varrho_{11}^{(0)}(Z, v_z) = 1$, and all other density matrix elements equal zero. In first order, we set

$$\tilde{\varrho}_{12}^{(1)}(Z, v_z) = \left[\tilde{\varrho}_{12}^{+}(v_z)e^{ikZ} + \tilde{\varrho}_{12}^{-}(v_z)e^{-ikZ}\right]. \tag{5.39}$$

Substituting the trial solution, equation (5.39), into equation (5.38c), with $[\varrho_{22}(Z, v_z) - \varrho_{11}(Z, v_z)]$ set equal to -1, we obtain

$$\begin{aligned}\frac{v_z \partial \tilde{\varrho}_{12}(Z, v_z)}{\partial Z} &= ikv_z\tilde{\varrho}_{12}^{+}(v_z)e^{ikZ} - ikv_z\tilde{\varrho}_{12}^{-}(v_z)e^{-ikZ} \\ &= -(\gamma - i\delta)\left[\tilde{\varrho}_{12}^{+}(v_z)e^{ikZ} + \tilde{\varrho}_{12}^{-}(v_z)e^{-ikZ}\right] \\ &\quad + i\chi_s^* \cos kZ.\end{aligned} \tag{5.40}$$

Equating coefficients of $e^{\pm ikZ}$, we find

$$\tilde{\varrho}_{12}^{+}(v_z) = \frac{1}{2}\frac{i\chi_s^*}{\gamma - i(\delta - kv_z)}, \tag{5.41a}$$

$$\tilde{\varrho}_{12}^{-}(v_z) = \frac{1}{2}\frac{i\chi_s^*}{\gamma - i(\delta + kv_z)}, \tag{5.41b}$$

and

$$\tilde{\varrho}_{12}(Z, v_z) = \frac{i\chi_s^*}{2}\left[\frac{e^{ikZ}}{\gamma - i(\delta - kv_z)} + \frac{e^{-ikZ}}{\gamma - i(\delta + kv_z)}\right]. \tag{5.42}$$

Last, for the force (5.33), we obtain

$$\mathbf{F}(Z, v_z) = \hbar k\hat{\mathbf{z}}\chi_s \sin kZ\tilde{\varrho}_{12}(Z, v_z) + \text{c.c.} \tag{5.43a}$$

$$\begin{aligned}&= \frac{1}{4}\hbar k\hat{\mathbf{z}}\,|\chi_s|^2\left\{\left[\frac{1}{\gamma - i(\delta + kv_z)} - \frac{1}{\gamma - i(\delta - kv_z)}\right] + \text{c.c.}\right\} \\ &+ \frac{1}{4}\hbar k\hat{\mathbf{z}}\,|\chi_s|^2\left\{\left[\frac{e^{2ikZ}}{\gamma - i(\delta - kv_z)} - \frac{e^{-2ikZ}}{\gamma - i(\delta + kv_z)}\right] + \text{c.c.}\right\}.\end{aligned} \tag{5.43b}$$

In the limit of large δ, the first line in equation (5.43b) is small and the second gives the gradient force, equation (5.37).

If we ask for the *spatially averaged force*, $\mathbf{F}(v_z) = \langle \mathbf{F}(Z, v_z)\rangle_Z$, then the second line in equation (5.43b) vanishes, and the first gives

$$\mathbf{F}(v_z) = \frac{\hbar k\gamma\hat{\mathbf{z}}\,|\chi_s|^2}{2}\left[\frac{1}{\gamma^2 + (\delta + kv_z)^2} - \frac{1}{\gamma^2 + (\delta - kv_z)^2}\right]. \tag{5.44}$$

This force is plotted in figure 5.3 as a function kv_z for $\delta > 0$. Since the force is positive for $v_z < 0$ and negative for $v_z > 0$, the force always tends to decrease the magnitude of the velocity: it is a *friction force*. Near $kv_z = 0$ (that is, for $kv_z \ll \gamma, |\delta|$)

$$\mathbf{F}(v_z) = -\hat{\mathbf{z}}\frac{2\hbar k^2 v_z\gamma\,|\chi_s|^2\,\delta}{(\gamma^2 + \delta^2)^2}. \tag{5.45}$$

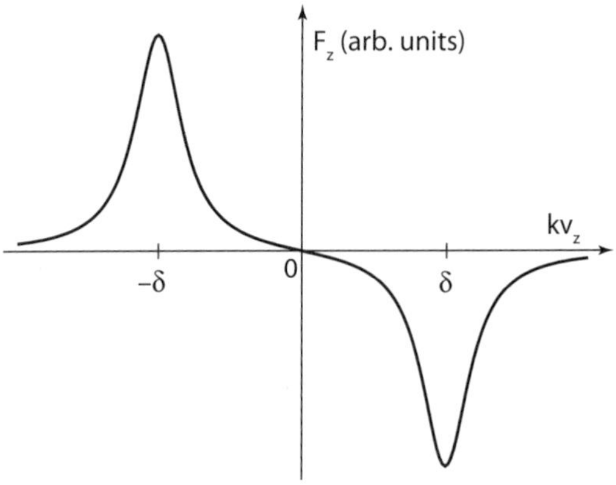

Figure 5.3. Force on an atom in a standing-wave optical field with positive δ. Since the force always opposes the motion, it is a friction force.

A vapor can be optically cooled if three, mutually perpendicular standing-wave fields are applied. In that case, the spatially averaged force is

$$\mathbf{F}(\mathbf{v}) = -\beta_f \mathbf{v}, \tag{5.46}$$

where the *friction coefficient* β_f is given by

$$\beta_f = \frac{2\hbar k^2 \gamma \left|\chi_s\right|^2 \delta}{(\gamma^2 + \delta^2)^2}. \tag{5.47}$$

The rate at which the energy $\mathcal{E}$ of atoms having velocity $\mathbf{v}$ is decreased is equal to

$$\frac{d}{dt}\mathcal{E} = \mathbf{F} \cdot \mathbf{v} = -\alpha_f \mathcal{E}, \tag{5.48}$$

where $\mathcal{E} = \frac{1}{2}Mv^2$ and

$$\alpha_f = \frac{4\hbar k^2 \gamma \left|\chi_s\right|^2 \delta}{M(\gamma^2 + \delta^2)^2} = \frac{2\beta_f}{M}. \tag{5.49}$$

The maximum friction force occurs when $\delta = \gamma/\sqrt{3}$.

The cooling does not go on forever. As a result of scattering into unoccupied vacuum modes, the atom gets random kicks that heat the vapor. To properly calculate the diffusion, we must quantize the center-of-mass motion. We can get a rough estimate of the diffusion by assuming that the atoms get a momentum kick equal to $\hbar k$ and acquire $(\hbar k)^2/2M$ of energy per kick so that the spatially averaged heating rate is

$$\left.\frac{d}{dt}\frac{p^2}{2M}\right|_{heating} \approx \frac{\hbar^2 k^2}{2M}\gamma_2 \langle \varrho_{22}(v_z)\rangle_Z. \tag{5.50}$$

In perturbation theory, the spatially averaged excited-state population $\langle \varrho_{22}(v_z)\rangle_Z$ is given by

$$\langle \varrho_{22}(v_z)\rangle_Z = \frac{\left|\chi_s\right|^2 \gamma}{2\gamma_2}\left[\frac{1}{\gamma^2 + (\delta + kv_z)^2} + \frac{1}{\gamma^2 + (\delta - kv_z)^2}\right], \tag{5.51}$$

which can be obtained by substituting equation (5.42) into equation (5.38b) and solving for $\varrho_{22}(Z, v_z)$. For $kv_z \ll \gamma, |\delta|$, and with $\gamma_2 = 2\gamma$, equation (5.51) reduces to

$$\langle \varrho_{22} \rangle_Z \approx \frac{|\chi_s|^2 / 2}{\gamma^2 + \delta^2}. \tag{5.52}$$

Therefore, one finds

$$\left. \frac{d\mathcal{E}}{dt} \right|_{heat} = \frac{\hbar^2 k^2}{4M} \frac{|\chi_s|^2 \gamma_2}{\gamma^2 + \delta^2} = \frac{2D}{M}, \tag{5.53}$$

where D is the *diffusion coefficient*, and

$$\frac{d}{dt}\mathcal{E} = -\frac{4\hbar k^2 \gamma |\chi_s|^2 \delta}{M(\gamma^2 + \delta^2)^2}\mathcal{E} + \frac{\hbar^2 k^2}{4M} \frac{|\chi_s|^2 \gamma_2}{\gamma^2 + \delta^2}. \tag{5.54}$$

In steady state and for the optimal detuning, $\delta = \gamma$, we find

$$\mathcal{E} = \frac{\hbar \gamma_2 (\gamma^2 + \delta^2)}{16\delta\gamma} \approx \hbar \frac{\gamma_2}{8} = \frac{1}{2} k_B T, \tag{5.55}$$

where T is the equilibrium temperature, and k_B is Boltzmann's constant. A more rigorous calculation includes fluctuations arising from the difference in the number of photons scattered by atoms from each traveling wave component of the standing-wave field—that is, contributions to diffusion from stimulated as well as spontaneous processes. With the inclusion of such stimulated processes, the final temperature increases to $T_D = 7\hbar\gamma_2/20k_B$, a result that is referred to as the *Doppler limit* of laser cooling [4].[2] For Na, one estimates that $T_D \approx 170\ \mu$K, which corresponds to $v \approx 25$ cm/s, and $kv = 2.7 \times 10^6\ \text{s}^{-1} \ll 6.3 \times 10^7\ \text{s}^{-1} = \gamma_2$. When experiments were done by Lett *et al.* [5] and Weiss *et al.* [6] to test this prediction, they found a *lower* temperature—for reasons we discuss later. For this discovery and the explanation of the cooling, William Phillips, Steven Chu, and Claude Cohen-Tannoudji shared the Nobel Prize in 1997.

5.4 Summary

In this chapter, we have seen how optical fields can be used to manipulate the center-of-mass motion of atoms. We derived equations that properly account for atomic motion and calculated the force on atoms produced by various field configurations. For the most part, we have considered large detunings or cold atoms, where Doppler shifts play a minimal role. We now return to a discussion of thermal vapors and see how *saturation spectroscopy* can be used to eliminate the Doppler width in spectral profiles. Before doing so, however, we take a slight detour and derive the so-called Maxwell-Bloch equations.

[2] The correct diffusion coefficient, $D = (7\hbar^2 k^2/20)(|\chi_s|^2 \gamma_2/(\gamma^2 + \delta^2))$ can be obtained using the formalism developed in the appendix in chapter 18. Many authors refer to $k_B T = \hbar\gamma_2/2$, instead of $k_B T = 7\hbar\gamma_2/20$, as the Doppler limit of laser cooling.

5.5 Appendix: Quantization of the Center-of-Mass Motion

As we often tell our students, atoms have never heard about wave packets. Wave packets are a construction of physicists. In other words, when we consider a vapor or an ensemble of atoms, we usually have a very limited amount of information. There is virtually *no* information about any individual atom's center-of-mass motion; rather, one has some general thermodynamical variables to describe the vapor. One is then at liberty to choose *any* wave packets for the atoms' center-of-mass motion that are consistent with these thermodynamic properties, even if they have no relation to what you might think would be a "reasonable" atomic wave packet. For example, suppose that a thermal vapor is characterized by a uniform density and a Maxwellian distribution of velocities. One can carry out calculations using plane-wave center-of-mass states with a (diagonal) thermal momentum distribution or using highly localized center-of-mass states, randomly distributed in space, in which *each* atom has the entire thermal momentum distribution [7]. Although neither of these descriptions seems physical, both models lead to the *same* results when considering atom–field interactions, since they correspond to the same density matrix for the atomic ensemble.

We now proceed to quantize the center-of-mass motion. Let $\mathbf{R}$ be the center-of-mass coordinate and $\mathbf{r}$ a relative electronic coordinate. The Hamiltonian is assumed to be of the form

$$\hat{H} = \frac{\hat{\mathbf{P}}^2}{2M} + \hat{H}_0(\mathbf{r}) + \hat{V}(\mathbf{r}, \mathbf{R}, t), \tag{5.56}$$

where $\hat{\mathbf{P}} = -i\hbar\nabla_{\mathbf{R}}$ is the momentum conjugate to $\mathbf{R}$. At this stage, all relaxation is neglected. A multitude of representations can be used to label density matrix elements. We consider four possible representations to characterize the center-of-mass motion: coordinate, momentum, sum and difference, and Wigner (coordinate-momentum) representations.

5.5.1 Coordinate Representation

If we expand the wave function as

$$\Psi(\mathbf{r}, \mathbf{R}, t) = \sum_{\alpha} A_\alpha(\mathbf{R}, t)\psi_\alpha(\mathbf{r}), \tag{5.57}$$

then the state amplitudes obey a Schrödinger-like equation

$$i\hbar\frac{\partial A_\alpha(\mathbf{R}, t)}{\partial t} = -\frac{\hbar^2}{2M}\nabla_R^2 A_\alpha(\mathbf{R}, t) + \sum_{\alpha'} V_{\alpha\alpha'}(\mathbf{R}, t)A_{\alpha'}(\mathbf{R}, t) + E_\alpha A_\alpha(\mathbf{R}, t), \tag{5.58}$$

where

$$V_{\alpha\alpha'}(\mathbf{R}, t) = \int d\mathbf{r}\psi_\alpha^*(\mathbf{r})\hat{V}(\mathbf{r}, \mathbf{R}, t)\psi_{\alpha'}(\mathbf{r}), \tag{5.59}$$

and E_α and $\psi_\alpha(\mathbf{r})$ are eigenenergies and eigenfunctions of $\hat{H}_0(\mathbf{r})$. Density matrix elements are defined by

$$\varrho_{\alpha\alpha'}(\mathbf{R}, \mathbf{R}', t) = A_\alpha(\mathbf{R}, t)A_{\alpha'}^*(\mathbf{R}', t). \tag{5.60}$$

Note that we must allow for nondiagonal values of the center-of-mass coordinates as well as those of the internal states. It then follows directly from equation (5.58) that

$$\begin{aligned} i\hbar\frac{\partial \varrho_{\alpha\alpha'}(\mathbf{R},\mathbf{R}',t)}{\partial t} &= (E_\alpha - E_{\alpha'})\varrho_{\alpha\alpha'}(\mathbf{R},\mathbf{R}',t) \\ &-(\hbar^2/2M)[\nabla_R^2\varrho_{\alpha\alpha'}(\mathbf{R},\mathbf{R}',t) - \nabla_{R'}^2\varrho_{\alpha\alpha'}(\mathbf{R},\mathbf{R}',t)] \\ &+\sum_{\alpha''}[V_{\alpha\alpha''}(\mathbf{R},t)\varrho_{\alpha''\alpha'}(\mathbf{R},\mathbf{R}',t) \\ &-\varrho_{\alpha\alpha''}(\mathbf{R},\mathbf{R}',t)V_{\alpha''\alpha'}(\mathbf{R}',t)]. \end{aligned} \tag{5.61}$$

This is a very complicated equation. In many problems, the atomic center-of-mass motion can be treated classically. Imagine that the atom can be described by a wave packet that is much larger than its average de Broglie wavelength λ_{dB} (to ensure no spreading), but much smaller than other characteristic lengths in the problem. Then we can take

$$\begin{aligned} V_{\alpha\alpha'}(\mathbf{R},t) &\approx V_{\alpha\alpha'}(\langle\mathbf{R}\rangle,t), \\ \varrho_{\alpha,\alpha'}(\mathbf{R},\mathbf{R}',t) &\approx \varrho_{\alpha,\alpha'}(\langle\mathbf{R}\rangle,\langle\mathbf{R}'\rangle,t), \end{aligned} \tag{5.62}$$

where $\langle\mathbf{R}\rangle$ is the average position of the wave packet.

The remaining terms in equation (5.61) are

$$\nabla_R\varrho_{\alpha,\alpha'}(\mathbf{R},\mathbf{R}',t) \quad \text{and} \quad \nabla_{R'}\varrho_{\alpha,\alpha'}(\mathbf{R},\mathbf{R}',t). \tag{5.63}$$

We can write

$$\begin{aligned} \nabla_R\varrho_{\alpha\alpha'}(\mathbf{R},\mathbf{R}',t) &= \nabla_R\left[A_\alpha(\mathbf{R},t)A^*_{\alpha'}(\mathbf{R}',t)\right] \\ &= \frac{1}{(2\pi\hbar)^{3/2}}\nabla_R\int d\mathbf{P}\Phi_\alpha(\mathbf{P},t)e^{\frac{i}{\hbar}\mathbf{P}\cdot\mathbf{R}}A^*_{\alpha'}(\mathbf{R}',t) \\ &= \frac{(i/\hbar)}{(2\pi\hbar)^{3/2}}\int d\mathbf{P}\,\mathbf{P}\Phi_\alpha(\mathbf{P},t)e^{\frac{i}{\hbar}\mathbf{P}\cdot\mathbf{R}}A^*_{\alpha'}(\mathbf{R}',t), \end{aligned} \tag{5.64}$$

where

$$\Phi_\alpha(\mathbf{P},t) = (2\pi\hbar)^{-3/2}\int d\mathbf{R}\; e^{-i\mathbf{P}\cdot\mathbf{R}/\hbar}A_\alpha(\mathbf{R},t) \tag{5.65}$$

is the wave function in momentum space. Since there is negligible spreading of the wave packet, the momentum of the wave packet is fairly well defined about some average value $\langle\mathbf{P}\rangle$. Setting $\mathbf{P} = \langle\mathbf{P}\rangle + (\mathbf{P} - \langle\mathbf{P}\rangle)$ and neglecting the contribution from the $(\mathbf{P} - \langle\mathbf{P}\rangle)$ term, we obtain

$$\nabla_R\varrho_{\alpha\alpha'}(\mathbf{R},\mathbf{R}',t) \sim \frac{i}{\hbar}\langle\mathbf{P}\rangle\varrho_{\alpha\alpha'}(\mathbf{R},\mathbf{R}',t). \tag{5.66}$$

Similarly,

$$\nabla_{R'}\varrho_{\alpha\alpha'}(\mathbf{R},\mathbf{R}',t) \sim -\frac{i}{\hbar}\langle\mathbf{P}\rangle\varrho_{\alpha\alpha'}(\mathbf{R},\mathbf{R}',t). \tag{5.67}$$

Therefore, setting $\mathbf{v} = \langle \mathbf{P} \rangle / M$ and rewriting $\mathbf{R} \equiv \langle \mathbf{R} \rangle$ as a classical variable, we can use these results to transform equation (5.61) into

$$\frac{\partial \varrho_{\alpha\alpha'}(\mathbf{R}, \mathbf{v}, t)}{\partial t} + \mathbf{v} \cdot \nabla \varrho_{\alpha\alpha'}(\mathbf{R}, \mathbf{v}, t) = -i\omega_{\alpha\alpha'}\varrho_{\alpha\alpha'}(\mathbf{R}, \mathbf{v}, t) - \frac{i}{\hbar}[\hat{V}(\mathbf{R}, t), \boldsymbol{\varrho}(\mathbf{R}, \mathbf{v}, t)]_{\alpha\alpha'}, \tag{5.68}$$

where $\omega_{\alpha\alpha'} = (E_\alpha - E_{\alpha'})/\hbar$, $\mathbf{R}$ and $\mathbf{v}$ are now classical variables giving the average position and velocity of the center of mass of the atomic wave packet, and the last term is the $\alpha\alpha'$ matrix element of the commutator. This is the basic equation we use for applications. Although the density matrix element $\varrho_{\alpha\alpha'}(\mathbf{R}, \mathbf{R}', t)$ has units of m^{-3}, we adopt a convention in which $\varrho_{\alpha\alpha'}(\mathbf{R}, \mathbf{v}, t)$ is dimensionless and corresponds to a single-particle density matrix element.

5.5.2 Momentum Representation

Density matrix elements in the momentum representation are defined in terms of the momentum state amplitude defined in equation (5.65). This amplitude obeys an integro-differential Schrödinger equation,

$$i\hbar\partial\Phi_\alpha(\mathbf{P}, t)/\partial t = (E_\alpha + E_P)\Phi_\alpha(\mathbf{P}, t) + (2\pi\hbar)^{-3/2}\sum_{\alpha'}\int d\mathbf{P}'\; V_{\alpha\alpha'}(\mathbf{P} - \mathbf{P}', t)\Phi_{\alpha'}(\mathbf{P}', t), \tag{5.69}$$

where

$$V_{\alpha\alpha'}(\mathbf{P}, t) = (2\pi\hbar)^{-3/2}\int d\mathbf{R}\; e^{-i\mathbf{P}\cdot\mathbf{R}/\hbar}\, V_{\alpha\alpha'}(\mathbf{R}, t), \tag{5.70}$$

and

$$E_P = \frac{P^2}{2M} \tag{5.71}$$

is the center-of-mass energy.

Defining density matrix elements as

$$\varrho_{\alpha\alpha'}(\mathbf{P}, \mathbf{P}', t) = \Phi_\alpha(\mathbf{P}, t)\Phi^*_{\alpha'}(\mathbf{P}', t), \tag{5.72}$$

we find from equation (5.69) that these elements obey

$$i\hbar\partial\varrho_{\alpha\alpha'}(\mathbf{P}, \mathbf{P}', t)/\partial t = (E_\alpha - E_{\alpha'})\varrho_{\alpha\alpha'}(\mathbf{P}, \mathbf{P}', t) + (E_P - E_{P'})\varrho_{\alpha\alpha'}(\mathbf{P}, \mathbf{P}', t) + \frac{1}{(2\pi\hbar)^{3/2}}\sum_{\alpha''}\int d\mathbf{P}''[V_{\alpha\alpha''}(\mathbf{P} - \mathbf{P}'', t)\varrho_{\alpha''\alpha'}(\mathbf{P}'', \mathbf{P}', t) - \varrho_{\alpha\alpha''}(\mathbf{P}, \mathbf{P}'', t)V_{\alpha''\alpha'}(\mathbf{P}'' - \mathbf{P}', t)]. \tag{5.73}$$

The momentum space representation is especially useful for considering the interaction of one or more plane waves with atoms, since the these fields transfer discrete momentum to the atoms. The density matrix element $\varrho_{\alpha\alpha'}(\mathbf{P}, \mathbf{P}', t)$ has units of $(\text{momentum})^{-3}$.

5.5.3 Sum and Difference Representation

A useful representation for problems involving spontaneous emission is one in which new variables

$$\mathbf{u} = \mathbf{R} - \mathbf{R}', \quad \mathbf{s} = (\mathbf{R} + \mathbf{R}')/2 \tag{5.74}$$

are defined such that density matrix elements

$$\varrho'_{\alpha\alpha'}(\mathbf{s}, \mathbf{u}, t) = A_\alpha\left(\mathbf{s} + \frac{\mathbf{u}}{2}, t\right) A^*_{\alpha'}\left(\mathbf{s} - \frac{\mathbf{u}}{2}, t\right) \tag{5.75}$$

satisfy the equation of motion

$$\begin{aligned} i\hbar\partial\varrho'_{\alpha\alpha'}(\mathbf{s}, \mathbf{u}, t)/\partial t &= (E_\alpha - E_{\alpha'})\varrho'_{\alpha\alpha'}(\mathbf{s}, \mathbf{u}, t) - (\hbar^2/M)(\nabla_s \cdot \nabla_u)\varrho'_{\alpha\alpha'}(\mathbf{s}, \mathbf{u}, t) \\ &+ \sum_{\alpha''}\left[V_{\alpha\alpha''}\left(\mathbf{s} + \frac{\mathbf{u}}{2}, t\right)\varrho'_{\alpha''\alpha'}(\mathbf{s}, \mathbf{u}, t)\right. \\ &\left. - \varrho'_{\alpha\alpha''}(\mathbf{s}, \mathbf{u}, t)\, V_{\alpha''\alpha'}\left(\mathbf{s} - \frac{\mathbf{u}}{2}, t\right)\right]. \end{aligned} \tag{5.76}$$

The primes distinguish $\varrho'_{\alpha\alpha'}(\mathbf{s}, \mathbf{u}, t)$ from $\varrho_{\alpha\alpha'}(\mathbf{R}, \mathbf{R}', t)$. In a classical limit, $\mathbf{u} \sim 0$ and $\mathbf{s} \sim \mathbf{R}$. The density matrix element $\varrho'_{\alpha\alpha'}(\mathbf{s}, \mathbf{u}, t)$ has units of m^{-3}.

5.5.4 Wigner Representation

In statistical mechanics, one defines a *phase-space distribution* that is a function of position and momentum. Since the position and momentum operators do not commute, it is impossible to define an analogous function in quantum mechanics. On the other hand, it *is* possible in quantum mechanics to define a class of functions that depend on position and momentum that, in certain limiting cases, may approximate a classical phase space distribution. The Wigner distribution can be defined by

$$\varrho_{\alpha\alpha'}(\mathbf{R}, \mathbf{P}, t) = (2\pi\hbar)^{-3}\int d\mathbf{u}\varrho'_{\alpha\alpha'}(\mathbf{R}, \mathbf{u}, t)e^{-i\mathbf{P}\cdot\mathbf{u}/\hbar} \tag{5.77a}$$

$$= (2\pi\hbar)^{-3}\int d\mathbf{u}\varrho_{\alpha\alpha'}\left(\mathbf{R} + \frac{\mathbf{u}}{2}, \mathbf{R} - \frac{\mathbf{u}}{2}, t\right)e^{-i\mathbf{P}\cdot\mathbf{u}/\hbar} \tag{5.77b}$$

$$= (2\pi\hbar)^{-3}\int d\mathbf{q}\varrho_{\alpha\alpha'}\left(\mathbf{P} + \frac{\mathbf{q}}{2}, \mathbf{P} - \frac{\mathbf{q}}{2}, t\right)e^{i\mathbf{R}\cdot\mathbf{q}/\hbar}. \tag{5.77c}$$

In this representation, **R** and **P** are *parameters*. Moreover, **P** is not the center-of-mass momentum of the wave packet nor is **R** the center-of-mass position of the wave packet, although they play the role of such quantities in most situations. Units of $\varrho_{\alpha\alpha'}(\mathbf{R}, \mathbf{P}, t)$ are volume$^{-3/2}$momentum$^{-3/2}$.

Using equations (5.76) and (5.77a), one can prove that $\varrho_{\alpha\alpha'}(\mathbf{R}, \mathbf{P}, t)$ satisfies the equation of motion

$$\begin{aligned} &i\hbar\{\partial\varrho_{\alpha\alpha'}(\mathbf{R}, \mathbf{P}, t)/\partial t + [(\mathbf{P}/M)\cdot\nabla_R]\varrho_{\alpha\alpha'}(\mathbf{R}, \mathbf{P}, t)\} \\ &= (E_\alpha - E_{\alpha'})\varrho_{\alpha\alpha'}(\mathbf{R}, \mathbf{P}, t) + \sum_{\alpha''}(2\pi\hbar)^{-3/2}\int d\mathbf{P}'\left[V_{\alpha\alpha''}(\mathbf{P}', t)\varrho_{\alpha''\alpha'}\left(\mathbf{R}, \mathbf{P} - \frac{\mathbf{P}'}{2}, t\right)\right. \\ &\left. - \varrho_{\alpha\alpha''}\left(\mathbf{R}, \mathbf{P} + \frac{\mathbf{P}'}{2}, t\right)V_{\alpha''\alpha'}(\mathbf{P}', t)\right]e^{i\mathbf{P}'\cdot\mathbf{R}/\hbar}. \end{aligned} \tag{5.78}$$

The Wigner function has some interesting and useful properties. Integrated over **P**, it yields

$$\int d\mathbf{P}\varrho_{\alpha\alpha'}(\mathbf{R},\mathbf{P},t) = \frac{1}{(2\pi\hbar)^3}\int d\mathbf{u}\int d\mathbf{P}\varrho_{\alpha\alpha'}\left(\mathbf{R}+\frac{\mathbf{u}}{2},\mathbf{R}-\frac{\mathbf{u}}{2},t\right)e^{-\frac{i}{\hbar}\mathbf{P}\cdot\mathbf{u}}$$
$$= \varrho_{\alpha\alpha'}(\mathbf{R},\mathbf{R},t) = A_\alpha(\mathbf{R},t)A^*_{\alpha'}(\mathbf{R},t), \tag{5.79}$$

and integrated over **R**,

$$\int d\mathbf{R}\varrho_{\alpha\alpha'}(\mathbf{R},\mathbf{P},t) = \frac{1}{(2\pi\hbar)^3}\int d\mathbf{q}\int d\mathbf{R}\varrho_{\alpha\alpha'}\left(\mathbf{P}+\frac{\mathbf{q}}{2},\mathbf{P}-\frac{\mathbf{q}}{2},t\right)e^{\frac{i}{\hbar}\mathbf{R}\cdot\mathbf{q}}$$
$$= \varrho_{\alpha\alpha'}(\mathbf{P},\mathbf{P},t) = \Phi_\alpha(\mathbf{P},t)\Phi^*_{\alpha'}(\mathbf{P},t). \tag{5.80}$$

For a single-state α, these quantities are simply the probability distributions in coordinate and momentum space, respectively. Thus, it appears that $\varrho_{\alpha\alpha}(\mathbf{R},\mathbf{P},t)$ is a phase-space distribution for the state α population. Although it resembles such a function in many cases, it is not a true phase-space distribution and can even take on negative values in certain instances. The negative values are a signature of the quantum nature of the state.

Perhaps one of the most useful properties of the Wigner function is that it allows you to calculate expectation values of operators in a simple manner. Consider an operator that is a function of position only, $\hat{O}(\mathbf{r},\mathbf{R})$. Then

$$\langle\hat{O}(t)\rangle = \int\Psi^*(\mathbf{r},\mathbf{R},t)\hat{O}(\mathbf{r},\mathbf{R})\Psi(\mathbf{r},\mathbf{R},t)\,d\mathbf{R}$$
$$= \sum_{\alpha\alpha'}\int A^*_\alpha(\mathbf{R},t)\,O_{\alpha\alpha'}(\mathbf{R})A_{\alpha'}(\mathbf{R},t)\,d\mathbf{R}, \tag{5.81}$$

where

$$O_{\alpha\alpha'}(\mathbf{R}) = \int\Psi^*_\alpha(\mathbf{r})\hat{O}(\mathbf{r},\mathbf{R})\Psi_{\alpha'}(\mathbf{r})\,d\mathbf{r} \tag{5.82}$$

is a *function* of R (not an operator). Since

$$A^*_\alpha(\mathbf{R},t)A_{\alpha'}(\mathbf{R},t) = \int d\mathbf{P}\varrho_{\alpha'\alpha}(\mathbf{R},\mathbf{P},t), \tag{5.83}$$

this can be rewritten as

$$\langle\hat{O}(t)\rangle = \sum_{\alpha\alpha'}\int d\mathbf{R}\int d\mathbf{P}\,O_{\alpha\alpha'}(\mathbf{R})\varrho_{\alpha'\alpha}(\mathbf{R},\mathbf{P},t). \tag{5.84}$$

Similarly, for an operator $\hat{O}(\mathbf{r},\mathbf{P})$, one finds

$$\langle\hat{O}(t)\rangle = \sum_{\alpha\alpha'}\int d\mathbf{R}\int d\mathbf{P}\,O_{\alpha\alpha'}(\mathbf{P})\varrho_{\alpha'\alpha}(\mathbf{R},\mathbf{P},t), \tag{5.85}$$

where

$$O_{\alpha\alpha'}(\mathbf{P}) = \int\Psi^*_\alpha(\mathbf{r})O(\mathbf{r},\mathbf{P})\Psi_{\alpha'}(\mathbf{r})\,d\mathbf{r} \tag{5.86}$$

can be considered as a function of **P**. Moreover, for any operator that is symmetric on the interchange of **R** and **P** such as $(\mathbf{R}\cdot\mathbf{P}+\mathbf{P}\cdot\mathbf{R})/2$, it is possible to show that (recall as an operator, $\mathbf{P}=-i\hbar\nabla_R$)

$$\langle \hat{O}(t)\rangle = \sum_{\alpha\alpha'}\int d\mathbf{R}\int d\mathbf{P}\, O_{\alpha\alpha'}(\mathbf{R},\mathbf{P})\varrho_{\alpha'\alpha}(\mathbf{R},\mathbf{P},t), \tag{5.87}$$

where $O(\mathbf{R},\mathbf{P})$ is now considered as a *classical function* of **R** and **P** and *not* an operator. Thus, the beauty of the Wigner distribution is that expectation values of operators can be calculated simply by treating both **R** and **P** as independent variables and $O(\mathbf{R},\mathbf{P})$ as a function rather than operator—that is, instead of using $(-i\hbar\mathbf{R}\cdot\nabla_R - i\hbar\nabla_R\cdot\mathbf{R})/2$, one uses $(\mathbf{R}\cdot\mathbf{P}+\mathbf{P}\cdot\mathbf{R})/2 = \mathbf{R}\cdot\mathbf{P}$ with the Wigner function.

For atom–plane-wave field interactions, where $V_{\alpha\alpha'}(\mathbf{R},t) \sim \exp[\pm i(\mathbf{k}\cdot\mathbf{R})]$, it follows from equation (5.70) that $V_{\alpha\alpha'}(\mathbf{P}',t) \sim \delta(\mathbf{P}'\pm\hbar\mathbf{k})$. If $\hbar k \ll |\langle\mathbf{P}\rangle|$, we can expand the integrand in equation (5.78) about $\mathbf{P}'=0$. To lowest order, we find

$$\begin{aligned} &i\hbar\{\partial\varrho_{\alpha\alpha'}(\mathbf{R},\mathbf{P},t)/\partial t + [(\mathbf{P}/M)\cdot\nabla_R]\varrho_{\alpha\alpha'}(\mathbf{R},\mathbf{P},t)\} \\ &= (E_\alpha - E_{\alpha'})\varrho_{\alpha\alpha'}(\mathbf{R},\mathbf{P},t) + [\hat{V}(\mathbf{R},t),\boldsymbol{\varrho}(\mathbf{R},\mathbf{P},t)]_{\alpha\alpha'}, \end{aligned} \tag{5.88}$$

which is identical in form to equation (5.68). Equation (5.88) is the lead term in an expansion of the Wigner function in a power series in $\hbar$.

For most problems of atom–field interactions, it is valid to treat the center-of-mass motion classically. However, whenever single-photon processes become important [such as diffusion of atoms resulting from spontaneous emission, or scattering from atoms having energies less than $\hbar^2k^2/(2M)$, where k is the optical field propagation constant], a quantized description of the center-of-mass motion is needed.

Problems

1. Plot the steady-state value of $\varrho_{11}(v_z)W_0(v_z)$ for an atom in a monochromatic plane-wave field. Show that there is a "hole" burned in this ground-state velocity distribution by the field.
2. Calculate the maximum force on an atom produced by a monochromatic, plane-wave field having half Rabi frequency $\chi/2\pi = 10$ MHz, given that $\gamma_2/2\pi = 10$ MHz and there are no collisions. Assume that $v_z = 200$ m/s, that the resonance wavelength is $\lambda_0 = 628$ nm, and that the field can be tuned within 1 GHz of resonance. Calculate the acceleration that this force produces for an atom having atomic mass 23.
3. For a 5-mW standing-wave laser field having a waist area of 4 mm^2, calculate the well depth of the ground-state potential produced by the field in units of the recoil energy $(\hbar^2k^2)/2M$ assuming a detuning of 3γ. Repeat the calculation for a FORT, in which the laser field has a power of 100 mW and is focused to a spot diameter of 20 μm. The detuning is $20\gamma_2$. Take $\gamma_2/2\pi = 6$ MHz, $\lambda = 780$ nm, $M = {}^{85}$Rb mass, and $(\mu_x)_{21} = -0.57ea_0$. Also calculate the

frequency spacing at the bottom of the wells assuming that the potentials can be approximated as harmonic in that region. Can atoms cooled to the Doppler limit of laser cooling be trapped in these potentials? Explain.

4. Consider a transition from a ground state having total angular momentum $J = 0$ to an excited state with $J = 1$. The atom is put in a magnetic field that varies as $\mathbf{B} = \beta Z\hat{\mathbf{z}}$ (β =positive constant) near $Z = 0$. Assume that the magnetic sublevels' energies vary as $E_m = E_2 + gmZ$, where g is a positive constant, and m is the magnetic quantum number. If σ_+ light is incident in the $+Z$ direction and σ_- light in the $-Z$ direction, show that an atom can be trapped in the Z direction if the field is detuned below the resonance frequency. Assume that the atoms are moving sufficiently slowly to neglect their Doppler shift. Note that this is the same detuning that will produce cooling for atoms located near $Z = 0$. This technique is the basis for the (magneto-optical trap)(MOT); see reference [8]. (σ_+ radiation drives $\Delta m = 1$ ground-to excited-state transitions, and σ_- radiation drives $\Delta m = -1$ ground-to excited-state transitions.)
5. In the *optical Stern-Gerlach experiment*, two plane-wave laser fields intersect at a small angle θ to form a standing-wave field having period $d = \lambda/\sin\theta \gg \lambda$. The field amplitude varies as $\cos(2\pi Z/d)$ in the Z direction. An atomic beam having diameter much less than d is incident perpendicular to the standing-wave field, and the atoms pass through the laser beam waist, which has diameter w_0. The longitudinal velocity of the beam is v, and the atoms enter the beam with $\varrho_{11} = 1$. If the field is resonant with the atomic transition, calculate the energy of the two semiclassical dressed states of the atom–field system and the maximum angular splitting of these states produced by the field. Assume that the atom–field interaction can be treated in an impulse approximation, and take $v = 1.76 \times 10^3$ m/s, atomic mass $= 4$, $w_0 = 39\,\mu$m, $d = 30\,\mu$m, $|\chi_s|/2\pi = 0.7$ GHz; see reference [9].
6. Repeat problem 5, but assume that the detuning $\delta/2\pi = 160$ MHz is sufficiently large that the atom stays adiabatically in the ground dressed state of the system. In this case, calculate the deflection of the atoms by the field as a function of Z.
7. For an off-resonance standing-wave field, what sign of delta is needed to have the force on the atoms attract atoms to the antinodes of the field? For this sign of delta, estimate the focal distance of one "lobe" of the field for atoms crossing the field having longitudinal de Broglie wavelength λ_B, assuming that the potential can be approximated as that of a harmonic oscillator. Evaluate this distance assuming $v = 5 \times 10^2$ m/s, atomic mass $= 85$, $w_0 = 30\ \mu$m, $\lambda = 780$ nm, $|\chi_s|/2\pi = 300$ MHz, and $\delta/2\pi = 10$ GHz.

8–9. Calculate the spatially averaged force for atoms having $kv \ll \gamma_2$ for two-level atoms interacting with a standing-wave field. To do this, solve the density matrix equations for $v = v_z = 0$, and then calculate a correction term linear in v. Show that in weak fields, your result reduces to equation (5.45). For strong fields, show that the sign of the force changes for a given detuning if one excludes a small region about $\delta = 0$ (i.e., for a given δ, if the force is a friction force for $\delta < 0$ for weak fields, it becomes an accelerating force for strong fields). Take $\gamma = \gamma_2/2$ and χ_s real.

10. Suppose that you want to solve equations (5.38a) to (5.38d) for arbitrary field strengths. For solutions that are periodic in Z, you can set

$$N = \varrho_{22} - \varrho_{11} = \sum_{n=-\infty}^{\infty} N_n e^{2inkZ},$$

where $N_{-n} = N_n^*$, and also set

$$\tilde{\varrho}_{21} = \sum_{n=-\infty}^{\infty} r_n e^{i(2n-1)kZ}.$$

Obtain equations that relate the Fourier coefficients N_n and r_n. By truncating the series and solving the resulting equations, you can obtain successive approximations to the exact solution in terms of continued fractions. From your equations, obtain a value for N to order χ_s^2. Take χ_s to be real.

In problems 11 to 22, it is assumed that an atom's time evolution is governed by the Hamiltonian

$$\hat{H} = \hat{\mathbf{P}}^2/(2M) + \hat{H}_0(\mathbf{r}) + \hat{V}(\mathbf{r}, \mathbf{R}, t),$$

where M is the atomic mass, $\hat{H}_0(\mathbf{r})$ is the free-atom Hamiltonian having eigenvalues E_α and eigenfunctions $\psi_\alpha(\mathbf{r})$, $\hat{\mathbf{P}}$ is the center-of-mass momentum operator for the atom, and $\hat{V}(\mathbf{r}, \mathbf{R}, t)$ is an interaction term describing the atom's interaction with external fields that can change the atom's internal state and center-of-mass momentum.

11. Derive equation (5.58).
12. Derive equation (5.61).
13. Derive equation (5.69).
14. Derive equation (5.73).
15. Derive equation (5.76).
16. Derive equation (5.78).
17. Prove explicitly that equation (5.87) is valid for the operator $\hat{\mathrm{O}}(\mathbf{R}, \mathbf{P}) = (\mathbf{R} \cdot \mathbf{P} + \mathbf{P} \cdot \mathbf{R})/2$ using $\mathbf{P} = -i\hbar\boldsymbol{\nabla}_R$.
18. Calculate the one-dimensional Wigner function for

$$\psi_1(x) = (2\pi a^2)^{-1/4} \exp(ip_0 x/\hbar) \exp[-(x/a)^2]$$

and

$$\psi_2(x) = (2\pi a^2)^{-1/4} \sqrt{\frac{e^{18}}{1+e^{18}}} \left\{\exp[-(x-3a)^2/a^2] + \exp[-(x+3a)^2/a^2]\right\}.$$

Use a three-dimensional plot to plot the Wigner function as a function of x and p for $a = 1$, $\hbar = 1$ and $p_0 = 2$. You will probably want to use a computer program to evaluate the integrals. Which of these wave functions is more classical in nature? Explain. The second wave function is referred to as a *Schrödinger cat state*.
19. For a free particle, use the Wigner distribution function (in one dimension) to evaluate $\langle p \rangle$, $\langle x \rangle$, $\langle \Delta p^2 \rangle$, and $\langle \Delta x^2 \rangle$ at time t in terms of their values at $t = 0$.

20. Prove Ehrenfest's theorem for a particle of mass M moving in a potential $V(\mathbf{R})$ using the Wigner distribution function.
21. The total density can be defined as

$$\varrho(\mathbf{R}, \mathbf{P}, t) = \sum_{\alpha} \varrho_{\alpha\alpha}(\mathbf{R}, \mathbf{P}, t).$$

Show that $\varrho(\mathbf{R}, \mathbf{P}, t)$ satisfies the equation of motion

$$i\hbar\, \{\partial\varrho(\mathbf{R}, \mathbf{P}, t)/\partial t + [(\mathbf{P}/M)\cdot\nabla_{\mathbf{R}}]\varrho(\mathbf{R}, \mathbf{P}, t)\} =$$
$$+\sum_{\alpha,\alpha'}(2\pi\hbar)^{-3/2}\int d\mathbf{P}'[V_{\alpha\alpha'}(\mathbf{P}', t)\varrho_{\alpha'\alpha}\left(\mathbf{R}, \mathbf{P} - \frac{\mathbf{P}'}{2}, t\right)$$
$$-\varrho_{\alpha\alpha'}\left(\mathbf{R}, \mathbf{P} + \frac{\mathbf{P}'}{2}, t\right) V_{\alpha'\alpha}(\mathbf{P},' t)]e^{i\mathbf{P}'\cdot\mathbf{R}/\hbar}.$$

Make an expansion of the integrand to first order in $\mathbf{P}'$ to obtain

$$\partial\varrho\ (\mathbf{R}, \mathbf{P}, t)/\partial t + [(\mathbf{P}/M)\cdot\nabla_{\mathbf{R}}]\varrho(\mathbf{R}, \mathbf{P}, t) =$$
$$+(i\hbar/2)\sum_{\alpha,\alpha'}[\nabla_{\mathbf{R}} V_{\alpha\alpha'}(\mathbf{R}, t)\cdot\nabla_{\mathbf{P}}\varrho_{\alpha'\alpha}(\mathbf{R}, \mathbf{P}, t)$$
$$+\nabla_{\mathbf{P}}\varrho_{\alpha\alpha'}(\mathbf{R}, \mathbf{P}, t)\cdot\nabla_{\mathbf{R}} V_{\alpha'\alpha}(\mathbf{R}, t)].$$

22. Now assume that the external potential is state-independent—that is,

$$V_{\alpha\alpha'}(\mathbf{R}, t) = V(\mathbf{R}, t)\delta_{\alpha,\alpha'}.$$

Prove that

$$i\hbar\{\partial\varrho(\mathbf{R}, \mathbf{P}, t)/\partial t + [(\mathbf{P}/M)\cdot\nabla_{\mathbf{R}}]\varrho(\mathbf{R}, \mathbf{P}, t)\} =$$
$$+(2\pi\hbar)^{-3/2}\int d\mathbf{P}'\ V(\mathbf{P}', t)\left[\varrho\left(\mathbf{R}, \mathbf{P} - \frac{\mathbf{P}'}{2}, t\right) - \varrho\left(\mathbf{R}, \mathbf{P} + \frac{\mathbf{P}'}{2}, t\right)\right] e^{i\mathbf{P}'\cdot\mathbf{R}/\hbar},$$

where $V(\mathbf{P}, t)$ is the Fourier transform of $V(\mathbf{R}, t)$. Make an expansion of the integrand to second order in $\mathbf{P}'$ to obtain

$$\partial\varrho(\mathbf{R}, \mathbf{P}, t)/\partial t + [(\mathbf{P}/M)\cdot\nabla_{\mathbf{R}}]\varrho(\mathbf{R}, \mathbf{P}, t) + [-\nabla_{\mathbf{R}} V(\mathbf{R}, t)]\cdot\nabla_{\mathbf{P}}\varrho(\mathbf{R}, \mathbf{P}, t) = 0,$$

and give an interpretation to your result. Note that corrections of order P'^2 do not contribute. In problem 21, the corrections of order P'^2 do not necessarily vanish and can serve as a source of diffusion for the distribution function.

References

[1] P. R. Berman, *Light scattering*, Contemporary Physics **49**, 313–330 (2008).

[2] T. W. Hänsch and A. L. Schawlow, *Cooling of gases by laser radiation*, Optics Communications **13**, 68–71 (1975).

[3] J. D. Miller, R. A. Cline, and D. J. Heinzen, *Far-off resonance optical trapping of atoms*, Physical Review A **47**, R4567–4570 (1993).

[4] D. J. Wineland and W. M. Itano, *Laser cooling of atoms*, Physical Review A **20**, 1521–1540 (1979).

[5] P. D. Lett, W. D. Phillips, S. L. Rolston, C. E. Tanner, R. N. Watts, and C. I. Westbrook, *Optical molasses,* Journal of the Optical Society of America B **6**, 2084–2107 (1989).

[6] D. S. Weiss, E. Riis, Y. Shevy, P. Jeffrey Ungar, and S. Chu, *Optical molasses and multilevel atoms: experiment,* Journal of the Optical Society of America B **6**, 2072–2083 (1989).

[7] R. Beach, S. R. Hartmann, and R. Friedberg, *Billiard-ball echo model*, Physical Review A **25**, 2658–2666 (1982).

[8] E. L. Raab, M. Prentiss, A. Cable, S. Chu, and D. E. Pritchard, *Trapping of neutral sodium atoms with radiation pressure*, Physical Review Letters **59**, 2631–2634 (1987).

[9] T. Sleator, T. Pfau, V. Balykin, O. Carnal, and J. Mlynek, *Experimental demonstration of the optical Stern-Gerlach effect*, Physical Review Letters **68**, 1996–1999 (1992).

Bibliography

A. Ashkin, *Applications of laser radiation pressure*, Science **210**, 1081–1088 (1980).

P. R. Berman, *Quantum-mechanical transport equation for atomic systems*, Physical Review A **5**, 927–939 (1972).

J. Dalibard and C. Cohen-Tannoudji, *Atomic motion in laser light: connection between semiclassical and quantum descriptions*, Journal of Physics B **18**, 1661–1683 (1985).

J. P. Gordon and A. Ashkin, *Motion of atoms in a radiation trap*, Physical Review A **21**, 1606–1617 (1980).

H. J. Metcalf and P. van der Straten, *Laser Cooling and Trapping* (Springer-Verlag, New York, 1999).

W. P. Schleich, *Quantum Optics in Phase Space* (Wiley-VCH, Berlin, 2001), chap. 20.

S. Stenholm, *Semiclassical expansion for systems with internal degrees of freedom,* Physical Review A **47**, 2523–2531 (1993).

———*The semiclassical theory of laser cooling,* Reviews of Modern Physics **58**, 699–739 (1986).

6

Maxwell-Bloch Equations

6.1 Wave Equation

Up until this point, we have looked at the modification of the atomic state vector produced by an external electromagnetic field; however, for the most part, we have ignored changes in the external field produced by the atoms. If we quantize the electromagnetic field, the total energy of the atoms plus field is conserved, guaranteeing that changes in the energy of the incident field are automatically included. Within the context of the semiclassical theory, however, we are restricted to a classical description of the incident field. Thus, any changes in the external field must be accounted for using Maxwell's equations. In this *Maxwell-Bloch* approach, the fields are described by Maxwell's equations, but the polarization that appears in Maxwell's equations is calculated quantum-mechanically in terms of the expectation value of the atomic dipole moment operator.

It is necessary to couple the quantum evolution equation for the atoms to Maxwell's equations for the fields in order to follow the evolution of the fields. To do this, we write Maxwell's equations but do not take $\mathbf{D} = \varepsilon_0\mathbf{E}$. Rather, we set $\mathbf{D} = \varepsilon_0\mathbf{E} + \mathbf{P}$ and calculate $\mathbf{P}$ quantum-mechanically as the expectation value of the atomic dipole moment operator. We take $\mathbf{B} = \mu_0\mathbf{H}$ and neglect any effects arising from magnetic polarization.

In the absence of free charges and currents, Maxwell's equations are

$$\nabla \cdot (\varepsilon_0\mathbf{E} + \mathbf{P}) = 0, \tag{6.1a}$$

$$\nabla \times \mathbf{E} = -\frac{\partial \mathbf{B}}{\partial t}, \tag{6.1b}$$

$$\nabla \times \mathbf{B} = \mu_0 \frac{\partial(\varepsilon_0\mathbf{E} + \mathbf{P})}{\partial t}, \tag{6.1c}$$

$$\nabla \cdot \mathbf{B} = 0, \tag{6.1d}$$

and the wave equation, equation (1.5), is

$$\nabla^2 \mathbf{E} - \frac{1}{c^2}\frac{\partial^2 \mathbf{E}}{\partial t^2} = \nabla(\nabla \cdot \mathbf{E}) + \frac{1}{c^2 \varepsilon_0}\frac{\partial^2 \mathbf{P}}{\partial t^2}. \tag{6.2}$$

With no free charges,

$$\nabla \cdot \mathbf{E} = \rho_b/\epsilon_0 = -\nabla \cdot \mathbf{P}/\epsilon_0, \tag{6.3}$$

where ρ_b is the bound charge density; this term vanishes in vacuum. In a medium, it can give rise to transverse effects such as self-trapping, self-focusing, and ring formation. We neglect any such transverse effects and consider essentially a plane wave limit in which the fields propagate in the $\hat{\mathbf{z}}$ direction and have polarization $\boldsymbol{\epsilon}$ in the XY plane, allowing us to set $\nabla \cdot \mathbf{P} \approx 0$ and to use

$$\frac{\partial^2 \mathbf{E}}{\partial Z^2} - \frac{1}{c^2}\frac{\partial^2 \mathbf{E}}{\partial t^2} = \frac{1}{c^2 \varepsilon_0}\frac{\partial^2 \mathbf{P}}{\partial t^2} \tag{6.4}$$

as the wave equation appropriate to our considerations. By setting $\nabla^2 \mathbf{E} \approx -\partial^2 \mathbf{E}/\partial Z^2$, we are neglecting any diffraction effects that would accompany a beam having finite radius a. Diffraction effects are minimal if $\lambda/a \ll 1$.

6.1.1 Pulse Propagation in a Linear Medium

To illustrate pulse propagation in a linear medium characterized by a frequency-dependent index of refraction, we first Fourier transform the positive frequency components of the fields using

$$\mathbf{E}^+(Z,t) = \frac{1}{\sqrt{2\pi}}\int_0^\infty \tilde{\mathbf{E}}^+(Z,\omega')e^{i[k'Z-\omega' t]}d\omega', \tag{6.5a}$$

$$\mathbf{P}^+(Z,t) = \frac{1}{\sqrt{2\pi}}\int_0^\infty \tilde{\mathbf{P}}^+(Z,\omega')e^{i[k'Z-\omega' t]}d\omega', \tag{6.5b}$$

where k' is a function of ω' to be determined. In section 6.2, we calculate the polarization $\mathbf{P}(Z,t)$ quantum-mechanically; here, we assume that the polarization $\tilde{\mathbf{P}}^+(Z,\omega)$ is related to the electric field $\tilde{\mathbf{E}}^+(Z,\omega)$ by

$$\tilde{\mathbf{P}}^+(Z,\omega) = \varepsilon_0 \chi(\omega)\tilde{\mathbf{E}}^+(Z,\omega) = \varepsilon_0\left[n^2(\omega)-1\right]\tilde{\mathbf{E}}^+(Z,\omega), \tag{6.6}$$

where $\chi(\omega)$ is the electric susceptibility, and $n(\omega)$ is the index of refraction, assumed to be both complex and frequency dependent. The total fields are given by

$$\mathbf{E}(Z,t) = \left[\mathbf{E}^+(Z,t) + \mathbf{E}^+(Z,t)^*\right], \tag{6.7a}$$

$$\mathbf{P}(Z,t) = \left[\mathbf{P}^+(Z,t) + \mathbf{P}^+(Z,t)^*\right]. \tag{6.7b}$$

Substituting equations (6.5) into equation (6.4) and using equation (6.6), we find that equation (6.5a) is a solution of equation (6.4) provided that

$$k'(\omega') = n(\omega')\frac{\omega'}{c}, \tag{6.8}$$

giving rise to a field amplitude

$$\mathbf{E}^{+}(Z,t)=\frac{1}{\sqrt{2\pi}}\int_0^{\infty}\tilde{\mathbf{E}}^{+}(Z,\omega')e^{-i\omega'[t-n(\omega')Z/c]}d\omega'. \tag{6.9}$$

It is now assumed that $\tilde{E}(Z,\omega)$ is a sharply peaked function about a central frequency $\omega'=\omega$. As a result, it is convenient to rewrite equation (6.9) as

$$\mathbf{E}^{+}(Z,t)\approx\frac{\hat{\boldsymbol{\epsilon}}}{2}e^{-i\omega[t-n(\omega)Z/c]}E(Z,t), \tag{6.10a}$$

$$\tilde{\mathbf{E}}^{+}(Z,\omega)=\frac{1}{2}\hat{\boldsymbol{\epsilon}}\tilde{E}(Z,\omega), \tag{6.10b}$$

where the complex pulse amplitude $E(Z,t)$ is given by

$$\begin{aligned}E(Z,t)&=\frac{1}{\sqrt{2\pi}}\int_0^{\infty}d\omega'\,\tilde{E}(Z,\omega')\exp\{-i\Delta t+i[\omega' n(\omega')-\omega n(\omega)]Z/c\}\\&=\frac{1}{\sqrt{2\pi}}\int_{-\omega}^{\infty}d\Delta\,\tilde{E}(Z,\omega+\Delta)\exp\{-i\Delta t+i[(\omega+\Delta)n(\omega+\Delta)-\omega n(\omega)]Z/c\}\\&\approx\frac{1}{\sqrt{2\pi}}\int_{-\infty}^{\infty}d\Delta\,\tilde{E}(Z,\omega+\Delta)\\&\quad\times\exp\{-i\Delta t+i[(\omega+\Delta)\,n(\omega+\Delta)-\omega n(\omega)]Z/c\}.\end{aligned} \tag{6.11}$$

We have set $\omega'=\omega+\Delta$ and extended the lower limit of the Δ integral to $-\infty$, since $\tilde{E}(Z,\omega+\Delta)$ is sharply peaked about $\Delta=0$. Expanding

$$(\omega+\Delta)\,n(\omega+\Delta)-\omega n(\omega)\approx\frac{d\,[\omega n(\omega)]}{d\omega}\Delta=\left[n(\omega)+\omega\frac{dn(\omega)}{d\omega}\right]\Delta, \tag{6.12}$$

correct to first order in Δ, we find that

$$\begin{aligned}E(Z,t)&\approx\frac{1}{\sqrt{2\pi}}\int_{-\infty}^{\infty}d\Delta\,\tilde{E}(Z,\omega+\Delta)\exp\left(-i\Delta\left\{t-\left[n(\omega)+\omega\frac{dn(\omega)}{d\omega}\right]\frac{Z}{c}\right\}\right)\\&=E\left\{0,t-\left[n(\omega)+\omega\frac{dn(\omega)}{d\omega}\right]\frac{Z}{c}\right\},\end{aligned} \tag{6.13}$$

or

$$\mathbf{E}^{+}(Z,t)\approx\frac{\hat{\boldsymbol{\epsilon}}}{2}e^{-i\omega[t-n(\omega)\frac{Z}{c}]}E\left\{0,t-\left[n(\omega)+\omega\frac{dn(\omega)}{d\omega}\right]\frac{Z}{c}\right\}. \tag{6.14}$$

For real $n(\omega)$, equation (6.14) implies that the field propagates without distortion or attenuation with a group velocity

$$v_g=\frac{c}{n(\omega)+\omega\frac{dn(\omega)}{d\omega}}, \tag{6.15}$$

and that the phase velocity is $v_{ph}=c/n(\omega)$. Of course, if we had included higher terms in the expansion (6.12), these terms would have resulted in distortion of the pulse shape as it propagates. These terms become important near resonance. If $n(\omega)$ is complex, the leading exponential $e^{-i\omega[t-n(\omega)Z/c]}$ results in attenuation {assuming that $\mathrm{Im}[n(\omega)] < 0\}$. The group velocity becomes complex, but the pulse propagation speed is still given by equation (6.15), if $\mathrm{Re}[n(\omega)]$ is substituted for $n(\omega)$.[1]

6.2 Maxwell-Bloch Equations

Rather than use equation (6.6) to relate the polarization to the field, we would like to calculate this polarization quantum-mechanically and obtain equations that allow us to find the fields in a self-consistent manner. We have seen already that the atomic response to an applied field can result in a nonlinear dependence of density matrix elements on the field amplitude. As a consequence, we expect that, in general, one finds a polarization that is a nonlinear function of the applied field amplitude.

The macroscopic polarization is defined as the average dipole moment per unit volume. As such, in quantum mechanics, it becomes the expectation value of the atomic electric dipole moment operator $\hat{\boldsymbol{\mu}}$, namely,

$$\begin{aligned}\langle \mathbf{P}(Z,t)\rangle &= \mathcal{N}\left[\boldsymbol{\mu}_{21}\langle \varrho_{12}(Z,t)\rangle + \boldsymbol{\mu}_{12}\langle \varrho_{21}(Z,t)\rangle\right] \\ &= \mathcal{N}\left[\boldsymbol{\mu}_{21}\langle \tilde{\varrho}_{12}(Z,t)\rangle e^{-i(kZ-\omega t)} + \boldsymbol{\mu}_{12}\langle \tilde{\varrho}_{21}(Z,t)\rangle\right] e^{i(kZ-\omega t)} \\ &= \left[\mathbf{P}^{+}(Z,t) + \mathbf{P}^{+}(Z,t)^{*}\right], \end{aligned} \tag{6.16}$$

where

$$\mathbf{P}^{+}(Z,t) = \mathcal{N}\boldsymbol{\mu}_{12}\langle \tilde{\varrho}_{21}(Z,t)\rangle e^{i(kZ-\omega t)}, \tag{6.17}$$

$$\langle \tilde{\varrho}_{ij}(\mathbf{R},t)\rangle = \int d\mathbf{v}\, W_0(\mathbf{v})\tilde{\varrho}_{ij}(\mathbf{R},\mathbf{v},t), \tag{6.18}$$

$\mathcal{N}$ is the density, and $W_0(\mathbf{v})$ is the velocity distribution. By combining equation (6.16) for the polarization, equations (5.2) for the evolution of density matrix elements, and the wave equation (6.4), one has a set of coupled partial differential equations that can be solved numerically given some initial conditions. Typically, one would suppose that an optical pulse is incident on a medium from vacuum and that the initial pulse field envelope is specified. The numerical solution of such a problem is far from trivial, since the integration step sizes must be taken to be much less than an optical period in time and much less than a wavelength in space. On the other hand, changes in the pulse envelope often occur on much slower time and longer length scales. To circumvent this problem, we often make a *slowly varying amplitude and phase approximation* (SVAPA).

[1] For complex $n(\omega)$, the point of stationary phase in the integrand of equation (6.11) occurs when

$$t = \frac{cZ}{n_r(\omega) + \omega\frac{dn_r(\omega)}{d\omega}},$$

where $n_r(\omega) = \mathrm{Re}[n(\omega)]$, implying that the group velocity is

$$v_g = c/\left[n_r(\omega) + \omega\frac{dn_r(\omega)}{d\omega}\right].$$

6.2.1 Slowly Varying Amplitude and Phase Approximation (SVAPA)

To proceed, we assume that the Fourier amplitudes $\tilde{\mathbf{E}}^+(Z, \omega')$ and $\tilde{\mathbf{P}}^+(Z, \omega')$ appearing in equations (6.5) are sharply peaked at the central field frequency ω. In that case, it is convenient to write the positive frequency components of the fields in the form

$$\mathbf{E}^+(Z, t) = \frac{1}{2}\hat{\mathbf{x}}E(Z, t)e^{i[kZ-\omega t]}, \tag{6.19a}$$

$$\tilde{\mathbf{E}}^+(Z, \omega) = \frac{1}{2}\hat{\mathbf{x}}\tilde{E}(Z, \omega), \tag{6.19b}$$

$$\mathbf{P}^+(Z, t) = \hat{\mathbf{x}}P(Z, t)e^{i[kZ-\omega t]}, \tag{6.19c}$$

$$\mathbf{P}^+(Z, \omega) = \hat{\mathbf{x}}P(Z, \omega), \tag{6.19d}$$

where $k = \omega/c$,

$$E(Z, t) = \frac{1}{\sqrt{2\pi}}\int_0^\infty \tilde{E}(Z, \omega')e^{i[(k'-k)Z-(\omega'-\omega)t]}d\omega', \tag{6.20a}$$

$$P(Z, t) = \frac{1}{\sqrt{2\pi}}\int_0^\infty \tilde{P}(Z, \omega')e^{i[(k'-k)Z-(\omega'-\omega)t]}d\omega', \tag{6.20b}$$

and we have taken the fields polarized in the $\hat{\mathbf{x}}$ direction. Equations (6.20) and (6.19) are formally identical to equations (6.5). Although both $\tilde{E}(Z, \omega')$ and $\tilde{P}(Z, \omega')$ are assumed to be sharply peaked at the central laser field frequency ω, it does not necessarily follow that $E(Z, t)$ and $P(Z, t)$ are slowly varying in space compared with $\lambda = 2\pi/k$ and slowly varying in time compared with ω^{-1}. We assume this to be the case for the moment, but you will see that this approximation can lead to an incorrect group velocity and index of refraction when the index of refraction of the medium differs considerably from unity.

The next step is to substitute equations (6.20) and (6.19) into the wave equation for $E(Z, t)$,

$$\frac{\partial^2 E(Z, t)}{\partial Z^2} - \frac{1}{c^2}\frac{\partial^2 E(Z, t)}{\partial t^2} = \frac{2}{c^2\varepsilon_0}\frac{\partial^2 P(Z, t)}{\partial t^2}, \tag{6.21}$$

and, owing to the slow variation of $E(Z, t)$ and $P(Z, t)$, to neglect terms such as $\partial^2 E(Z, t)/\partial t^2$, $\partial^2 E(Z, t)/\partial Z^2$, $\partial P(Z, t)/\partial t$, and $\partial^2 P(Z, t)/\partial t^2$. Thus, we approximate

$$\frac{\partial\left[E(Z, t)e^{i(kZ-\omega t)}\right]}{\partial Z} = \left[ikE(Z, t) + \frac{\partial E(Z, t)}{\partial Z}\right]e^{i(kZ-\omega t)}, \tag{6.22a}$$

$$\frac{\partial^2\left[E(Z, t)e^{i(kZ-\omega t)}\right]}{\partial Z^2} \approx \left[-k^2E(Z, t) + 2ik\frac{\partial E(Z, t)}{\partial Z}\right]e^{i(kZ-\omega t)}, \tag{6.22b}$$

and, similarly,

$$\frac{\partial^2\left[E(Z, t)e^{i(kZ-\omega t)}\right]}{\partial t^2} \approx \left[-\omega^2E(Z, t) - 2i\omega\frac{\partial E(Z, t)}{\partial t}\right]e^{i(kZ-\omega t)}, \tag{6.23a}$$

$$\frac{\partial^2\left[P(Z, t)e^{i(kZ-\omega t)}\right]}{\partial t^2} \approx -\omega^2P(Z, t)e^{i(kZ-\omega t)}. \tag{6.23b}$$

(We do not keep the first derivative in the polarization since the zero-order terms do not cancel as they do for the electric field amplitude, but see the appendix and problem 5 for a further discussion of this point.)

Substituting these results into the wave equation (6.21) and using the fact that $\omega = kc$, we obtain

$$\left(\frac{\partial}{\partial Z} + \frac{1}{c}\frac{\partial}{\partial t}\right) E(Z,t) = \frac{ik}{\varepsilon_0} P(Z,t). \tag{6.24}$$

From equations (6.16) to (6.18) and (6.19b), it follows that $P(Z,t) = \mathcal{N}(\mu_x)_{12} \times \langle\tilde{\varrho}_{21}(Z,t)\rangle$, so equation (6.24) can be rewritten in the form

$$\left(\frac{\partial}{\partial Z} + \frac{1}{c}\frac{\partial}{\partial t}\right) E(Z,t) = \frac{i\mathcal{N}k(\mu_x)_{12}}{\varepsilon_0}\langle\tilde{\varrho}_{21}(Z,t)\rangle, \tag{6.25}$$

where $(\mu_x)_{12}$ is the x component of $\boldsymbol{\mu}_{12}$, and $\langle\tilde{\varrho}_{21}(Z,t)\rangle$ is the velocity average defined by equation (6.18). Thus, we have accomplished our task of relating the evolution of the complex electric field amplitude to density matrix elements that can be evaluated using the density matrix or Bloch equations. Equation (6.25), together with the density matrix or Bloch equations, are referred to as the Maxwell-Bloch equations in the slowly varying amplitude and phase approximation. In the limit that a steady-state field distribution has been reached as in the absorption problem discussed in chapter 4, equation (6.25) reduces to

$$\frac{\partial E(Z)}{\partial Z} = \frac{i\mathcal{N}k(\mu_x)_{12}}{\varepsilon_0}\langle\tilde{\varrho}_{21}(Z)\rangle. \tag{6.26}$$

As we alluded to earlier, there is a problem with this approach when the index of refraction n differs appreciably from unity. It is not difficult to understand why this is the case. In a dielectric medium, $\mathrm{E}^{+}(Z,t)$ varies as $\exp[inkZ]$ and is proportional to $E(Z,t)\exp(ikZ)$; as a consequence, $E(Z,t)$ varies as $\exp[ik(n-1)Z]$, implying that $\partial^2 E(Z,t)/\partial Z^2$ is no longer negligible compared with $k\partial E(Z,t)/\partial Z$. An alternative approach to the slowly varying amplitude and phase approximation that yields the correct group velocity and index of refraction is given in the appendix. Thus, equation (6.25) must be used with some caution. In general, it is valid for a medium, such as a dilute vapor, in which the index of refraction is approximately equal to unity.

6.3 Linear Absorption and Dispersion—Stationary Atoms

We now return to the problem of linear absorption by an atomic vapor as a simple application of the Maxwell-Bloch formalism. Recall that, by absorption, we really mean the loss of intensity of the beam as it traverses the medium—in steady state, the radiation is scattered rather than absorbed. We consider the steady-state distribution produced by a quasi-monochromatic field in a uniform medium of *stationary* two-level atoms, for which $\langle\tilde{\varrho}_{21}(Z)\rangle = \tilde{\varrho}_{21}(Z)$, since no velocity average is needed. From the density matrix equations (5.2), the steady-state value of $\tilde{\varrho}_{21}$ to lowest order in the applied field is given by

$$\tilde{\varrho}_{21}(Z) = \frac{-i\chi(Z)}{\gamma + i\delta}, \tag{6.27}$$

where $\chi(Z) = -(\mu_x)_{21}\, E(Z)/2\hbar$. Therefore, from equation (6.26), the envelope function $E(Z)$ can be obtained by using

$$\frac{\partial E(Z)}{\partial Z} = -\frac{k\mathcal{N}|(\mu_x)_{12}|^2 E(Z)}{2\hbar\varepsilon_0(\gamma + i\delta)} = -\frac{\alpha}{2}E(Z), \tag{6.28}$$

where

$$\alpha = \alpha_r + i\alpha_i = \frac{k\mathcal{N}|(\mu_x)_{12}|^2}{\hbar\varepsilon_0(\gamma^2 + \delta^2)}(\gamma - i\delta) \tag{6.29}$$

is a complex absorption coefficient.

From equation (6.28), we find

$$E(Z) = E(0)e^{-\alpha Z/2}, \tag{6.30}$$

which implies that

$$\mathbf{E}(Z, t) = \frac{1}{2}\hat{\mathbf{x}}E(0)e^{i(k-\alpha_i/2)Z}e^{-\alpha_r Z/2}e^{-i\omega t} + \text{c.c.} \tag{6.31}$$

From this equation and equation (6.29), we see that

$$k - \alpha_i/2 = k\left[1 + \frac{\mathcal{N}|(\mu_x)_{12}|^2\delta}{2\hbar\varepsilon_0(\gamma^2 + \delta^2)}\right] \equiv nk, \tag{6.32}$$

implying that the index of refraction n is equal to[2]

$$n = 1 + \frac{\mathcal{N}|(\mu_x)_{12}|^2\delta}{2\hbar\varepsilon_0(\gamma^2 + \delta^2)}. \tag{6.33}$$

In addition, the absorption coefficient

$$\alpha_r = \frac{\mathcal{N}|(\mu_x)_{12}|^2 k}{\hbar\varepsilon_0\gamma}\frac{\gamma^2}{(\gamma^2 + \delta^2)} \tag{6.34}$$

agrees with our previous result, equation (4.22), in the limit of weak fields. The index of refraction is plotted in figure 6.1 as a function of $-\delta = \omega - \omega_0$. For most of the range of δ, the index increases with increasing ω ($dn/d\omega > 0$), and this is referred to as *normal dispersion*. In the region about the resonance, $dn/d\omega < 0$, and this region is referred to as *anomalous dispersion*.

One aspect of absorption that can be generalized to a number of other processes involving the exchange of energy between different field modes is the relationship between the field amplitude and the coherence generated in the medium by the field. In the case of linear absorption, the field amplitude is

$$\mathbf{E}^+(Z, t) = \frac{1}{2}\hat{\mathbf{x}}E(Z)e^{ikZ-i\omega t}, \tag{6.35}$$

[2] This equation is valid only if $k|\alpha| \ll 1$; the more general result,

$$n = \left[1 + \frac{\mathcal{N}|(\mu_x)_{12}|^2\delta}{\hbar\varepsilon_0(\gamma^2 + \delta^2)}\right]^{1/2},$$

is derived in the appendix.

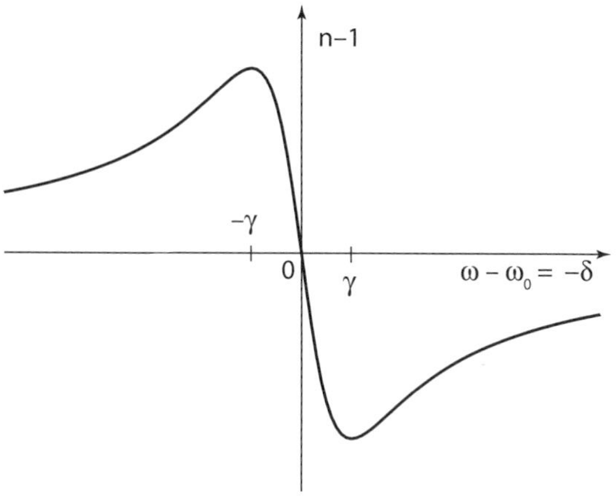

Figure 6.1. Index of refraction in the vicinity of a resonance as a function of $-\delta = \omega - \omega_0$.

while the associated dipole coherence is

$$\varrho_{21}(Z,t) = -\frac{i\chi(Z)}{\gamma + i\delta} e^{ikZ-i\omega t} = -\frac{|\chi(Z)|\, e^{ikZ-i\omega t+i\phi+i\beta}}{\sqrt{\gamma^2+\delta^2}}, \tag{6.36}$$

where

$$\tan\beta = \frac{\gamma}{\delta}, \tag{6.37}$$

and we have set $E(Z) = |E(Z)|\, e^{i\phi}$. The phase difference β between the dipole coherence and the applied field allows the medium to extract energy from the field. The power delivered to an atom by the field is equal to the scalar product of the force $\mathbf{F}$ on the electron in the atom with its velocity $\langle \dot{\mathbf{r}} \rangle$. As a consequence, the cycle-averaged power per unit volume $[dW(Z)/dt]$ provided by the field to the medium is given by

$$\begin{aligned}
\frac{dW(Z)}{dt} &= \overline{\mathbf{F}(Z) \cdot \frac{d\langle \mathbf{r}(Z) \rangle}{dt}} \\
&= \left[-\frac{1}{2} eE(Z)\right] [i\omega x_{21}\tilde{\varrho}_{12}(Z)] + \text{c.c.} \\
&= -\frac{i}{2}\omega e E(Z) x_{21} \frac{i\chi^*(Z)}{\gamma - i\delta} + \text{c.c.} \\
&= \hbar\omega \frac{2\gamma |\chi(Z)|^2}{\gamma^2+\delta^2} = \hbar\omega\gamma_2\varrho_{22}(Z),
\end{aligned} \tag{6.38}$$

which provides some justification for equation (4.15). On the other hand, we can also write this result as

$$\frac{dW(Z)}{dt} = \frac{2\hbar\omega|\chi(Z)|^2}{\sqrt{\gamma^2+\delta^2}} \sin\beta, \tag{6.39}$$

with $0 \le \beta \le \pi$. If there is no relaxation ($\gamma = 0$), then $\beta = 0$ or π and, on average, the field does no work on the medium. In steady-state when $\gamma \neq 0$, the energy given to the medium by the field is counterbalanced by the energy lost by the medium to "friction," resulting from the damping produced by the vacuum field. The overall process is one of scattering of the incident field into unoccupied modes of the vacuum field, with the atoms serving as intermediaries.

6.4 Linear Pulse Propagation

It is clear from equations (6.30) and (6.32) that the validity condition for the SVAPA, $[\partial E(Z)/\partial Z] \ll kE(Z)$, can be satisfied only if $k\,|\alpha| \sim (n-1) \ll 1$, a condition that is equivalent to requiring that the index of refraction of the medium is approximately equal to unity. To test the validity of the slowly varying amplitude and phase approximation, equation (6.25), we can consider linear pulse propagation in the adiabatic following limit for a medium of stationary atoms. Recall that the adiabatic following limit is one where the field varies slowly in a time of order $|\gamma + i\delta|^{-1}$. In the linear regime, one sets $\varrho_{11} = 1$ and calculates $\tilde{\varrho}_{21}(Z, t)$ from [see equation (5.9d)]

$$\partial\tilde{\varrho}_{21}(Z,t)/\partial t = -(\gamma + i\delta)\,\tilde{\varrho}_{21}(Z,t) - i\chi(Z,t), \tag{6.40}$$

where $\chi(Z,t) = -(\mu_x)_{21}\,E(Z,t)/2\hbar$. Since the atoms are stationary, we need not concern ourselves with any velocity averaging. The solution of equation (6.40) in the adiabatic following limit is

$$\begin{aligned}\tilde{\varrho}_{21}(Z,t) &= -i\int_{-\infty}^{t} \chi(Z,t')e^{-(\gamma+i\delta)(t-t')}dt' \\ &\approx -\frac{i\chi(Z,t)}{\gamma+i\delta} + \frac{i\partial\chi(Z,t)/\partial t}{(\gamma+i\delta)^2} - \cdots \\ &= \frac{i(\mu_x)_{21}}{2\hbar(\gamma+i\delta)}\left[E(Z,t) - \frac{\partial E(Z,t)/\partial t}{(\gamma+i\delta)} + \cdots\right],\end{aligned} \tag{6.41}$$

where we have generated the series using integration by parts.

Substituting this result into equation (6.25), we find

$$\frac{\partial E(Z,t)}{\partial Z} + \frac{1}{c}\frac{\partial E(Z,t)}{\partial t} = -\frac{\alpha}{2}\left[E(Z,t) - \frac{\partial E(Z,t)/\partial t}{(\gamma+i\delta)}\right], \tag{6.42}$$

where α is given by equation (6.29). We define a complex index of refraction by

$$\begin{aligned}n_c(\omega) &= 1 + i\alpha/2k \\ &= 1 + \frac{i\mathcal{N}|(\mu_x)_{12}|^2}{2\hbar\varepsilon_0\,[\gamma + i(\omega_0 - \omega)]},\end{aligned} \tag{6.43}$$

where $k = \omega/c$. It then follows that

$$\begin{aligned}\frac{dn_c}{d\omega} &= \frac{i\mathcal{N}|(\mu_x)_{12}|^2}{2\hbar\varepsilon_0\,(\gamma+i\delta)}\frac{i}{(\gamma+i\delta)} = \frac{i\,[n_c(\omega)-1]}{(\gamma+i\delta)} \\ &= \frac{i}{(\gamma+i\delta)}\frac{i\alpha}{2k},\end{aligned} \tag{6.44}$$

or

$$\frac{\alpha}{2(\gamma+i\delta)} = -\frac{\omega}{c}\frac{dn_c}{d\omega}, \tag{6.45}$$

which, when substituted into equation (6.42), yields

$$\frac{\partial E(Z,t)}{\partial Z} + \frac{1}{c}\left(1 + \omega\frac{dn_c}{d\omega}\right)\frac{\partial E(Z,t)}{\partial t} = -\frac{\alpha}{2}E(Z,t). \tag{6.46}$$

This equation implies that the group velocity is equal to $c/[1 + \omega(dn/d\omega)]$, whereas the expected result is $c/[n + \omega(dn/d\omega)]$, where $n = \mathrm{Re}\, n_c$ is the index of refraction. Thus, this result shows that, for this form of the slowly varying amplitude and phase approximation to be valid, one must have $n \approx 1$, as we have pointed out previously.

6.5 Other Problems with the Maxwell-Bloch Equations

The Maxwell-Bloch equations are used successfully to model many atom–field systems. We have already noted some of their limitations, related mainly to approximations implicit in going over to the slowly varying amplitude and phase approximation. However, even if we do not make the slowly varying amplitude and phase approximation, there must still be something rotten (or at least a little spoiled) in the state of Denmark.

A simple example helps to illustrate what is missing. Imagine that a field propagates in a medium that can be characterized by a *real* index of refraction. From the quantum point of view, this situation corresponds to propagation of a field whose frequency is far off resonance from an atomic transition, such that the imaginary part of the index of refraction (which varies as δ^{-2}) can be neglected. Moreover, let us neglect dispersion. In this limit, the field propagates through the medium without loss or distortion. But this is an approximate result only, since there is a small, but finite, loss owing to scattering. The Maxwell-Bloch equations properly account for this loss insofar as the amplitude of the field is diminished as the field propagates in the medium, owing to the small but nonvanishing imaginary part of the index of refraction. On the other hand, in the approach followed in this chapter, these equations do *not* allow one to calculate the field that is *scattered* by the atoms. For conservation of energy, this scattered field must exist—it is nothing more than Rayleigh scattering.

What is missing from a conventional derivation of the Maxwell-Bloch equations is the role of *fluctuations*. In some sense, we have violated one of the maxims of quantum mechanics—*sum over final states and average over initial states*. Normally, the average over particle position and frequency (velocity) is carried out *after* one obtains expressions for expectation values of quantum-mechanical operators. In deriving the Maxwell-Bloch equations, we perform this macroscopic average directly in the field *amplitude* equations, whereas we should wait to carry out the average until we calculate field *intensities*. For example, the field radiated by a one-dimensional array of N dipole oscillators in a direction transverse to the line of the oscillators is proportional to

$$S = \sum_{j=1}^{N} e^{ikX_j}, \tag{6.47}$$

where X_j is the position of the jth oscillator. If we average this over random positions of the oscillators, then $\langle S \rangle = 0$. On the other hand, the average field *intensity* is proportional to

$$\langle |S|^2 \rangle = \left\langle \sum_{j,j'=1}^{N} e^{ik(X_j - X_{j'})} \right\rangle = N + \left\langle \sum_{j,j' \neq j}^{N} e^{ik(X_j - X_{j'})} \right\rangle = N. \tag{6.48}$$

The terms with $j = j'$ give rise to a signal proportional to N. It is precisely such a contribution that is responsible for Rayleigh scattering.

Thus, the Maxwell-Bloch treatment is reliable only if fluctuations in both particle position and frequency can be neglected. Although such an approximation is not valid for the scattered field, in general, it is often valid for the fields that are incident on the medium. For most of the problems considered in this text, the fields that propagate in the medium can be taken to have a cylindrical cross-sectional area equal to A. To neglect fluctuations in particle position for such fields, it is necessary that we break the medium into a number of slices in the direction of propagation, each slice having a length that is much less than a wavelength. Then, each atom in the slice sees the same phase of the incident field. As a consequence, fluctuations in particle position are negligible if the number of atoms in such a slice is much greater than unity,

$$\mathcal{N} A\lambda \gg 1, \tag{6.49}$$

where $\mathcal{N}$ is the atomic density. This condition is much less severe than the one that is often imposed for a macroscopic description of Maxwell's equations (the number of atoms in a sphere whose radius is less than λ be much greater than unity). For a beam having cross-sectional area of 1 cm^2, inequality (6.49) is satisfied for densities $\mathcal{N} \gg 10^{10}$ atoms/m^3 $= 10^6$ atoms/cm^3, a condition that is satisfied even in dilute vapors. In general, fluctuations in atomic velocity or frequency can be neglected as well, unless we look at the contributions from only a few of the atoms within the inhomogeneous width associated with the transition (as in single-molecule spectroscopy). The message to take away is that the Maxwell-Bloch equations can often be used to describe propagation of fields in both vapors and solids, but they do not provide a complete description for the scattered fields.

6.6 Summary

In this chapter, we derived the Maxwell-Bloch equations and discussed some applications. Although there are problems associated with the Maxwell-Bloch equations and with the slowly varying amplitude and phase approximation (SVAPA), there is a wide range of problems for which Maxwell-Bloch equations in the slowly varying amplitude and phase approximation retain a high degree of validity. We return to these equations often throughout the text.

6.7 Appendix: Slowly Varying Amplitude and Phase Approximation—Part II

For detunings $\omega_0 \gg |\delta| \gg \gamma$, the index of refraction for a dilute vapor varies as

$$n = 1 + \frac{\mathcal{N} |(\mu_x)_{12}|^2}{2\hbar\varepsilon_0\delta}, \tag{6.50}$$

and $\omega\,(dn/d\omega) \gg n$; that is, the dominant contribution to the group velocity is a result of dispersion in the medium. On the other hand, for material such as glass or water, the dispersion is of secondary importance in the visible part of the spectrum, since the frequency of optical radiation is far from any electronic transition in these materials (the RWA cannot be used in this case); as a consequence, the group velocity in glass or water is determined predominantly by their index of refraction at some arbitrary frequency in the visible. As we have seen in equation (6.46), the treatment in section 6.4 led to a group velocity $v_g = c\,[1 + \omega\,(dn/d\omega)]$ instead of $v_g = c\,[n + \omega\,(dn/d\omega)]$. Moreover, the expression for the index of refraction (6.33) is valid only for $n \approx 1$. It would be nice to have a formulation of the slowly varying amplitude and phase approximation that leads to the correct group velocity in the absence of dispersion and gives the correct index of refraction when $n > 1$.

One possible error in the approach we used is linked to the fact that we approximated

$$\frac{\partial^2\left[P(Z,t)e^{i(kZ-\omega t)}\right]}{\partial t^2} \approx -\omega^2 P(Z,t)e^{i(kZ-\omega t)}, \tag{6.51}$$

equation (6.23b). If, instead, we keep the next leading term,

$$\frac{\partial^2\left[P(Z,t)e^{i(kZ-\omega t)}\right]}{\partial t^2} \approx \left[-\omega^2 P(Z,t) - 2i\omega\frac{\partial P(Z,t)}{\partial t}\right] e^{i(kZ-\omega t)}, \tag{6.52}$$

it is not difficult to show (see problem 5) that the group velocity is replaced by $v_g = c/n^2$ in the absence of dispersion. Thus, although the result is changed, it is still wrong.

It *is* possible to modify the slowly varying amplitude and phase approximation in a manner that leads to the correct pulse propagation speed and a correct index of refraction. To do so, we must specify the phase of the field explicitly and make sure that any corrections to this phase are much less than unity. If this is accomplished, the validity condition for the SVAPA becomes $k\alpha_r \ll 1$ instead of $k\,|\alpha| \ll 1$; for detunings $|\delta| \gg \gamma$, $\alpha_r/\,|\alpha| = \gamma/|\delta|$, and the SVAPA can be valid in cases where the index of refraction deviates from unity.

To arrive at this alternative form for the SVAPA, we write the field and polarization as

$$\mathbf{E}^+(Z,t) = \frac{1}{2}\hat{\mathbf{x}}|E(Z,t)|e^{i[kZ-\omega t+\phi(Z,t)]}, \tag{6.53a}$$

$$\mathbf{P}^+(Z,t) = \hat{\mathbf{x}}P(Z,t)e^{i[kZ-\omega t+\phi(Z,t)]}, \tag{6.53b}$$

where $k = \omega/c$, $\phi(Z,t)$ is real, but $P(Z,t)$ can still be complex. In fact, equations (6.53) in some sense correspond to the second definition of the field interaction representation. It is convenient to write

$$\begin{aligned} P(Z,t) &= \mathcal{N}\,(\mu_x)_{12}\,\langle\tilde{\varrho}_{21}(Z,t)\rangle \\ &= \mathcal{N}\,(\mu_x)_{12}\,[\langle u(Z,t)\rangle - i\langle v(Z,t)\rangle]/2, \end{aligned} \tag{6.54}$$

where $u(Z,t)$ and $v(Z,t)$ are components of the Bloch vector. For simplicity, $(\mu_x)_{12}$ is taken to be real.

When expressions (6.53) and (6.54) are substituted into the wave equation and second-order spatial and temporal derivatives of $|E(Z,t)|$, $\phi(Z,t)$, and $P(Z,t)$ are neglected, one obtains, in analogy with equations (6.22) and (6.23), but using equation (6.52) instead of equation (6.23b),

$$\begin{aligned}
&-k\frac{\partial\phi(Z,t)}{\partial z}|E(Z,t)|+i\left[k+\frac{\partial\phi(Z,t)}{\partial Z}\right]\frac{\partial|E(Z,t)|}{\partial Z}-\frac{|E(Z,t)|}{2}\left[\frac{\partial\phi(Z,t)}{\partial Z}\right]^2\\
&+\frac{k}{c}\frac{\partial\phi(Z,t)}{\partial t}|E(Z,t)|+\frac{i}{c}\left[k-\frac{1}{c}\frac{\partial\phi(Z,t)}{\partial t}\right]\frac{\partial|E(Z,t)|}{\partial t}+\frac{|E(Z,t)|}{2}\left[\frac{\partial\phi(Z,t)}{\partial t}\right]^2\\
&=-\frac{\mathcal{N}k^2(\mu_x)_{12}}{2\epsilon_0}\left[\langle u(Z,t)\rangle-i\langle v(Z,t)\rangle\right]-\frac{ik\mathcal{N}k^2(\mu_x)_{12}}{c\epsilon_0}\frac{\partial\left[\langle u(Z,t)\rangle-i\langle v(Z,t)\rangle\right]}{\partial t}.
\end{aligned} \tag{6.55}$$

The main difference between equations (6.55) and (6.25), aside from the inclusion of equation (6.52), is that we keep terms of the form $[\partial\phi(Z,t)/\partial t][\partial|E(Z,t)|/\partial t]$ and $[\partial\phi(Z,t)/\partial Z][\partial|E(Z,t)|/\partial Z]$, as well as terms varying as $[\partial\phi(Z,t)/\partial Z]^2$ and $[\partial\phi(Z,t)/\partial t]^2$; such terms are not included in equation (6.25).

Equating real and imaginary parts in equation (6.55), we find

$$\begin{aligned}
&\left[-\frac{1}{2k}\left(\frac{\partial\phi(Z,t)}{\partial Z}\right)^2+\frac{\partial\phi(Z,t)}{\partial Z}+\frac{1}{c}\frac{\partial\phi(Z,t)}{\partial t}\right]|E(Z,t)|\\
&\qquad=\frac{\mathcal{N}k(\mu_x)_{12}}{2\epsilon_0}\langle u(Z,t)\rangle+\frac{\mathcal{N}(\mu_x)_{12}}{c\epsilon_0}\frac{\partial\langle v(Z,t)\rangle}{\partial t},
\end{aligned} \tag{6.56a}$$

$$\begin{aligned}
&\left[k+\frac{\partial\phi(Z,t)}{\partial Z}\right]\frac{\partial|E(Z,t)|}{\partial Z}+\frac{1}{c}\left[k-\frac{1}{c}\frac{\partial\phi(Z,t)}{\partial t}\right]\frac{\partial|E(Z,t)|}{\partial t}\\
&\qquad=\frac{\mathcal{N}k^2(\mu_x)_{12}}{2\epsilon_0}\langle v(Z,t)\rangle-\frac{\mathcal{N}k(\mu_x)_{12}}{c\epsilon_0}\frac{\partial\langle u(Z,t)\rangle}{\partial t}.
\end{aligned} \tag{6.56b}$$

Equations (6.56) represent an alternative form of Maxwell's equations in the slowly varying amplitude and phase approximation that is valid when the index of refraction of the medium departs from unity. They should be combined with equations (3.98) for density matrix elements. In general, these combined Maxwell-Bloch equations must be solved numerically.

To show that equations (6.56) lead to the correct index of refraction and group velocity, we consider two limiting cases. First, we solve the equations in steady state for a weak input field and stationary atoms. If all time derivatives in equations (6.56) are set equal to zero, we obtain

$$\left[-\frac{1}{2k}\left(\frac{\partial\phi(Z)}{\partial Z}\right)^2+\frac{\partial\phi(Z)}{\partial Z}\right]|E(Z)|=\frac{\mathcal{N}k(\mu_x)_{12}}{2\epsilon_0}u(Z), \tag{6.57a}$$

$$\left[k+\frac{\partial\phi(Z)}{\partial Z}\right]\frac{\partial|E(Z)|}{\partial Z}=\frac{\mathcal{N}k^2(\mu_x)_{12}}{2\epsilon_0}v(Z). \tag{6.57b}$$

We look for a self-consistent solution in which

$$\phi(Z)=k\beta Z, \tag{6.58}$$

substitute this into equations (6.57), and find

$$-\frac{\beta^2}{2}+\beta=\frac{\mathcal{N}(\mu_x)_{12}}{2\epsilon_0}\frac{u(Z)}{|E(Z)|}=\frac{\mathcal{N}(\mu_x)_{12}^2}{2\hbar\epsilon_0}\frac{\delta}{\gamma^2+\delta^2}=-\frac{\alpha_i}{2k}, \tag{6.59a}$$

$$(1+\beta)\frac{\partial|E(Z)|}{\partial Z}=\frac{\mathcal{N}k(\mu_x)_{12}}{2\epsilon_0}v(Z)=-\frac{\mathcal{N}k(\mu_x)_{12}^2}{2\hbar\epsilon_0}\frac{\gamma|E(Z)|}{\gamma^2+\delta^2}=-\frac{\alpha_r}{2}|E(Z)|, \tag{6.59b}$$

where α_i and α_r are defined by equation (6.29).

The solutions for β and $|E(Z)|$ are

$$\beta=-1+\sqrt{1+\alpha_i/k}, \tag{6.60a}$$

$$|E(Z)|=|E(0)|\exp\left[-\frac{\alpha_r Z}{2(1+\beta)}\right], \tag{6.60b}$$

which, when substituted into equation (6.53a), yields

$$\mathbf{E}^+(Z,t)=\frac{1}{2}\hat{\mathbf{x}}|E(0)|\exp\left\{-\frac{\alpha_r Z}{2n(\omega)}+i[n(\omega)kZ-\omega t]\right\}, \tag{6.61}$$

where the (real) index of refraction is given by

$$n(\omega)=\sqrt{1+\alpha_i/k}=1+\beta. \tag{6.62}$$

Equation (6.61) is valid provided that $k\alpha_r \ll 1$ and agrees with equation (6.31) in the limit that $k|\alpha| \ll 1$.

Next, we solve the equations in the adiabatic following limit for a weak input field and stationary atoms. To lowest order, the solution for the phase is still given by equations (6.58) and (6.62). When these equations are substituted into equation (6.56b), we can obtain

$$\begin{aligned}n(\omega)\frac{\partial|E(Z,t)|}{\partial Z}+\frac{1}{c}\frac{\partial|E(Z,t)|}{\partial t}&=\frac{\mathcal{N}k(\mu_x)_{12}}{2\epsilon_0}v(Z,t)\\&\quad-\frac{\mathcal{N}(\mu_x)_{12}}{c\epsilon_0}\frac{\partial u(Z,t)}{\partial t}.\end{aligned} \tag{6.63}$$

The next step is use equation (6.41) for $\tilde{\varrho}_{21}(Z,t)$ to write

$$\begin{aligned}v(Z,t)&=-2\mathrm{Im}\tilde{\varrho}_{21}(Z,t)\\&=-\frac{(\mu_x)_{21}\gamma|E(Z,t)|}{\hbar(\gamma^2+\delta^2)}+\frac{(\mu_x)_{21}(\gamma^2-\delta^2)}{\hbar(\gamma^2+\delta^2)}\frac{\partial|E(Z,t)|}{\partial t}\end{aligned} \tag{6.64}$$

and

$$\begin{aligned}u(Z,t)&=2\mathrm{Re}\,\tilde{\varrho}_{21}(Z,t)\\&\approx\frac{(\mu_x)_{21}\delta|E(Z,t)|}{\hbar(\gamma^2+\delta^2)}.\end{aligned} \tag{6.65}$$

When equations (6.64) and (6.65) are inserted into equation (6.63), equation (6.29) is used, and equation (6.62) is written in the form

$$n^2(\omega)=1+\frac{\mathcal{N}|(\mu_x)_{12}|^2\delta}{\hbar\varepsilon_0(\gamma^2+\delta^2)}, \tag{6.66}$$

or, equivalently, as

$$2n(\omega)\frac{dn(\omega)}{d\omega} = -\frac{\mathcal{N}|(\mu_x)_{12}|^2(\gamma^2-\delta^2)}{\hbar\varepsilon_0(\gamma^2+\delta^2)^2}, \tag{6.67}$$

we find

$$\begin{aligned} n(\omega)\frac{\partial|E(Z,t)|}{\partial Z} + \frac{1}{c}\frac{\partial|E(Z,t)|}{\partial t} &= -\frac{\alpha_r}{2}|E(Z,t)| \\ &\quad -\frac{\omega n(\omega)}{c}\frac{dn(\omega)}{d\omega}\frac{\partial|E(Z,t)|}{\partial t} - \frac{\left[n^2(\omega)-1\right]}{c}\frac{\partial|E(Z,t)|}{\partial t}, \end{aligned} \tag{6.68}$$

which can be rewritten as

$$\frac{\partial|E(Z,t)|}{\partial Z} + \frac{1}{c}\left[n(\omega)+\omega\frac{dn(\omega)}{d\omega}\right]\frac{\partial|E(Z,t)|}{\partial t} = -\frac{\alpha_r}{2n(\omega)}|E(Z,t)|. \tag{6.69}$$

Equation (6.69) gives the correct group velocity.

Problems

1. Determine the condition that allows one to neglect subsequent terms in the expansion (6.12) for an index of refraction given by equation (6.43).
2. For a density of 10^{10} atoms/cm^3, and reasonable atomic parameters ($\bar{\lambda} = \lambda/2\pi = 10^{-5}$ cm, $\gamma = 10^8\ \mathrm{s}^{-1}$; recall that $\alpha_0 = 6\pi\bar{\lambda}^2\mathcal{N}$), estimate the contributions to the group velocity from $n-1$ and $\omega(dn/d\omega)$ if $\delta = 100\gamma$.
3. For the parameters of problem 2, solve equation (6.25) numerically. Take your initial pulse envelope as $\chi(Z,0) = \chi_0\exp[-(10\gamma z/c)^2]$, and show that it is the $\omega(dn/d\omega)$ term that modifies the group velocity. (The signature of this term is a δ^{-2} dependence.) Assume a weak incident field (Rabi frequency much less than decay rates) so that $\tilde{\varrho}_{21}(Z,t)$ is determined from

$$d\tilde{\varrho}_{21}(Z,t)/dt = -(\gamma+i\delta)\tilde{\varrho}_{21}(Z,t) - i\chi(Z,t).$$

   You might want to reexpress equation (6.25) in terms of the Rabi frequency.
4. In deriving equation (6.25), keep the next-order correction to $\partial^2 P(Z,t)/\partial t^2$—that is, take

$$\frac{\partial^2 P(Z,t)}{\partial t^2} \approx \left[-\omega^2 P(Z,t) - 2i\omega\frac{\partial P(Z,t)}{\partial t}\right]e^{i[kZ-\omega t]}.$$

   Neglecting dispersion, show that this modifies the result (6.46) obtained in the text and leads to a group velocity c/n^2 for $n \approx 1$. Thus, additional terms must be kept if the slowly varying amplitude and phase approximation is to reproduce the correct group velocity.
5. Consider a transition for which $\omega_0 \gg \omega \gg \gamma$ and ω is an optical frequency. Without making the RWA, show that equation (6.66) for the index of

refraction should be replaced by

$$n(\omega) = \sqrt{1 + \frac{\mathcal{N}\,(\mu_x)_{12}^2}{\hbar\epsilon_0}\frac{2\omega_0}{\omega_0^2 - \omega^2}}$$

$$= \sqrt{1 + 6\pi\mathcal{N}\bar{\lambda}^3\,\frac{2\gamma\omega_0}{\omega_0^2 - \omega^2}},$$

where $\bar{\lambda} = 2\pi/\omega$. [For off-resonance excitation, one should replace $\bar{\lambda}_0$ by $\bar{\lambda}$ in equation (4.28).] Assuming that $\mathcal{N}\bar{\lambda}^3\gamma$ is the same as for typical optical transitions, estimate the density for which the index of refraction deviates significantly from unity. In the limit that $\omega_0 \gg \omega$, compare the values of $n(\omega)$ and $\omega dn(\omega)/d\omega$ to show that dispersion provides only a small correction to the group velocity. For water, $n \approx 1.34$ at 550 nm and the first electronic transition is at $\lambda \approx 166$ nm; estimate the dispersion associated with this transition, and show that it leads to a value that is consistent with the experimental value of the dispersion of water, which is approximately 2% over the visible range.

Bibliography

L. Allen and J. H. Eberly, *Optical Resonance and Two-Level Atoms* (Wiley, New York, 1985), chaps. 1, 4, and 5.

W. E. Lamb Jr., *Theory of an optical maser*, Physical Review **134**, A1429–A1450 (1964).

P. W. Milonni and J. H. Eberly, *Laser Physics* (Wiley, Hoboken, NJ, 2009), secs. 8.3 and 8.4.

B. W. Shore, *The Theory of Coherent Excitation*, vol. 1 (Wiley-Interscience, New York, 1990), chap. 12.

A. E. Siegman, *Lasers* (University Science Books, Mill Valley, CA, 1986), sec. 24.4.

7

Two-Level Atoms in Two or More Fields:

Introduction to Saturation Spectroscopy

Shortly after the discovery of the gas laser, it became clear that the principles underlying its operation also had important implications for the spectroscopy of atomic vapors. We have already seen that, for thermal vapors, the factor that limits the resolution achievable in linear spectroscopy is the inhomogeneous broadening produced by Doppler broadening. We have also noted that this is not an intrinsic limitation to the resolution, since it could be eliminated by using cold atoms. With the development of the gas laser, it became evident that *nonlinear* or *saturation* spectroscopy could be used to eliminate the Doppler broadening, even in thermal vapors. The price one pays is that only a small fraction of the atoms in the vapor cell contribute to the observed signals. The use of saturation effects revolutionized the spectroscopy of vapors, liquids, and solids and is often referred to as *laser spectroscopy* since it requires the use of moderately intense, quasi-single-mode sources. Aside from its use in providing a probe of matter, laser spectroscopy has important applications in metrology; laser techniques can be used to stabilize resonant cavities or "slave" lasers to a narrow atomic or molecular transition frequency. In this chapter, we examine the basic concepts needed to understand the way in which laser spectroscopy can be used to achieve these results.

7.1 Two-Level Atoms and *N* Fields—Third-Order Perturbation Theory

We have seen that it is possible to obtain a steady-state solution of the density matrix equations in the field interaction representation when a single, plane-wave, monochromatic field interacts with an ensemble of two-level atoms. If we add a second field having a different frequency (or even a different direction in the case of moving atoms), the problem becomes significantly more difficult to solve. This may seem surprising at first. For a harmonic oscillator, one simply uses superposition to account for the presence of a number of fields having different frequencies. The superposition principle applies because the harmonic oscillator is an intrinsically linear device. As we have stressed on several occasions, an atom is not a harmonic oscillator; it is intrinsically a nonlinear device. It is just this nonlinearity that we exploit in laser spectroscopy.

Since the atomic response to the field is nonlinear, a superposition approach is doomed to failure. In fact, it is not difficult to see why. To lowest order in the applied fields, the atomic dipoles respond at all the frequencies of the applied fields, just as in the linear case. However, to second order in the fields, there are changes in state populations that depend on the beat frequencies of each pair of fields. In higher order, many different combinations of the field frequencies enter. As a consequence, the only way to solve the problem of a two-level atom interacting with two or more fields of arbitrary intensity having different frequencies is to solve it numerically. Analytic solutions can be obtained using perturbation theory or in the limit that only one of the fields is intense. We look first at a perturbative solution and defer the case of one strong and one weak field to appendix A.

We assume that the total incident field can be written as the sum of N separate fields,

$$\mathbf{E}(\mathbf{R}, t) = \frac{1}{2} \sum_{\mu=1}^{N} \hat{\boldsymbol{\epsilon}}_\mu E_\mu e^{i\mathbf{k}_\mu \cdot \mathbf{R} - i\omega_\mu t} + \text{c.c.}, \tag{7.1}$$

where $\hat{\boldsymbol{\epsilon}}_\mu$ is the polarization of field μ and $k_\mu = \omega_\mu / c$. Although not indicated explicitly, the complex field amplitudes E_μ can be slowly varying functions of space and time. It is of some interest to calculate the total field intensity

$$\left\langle |\mathbf{E}(\mathbf{R}, t)|^2 \right\rangle = \frac{1}{2} \sum_{\mu,\nu=1}^{N} (\hat{\boldsymbol{\epsilon}}_\mu \cdot \hat{\boldsymbol{\epsilon}}_\nu) |E_\mu E_\nu| \cos(\mathbf{k}_{\mu\nu} \cdot \mathbf{R} + \delta_{\mu\nu} t + \varphi_{\mu\nu}), \tag{7.2}$$

where

$$\mathbf{k}_{\mu\nu} = \mathbf{k}_\mu - \mathbf{k}_\nu, \tag{7.3a}$$

$$\delta_\mu = \omega_0 - \omega_\mu, \tag{7.3b}$$

$$\delta_{\mu\nu} = \delta_\mu - \delta_\nu = \omega_\nu - \omega_\mu, \tag{7.3c}$$

$$\varphi_{\mu\nu} = \varphi_\mu - \varphi_\nu, \tag{7.3d}$$

$$E_\mu = |E_\mu| e^{i\varphi_\mu}, \tag{7.3e}$$

ω_0 is the transition frequency, and the brackets refer to a time average over an optical period. We see that the field intensity consists of a constant background from each field separately ($\mu = \nu$) and standing-wave patterns resulting from the interference of the different fields. The standing-wave components are not stationary; they move with speed $\delta_{\mu\nu} / k_{\mu\nu}$. As we shall see, these external fields give rise to moving atomic population gratings, and the relative phase of the population gratings and the field standing-wave pattern is an important parameter in laser cooling schemes.

To calculate the atomic response to the fields, it is convenient to work in the normal interaction representation, since there are many field frequencies but only a single atomic frequency. Using the fact that $\varrho_{11}(\mathbf{R}, \mathbf{v}, t) + \varrho_{22}(\mathbf{R}, \mathbf{v}, t) = 1$, one finds that the appropriate equations in the interaction representation are

$$\frac{\partial \varrho_{12}^I(\mathbf{R}, \mathbf{v}, t)}{\partial t} + \mathbf{v} \cdot \nabla \varrho_{12}^I(\mathbf{R}, \mathbf{v}, t) = -\sum_{\mu=1}^{N} i\chi_\mu^* e^{-i\mathbf{k}_\mu \cdot \mathbf{R} - i\delta_\mu t} \left[2\varrho_{22}(\mathbf{R}, \mathbf{v}, t) - 1\right] - \gamma \varrho_{12}^I(\mathbf{R}, \mathbf{v}, t), \tag{7.4a}$$

$$\frac{\partial \varrho_{22}(\mathbf{R}, \mathbf{v}, t)}{\partial t} + \mathbf{v} \cdot \nabla \varrho_{22}(\mathbf{R}, \mathbf{v}, t) = -\sum_{\mu=1}^{N} \left[i\chi_\mu^* e^{-i\mathbf{k}_\mu \cdot \mathbf{R} - i\delta_\mu t} \varrho_{21}^I(\mathbf{R}, \mathbf{v}, t) - i\chi_\mu e^{i\mathbf{k}_\mu \cdot \mathbf{R} + i\delta_\mu t} \varrho_{12}^I(\mathbf{R}, \mathbf{v}, t) \right] - \gamma_2 \varrho_{22}(\mathbf{R}, \mathbf{v}, t), \tag{7.4b}$$

$$\varrho_{21}^I(\mathbf{R}, \mathbf{v}, t) = \left[\varrho_{12}^I(\mathbf{R}, \mathbf{v}, t)\right]^*, \tag{7.4c}$$

$$\varrho_{11}(\mathbf{R}, \mathbf{v}, t) = 1 - \varrho_{22}(\mathbf{R}, \mathbf{v}, t), \tag{7.4d}$$

where

$$\chi_\mu = -\left(\boldsymbol{\mu}_{21} \cdot \boldsymbol{\epsilon}_\mu\right) E_\mu / 2\hbar, \tag{7.5}$$

and $\delta_\mu = \omega_0 - \omega_\mu$. These equations can be solved as a perturbation series in powers of the external field amplitudes. We are interested in this chapter only in the steady-state response—that is, values of the density matrix elements once all transients have died away. Recall that $\varrho_{\alpha\alpha'}(\mathbf{R}, \mathbf{v}, t)$ is a single-particle density matrix element. The phase-space density matrix elements are given by

$$\mathcal{R}_{\alpha\alpha'}(\mathbf{R}, \mathbf{v}, t) = \mathcal{N}(\mathbf{R}) W_0(\mathbf{v}) \varrho_{\alpha\alpha'}(\mathbf{R}, \mathbf{v}, t), \tag{7.6}$$

where $\mathcal{N}(\mathbf{R})$ is the spatial density, and $W_0(\mathbf{v})$ is the velocity distribution in the absence of any applied fields.

7.1.1 Zeroth Order

To zeroth order in the applied fields, the atomic density matrix elements relax to their equilibrium values, $\varrho_{11}(\mathbf{R}, \mathbf{v}, t) = 1$, $\varrho_{21}^I(\mathbf{R}, \mathbf{v}, t) = \varrho_{12}^I(\mathbf{R}, \mathbf{v}, t) = \varrho_{22}(\mathbf{R}, \mathbf{v}, t) = 0$.

7.1.2 First Order

To first order in the applied fields, we assume a solution of the form

$$\varrho_{12}^{I(1)}(\mathbf{R}, \mathbf{v}, t) = \sum_{\mu=1}^{N} \varrho_{12}(\mathbf{v}; \mu) e^{-i\mathbf{k}_\mu \cdot \mathbf{R} - i\delta_\mu t}. \tag{7.7}$$

Substituting this expression into equation (7.4a) and equating coefficients of $e^{-i\mathbf{k}_\mu \cdot \mathbf{R} - i\delta_\mu t}$, we find that

$$\varrho_{12}(\mathbf{v}; \mu) = \frac{i\chi_\mu^*}{\gamma - i\left(\delta_\mu + \mathbf{k}_\mu \cdot \mathbf{v}\right)}. \tag{7.8}$$

To first order, the response is the sum of contributions from each field; the response is linear, and a superposition principle holds.

7.1.3 Second Order

To second order in the applied fields, we assume a solution of the form

$$\varrho_{22}^{(2)}(\mathbf{R}, \mathbf{v}, t) = \sum_{\mu,\nu=1}^{N} \varrho_{22}(\mathbf{v}; \mu, \nu) e^{-i\left(\mathbf{k}_{\mu\nu}\cdot\mathbf{R}+\delta_{\mu\nu}t\right)}. \tag{7.9}$$

Substituting this expression and the solution [equations (7.8) and (7.7)] into equation (7.4b) and equating coefficients of $e^{-i\left(\mathbf{k}_{\mu\nu}\cdot\mathbf{R}+\delta_{\mu\nu}t\right)}$, we find that (after the substitution, it helps to interchange μ and ν in the second sum)

$$\begin{aligned}
\varrho_{22}(\mathbf{v}; \mu, \nu) &= \frac{\chi_\mu^* \chi_\nu}{\gamma_2 - i\left(\delta_{\mu\nu} + \mathbf{k}_{\mu\nu}\cdot\mathbf{v}\right)} \left[\frac{1}{\gamma + i(\delta_\nu + \mathbf{k}_\nu\cdot\mathbf{v})} + \frac{1}{\gamma - i\left(\delta_\mu + \mathbf{k}_\mu\cdot\mathbf{v}\right)}\right] \\
&= \frac{\chi_\mu^* \chi_\nu}{\gamma_2 - i\left(\delta_{\mu\nu} + \mathbf{k}_{\mu\nu}\cdot\mathbf{v}\right)} \left\{\frac{2\gamma - i\left(\delta_{\mu\nu} + \mathbf{k}_{\mu\nu}\cdot\mathbf{v}\right)}{\left[\gamma + i(\delta_\nu + \mathbf{k}_\nu\cdot\mathbf{v})\right]\left[\gamma - i\left(\delta_\mu + \mathbf{k}_\mu\cdot\mathbf{v}\right)\right]}\right\} \\
&= \frac{\chi_\mu^* \chi_\nu}{\left[\gamma + i(\delta_\nu + \mathbf{k}_\nu\cdot\mathbf{v})\right]\left[\gamma - i\left(\delta_\mu + \mathbf{k}_\mu\cdot\mathbf{v}\right)\right]} \\
&\quad \times \left[1 + \frac{2\Gamma}{\gamma_2 - i\left(\delta_{\mu\nu} + \mathbf{k}_{\mu\nu}\cdot\mathbf{v}\right)}\right],
\end{aligned} \tag{7.10}$$

where we used the fact that $(2\gamma - \gamma_2)$ is equal to the collision rate Γ, since $\gamma = \gamma_2/2 + \Gamma$.

If we write

$$\varrho_{22}(\mathbf{v}; \mu, \nu) = \left|\varrho_{22}(\mathbf{v}; \mu, \nu)\right| e^{i\alpha(\mu,\nu)}, \tag{7.11}$$

then

$$\varrho_{22}^{(2)}(\mathbf{R}, \mathbf{v}, t) = \sum_{\mu,\nu=1}^{N} \left|\varrho_{22}(\mathbf{v}; \mu, \nu)\right| \cos\left[\mathbf{k}_{\mu\nu}\cdot\mathbf{R} + \delta_{\mu\nu}t - \alpha(\mu, \nu)\right]. \tag{7.12}$$

The population consists of a constant background and a sum of moving gratings, which, in general, are shifted in phase from the moving standing waves in the incident field intensity.

7.1.4 Third Order

To third order in the applied fields, we assume a solution of the form

$$\varrho_{12}^{I(3)}(\mathbf{R}, \mathbf{v}, t) = \sum_{\mu,\nu,\sigma=1}^{N} \varrho_{12}(\mathbf{v}; \mu, \nu, \sigma) e^{-i\left[(\mathbf{k}_\mu - \mathbf{k}_\nu + \mathbf{k}_\sigma)\cdot\mathbf{R} + (\delta_\mu - \delta_\nu + \delta_\sigma)t\right]}. \tag{7.13}$$

Substituting this expression and the second-order solution (7.9) for $\varrho_{22}(\mathbf{R}, \mathbf{v}, t)$ into equation (7.4a) and equating coefficients of $e^{-i\left[(\mathbf{k}_\mu-\mathbf{k}_\nu+\mathbf{k}_\sigma)\cdot\mathbf{R}+(\delta_\mu-\delta_\nu+\delta_\sigma)t\right]}$, we find that

$$\varrho_{12}^{I(3)}(\mathbf{R}, \mathbf{v}, t) = -2i \sum_{\mu,\nu,\sigma=1}^{N} \chi_\mu^* \chi_\nu \chi_\sigma^* e^{-i(\mathbf{k}_s\cdot\mathbf{R}+\delta_s t)}$$
$$\times \frac{1}{[\gamma - i\delta_s(\mathbf{v})]\left[\gamma - i\delta_\mu(\mathbf{v})\right][\gamma + i\delta_\nu(\mathbf{v})]}\left[1 + \frac{2\Gamma}{\gamma_2 - i\delta_{\mu\nu}(\mathbf{v})}\right], \tag{7.14}$$

where

$$\delta_\mu(\mathbf{v}) = \delta_\mu + \mathbf{k}_\mu \cdot \mathbf{v}, \tag{7.15}$$

$$\delta_{\mu\nu}(\mathbf{v}) = \delta_\mu(\mathbf{v}) - \delta_\nu(\mathbf{v}) = \omega_\nu - \omega_\mu + \left(\mathbf{k}_\mu - \mathbf{k}_\nu\right)\cdot \mathbf{v}, \tag{7.16}$$

$$\delta_s(\mathbf{v}) = \delta_\mu(\mathbf{v}) - \delta_\nu(\mathbf{v}) + \delta_\sigma(\mathbf{v}), \tag{7.17}$$

$$\delta_s = \delta_\mu - \delta_\nu + \delta_\sigma, \tag{7.18}$$

$$\mathbf{k}_s = \mathbf{k}_\mu - \mathbf{k}_\nu + \mathbf{k}_\sigma. \tag{7.19}$$

Each possible value of μ, ν, σ corresponds to a contribution to the third-order polarization; for N fields, there are N^3 such contributions.

Equation (7.14) is a very rich expression, containing many secrets and mysteries. It can be used to analyze a multitude of problems. First, we note that if $\mu = \nu$, then $\mathbf{k}_s = \mathbf{k}_\sigma$ and the corresponding term in equation (7.14) varies as $e^{-i(\mathbf{k}_\sigma\cdot\mathbf{R}+\delta_\sigma t)}$, which implies that $\varrho_{12}^{(3)}(\mathbf{R}, \mathbf{v}, t)$ varies as $e^{-i(\mathbf{k}_\sigma\cdot\mathbf{R}-\omega_\sigma t)}$. In other words, this contribution corresponds to a modification of field σ. Similarly, if $\sigma = \nu$, then $\mathbf{k}_s = \mathbf{k}_\mu$ and the corresponding term in equation (7.14) corresponds to a modification of field μ. On the other hand, if neither of these conditions hold, one finds that many more frequencies and field vectors are present in the atomic response than were present in the initial fields. This is a consequence of the nonlinear nature of the interaction. For example, if three fields are incident, then $\varrho_{12}^{(3)}(\mathbf{R}, \mathbf{v}, t)$ varies as $e^{-i(\mathbf{k}_s\cdot\mathbf{R}-\omega_s t)}$, where

$$\begin{aligned}\omega_s &= \omega_0 - \left(\delta_\mu - \delta_\nu + \delta_\sigma\right)\\ &= \omega_\mu - \omega_\nu + \omega_\sigma\end{aligned} \tag{7.20}$$

can take on the following values:

$$\begin{aligned}\omega_s = {}&\omega_1, \omega_2, \omega_3, \omega_1 - \omega_2 + \omega_3, \omega_1 - \omega_3 + \omega_2, \omega_2 - \omega_1 + \omega_3, 2\omega_1 - \omega_2,\\ &2\omega_1 - \omega_3, 2\omega_2 - \omega_1, 2\omega_2 - \omega_3, 2\omega_3 - \omega_1, 2\omega_3 - \omega_2,\end{aligned} \tag{7.21}$$

along with the corresponding values of $\mathbf{k}_s$. The first three frequencies correspond to frequencies of the applied fields, but all the rest correspond to *new* field frequencies generated in the medium. As this example shows, although there are N^3 contributions to the third-order polarization, the number of distinct frequencies, $\omega_s = \omega_\mu - \omega_\nu + \omega_\sigma$, appearing in the third-order polarization can be significantly less than N^3.

Four-wave mixing refers to a process in which three fields give rise to a new field in the medium. The new field characterized by $(\mathbf{k}_s, \omega_s)$ propagates in the medium only if $|k_s - \omega_s/c|\, L \ll 1$, a condition referred to as *phase matching*. Both four-wave mixing and phase matching are discussed in more detail in appendix B, including a *phase conjugate* geometry that enables one to reduce or eliminate phase distortion in the generated signal.

The next feature to study in equation (7.14) is the position of any resonances that occur as the frequency of each field is varied. In the case of stationary atoms, we see that there are resonances whenever $\delta_\mu = 0$—that is, when each field frequency is resonant with the atomic transition frequency. Moreover, there is also a resonance at $\delta_s = 0$, when the new field frequencies equal the transition frequency. Last, there is an additional "resonance" when $\delta_{\mu\nu} = \delta_\mu - \delta_\nu = \omega_\nu - \omega_\mu = 0$. The term "resonance" appears in quotation marks here, since this resonance has nothing to do with the atomic transition frequency and is a resonance between different applied field frequencies. You will see in the next chapter that this is closely related to a *two-photon* resonance condition. Also, it is interesting to note that this term vanishes in the absence of collisions ($\gamma = \gamma_2/2$). As such, these resonances have been referred to as *pressure-induced extra resonances* [1].

7.2 *N*=2: Saturation Spectroscopy for Stationary Atoms

An important case of practical interest is one in which there are two external fields, $N=2$. Moreover, we can gain considerable insight into this problem by considering the case of stationary atoms in the absence of collisions, $\gamma = \gamma_2/2$. In this limit, it follows from equations (7.9) and (7.10) that

$$\varrho_{22}^{(2)}(\mathbf{R},t) = \frac{|\chi_1|^2}{\gamma^2+\delta_1^2} + \frac{|\chi_2|^2}{\gamma^2+\delta_2^2} + \left[\frac{\chi_1^*\chi_2 e^{-i(\mathbf{k}_{12}\cdot\mathbf{R}+\delta_{12}t)}}{(\gamma+i\delta_1)(\gamma-i\delta_2)} + \frac{\chi_2^*\chi_1 e^{i(\mathbf{k}_{12}\cdot\mathbf{R}+\delta_{12}t)}}{(\gamma+i\delta_2)(\gamma-i\delta_1)}\right]. \tag{7.22}$$

The first two terms represent absorption (i.e., scattering) of each of the fields, resulting in an excited-state population that leads to a corresponding *decrease* in the absorption of the fields in third order. This is the origin of the term *saturation spectroscopy,* since the field absorption is reduced by the saturating effects of the field. However, we see that there is an *additional* term in equation (7.22) that is of the form $A\cos(\mathbf{k}_{12}\cdot\mathbf{R}+\delta_{12}t+\phi)$, where A and ϕ are parameters that depend on χ_μ, δ_μ, and γ. As such, this term corresponds to a *population grating* produced by the interference of the two fields. In third order, this can result in a coherent exchange of energy between the two fields.

We now want to calculate the atomic response to third order in the fields. To do so, we assume that the first field is a *pump* field, having a higher intensity than the second field, which is referred to as a *probe* field. The idea is to calculate the atomic response

$$\varrho_{12}^{I}(\mathbf{R},t) \approx \varrho_{12}^{I(1)}(\mathbf{R},t) + \varrho_{12}^{I(3)}(\mathbf{R},t) \tag{7.23}$$

that oscillates at the probe field frequency. The first-order response is given by equation (7.8). From equation (7.14), one finds that there are eight contributions to the third-order polarization when $N=2$. These are listed schematically in table 7.1.

TABLE 7.1
Values of μ, ν, and σ contributing to the third-order response $\varrho_{12}^{I(3)}(\mathbf{R}, \mathbf{v}, t)$.

μ, ν, σ	Rabi frequencies	Frequency
1, 1, 1	$\lvert\chi_1\rvert^2\chi_1^*$	ω_1
1, 1, 2	$\lvert\chi_1\rvert^2\chi_2^*$	ω_2
1, 2, 1	$(\chi_1^*)^2\chi_2$	$2\omega_1 - \omega_2$
1, 2, 2	$\lvert\chi_2\rvert^2\chi_1^*$	ω_1
2, 1, 1	$\lvert\chi_1\rvert^2\chi_2^*$	ω_2
2, 1, 2	$(\chi_2^*)^2\chi_1$	$2\omega_2 - \omega_1$
2, 2, 1	$\lvert\chi_2\rvert^2\chi_1^*$	ω_1
2, 2, 2	$\lvert\chi_2\rvert^2\chi_2^*$	ω_2

Since we want the third-order response that is at the probe frequency and is linear in the probe field Rabi frequency, the only contributing terms are $(\mu, \nu, \sigma) = (1, 1, 2), (2, 1, 1)$. Such terms provide us with the pump field–induced modifications of the linear absorption that are of order $|\chi_1|^2$.

Including the linear and nonlinear contributions from equations (7.8) and (7.14), respectively, we find that

$$\varrho_{12}(\mathbf{R}, t) = \varrho_{12}^I(\mathbf{R}, t)e^{i\omega_0 t} = \frac{i\chi_2^* e^{-i(\mathbf{k}_2\cdot\mathbf{R}-\omega_2 t)}}{\gamma - i\delta_2} \times \left\{1 - 2|\chi_1|^2\left[\frac{1}{\gamma^2+\delta_1^2} + \frac{1}{(\gamma - i\delta_2)(\gamma + i\delta_1)}\right]\right\}, \tag{7.24}$$

where the contributions $(\mu, \nu, \sigma) = (1, 1, 2), (2, 1, 1)$ have been used in equation (7.14). Equation (7.24) represents the atomic response to third order in the applied field amplitudes that is at the probe frequency and linear in the probe field Rabi frequency.

The first term in the square brackets always results in reduced absorption of the probe field, as a result of "saturation" by the pump field. In other words, this contribution corresponds to the pump field producing an excited-state population having a Lorenztian distribution as a function of frequency that results in reduced absorption for the probe field. However, the second term in square brackets in equation (7.24) can result in either decreased or increased probe field absorption. It is a *coherence* or *grating* term that can be attributed to scattering of the pump field from the spatial grating created by the pump and probe fields in second order.

In the limit that $|\delta_1| \gg \gamma$, equation (7.24) reduces to

$$\begin{aligned}\varrho_{12}(\mathbf{R}, t) &\approx \frac{i\chi_2^* e^{-i(\mathbf{k}_2\cdot\mathbf{R}-\omega_2 t)}}{\gamma - i\delta_2}\left\{1 - 2|\chi_1|^2\left[\frac{1}{(\gamma - i\delta_2)(i\delta_1)}\right]\right\} \\ &\approx \frac{i\chi_2^* e^{-i(\mathbf{k}_2\cdot\mathbf{R}-\omega_2 t)}}{(\gamma - i\delta_2)\left\{1 + 2|\chi_1|^2\left[\frac{1}{(\gamma - i\delta_2)(i\delta_1)}\right]\right\}} \\ &= \frac{i\chi_2^* e^{-i(\mathbf{k}_2\cdot\mathbf{R}-\omega_2 t)}}{\left[\gamma - i\left(\delta_2 + \frac{2|\chi_1|^2}{\delta_1}\right)\right]}.\end{aligned} \tag{7.25}$$

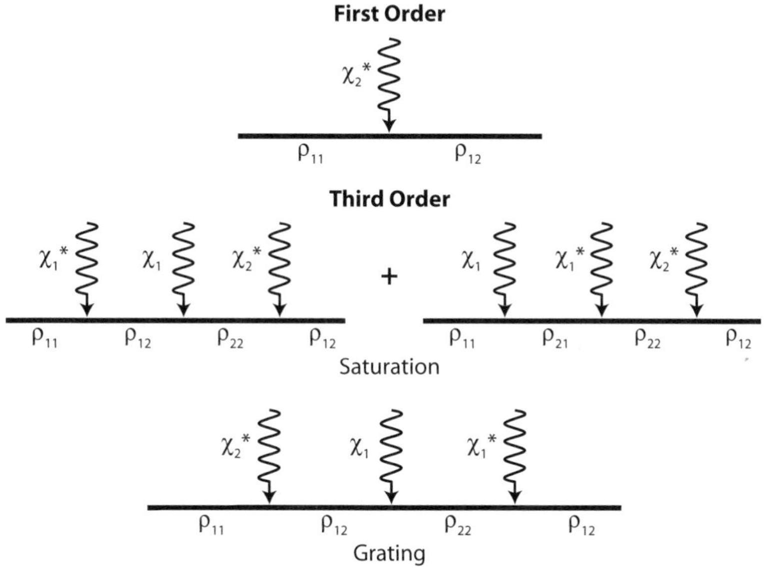

Figure 7.1. "Time-ordered" diagrams that schematically represent the contributions to ϱ_{12} to first order in the probe field and up to second order in the pump field.

Since $\delta_2 = \omega_0 - \omega_2$ has been replaced by $[(\omega_0 + 2|\chi_1|^2/\delta_1) - \omega_2]$, the net result of the pump field is to produce a light shift of the transition frequency by an amount $2|\chi_1|^2/\delta_1$; the ground state is lowered by $|\chi_1|^2/\delta_1$, and the excited state is raised by the same amount.

On the other hand, we can consider the case where both fields have the same frequency ($\delta_1 = \delta_2 = \delta$) and are distinguished by their directions of propagation. In that limit, the contributions from the saturation and grating terms are identical, and equation (7.24) reduces to

$$\varrho_{12}(\mathbf{R}, t) = \varrho_{12}^I(\mathbf{R}, t)e^{i\omega_0 t} = \frac{i\chi_2^* e^{-i(\mathbf{k}_2\cdot\mathbf{R} - \omega_2 t)}}{\gamma - i\delta} \times \left(1 - \frac{4|\chi_1|^2}{\gamma^2 + \delta^2}\right). \tag{7.26}$$

The nonlinear absorption profile is the square of a Lorentzian having width (FWHM) of $2(\sqrt{2} - 1)^{1/2}\gamma = 1.29\gamma$, which is narrower than the 2γ width of the linear absorption profile.

Even though we are working in a steady-state limit, one can view the saturation and grating contributions in terms of the time-ordered diagrams shown in figure 7.1. The diagrams indicate how the field interactions, represented by wavy arrows, modify the atomic density matrix elements. In first order, the probe field E_2^* acts to produce ϱ_{12}. In third order, there are two contributions to ϱ_{12}; the saturation term involves the creation of unmodulated excited-state population by the pump field followed by "absorption" of the probe field, while the grating term involves the creation of a spatially modulated population by the probe and pump fields, followed by scattering of the pump field from this grating. Thus, the conventional picture that the pump field saturates the transition and reduces the probe field absorption does not fully encompass the underlying physical processes that are present. The contributions represented schematically in figure 7.1 correspond simply to terms that arise in a perturbative solution of equations (7.4) for ϱ_{12}^{I} that are of order χ_2^* or $|\chi_1|^2\,\chi_2^*$.

7.3 *N*=2: Saturation Spectroscopy for Moving Atoms in Counterpropagating Fields—Hole Burning

Of greater interest in saturation spectroscopy is the response of atoms in a thermal vapor. A convenient field geometry to explore the saturation spectroscopy of moving atoms is one in which two counterpropagating fields are applied to the medium. Experimentally, this is often achieved by reflecting a portion of a pump field back into the medium, the reflected field constituting the probe field whose absorption profile we want to monitor. The two fields have the same frequency $\omega_1 = \omega_2 = \omega$, so

$$\begin{aligned}\delta_1 &= \omega_0 - \omega = \delta_2 \equiv \delta,\\ \delta_{21} &= \omega_1 - \omega_2 = 0,\end{aligned} \tag{7.27}$$

where subscript $_2$ refers to the probe field, and subscript $_1$ refers to the pump field. The propagation vectors for the two fields are

$$\mathbf{k}_1 = -\mathbf{k}_2 \equiv \mathbf{k}, \tag{7.28}$$

implying that

$$\begin{aligned}\delta_1(\mathbf{v}) &= \delta + \mathbf{k}\cdot\mathbf{v},\\ \delta_2(\mathbf{v}) &= \delta - \mathbf{k}\cdot\mathbf{v}.\end{aligned} \tag{7.29}$$

We do not set $\gamma = \gamma_2/2$, allowing for collision effects.

7.3.1 Hole Burning and Atomic Population Gratings

To second order in the applied fields, the upper-state population, calculated using equations (7.9) and (7.10), is

$$\begin{aligned}\varrho_{22}^{(2)}(\mathbf{R},\mathbf{v}) =& \frac{(2\gamma/\gamma_2)\,|\chi_1|^2}{\gamma^2 + (\delta+\mathbf{k}\cdot\mathbf{v})^2} + \frac{(2\gamma/\gamma_2)\,|\chi_2|^2}{\gamma^2 + (\delta-\mathbf{k}\cdot\mathbf{v})^2}\\ &+ \left\{\begin{array}{l} \frac{\chi_1^*\chi_2 e^{-2i\mathbf{k}\cdot\mathbf{R}}}{[\gamma+i(\delta-\mathbf{k}\cdot\mathbf{v})][\gamma-i(\delta+\mathbf{k}\cdot\mathbf{v})]}\left(1+\frac{2\Gamma}{\gamma_2-2i\mathbf{k}\cdot\mathbf{v}}\right)\\ +\frac{\chi_2^*\chi_1 e^{2i\mathbf{k}\cdot\mathbf{R}}}{[\gamma+i(\delta+\mathbf{k}\cdot\mathbf{v})][\gamma-i(\delta-\mathbf{k}\cdot\mathbf{v})]}\left(1+\frac{2\Gamma}{\gamma_2+2i\mathbf{k}\cdot\mathbf{v}}\right)\end{array}\right\}.\end{aligned} \tag{7.30}$$

There are some interesting implications of these results. For the *spatially averaged* excited-state population, denoted by $\langle\varrho_{22}^{(2)}(\mathbf{v})\rangle$, the last two terms vanish, and we find

$$\left\langle \varrho_{22}^{(2)}(\mathbf{v}) \right\rangle = \frac{(2\gamma/\gamma_2)\,|\chi_1|^2}{\gamma^2 + (\delta + \mathbf{k}\cdot\mathbf{v})^2} + \frac{(2\gamma/\gamma_2)\,|\chi_2|^2}{\gamma^2 + (\delta - \mathbf{k}\cdot\mathbf{v})^2}$$
$$\approx \frac{(2\gamma/\gamma_2)\,|\chi_1|^2}{\gamma^2 + (\delta + \mathbf{k}\cdot\mathbf{v})^2}, \tag{7.31}$$

with the second line of the equation valid when $|\chi_1| \gg |\chi_2|$. To understand the manner in which the fields modify the atomic state populations, it is convenient to introduce the spatially averaged, excited-state population density (in velocity space) defined by

$$\mathcal{R}_{22}(\mathbf{v}) = \int d\mathbf{R}\, \mathcal{R}_{22}(\mathbf{R}, \mathbf{v}) = W_0(\mathbf{v}) \left\langle \varrho_{22}^{(2)}(\mathbf{v}) \right\rangle. \tag{7.32}$$

At low pressures and ambient vapor cell temperatures, one encounters situations in which

$$|\delta|, \gamma \ll ku, \tag{7.33}$$

where u is the most probable atomic speed. Condition (7.33) is referred to as the *Doppler limit*, since the Doppler width provides the dominant contribution to the linear absorption line width. In the Doppler limit, there is maximum excitation of spatially averaged state 2 population when $\mathbf{k}\cdot\mathbf{v} = -\delta$, since only those atoms having $\mathbf{k}\cdot\mathbf{v} = -\delta$ are resonant with the field. In figure 7.2, we plot the spatially averaged, ground-state velocity distribution

$$\mathcal{R}_{11}(\mathbf{v}) = W_0(\mathbf{v}) \left[1 - \left\langle \varrho_{22}^{(2)}(\mathbf{v}) \right\rangle\right] \tag{7.34}$$

as a function $\mathbf{k}\cdot\mathbf{v}$. One sees that there is a "hole" of width 2γ "burned" in the ground-state population density centered at $\mathbf{k}\cdot\mathbf{v} = -\delta$. This corresponds to the famous *hole-burning model* introduced by Bennett [2] in the late 1960s. Whenever a quasi-monochromatic field interacts with an ensemble of atoms having a broad (compared with the natural width), *inhomogeneous* distribution of transition frequencies, hole burning can occur. In solids, there is often a large inhomogeneous width introduced by strain in the medium; as such, hole burning occurs in solids as well as thermal vapors.

The last two terms in equation (7.30) correspond to a stationary grating created by the pump and probe fields that contributes mainly in the region about $\mathbf{k}\cdot\mathbf{v} = 0$. As you will see, this term contributes negligibly in the Doppler limit, but would contribute, for example, if $\delta > ku$. It will prove instructive to write the grating term as

$$G(\mathbf{R},\mathbf{v}) = \left\{ \begin{array}{l} \frac{\chi_1^* \chi_2 e^{-2i\mathbf{k}\cdot\mathbf{R}}}{[\gamma + i(\delta - \mathbf{k}\cdot\mathbf{v})][\gamma - i(\delta + \mathbf{k}\cdot\mathbf{v})]} \left(1 + \frac{2\Gamma}{\gamma_2 - 2i\mathbf{k}\cdot\mathbf{v}}\right) \\ + \frac{\chi_2^* \chi_1 e^{2i\mathbf{k}\cdot\mathbf{R}}}{[\gamma + i(\delta + \mathbf{k}\cdot\mathbf{v})][\gamma - i(\delta - \mathbf{k}\cdot\mathbf{v})]} \left(1 + \frac{2\Gamma}{\gamma_2 + 2i\mathbf{k}\cdot\mathbf{v}}\right) \end{array} \right\}$$
$$= |G(\mathbf{R},\mathbf{v})| \cos(2\mathbf{k}\cdot\mathbf{R} + \varphi_{12} + \xi), \tag{7.35}$$

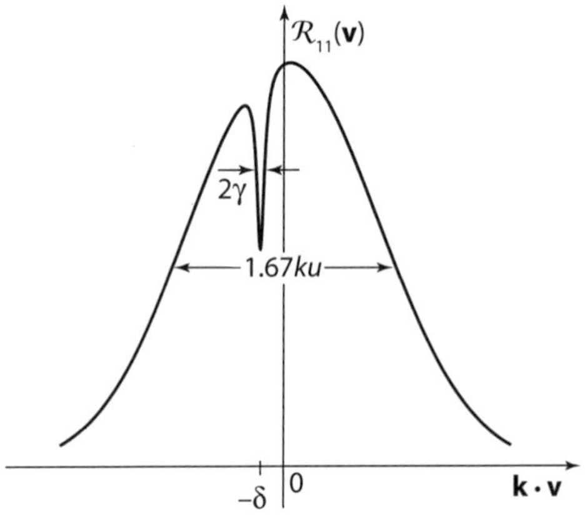

Figure 7.2. Ground-state population density $\mathcal{R}_{11}(\mathbf{v})$ as a function of $\mathbf{k}\cdot\mathbf{v}$. A hole is "burned" for $\mathbf{k}\cdot\mathbf{v} = -\delta$.

where

$$|G(\mathbf{R},\mathbf{v})| = 2\left|\frac{\chi_1\chi_2^*}{[\gamma + i(\delta + \mathbf{k}\cdot\mathbf{v})]\,[\gamma - i(\delta - \mathbf{k}\cdot\mathbf{v})]}\left(1 + \frac{2\Gamma}{\gamma_2 + 2i\mathbf{k}\cdot\mathbf{v}}\right)\right|, \tag{7.36}$$

$$\xi = \mathrm{Arg}\left\{\frac{1}{[\gamma + i(\delta + \mathbf{k}\cdot\mathbf{v})]\,[\gamma - i(\delta - \mathbf{k}\cdot\mathbf{v})]}\left(1 + \frac{2\Gamma}{\gamma_2 + 2i\mathbf{k}\cdot\mathbf{v}}\right)\right\}, \tag{7.37}$$

and φ_{12} is the relative phase of fields 1 and 2 appearing in equation (7.2). In this form, one can see that the atomic population grating is shifted by ξ from the standing-wave field intensity given in equation (7.2). By writing Maxwell's equations for the amplitudes of the two traveling wave components of the standing-wave field, one can show that, provided that $\xi \neq 0$, this phase shift can result in energy exchange between the two fields.

7.3.2 Probe Field Absorption

We now calculate the linear probe field absorption (remember that actually we are calculating the loss in the probe field intensity that results from *scattering* by the medium), including modifications to the absorption that are of second order in the pump field amplitude. As earlier, we write

$$\varrho_{12}^{I}(\mathbf{R},\mathbf{v},t) \approx \varrho_{12}^{I(1)}(\mathbf{R},\mathbf{v},t) + \varrho_{12}^{I(3)}(\mathbf{R},\mathbf{v},t) \tag{7.38}$$

and look for the contributions that are first order in the probe field Rabi frequency χ_2^*. The $\varrho_{12}^{I(1)}(\mathbf{R},\mathbf{v},t)$ term represents the linear response of the medium to the probe field, while the $\varrho_{12}^{I(3)}(\mathbf{R},\mathbf{v},t)$ term gives the modified response resulting from the saturating effects of the pump field. From equations (7.7), (7.8), (7.14), (7.27), and

(7.28), we find

$$\begin{aligned}\varrho_{12}(\mathbf{R},\mathbf{v},t) &= \varrho_{12}^{I}(\mathbf{R},\mathbf{v},t)e^{i\omega_0 t}\\ &= \frac{i\chi_2^* e^{-i(-\mathbf{k}\cdot\mathbf{R}-\omega t)}}{\gamma - i(\delta - \mathbf{k}\cdot\mathbf{v})}\left\{1 - 2|\chi_1|^2\left[\frac{2\gamma}{\gamma_2}\frac{1}{\gamma^2+(\delta+\mathbf{k}\cdot\mathbf{v})^2}\right.\right.\\ &\quad \left.\left.+\frac{1}{[\gamma - i(\delta-\mathbf{k}\cdot\mathbf{v})][\gamma + i(\delta+\mathbf{k}\cdot\mathbf{v})]}\left(1+\frac{2\Gamma}{\gamma_2+2i\mathbf{k}\cdot\mathbf{v}}\right)\right]\right\},\end{aligned} \tag{7.39}$$

where the contributions $(\mu, \nu, \sigma) = (1, 1, 2), (2, 1, 1)$ have been used in calculating the third-order contribution, since these lead to a response at the probe field frequency that is linear in χ_2^*.

We are now able to calculate the probe field absorption using equation (7.39) and the Maxwell-Bloch equations. Let us take $\mathbf{k}$ along the Z axis. The steady-state Maxwell-Bloch equation (6.25) for the probe field amplitude is given by

$$\frac{\partial E_2(Z)}{\partial Z} = -\frac{i\mathcal{N}k(\mu_x)_{12}}{\varepsilon_0}\langle\tilde{\varrho}_{21}(Z)\rangle, \tag{7.40}$$

where

$$\langle\tilde{\varrho}_{21}(Z)\rangle = \left\langle\tilde{\varrho}_{21}^{(1)}(Z)\right\rangle + \left\langle\tilde{\varrho}_{21}^{(3)}(Z)\right\rangle, \tag{7.41}$$

and, from equations (6.18) and (7.39),

$$\left\langle\tilde{\varrho}_{21}^{(1)}(Z)\right\rangle = -\int_{-\infty}^{\infty} dv_z \frac{i\chi_2(Z)W_0(v_z)}{\gamma + i(\delta - kv_z)}, \tag{7.42}$$

and

$$\begin{aligned}\left\langle\tilde{\varrho}_{21}^{(3)}(Z)\right\rangle &= \frac{4\gamma}{\gamma_2}|\chi_1|^2\int_{-\infty}^{\infty} dv_z \frac{i\chi_2(Z)W_0(v_z)}{\gamma + i(\delta - kv_z)}\frac{1}{\gamma^2 + (\delta + kv_z)^2}\\ &\quad +2|\chi_1|^2\int_{-\infty}^{\infty} dv_z \frac{i\chi_2(Z)W_0(v_z)}{\gamma + i(\delta - kv_z)}\\ &\quad \times\frac{1}{[\gamma + i(\delta - kv_z)][\gamma - i(\delta + kv_z)]}\left(1+\frac{2\Gamma}{\gamma_2 - 2ikv_z}\right),\end{aligned} \tag{7.43}$$

with

$$W_0(v_z) = \frac{1}{\sqrt{\pi}u}e^{-v_z^2/u^2}. \tag{7.44}$$

The minus sign appears in equation (7.40) since the probe field propagates in the $-\hat{\mathbf{z}}$ direction. We concentrate only on the nonlinear, third-order contribution, which depends on the presence of field 1. This term can be isolated by modulating field 1 and detecting the contribution to field 2's absorption that is modulated.

Keeping only the third-order contribution (7.43) in equation (7.40), we find

$$\frac{\partial E_2(Z)}{\partial Z} = \frac{\alpha_s}{2}E_2(Z) + \frac{\alpha_{12}}{2}E_2(Z), \tag{7.45}$$

where

$$\alpha_s = -2\alpha_0\gamma|\chi_1|^2 \int dv_z \frac{W_0(v_z)}{[\gamma + i(\delta - kv_z)]\left[\gamma^2 + (\delta + kv_z)^2\right]}, \tag{7.46}$$

$$\begin{aligned}\alpha_{12} = &-\alpha_0\gamma_2|\chi_1|^2 \int_{-\infty}^{\infty} dv_z \frac{W_0(v_z)}{\gamma + i(\delta - kv_z)} \\ &\times \frac{1}{[\gamma + i(\delta - kv_z)]\,[\gamma - i(\delta + kv_z)]}\left(1 + \frac{2\Gamma}{\gamma_2 - 2ikv_z}\right),\end{aligned} \tag{7.47}$$

and definition (4.23) for α_0 has been used. There are two contributions to $\partial E_2^+/\partial Z$. The first is the saturation term and the second is the grating or *back-scattering* term.

7.3.2.1 Saturation term

The integral in equation (7.46) can be expressed exactly in terms of plasma dispersion functions if the integrand is written in terms of partial fractions. In the Doppler limit defined by equation (7.33), an approximate expression can be obtained if the Gaussian is evaluated at $v_z = \delta/k$. In that case, we evaluate the integral using contour integration to arrive at

$$\mathrm{Re}\,\alpha_s \sim -\alpha_0 \frac{\sqrt{\pi}\gamma}{ku} e^{-\delta^2/k^2u^2} \frac{|\chi_1|^2}{\gamma^2 + \delta^2}. \tag{7.48}$$

Since $\mathrm{Re}\,\alpha_s$ is negative, this term corresponds to reduced absorption as the probe propagates in the $-\hat{\mathbf{z}}$ direction. This nonlinear absorption profile is a Lorentzian centered at $\delta = 0$ having width (FWHM) 2γ—the Doppler width is gone! This contribution has a simple physical interpretation. The pump field excites only those atoms having $|kv_z + \delta| \lesssim \gamma$ or $|v_z + \delta/k| \lesssim \gamma/k$. The second field interacts with these atoms only if $|kv_z + \delta| \lesssim \gamma$, or $|2\delta| \lesssim 2\gamma$. Therefore, only if $|\delta| \lesssim \gamma$ is there a contribution from the saturation term.

If this term is combined with the linear absorption term, obtained from equations (7.42) and (7.40) as

$$\alpha^{(1)} = \alpha_0 \frac{\gamma_2}{2} \int_{-\infty}^{\infty} dv_z \frac{W_0(v_z)}{\gamma + i(\delta - kv_z)}, \tag{7.49}$$

one gets a net absorption coefficient in the Doppler limit that is given by

$$\mathrm{Re}\left(\alpha^{(1)} + \alpha_s\right) = \alpha_0 \frac{\sqrt{\pi}\gamma}{ku} e^{-\delta^2/k^2u^2} \frac{\gamma_2}{2\gamma}\left[1 - \frac{|\chi_1|^2\,(2\gamma/\gamma_2)}{\gamma^2 + \delta^2}\right]. \tag{7.50}$$

This absorption coefficient is plotted in figure 7.3 as a function of δ. (Recall that the pump and probe fields have the same frequency, so as δ is varied, *both* the pump and probe field frequencies are changing.)

One sees a dip in the center of the profile having width 2γ, which is often referred to as a *Lamb dip*, after Willis Lamb, who predicted a similar dip in laser power as a function of cavity detuning in his theory of a gas laser [3]. This dip has important experimental consequences. It can be used to lock laser fields or resonant cavities to a specific atomic frequency. If we keep the laser or cavity frequency at the center of

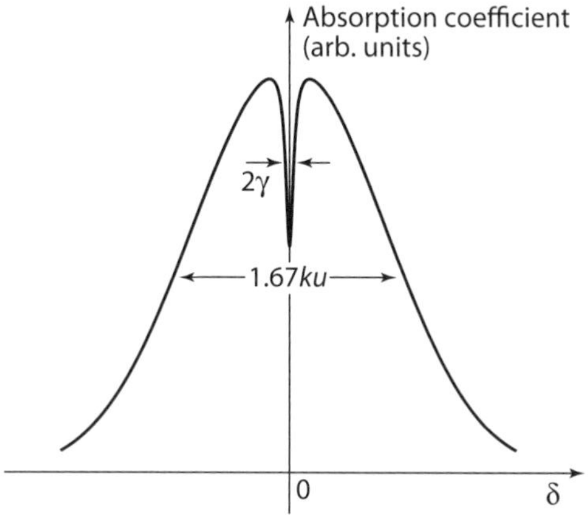

Figure 7.3. Absorption coefficient (in arbitrary units) as a function of detuning in saturated absorption.

the absorption profile, we can be certain that the frequency is locked to the atomic transition frequency.

7.3.2.2 Back-scattering term

In the Doppler limit, the back-scattering term α_{12} given by equation (7.47) vanishes, since all the poles in the contour integration are in the same half plane. The grating term is unimportant, since the atoms move several wavelengths in their excited-state lifetime; as such, the contribution from the grating term is "washed out" on integrating over velocities.

On the other hand, for detunings $|\delta| \gg ku$, the grating term cannot be neglected. If $|\delta| \gg ku$, *all* the atoms are off resonance, and the relative contributions of the saturation and grating terms are comparable.

7.4 Saturation Spectroscopy in Inhomogeneously Broadened Solids

We have already alluded to the fact that saturation spectroscopy can be used to probe solids as well as thermal vapors. Optical transitions in solids are often characterized by large inhomogeneous widths resulting from interactions between the host medium and the atoms in the medium that interact with the optical fields. Of course, the atoms are not in motion in a solid, but owing to the inhomogeneous width, the same principles of hole burning and saturation spectroscopy that we encountered for vapors applies equally well.

In other words, the transition frequency ω_0 is different for atoms located at different positions in a solid. We assume that there is some average transition frequency $\bar{\omega}_0$ and that the distribution $W_f(\Delta)$ of displacements Δ about this average frequency is given by

$$W_f(\Delta) = \frac{1}{\sqrt{\pi}\sigma_w} e^{-(\Delta/\sigma_w)^2}, \tag{7.51}$$

where σ_w characterizes the width of the inhomogeneous distribution. In some sense, the distribution of detunings $\delta(\Delta)$ replaces the Doppler distribution that is relevant for vapors. In terms of the transition frequency ω_0 and an applied field frequency ω,

$$\delta(\Delta) = \omega_0 - \omega = \delta_0 + \Delta, \tag{7.52}$$

where

$$\delta_0 = \bar{\omega}_0 - \omega, \qquad \Delta = \omega_0 - \bar{\omega}_0. \tag{7.53}$$

As a consequence, equation (7.14) is replaced by

$$\varrho_{12}^{I(3)}(\mathbf{R}, \Delta, t) = -2i \sum_{\mu,\nu,\sigma=1}^{N} \chi_\mu^* \chi_\nu \chi_\sigma^* e^{-i[\mathbf{k}_s \cdot \mathbf{R} + \delta_s t]} \times \frac{1}{[\gamma - i\delta_s(\Delta)]\,[\gamma - i\delta_\mu(\Delta)]\,[\gamma + i\delta_\nu(\Delta)]} \left(1 + \frac{2\Gamma}{\gamma_2 - i\delta_{\mu\nu}}\right), \tag{7.54}$$

where

$$\delta_\mu(\Delta) = \omega_0 - \omega_\mu = \bar{\omega}_0 - \omega_\mu + \Delta, \tag{7.55}$$

$$\delta_{\mu\nu} = \omega_\nu - \omega_\mu, \tag{7.56}$$

$$\delta_s(\Delta) = \delta_\mu(\Delta) - \delta_\nu(\Delta) + \delta_\sigma(\Delta). \tag{7.57}$$

There are a few fundamental differences between equations (7.14) and (7.54). In solids, the *directions* of the incident fields are irrelevant, since there is no Doppler effect. As a consequence, there is no washing out of the grating term for counterpropagating fields; the grating and absorption terms contribute on equal footing, regardless of the directions of the fields. The last factor in equation (7.54) depends only on the field frequencies and not on the atomic frequencies. As a consequence, this "collision"-induced resonance[1] cannot vanish as a result of averaging over the inhomogeneous distribution of transition frequencies as it does for a thermal vapor and counterpropagating fields [4].

Moreover, it is no longer possible to get a narrow line shape using pump and probe fields having the same frequency. In the case of a thermal vapor with counterpropagating fields, each moving atom is exposed to *two* field frequencies, owing to the Doppler effect. Saturation spectroscopy relied on the fact that both fields were resonant only for atoms having $|\mathbf{k} \cdot \mathbf{v}| \lesssim \gamma$—in other words, only those atoms in a narrow frequency range interact with *both* fields. In solids, since there is no Doppler effect, atoms over the entire inhomogeneous width interact with both fields if they have the same frequency. To obtain a narrow, homogeneously broadened line, one must fix the pump field frequency and vary the probe field frequency. As the probe frequency is tuned, atoms are brought into resonance with both fields only when the probe frequency equals the pump frequency. In this manner, one recovers all the features of saturation spectroscopy.

[1] Collisions per se are absent in a solid, but a nonvanishing Γ can result from dephasing processes produced by the host medium.

7.5 Summary

Several important concepts were introduced in this chapter. We have seen that it can be advantageous to use one field to "pump" or prepare atoms and a second field to probe these atoms. The pump field can modify the velocity distribution associated with atomic state populations. In saturation spectroscopy, the pump field excites a narrow velocity class of excited atoms that are probed by a second field. As a consequence, it is possible to eliminate the Doppler width in the probe absorption spectrum by using only these selected atoms. The term *saturation spectroscopy* can be somewhat misleading. We have seen that there is another physical process that enters when one drives atomic ensembles with two or more fields. The fields can create population gratings off which any of the fields can scatter. This coherent scattering process can lead to four-wave mixing signals and the coherent exchange of energy between the fields. In the next chapter, we turn our attention to *three-level* atoms to explore new types of phenomena that can arise in such systems.

7.6 Appendix A: Saturation Spectroscopy—Stationary Atoms in One Strong and One Weak Field

Consider a pump–probe experiment in which two fields drive two-level transitions in an ensemble of *stationary* atoms. The field frequencies are $\omega_1 = \omega$ and $\omega_2 = \omega + \Delta$. We want to calculate the probe field (field 2) absorption to *all* orders in the pump field (field 1) intensity. The probe field absorption spectrum exhibits many interesting features not found in the perturbative solution that was considered earlier. You are asked to explore some of these features in the problems. In this appendix, we simply outline the approach needed to solve this problem.

Since the first field is strong, it makes sense to work in the field interaction representation for the first field. In the rotating-wave approximation (RWA), the equations to be solved are

$$\frac{\partial \tilde{\varrho}_{12}}{\partial t} = -i\chi_1^*(2\varrho_{22} - 1) - i\chi_2^* e^{i\Delta t}(2\varrho_{22} - 1) - (\gamma - i\delta_1)\tilde{\varrho}_{12}, \tag{7.58a}$$

$$\frac{\partial \varrho_{22}}{\partial t} = i\chi_1^*\tilde{\varrho}_{21} - i\chi_1\tilde{\varrho}_{12} + i\chi_2^* e^{i\Delta t}\tilde{\varrho}_{21} - i\chi_2 e^{-i\Delta t}\tilde{\varrho}_{12} - \gamma_2\varrho_{22}, \tag{7.58b}$$

$$\tilde{\varrho}_{21} = [\tilde{\varrho}_{12}]^*, \tag{7.58c}$$

$$\varrho_{11} = 1 - \varrho_{22}, \tag{7.58d}$$

$$\Delta = \omega_2 - \omega_1 = \delta_1 - \delta_2, \tag{7.58e}$$

where any spatial phase factors can be included in the χ's. We seek a steady-state solution (i.e., a solution once all transients have died away) that is first order in χ_2 but all orders in χ_1. We have already obtained the solution that is zeroth order in χ_2 [see equation (4.11)], for which density matrix elements are denoted by $\tilde{\varrho}_{ij}^{(0)}$.

We try a solution

$$\tilde{\varrho}_{ij} = \tilde{\varrho}_{ij}^{(0)} + \tilde{\varrho}_{ij}^{(1)}, \tag{7.59}$$

where $\tilde{\varrho}_{ij}^{(1)}$ is first order in χ_2 or χ_2^*, substitute the trial solution in the original differential equations, and keep only terms linear in the probe field amplitude to obtain

$$\frac{\partial \tilde{\varrho}_{12}^{(1)}}{\partial t} = -2i\chi_1^* \varrho_{22}^{(1)} - i\chi_2^* e^{i\Delta t}\left[2\varrho_{22}^{(0)} - 1\right] - (\gamma - i\delta_1)\tilde{\varrho}_{12}^{(1)}, \tag{7.60a}$$

$$\frac{\partial \varrho_{22}^{(1)}}{\partial t} = i\chi_1^* \tilde{\varrho}_{21}^{(1)} - i\chi_1 \tilde{\varrho}_{12}^{(1)} + i\chi_2^* e^{i\Delta t}\tilde{\varrho}_{21}^{(0)} - i\chi_2 e^{-i\Delta t}\tilde{\varrho}_{12}^{(0)} - \gamma_2 \varrho_{22}^{(1)}, \tag{7.60b}$$

$$\tilde{\varrho}_{21} = [\tilde{\varrho}_{12}]^*, \tag{7.60c}$$

$$\varrho_{11}^{(1)} = -\varrho_{22}^{(1)}. \tag{7.60d}$$

To solve these equations, we guess a solution of the form

$$\tilde{\varrho}_{12}^{(1)} = a_{12} + b_{12}e^{i\Delta t} + c_{12}e^{-i\Delta t}, \tag{7.61a}$$

$$\tilde{\varrho}_{22}^{(1)} = a_{22} + b_{22}e^{i\Delta t} + c_{22}e^{-i\Delta t}, \tag{7.61b}$$

substitute it into the differential equations, and equate constant coefficients and coefficients of $e^{\pm i\Delta t}$ to obtain

$$\gamma_2 a_{22} - i\chi_1^* a_{12}^* + i\chi_1 a_{12} = 0, \tag{7.62a}$$

$$(\gamma - i\delta_1) a_{12} + 2i\chi_1^* a_{22} = 0, \tag{7.62b}$$

$$(\gamma_2 + i\Delta) b_{22} - i\chi_1^* c_{12}^* + i\chi_1 b_{12} = i\chi_2^* \tilde{\varrho}_{21}^{(0)}, \tag{7.62c}$$

$$[\gamma + i(\Delta - \delta_1)] b_{12} + 2i\chi_1^* b_{22} = -i\chi_2^*\left[2\varrho_{22}^{(0)} - 1\right], \tag{7.62d}$$

$$[\gamma + i(\Delta + \delta_1)] c_{12}^* - 2i\chi_1 c_{22} = 0. \tag{7.62e}$$

Before solving these equations, it is useful to note that

$$\begin{aligned}\varrho_{12}^{(1)} &= \tilde{\varrho}_{12}^{(1)} e^{i\omega_1 t} = \left(a_{12} + b_{12}e^{i\Delta t} + c_{12}e^{-i\Delta t}\right)e^{i\omega_1 t} \\ &= a_{12}e^{i\omega_1 t} + b_{12}e^{i\omega_2 t} + c_{12}e^{i(2\omega_1 - \omega_2)t}.\end{aligned} \tag{7.63}$$

Thus, the a_{12} term is a modification of the pump field, the b_{12} term gives the probe field absorption, which is what is being sought here, and the c_{12} term is a new field generated via four-wave mixing.

It is simple to solve equations (7.62) analytically on your own or using a computer. By either method, one obtains $a_{22} = a_{12} = 0$ and

$$b_{12} = \frac{-i\chi_2^* N^{(0)}}{\mu_{12}'}\left[1 - \frac{2|\chi_1|^2}{\mu_{12}'\mu_{21}^p(1+B)}\left(1 + \frac{2\Gamma}{\mu_2}\right)\right], \tag{7.64}$$

where

$$\mu_2 = \gamma_2 + i\Delta, \tag{7.65a}$$

$$\mu_{12} = \gamma + i\Delta, \tag{7.65b}$$

$$\mu'_{12} = \gamma - i\delta_2 = \gamma + i(\Delta - \delta_1), \tag{7.65c}$$

$$\bar{\mu}_{12} = \gamma + i(\Delta + \delta_1), \tag{7.65d}$$

$$\mu^p_{21} = \gamma + i\delta_1, \tag{7.65e}$$

$$B = \frac{4|\chi_1|^2\mu_{12}}{\mu'_{12}\bar{\mu}_{12}\mu_2}, \tag{7.65f}$$

$$N^{(0)} = \left[\varrho^{(0)}_{22} - \varrho^{(0)}_{11}\right] = -\left[1 + \frac{4(\gamma/\gamma_2)|\chi_1|^2}{\gamma^2 + \delta_1^2}\right]^{-1}. \tag{7.65g}$$

Although this is a complicated expression, it is easy to program on a computer. The probe absorption is proportional to $\mathrm{Im}(b_{12}/\chi_2^*)$. When $\delta_1 = 0$ and $|\chi_1| \gg \gamma$, the pump field almost totally bleaches the medium, and there is very little probe absorption or gain, but owing to interesting interference phenomena, there can be *either* absorption or gain for the probe field, depending on the value of Δ. For $|\delta_1| \gg \gamma, |\chi_1|$, there are resonances at $\Delta = 0, \pm\delta_1$. The resonance at $\Delta = -\delta_1$ $(2\omega_1 - \omega_2 = \omega_0)$ is a *gain* term that results, in lowest order, from the absorption of two pump photons and the emission of one probe photon. There is a dispersion-like resonance centered at $\Delta = 0$ that is a little more difficult to explain [5].

Although the probe absorption has been calculated for stationary atoms, it is relatively simple to generalize equation (7.64) to the case of a thermal vapor excited by these fields. One need make only the replacements

$$\begin{aligned} \Delta &\rightarrow \Delta - \mathbf{k}_{21}\cdot\mathbf{v}, \\ \delta_j &\rightarrow \Delta + \mathbf{k}_j\cdot\mathbf{v} \end{aligned} \tag{7.66}$$

in equations (7.65). The velocity integration in this case must be carried out numerically.

7.7 Appendix B: Four-Wave Mixing

Let us consider three fields incident on a vapor, as shown in figure 7.4. Equation (7.14) gives the coherence $\varrho^{I(3)}_{12}(\mathbf{R}, \mathbf{v}, t)$ to third order in the applied fields. In this section, we consider only two terms in the sum, those for which

$$\mathbf{k}_s = \mathbf{k}_1 - \mathbf{k}_3 + \mathbf{k}_2. \tag{7.67}$$

In terms of the summation indices in equation (7.14), this value of $\mathbf{k}_s$ corresponds to $(\mu, \nu, \sigma) = (1, 3, 2), (2, 3, 1)$. For these terms, $\varrho^{I(3)}_{21}(\mathbf{R}, \mathbf{v}, t)$ is proportional to $\chi_1\chi_2\chi_3^*$. This leads to a polarization contribution

$$P \sim \tilde{\varrho}_{21}(\mathbf{R}, \mathbf{v})e^{i\mathbf{k}_s\cdot\mathbf{R} - i\omega_s t}, \tag{7.68}$$

where $\mathbf{k}_s = \mathbf{k}_1 + \mathbf{k}_2 - \mathbf{k}_3$, $\omega_s = \omega_1 + \omega_2 - \omega_3$.

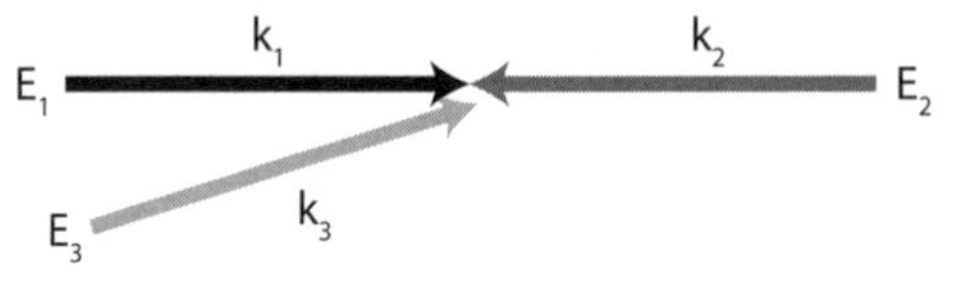

Figure 7.4. Field geometry for four-wave mixing.

Does this polarization give rise to a *new* field in the medium? To answer this question, we return to the wave equation (6.21) but do not take $k_s = \omega_s/c$; however, we still assume that $|(k_s^2 - \frac{\omega_s^2}{c^2})/k_s|\lambda \ll 1$ to ensure that the SVAPA is valid. With this assumption, equation (6.21) is replaced by

$$\frac{\partial E_s(Z)}{\partial Z} - \frac{i}{2k_s}\left(k_s^2 - \frac{\omega_s^2}{c^2}\right)E_s(Z) = \frac{ik_s\mathcal{N}\mu_{12}\langle\tilde{\varrho}_{21}(Z)\rangle}{\varepsilon_0}, \tag{7.69}$$

where

$$\begin{aligned}\tilde{\varrho}_{21}^{(3)}(Z) = 2i\int d\mathbf{v}\,W_0(\mathbf{v})\chi_1(Z)\chi_2(Z)\chi_3^*(Z) \sum_{\mu,\nu,\sigma=(1,3,2),(2,3,1)} \\ \times\frac{1}{[\gamma + i\delta_s(\mathbf{v})]\,[\gamma + i\delta_\mu(\mathbf{v})]\,[\gamma - i\delta_\nu(\mathbf{v})]} \\ \times\left[1 + \frac{2\Gamma}{\gamma_2 + i\delta_{\mu\nu}(\mathbf{v})}\right]\end{aligned} \tag{7.70}$$

and Z is taken to be in the direction of $\mathbf{k}_s$. There can be some Z dependence in each of the field amplitudes and Rabi frequencies, reflecting loss or gain for the fields as they propagate in the medium.

If $k_s \neq \omega_s/c$, $E_s(Z)$ oscillates as a function of Z. If we write

$$\frac{1}{2k_s}\left(k_s^2 - \frac{\omega_s^2}{c^2}\right) = \Delta k_s, \tag{7.71}$$

the field does not build up if $\Delta k_s L \gg 1$, where L is the length of the sample. Thus, one must try to adjust things so that $k_s = \omega_s/c$. This is called *phase matching*. Phase matching can be achieved in some cases by using the dispersive properties of the index of refraction. One set of conditions that guarantees phase matching is $\omega_1 = \omega_2 = \omega_3$, $\mathbf{k}_1 = -\mathbf{k}_2$, and $k_3 = k_1$. In this case, $\mathbf{k}_s = -\mathbf{k}_3$ and $\omega_s = \omega_3$, so that the new field propagates in a direction opposite to field 3. This is called a *phase conjugate geometry*, since the generated wave varies as χ_3^*. If we send a probe χ_3 into the medium, the four-wave mixing signal is generated in the $-\mathbf{k}_3$ direction and "undoes" any phase distortion produced on the χ_3 field by the medium! The classic demonstration of phase-conjugate four-wave mixing involves sending a probe signal through frosted glass and having the phase conjugate signal return through the glass, producing an undistorted replica of the input probe [6]. More practical applications could involve the removal of phase distortions of the atmosphere when a laser probe field is used to monitor pollutants. Phase conjugate mirrors also can be formed using counterpropagating pump beams. A probe beam incident perpendicular to these beams will be reflected in a distortionless manner; moreover, you can never "burn" such a mirror and can even have gain on the reflected beam!

If fields 1 and 2 are strong, and if depletion of the pump fields can be neglected, then the coupled equations for the probe and signal field are of the form

$$\frac{\partial E_s(Z)}{\partial Z} = AE_3^*(Z) - \alpha_s^{(1)} E_s(Z), \tag{7.72a}$$

$$\frac{\partial E_3^*(Z)}{\partial Z} = BE_s(Z) - \alpha_3^{(1)} E_3(Z), \tag{7.72b}$$

where A and B are proportional to the pump field intensity, and we have added in terms representing the linear absorption of these fields. These equations can be solved, subject to the conditions $E_3^*(Z = 0) = \text{constant}$ and $E_s(Z = L) = 0$ (see reference [7]).

Last, we note that the two contributions in the sum in equation (7.70) can be given a simple physical interpretation. The term with $(\mu, \nu, \sigma) = (1, 3, 2)$ corresponds to fields 1 and 3, forming a grating off which field 2 scatters in the $\mathbf{k}_s$ direction, while the term with $(\mu, \nu, \sigma) = (1, 3, 2)$ corresponds to fields 2 and 3, forming a grating off which field 1 scatters in the $\mathbf{k}_s$ direction. In the limit that $\mathbf{k}_1$ and $\mathbf{k}_3$ subtend a small angle, both terms $(\mu, \nu, \sigma) = (1, 3, 2), (2, 3, 1)$ contribute for stationary atoms or for $\delta \gg ku$, since the velocity average is unimportant in these limits. On the other hand, if one is in the Doppler limit, only the "forward grating," $(\mu, \nu, \sigma) = (1, 3, 2)$, is nonvanishing when the velocity average is taken.

Problems

1. **Kerr effect.** Extend the calculation of linear absorption for stationary atoms given in the text to include the lowest order nonlinear corrections to the polarization—that is, include terms in the polarization up to order χ^3. Calculate the index of refraction to order χ^2, and show that if the incident beam has a Gaussian transverse profile, it is possible for the medium to focus the laser field. This is referred to as *self-focusing* and occurs if the index effect is sufficiently large to overcome diffraction effects.
2. Derive equations (7.10) and (7.14).
3. Consider a pump–probe experiment in which two fields drive a two-level transition in an atomic beam. The propagation vectors of both fields are perpendicular to the beam such that $\mathbf{k}_1 \cdot \mathbf{v} = \mathbf{k}_2 \cdot \mathbf{v} = 0$. The field frequencies are $\omega_1 = \omega$ and $\omega_2 = \omega + \Delta$, and the detuning of the first field is $\delta_1 = \omega_0 - \omega_1$. Calculate the steady-state value (*steady state* means that all transients have died away) of ϱ_{12} to order $|\chi_1|^2\chi_2^*$. Consider the limit in which $|\delta_1| \gg \gamma$. To this order in $|\chi_1|^2\chi_2^*$, ϱ_{12} varies as $\tilde{\varrho}_{12}\exp[-i\mathbf{k}_2 \cdot \mathbf{R} + i\omega_2 t]$. Show that there is a dispersive-like structure in the probe absorption spectrum [proportional to $\text{Im}(\tilde{\varrho}_{12}/\chi_2^*)$] centered at $\Delta = 0$ and that the linear absorption resonance at $\delta_2 = 0$ $(\Delta = -\delta_1)$ is slightly pulled toward $\Delta = 0$. Moreover, show that if $\gamma = \gamma_2/2$ (as it would if the only relaxation was produced by spontaneous decay), the dispersion-like resonance vanishes. The appearance of the resonance at $\Delta = 0$ when collisions are present (so that $\gamma_2 \neq 2\gamma$) has been referred to as *pressure-induced extra resonances* (see reference [1]).

4–5. Now repeat the calculation of problem 3 to all orders in χ_1 and to first order in χ_2^*. To do this, expand each density matrix element as $\tilde{\varrho}_{ij} = \tilde{\varrho}_{ij}^{(0)} + \tilde{\varrho}_{ij}^{(1)}$, where $\tilde{\varrho}_{ij}^{(0)}$ is zeroth order in χ_2^*, and $\tilde{\varrho}_{ij}^{(1)}$ is first order in χ_2^*—that is, derive equation (7.64). Plot $\mathrm{Im}(b_{12}/\chi_2^*)$ as a function of Δ for $\delta_1 = 0$, $|\chi_1| \gg \gamma$ and for $|\delta_1| \gg \gamma, |\chi_1|$. In the latter case, show that there are resonances at $\Delta = 0, \pm\delta_1$. The absorption at $\Delta = -\delta_1$ and gain at $\Delta = \delta_1$ can be explained easily in terms of semiclassical dressed states if you care to try.

6. Consider two counterpropagating fields incident on a vapor. Take both fields polarized in the $\hat{\mathbf{x}}$ direction so that the total field component in this direction is

$$E = E_+ \cos(kZ - \omega t) + E_- \cos(kZ + \omega t), \quad E_\pm \text{ real.}$$

Show that the time-averaged intensity of this field has a spatially varying part that varies as $E_+E_- \cos(2kZ)$. These fields interact with an ensemble of two-level atoms. To second order in the field amplitudes, show that the population difference $\varrho_{22} - \varrho_{11}$ has a spatially varying part that can be written as $A\cos(2kZ+\xi)$, where A and ξ are real. Calculate A and ξ as a function of kv_z, and show that this spatial grating is phase shifted from the incident spatial variation of the field intensity by an amount ξ that depends on the atomic velocity.

7–8. For the field of problem 6, the population difference $N = \varrho_{22} - \varrho_{11}$ can be expanded as $N = \varrho_{22} - \varrho_{11} = \sum_{n=-\infty}^{\infty} N_n \exp(2inkZ)$, where $N_n = N_{-n}^*$. Derive coupled Maxwell-Bloch equations for the field amplitudes $E_\pm$, and show that they depend only on N_0 and $N_{\pm 1}$. Writing $N_1 = Ae^{i(\xi+\phi)}$, where A and ξ are real and $\phi = \mathrm{Arg}(\chi_+) - \mathrm{Arg}(\chi_-)$, determine the conditions on ξ and δ for there to be an amplification of field E_+ at the expense of E_-. (Assume that $|\delta| \gg kv, \gamma$ throughout this problem.) To second order in the field amplitudes, show that there can be gain for one of the fields if $\gamma \neq \gamma_2/2$, but not if $\gamma = \gamma_2/2$. Using this result, prove that the atoms are cooled for $\delta < 0$ as a result of the back-scattering of the fields if $\gamma \neq \gamma_2/2$.

9. Suppose that you have a ground-state density of stationary atoms given by $\varrho_{11}(Z) = \cos^2(kZ)$ in a sample of length L. A plane wave is incident in the positive Z direction. Show that there is a propagating wave that is generated in the sample that is directed in the $-Z$ direction. (Hint: Use the Maxwell-Bloch equations.) Neglecting absorption, calculate the amplitude of the wave that exits the *entrance* face of the sample, assuming that the density in the sample is equal to $\mathcal{N}$.

10. Evaluate the integrals (7.46) and (7.47) exactly in terms of the plasma dispersion or complex error function. By evaluating these expressions, show that in the Doppler limit (7.33), the results agree with those derived in the text. Neglect collisions in the grating term.

11. In the Doppler limit, calculate the contribution to the probe absorption signal that is first order in the pump field intensity and when the pump field has frequency ω_1 and the probe field frequency is varied. Assume that the fields are counterpropagating and that the grating term still vanishes. Show that this contribution to the absorption coefficient is a minimum (maximum negative absorption) when $\omega_2 = 2\omega_0 - \omega_1$, but that the width of the resonance is 4γ rather than 2γ, as we found in saturated absorption with a standing-wave field of a given frequency. Interpret the result in terms of a simple hole-burning model.

12. Repeat problem 11 for a solid. Take $\gamma = \gamma_2/2$, and show that the grating term is nonvanishing when an integral over the inhomogeneous frequency distribution is taken. Moreover, show that the grating term can have a sign opposite to that of the saturation term for some detunings.
13. Show that, in the Doppler limit, only one of the terms in equation (7.70) contributes, and explain this result in terms of the gratings that are formed in the medium. Assume that fields 1 and 3 are nearly copropagating. On the other hand, show that for either stationary atoms, or for detunings much larger than the Doppler widths, both terms in the sum contribute. Neglect collisions.

References

[1] For a review, see L. Rothberg, *Dephasing-induced coherence phenomena*, in *Progress in Optics* XXIV, edited by E. Wolf (Elsevier Scientific, Amsterdam, 1987), pp. 39–101.

[2] W. R. Bennett Jr., *Hole burning effects in a He-Ne optical maser*, Physical Review **126**, 580–593 (1962).

[3] W. E. Lamb Jr., *Theory of an optical maser*, Physical Review **134**, A1429–A1450 (1964).

[4] D. G. Steel and S. C. Rand, *Ultranarrow nonlinear optical resonances in solids*, Physical Review Letters **55**, 2285–2288 (1985).

[5] G. Grynberg and C. Cohen-Tannoudji, *Central resonance of the Mollow absorption spectrum: physical origin of gain without population inversion*, Optics Communications **96**, 150–163 (2000); P. R. Berman and G. Khitrova, *Theory of pump-probe spectroscopy using an amplitude approach*, Optics Communications **179**, 19–27 (2000).

[6] For a review, see D. L. Pepper, in *Laser Handbook*, vol. 4, edited by M. I. Stich and M. Bass (Elsevier Science, Amsterdam, 1985), pp. 333–485; D. L. Pepper, *Applications of optical phase conjugation*, Scientific American **254**, 74–83 (1986); R. L. Abrams, J. F. Lam, R. C. Lind, D. G. Steel, and P. F. Liao, *Phase conjugation and high-resolution spectroscopy by resonant degenerate four-wave mixing*, in *Optical Phase Conjugation* (Academic Press, New York, 1983), pp. 211–284.

[7] R. L. Abrams and R. L. Lind, *Degenerate four-wave mixing in absorbing media*, Optics Letters **2**, 94–96 (1978); **3**, 205, errata (1978).

Bibliography

In addition to the books listed in chapter 1, the following is a nonexhaustive list of some early articles on nonlinear spectroscopy of two-level atoms:

G. S. Agarwal, *Absorption spectra of strongly driven two-level atoms*, Physical Review A **19**, 923–924 (1979).

E. V. Baklanov and V. P. Chebotaev, *Effects of fields in resonant interaction of opposing waves in a gas I*, Soviet Physics–JETP **33**, 300–308 (1971).

———, *Resonance interaction of unidirectional waves in gases*, Soviet Physics–JETP **34**, 490–494 (1972).

R. W. Boyd and S. Mukamel, *Origin of spectral holes in pump-probe studies of homogenously broadened lines*, Physical Review A **29**, 1973–1983 (1984).

V. P. Chebotaev and V. S. Letokhov, *Nonlinear narrow optical resonances induced by laser radiation,* Progress in Quantum Electronics **4**, 111–206 (1975).

M. Gruneisen, K. Macdonald, and R. W. Boyd, *Induced gain and modified absorption of a weak probe beam in a strongly driven sodium vapor*, Journal of the Optical Society of America B **5**, 123–128 (1988).

J. L. Hall, *The lineshape problem in laser saturated molecular absorption*, in *Lectures in Theoretical Physics XII*, edited by K. T. Mahahthappa and W. E. Brittin (Gordon and Breach, New York, 1970), pp. 161–210.

S. Haroche and F. Hartmann, *Theory of saturated-absorption line shapes*, Physical Review A **6**, 1280–1300 (1972).

L. W. Hillman, R. W. Boyd, J. Krasinski, and C. R. Stroud Jr., *Observation of a spectral hole due to population oscillations in a homogenously broadened optical absorption line*, Optics Communications **45**, 416–419 (1983).

G. Khitrova, P. R. Berman, and M. Sargent III, *Theory of pump-probe spectroscopy*, Journal of the Optical Society of America B **5**, 160–170 (1988).

B. R. Mollow, *Stimulated emission and absorption near resonance for driven systems*, Physical Review A **5**, 2217–2222 (1972).

M. Sargent III, *Spectroscopic techniques based on Lamb's laser theory*, Physics Reports **43**, 223–265 (1978).

M. Sargent III and P. E. Toschek, *Unidirectional saturation spectroscopy II: general lifetimes, interpretations, and analogies,* Applied Physics **11**, 107–120 (1976).

J. H. Shirley, *Semiclassical theory of saturated absorption in gases,* Physical Review A **8**, 347–368 (1973).

S. Stenholm, *Theoretical foundations of laser spectroscopy*, Physics Reports **43**, 151–221 (1978).

Yu. A. Vdovin, V. M. Ermachenko, A. I. Popov, and E. D. Protsenko, *Observation of fine structure within the limits of homogenous line width*, JETP Letters **15**, 282–284 (1972).

F. Y. Wu, S. Ezekiel, M. Ducloy, and B. R. Mollow, *Observation of amplification in a strongly driven two-level atomic system at optical frequencies*, Physical Review Letters **38**, 1077–1080 (1977).

For discussions of four-wave mixing, including additional references, see:

R. L. Abrams, J. F. Lam, R. C. Lind, D. G. Steel, and P. F. Liao, *Phase conjugation and high-resolution spectroscopy by resonant degenerate four-wave mixing*, in *Optical Phase Conjugation* (Academic Press, New York, 1983), pp. 211–284.

P. Meystre and M. Sargent III, *Elements of Quantum Optics*, 3rd ed. (Springer-Verlag, Berlin, 1998), chap. 10.

Y. R. Shen, *The Principles of Nonlinear Optics* (Wiley-Interscience, New York, 1984), chaps. 14 and 15.

8

Three-Level Atoms:

Applications to Nonlinear Spectroscopy—Open Quantum Systems

We now broaden our horizons and add another level to our atom. As you will see, even for the limited examples presented in this and the next chapter, the addition of a third level opens new vistas. New and important atom–field dynamics appear when a three-level atom interacts with two or more optical fields. Often, one field drives a given transition in the atom to prepare or *dress* the atom, while a second field drives a coupled transition to exploit the state of the atom that has been prepared by the first field.

8.1 Hamiltonian for Λ, V, and Cascade Systems

The prototypical system consists of a three-level atom interacting with two traveling wave fields. This system has received a great deal of attention in recent years, owing to applications involving adiabatic transfer of population, lasing without inversion, dark states, coherent population trapping, laser cooling, and slow light. The original interest in such systems was for nonlinear spectroscopy—elimination of the Doppler width [1, 2].

There are, in effect, three types of three-level schemes, and these are shown in figure 8.1. We refer to these level schemes as the cascade, V, and Λ configurations, respectively. Two fields,

$$\mathbf{E}(\mathbf{R}, t) = \frac{1}{2}\hat{\boldsymbol{\epsilon}}\left[E(\mathbf{R}, t)e^{i(\mathbf{k}\cdot\mathbf{R}-\omega t)} + E(\mathbf{R}, t)^* e^{-i(\mathbf{k}\cdot\mathbf{R}-\omega t)}\right], \tag{8.1a}$$

$$\mathbf{E}'(\mathbf{R}, t) = \frac{1}{2}\hat{\boldsymbol{\epsilon}}'\left[E'(\mathbf{R}, t)e^{i(\mathbf{k}'\cdot\mathbf{R}-\omega' t)} + E'(\mathbf{R}, t)^* e^{-i(\mathbf{k}'\cdot\mathbf{R}-\omega' t)}\right], \tag{8.1b}$$

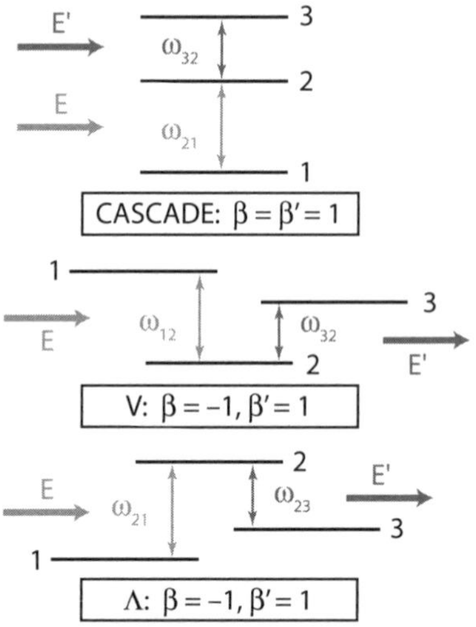

Figure 8.1. Three-level systems: cascade, V, and Λ. The quantities β and β' are defined in section 8.1.3.

are incident on an ensemble of such three-level atoms. As in the two-level case, the complex field amplitudes $E(\mathbf{R}, t)$ and $E'(\mathbf{R}, t)$ vary slowly in space over a distance of a wavelength and slowly in time over an optical period.

We assume that field $\mathbf{E}$ drives only the 1–2 transition and field $\mathbf{E}'$ only the 2–3 transition. States 1 and 3 have the same parity, which is opposite to that of state 2. The energy of level 2 is arbitrarily taken equal to zero in order to give the equations for all the level configurations a similar structure.

The assumption that each field drives a single transition must be given some justification. Clearly, if the 1–2 and 2–3 transition frequencies differ by an amount that is much greater than all decay rates, detunings, and Rabi frequencies in the problem, and if field $\mathbf{E}$ is nearly resonant with the 1–2 transition and $\mathbf{E}'$ with 2–3 transition, then this approximation is valid. Alternatively, given the selection rules for the transitions, it may be possible to choose the field polarizations to ensure that each field drives a single transition. For example, in the Λ scheme, suppose that the transition frequencies are identical, but that states 1 and 3 have $J = 1, m = \mp 1$, respectively, while state 2 has $J = 0$. By choosing σ_+ circular polarization for field $\mathbf{E}$ and σ_- circular polarization for field $\mathbf{E}'$, then field $\mathbf{E}$ drives only the 1–2 transition and field $\mathbf{E}'$ only the 2–3 transition.

As in the two-level case, let us consider first a stationary atom at position $\mathbf{R}$ interacting with the two fields. It is assumed that the rotating-wave approximation (RWA) is valid for both fields. However, the RWA brings in different terms for the three systems, depending on whether levels 1 and 3 lie above or below level 2 in energy. In all cases, the interaction Hamiltonian is

$$\hat{V}(\mathbf{R}, t) = -\hat{\boldsymbol{\mu}} \cdot \left[\mathbf{E}(\mathbf{R}, t) + \mathbf{E}'(\mathbf{R}, t) \right], \tag{8.2}$$

and owing to the parity of the levels, $V_{13} = V_{31} = V_{11} = V_{33} = V_{22} = 0$.

8.1.1 Cascade Configuration

In the Schrödinger, interaction, and field interaction representations, for the cascade-level configuration, we set

$$|\Psi(t)\rangle = a_1(t)|1\rangle + a_2(t)|2\rangle + a_3(t)|3\rangle \quad \text{Schrödinger,} \tag{8.3a}$$

$$|\Psi(t)\rangle = c_1(t)e^{i\omega_{21}t}|1\rangle + c_2(t)|2\rangle + c_3(t)e^{-i\omega_{32}t}|3\rangle \quad \text{interaction,} \tag{8.3b}$$

$$|\Psi(t)\rangle = \tilde{c}_1(t)e^{i\omega t}|1\rangle + \tilde{c}_2(t)|2\rangle + \tilde{c}_3(t)e^{-i\omega' t}|3\rangle \quad \text{field interaction,} \tag{8.3c}$$

where

$$\omega_{21} = \frac{E_2 - E_1}{\hbar} = -\frac{E_1}{\hbar} > 0, \tag{8.4a}$$

$$\omega_{32} = \frac{E_3 - E_2}{\hbar} = \frac{E_3}{\hbar}. \tag{8.4b}$$

We could also modify the field interaction representation to include the phases of the applied fields, if that proves to be convenient. Moreover, one can introduce semiclassical dressed states, but they are rather complicated except in the case that the detunings for both fields are equal in magnitude but opposite in sign [3]. We discuss three-level dressed states in the context of dark states and adiabatic following in chapter 9.

Since each field drives a single transition, it is most convenient to work in the field interaction representation. For the interaction potential (8.2) and the fields (8.1), it follows from Schrödinger equation with the state vector (8.3c) that the state amplitudes evolve as

$$i\hbar\frac{d\tilde{\mathbf{c}}}{dt} = \tilde{\mathbf{H}}_c\tilde{\mathbf{c}}, \tag{8.5}$$

where

$$\tilde{\mathbf{H}}_c = \hbar\begin{pmatrix} -\delta_c & \chi_c^* e^{-i\mathbf{k}\cdot\mathbf{R}} & 0 \\ \chi_c e^{i\mathbf{k}\cdot\mathbf{R}} & 0 & \chi_c'^* e^{-i\mathbf{k}'\cdot\mathbf{R}} \\ 0 & \chi_c' e^{i\mathbf{k}'\cdot\mathbf{R}} & \delta_c' \end{pmatrix}, \tag{8.6}$$

$$\chi_c = -\frac{\boldsymbol{\mu}_{21}\cdot\hat{\boldsymbol{\epsilon}}E(\mathbf{R},t)}{2\hbar}, \tag{8.7a}$$

$$\chi_c' = -\frac{\boldsymbol{\mu}_{32}\cdot\hat{\boldsymbol{\epsilon}}'E'(\mathbf{R},t)}{2\hbar}, \tag{8.7b}$$

$$\delta_c = \omega_{21} - \omega, \tag{8.8a}$$

$$\delta_c' = \omega_{32} - \omega', \tag{8.8b}$$

and the c subscript stands for *cascade*. Although not indicated explicitly, the Rabi frequencies can be slowly varying functions of space and time. In component form, equation (8.5) is

$$\dot{\tilde{c}}_1 = i\delta_c\tilde{c}_1 - i\chi_c^* e^{-i\mathbf{k}\cdot\mathbf{R}}\tilde{c}_2, \tag{8.9a}$$

$$\dot{\tilde{c}}_2 = -i\chi_c e^{i\mathbf{k}\cdot\mathbf{R}}\tilde{c}_1 - i\chi_c'^* e^{-i\mathbf{k}'\cdot\mathbf{R}}\tilde{c}_3, \tag{8.9b}$$

$$\dot{\tilde{c}}_3 = -i\delta_c'\tilde{c}_3 - i\chi_c' e^{i\mathbf{k}'\cdot\mathbf{R}}\tilde{c}_2. \tag{8.9c}$$

8.1.2 V and Λ Configurations

In a similar manner, for the V and Λ configurations, we write

$$|\Psi(t)\rangle = c_1(t)e^{-i\omega_{12}t}|1\rangle + c_2(t)|2\rangle + c_3(t)e^{-i\omega_{32}t}|3\rangle \quad \text{interaction}, \tag{8.10a}$$

$$|\Psi(t)\rangle = \tilde{c}_1(t)e^{-i\omega t}|1\rangle + \tilde{c}_2(t)|2\rangle + \tilde{c}_3(t)e^{-i\omega' t}|3\rangle \quad \text{field interaction}, \tag{8.10b}$$

and

$$|\Psi(t)\rangle = c_1(t)e^{i\omega_{21}t}|1\rangle + c_2(t)|2\rangle + c_3(t)e^{i\omega_{23}t}|3\rangle \quad \text{interaction}, \tag{8.11a}$$

$$|\Psi(t)\rangle = \tilde{c}_1(t)e^{i\omega t}|1\rangle + \tilde{c}_2(t)|2\rangle + \tilde{c}_3(t)e^{i\omega' t}|3\rangle \quad \text{field interaction}, \tag{8.11b}$$

respectively.

In the field interaction representation, the corresponding Hamiltonians for the V and Λ configurations are

$$\tilde{\mathbf{H}}_V = \hbar\begin{pmatrix} \delta_V & \chi_V e^{i\mathbf{k}\cdot\mathbf{R}} & 0 \\ \chi_V^* e^{-i\mathbf{k}\cdot\mathbf{R}} & 0 & \chi_V'^* e^{-i\mathbf{k}'\cdot\mathbf{R}} \\ 0 & \chi_V' e^{i\mathbf{k}'\cdot\mathbf{R}} & \delta_V' \end{pmatrix}, \tag{8.12a}$$

$$\tilde{\mathbf{H}}_\Lambda = \hbar\begin{pmatrix} -\delta_\Lambda & \chi_\Lambda^* e^{-i\mathbf{k}\cdot\mathbf{R}} & 0 \\ \chi_\Lambda e^{i\mathbf{k}\cdot\mathbf{R}} & 0 & \chi_\Lambda' e^{i\mathbf{k}'\cdot\mathbf{R}} \\ 0 & \chi_\Lambda'^* e^{-i\mathbf{k}'\cdot\mathbf{R}} & -\delta_\Lambda' \end{pmatrix}, \tag{8.12b}$$

respectively, where

$$\chi_V = -\frac{\boldsymbol{\mu}_{12}\cdot\hat{\boldsymbol{\epsilon}}E(\mathbf{R},t)}{2\hbar}, \tag{8.13a}$$

$$\chi_V' = -\frac{\boldsymbol{\mu}_{32}\cdot\hat{\boldsymbol{\epsilon}}'E'(\mathbf{R},t)}{2\hbar}, \tag{8.13b}$$

$$\chi_\Lambda = -\frac{\boldsymbol{\mu}_{21}\cdot\hat{\boldsymbol{\epsilon}}E(\mathbf{R},t)}{2\hbar}, \tag{8.13c}$$

$$\chi_\Lambda' = -\frac{\boldsymbol{\mu}_{23}\cdot\hat{\boldsymbol{\epsilon}}'E'(\mathbf{R},t)}{2\hbar}, \tag{8.13d}$$

and

$$\delta_V = \omega_{12} - \omega, \tag{8.14a}$$

$$\delta_V' = \omega_{32} - \omega', \tag{8.14b}$$

$$\delta_\Lambda = \omega_{21} - \omega, \tag{8.14c}$$

$$\delta_\Lambda' = \omega_{23} - \omega'. \tag{8.14d}$$

8.1.3 All Configurations

It is possible to treat all level configurations with a single notation by defining

$$\chi_\beta = -\frac{\boldsymbol{\mu}_{21}\cdot\hat{\boldsymbol{\epsilon}}E(\mathbf{R},t)}{2\hbar} \quad \text{for } \beta = 1, \tag{8.15a}$$

$$\chi_\beta = \left[-\frac{\boldsymbol{\mu}_{12}\cdot\hat{\boldsymbol{\epsilon}}E(\mathbf{R},t)}{2\hbar}\right]^* \quad \text{for } \beta = -1, \tag{8.15b}$$

$$\chi'_{\beta'} = -\frac{\boldsymbol{\mu}_{32}\cdot\hat{\boldsymbol{\epsilon}}'E'(\mathbf{R},t)}{2\hbar} \quad \text{for } \beta' = 1, \tag{8.15c}$$

$$\chi'_{\beta'} = \left[-\frac{\boldsymbol{\mu}_{23}\cdot\hat{\boldsymbol{\epsilon}}'E'(\mathbf{R},t)}{2\hbar}\right]^* \quad \text{for } \beta' = -1, \tag{8.15d}$$

and

$$\delta_\beta = \omega_{21} - \omega \quad \text{for } \beta = 1, \tag{8.16a}$$

$$\delta_\beta = \omega_{12} - \omega \quad \text{for } \beta = -1, \tag{8.16b}$$

$$\delta'_{\beta'} = \omega_{32} - \omega' \quad \text{for } \beta' = 1, \tag{8.16c}$$

$$\delta'_{\beta'} = \omega_{23} - \omega' \quad \text{for } \beta' = -1, \tag{8.16d}$$

where the β's correspond to the different configurations as follows:

$$\beta = \beta' = 1 \quad \text{cascade}, \tag{8.17a}$$

$$\beta = -1, \quad \beta' = 1 \quad V, \tag{8.17b}$$

$$\beta = 1, \quad \beta' = -1 \quad \Lambda. \tag{8.17c}$$

For example, the Hamiltonian is now given by

$$\tilde{\mathrm{H}} = \hbar \begin{pmatrix} -\beta\delta_\beta & \chi_\beta^* e^{-i\mathbf{k}\cdot\mathbf{R}} & 0 \\ \chi_\beta e^{i\mathbf{k}\cdot\mathbf{R}} & 0 & \chi_{\beta'}^{\prime *} e^{-i\mathbf{k}'\cdot\mathbf{R}} \\ 0 & \chi'_{\beta'} e^{i\mathbf{k}'\cdot\mathbf{R}} & \beta'\delta'_{\beta'} \end{pmatrix}, \tag{8.18}$$

valid for any configuration.

8.2 Density Matrix Equations in the Field Interaction Representation

As in the two-level problem, we can introduce atomic motion in a simple fashion if this motion is treated classically. Setting

$$\varrho_{12}(\mathbf{R},\mathbf{v},t) = \tilde{\varrho}_{12}(\mathbf{R},\mathbf{v},t)e^{-i\beta(\mathbf{k}\cdot\mathbf{R}-\omega t)}, \tag{8.19a}$$

$$\varrho_{23}(\mathbf{R},\mathbf{v},t) = \tilde{\varrho}_{23}(\mathbf{R},\mathbf{v},t)e^{-i\beta'(\mathbf{k}'\cdot\mathbf{R}-\omega' t)}, \tag{8.19b}$$

$$\varrho_{13}(\mathbf{R},\mathbf{v},t) = \tilde{\varrho}_{13}(\mathbf{R},\mathbf{v},t)e^{-i\beta(\mathbf{k}\cdot\mathbf{R}-\omega t)-i\beta'(\mathbf{k}'\cdot\mathbf{R}-\omega' t)}, \tag{8.19c}$$

using the Hamiltonian (8.18), and adding in relaxation terms, we obtain the following density matrix equations:

$$\frac{\partial \varrho_{11}}{\partial t} + \mathbf{v} \cdot \nabla \varrho_{11} = -i\chi_\beta^* \tilde{\varrho}_{21} + i\chi_\beta \tilde{\varrho}_{12} - \gamma_1 \varrho_{11} + \gamma_{2,1}\varrho_{22}, \tag{8.20a}$$

$$\frac{\partial \varrho_{22}}{\partial t} + \mathbf{v} \cdot \nabla \varrho_{22} = i\chi_\beta^* \tilde{\varrho}_{21} - i\chi_\beta \tilde{\varrho}_{12} - i\chi_{\beta'}^{\prime *} \tilde{\varrho}_{32} + i\chi_{\beta'}^{\prime} \tilde{\varrho}_{23}$$
$$-\gamma_2 \varrho_{22} + \gamma_{1,2}\varrho_{11} + \gamma_{3,2}\varrho_{33}, \tag{8.20b}$$

$$\frac{\partial \varrho_{33}}{\partial t} + \mathbf{v} \cdot \nabla \varrho_{33} = i\chi_{\beta'}^{\prime *} \tilde{\varrho}_{32} - i\chi_{\beta'}^{\prime} \tilde{\varrho}_{23} - \gamma_3 \varrho_{33} + \gamma_{2,3}\varrho_{22}, \tag{8.20c}$$

$$\frac{\partial \tilde{\varrho}_{12}}{\partial t} + \mathbf{v} \cdot \nabla \tilde{\varrho}_{12} = -i\chi_\beta^* (\varrho_{22} - \varrho_{11}) + i\chi_{\beta'}^{\prime} \tilde{\varrho}_{13}$$
$$- \left[\gamma_{12} - i\beta\delta_\beta(\mathbf{v})\right] \tilde{\varrho}_{12}, \tag{8.20d}$$

$$\frac{\partial \tilde{\varrho}_{13}}{\partial t} + \mathbf{v} \cdot \nabla \tilde{\varrho}_{13} = -i\chi_\beta^* \tilde{\varrho}_{23} + i\chi_{\beta'}^{\prime *} \tilde{\varrho}_{12}$$
$$- \left\{\gamma_{13} - i\left[\beta\delta_\beta(\mathbf{v}) + \beta'\delta'_{\beta'}(\mathbf{v})\right]\right\} \tilde{\varrho}_{13}, \tag{8.20e}$$

$$\frac{\partial \tilde{\varrho}_{23}}{\partial t} + \mathbf{v} \cdot \nabla \tilde{\varrho}_{23} = -i\chi_{\beta'}^{\prime *} (\varrho_{33} - \varrho_{22}) - i\chi_\beta \tilde{\varrho}_{13}$$
$$- \left[\gamma_{23} - i\beta'\delta'_{\beta'}(\mathbf{v})\right] \tilde{\varrho}_{23}, \tag{8.20f}$$

$$\tilde{\varrho}_{ij} = \left(\tilde{\varrho}_{ji}\right)^*, \tag{8.20g}$$

where it is understood that the Rabi frequencies can be slowly varying functions of space and time. The detunings appearing in these equations are defined as

$$\delta_\beta(\mathbf{v}) = \delta_\beta + \beta \mathbf{k} \cdot \mathbf{v}, \tag{8.21a}$$

$$\delta'_{\beta'}(\mathbf{v}) = \delta'_\beta + \beta' \mathbf{k}' \cdot \mathbf{v}, \tag{8.21b}$$

and the decay rates as

$$\gamma_{12} = (\gamma_1 + \gamma_2)/2 + \Gamma_{12}, \tag{8.22a}$$

$$\gamma_{23} = (\gamma_2 + \gamma_3)/2 + \Gamma_{23}, \tag{8.22b}$$

$$\gamma_{13} = (\gamma_1 + \gamma_3)/2 + \Gamma_{13}, \tag{8.22c}$$

where the Γ_{ij} are collisional dephasing rates associated with the ij transition. The decay rate $\gamma_{i,j}$ appearing in equations (8.20) is the *partial spontaneous decay rate* from level i to j (not to be confused with the dephasing rates γ_{ij}). It is important to remember that some of the $\gamma_{i,j}$'s and γ_i's vanish in certain configurations—namely,

$$\gamma_1 = \gamma_{2,3} = \gamma_{1,2} = 0, \quad \gamma_{2,1} = \gamma_2, \quad \gamma_{3,2} = \gamma_3 \quad \text{cascade}, \tag{8.23a}$$

$$\gamma_2 = \gamma_{2,3} = \gamma_{2,1} = 0, \quad \gamma_{1,2} = \gamma_1, \quad \gamma_{3,2} = \gamma_3 \quad \text{V}, \tag{8.23b}$$

$$\gamma_1 = \gamma_3 = \gamma_{1,2} = \gamma_{3,2} = 0, \quad \gamma_{2,1} + \gamma_{2,3} = \gamma_2 \quad \Lambda. \tag{8.23c}$$

In all cases, total population is conserved,

$$\varrho_{11} + \varrho_{22} + \varrho_{33} = 1. \tag{8.24}$$

It might seem that the collision rates, Γ_{12}, Γ_{23}, and Γ_{13}, are independent parameters, but this is *not* the case. In fact, putting in arbitrary values for these parameters can lead to negative populations [4]! The interdependence of the collision rates can be seen using a simple example. If the collisional interaction is the same for levels 2 and 3, then $\Gamma_{23} = 0$ and $\Gamma_{13} = \Gamma_{12}$, since the collision-induced phase shifts are identical for the state amplitudes 2 and 3.

8.3 Steady-State Solutions—Nonlinear Spectroscopy

Equations (8.20) must be solved numerically when the Rabi frequencies are functions of time. In the case of constant Rabi frequencies, however, one can solve these equations in steady state by setting the entire left-hand side of the equations equal to zero. One is faced with solving eight linear equations for the density matrix elements since one of the populations can be eliminated using condition (8.24). Symbolic computer programs such as Mathematica have no problem in obtaining analytic expressions for the steady-state density matrix elements, but they are too lengthy to display here. Moreover, any averaging over the velocity distribution must still be carried out numerically.

Once solutions are obtained, the field absorption coefficients can be obtained from the steady-state solutions of the Maxwell-Bloch equations (6.25), assuming that both fields propagate parallel to the Z axis. In that case, for fields propagating in the $\hat{\mathbf{z}}$ direction, one finds

$$\frac{\partial E(Z)}{\partial Z} = \frac{i\mathcal{N}k\mu_{12}}{\varepsilon_0}\langle\tilde{\varrho}_{21}(Z)\rangle \quad \text{cascade, } \Lambda, \tag{8.25a}$$

$$\frac{\partial E(Z)}{\partial Z} = \frac{i\mathcal{N}k\mu_{21}}{\varepsilon_0}\langle\tilde{\varrho}_{12}(Z)\rangle \quad V, \tag{8.25b}$$

for field E with $\mu_{ij} = \boldsymbol{\mu}_{ij} \cdot \hat{\boldsymbol{\epsilon}}$, and

$$\frac{\partial E'(Z)}{\partial Z} = \frac{i\mathcal{N}k\mu_{23}}{\varepsilon_0}\langle\tilde{\varrho}_{32}(Z)\rangle \quad \text{cascade, } V, \tag{8.26a}$$

$$\frac{\partial E'(Z)}{\partial Z} = \frac{i\mathcal{N}k\mu_{32}}{\varepsilon_0}\langle\tilde{\varrho}_{23}(Z)\rangle \quad \Lambda, \tag{8.26b}$$

for field E' with $\mu_{ij} = \boldsymbol{\mu}_{ij} \cdot \hat{\boldsymbol{\epsilon}}'$. For fields propagating in the $-\hat{\mathbf{z}}$ direction, the signs in these equations should be changed. We will refer to field E as the *pump* field and field E' as the *probe* field, although we have not yet imposed any restrictions on the strengths of these fields.

In this section, we calculate the probe field E' absorption coefficient for the cascade configuration in lowest order perturbation theory. The probe field absorption coefficient is proportional to the imaginary part of $\tilde{\varrho}_{32}$, which, in turn, is proportional to ϱ_{33} in steady state [see equation (8.20c)]. We leave the case of the V and Λ configurations to the problems, although the Λ configuration is discussed in some detail in chapter 9. Both fields propagate parallel to the Z axis, but the fields can be co- or counterpropagating. As you will see, the resulting excitation probability offers a new nonlinear spectroscopic tool for obtaining Doppler-free spectra in vapors.

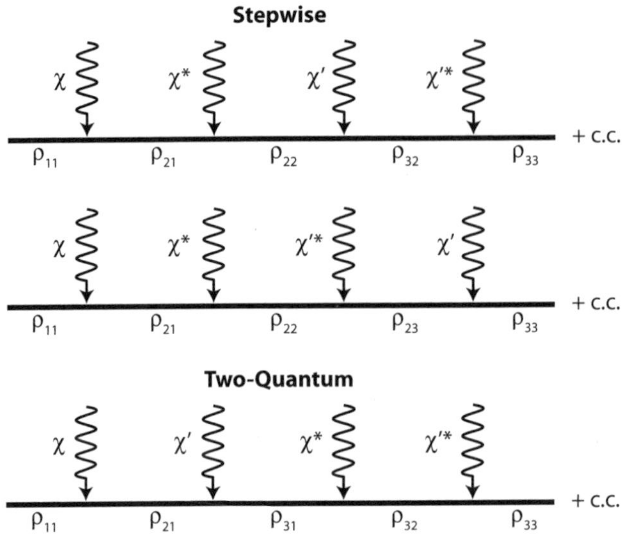

Figure 8.2. Schematic representation of the stepwise and two-quantum perturbative chains that lead to ϱ_{33}.

The starting point of the calculation is the zeroth-order solution, with $\varrho_{11} = 1$ and all other density matrix elements equal to zero. We need to calculate ϱ_{33} to order $|\chi\chi'|^2$. As in the case of a two-level atom driven by two fields, we can represent the perturbation chains leading to ϱ_{33} using the "time-ordered" diagrams shown in figure 8.2. There are two-classes of chains that are discussed in the following, *step-wise* and *two-quantum*. Any single chain involves a sequence of calculations using the steady-state solutions of equations (8.20). For example, the top-left chain in figure 8.2, $\varrho_{11} \rightarrow \varrho_{21} \rightarrow \varrho_{22} \rightarrow \varrho_{32} \rightarrow \varrho_{33}$, involves starting with $\varrho_{11} = 1$, calculating

$$\tilde{\varrho}_{21} = \frac{-i\chi}{\gamma_{12} + i\delta(\mathbf{v})} \tag{8.27}$$

from (the complex conjugate of) equation (8.20d),

$$\varrho_{22} = \frac{i\chi^*\tilde{\varrho}_{21}}{\gamma_2} = \frac{1}{\gamma_2}\frac{|\chi|^2}{\gamma_{12} + i\delta(\mathbf{v})} \tag{8.28}$$

from equation (8.20b),

$$\tilde{\varrho}_{32} = \frac{-i\chi'}{\gamma_{23} + i\delta'(\mathbf{v})}\varrho_{22} = \frac{1}{\gamma_2}\frac{-i\chi'}{\gamma_{23} + i\delta'(\mathbf{v})}\frac{|\chi|^2}{\gamma_{12} + i\delta(\mathbf{v})} \tag{8.29}$$

from (the complex conjugate of) equation (8.20f), and last,

$$\varrho_{33} = \frac{i\chi'^*\tilde{\varrho}_{32}}{\gamma_3} = \frac{1}{\gamma_3\gamma_2}\frac{|\chi'|^2}{\gamma_{23} + i\delta'(\mathbf{v})}\frac{|\chi|^2}{\gamma_{12} + i\delta(\mathbf{v})} \tag{8.30}$$

from equation (8.20c). In these expressions,

$$\chi = -\frac{\boldsymbol{\mu}_{21}\cdot\hat{\boldsymbol{\epsilon}}E(\mathbf{R},t)}{2\hbar}, \quad \chi' = -\frac{\boldsymbol{\mu}_{32}\cdot\hat{\boldsymbol{\epsilon}}'E'(\mathbf{R},t)}{2\hbar}, \tag{8.31a}$$

$$\delta(\mathbf{v}) = \omega_{21} - \omega + \mathbf{k}\cdot\mathbf{v}, \quad \delta'(\mathbf{v}) = \omega_{32} - \omega' + \mathbf{k}'\cdot\mathbf{v}. \tag{8.31b}$$

Other chains are calculated in a similar manner. In fact, it is easy to establish "rules" for calculating the contributions from these chains, if one wants. Note that the populations ϱ_{22} and ϱ_{33} appearing in equations (8.28) and (8.30), respectively, are not real—they represent the *partial* contributions to these populations from this specific chain. Of course, when one sums the contributions from all the step-wise chains, ϱ_{22} and ϱ_{33} will be real and positive.

The step-wise (SW) contribution involves the sequential or step-wise absorption of each field by the atoms and is characterized by the presence of the intermediate-state population ϱ_{22}. Summing up the contributions from the four chains

$$\begin{aligned}
&\varrho_{11} \to \varrho_{21} \to \varrho_{22} \to \varrho_{32} \to \varrho_{33},\\
&\varrho_{11} \to \varrho_{12} \to \varrho_{22} \to \varrho_{23} \to \varrho_{33},\\
&\varrho_{11} \to \varrho_{21} \to \varrho_{22} \to \varrho_{23} \to \varrho_{33},\\
&\varrho_{11} \to \varrho_{12} \to \varrho_{22} \to \varrho_{32} \to \varrho_{33},
\end{aligned} \tag{8.32}$$

of the stepwise contribution represented in figure 8.2, we find

$$\begin{aligned}
\varrho_{33}^{\mathrm{SW}}(\mathbf{v}) &= \frac{-i\chi}{\gamma_{12}+i\delta(\mathbf{v})}\frac{i\chi^*}{\gamma_2}\frac{-i\chi'}{\gamma_{23}+i\delta'(\mathbf{v})}\frac{i\chi'^*}{\gamma_3} + \text{c.c.}\\
&\quad + \frac{i\chi^*}{\gamma_{12}-i\delta(\mathbf{v})}\frac{-i\chi}{\gamma_2}\frac{-i\chi'}{\gamma_{23}+i\delta'(\mathbf{v})}\frac{i\chi'^*}{\gamma_3} + \text{c.c.}\\
&= \frac{4|\chi\chi'|^2}{\gamma_2\gamma_3}\frac{\gamma_{12}}{\gamma_{12}^2+(\delta+\mathbf{k}\cdot\mathbf{v})^2}\frac{\gamma_{23}}{\gamma_{23}^2+(\delta'+\mathbf{k}'\cdot\mathbf{v})^2}.
\end{aligned} \tag{8.33}$$

As might have been expected, this contribution is just the product of the Lorentzian absorption profile of each field acting on its respective transition. The first two lines of equation (8.33) are written in a manner that illustrates how figure 8.2 can be used to write the perturbative result by application of some simple rules.

The two-quantum (TQ) chains do *not* involve the intermediate-state population; rather, the coherence ϱ_{31} or ϱ_{13} appears in these chains. We refer to these as two-quantum chains since they are closely related to the type of contribution one encounters in two-photon transitions (see section 8.4); moreover, this chain is nonvanishing only if both fields are present *simultaneously*. Of course, we have assumed both fields to be present simultaneously, but had we used spatially separated pulsed fields, the two-quantum contribution would vanish, whereas the step-wise terms could contribute.

The left-hand diagram of the two-quantum chain, $\varrho_{11} \to \varrho_{21} \to \varrho_{31} \to \varrho_{32} \to \varrho_{33}$, can be evaluated by starting with $\varrho_{11} = 1$, calculating

$$\tilde{\varrho}_{21} = \frac{-i\chi}{\gamma_{12}+i\delta(\mathbf{v})} \tag{8.34}$$

from (the complex conjugate of) equation (8.20d),

$$\tilde{\varrho}_{31} = \frac{-i\chi'\tilde{\varrho}_{21}}{\gamma_{13} + i\left[\delta(\mathbf{v}) + \delta'(\mathbf{v})\right]} \tag{8.35}$$

from (the complex conjugate) of equation (8.20e),

$$\tilde{\varrho}_{32} = \frac{i\chi^*}{\gamma_{23} + i\delta'(\mathbf{v})}\tilde{\varrho}_{31} \tag{8.36}$$

from (the complex conjugate of) equation (8.20f), and last,

$$\varrho_{33} = \frac{i\chi'^*\tilde{\varrho}_{32}}{\gamma_3} \tag{8.37}$$

from equation (8.20c).

Summing the two chains

$$\begin{aligned} &\varrho_{11} \to \varrho_{21} \to \varrho_{31} \to \varrho_{32} \to \varrho_{33}, \\ &\varrho_{11} \to \varrho_{12} \to \varrho_{13} \to \varrho_{23} \to \varrho_{33}, \end{aligned} \tag{8.38}$$

of the two-quantum contribution shown in figure 8.2, we obtain

$$\begin{aligned} \varrho_{33}^{\mathrm{TQ}}(\mathbf{v}) &= \frac{-i\chi}{\gamma_{12} + i\delta(\mathbf{v})} \frac{-i\chi'}{\gamma_{13} + i\left[\delta(\mathbf{v}) + \delta'(\mathbf{v})\right]} \\ &\quad \times \frac{i\chi^*}{\gamma_{23} + i\delta'(\mathbf{v})} \frac{i\chi'^*}{\gamma_3} + \text{c.c.}, \\ &= \frac{|\chi\chi'|^2}{\gamma_3} \frac{1}{\gamma_{12} + i(\delta + \mathbf{k}\cdot\mathbf{v})} \frac{1}{\gamma_{23} + i(\delta' + \mathbf{k}'\cdot\mathbf{v})} \\ &\quad \times \frac{1}{\gamma_{13} + i\left[\delta + \delta' + (\mathbf{k} + \mathbf{k}')\cdot\mathbf{v}\right]} + \text{c.c.} \end{aligned} \tag{8.39}$$

If we look at equations (8.33) and (8.39), it would appear that, for fixed δ, there are resonances when $\delta' = 0$ or $\delta + \delta' = 0$ (for $\mathbf{v} = 0$). However, if all the collision rates vanish, we can combine terms to obtain

$$\begin{aligned} \varrho_{33}(\mathbf{v}) &= \varrho_{33}^{\mathrm{SW}}(\mathbf{v}) + \varrho_{33}^{\mathrm{TQ}}(\mathbf{v}) \\ &= \frac{|\chi\chi'|^2}{\left[(\gamma_2/2)^2 + (\delta + \mathbf{k}\cdot\mathbf{v})^2\right]\left\{(\gamma_3/2)^2 + \left[\delta + \delta' + (\mathbf{k} + \mathbf{k}')\cdot\mathbf{v}\right]^2\right\}}. \end{aligned} \tag{8.40}$$

The resonance at $\delta' = 0$ has vanished! There is a simple physical interpretation of this result. In lowest order perturbation theory, the transition from state 1 to 3 is really a two-photon process, provided that there are no collisions or other mechanism to give energy to or take energy from the system. As such, the width associated with this transition depends only on the widths of the initial and final states, and not on the width of the intermediate-state 2.

To get the line shape

$$I = \int d\mathbf{v}\varrho_{33}(\mathbf{v})W_0(\mathbf{v}) \tag{8.41}$$

as a function of δ' for fixed δ, we look at the case

$$\mathbf{k} = k\hat{\mathbf{z}}, \quad \mathbf{k}' = \eta k'\hat{\mathbf{z}}, \tag{8.42}$$

where $\eta = 1$ for copropagating fields and $\eta = -1$ for counterpropagating fields. All the integrals contributing to the signal can be expressed in terms of plasma dispersion functions [5], but we look at two limiting cases for which simplified expressions can be obtained. We consider first the case of stationary atoms and second the Doppler limit, $|\delta|, \gamma_{12} \ll ku$; $|\delta'|, \gamma_{23} \ll k'u$.

8.3.1 Stationary Atoms

In this case, we can combine equations (8.33) and (8.39), use equations (8.22), and do a little algebra to obtain

$$\begin{aligned} I = {} & \frac{|\chi\chi'|^2}{\gamma_3} \frac{1}{\gamma_{12}^2 + \delta^2} \left\{ \frac{2\gamma_{13}}{\left[\gamma_{13}^2 + (\delta + \delta')^2\right]} + \frac{2\Gamma_{12}}{\gamma_2} \frac{2\gamma_{23}}{\gamma_{23}^2 + \delta'^2} \right\} \\ & + \frac{|\chi\chi'|^2}{\gamma_3} \frac{(\Gamma_{13} + \Gamma_{12} - \Gamma_{23})}{\gamma_{12}^2 + \delta^2} \left\{ \begin{array}{c} \frac{1}{[\gamma_{13} - i(\delta + \delta')]} \frac{1}{(\gamma_{23} - i\delta')} \\ + \text{c.c.} \end{array} \right\}. \end{aligned} \tag{8.43}$$

We have allowed for collisions to occur even though we assume that the atoms are stationary. In effect, instead of being applicable to a vapor, equation (8.43) more realistically models atoms in a solid, where the Γ_{ij}'s are *dephasing* rates produced by the host medium. In the absence of dephasing, there is a single resonance centered at $\delta' = -\delta$. Dephasing introduces a new resonance centered at $\delta' = 0$ via the term proportional to $2\Gamma_{12}/\gamma_2$ in equation (8.43)—in effect, the dephasing provides the energy mismatch $\hbar\delta$ between the incident field **E** frequency ω and the 1–2 transition frequency ω_{21} to allow for excitation of level 2. This term is also present for vapors in the large detuning limit, $|\delta| \gg \gamma_{12}$. The last term in equation (8.43) is small in the vicinity of the resonances in the limit that $|\delta| \gg \gamma_{12}$. In the limit that $|\delta| \gg \gamma_{12}$, the two-quantum and *dephasing-* or *collision-induced* resonances are resolved (see figure 8.3) [6].

For this relatively simple system, we have a pretty interesting result. Dephasing results in a *redistribution* of the probe absorption profile. The ratio of the amplitude of the dephasing-induced and two-quantum components is $(2\Gamma_{12}/\gamma_2)(\gamma_{13}/\gamma_{23})$, which grows with increasing dephasing.

8.3.2 Moving Atoms: Doppler Limit

In the Doppler limit,

$$|\delta| < ku, \quad \gamma_{12} \ll ku, \quad |\delta'| < k'u, \quad \gamma_{23} \ll k'u, \tag{8.44}$$

the various integrals can be evaluated using contour integration. The results depend critically on whether the fields are co- or counterpropagating. It is assumed that the velocity distribution is the Gaussian (4.41).

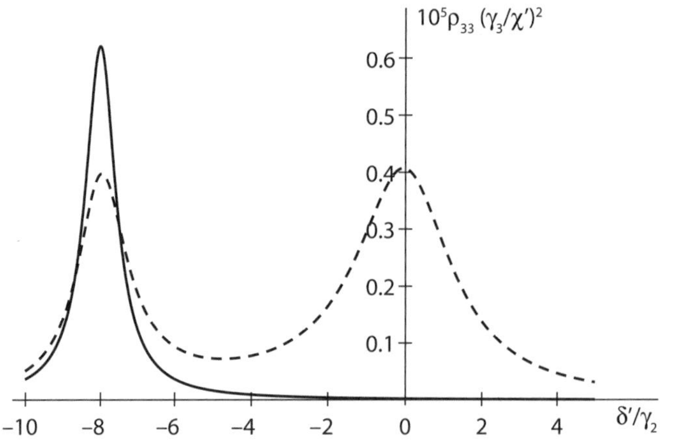

Figure 8.3. Graph of the probe absorption profile as a function of probe detuning δ'/γ_2. The quantity plotted is $10^5\varrho_{33}(\gamma_3/\chi')^2$, which is proportional to the probe absorption coefficient. In this and other figures in this chapter, all frequencies are taken in units of γ_2; in other words, we set $\gamma_2 = 1$. In this figure, $\gamma_3 = 1$, $\delta = 8$, and $\chi = 0.01$. The solid curve represents no dephasing, and the dashed curve represents dephasing with $\Gamma_{12} = 1$, $\Gamma_{23} = 0.5$, $\Gamma_{13} = 0.3$. There is an extra resonance in the presence of dephasing.

8.3.2.1 Stepwise contribution

The stepwise contribution, obtained using equation (8.33) is

$$\begin{aligned} I^{\mathrm{SW}} &= \frac{4|\chi\chi'|^2}{\gamma_2\gamma_3}\frac{1}{\sqrt{\pi}u}\int_{-\infty}^{\infty} dv_z e^{-v_z^2/u^2}\frac{\gamma_{12}}{\left[\gamma_{12}^2+(\delta+kv_z)^2\right]}\frac{\gamma_{23}}{\left[\gamma_{23}^2+(\delta'+\eta kv_z)^2\right]} \\ &\approx \frac{4\sqrt{\pi}|\chi\chi'|^2}{\gamma_2\gamma_3}\frac{e^{-\delta^2/k^2u^2}}{ku}\frac{\gamma_a}{\gamma_a^2+\delta_a^2}, \end{aligned} \tag{8.45}$$

where

$$\gamma_a = \gamma_{23} + \frac{k'}{k}\gamma_{12}, \tag{8.46a}$$

$$\delta_a = \delta' - \eta\frac{k'}{k}\delta. \tag{8.46b}$$

There is a simple interpretation of this result. The first field excites atoms having velocity $|kv_z + \delta| \lesssim \gamma_{12}$ or $|v_z + \delta/k| \lesssim \gamma_{12}/k$. The second field is resonant if $|\delta' + \eta k' v_z| \lesssim \gamma_{23}$. Substituting the value of v_z determined by the first field, we find the resonance condition

$$\left|\delta' - \eta\frac{k'}{k}\delta\right| \lesssim \gamma_{23} + \frac{k'}{k}\gamma_{12}, \tag{8.47}$$

in agreement with equations (8.45) and (8.46).

8.3.2.2 *Two-quantum contribution*

For co-propagating waves, $\eta = 1$, the integral of the TQ term, equation (8.39), does not contribute since all the poles are in the same half-plane. For counterpropagating waves, $\eta = -1$, there is a contribution. Using equation (8.39), we find

$$\begin{aligned} I^{\mathrm{TQ}} &= \frac{|\chi\chi'|^2}{\gamma_3}\frac{1}{\sqrt{\pi}u}\int_{-\infty}^{\infty} dv_z e^{-v_z^2/u^2}\frac{1}{\gamma_{12}+i(\delta+kv_z)}\frac{1}{\gamma_{23}+i(\delta'+\eta k' v_z)} \\ &\quad\times\frac{1}{\gamma_{13}+i[\delta+\delta'+(k+\eta k')\,v_z]}+\text{c.c.} \\ &= \frac{4\sqrt{\pi}|\chi\chi'|^2}{\gamma_3}\frac{e^{-\delta^2/k^2u^2}}{ku}\frac{\gamma_a\gamma_b-\delta_a\delta_b}{\left(\gamma_a^2+\delta_a^2\right)\left(\gamma_b^2+\delta_b^2\right)}\Theta(-\eta), \end{aligned} \tag{8.48}$$

where

$$\gamma_b = \left(\gamma_{13}+\frac{k-k'}{k'}\gamma_{23}\right)\Theta(k-k')+\left(\gamma_{13}+\frac{k'-k}{k}\gamma_{12}\right)\Theta(k'-k), \tag{8.49a}$$

$$\delta_b = \left(\delta+\frac{k}{k'}\delta'\right)\Theta(k-k')+\left(\delta'+\frac{k'}{k}\delta\right)\Theta(k'-k), \tag{8.49b}$$

and $\Theta(x)$ is the Heaviside step function defined by

$$\Theta(x) = \begin{cases} 1 & \text{for } x \geq 0 \\ 0 & \text{for } x < 0 \end{cases}. \tag{8.50}$$

By multiplying equation (8.48) by γ_2/γ_2 and carrying out some mathematical manipulations, we can recast equation (8.48) into the suggestive form

$$I^{\mathrm{TQ}} = \frac{4\sqrt{\pi}|\chi\chi'|^2}{\gamma_2\gamma_3}\frac{e^{-\delta^2/k^2u^2}}{ku}\Theta(-\eta)\left[\begin{array}{c} -\frac{\gamma_a}{\gamma_a^2+\delta_a^2}+\frac{\gamma_b}{\gamma_b^2+\delta_b^2} \\ +\frac{\gamma_a\gamma_b-\delta_a\delta_b}{\left(\gamma_a^2+\delta_a^2\right)\left(\gamma_b^2+\delta_b^2\right)}\Gamma_c \end{array}\right], \tag{8.51}$$

where

$$\Gamma_c = \Gamma_{13}-\Gamma_{12}-\Gamma_{23}. \tag{8.52}$$

8.3.2.3 *Total line shape*

Combining equations (8.45) and (8.51), we find

$$I = \frac{4\sqrt{\pi}|\chi\chi'|^2}{\gamma_2\gamma_3}\frac{e^{-\delta^2/k^2u^2}}{ku}\left[\begin{array}{c} \frac{\gamma_a}{\gamma_a^2+\delta_a^2}\Theta(\eta)+\frac{\gamma_b}{\gamma_b^2+\delta_b^2}\Theta(-\eta) \\ +\frac{\gamma_a\gamma_b-\delta_a\delta_b}{\left(\gamma_a^2+\delta_a^2\right)\left(\gamma_b^2+\delta_b^2\right)}\Gamma_c\Theta(-\eta) \end{array}\right]. \tag{8.53}$$

The third term is a collision-induced term that can contribute significantly only if

$$\left|\left(\frac{k'}{k}-\frac{k}{k'}\right)\delta\right| \lesssim \gamma_a, \gamma_b. \tag{8.54}$$

Aside from this collision-induced term, the line shape is a Lorentzian, centered at

$$\delta' - \eta\frac{k'}{k}\delta = 0, \tag{8.55}$$

having width (FWHM; in the absence of collisions)

$$2\gamma = \frac{|k + \eta k'|}{k}\gamma_2 + \gamma_3. \tag{8.56}$$

The probe absorption profile (8.53) is a Doppler-free line shape, arising from the fact that only the velocity subgroup of atoms resonant with the pump field contributes significantly to the probe absorption. For counterpropagating waves, the width of the resonance is diminished owing to Doppler phase cancellation of the two-quantum processes. This narrowing is related to two-photon Doppler-free spectroscopy (see section 8.4). For more details and the general result, including collision shifts and other collision effects, see reference [7].

8.4 Autler-Townes Splitting

With increasing pump or probe field intensities, the perturbation solution eventually fails. Using symbolic computer programs, one can obtain analytic solutions for $\tilde{\varrho}_{21}(\mathbf{v})$ and $\tilde{\varrho}_{32}(\mathbf{v})$. In general, one must integrate these expressions numerically over the velocity distribution and then insert the results into equations (8.25a) or (8.26a) to find the pump or probe field absorption. In this section, we consider stationary atoms only. The situation becomes fairly complicated for moving atoms, but there exist extensive discussions of this problem in the literature (see reference [8]). It is a fairly simple matter to use computer solutions for $\tilde{\varrho}_{32}$ and to plot the probe absorption coefficient

$$\alpha' = 2\,\mathrm{Im}\left[\frac{k\mu_{23}}{\varepsilon_0}\left\langle\frac{\tilde{\varrho}_{32}(Z)}{E'(Z)}\right\rangle\right] \tag{8.57}$$

as a function of δ'. One could equally well graph ϱ_{33} as a function of δ', since α' is proportional to ϱ_{33}. One such plot is shown in figure 8.4 and illustrates what is known as the *Autler-Townes splitting*, originally observed on microwave molecular transitions [9].

To calculate the Autler-Townes splitting, we assume that the probe field is weak but that the pump field has arbitrary intensity. To calculate $\tilde{\varrho}_{32}$, it is easiest to take the appropriate limit of a computer-generated solution, but we offer the more traditional approach. We need to calculate $\tilde{\varrho}_{32}$ to first order in χ'. First, we calculate $\varrho_{11}^{(0)}, \varrho_{22}^{(0)}, \tilde{\varrho}_{12}^{(0)}$, and $\tilde{\varrho}_{21}^{(0)}$ to zeroth order in χ'. These are simply the steady-state values of the two-level problem with one field that we have calculated previously and are given in equations (4.11).

To calculate $\tilde{\varrho}_{32}$, we set $\tilde{\varrho}_{32} = i\chi' R_{32}$ and $\tilde{\varrho}_{31} = i\chi' R_{31}$, substitute these expressions in equations (8.20), keep only terms linear in χ', and equate coefficients of χ' to obtain the following coupled equations for the steady-state values of R_{32} and R_{31}:

$$\begin{aligned}\left[\gamma_{23} + i\delta'\right] R_{32} &= -\varrho_{22}^{(0)} + i\chi^* R_{31},\\ \left[\gamma_{13} + i\left(\delta + \delta'\right)\right] R_{31} &= -\tilde{\varrho}_{21}^{(0)} + i\chi R_{32}.\end{aligned} \tag{8.58}$$

Note that the pump field Rabi frequency enters into these equations; in some sense, the presence of the Rabi frequency reflects processes of the type shown in the

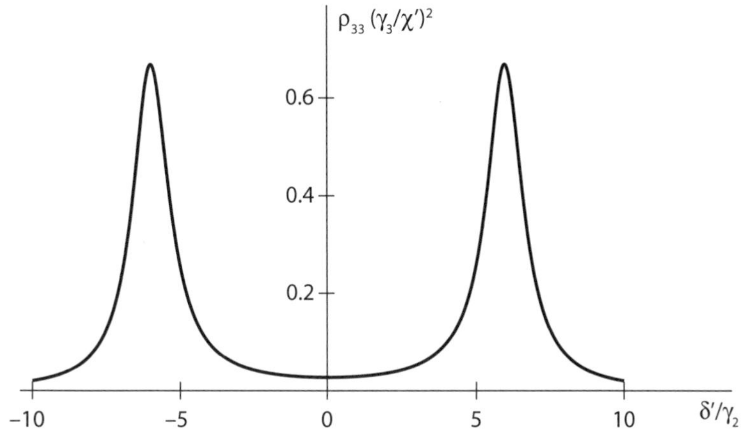

Figure 8.4. Autler-Townes splitting: probe absorption profile for stationary atoms with no dephasing and $\gamma_3 = 1, \delta = 0$, and $\chi = 6$ (all in units of γ_2). The quantity plotted is $\varrho_{33}(\gamma_3/\chi')^2$, which is proportional to the probe absorption coefficient.

two-quantum contribution of figure 8.2, in which the probe field acts before some of the pump field interactions. We can solve equations (8.58) to obtain

$$\tilde{\varrho}_{32} = i\chi' R_{32} = -\frac{i\chi' \left\{ [\gamma_{13} + i(\delta + \delta')]\, \varrho_{22}^{(0)} + i\chi^* \tilde{\varrho}_{21}(0) \right\}}{(\gamma_{23} + i\delta')\,[\gamma_{13} + i(\delta + \delta')] + |\chi|^2}. \tag{8.59}$$

When $\Omega = (\delta^2 + 4|\chi|^2)^{1/2} \gg \gamma_{23}, \gamma_{13}$, there are *two* resolved resonances that are centered at

$$\delta' = -\frac{1}{2}(\delta \pm \sqrt{\delta^2 + 4|\chi|^2}) = -\frac{1}{2}(\delta \pm \Omega). \tag{8.60}$$

The 2Ω separation of these resonances is referred to as the Autler-Townes splitting.

If the pump field is on resonance ($\delta = 0$) and if $|\chi| \gg \gamma_{12}$, then $\varrho_{22}^{(0)} \sim 1/2$, $i\chi^* \tilde{\varrho}_{21}(0) \sim \gamma_2/4$, and

$$\tilde{\varrho}_{32} = -\frac{1}{2} \frac{i\chi' (\gamma_{13} + \gamma_2/2 + i\delta')}{(\gamma_{23} + i\delta')(\gamma_{13} + i\delta') + |\chi|^2}. \tag{8.61}$$

If $|\chi| \gg \gamma_{23}, \gamma_{13}$, the resonances occur symmetrically, centered at $\delta' = \pm|\chi| = \pm|\Omega_0|/2$ having width $(\gamma_{13} + \gamma_{23})$. The splitting between the peaks of the resonances can be used as a measure of the pump field strength.

The Autler-Townes splitting has a simple interpretation in terms of the semiclassical dressed states. In fact, one can obtain the positions, widths, and relative amplitudes of the resonances directly using the dressed states, provided that the secular approximation is valid. The semiclassical dressed-state approach to this problem is an ideal method for elucidating the physics. We neglect collisions, which complicates matters somewhat, and take the Rabi frequencies to be real. The "bare" and dressed states are shown in figure 8.5. The strong pump field dresses only states $|\tilde{1}(t)\rangle$ and $|\tilde{2}(t)\rangle$.

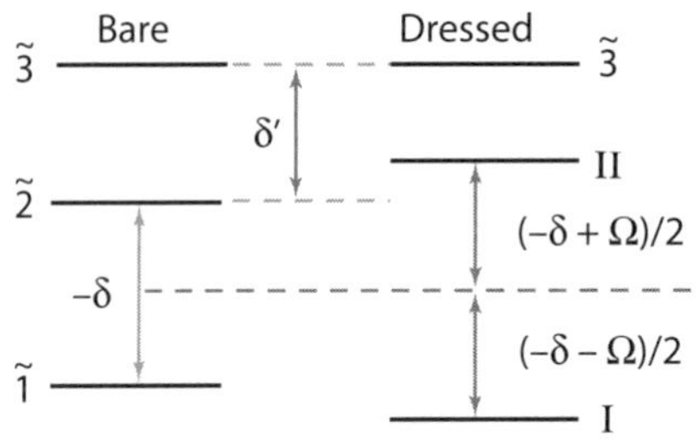

Figure 8.5. Levels for the cascade configuration in the "bare" (field-interaction) and dressed bases.

One can use figure 8.5 to obtain the line profile by inspection; formal justification for this procedure is given later in this section. In the secular approximation, $\Omega \gg \gamma_{12}$, we can neglect the coherence $\varrho_{I,II}$. As such, the probe absorption profile can be considered as being composed of the sum of two *independent* profiles, $\varrho_{33}(I)$ and $\varrho_{33}(II)$, respectively, that depend only on the dressed-state populations ($\varrho_{I,I}$ and $\varrho_{II,II}$), the square of the transition amplitudes from each dressed state to state 3 (proportional to $|\chi'_{3,I}|^2$ and $|\chi'_{3,II}|^2$, respectively), and the decay rates (γ_I and γ_{II}) of the dressed states. In terms of these quantities, the individual line profiles are the Lorentzians

$$\varrho_{33}(I) = \frac{|\chi'_{3,I}|^2}{\gamma_3} \frac{2\gamma_{I3}}{\gamma_{I3}^2 + \left[\delta' + \frac{1}{2}(\delta + \Omega)\right]^2} \varrho_{I,I}, \tag{8.62a}$$

$$\varrho_{33}(II) = \frac{|\chi'_{3,II}|^2}{\gamma_3} \frac{2\gamma_{II3}}{\gamma_{II3}^2 + \left[\delta' + \frac{1}{2}(\delta - \Omega)\right]^2} \varrho_{II,II}. \tag{8.62b}$$

The various quantities appearing in these equations are determined as follows. The dressed-state populations are given by the solutions of equations (3.63) with $\Gamma = 0$, namely,

$$\varrho_{I,I} = \frac{\cos^4\theta}{\sin^4\theta + \cos^4\theta}, \qquad \varrho_{II,II} = \frac{\sin^4\theta}{\sin^4\theta + \cos^4\theta}, \tag{8.63}$$

where $\tan 2\theta = |\Omega_0|/\delta$ is the dressing angle. Since

$$|I\rangle = \cos\theta|\tilde{1}\rangle - \sin\theta|\tilde{2}\rangle, \tag{8.64}$$

$$|II\rangle = \sin\theta|\tilde{1}\rangle + \cos\theta|\tilde{2}\rangle, \tag{8.65}$$

it follows that the Rabi frequencies are given by

$$\chi'_{3,I} = -\frac{\hat{\epsilon}' \cdot E'}{2\hbar}\langle\tilde{3}|\hat{\mu}(\cos\theta|\tilde{1}\rangle - \sin\theta|\tilde{2}\rangle) = -\sin\theta\chi', \tag{8.66a}$$

$$\chi'_{3,II} = -\frac{\hat{\epsilon}' \cdot E'}{2\hbar}\langle\tilde{3}|\hat{\mu}(\sin\theta|\tilde{1}\rangle + \cos\theta|\tilde{2}\rangle) = \cos\theta\chi'. \tag{8.66b}$$

The decay rates of states $|I\rangle$ and $|II\rangle$ are proportional to the absolute square of the matrix element of the dipole moment operator between states $|I\rangle$ or $|II\rangle$ and state $|\tilde{1}\rangle$, respectively. As such, $\gamma_I = \sin^2\theta\gamma_2$, $\gamma_{II} = \cos^2\theta\gamma_2$, and

$$\gamma_{I3} = \frac{\gamma_I + \gamma_3}{2} = \frac{\gamma_2 \sin^2\theta + \gamma_3}{2}, \tag{8.67a}$$

$$\gamma_{II3} = \frac{\gamma_{II} + \gamma_3}{2} = \frac{\gamma_2 \cos^2\theta + \gamma_3}{2}. \tag{8.67b}$$

When these values are substituted into equation (8.62), we find

$$\varrho_{33}(I) = 2\frac{|\chi'|^2}{\gamma_3}\frac{\gamma_{I3}}{\gamma_{I3}^2 + \left[\delta' + \frac{1}{2}(\delta+\Omega)\right]^2}\frac{\sin^2\theta\cos^4\theta}{\sin^4\theta + \cos^4\theta}, \tag{8.68a}$$

$$\varrho_{33}(II) = 2\frac{|\chi'|^2}{\gamma_3}\frac{\gamma_{II3}}{\gamma_{II3}^2 + \left[\delta' + \frac{1}{2}(\delta-\Omega)\right]^2}\frac{\sin^4\theta\cos^2\theta}{\sin^4\theta + \cos^4\theta}, \tag{8.68b}$$

from which it follows that the relative amplitude of the two resonances is

$$\frac{\varrho_{33}(II)}{\varrho_{33}(I)} = \frac{\cos^2\theta}{\sin^2\theta}\left(\frac{\gamma_2\sin^2\theta + \gamma_3}{\gamma_2\cos^2\theta + \gamma_3}\right). \tag{8.69}$$

For zero pump field detuning, $\theta = \pi/4$, $\varrho_{33}(II)/\varrho_{33}(I) = 1$, and the FWHM of each resonance is $[(\gamma_2/2) + \gamma_3]$. The reason that the width has a factor of $\gamma_2/2$ is that only half of each dressed state contributes to the decay. Note that the dressed-state approach has greatly simplified the problem. In the secular approximation, the absorption profile can be written by inspection!

This approach can be justified formally. To do so, we transform the Hamiltonian (8.6) into the dressed-state basis using

$$\tilde{\mathbf{H}}_D = \mathbf{T}\tilde{\mathbf{H}}_c\mathbf{T}^\dagger = \hbar\begin{pmatrix} -\frac{1}{2}(\delta+\Omega) & 0 & \chi'_{3,I} \\ 0 & \frac{1}{2}(\Omega-\delta) & \chi'_{3,II} \\ \chi'_{3,I} & \chi'_{3,II} & \delta' \end{pmatrix}, \tag{8.70}$$

where

$$\mathbf{T} = \begin{pmatrix} \cos\theta & -\sin\theta & 0 \\ \sin\theta & \cos\theta & 0 \\ 0 & 0 & 1 \end{pmatrix}, \tag{8.71}$$

we have set $\mathbf{R} = 0$ and used equation (8.66). Including decay (see problem 14), the relevant density matrix equations for the Hamiltonian (8.70) in the secular approximation and to first order in χ' are

$$\varrho_{33} = \frac{i\chi_{3,II}\left(\tilde{\varrho}_{3,II} - \tilde{\varrho}_{II,3}\right)}{\gamma_3} + \frac{i\chi'_{3,I}\left(\tilde{\varrho}_{3,I} - \tilde{\varrho}_{I,3}\right)}{\gamma_3}, \tag{8.72a}$$

$$\frac{\partial\tilde{\varrho}_{I,3}}{\partial t} = i\chi'_{3,I}\varrho_{I,I} - \left\{\gamma_{I3} - i\left[\delta' + \frac{1}{2}(\delta+\Omega)\right]\right\}\tilde{\varrho}_{I,3}, \tag{8.72b}$$

$$\frac{\partial\tilde{\varrho}_{II,3}}{\partial t} = i\chi'_{3,II}\varrho_{II,II} - \left\{\gamma_{II3} - i\left[\delta' + \frac{1}{2}(\delta-\Omega)\right]\right\}\tilde{\varrho}_{II,3}. \tag{8.72c}$$

Solving in steady state using the values (8.63), we obtain the solution (8.68).

8.5 Two-Photon Spectroscopy

When the pump field is far off resonance, $|\delta| \gg \gamma_{12}, ku$, it is still possible to resonantly drive transitions between levels 1 and 3 in the cascade and Λ configurations. (It is also possible in the V configuration if one starts in levels 1 or 3 and decay is negligible.) In this limit, two-quantum processes are dominant, and one speaks of *two-photon* transitions. For the sake of definiteness, we consider the cascade configuration. If $|\delta| \gg \gamma_{12}, ku$, but $\delta + \delta' \approx 0$, it is possible to adiabatically eliminate state $|2\rangle$ from the calculation and get an effective two-level problem involving levels 1 and 3. Although this limit can be derived from our three-level equations, it is instructive to rederive the results so that they can be generalized to very large detunings from a *number* of intermediate states.

Thus, we consider two fields driving a transition between states 1 and 3 having the same parity (see figure 8.6). There are other states m far off-resonance with the incident fields that have parity opposite to states 1 and 3. The states m need not lie between levels 1 and 3 and can even include continuum states. We change the notation for the fields slightly and take

$$\mathbf{E}(\mathbf{R}, t) = \frac{1}{2}\hat{\boldsymbol{\epsilon}}_1 \left[\mathcal{E}_1(\mathbf{R}, t)e^{i(\mathbf{k}_1\cdot\mathbf{R}-\omega_1 t)} + \text{c.c.}\right], \tag{8.73a}$$

$$\mathbf{E}'(\mathbf{R}, t) = \frac{1}{2}\hat{\boldsymbol{\epsilon}}_2 \left[\mathcal{E}_2(\mathbf{R}, t)e^{i(\mathbf{k}_2\cdot\mathbf{R}-\omega_2 t)} + \text{c.c.}\right], \tag{8.73b}$$

where $\omega_1 = \omega$ and $\omega_2 = \omega'$.

In the interaction representation, *but without making the RWA and allowing each field to drive each transition*, the equations for the probability amplitudes for an atom at the origin, in the absence of relaxation, are

$$\dot{c}_1 = -i\sum_{j=1}^{2}\sum_m \left[\chi_{1m}^{(j)}e^{-i(\omega_j-\omega_{1m})t} + \chi_{1m}^{(j)*}e^{i(\omega_j+\omega_{1m})t}\right]c_m, \tag{8.74a}$$

$$\dot{c}_3 = -i\sum_{j=1}^{2}\sum_m \left[\chi_{3m}^{(j)}e^{-i(\omega_j-\omega_{3m})t} + \chi_{3m}^{(j)*}e^{i(\omega_j+\omega_{3m})t}\right]c_m, \tag{8.74b}$$

$$\begin{aligned}\dot{c}_m = -i\sum_{j=1}^{2}\Big\{&\left[\chi_{m1}^{(j)}e^{-i(\omega_j-\omega_{m1})t} + \chi_{m1}^{(j)*}e^{i(\omega_j+\omega_{m1})t}\right]c_1 \\ &+ \left[\chi_{m3}^{(j)}e^{-i(\omega_j-\omega_{m3})t} + \chi_{m3}^{(j)*}e^{i(\omega_j+\omega_{m3})t}\right]c_3\Big\},\end{aligned} \tag{8.74c}$$

where

$$\chi_{mn}^{(j)} = -\frac{1}{2\hbar}\langle m|\hat{\boldsymbol{\mu}}\cdot\hat{\boldsymbol{\epsilon}}_j|n\rangle\mathcal{E}_j, \tag{8.75}$$

$\omega_{mn} = (E_m - E_n)/\hbar$, and E_m is the energy of state m. It is assumed that $\langle m|\hat{\boldsymbol{\mu}}\cdot\hat{\boldsymbol{\epsilon}}_j|n\rangle$ is real. The reason we do not make the RWA and that we allow each field to drive *both* transitions is that the frequency detunings may be sufficiently large for both transitions to invalidate the assumption that each field drives a single transition.

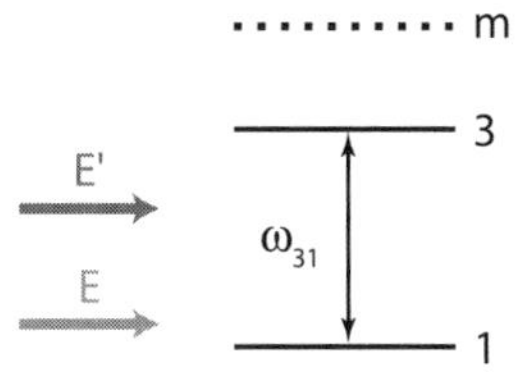

Figure 8.6. Two-photon transitions can proceed from state 1 to state 3 via intermediate states m.

If the fields are turned on slowly compared with detunings to states m, we can integrate equation (8.74c) by parts and keep the lead term to obtain

$$c_m \approx -\sum_{j=1}^{2}\left\{\left[-\frac{\chi_{m1}^{(j)}e^{-i(\omega_j-\omega_{m1})t}}{\omega_j-\omega_{m1}}+\frac{\chi_{m1}^{(j)*}e^{i(\omega_j+\omega_{m1})t}}{\omega_j+\omega_{m1}}\right]c_1\right.$$

$$\left.+\left[-\frac{\chi_{m3}^{(j)}e^{-i(\omega_j-\omega_{m3})t}}{\omega_j-\omega_{m3}}+\frac{\chi_{m3}^{(j)*}e^{i(\omega_j+\omega_{m3})t}}{\omega_j+\omega_{m3}}\right]c_3\right\}. \qquad (8.76)$$

Substituting equation (8.76) into equations (8.74a) and (8.74b), and keeping only slowly varying terms, those that are constant or vary as $e^{\pm i(\omega+\omega'-\omega_{31})t}$, we obtain

$$\dot{c}_1 = -is_1c_1 - i\chi_{tq}^{*}e^{-i(\delta+\delta')t}c_3, \qquad (8.77a)$$

$$\dot{c}_3 = -is_3c_3 - i\chi_{tq}e^{i(\delta+\delta')t}c_1, \qquad (8.77b)$$

where

$$s_1 = -\sum_{j=1}^{2}\sum_m\left[\frac{|\chi_{1m}^{(j)}|^2}{\omega_j-\omega_{m1}}-\frac{|\chi_{1m}^{(j)}|^2}{\omega_j+\omega_{m1}}\right] = -2\sum_{j=1}^{2}\sum_m\frac{|\chi_{1m}^{(j)}|^2\omega_{m1}}{\omega_j^2-\omega_{m1}^2}, \qquad (8.78a)$$

$$s_3 = -\sum_{j=1}^{2}\sum_m\left[\frac{|\chi_{3m}^{(j)}|^2}{\omega_j-\omega_{m3}}-\frac{|\chi_{3m}^{(j)}|^2}{\omega_j+\omega_{m3}}\right] = -2\sum_{j=1}^{2}\sum_m\frac{|\chi_{3m}^{(j)}|^2\omega_{m3}}{\omega_j^2-\omega_{m3}^2}, \qquad (8.78b)$$

$$\chi_{tq} = -\sum_m\left[\frac{\chi_{3m}^{(2)}\chi_{m1}^{(1)}}{\omega_1-\omega_{m1}}+\frac{\chi_{3m}^{(1)}\chi_{m1}^{(2)}}{\omega_2-\omega_{m1}}\right], \qquad (8.78c)$$

and $(\delta+\delta') = \omega_{31}-(\omega+\omega')$. The fields produce a *light shift* in each of the levels represented by the s_1 and s_3 terms, as well as two-quantum coupling represented by the χ_{tq} terms. The problem has been reduced to an effective two-level problem plus the light shifts. Therefore, the equations are the same as in the two-level case, and decay terms can be added into the density matrix equations. In this manner, we obtain

$$\frac{\partial \varrho_{11}}{\partial t} + \mathbf{v} \cdot \nabla \varrho_{11} = -i\chi_{tq}^{*}\tilde{\varrho}_{31} + i\chi_{tq}\tilde{\varrho}_{13}, \tag{8.79a}$$

$$\frac{\partial \varrho_{33}}{\partial t} + \mathbf{v} \cdot \nabla \varrho_{33} = i\chi_{tq}^{*}\tilde{\varrho}_{31} - i\chi_{tq}\tilde{\varrho}_{13} - \gamma_3\varrho_{33}, \tag{8.79b}$$

$$\frac{\partial \tilde{\varrho}_{13}}{\partial t} + \mathbf{v} \cdot \nabla \tilde{\varrho}_{13} = -i\chi_{tq}^{*}(\varrho_{33} - \varrho_{11}) - \gamma_{13}\tilde{\varrho}_{13}$$

$$+ i\left[\delta + \delta' + (\mathbf{k} + \mathbf{k}') \cdot \mathbf{v} - s_1 + s_3\right]\tilde{\varrho}_{13}, \tag{8.79c}$$

$$\tilde{\varrho}_{31} = \left(\tilde{\varrho}_{13}\right)^{*}. \tag{8.79d}$$

An *in term* must also be included in the equation for $\partial\varrho_{11}/\partial t$ to allow for repopulation of this state, but it generally involves a cascaded emission through intermediate states not shown in figure 8.6.

The two-photon absorption line shape has a Doppler width determined by $|\mathbf{k}+\mathbf{k}'|u$. If $\mathbf{k} = -\mathbf{k}'$, that is, for equal frequencies and oppositely traveling waves, the Doppler effect is totally suppressed, and the resonance width is $2\gamma_{13}$. The resultant line shape is referred to as a *two-photon Doppler-free profile* [10]. In contrast to saturation spectroscopy, the entire velocity ensemble contributes to the line shape. The price you pay is that the transition strength is small, since the intermediate state is far-detuned.[1]

For cw lasers, if $\chi/2\pi \approx 10^8$ Hz, the light shifts and two-photon coupling strength is of order 10 Hz if $(\omega_j - \omega_{m1})/2\pi \approx 10^{15}$ Hz, but could be as large as 1.0 MHz if $(\omega_j - \omega_{m1})/2\pi \approx 10$ GHz (nearly resonant intermediate state).

Some caution must be taken if equations (8.79) are used for a Λ rather than cascade system. The problem is that γ_{13} is equal to zero for a Λ scheme in which levels 1 and 3 are ground-state sublevels. In that case, it is necessary to include higher order terms in the $\partial\tilde{\varrho}_{13}/\partial t$ equation that correspond to off-resonant scattering of the incident fields. Such terms are included in chapter 9.

8.6 Open versus Closed Quantum Systems

It is sometimes necessary to consider *open quantum systems*. For example, in our three-level problem in the Λ configuration, imagine that only field $\mathbf{E}$ is present. Eventually, all the population decays to state 3. If we take as our quantum system levels 1 and 2 only, there is leakage out of the system, and population is no longer conserved for the two-level subsystem. In this case, if the vapor is not replenished with new atoms in either states 1 or 2, the only steady state for the system is one in which all density matrix elements are equal to zero. Atoms in gas lasers

[1] It was assumed implicitly in this section that the frequency of each field *separately* was not approximately equal to $\omega_{31}/2$ so that a single field could not drive two-photon transitions on its own. This is no longer the case if we take $\omega = \omega' \approx \omega_{31}/2$, as in the case of two-photon Doppler-free spectroscopy; however, there is a large Doppler width $\sim 2ku$ associated with the two-photon transitions from each field separately so that the relative contribution of these terms is small.

provide a good example of an open quantum system. The two atomic levels between which lasing takes place are excited states that must be pumped incoherently via an electric discharge or collisions with metastable atoms. To achieve a steady state in open quantum systems, there must be some *incoherent* pumping of the levels. Just as collisions led to new resonances in nonlinear spectroscopy, relaxation rates in open quantum systems can produce similar effects. For example, in the cascade configuration for stationary atoms, if level 1 has some associated decay rate, there are always resonances centered at $\delta' = -\delta$ and $\delta' = 0$, even in the absence of collisions.

There is another new feature that arises in open quantum systems. For a *closed* two-level system, ϱ_{11} can be eliminated using the fact that $\varrho_{11} = 1 - \varrho_{22}$. In the remaining equations, only the decay rates γ_2 and γ_{12} appear. As a consequence, for relaxation effects to be discernible, they must occur on a timescale of order γ_2^{-1}. On the other hand, in open systems, there is a new timescale, γ_1^{-1}, that enters. If $\gamma_1^{-1} \gg \gamma_2^{-1}$, as is often the case, the sensitivity to relaxation effects, such as ground-state velocity-changing collisions, can be increased significantly. Examples are given in chapter 10.

A *qualitatively different* type of open quantum system results from atoms leaving the interaction volume as a consequence of their motion. This loss rate is *state-independent*; as such, it provides an overall increase in line widths known as *transit-time broadening*. Transit-time broadening does not lead to *additional* resonances in nonlinear spectroscopic line shapes.

Open quantum systems can be modeled easily by going over to ensemble density matrix elements $\mathcal{R}_{ij}(\mathbf{R}, \mathbf{v}, t)$ and adding incoherent pumping terms $\Lambda_i(\mathbf{R}, \mathbf{v})$ and additional loss terms $-\Gamma_i \mathcal{R}_{ii}(\mathbf{R}, \mathbf{v}, t)$ to the population evolution equations for $\dot{\mathcal{R}}_{ii}$.

8.7 Summary

We have introduced some of the basic concepts involved in the interaction of three-level atoms with two fields. The fields drive transitions that share a common level. We have seen how this scheme can be used to eliminate the Doppler width in spectral profiles. Moreover, we have seen that there are two, qualitatively different processes that enter, step-wise and two-quantum. In the next chapter, we continue our study of three-level systems, but concentrate on the Λ configuration. We will see how the Λ configuration differs from the other configurations in one important aspect, and how this characteristic can be used to generate "slow light."

Problems

1. Derive equations (8.33) and (8.39).
2. Derive equation (8.40).
3. Derive equation (8.43).
4. The step-wise and two-quantum contributions to probe absorption in perturbation theory are given in equations (8.45) and (8.48). Assume that $\mathbf{k} = k\hat{\mathbf{z}}$ and $\mathbf{k} = \eta k'\hat{\mathbf{z}}$, with $\eta = \pm 1$. Show that, without taking the Doppler limit, the

velocity integrated step-wise contribution can be written in terms of

$$I_1(\mu_1, \mu_2, \eta_2/\eta_1) = \frac{1}{\sqrt{\pi}u}\int_{-\infty}^{\infty} dv_z e^{-v_z^2/u^2} \frac{1}{(\mu_1 + \eta_1 v_z/u)(\mu_2 + \eta_2 v_z/u)}$$

and the velocity integrated two-quantum contribution in terms of

$$I_2(\mu_1, \mu_2, \mu_3, \eta_2/\eta_1, \eta_3/\eta_1) = \frac{1}{\sqrt{\pi}u}\int_{-\infty}^{\infty} dv_z e^{-v_z^2/u^2}$$

$$\times \frac{1}{(\mu_1 + \eta_1 v_z/u)(\mu_2 + \eta_2 v_z/u)(\mu_3 + \eta_3 v_z/u)}.$$

5. If the integrals in problem 4 are written such that $\text{Im}\,\mu_i > 0$, then prove that

$$I_1(\mu_1, \mu_2, \eta) = -M(\mu_1, \mu_2, \eta)\,[Z(\mu_1) - \eta Z(\mu_2)],$$

$$I_2(\mu_1, \mu_2, \mu_3, \eta, \eta') = M(\mu_1, \mu_2, \eta)\left[I_1(\mu_1, \mu_3, \eta') - \eta I_1(\mu_1, \mu_2, \eta'/\eta)\right],$$

where

$$M(\mu_1, \mu_2, \eta) = \frac{1}{\mu_2 - \eta\mu_1},$$

and

$$Z(\mu) = i\sqrt{\pi}\,[w(\mu)] = -\frac{1}{\sqrt{\pi}u}\int_{-\infty}^{\infty} dv_z e^{-v_z^2/u^2} \frac{1}{(\mu \pm v_z/u)},$$

with $\text{Im}\,\mu > 0$.

6. Use a symbolic program to obtain steady-state solutions to equations (8.20) for both the cascade and Λ configurations for stationary atoms in the absence of collisions. Plot the probe absorption, $\text{Im}\,\tilde{\varrho}_{23}/\chi'$ as a function of δ' for $\delta = 0$, $\gamma_3 = 1$, $\chi = 4$, $\chi' = 0.1$, and $\delta = 0$, $\gamma_3 = 1$, $\chi = 0.3$, $\chi' = 0.1$, with all frequencies in terms of γ_2, which is arbitrarily set equal to 1. Show that in the case of the Λ configuration, the probe absorption vanishes identically at $\delta' = 0$. This is a general result for the Λ configuration—the probe absorption vanishes in steady state whenever $\delta' = \delta$.
7. Use a symbolic program to obtain a steady-state solution to equations (8.20) for the cascade configuration for stationary atoms. Expand the expression for $\tilde{\varrho}_{32}$ to first order in χ', and show that it agrees with equation (8.59).
8. Consider the case of two-photon Doppler-free spectroscopy and calculate the steady-state excited-state probability to lowest order in the applied fields, including terms that involve absorption of two photons from each field separately. Take $\delta = \delta'$ and $\mathbf{k} = -\mathbf{k}' = -k\hat{\mathbf{z}}$. In the Doppler limit, show that the contribution from these terms is much smaller than the Doppler-free terms if the fields are tuned to resonance, but that the two-photon absorption from the individual fields provides a broad pedestal to the absorption signal.
9. Compare the "exact" solution $\varrho_{33}/|\chi'|^2 = -2\chi'^*\text{Im}(\tilde{\varrho}_{32})/|\chi'|^2$ with $\tilde{\varrho}_{32}$ given by equation (8.59), with the dressed-state solution

$$\varrho_{33}/|\chi'|^2 = [\varrho_{33}(I) + \varrho_{33}(II)]\,/|\chi'|^2,$$

with $\varrho_{33}(I)$ and $\varrho_{33}(II)$ given by equations (8.68) for $\delta = 0$, $\gamma_2 = \gamma_3 = 1$, $\chi = 4$ and for $\delta = 6$, $\gamma_2 = \gamma_3 = 1$, $\chi = 1$. Plot the exact and dressed-state solutions in these cases. Neglect collisions.

10. Consider the V and Λ three-level schemes. In the V scheme, assume that all the population is in state 2 (the "point" of the V) in the absence of the fields, and that both states 1 and 3 decay to state 2. In the Λ scheme, assume that states 1 and 3 each have some equilibrium population in the absence of the applied fields, and that state 2 decays to both states 1 and 3 ($\gamma_{2,1}+\gamma_{2,3}=\gamma_2$). Draw perturbation chains similar to those in figure 8.2 that lead to a contribution to probe absorption $\tilde{\varrho}_{32}$ that is of order $|\chi|^2\chi'$ for the V scheme and $|\chi|^2\chi'^*$ for the Λ scheme. No calculations are required in this problem—just draw the perturbation chains, but be sure to include contributions resulting from spontaneous decay. Suppose that the field χ is detuned by a large amount ($\delta \gg \gamma, ku$) from resonance. Which of the perturbation chains will dominate? Why? In this case, the nonlinear contribution to the probe absorption can be calculated as a dispersion curve centered at $\delta' = 0$. How do you interpret this result?
11. Neglecting collisions, show that, to within a factor k/k', the term in brackets in equation (8.53) can be written as

$$\frac{\bar{\gamma}}{\bar{\gamma}^2+\bar{\delta}^2},$$

where

$$\bar{\gamma} = \left|\frac{k'+\eta k}{k}\right|\frac{\gamma_2}{2}+\frac{\gamma_3}{2},$$

$$\bar{\delta} = \delta' - \eta\frac{k'}{k}\delta.$$

Why is the width less for counterpropagating than for copropagating waves?
12. Consider the case of counterpropagating fields and moving atoms in the cascade configuration. Take the limit that $\delta \gg ku$, but that all the decay rates are less than ku. Show that the probe absorption consists of a single Gaussian in the absence of collisions and two Gaussians when collisions are present. Interpret your results.
13. Given that the relaxation of density matrix elements is $\dot{\varrho}_{11}=0$, $d\tilde{\varrho}_{23}/dt = -\left[(\gamma_2+\gamma_3)/2\right]\tilde{\varrho}_{23}$, $d\tilde{\varrho}_{13}/dt = -(\gamma_3/2)\,\tilde{\varrho}_{13}$, obtain the corresponding equations for the dressed-state density matrix elements $\dot{\varrho}_{I,3}$ and $\dot{\varrho}_{II,3}$, and show that the result differs from that used in equation (8.72) since $\dot{\varrho}_{I,3}$ is coupled to $\varrho_{II,3}$ and $\dot{\varrho}_{II,3}$ is coupled to $\varrho_{I,3}$. Why were these terms ignored?
14. To see the role that an open system can play, reconsider pump–probe spectroscopy on a stationary, two-level atom, but add terms $-\gamma_1\varrho_{11}+\Lambda_1$ to the equation for $\dot{\varrho}_{11}$. To zeroth order in the applied fields, the solution for ϱ_{11} is $\varrho_{11}^{(0)} = \Lambda_1/\gamma_1$. Calculate the population difference $\varrho_{22}-\varrho_{11}$ to second order in the applied fields, and show that there is a new, dispersion-like resonance having width $2\gamma_1$ centered at $\omega' = \omega$, even in the absence of dephasing, $\Gamma_{12}=0$. Is this a "subnatural" resonance? That is, does this resonance allow you to determine the transition frequency in the atom with a resolution less than γ_{12}? Explain.

References

[1] M. S. Feld and A. Javan, *Laser-induced line narrowing effects in coupled Doppler-broadened transitions*, Physical Review **177**, 540–562 (1969).

[2] Th. Hänsch and P. Toschek, *Theory of a three-level gas amplifier*, Zeitschrift für Physik **236**, 213–244 (1970).

[3] P. R. Berman and R. Salomaa, *Comparison between dressed-atom and bare-atom pictures in laser spectroscopy*, Physical Review A **25**, 2667–2692 (1982).

[4] P. R. Berman and R. C. O'Connell, *Constraints on dephasing widths and shifts in three-level quantum systems*, Physical Review A **71**, 022501, 1–5 (2005).

[5] P. R. Berman, *Study of collisions by laser spectroscopy*, in *Advances in Atomic, Molecular and Optical Physics*, edited by D. R. Bates and B. Bederson, vol. 13 (Academic Press, New York, 1977), pp. 57–112.

[6] P. F. Liao, J. E. Bjorkholm, and P. R. Berman, *Study of collisional redistribution using two-photon absorption with a nearly resonant intermediate state*, Physical Review A **20**, 1489–1494 (1979).

[7] P. R. Berman, *Effects of collisions on linear and nonlinear spectroscopic line shapes*, Physics Reports **43**, 101–149 (1978).

[8] R. Salomaa and S. Stenholm, *Two-photon spectroscopy: effects of a resonant intermediate state*, Journal of Physics B **8**, 1795–1805 (1975); R. Salomaa and S. Stenholm, *Two-photon spectroscopy II: effects of residual Doppler broadening*, Journal of Physics B **9**, 1221–1235 (1976); R. Salomaa and S. Stenholm, *Two-photon spectroscopy III: general strong-field aspects*, Journal of Physics B **10**, 3005–3021 (1977); R. Salomaa, *Occurrence of split spectra in Doppler-free two-photon spectroscopy*, Physica Scripta **15**, 251–258 (1977).

[9] S. H. Autler and C. H. Townes, *Stark effect in rapidly varying fields*, Physical Review **100**, 703–722 (1955).

[10] B. Cagnac, G. Grynberg, and F. Biraben, *Spectroscopie d'absorption multiphonique sans effet Doppler*, Journal de Physique **34**, 845–858 (1973); F. Biraben, B. Cagnac, and G. Grynberg, *Experimental evidence of two-photon transition without Doppler broadening*, Physical Review Letters **32**, 643–645 (1974); T. W. Hänsch, S. A. Lee, R. Wallenstein, and C. Wieman, *Doppler-free two photon spectroscopy of hydrogen 1S-2S*, Physical Review Letters **34**, 307–309 (1975).

Bibliography

Three-level saturation spectroscopy is discussed in the books on laser spectroscopy listed in chapter 1. In addition, the following is a representative list of some early articles in this field:

I. M. Beterov, Yu. A. Matyugin, and V. P. Chebotaev, *Spectroscopy of two-quantum transitions in a gas near resonances*, Soviet Physics–JETP **37**, 756–763 (1973).

M. S. Feld and A. Javan, *Laser-induced line narrowing effects in coupled Doppler-broadened transitions*, Physical Review **177**, 540–562 (1969).

Th. Hänsch and P. Toschek, *Theory of a three-level gas amplifier*, Zeitschrift für Physik **236**, 213–244 (1970).

G. E. Notkin, S. G. Rautian, and A. A. Feoktistov, *Contribution to the theory of spontaneous emission from atoms in external fields*, Soviet Physics–JETP **25**, 1112–1121 (1967).

T. Ya. Popova, A. K. Popov, S. G. Rautian, and R. I. Sokolovskiĭ, *Nonlinear interference effects in emission, absorption, and generation spectra*, Soviet Physics–JETP **30**, 466–472 (1970).

C. Selsart and J.-C. Keller, *Absorption line narrowing in a three-level system of neon under interaction with two quasi-resonant fields*, Optics Communications **15**, 91–94 (1975).

R. M. Whitley and C. R. Stroud Jr., *Double optical resonance*, Physical Review A **14**, 1498–1513 (1976).

9

Three-Level Λ Atoms:

Dark States, Adiabatic Following, and Slow Light

In the previous chapter, we concentrated mainly on the cascade three-level system to illustrate several features of saturation spectroscopy. In this chapter, we examine the Λ configuration in more detail. A critical feature of a closed Λ configuration occurs when levels 1 and 3 are stable or metastable. As you will see, this feature can be exploited to produce such effects as stimulated Raman adiabatic passage and slow light. Moreover, the stability of levels 1 and 3 plays a role in several schemes involving the storage of quantum information. In such schemes, one attempts to transfer the quantum information contained in an incoming, quantized optical field to a superposition state in the atomic ensemble. A method for achieving this goal is discussed in chapter 21. To simplify matters, we neglect atomic motion and, unless noted otherwise, also neglect any collisional or other dephasing processes. In the appendix, we calculate the force exerted by the fields on atoms in the Λ configuration.

9.1 Dark States

There is a novel feature of three-level systems that can occur when $\beta\delta_\beta = -\beta'\delta_{\beta'}$. If we return to the Hamiltonian (8.18) for a single atom located at $\mathbf{R} = 0$, we find that the state amplitudes evolve as

$$\dot{\tilde{c}}_1 = i\beta\delta_\beta\tilde{c}_1 - i\chi_\beta^*(t)\tilde{c}_2, \tag{9.1a}$$

$$\dot{\tilde{c}}_2 = -i\chi_\beta(t)\tilde{c}_1 - i\chi_{\beta'}^{\prime *}(t)\tilde{c}_3, \tag{9.1b}$$

$$\dot{\tilde{c}}_3 = -i\beta'\delta'_{\beta'}\tilde{c}_3 - i\chi'_{\beta'}(t)\tilde{c}_2\,. \tag{9.1c}$$

Let us introduce a new state amplitude defined by

$$c_D = \chi'_{\beta'}\tilde{c}_1 - \chi_\beta^*\tilde{c}_3. \tag{9.2}$$

Provided that

$$\beta\delta_\beta = -\beta'\delta'_{\beta'}, \tag{9.3}$$

it follows immediately that, for *constant* Rabi frequencies,

$$\dot{c}_D = \chi'_{\beta'}\dot{\tilde{c}}_1 - \chi^*_\beta\dot{\tilde{c}}_3 = i\beta\delta_\beta\left(\chi'_{\beta'}\tilde{c}_1 - \chi^*_\beta\tilde{c}_3\right) = i\beta\delta_\beta c_D. \tag{9.4}$$

The state amplitude c_D is completely decoupled from the applied fields! As such, c_D is referred to as a *dark-state amplitude* [1]. For simplicity, we take both χ_β and $\chi'_{\beta'}$ *real and positive*, unless indicated otherwise.

Of course, as a result of relaxation, the dark-state amplitude decays, *unless both* states 1 and 3 are stable or metastable and the dephasing or collision rate $\Gamma_{13} = 0$. In the Λ configuration, it is sometimes possible to choose states 1 and 3 as different states of the ground-state manifold of levels. In this case, an atom prepared in the state

$$|D\rangle = \frac{\chi'\,|1\rangle - \chi\,|3\rangle}{\sqrt{\chi^2 + \chi'^2}} \tag{9.5}$$

at $t = 0$ stays there forever since it does not interact with the incident fields—it is truly a dark state, *provided that*

$$\delta = \omega_{21} - \omega = \delta' = \omega_{23} - \omega' \tag{9.6}$$

and $\Gamma_{13} = 0$. For symmetry purposes, it is convenient to define the Rabi frequencies χ and χ' by

$$\chi = \chi_{\beta=1} = -\frac{\boldsymbol{\mu}_{21}\cdot\hat{\boldsymbol{\epsilon}}E}{2\hbar}, \tag{9.7a}$$

$$\chi' = \left(\chi_{\beta'=-1}\right)^* = -\frac{\boldsymbol{\mu}_{23}\cdot\hat{\boldsymbol{\epsilon}}'E'}{2\hbar}, \tag{9.7b}$$

where E and E' are the complex field amplitudes.

We limit our discussion to the Λ configuration in which states 1 and 3 are stable ground-state sublevels, since the dark states for the other configurations decay as a result of spontaneous emission. Why is state (9.5) dark when $\delta = \delta'$? In effect, the selection rules are such that the "polarization" of the applied fields cannot drive transitions from state $|D\rangle$ to state $|2\rangle$. This is most easily seen if we take levels 1 and 3 to be the $m = \mp 1$ degenerate magnetic sublevels of a $J = 1$ ground state, level 2 to be a $J = 0$ excited state, and fields **E** and **E**′ to have equal amplitudes and σ_+ and σ_- polarizations, respectively.[1] In effect, the incident fields can be considered as a *single* field that is linearly polarized in the y direction. Instead of using the normal magnetic state basis $|m\rangle$ for the ground-state sublevels, one could define an x, y, z ground-state basis in which

$$|x\rangle = -(|1\rangle - |-1\rangle)/\sqrt{2}, \quad |y\rangle = i(|1\rangle + |-1\rangle)/\sqrt{2}, \quad |z\rangle = |0\rangle.$$

[1] As discussed in detail in the appendix in chapter 16, $\sigma_\pm$ polarized fields drive $\Delta m = \pm 1$ transitions on absorption. The unit vector in the y direction can be written as $\hat{\mathbf{y}} = i(\hat{\boldsymbol{\epsilon}}_+ + \hat{\boldsymbol{\epsilon}}_-)/\sqrt{2}$, where $\hat{\boldsymbol{\epsilon}}_\pm = \mp(\hat{\mathbf{x}} \pm i\hat{\mathbf{y}})/\sqrt{2}$ are unit vectors for $\sigma_\pm$ polarization.

For an incident field that is y polarized, both $|x\rangle$ and $|z\rangle$ are dark states, since they are not coupled to the excited state by the field.

This simple example suggests that it is useful to redefine our basis vectors as

$$|D(t)\rangle\rangle = c|\tilde{1}(t)\rangle - s|\tilde{3}(t)\rangle, \tag{9.8a}$$

$$|B(t)\rangle\rangle = s|\tilde{1}(t)\rangle + c|\tilde{3}(t)\rangle, \tag{9.8b}$$

$$|\tilde{2}(t)\rangle\rangle = |\tilde{2}(t)\rangle, \tag{9.8c}$$

where

$$c = \frac{\chi'}{X} = \cos\Theta, \quad s = \frac{\chi}{X} = \sin\Theta,$$
$$X = \sqrt{\chi^2 + \chi'^2}, \quad \tan\Theta = \frac{\chi}{\chi'}. \tag{9.9}$$

In terms of these new basis kets, and for $\delta = \delta'$, the Hamiltonian (8.12b) is transformed to

$$\mathbf{H}_{dark} = \hbar \begin{pmatrix} -\delta & 0 & 0 \\ 0 & 0 & X \\ 0 & X & -\delta \end{pmatrix}, \tag{9.10}$$

where the states are ordered as $|D(t)\rangle\rangle$, $|\tilde{2}(t)\rangle\rangle$, $|B(t)\rangle\rangle$. State $|D(t)\rangle\rangle$ is decoupled from the fields. From this point onward, we drop the explicit time dependence in the kets. You should remember, however, that a dark state is *not* an energy eigenstate of the original Hamiltonian, since the dark state $|D(t)\rangle\rangle$ given in equation (9.8a) is a linear superposition of two states $|1\rangle$ and $|3\rangle$ having a relative phase that oscillates as a function of time if states $|1\rangle$ and $|3\rangle$ have different eigenenergies.

To obtain the density matrix equations, we use $i\hbar\dot{\varrho} = [\mathbf{H}_{dark}, \varrho]$ and add in relaxation terms. Including relaxation is done most easily by first using equation (9.8) to show that density matrix elements in the two bases are related by

$$\varrho_{DD} = c^2\varrho_{11} + s^2\varrho_{33} - sc(\tilde{\varrho}_{13} + \tilde{\varrho}_{31}), \tag{9.11a}$$

$$\varrho_{BB} = c^2\varrho_{33} + s^2\varrho_{11} + sc(\tilde{\varrho}_{13} + \tilde{\varrho}_{31}), \tag{9.11b}$$

$$\varrho_{DB} = c^2\tilde{\varrho}_{13} - s^2\tilde{\varrho}_{31} + sc(\varrho_{11} - \varrho_{33}), \tag{9.11c}$$

$$\varrho_{D2} = c\tilde{\varrho}_{12} - s\tilde{\varrho}_{32}, \tag{9.11d}$$

$$\varrho_{B2} = c\tilde{\varrho}_{32} + s\tilde{\varrho}_{12}, \tag{9.11e}$$

along with $\varrho_{ji} = \varrho_{ij}^*$, and the inverse relationship, obtained by letting s go to $-s$. It then follows from equation (9.11a), for example, that

$$\begin{aligned}\dot{\varrho}_{DD}|_{\text{relaxation}} &= c^2\dot{\varrho}_{11} + s^2\dot{\varrho}_{33} - sc(\dot{\tilde{\varrho}}_{13} + \dot{\tilde{\varrho}}_{31}) \\ &= \gamma_{2,1}c^2\varrho_{22} + \gamma_{2,3}s^2\varrho_{22} + sc\Gamma_{13}(\tilde{\varrho}_{13} + \tilde{\varrho}_{31}) \\ &= \gamma_{2,1}c^2\varrho_{22} + \gamma_{2,3}s^2\varrho_{22} \\ &\quad + sc\Gamma_{13}[2sc(\varrho_{BB} - \varrho_{DD}) + (c^2 - s^2)(\varrho_{BD} + \varrho_{DB})].\end{aligned} \tag{9.12}$$

Following a similar procedure for all density matrix elements, and neglecting collisions or dephasing ($\Gamma_{13} = 0$), we find

$$\dot{\varrho}_{DD} = \left(c^2\gamma_{2,1} + s^2\gamma_{2,3}\right)\varrho_{22}, \tag{9.13a}$$

$$\dot{\varrho}_{BB} = iX(\varrho_{B2} - \varrho_{2B}) + \left(s^2\gamma_{2,1} + c^2\gamma_{2,3}\right)\varrho_{22}, \tag{9.13b}$$

$$\dot{\varrho}_{22} = -iX(\varrho_{B2} - \varrho_{2B}) - \gamma_2\varrho_{22}, \tag{9.13c}$$

$$\dot{\varrho}_{DB} = iX\varrho_{D2} - sc(\gamma_{2,3} - \gamma_{2,1})\varrho_{22}, \tag{9.13d}$$

$$\dot{\varrho}_{D2} = iX\varrho_{DB} - \left(\frac{\gamma_2}{2} - i\delta\right)\varrho_{D2}, \tag{9.13e}$$

$$\dot{\varrho}_{B2} = iX(\varrho_{BB} - \varrho_{22}) - \left(\frac{\gamma_2}{2} - i\delta\right)\varrho_{B2}, \tag{9.13f}$$

$$\varrho_{ji} = \varrho_{ij}^*. \tag{9.13g}$$

The only steady-state solution of these equations is $\varrho_{DD} = 1$, with all other density matrix elements equal to zero. In the original basis, this corresponds to

$$\tilde{\varrho}_{13} = \tilde{\varrho}_{31} = -sc = -\frac{\chi\chi'}{\chi^2 + \chi'^2},$$

$$\varrho_{11} = c^2 = \frac{\chi'^2}{\chi^2 + \chi'^2}, \qquad \varrho_{33} = s^2 = \frac{\chi^2}{\chi^2 + \chi'^2}, \tag{9.14}$$

with all other density matrix elements equal to zero. The system always evolves to a dark state with a rate of order $\gamma_2\varrho_{22}$. It is also clear from equation (9.12) that had we included collisions or dephasing, states $|B\rangle$ and $|D\rangle$ would be coupled by Γ_{13}. Thus, although state $|D\rangle$ is still a dark state for the system when $\Gamma_{13} \neq 0$, it is no longer a *steady state* to which the atom evolves.

In the limit that the probe field is weak, it is possible to get a rather simple expression for the probe field absorption. The density matrix equations for the Λ configuration in the field interaction basis are

$$\dot{\varrho}_{11} = i\left(\chi\tilde{\varrho}_{12} - \chi^*\tilde{\varrho}_{21}\right) + \gamma_{2,1}\varrho_{22}, \tag{9.15a}$$

$$\dot{\varrho}_{33} = i\left(\chi'\tilde{\varrho}_{32} - \chi'^*\tilde{\varrho}_{23}\right) + \gamma_{2,3}\varrho_{22}, \tag{9.15b}$$

$$\dot{\varrho}_{22} = -i\left(\chi\tilde{\varrho}_{12} - \chi^*\tilde{\varrho}_{21}\right) - i\left(\chi'\tilde{\varrho}_{32} - \chi'^*\tilde{\varrho}_{23}\right) - \gamma_2\varrho_{22}, \tag{9.15c}$$

$$\dot{\tilde{\varrho}}_{13} = i\chi'\tilde{\varrho}_{12} - i\chi^*\tilde{\varrho}_{23} - \left[\Gamma_{13} - i(\delta - \delta')\right]\tilde{\varrho}_{13}, \tag{9.15d}$$

$$\dot{\tilde{\varrho}}_{12} = i\chi'^*\tilde{\varrho}_{13} - i\chi^*(\varrho_{22} - \varrho_{11}) - (\gamma_{12} - i\delta)\tilde{\varrho}_{12}, \tag{9.15e}$$

$$\dot{\tilde{\varrho}}_{32} = i\chi^*\tilde{\varrho}_{31} - i\chi'^*(\varrho_{22} - \varrho_{33}) - (\gamma_{32} - i\delta')\tilde{\varrho}_{32}, \tag{9.15f}$$

$$\varrho_{ji} = \varrho_{ij}^*, \tag{9.15g}$$

where we have included collisional or dephasing terms $[\gamma_{ij} = (\gamma_i + \gamma_i)/2 + \Gamma_{ij}]$ and have not set $\delta = \delta'$. For reasons that will become apparent, we now consider field $\mathbf{E}'$ to be the pump field and field $\mathbf{E}$ to be the probe field. Following a procedure similar to that which led to equation (8.59) or using the appropriate limit of a computer solution of equations (9.15), we find to lowest order in χ that, in steady state,

$$\tilde{\varrho}_{21} = -\chi \frac{i\Gamma_{13} - (\delta - \delta')}{i(\delta - \delta')\gamma_{12} + \gamma_{12}\Gamma_{13} - \delta(\delta - \delta') + i\Gamma_{13}\delta + \chi'^2}. \tag{9.16}$$

As expected, if $[i\Gamma_{13} - (\delta - \delta')] = 0$, there is no steady-state absorption.

9.2 Adiabatic Following—Stimulated Raman Adiabatic Passage

Since dark states involve a coherent superposition of two, stable internal states, they might be useful for storing quantum information. In quantum information storage schemes, one must somehow transfer optical coherence to atomic state coherence without loss of fidelity. Allowing atoms to evolve to dark states as a result of relaxation invariably results in a loss of information. Moreover, it takes a relatively long time, since one must wait a time at least of order γ_2^{-1} for the atoms to decay to the dark state. This does not mean that dark states are not useful in quantum information schemes, only that they must be used in another fashion. In this section, we describe one such scheme, stimulated Raman adiabatic passage, or STIRAP [2]. The STIRAP method extends our previous calculations of semiclassical dressed states to three-level atoms. STIRAP allows one to convert an initial state population ϱ_{11} to an arbitrary superposition of states 1 and 3 in a time much shorter than γ_2^{-1}. Moreover, the conversion is made with the population of level 2 remaining negligibly small—as such, spontaneous decay plays only a minimal role in STIRAP.

We start with the Hamiltonian for a Λ three-level configuration with $\delta = \delta'$,

$$\tilde{\mathbf{H}}(t) = \hbar \begin{pmatrix} 0 & \chi(t) & 0 \\ \chi(t) & \delta & \chi'(t) \\ 0 & \chi'(t) & 0 \end{pmatrix}, \tag{9.17}$$

where the Rabi frequencies are assumed to be real but can be functions of time. For symmetry purposes, we have changed the zero of energy of the Hamiltonian (8.18) by adding an energy $\hbar\delta$ to each of the states. In dealing with STIRAP, one calculates instantaneous eigenstates [semiclassical dressed states of $\tilde{\mathbf{H}}(t)$] of the three-level system and then sees how they are coupled by the nonadiabatic terms and the decay rates. The eigenvalues of $\tilde{\mathbf{H}}(t)$ are

$$\omega_D = 0, \tag{9.18a}$$

$$\omega_{A,B} = \frac{\delta}{2} \mp \sqrt{X^2 + \frac{\delta^2}{4}}, \tag{9.18b}$$

and the corresponding eigenkets are

$$|D\rangle = -c|\tilde{1}\rangle + s|\tilde{3}\rangle, \tag{9.19a}$$

$$|A\rangle = (s'|\tilde{1}\rangle + c'|\tilde{3}\rangle + |\tilde{2}\rangle)/\sqrt{1 + s'^2 + c'^2}, \tag{9.19b}$$

$$|B\rangle = (s''|\tilde{1}\rangle + c''|\tilde{3}\rangle + |\tilde{2}\rangle)/\sqrt{1 + s''^2 + c''^2}, \tag{9.19c}$$

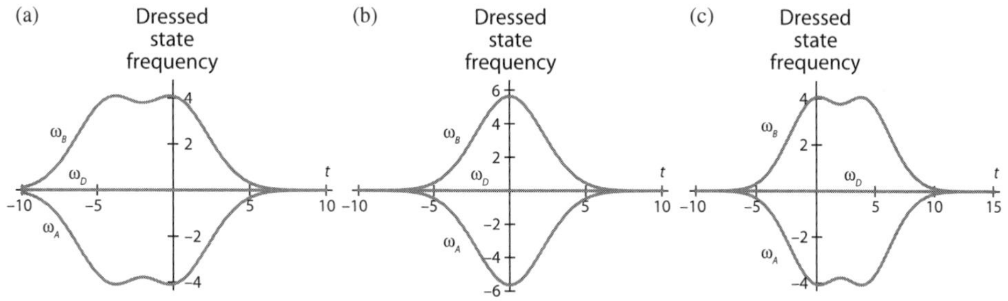

Figure 9.1. Dressed-state frequencies in STIRAP for $\delta = 0$ and Gaussian pulses: (a) pulse χ' precedes pulse χ; (b) pulses χ' and χ are simultaneous; (c) pulse χ' follows pulse χ. The graphs are drawn using $\chi(t) = 4\exp(-t^2/10)$ and $\chi'(t) = 4\exp[-(t-\tau)^2/10]$, with $\tau = -4, 0, 4$.

where

$$c = \chi'/X, \quad s = \chi/X, \qquad \tan\Theta = \chi/\chi', \tag{9.20a}$$

$$c' = \frac{\chi'}{\frac{\delta}{2} - \sqrt{X^2 + \frac{\delta^2}{4}}}, \quad s' = \frac{\chi}{\frac{\delta}{2} - \sqrt{X^2 + \frac{\delta^2}{4}}}, \tag{9.20b}$$

$$c'' = \frac{\chi'}{\frac{\delta}{2} + \sqrt{X^2 + \frac{\delta^2}{4}}}, \quad s'' = \frac{\chi}{\frac{\delta}{2} + \sqrt{X^2 + \frac{\delta^2}{4}}}. \tag{9.20c}$$

Although not indicated explicitly, all these quantities are functions of time for pulsed fields. Note, however, that $\dot{\Theta} = 0$ and an atom that is prepared in a dark state remains there, if the time dependence of both fields is identical. Transition frequencies between the states (9.19) are defined by $\omega_{\alpha\beta} = \omega_\alpha - \omega_\beta$, where the ω_α are given by equation (9.18) and α and β can take on any of the values D, A, B.

For a resonant field, $\delta = 0$, we find

$$\omega_D = 0, \tag{9.21a}$$

$$\omega_{A,B} = \mp X, \tag{9.21b}$$

and

$$|D\rangle = -c|\tilde{1}\rangle + s|\tilde{3}\rangle, \tag{9.22a}$$

$$|A\rangle = -(s|\tilde{1}\rangle + c|\tilde{3}\rangle - |\tilde{2}\rangle)/\sqrt{2}, \tag{9.22b}$$

$$|B\rangle = (s|\tilde{1}\rangle + c|\tilde{3}\rangle + |\tilde{2}\rangle)/\sqrt{2}. \tag{9.22c}$$

The eigenvalues are plotted in figure 9.1 as a function of time for field pulses having a Gaussian envelope, for $\delta = 0$, and with different relative delay between the pulses.

It is important to note that spontaneous emission does not perturb the dark state, since state $|D\rangle$ does not contain an admixture of decaying state $|\tilde{2}\rangle$—the only loss state $|D\rangle$ experiences arises from nonadiabatic contributions. An atom that is prepared in this dark state stays there forever, unless nonadiabatic conditions result in a transfer out of this state.

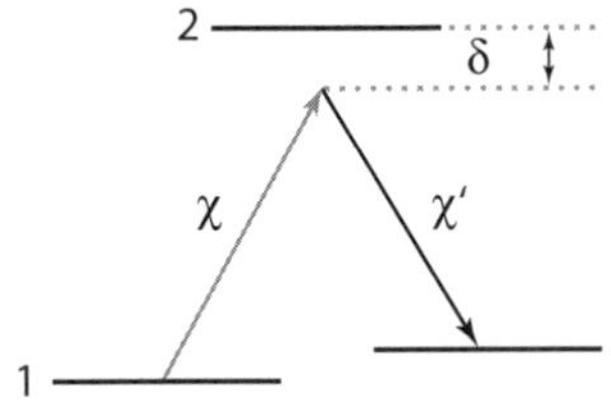

Figure 9.2. Three-level Λ scheme, with $\delta = \delta'$.

The idea behind STIRAP is to arrange the fields to ensure that atoms remain in dressed state $|D\rangle$ (having eigenvalue 0) for the entire atom–field interaction. One starts by placing all the atoms in state $|\tilde{1}\rangle$ when both fields are off. The level scheme is shown in figure 9.2.

Stage 1. The atoms are prepared in state $|\tilde{1}\rangle$. Since $s' = s'' = 0$, this initial condition corresponds to preparation in dressed state $|D\rangle$ [see equations (9.19)]. In stage 1, the field E′ is turned on, and the atoms stay in dressed state $|D\rangle = |\tilde{1}\rangle$, irrespective of how slowly or rapidly the field is turned on (since state $|\tilde{1}\rangle$ is not coupled to field E′). Clearly, in stage 1, relaxation cannot be playing any role, since the atoms remain in bare state $|\tilde{1}\rangle$ during this stage.

Stage 2. When the Rabi frequency χ' has been ramped up to a sufficiently large value, one can turn on field E sufficiently slowly to ensure that the atom stays in dressed state $|D\rangle$. Typically, one must turn on the field in a time that is long compared with ω_{DA}^{-1} and ω_{BD}^{-1}, which are of order $1/\chi'$ if $\delta = 0$. Relaxation does not couple population out of this dressed state. In terms of figure 9.1, the probe field can be turned on once the frequency separations ω_{DA} and ω_{BD} are sufficiently large.

Stage 3. In stage 3, one can manipulate both fields at will, making sure to change the fields in a time that is long compared with ω_{DA}^{-1} and ω_{BD}^{-1} (which are of order $1/X$ if $\delta = 0$) in order to maintain adiabaticity. For example, if one turns *off* field E′ adiabatically in this stage, population is transferred totally to level 3 from level 1, since state $|D\rangle$ adiabatically evolves into state $|\tilde{3}\rangle$ in this limit. In contrast to two-level adiabatic switching, relaxation does not degrade this result, assuming, as we have, that states $|1\rangle$ and $|3\rangle$ are stable; moreover, we can use *resonant* fields for STIRAP adiabatic switching, whereas a chirped, *detuned* field was used for two-level, adiabatic switching. Alternatively, to arrive at some selected value of $\tilde{\varrho}_{13}$, one could then turn off both fields suddenly (in a time much less than X^{-1}) when $\tilde{\varrho}_{13}$ reaches the selected value.

Adiabatic switching and control via STIRAP are illustrated in figure 9.3, obtained from a numerical solution of equations (9.15) with $\varrho_{11}(-\infty) = 1$. In figures 9.3(a) and 9.3(b),

$$\chi(t) = 10e^{-(t-.5)^2}, \quad \chi'(t) = 10e^{-(t+.5)^2},$$

where time is measured in units of γ_2^{-1} and frequency in units of γ_2. It is seen that switching from level 1 to 3 is almost perfect and that the population of level 2 remains small throughout the entire STIRAP pulse sequence. In figure 9.3(c),

$$\chi(t) = 10e^{-(t-.5)^2}\Theta(-t), \quad \chi'(t) = 10e^{-(t+.5)^2}\Theta(-t),$$

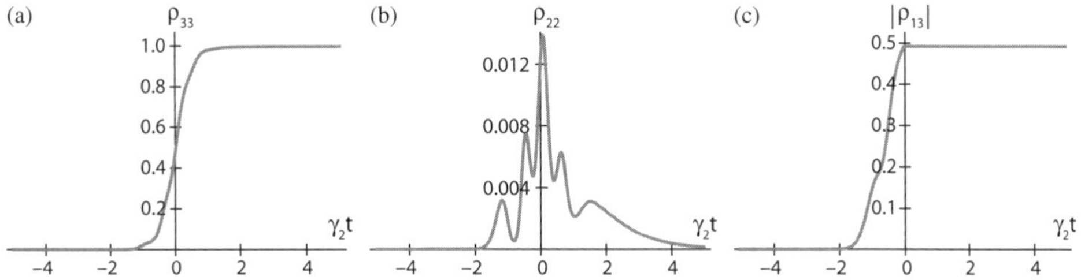

Figure 9.3. Adiabatic switching and control via STIRAP. If pulse $\chi(t)$ follows pulse $\chi'(t)$, it is possible to adiabatically switch population from level 1 to level 3 [plot (a)] while minimally populating level 2 [plot (b)]. With the same sequence but cutting off both pulses when they overlap, it is possible to optimize the coherence ρ_{13} [plot (c)].

where $\Theta(t)$ is a Heaviside function (not to be confused with the angle Θ). For this pulse sequence in which both fields are turned off suddenly, the absolute value of the ground-state coherence ϱ_{13} is optimized.

For reference purposes, we write the equations for density matrix elements in the semiclassical dressed-state basis when $\delta = \delta' = 0$, for equal branching ratios, $\gamma_{2,1} = \gamma_{2,3} = \gamma_2/2$. The transformation from the bare to the dressed-state basis, obtained using equations (9.22), is given by

$$\begin{pmatrix} \varrho_{DD} \\ \varrho_{DA} \\ \varrho_{DB} \\ \varrho_{AD} \\ \varrho_{AA} \\ \varrho_{AB} \\ \varrho_{BD} \\ \varrho_{BA} \\ \varrho_{BB} \end{pmatrix} = \begin{pmatrix} c^2 & s^2 & -sc & -sc & 0 & 0 & 0 & 0 & 0 \\ \frac{sc}{\sqrt{2}} & -\frac{sc}{\sqrt{2}} & \frac{c^2}{\sqrt{2}} & -\frac{s^2}{\sqrt{2}} & -\frac{c}{\sqrt{2}} & 0 & \frac{s}{\sqrt{2}} & 0 & 0 \\ -\frac{sc}{\sqrt{2}} & \frac{sc}{\sqrt{2}} & -\frac{c^2}{\sqrt{2}} & \frac{s^2}{\sqrt{2}} & -\frac{c}{\sqrt{2}} & 0 & \frac{s}{\sqrt{2}} & 0 & 0 \\ \frac{sc}{\sqrt{2}} & -\frac{sc}{\sqrt{2}} & -\frac{s^2}{\sqrt{2}} & \frac{c^2}{\sqrt{2}} & 0 & -\frac{c}{\sqrt{2}} & 0 & \frac{s}{\sqrt{2}} & 0 \\ \frac{s^2}{2} & \frac{c^2}{2} & \frac{sc}{2} & \frac{sc}{2} & -\frac{s}{2} & -\frac{s}{2} & -\frac{c}{2} & -\frac{c}{2} & \frac{1}{2} \\ -\frac{s^2}{2} & -\frac{c^2}{2} & -\frac{sc}{2} & -\frac{sc}{2} & -\frac{s}{2} & \frac{s}{2} & -\frac{c}{2} & \frac{c}{2} & \frac{1}{2} \\ -\frac{sc}{\sqrt{2}} & \frac{sc}{\sqrt{2}} & \frac{s^2}{\sqrt{2}} & -\frac{c^2}{\sqrt{2}} & 0 & -\frac{c}{\sqrt{2}} & 0 & \frac{s}{\sqrt{2}} & 0 \\ -\frac{s^2}{2} & -\frac{c^2}{2} & -\frac{sc}{2} & -\frac{sc}{2} & \frac{s}{2} & -\frac{s}{2} & \frac{c}{2} & -\frac{c}{2} & \frac{1}{2} \\ \frac{s^2}{2} & \frac{c^2}{2} & \frac{sc}{2} & \frac{sc}{2} & \frac{s}{2} & \frac{s}{2} & \frac{c}{2} & \frac{c}{2} & \frac{1}{2} \end{pmatrix} \begin{pmatrix} \varrho_{11} \\ \varrho_{33} \\ \tilde{\varrho}_{13} \\ \tilde{\varrho}_{31} \\ \tilde{\varrho}_{12} \\ \tilde{\varrho}_{21} \\ \tilde{\varrho}_{32} \\ \tilde{\varrho}_{23} \\ \varrho_{22} \end{pmatrix}, \tag{9.23}$$

with the inverse transformation having $s \to -s$. Using equations for the time evolution of these density matrix elements (which we do not write explicitly), it is possible to show that ϱ_{22} remains small as field **E** is turned on, provided that $\dot{\Theta}^2/\chi'^2 \ll 1$ and $|\dot{\Theta}\gamma_2/\chi'^2| \ll 1$, where Θ is defined in equation (9.9). For nonzero detuning $|\delta| \gg X = (\chi^2 + \chi'^2)^{1/2}$, the adiabaticity condition becomes more severe, since the separation $\omega_{DA} = [X^2 + (\delta^2/4)]^{1/2} - \delta/2 \sim X^2/|\delta| \ll X$.

9.3 Slow Light

There has been a great deal of excitement connected with the use of the nonlinear properties of atomic ensembles to decrease the group velocity of light as it

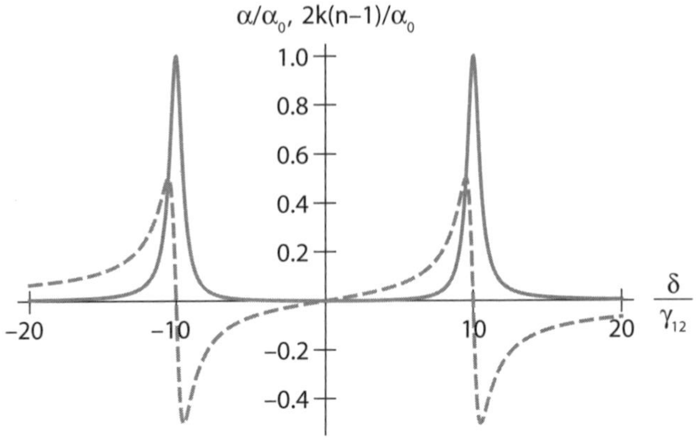

Figure 9.4. Graphs of the absorption coefficient (solid line) in units of α_0 and $(n-1)$ (dashed line) in units of $2k/\alpha_0$ as a function of δ/γ_{12} for $\chi'/\gamma_{12} = 10$.

propagates in a medium [3]. It is not simply the fact that one can slow the group velocity that has aroused interest in this subject, but the ability to transfer information carried by the optical fields to atomic state coherence. The slow light propagates with minimal distortion in a medium whose index of refraction is approximately equal to unity over the bandwidth of the pulse. This is remarkable since the medium is optically dense at optical frequencies but has been rendered transparent by the application of a control field. Using the concepts of dark states and adiabatic following introduced in the previous sections, one can gain an understanding of the underlying physics associated with slow light generation.

The physical system is identical to the one we considered in discussing STIRAP. As in STIRAP, the first step is to prepare all the atoms in the medium in state $|1\rangle$ and send in the field $\mathbf{E}'$. Field $\mathbf{E}'$ is chosen to be sufficiently intense to create a *frequency window,* enabling field $\mathbf{E}$ to enter the medium without reflection. To understand how this frequency window is formed, recall that the absorption of probe field $\mathbf{E}$ is governed by density matrix element

$$\tilde{\varrho}_{21} = \frac{\chi\delta}{-\delta^2 + i\gamma_{12}\delta + \chi'^2} \tag{9.24}$$

given in equation (9.16), where we have set $\delta' = 0$ and neglect collisions and dephasing ($\gamma_{12} = \gamma_2/2$, $\Gamma_{13} = 0$). Following the same procedure used in section 6.3, we can use the Maxwell-Bloch equations to show that the absorption coefficient and index of refraction for the probe field are [see equations (6.25), (6.33), (6.34), (4.23), and (6.66)]

$$\alpha = \alpha_0 \frac{\gamma_{12}^2\delta^2}{\left(\chi'^2 - \delta^2\right)^2 + (\gamma_{12}\delta)^2} \tag{9.25}$$

and

$$n^2 = 1 + \frac{\alpha_0}{k}\frac{\gamma_{12}\delta\left(\chi'^2 - \delta^2\right)}{\left(\chi'^2 - \delta^2\right)^2 + (\gamma_{12}\delta)^2}, \tag{9.26}$$

respectively. These quantities are plotted in figure 9.4. As you can see, the index of refraction is approximately equal to unity, and there is negligible absorption in the vicinity of resonance, $\delta = 0$; however, $dn/d\omega$ can be large in this region, resulting in a reduced group velocity for the probe field.

Let us assume that the probe field pulse has a frequency bandwidth $\Delta\omega$ centered about the 1–2 transition frequency. As such, the probe detuning satisfies $|\delta| \leq \Delta\omega$. If

$$\left(\gamma_{12}\Delta\omega/\chi'^2\right)^2 \ll 1, \tag{9.27}$$

there is negligible probe absorption over distances of length $\alpha_0^{-1}(\chi'^2/\gamma_{12}\Delta\omega)^2 \gg \alpha_0^{-1}$. In other words, as long as

$$\Delta\omega \ll \frac{\chi'^2}{\gamma_{12}\sqrt{\alpha_0 L}}, \tag{9.28}$$

where L is the sample length, there is negligible probe absorption. Moreover, if

$$(\alpha_0/2k)\left(\gamma_{12}\Delta\omega/\chi'^2\right) \sim \mathcal{N}\lambda^3\left(\gamma_{12}\Delta\omega/\chi'^2\right) \ll 1, \tag{9.29}$$

there is negligible reflection as the probe field enters the medium since $n \approx 1$. The presence of a strong pump field in this three-level system has rendered the medium transparent, a process referred to as *electromagnetic-induced transparency* or EIT [4].

To obtain expressions for the field amplitudes as they propagate in the medium, we combine the equations for the three-level system (9.15) with Maxwell's equations for the fields. In the slowly varying amplitude and phase approximation, it follows from equations (6.25) and (9.7) that

$$\left(\frac{\partial}{\partial Z} + \frac{1}{c}\frac{\partial}{\partial t}\right)\chi = -i\zeta\tilde{\varrho}_{21}, \tag{9.30a}$$

$$\left(\frac{\partial}{\partial Z} + \frac{1}{c}\frac{\partial}{\partial t}\right)\chi' = -i\zeta\tilde{\varrho}_{23}, \tag{9.30b}$$

where

$$\zeta = 3\lambda^2\mathcal{N}\gamma_2/8\pi, \tag{9.31}$$

$\mathcal{N}$ is the atomic density, and $\tilde{\varrho}_{21}$ and $\tilde{\varrho}_{23}$ are single-particle density matrix elements. The equations have been written in terms of the Rabi frequencies instead of the field amplitudes, and it has been assumed that the wavelengths of the two transitions are approximately equal. The key point here is that even though $\tilde{\varrho}_{21}$ can be very small (it is identically zero for exact EIT), when multiplied by ζ, it can lead to complete "absorption" of the probe field for sufficiently high density $\mathcal{N}$. Thus, *each* atom has $\varrho_{33} \approx 0$ for a weak probe field, but the *ensemble* of atoms can still result in the probe field being "absorbed."

In what follows, we assume always that the Rabi frequency χ' is controlled externally so that we need not be concerned with its propagation equation. This may not be strictly true if its field intensity goes to zero, but we will be content to restrict our analysis to equation (9.30a). To account for pump propagation effects, one would need to solve the set of coupled equations (9.30).

It may helpful to give a qualitative description of the atom–field interaction before giving the mathematical analysis. The atoms are prepared in level 1, and the pump field is applied on the 2–3 transition. If condition (9.29) holds, the probe field enters the medium without reflection but is *compressed spatially* as it enters the medium and propagates without loss and with reduced group velocity. This situation corresponds *approximately* to a dark state for the atoms, with slight corrections that allow for propagation effects. The pulse excites a ground-state coherence in the atoms that adiabatically follows the pulse amplitude; following the pulse, the atoms are left in their initial state. Thus, any "information" is always stored as a coherence in the atoms that overlap the probe pulse (even as it goes to zero). In the next stage, one reduces or turns off the pump field. As you will see, this can be done fairly rapidly. At this point, the fields in the medium vanish, but the atoms are left with coherence that can "store" quantum information that was encoded in the probe pulse. The pump field is then turned back on, and the probe field can be restored, eventually decompressing as it leaves the medium.

The way to proceed analytically is to solve the density matrix equations in the limit of a weak probe field with $\varrho_{11} \approx 1$ and $\varrho_{22} \approx \varrho_{33} \approx 0$ (reference [5]). We neglect collisions and set $\delta' = 0$ and $|\delta| \ll \chi'$. From equations (9.15d) and (9.15e), we find

$$\tilde{\varrho}_{21} \approx \frac{i}{\chi'}\frac{\partial \tilde{\varrho}_{31}}{\partial t} \tag{9.32}$$

and

$$\begin{aligned}\tilde{\varrho}_{31} &\approx \frac{i}{\chi'}\left[\frac{\partial \tilde{\varrho}_{21}}{\partial t} + i\chi + \left(\frac{\gamma_2}{2} + i\delta\right)\tilde{\varrho}_{21}\right] \\ &= -\frac{\chi}{\chi'} - \frac{1}{\chi'^2}\left[\left(\frac{\gamma_2}{2} + i\delta\right)\frac{\partial \tilde{\varrho}_{31}}{\partial t} + \frac{\partial^2 \tilde{\varrho}_{31}}{\partial t^2}\right].\end{aligned} \tag{9.33}$$

Note that the lead term in this expression is just the value of $\tilde{\varrho}_{31}$ for a dark state (9.14) (when $\chi \ll \chi'$) and that the remainder is a correction resulting from field propagation.

If we neglect the correction to $\tilde{\varrho}_{31}$ from the time derivative terms and substitute equations (9.33) and (9.32) into equation (9.30a), assuming that χ' is constant, we can obtain

$$\left(\frac{\partial}{\partial Z} + \frac{1}{c}\frac{\partial}{\partial t}\right)\chi = -\frac{\zeta}{\chi'^2}\frac{\partial \chi}{\partial t}, \tag{9.34}$$

or

$$\left[\frac{\partial}{\partial Z} + \left(\frac{1}{c} + \frac{\zeta}{\chi'^2}\right)\frac{\partial}{\partial t}\right]\chi = 0. \tag{9.35}$$

The probe pulse propagates *without distortion* with a group velocity

$$v_g = \frac{\chi'^2}{\zeta c + \chi'^2}c. \tag{9.36}$$

The pulse is compressed spatially by a factor v_g/c. If $\zeta c \gg \chi'^2$, the group velocity is much less than c.

It is also possible to define a *polariton* state (combined state of the atomic coherence and probe field amplitude) as

$$\Psi(Z,t) = \frac{\cos\xi}{\sqrt{\zeta c}}\chi(Z,t) - \sin\xi\,\tilde{\varrho}_{31}(Z,t), \tag{9.37}$$

along with a *bright* state

$$\Phi(Z,t) = \frac{\sin\xi}{\sqrt{\zeta c}}\chi(Z,t) + \cos\xi\,\tilde{\varrho}_{31}(Z,t), \tag{9.38}$$

where

$$\tan\xi = \frac{\sqrt{\zeta c}}{\chi'}, \qquad \cos\xi = \frac{\chi'}{\sqrt{\zeta c + \chi'^2}}, \qquad \sin\xi = \frac{\sqrt{\zeta c}}{\sqrt{\zeta c + \chi'^2}}. \tag{9.39}$$

It then follows, using the approximate expressions (9.32) and (9.33) for $\tilde{\varrho}_{21}$ and $\tilde{\varrho}_{31}$, that

$$\left(\frac{\partial}{\partial Z} + \frac{\sec^2\xi}{c}\frac{\partial}{\partial t}\right)\Psi = -\frac{\dot{\xi}}{\cos^2\xi}\Phi - \tan\xi\frac{\partial\Phi}{\partial Z}, \tag{9.40a}$$

$$\begin{aligned}\Phi &= \frac{\sin\xi}{\zeta c}\left[\frac{\partial}{\partial t} + (\gamma - i\delta)\right]\left(\tan\xi\frac{\partial}{\partial t}\right)(\sin\xi\Psi - \cos\xi\Phi)\\ &\approx \frac{\cos^2\xi}{\chi'^2}\left[\frac{\partial}{\partial t} + (\gamma - i\delta)\right]\frac{\partial}{\partial t}(\sin\xi\Psi - \cos\xi\Phi)\end{aligned} \tag{9.40b}$$

for χ' constant.

In the limit that $\tilde{\varrho}_{31}(Z,t) \approx -\chi/\chi'$, we can use equations (9.37) to (9.39) to obtain the solution

$$\Phi(Z,t) = 0, \qquad \Psi(Z,t) = \frac{\chi(Z,t)}{\chi'}\frac{\sqrt{\zeta c}}{\sqrt{\zeta c + \chi'^2}}\left(1 + \frac{\chi'^2}{\zeta c}\right). \tag{9.41}$$

Moreover, in this limit, it follows from equation (9.40a) that $\Psi(Z,t)$ propagates with the group velocity $v_g = c\cos^2\xi = \chi'^2 c/(\zeta c + \chi'^2)$, in agreement with equation (9.36). Under these conditions, Ψ represents a *soliton*, a wave that propagates without distortion. The soliton state (9.41) reflects the fact that excitation is shared between the probe field and the atomic state coherence, although most of the state Ψ is always in atomic state coherence when $\zeta c \gg \chi'^2$.

In order to store the probe field information in the atomic state coherence and eventually restore it, one must turn the pump field off and on. The key point is to turn off the pump field while leaving $\Phi(Z,t) \approx 0$. This is accomplished if one turns off the field in a time that is much longer than $1/\zeta c$; the system remains approximately in the soliton state with all of the state transferred to atomic coherence. However, even if the field is turned off and on *instantaneously*, there is a correction to the field only of order $\sqrt{\chi}/\zeta c$. This occurs because most of the soliton state is already in the atomic state coherence in this limit, and turning off the field just produces $\Phi(Z,t) \approx \sqrt{\chi}/\zeta c$. When the pump field is restored, the probe field is restored (with corrections of order $\sqrt{\chi}/\zeta c$) and exits the medium. All this works

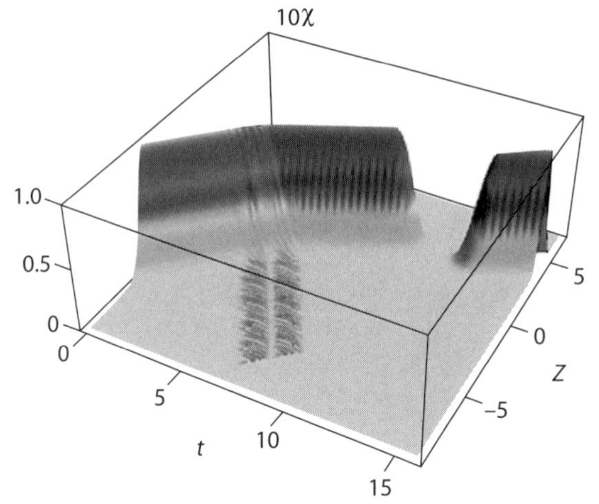

Figure 9.5. Field amplitude for a wave that enters a medium and slows its velocity by half before the control field is turned off. The field is not shown leaving the medium.

only when the probe pulse is adiabatically converted to the soliton state without loss as it enters the medium, is stored totally in the medium as ground-state coherence between levels 1 and 3, and is restored adiabatically from the soliton state to the initial input state as the control field is turned back on. These conditions can be satisfied if both the bandwidth condition (9.29) holds and the medium is optically dense, $\alpha_0 L \gg 1$.

In figure 9.5, we show a numerical solution of the Maxwell-Bloch equations for the probe field amplitude $\chi(Z, t)$. As can be seen, the probe field slows as it enters the medium, diminishes to zero when the pump field is turned off suddenly, and is restored when the pump field is turned back on. The corresponding plot of $-\tilde{\varrho}_{31}$ is shown in figure 9.6.

The graphs are drawn in dimensionless units with $c = 1$, $\delta = \delta' = 0$, $\gamma_2 = 2\gamma_{2,1} = 2\gamma_{3,1} = 1$, $\zeta = 400$, $\chi(Z, 0) = 0.1\exp[-(Z+4)^2]$, $\chi'(t) = 20$ for ($t < 9$ or $t > 14$), and $\chi'(t) = 20\left\{\exp[-(t-9)^2] + \exp[-(t-14)^2]\right\}$ for $9 \leq t \leq 14$. For these values of the parameters, it follows from equations (9.35) and (9.36) that $v_g/c = 1/2$ and that the pulse is compressed spatially by a factor of two as it enters the medium. Moreover, for ($t < 9$ or $t > 14$), when the control pulse is on, the maximum value of $-\tilde{\varrho}_{31}$ predicted by equation (9.33) is $-\tilde{\varrho}_{31} \approx 0.1/20 = 0.005$ and occurs at points corresponding to the peak of the input pulse in the medium. These features can be seen in figures 9.5 and 9.6.

9.4 Effective Two-State Problem for the Λ Configuration

You saw in the previous chapter that one can transform the three-level problem into an effective two-level problem when adiabatic elimination of state $|2\rangle$ is justified. In dealing with the Λ configuration, we have to be a little more careful, since states $|1\rangle$ and $|3\rangle$ are stable. The resulting equations that are to be derived have

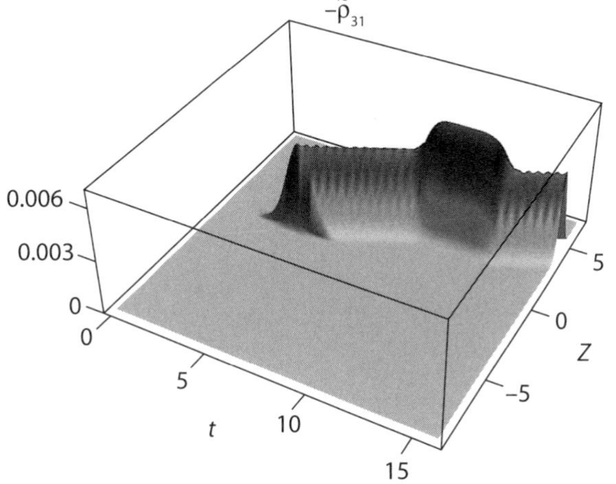

Figure 9.6. The value of $-\mathrm{Re}\,\tilde{\varrho}_{31}$ for the parameters of figure 9.5. When the pump field is turned off, the coherence increases in the medium.

many applications in quantum information. We neglect collisions or dephasing ($\gamma_{12} = \gamma_{32} = \gamma_2/2$; $\gamma_{13} = 0$) and assume that the applied fields vary in a time that is slow compared with $|\delta|^{-1}$. It is assumed that $|\delta| \gg \gamma_2$ and that $|\delta - \delta'| \ll |\delta|$, since these are the conditions needed to get an effective two-level problem, but we keep all terms to order $\delta^{-2} \approx (\delta')^{-2} \approx (\delta\delta')^{-1}$ to include relaxation of ground-state coherence produced by scattering of the fields into previously unoccupied vacuum field modes and terms of order $(|\chi|^2/\delta - |\chi'|^2/\delta')$ to include the light shifts of the levels.

We start with equations (9.15). In the limit that the fields vary slowly compared with $|\delta|^{-1}$, we can solve equations (9.15e) and (9.15f) quasistatically for the optical coherence (neglecting the ϱ_{22} terms that are of higher order) to obtain

$$\tilde{\varrho}_{12} = \frac{i\chi'^*\tilde{\varrho}_{13} + i\chi^*\varrho_{11}}{\frac{\gamma_2}{2} - i\delta}, \tag{9.42a}$$

$$\tilde{\varrho}_{32} = \frac{i\chi^*\tilde{\varrho}_{31} + i\chi'^*\varrho_{33}}{\frac{\gamma_2}{2} - i\delta'}. \tag{9.42b}$$

These values are substituted into equation (9.15c), which is solved quasistatically to give

$$\varrho_{22} = \frac{-i(\chi\tilde{\varrho}_{12} - \chi^*\tilde{\varrho}_{21}) - i(\chi'\tilde{\varrho}_{32} - \chi'^*\tilde{\varrho}_{23})}{\gamma_2}$$

$$\approx \frac{1}{\delta^2}\{|\chi|^2\varrho_{11} + |\chi'|^2\varrho_{33} + [\chi\chi'^*\tilde{\varrho}_{13} + (\chi\chi'^*)^*\tilde{\varrho}_{31}]\}, \tag{9.43}$$

where we have used the fact that $\delta^{-2} \approx (\delta')^{-2} \approx (\delta\delta')^{-1}$. Last, equations (9.42) and (9.43) are substituted into equations (9.15a) and (9.15d) to arrive at

$$\dot{\varrho}_{11} \approx -\frac{\gamma_{2,3}|\chi|^2}{\delta^2}\varrho_{11} + \frac{\gamma_{2,1}|\chi'|^2}{\delta^2}\varrho_{33} + i(\chi_{tq}\tilde{\varrho}_{13} - \chi_{tq}^*\tilde{\varrho}_{31}) + \frac{(\gamma_{2,3} - \gamma_{2,1})}{2\delta}(\chi_{tq}\tilde{\varrho}_{13} + \chi_{tq}^*\tilde{\varrho}_{31}), \tag{9.44a}$$

$$\dot{\tilde{\varrho}}_{13} = i(\delta - \delta')\tilde{\varrho}_{13} - i\chi_{tq}^*(\varrho_{33} - \varrho_{11}) + \frac{\gamma_2\chi_{tq}^*}{2\delta}(\varrho_{33} + \varrho_{11}) - \frac{\gamma_2(|\chi|^2 + |\chi'|^2)}{2\delta^2}\tilde{\varrho}_{13} + i\left(\frac{|\chi|^2}{\delta'} - \frac{|\chi'|^2}{\delta}\right)\tilde{\varrho}_{13}, \tag{9.44b}$$

$$\varrho_{33} = 1 - \varrho_{11}, \tag{9.44c}$$

where

$$\chi_{tq} = -\frac{\chi\chi'^*}{\delta}, \tag{9.45}$$

the relationship $(\gamma_{2,1} + \gamma_{2,3}) = \gamma_2$ has been used, and terms of order $(\gamma_2/\delta)^2$ and $|\delta - \delta'|/|\delta|$ have been neglected. These equations differ from equations (8.79) for the cascade system in that the entire contribution to the decay of the 1–3 state coherence results from off-resonant scattering of the input fields.

Since $(\varrho_{11} + \varrho_{33}) \approx 1$, there is a *driving* term proportional to $\gamma_2\chi_{tq}^*/2\delta$ for the ground-state coherence. This term actually corresponds to a coherent pumping of $\tilde{\varrho}_{13}$ by the fields. Note the difference between the dark-state basis and the original state basis pictures. In the dark-state picture, spontaneous emission populates the dark state that is decoupled from the field—the dark state is a linear combination of states $|1\rangle$ and $|3\rangle$, resulting in a coherence between these states. On the other hand, in the bare-state picture, the steady-state ground-state coherence results from an equilibrium that is reached between the coherent pumping of that coherence by the input fields and the decay of the coherence resulting from spontaneous emission (scattering of the fields into unoccupied vacuum field modes).

The term $i[(|\chi|^2/\delta') - (|\chi'|^2/\delta)]\tilde{\varrho}_{13}$ in equation (9.44c) corresponds to a light shift. You might have expected that the light shift term would be $i[(|\chi|^2/\delta) - (|\chi'|^2/\delta')]\tilde{\varrho}_{13}$, reflecting the light shifts from each field separately, but you can verify that equation (9.44c) gives the correct result by solving the full, three-state problem. Had we neglected terms of order $|\delta|^{-2}$, there would not be a steady-state solution. In that limit, the adiabatic elimination procedure has to be modified somewhat, and the light shifts are as expected. Details are relegated to the problems.

9.5 Summary

We have seen that the Λ configuration brings into play new features not found in the cascade configuration. It is the stability of levels 1 and 3 that leads to new possibilities for slow light production and the storage of optical signal in atomic state coherence. We now turn our attention toward coherent transient spectroscopy.

9.6 Appendix: Force on an Atom in the Λ Configuration

We can generalize our result on the force on a two-level atom to the three-level atom with minimal difficulty. In doing so, we gain additional insight into the friction force. The obvious generalization of equation (5.15) for the time-averaged friction force to a three-level atom is

$$\mathbf{F} = -\hbar[i\chi\mathbf{k}(\tilde{\varrho}_{12} - \tilde{\varrho}_{21}) + i\chi'\mathbf{k}'(\tilde{\varrho}_{32} - \tilde{\varrho}_{23})]. \tag{9.46}$$

It then follows from equations (9.15a) and (9.15b) that

$$\mathbf{F} = \hbar[(\mathbf{k}\gamma_{2,1} + \mathbf{k}'\gamma_{2,3})\varrho_{22} - (\mathbf{k}\dot{\varrho}_{11} + \mathbf{k}'\dot{\varrho}_{33})]. \tag{9.47}$$

In steady state, the averaged friction force vanishes regardless of the detunings δ, δ', and regardless of the ratio of the Rabi frequencies associated with the coupled transitions provided that

$$\mathbf{k}\gamma_{2,1} + \mathbf{k}'\gamma_{2,3} = 0. \tag{9.48}$$

For counterpropagating fields and $\mathbf{k} \approx -\mathbf{k}'$, the friction force vanishes for equal branching ratios, $\gamma_{2,1} = \gamma_{2,3}$ (see reference [6]). Even if condition (9.48) is not satisfied, the steady-state, averaged friction force vanishes provided that $\delta = \delta'$, since there is a dark state in this limit and $\varrho_{22} = 0$ in steady state.

Equation (9.47) can be given a simple physical interpretation. In steady state, the second term vanishes, and the force arises solely from scattered radiation. Each photon scattered on the 2–1 transition involves a loss of one photon from field E, and each photon scattered on the 3–2 transition involves a loss of one photon from field E', with a corresponding force on the atoms that is proportional to the scattering rates $\mathbf{k}\gamma_{2,1}$ and $\mathbf{k}'\gamma_{3,2}$, respectively. In the transient regime, there is an additional contribution to the force, represented by the second term in equation (9.46). This term corresponds to a stimulated exchange of momentum between the fields resulting in a corresponding change in momentum of the atoms.

The result (9.47) is valid for all atom–field detunings. In the adiabatic limit of section 9.4, we can show that the vanishing of the friction force results from the cancellation of two contributions. From equations (9.46) and (9.42), it follows that in steady state and for real Rabi frequencies,

$$\begin{aligned}\mathbf{F} = \hbar\gamma_2 &\left(\frac{\mathbf{k}\chi^2\varrho_{11}}{\frac{\gamma_2^2}{4} + \delta^2} + \frac{\mathbf{k}'\chi'^2\varrho_{33}}{\frac{\gamma_2^2}{4} + \delta'^2}\right) \\ &+ \hbar\chi\chi'\left[\mathbf{k}\left(\frac{\tilde{\varrho}_{13}}{\frac{\gamma_2}{2} - i\delta} + \frac{\tilde{\varrho}_{31}}{\frac{\gamma_2}{2} + i\delta}\right) + \mathbf{k}'\left(\frac{\tilde{\varrho}_{31}}{\frac{\gamma_2}{2} - i\delta'} + \frac{\tilde{\varrho}_{13}}{\frac{\gamma_2}{2} + i\delta'}\right)\right].\end{aligned} \tag{9.49}$$

Under dark-state conditions (9.14), this force vanishes. When condition (9.48) holds, the force also vanishes as a result of the cancellation of population and coherence terms, each of which separately is nonvanishing.

Problems

1–2. Consider the three-level problem, neglecting decay and for resonant fields. In dimensionless variables, the equations for the state amplitudes in the interaction picture are

$$
\begin{aligned}
dc_1/d\tau &= -iac_2, \\
dc_2/d\tau &= -iac_1 - ibc_3, \\
dc_3/d\tau &= -ibc_2,
\end{aligned}
$$

where $a = \chi T$, $b = \chi' T$, and time is measured in units of $\tau = t/T$. Take a and b to be Gaussian functions of time:

$$
\begin{aligned}
a &= (1/\sqrt{\pi})a_0 \exp(-\tau^2), \\
b &= (1/\sqrt{\pi})b_0 \exp[-(\tau - \tau_0)^2],
\end{aligned}
$$

where τ_0 is a delay that can be positive or negative. Consider the limit in which $a_0 \gg 1$ and $b_0 \gg 1$. Use computer solutions to show that if at $t = -\infty$, $c_1 = 1$, it is possible to maximize the state 3 population at $\tau = \infty$ by choosing τ_0 to be negative (*counterintuitive pulse sequence*).

To be specific, look at $a_0 = b_0 = 9.5, 10, 11$ for $\tau_0 = 0.75, 0, -0.75$. Show that if $\tau_0 = -0.75$, one maximizes the state 3 population at $\tau = \infty$ in a fairly robust manner.

3. Derive the equivalent of equations (9.13), including collision or dephasing terms, and without the assumption that $\delta = \delta'$. Show that they are given by

$$
\begin{aligned}
\dot{\varrho}_{DD} &= (\gamma_{2,1}c^2 + \gamma_{2,3}s^2)\varrho_{22} - 2s^2c^2\Gamma_{13}(\varrho_{DD} - \varrho_{BB}) \\
&\quad + sc[\Gamma_{13}(c^2 - s^2)(\varrho_{DB} + \varrho_{BD}) - i(\delta - \delta')(\varrho_{DB} - \varrho_{BD})], \\
\dot{\varrho}_{BB} &= iX(\varrho_{B2} - \varrho_{2B}) + (\gamma_{2,1}s^2 + \gamma_{2,3}c^2)\varrho_{22} - 2s^2c^2\Gamma_{13}(\varrho_{BB} - \varrho_{DD}) \\
&\quad + sc[-\Gamma_{13}(c^2 - s^2)(\varrho_{DB} + \varrho_{BD}) + i(\delta - \delta')(\varrho_{DB} - \varrho_{BD})], \\
\dot{\varrho}_{22} &= -iX(\varrho_{B2} - \varrho_{2B}) - \gamma_2\varrho_{22}, \\
\dot{\varrho}_{DB} &= iX\varrho_{D2} + sc[\Gamma_{13}(c^2 - s^2)(\varrho_{DD} - \varrho_{BB}) - i(\delta - \delta')(\varrho_{DD} - \varrho_{BB})] \\
&\quad - [\Gamma_{13}(c^4 + s^4) + i(\delta - \delta')(c^2 - s^2)]\varrho_{DB} \\
&\quad + 2s^2c^2\Gamma_{13}\varrho_{BB} - sc(\gamma_{2,3} - \gamma_{2,1})\varrho_{22}, \\
\dot{\varrho}_{D2} &= iX\varrho_{DB} - [\gamma_{32}s^2 + \gamma_{12}c^2 - i(\delta c^2 + \delta' s^2)]\varrho_{D2} \\
&\quad - sc[(\gamma_{12} - \gamma_{32}) - i(\delta - \delta')]\varrho_{B2}, \\
\dot{\varrho}_{B2} &= iX(\varrho_{BB} - \varrho_{22}) - [\gamma_{32}c^2 + \gamma_{12}s^2 - i(\delta s^2 + \delta' c^2)]\varrho_{B2} \\
&\quad - sc[(\gamma_{12} - \gamma_{32}) + i(\delta - \delta')]\varrho_{D2}, \\
\varrho_{ji} &= \varrho_{ij}^*.
\end{aligned}
$$

Show that if $\Gamma_{13} \neq 0$, the dark state is no longer a stationary state of the system.

4. Solve equations (9.15) numerically in dimensionless units, where $\gamma_2 = 1$, $\chi = 4$, $\chi' = 2$, $\gamma_{2,1} = \gamma_{2,3} = 1/2$, and $\delta = \delta' = 5$, for $\varrho_{11}(0) = \varrho_{33}(0) = 1/2$ and all other $\varrho_{ij}(0) = 0$. In particular, plot ϱ_{22} and ϱ_{33} as a function of time, and show that they evolve to their appropriate dark-state values.
5. Repeat problem 4, but take $\Gamma_{13} = 0.2$. Show that in this case, the system evolves to a steady state that is not a dark state. Repeat problem 4, but take $\delta = 2\delta' = 10$. Show that in this case, the system evolves to a steady state that is not a dark state.
6. With $|\delta - \delta'|/|\delta| \ll 1$, derive equations (9.44). Show that all time derivatives vanish when the steady-state values of density matrix elements corresponding to a dark state, equations (9.14), are used.
7. Consider pump–probe spectroscopy of a two-level atom (see appendix A in chapter 7). For a detuning $|\delta| \gg \gamma$, to order χ^2, there is a dispersion-like structure centered near $\delta' = \delta$. Prove that there is no absorption at $\delta' = \delta$, but that the index of refraction of the medium is enhanced at this detuning. Calculate the enhancement factor for a medium having density $\mathcal{N}$.
8. Show that the light shifts predicted in equation (9.44) are correct by considering the steady-state solution of the three-level problem (obtained by computer) with $\chi = 20$, $\delta = 100$, and $\chi' = 0.1$ in some dimensionless units. Show that the maximum of $|\operatorname{Im}\tilde{\varrho}_{23}|$ occurs at $\delta' \approx \delta + 3.85$, which corresponds to $\delta + \chi^2/\delta' = 103.85$ and not $\delta + \chi^2/\delta = 104.0$.
9. Solve the *amplitude* equations for the three-level problem in the normal interaction representation by adiabatically eliminating state 2, assuming large detunings, and neglecting all decay. Show that the corresponding density matrix equations are

$$\dot{\varrho}_{11} = -i\left(\frac{\chi^*\chi'}{\delta}\tilde{\varrho}_{13} - \frac{\chi\chi'^*}{\delta'}\tilde{\varrho}_{31}\right),$$

$$\dot{\tilde{\varrho}}_{13} = i(\delta - \delta')\tilde{\varrho}_{13} - i\left(\frac{|\chi|^2}{\delta} - \frac{|\chi'|^2}{\delta'}\right)\tilde{\varrho}_{13} + i\left(\frac{\chi^*\chi'}{\delta}\varrho_{33} - \frac{\chi^*\chi'}{\delta'}\varrho_{11}\right),$$

$$\varrho_{33} = 1 - \varrho_{11}.$$

The light shifts are now at their expected values, but there is no steady-state solution to the equations. As such, the procedure used to arrive at equations (9.42) is not quite correct. The first term in equation (9.42a) arises from

$$\tilde{\varrho}_{12} = i\chi'^* \int_{-\infty}^{t} dt' \tilde{\varrho}_{13}(t') e^{-(\gamma_{12} - i\delta)(t-t')}.$$

When decay of $\tilde{\varrho}_{13}(t')$ resulting from spontaneous emission was included, $\tilde{\varrho}_{13}$ reaches a constant value and can be removed from the integral. In this case, however, no steady state is reached, and we must write $\tilde{\varrho}_{13}(t') \approx \tilde{\varrho}_{13}(t)e^{-i(\delta-\delta')(t-t')}$—with this substitution, the light shift resulting from this term varies as $|\chi|^2/\delta$ and not $|\chi|^2/\delta'$.
10. Using the steady-state solution of equations (9.44), prove that the force (9.49) vanishes when condition (9.48) holds.

11. An interesting limit of equations (9.50) is one where we artificially set $\gamma_{2,3}$, $\gamma_{2,1}$, and γ_2 equal to zero in equation (9.50a), keep $\Gamma_{13} = 0$, but allow for a collision rate $\Gamma_{12} = \Gamma_{32} \neq 0$. In this limit, show that the steady-state solution is $\varrho_{DD} = \varrho_{DD}(t=0)$ and $\tilde{\varrho}_{B2} = \tilde{\varrho}_{D2} = \tilde{\varrho}_{DB} = 0$; $\rho_{BB} = \varrho_{22} = [1 - \varrho_{DD}(0)]/2$. This implies that, in steady state, $\tilde{\varrho}_{13} = sc\,[1 - 3\varrho_{DD}(0)]/2$. The steady-state values depend on the initial population of the dark state. For example, if we start in level 1 with equal Rabi frequencies, then $\tilde{\varrho}_{13} = -1/8$ in steady state. Thus, one can anticipate that there are *two* timescales in the problem when collisions are present, provided that $\Gamma_{13} = 0$ and $\Gamma_{12} = \Gamma_{32} \gg \gamma_2$. It is possible to approach the quasi-steady-state values associated with $\gamma_{2,3} = \gamma_{2,1} = 0$ with a rate of order $4\Gamma_{12}X^2/(\Gamma_{12}^2 + \delta^2)$, and then approach the true steady-state values with a rate of order $\gamma_2\varrho_{22}$. To see this, plot $\varrho_{33}(t)$ for $\varrho_{11}(0) = 1$, $\chi = \chi' = 50$, $\delta = \delta' = 50$, $\Gamma_{12} = \Gamma_{32} = 60$, $\Gamma_{13} = 0$, in units where $\gamma_2 = 2\gamma_{2,3} = 2\gamma_{2,1} = 1$.
12. When impurity ions are embedded in solids, they can have optical transitions with lifetimes of several hours. Such systems have large inhomogeneous broadening. Suppose that in a preparatory phase, a spectral hole was burned in the frequency distribution such that the distribution of frequencies is given by

$$W(\Delta) \approx \begin{cases} 0, & -\Delta_0 \le \Delta \le \Delta_0, \\ \frac{1}{2\Delta}, & -\Delta_w \le \Delta \le -\Delta_0, \\ \frac{1}{2\Delta}, & \Delta_0 \le \Delta \le \Delta_w, \end{cases}$$

where $\Delta_w \gg \Delta_0 \gg \gamma$. A pulse is incident whose bandwidth is contained in the frequency window $-\Delta_0 \ll \Delta \ll \Delta_0$, so that it interacts only *off-resonantly* with atoms outside this frequency window. Use equations (6.42) [with the average taken over $W(\Delta)$] and equation (6.41) to show that this pulse travels with reduced group velocity. (Thanks to J.-L. LeGouët of Laboratoire Aimé Cotton for this example.)

References

[1] E. Arimondo and G. Orriols, *Non-absorbing atomic coherences by coherent 2-photon transitions in a 3-level optical pumping*, Lettere al Nuovo Cimento **17**, 333–338 (1976); R. Gray, R. M. Whitley, and C. R. Stroud Jr., *Coherent trapping of atomic populations*, Optics Letters **3**, 218–220 (1976).

[2] J. Oreg, F. T. Hioe, and J. H. Eberly, *Adiabatic following in multilevel systems*, Physical Review A **29**, 690–697 (1985); U. Gaubatz, P. Rudecki, S. Schiemann, and K. Bergmann, *Population transfer between molecular vibrational levels by stimulated Raman scattering with partially overlapping laser fields: a new concept and experimental results*, Journal of Chemical Physics **92**, 5363–5376 (1990).

[3] D. Grischkowsky, *Adiabatic following and slow optical pulse propagation in rubidium vapor*, Physical Review A **7**, 2096–2102 (1973).

[4] K.-J. Boller, A. Imamoğlu, and S. E. Harris, *Observation of electromagnetically induced transparency*, Physical Review Letters **66**, 2593–2596 (1991).

[5] M. Fleischhauer and M. D. Lukin, *Quantum memory for photons: dark-state polaritons*, Physical Review A **65**, 022314, 1–12 (2002).

[6] F. Papoff, F. Mauri, and E. Arimondo, *Transient velocity-selective coherent population trapping in one dimension*, Journal of the Optical Society of America B **9**, 321–331 (1992).

Bibliography

There are probably more than a thousand articles on the subject matter in this chapter. We list only a few articles and reviews for each topic.

Three-Level Λ Systems

E. Arimondo, *Coherent population trapping in laser spectroscopy*, in *Progress in Optics*, edited by E. Wolf, vol. 35 (Elsevier, Amsterdam, 1996), pp. 257–354. This is a review article.

E. Arimondo and G. Orriols, *Non-absorbing atomic coherences by coherent 2-photon transitions in a 3-level optical pumping*, Lettere al Nuovo Cimento **17**, 333–338 (1976).

J. H. Eberly and V. V. Kozlov, *Wave equation for dark coherence in three-level media*, Physical Review Letters **88**, 243604, 1–4 (2002).

H. Friedmann and A. D. Wilson-Gordon, *Nonlinear refractive index near points of zero absorption and in the dead zone*, Physical Review A **52**, 4070–4077 (1995).

H. R. Gray, R. M. Whitley, and C. R. Stroud Jr., *Coherent trapping of atomic populations*, Optics Letters **3**, 218–220 (1978).

M. Löffler, D. E. Nikinov, O. A. Kocharovskaya, and M. O. Scully, *Strong field index enhancement via selective population of dressed states*, Physical Review A **56**, 5014–5021 (1997).

U. Rathe, M. Fleishhauer, S.-Y. Zhu, T. W. Hänsch, and M. O. Scully, *Nonlinear theory of index enhancement via quantum coherence and interference*, Physical Review A **47**, 4994–5002 (1993).

M. O. Scully, *Enhancement of the index of refraction via quantum coherence*, Physical Review Letters **67**, 1855–1858 (1991).

D. D. Yavuz, *Refractive index enhancement in a far-off resonant atomic system*, Physical Review Letters **95**, 223601, 1–4 (2005).

A. S. Zibrov, M. D. Lukin, L. Hollberg, D. E. Nikonov, M. O. Scully, H. G. Robinson, and V. L. Velichansky, *Experimental demonstration of enhanced index of refraction via quantum coherence in Rb*, Physical Review Letters **76**, 3935–3938 (1996).

STIRAP

U. Gaubatz, P. Rudecki, S. Schiemann, and K. Bergmann, *Population transfer between molecular vibrational levels by stimulated Raman scattering with partially overlapping laser fields: a new concept and experimental results*, Journal of Chemical Physics **92**, 5363–5376 (1990).

J. Oreg, F. T. Hioe, and J. H. Eberly, *Adiabatic following in multilevel systems*, Physical Review A **29**, 690–697 (1985).

B. W. Shore, K. Bergmann, J. Oreg, and S. Rosenwaks, *Multilevel adiabatic population transfer*, Physical Review A **44**, 7442–7447 (1991).

N. V. Vitanov, M. Fleischhauer, B. W. Shore, and K. Bergmann, *Coherent manipulation of atoms and molecules by sequential laser pulses*, in *Advances in Atomic, Molecular, and Optical Physics,* edited by B. Bederson and H. Walther, vol. 46 (Academic Press, New York, 2001), pp. 55–190. This is a comprehensive review article with extensive references.

EIT

K.-J. Boller, A. Imamoğlu, and S. E. Harris, *Observation of electromagnetically induced transparency*, Physical Review Letters **66**, 2593–2596 (1991).

M. Fleischauer, A. Imamoğlu, and J. Marangos, *Electromagnetically induced transparency: optics in coherent media*, Reviews of Modern Physics **77**, 633–673 (2003). This is a review article with extensive references.

M. D. Lukin, *Trapping and manipulating photon states in atomic ensembles*, Reviews of Modern Physics **75**, 457–472 (2003). This is a "colloquium" review.

Slow Light

R. W. Boyd and D. J. Gauthier, *"Slow" and "fast" light*, in *Progress in Optics*, edited by E. Wolf, vol. 43 (Elsevier, Amsterdam, 2002), pp. 497–530. This is a review article.

D. Budker, D. F. Kimball, S. M. Rochester, and V. V. Yashchuk, *Nonlinear magneto-optics reduced group velocity of light in atomic vapor with slow ground state relaxation*, Physical Review Letters **83**, 1767–1770 (1999).

M. Fleischhauer and M. D. Lukin, *Quantum memory for photons: dark-state polaritons*, Physical Review A **65**, 022314, 1–12 (2002).

D. Grischkowsky, *Adiabatic following and slow optical pulse propagation in rubidium vapor*, Physical Review A **7**, 2096–2102 (1973).

L. V. Hau, S. E. Harris, Z. Dutton, and C. H. Behroozi, *Light speed reduction to 17 metres per second in an ultracold atomic gas*, Nature **397**, 594–598 (1999).

M. M. Kash, V. A. Sautenkov, A. S. Zibrov, L. Hollberg, G. W. Welch, M. D. Lukin, Y. Rostovtsev, E. S. Fry, and M. O. Scully, *Ultraslow group velocity and enhanced nonlinear optical effects in a coherently driven hot atomic gas*, Physical Review Letters **82**, 5229–5332 (1999).

C. Liu, Z. Dutton, C. H. Behroozi, and L. V. Hau, *Observation of coherent optical information storage in an atomic medium using halted light pulses*, Nature **409**, 490–493 (2001).

J. J. Longdell, E. Fraval, M. J. Sellars, and N. B. Manson, *Stopped light with storage times greater than one second using electromagnetically induced transparency in a solid*, Physical Review Letters **95**, 063601, 1–4 (2005).

P. W. Milonni, *Controlling the speed of light pulses*, Journal of Physics B **35**, R31–R56 (2002). This is a review article.

———, *Fast Light, Slow Light, and Left-Handed Light* (Taylor and Francis Group, New York, 2005).

D. F. Phillips, A. Fleischhauer, A. Mair, R. L. Walsworth, and M. D. Lukin, *Storage of light in atomic vapor*, Physical Review Letters **86**, 783–786 (2001).

O. Schmidt, R. Wynands, Z. Hussein, and D. Meschede, *Steep dispersion and group velocity below c/3000 in coherent population trapping*, Physical Review A **53**, R27–R30 (1996).

A. V. Turukhin, V. S. Sudarshanan, M. S. Shahriar, J. A. Musser, B. S. Ham, and P. R. Hemmer, *Observation of ultraslow and stored light pulses in a solid*, Physical Review Letters **88**, 023602, 1–4 (2002).

R. G. Ulbrich and G. W. Fehrenbach, *Polariton wave packet propagation in the exciton resonance of a semiconductor*, Physical Review Letters **43**, 963–966 (1979).

10

Coherent Transients

Optical spectroscopic methods fall into two broad categories: *continuous-wave* (cw) or *stationary* spectroscopy, and *time-dependent* or *transient* spectroscopy. To this point, we have considered cw spectroscopy, in which the absorption or emission line shapes are measured as a function of the frequency of a probe field. In *coherent transient optical spectroscopy*, pulsed optical fields are used to create atomic state populations or coherence between atomic states. Following the excitation, the time evolution of the atoms is monitored. By using different pulse sequences, it is possible to extract both transition frequencies and relaxation rates from the coherent transient signals. Whether transient spectroscopy offers distinct advantages over cw spectroscopy depends on a number of factors, such as signal to noise and the reliability of line shape formulas [1].

In describing the coherent transient spectroscopy of two-level quantum systems, one often uses new labels for the relaxation rates that were already encountered in cw spectroscopy. The quantity γ_2 is referred to as the *longitudinal relaxation rate* and $T_1 = \gamma_2^{-1}$ as the *longitudinal relaxation time*. The rate γ at which $\tilde{\varrho}_{12}$ decays is referred to as the *transverse relaxation rate* and $T_2 = \gamma^{-1}$ as the *transverse relaxation time*. If there is inhomogeneous broadening in the system, one often associates a relaxation time $T_2^* = (\Delta\omega_0)^{-1}$ with the inhomogeneous broadening, where $\Delta\omega_0$ is the inhomogeneous frequency width characterizing the atomic ensemble. In the case of purely radiative broadening (spontaneous emission), $\gamma_2 = 2\gamma$ and $T_1 = T_2/2$. For open quantum systems, an additional relaxation rate, related to ground-state decay, must also be specified.

10.1 Coherent Transient Signals

The underlying principle of coherent transient spectroscopy is quite simple. Optical fields are used to create a phased array of atomic dipoles that then radiate a coherent signal. In a typical experiment, one applies one or more "short" radiation pulses to an atomic sample and monitors the radiation field emitted by the atoms following the pulses. A short pulse is one satisfying

$$|\delta|\tau, ku\tau, \sigma_w\tau, \gamma\tau, \gamma_2\tau \ll 1, \tag{10.1}$$

where τ is the pulse duration. Conditions (10.1) allow one to neglect any effects of detuning (including Doppler shifts or other types of inhomogeneous broadening) or relaxation *during* the pulse's action. The change in the atomic density matrix following each pulse is determined solely by the *pulse area* defined as

$$A = \int_{-\infty}^{\infty} \Omega_0(t)\, dt, \tag{10.2}$$

where we take the Rabi frequency $\Omega_0(t) = 2\chi(t)$ to be real and positive. Moreover, the participation of all the atoms is ensured by the large bandwidth of the short excitation pulses. Recall that in two-photon Doppler-free spectroscopy, we also had full participation of the atoms, but at a cost of small signal amplitude, owing to a large detuning from the intermediate state.

To calculate a coherent transient signal, one must piece together intervals in which the fields are applied, with intervals of *free evolution*—that is, evolution in the absence of the fields. There is a question as to the most convenient representation to use for the calculations. On one hand, the field interaction representation might seem the best. This allows for a simple pictorial representation of both the interaction and free zones using the Bloch vector. The major downside of this representation is that it is usually defined with respect to a single **k** vector. If one uses more than one interaction zone with fields having different **k** vectors, she has to be careful to redefine the field interaction representation for each zone. This problem is avoided if the standard interaction representation is used; from a computational viewpoint, therefore, the standard interaction representation might be preferred. In our discussion of coherent transients, we give pictorial representations of coherent transients using the Bloch vector picture, but base calculations on the standard interaction representation. In appendix A, various transfer matrices are given in terms of Bloch variables and field interaction density matrix elements.

When conditions (10.1) hold, it is possible to use an amplitude picture *during* the pulse, since the decay terms can be neglected on the timescale of the pulse. In the standard interaction representation, the state amplitudes during a pulse evolve according to

$$\dot{c}_1 = -i\chi(t)e^{-i\mathbf{k}\cdot\mathbf{R}_a(t)-i\delta t}c_2, \tag{10.3a}$$

$$\dot{c}_2 = -i\chi(t)e^{i\mathbf{k}\cdot\mathbf{R}_a(t)+i\delta t}c_1, \tag{10.3b}$$

where $\mathbf{R}_a(t)$ is the position of an atom at time t, and decay has been neglected. For a pulse centered at $t = T_j$, the inequalities $ku\tau, |\delta|\tau \ll 1$ guarantee that an atom's position is essentially frozen at $\mathbf{R}_a(T_j)$ during the pulse, as is the phase factor, $e^{\pm i\delta t} = e^{\pm i\delta T_j}$. Moreover, if an atom is going to be at position $\mathbf{R}$ at time t, then

$$\mathbf{R}_a(T_j) = \mathbf{R} - \mathbf{v}(t - T_j), \tag{10.4}$$

where $\mathbf{v}$ is the atomic velocity, provided that any changes in atomic velocity resulting from collisions are ignored. In effect, we have used a "trick" to relate the position of an atom at time T_j to the position $\mathbf{R}$ at some later time t at which the signal is measured.

As a consequence of these approximations, equations (10.3) can be replaced by

$$\dot{c}_1 = -i\chi_j(t)e^{-i\mathbf{k}\cdot\mathbf{R}_a(T_j)-i\delta T_j}c_2, \tag{10.5a}$$

$$\dot{c}_2 = -i\chi_j(t)e^{i\mathbf{k}\cdot\mathbf{R}_a(T_j)+i\delta T_j}c_1, \tag{10.5b}$$

during a pulse centered at $t = T_j$ having Rabi frequency $\Omega_j(t) = 2\chi_j(t)$. It is a simple matter to integrate these equations to obtain

$$c_1^+ = \cos\left(A_j/2\right)\, c_1^- - ie^{-i\mathbf{k}\cdot\mathbf{R}_a(T_j)-i\delta T_j}\sin\left(A_j/2\right)\, c_2^-, \tag{10.6a}$$

$$c_2^+ = \cos\left(A_j/2\right)\, c_2^- - ie^{i\mathbf{k}\cdot\mathbf{R}_a(T_j)+i\delta T_j}\sin\left(A_j/2\right)\, c_1^-, \tag{10.6b}$$

where $\pm$ refer to times immediately preceding and following the pulse having area A_j.

Using this result, one can construct density matrix elements, $(\varrho_{ij}^I)^+ = c_i^+ c_j^{+*}$, following the pulse in terms of those, $(\varrho_{ij}^I)^- = c_i^- c_j^{-*}$, before the pulse as

$$\begin{pmatrix} \varrho_{11}^I \\ \varrho_{22}^I \\ \varrho_{12}^I \\ \varrho_{21}^I \end{pmatrix}^+ = \mathbf{U}(T_j)\begin{pmatrix} \varrho_{11}^I \\ \varrho_{22}^I \\ \varrho_{12}^I \\ \varrho_{21}^I \end{pmatrix}^-, \tag{10.7}$$

where

$$\mathbf{U}(T_j) =$$

$$\frac{1}{2}\begin{pmatrix} 1+\cos A_j & 1-\cos A_j & i\sin A_j\, e^{i\Phi(T)} & -i\sin A_j\, e^{-i\Phi(T_j)} \\ 1-\cos A_j & 1+\cos A_j & -i\sin A_j\, e^{i\Phi(T)} & i\sin A_j\, e^{-i\Phi(T_j)} \\ i\sin A_j\, e^{-i\Phi(T_j)} & -i\sin A_j\, e^{-i\Phi(T_j)} & 1+\cos A_j & \left(1-\cos A_j\right)\, e^{-2i\Phi(T_j)} \\ -i\sin A_j\, e^{i\Phi(T_j)} & i\sin A_j\, e^{i\Phi(T_j)} & (1-\cos A_j)\, e^{2i\Phi(T_j)} & 1+\cos A_j \end{pmatrix}, \tag{10.8}$$

$$\begin{aligned} \Phi(T_j) &= \mathbf{k}_j\cdot\mathbf{R}(T_j) + \delta T_j, \\ \mathbf{R}(T_j) &= \mathbf{R} - \mathbf{v}(t - T_j), \end{aligned} \tag{10.9}$$

$\mathbf{v}$ is the atomic velocity, and $\mathbf{R}(T_j) \equiv \mathbf{R}_a(T_j)$. Atoms are modeled as two-level quantum systems, and any changes in velocity resulting from collisions have been ignored.

Between the pulses, or following the last pulse, the atoms evolve freely in the absence of any applied fields, with

$$\dot{\varrho}_{11}^I = \gamma_{2,1}\varrho_{22}^I, \quad \dot{\varrho}_{22}^I = -\gamma_2\varrho_{22}^I, \quad \dot{\varrho}_{12}^I = -\gamma\varrho_{12}^I, \quad \dot{\varrho}_{21}^I = -\gamma\varrho_{21}^I, \tag{10.10}$$

where we have included for the possibility of an open system by allowing the repopulation of state 1 to occur at a rate that is different from the rate of population loss from state 2. The solution of equation (10.10) is

$$\begin{pmatrix} \varrho_{11}^I(t) \\ \varrho_{22}^I(t) \\ \varrho_{12}^I(t) \\ \varrho_{21}^I(t) \end{pmatrix} = \mathbf{F}(t-t_0) \begin{pmatrix} \varrho_{11}^I(t_0) \\ \varrho_{22}^I(t_0) \\ \varrho_{12}^I(t_0) \\ \varrho_{21}^I(t_0) \end{pmatrix}, \tag{10.11}$$

where

$$\mathbf{F}(\tau) = \begin{pmatrix} 1 & \frac{\gamma_{2,1}}{\gamma_2}\left(1-e^{-\gamma_2\tau}\right) & 0 & 0 \\ 0 & e^{-\gamma_2\tau} & 0 & 0 \\ 0 & 0 & e^{-\gamma\tau} & 0 \\ 0 & 0 & 0 & e^{-\gamma\tau} \end{pmatrix}. \tag{10.12}$$

A coherent transient signal is constructed by piecing together field interaction zones and free evolution periods. For example, for three pulses,

$$\boldsymbol{\varrho}^I(t) = \mathbf{F}(t-T_3)\mathbf{U}(T_3)\mathbf{F}(T_3-T_2)\mathbf{U}(T_2)\mathbf{F}(T_2-T_1)\mathbf{U}(T_1)\boldsymbol{\varrho}^I(0), \tag{10.13}$$

where $\boldsymbol{\varrho}^I$ is considered as a column vector. Solutions of this nature are easily incorporated into a computer program.

Once $\varrho_{21}^I(\mathbf{R}, \mathbf{v}, t)$ has been obtained in this fashion, we use the Maxwell-Bloch equations (6.25) to calculate the signal radiated by the sample. This can actually get to be a bit confusing, since the field interaction representation was used in deriving the Maxwell-Bloch equations. In that derivation, we set $\varrho_{21}(\mathbf{R}, t) = \tilde{\varrho}_{21}e^{i\mathbf{k}_s\cdot\mathbf{R}-i\omega_s t}$ for a signal field having propagation vector $\mathbf{k}_s$ and frequency ω_s and found that phase matching occurs when $k_s = \omega_s/c$. The key ingredient then is to get an expression for

$$\varrho_{21}(\mathbf{R}, \mathbf{v}, t) = \varrho_{21}^I(\mathbf{R}, \mathbf{v}, t)e^{-i\omega_0 t} = \tilde{\varrho}_{21}e^{i\mathbf{k}_s\cdot\mathbf{R}-i\omega_s t} \tag{10.14}$$

and to identify both $\mathbf{k}_s$ and ω_s in this expression.

We are now in a position to calculate the radiated signal. In all examples, it is assumed that all fields, including the signal field, are polarized approximately in the $\hat{\mathbf{x}}$ direction and propagate approximately in the $\pm\hat{\mathbf{z}}$ direction. Moreover, it is assumed that phase matching is at least approximately satisfied and that the duration of the signal pulse, typically of order γ^{-1}, is larger than the sample length divided by the speed of light (quasi-steady-state regime). In this limit, the Maxwell-Bloch equations (6.25) for the signal field amplitude $E_s(Z, t)$ can be written as

$$\frac{\partial E_s(Z,t)}{\partial Z} = \frac{ik_s\mathcal{N}(\mu_x)_{12}}{\epsilon_0}\left\langle \varrho_{21}^I(Z,t)e^{-i\delta t}\right\rangle e^{-i\mathbf{k}_s\cdot\mathbf{R}}. \tag{10.15}$$

The average is over the velocity distribution or a distribution of ω_0's in the case of inhomogeneous broadening in a solid. If the sample is optically thin, the average power exiting a sample of length L and cross-sectional area σ is given by

$$\mathcal{P}(L,t) = \frac{1}{2}\epsilon_0 c\sigma \left|\frac{k_s(\mu_x)_{12}\mathcal{N}L}{\epsilon_0}\right|^2 \left|\left\langle \varrho_{21}^I(t)e^{-i\delta t}\right\rangle\right|^2, \tag{10.16}$$

or, making use of equation (4.27),

$$\mathcal{P}(L,t) = \frac{3\gamma_2}{8\pi} N\hbar\omega_0 \left(\mathcal{N}\lambda_0^2 L\right) \left|\left\langle \varrho_{21}^I(t) e^{-i\delta t}\right\rangle\right|^2, \tag{10.17}$$

where N is the total number of atoms in the sample, $\lambda_0 = 2\pi/k_0$, and it has been assumed that $k_s \approx k_0 = \omega_0/c$. One could extract the (complex) field amplitude rather than the field intensity by heterodyning the output with a reference field.

Several illustrative examples of coherent transient phenomena are presented, applicable to inhomogeneously broadened vapors and solids. The Maxwell-Bloch formalism is used to analyze free polarization decay, photon echoes, stimulated photon echoes, optical Ramsey fringes, and frequency combs. Although the examples considered are fairly simple, it should be appreciated that much more elaborate coherent transient techniques are applied routinely to probe complex materials such as liquids and solids. A method for obtaining optical Ramsey fringe using an atomic beam passing through several spatially separated field zones is discussed briefly in appendix B.

10.2 Free Polarization Decay

As a first application of the Maxwell-Bloch equations, we consider *free polarization decay* (FPD), which is the analogue of *free induction decay* (FID) in *nuclear magnetic resonance* (NMR). The basic idea behind FPD is very simple. An external field pulse is applied to an ensemble of atoms. The field creates a phased array of atomic dipoles that radiate coherently in the direction of the incident applied field. The decay of the FPD signal provides information about the transverse relaxation times. In this and all future examples, it is assumed that the pulse amplitude is unchanged as the pulse propagates through the medium, $\Omega_0(Z,t) = \Omega_0(Z-ct,0)$. We discuss both homogeneous and inhomogeneous broadening.

10.2.1 Homogeneous Broadening

A short pulse is applied at $T = 0$ having propagation vector $\mathbf{k}$ and frequency ω, with $k = \omega/c$. Condition (10.1) allows us to neglect any effects of detuning or relaxation during the pulse's action. Moreover, we assume that the sample length L is sufficiently small that

$$|\delta|\,(L/c),\; ku(L/c),\; \gamma(L/c),\; \gamma_2(L/c) \ll 1, \tag{10.18}$$

allowing us to neglect any atomic state evolution resulting from relaxation or atom–field detuning for the time it takes the excitation pulse to pass through the sample. Before the pulse arrives, each atom is in its ground state, implying that the components of the Bloch vector are $u = v = 0$, $w = -1$; that is, the Bloch vector points down [see figure 10.1(a)].

During the pulse, the pseudofield vector can be approximated as $\boldsymbol{\Omega}(t) = [\Omega_0(t), 0, 0]$, owing to condition (10.1). The Bloch vector precesses in the wv plane and, for a $\pi/2$ pulse, is aligned along the v axis just after the pulse. *Following* the pulse, the pseudofield vector is $\boldsymbol{\Omega} = (0, 0, \delta)$, and the Bloch vector precesses about the w axis as it decays.

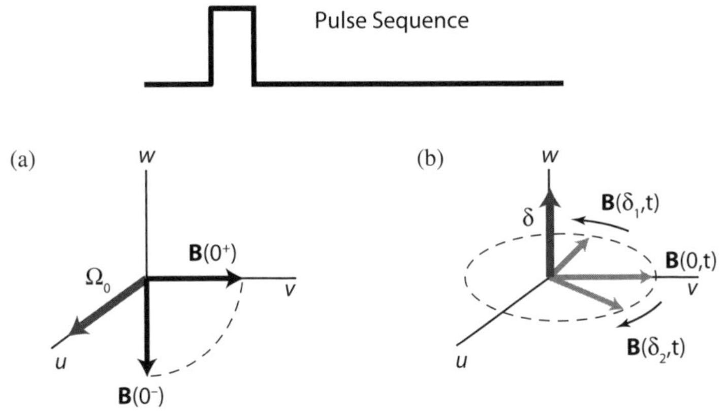

Figure 10.1. Pulse sequence for free polarization decay and associated Bloch diagrams. (a) A $\pi/2$ pulse excites atomic coherence. (b) The Bloch vector precesses about the w axis as it decays. Shown are the Bloch vectors corresponding to three different detunings at time t.

From this result, we can draw two conclusions. First, since the applied field is off when the atoms radiate, radiation is emitted at the natural frequency ω_0. This conclusion follows formally from equations (10.7), (10.11), and (10.13), which can be used to write

$$\boldsymbol{\varrho}^I(t) = \mathbf{F}(t)\mathbf{U}(0)\boldsymbol{\varrho}^I(0), \tag{10.19}$$

from which it follows that

$$\varrho_{21}^I(t) = -i\frac{\sin A}{2}e^{i\mathbf{k}\cdot\mathbf{R}}e^{-\gamma t} \tag{10.20}$$

and

$$\varrho_{21}(\mathbf{R}, t) = -i\frac{\sin A}{2}e^{i\mathbf{k}\cdot\mathbf{R}}e^{-\gamma t}e^{-i\omega_0 t}. \tag{10.21}$$

It is clear that this density matrix element gives rise to a phase-matched signal in the direction of the incident propagation vector $\mathbf{k}_s = \mathbf{k}$ and to both a polarization and signal field that oscillate at the atomic frequency ω_0.

Combining equations (10.17) and (10.20), we find that the FPD power exiting the sample is equal to

$$\mathcal{P}(L, t) = \frac{3\gamma_2}{8\pi}N\hbar\omega_0\left(\mathcal{N}\lambda_0^2 L\right)e^{-2\gamma t}\frac{\sin^2 A}{4}. \tag{10.22}$$

The signal is maximal for a pulse area of $\pi/2$. A measure of the output power as a function of time following the excitation pulse enables one to obtain a value for the transverse relaxation rate γ.

As in all coherent transients of this nature, the signal varies as the square of the density. The incident field pulse creates a phased array of atomic dipoles (similar to a phased array of antennas). The fields radiated by the dipoles interfere constructively in the "phase-matched" direction (in this case, the direction of the incident field pulse), so that the signal emitted in this direction is proportional to the square of the number of atoms in the volume. This type of cooperative emission is to be distinguished from cooperative effects in *spontaneous decay* from an ensemble of atoms (superradiance). Cooperative spontaneous decay is a purely quantum effect—the atoms never acquire a dipole moment. For a pencil-like sample, the neglect of cooperative decay effects is based on the assumption that $\mathcal{N}L/k^2 \ll 1$ (see references [2–5]).[1] In this limit, the coherent radiated field energy, obtained from equation (10.17), is always much less than the energy originally stored in the sample. Most of the energy originally stored in the sample is lost via incoherent emission (spontaneous decay).

10.2.2 Inhomogeneous Broadening

Often, the atoms or molecules are characterized by an inhomogeneous distribution of frequencies—that is, there can be a distribution of transition frequencies distributed about a central frequency $\bar{\omega}_0$. In a solid, this can occur as a result of different strains in the host medium, as discussed in section 7.3. In a vapor, the velocity distribution of the atoms results in a distribution of atomic transition frequencies, when viewed in the laboratory frame. To discuss both solids and vapors in the same context, we define

$$\begin{aligned}\tilde{\delta} &\equiv \delta(\omega_0, \mathbf{k}, \mathbf{v}) = \omega_0 - \omega + \mathbf{k}\cdot\mathbf{v} \\ &= \delta + \mathbf{k}\cdot\mathbf{v} = \delta_0 + \Delta + \mathbf{k}\cdot\mathbf{v},\end{aligned} \tag{10.23}$$

where

$$\delta = \omega_0 - \omega, \qquad \delta_0 = \bar{\omega}_0 - \omega, \qquad \Delta = \omega_0 - \bar{\omega}_0. \tag{10.24}$$

For a solid, there is an inhomogeneous frequency distribution, $W_f(\Delta)$, which, for the sake of definiteness, we take as the Gaussian

$$W_f(\Delta) = \frac{1}{\sqrt{\pi}\sigma_w} e^{-(\Delta/\sigma_w)^2}, \tag{10.25}$$

where σ_w characterizes the width of the inhomogeneous distribution. In a vapor, $\Delta = \sigma_w = 0$, but there is a Maxwellian velocity distribution

$$W_0(\mathbf{v}) = \frac{1}{\left(\pi u^2\right)^{3/2}} e^{-(v/u)^2}, \tag{10.26}$$

where u is the most probable atomic speed. For moving atoms, we must include the velocity-dependent terms in the phase factors $e^{\pm i\mathbf{k}\cdot\mathbf{R}(T_j)} = e^{\pm i\mathbf{k}\cdot[\mathbf{R}-\mathbf{v}(t-T_j)]}$ that appear in equation (10.7).

[1] In the case of inhomogeneously broadened media, the condition is $\mathcal{N}L/k^2 \ll 1$ is replaced by $\mathcal{N}L/k^2\,(\gamma_2/\Delta\omega) \ll 1$, where $\Delta\omega$ is the inhomogeneous width. Cooperative effects of this nature also occur for classical dipoles; the radiative damping of any one dipole is modified by the fields radiated by the other dipoles.

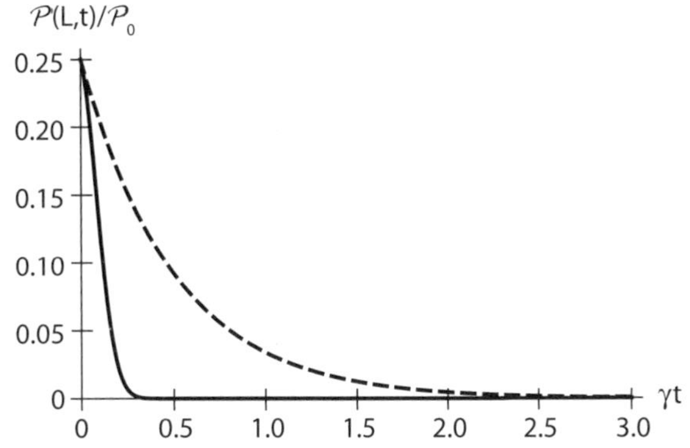

Figure 10.2. Free polarization decay signal for stationary (dashed curve) and inhomogeneously broadened (solid curve) atoms.

The net effect of including inhomogeneous broadening is that equation (10.22) must be replaced by

$$\begin{aligned}\mathcal{P}(L,t) &= \mathcal{P}_0 \left|\langle \varrho_{21}^I(t)e^{-i\delta t}\rangle\right|^2 \\ &= \mathcal{P}_0 \frac{\sin^2 A}{4}\left|\int d\mathbf{v}\, W_0(\mathbf{v})\int d\Delta\, W_f(\Delta)e^{-[\gamma+i(\Delta+\mathbf{k}\cdot\mathbf{v})]t}\right|^2 \\ &= \mathcal{P}_0 \frac{\sin^2 A}{4} e^{-2\gamma t}e^{-(\sigma_w t)^2/2}e^{-(kut)^2/2},\end{aligned} \tag{10.27}$$

where

$$\mathcal{P}_0 = \frac{3\gamma_2}{8\pi}N\hbar\bar{\omega}_0(\mathcal{N}\bar{\lambda}_0^2 L), \tag{10.28}$$

$\bar{\lambda}_0 = 2\pi c/\bar{\omega}_0$, and we used the fact that

$$\begin{aligned}\varrho_{21}^I(t)e^{-i\delta t} &= -i\frac{\sin A}{2}e^{i\mathbf{k}\cdot\mathbf{R}(0)}e^{-\gamma t}e^{-i(\delta_0+\Delta)t} \\ &= -i\frac{\sin A}{2}e^{i\mathbf{k}\cdot(\mathbf{R}-\mathbf{v}t)}e^{-\gamma t}e^{-i(\delta_0+\Delta)t}.\end{aligned} \tag{10.29}$$

In a solid, $u \sim 0$; in a vapor, $\sigma_w \sim 0$.

If $\sigma_w \gg \gamma$ (solids) or $ku \gg \gamma$ (vapors), the signal decays mainly as a result of inhomogeneous broadening. To use a Bloch vector picture to model FPD in an inhomogeneously broadened system, we must remember that *between* pulses, the effective detuning appearing in the field interaction representation is $\delta(\omega_0, \mathbf{k}, \mathbf{v})$. Bloch vectors corresponding to different frequencies precess about the w axis at different rates, implying that the optical dipoles created by the pulse lose their relative phase in a time of order $T_2^* = \sigma_w^{-1}$ (solids) or $(ku)^{-1}$ (vapors) [see figure 10.1(b)]. The FPD signal can be used to measure T_2^*, which can be viewed as an inhomogeneous, transverse relaxation time. In a room temperature vapor, ku/γ is typically of order of 100. In a solid, σ_w/γ can be orders of magnitude larger. In figure 10.2, the FPD signal $\mathcal{P}(L,t)/\mathcal{P}_0$ is plotted as a function of γt for a pulse area $A = \pi/2$, $\sigma_w = 0$, and $k_1 u/\gamma = 0$ (dashed curve—stationary atoms) and $k_1 u/\gamma = 10$ (solid curve—inhomogeneously broadened sample).

The rapid decay of the FPD signal in inhomogeneously broadened media has a simple physical explanation. The pulse excites a phased array of atomic dipoles that begins to radiate coherently. However, these atoms radiate at different frequencies as seen in the laboratory frame. After a time on the order of the inverse of the width of the inhomogeneous frequency distribution, the dipoles are out of phase, and while each atom radiates coherently, there is no longer any constructive interference from the ensemble of atoms.

10.3 Photon Echo

Although the FPD signal produced by short excitation pulses decays in a time of order T_2^*, the coherence of *individual* atoms decays in a much longer time, T_2. The question arises as to whether it is possible to bring all the atomic dipoles back into phase so that they can radiate a coherent signal. The *photon echo* accomplishes this goal, although it has very little to do with either photons or echoes, but it does sound good. The photon echo, first observed in ruby by Kurnit *et al.* [6], is the optical analogue of the spin echo in NMR [7]. A pulse having propagation vector $\mathbf{k}_1$ and frequency ω is applied at $t = 0$, and a second pulse having propagation vector $\mathbf{k}_2 = \mathbf{k}_1$ is applied at $t = T_{21}$. The echo is radiated at time $t = 2T_{21}$ in a direction $\mathbf{k}_s = \mathbf{k}_1$. There are many ways to explain echo formation and some of these are indicated in the following.

In the Bloch vector picture, a $\pi/2$ pulse excites the optical dipoles at $t = 0$, bringing the Bloch vector along the v axis [figure 10.3(a)]. The Bloch vector then begins to precess in the uv plane at a rate equal to the atom–field detuning. In an inhomogeneously broadened medium, the Bloch vectors associated with different atoms precess at different rates and dephase relative to each other in a time T_2^* [figure 10.3(b)]. The dipole coherence is not lost, however. If at time T_{21} a π pulse is applied, the net effect of the pulse is to cause a reflection about the uw plane [figure 10.3(c)]. As the atoms continue to precess at different rates [figure 10.3(d)], the rates are such that the Bloch vectors for *all* the atoms become aligned with the $-v$ axis at time $t = 2T_{21}$ and the "echo" signal is emitted [figure 10.3(e)].

It is not necessary that the pulse areas be equal to $\pi/2$ and π, although these areas lead to a maximal signal. What *is* necessary is that the second pulse produce at least a partial reflection about the uw plane. This reflection takes the Bloch vector components $u + iv$ into $u - iv$ or, equivalently, takes density matrix element $\tilde{\varrho}_{12}$ into $\tilde{\varrho}_{21}$. Since $\tilde{\varrho}_{12}$ and $\tilde{\varrho}_{21}$ are related to the real and imaginary parts of the average dipole moment operator, the second pulse must couple these real and imaginary parts. *Such coupling is impossible for a linear atom–field interaction.* Thus, by its very nature, the photon echo can occur only when a nonlinear atomic response is present.

An alternative way to picture echo formation is to use phase or double-sided Feynman diagrams [8] that keep track of the relative phase of the different dipoles. Diagrams similar to those indicated in figure 10.4 were introduced by Beach, Hartmann, and Friedberg in the context of a "billiard ball echo model" [9] and have been used extensively in theories of atom interferometry. Each line represents a field amplitude. The abscissa is time, and the ordinate is the *phase* associated with the amplitude. As in the Bloch vector picture, the phases represented in this diagram

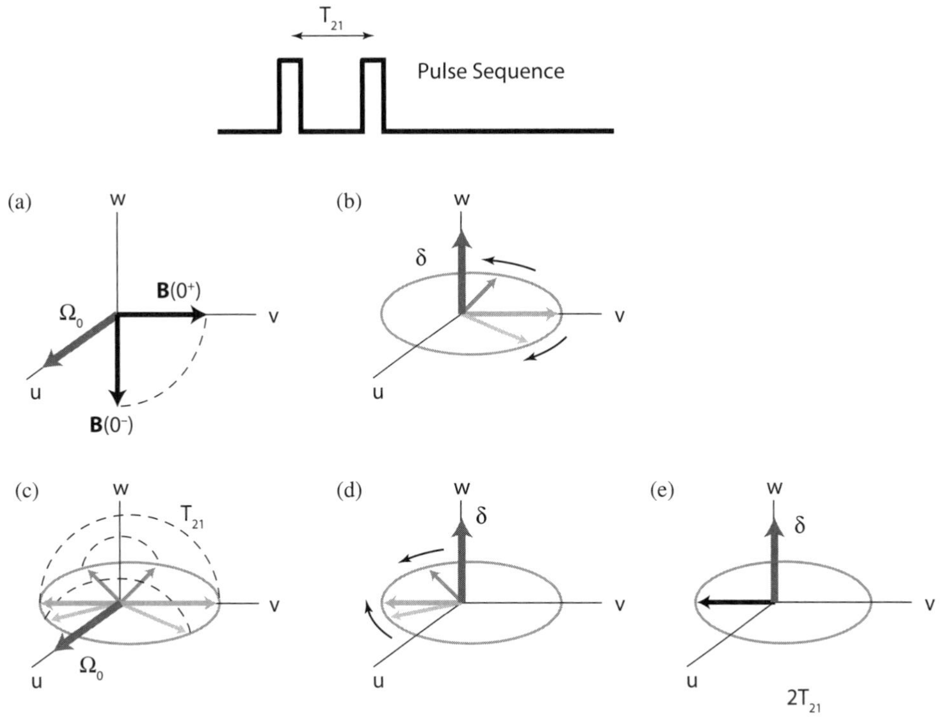

Figure 10.3. Pulse sequence and Bloch vector picture of the two-pulse photon echo.

are those associated with the *field interaction representation*. In the absence of any interactions, it follows from equations (2.99) that the phase associated with the state 1 amplitude in the field interaction representation is $\tilde{\delta}t/2$ and that associated with the state 2 amplitude is $-\tilde{\delta}t/2$, where $\tilde{\delta}$ is given by equation (10.23) with $\mathbf{k} = \mathbf{k}_1$. In these diagrams, the phase of each amplitude is displaced by $-\tilde{\delta}t/2$ so that the state 1 amplitude evolves without any phase change and the state 2 amplitude evolves with a phase equal to $-\tilde{\delta}t$.

Suppose that a pulse is applied at $t = T_j$ that drives transitions between states 1 and 2. When state 1 is converted to state 2, the state 2 amplitude acquires a phase factor $e^{-i\tilde{\delta}(t-T_j)}$ following the pulse. When state 2 is converted to state 1, the state 1 amplitude acquires a phase factor $e^{i\tilde{\delta}(t-T_j)}$ following the pulse.

The concepts are best illustrated by making reference to figure 10.4, which is the phase diagram for the photon echo. An atom starts in state 1 at $t = 0$. At $t = 0$, a pulse is applied that creates a coherent superposition of states 1 and 2. The pulse duration is sufficiently small that it appears to be instantaneous in this and subsequent diagrams. The phase associated with the state 1 amplitude does not change following the pulse, but that associated with the state 2 amplitude does, varying as $-\tilde{\delta}t = -(\delta_0 + \Delta + \mathbf{k}_1 \cdot \mathbf{v})t$ following the pulse, as seen in the figure. The slope of this line differs for atoms having different Δ or $\mathbf{v}$. A vertical cut establishes the density matrix element of interest, and the vertical distance between the two amplitudes is a measure of the *relative* phase of the amplitudes, in the field

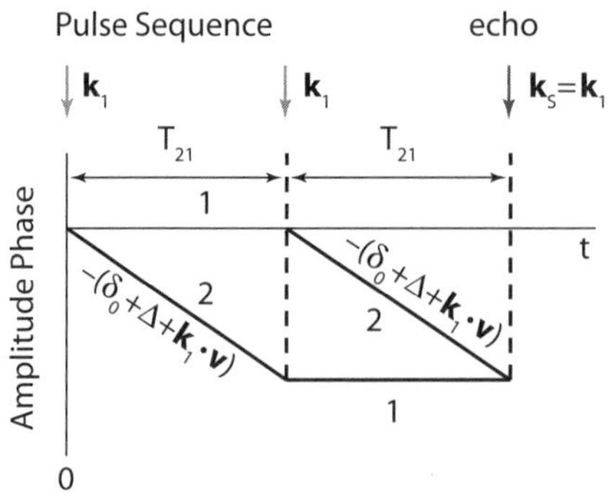

Figure 10.4. Phase diagram for the two-pulse photon echo.

interaction representation. For example, between $t = 0$ and $t = T_{21}$ in figure 10.4, the density matrix element $\tilde{\varrho}_{12}$ has been created with relative phase

$$\phi = 0 - \left(-\tilde{\delta}t\right) = \tilde{\delta}t = (\delta_0 + \Delta + \mathbf{k}_1 \cdot \mathbf{v})\,t, \tag{10.30}$$

which grows with increasing t.

The application of a second pulse at $t = T_{21}$ converts $\tilde{\varrho}_{12}$ to $\tilde{\varrho}_{21}$. This is seen in the figure by the change that occurs for *both* state amplitudes at time T_{21}. State 1 is converted to state 2, which acquires a phase $-(\delta_0 + \Delta + \mathbf{k}_1 \cdot \mathbf{v})(t - T_{21})$ following the pulse. Simultaneously state 2 is converted back to state 1, which has no further phase accumulation—that is, the state 1 amplitude acquires a phase factor $e^{i\tilde{\delta}(t-T_{21})}$ that combines with the phase factor $e^{-i\tilde{\delta}t}$ produced by the first pulse to result in a *constant*, net phase factor, $e^{-i\tilde{\delta}T_{21}}$, for the state 1 amplitude following the second pulse. What is not shown in the diagram are the weighting functions for conversion of the state amplitudes, which are proportional to the sin or cos of the pulse area.

One finds significant contributions to the dipole coherence at a given time *only* when the relative phase is the *same* for all the optical dipoles at that time. Owing to inhomogeneous broadening, the slopes of the slanted lines in figure 10.4 would *differ* for different atoms in both solids and vapors. On averaging over an inhomogeneous frequency distribution, the average dipole coherence vanishes, except at times near *crossings* of the state amplitudes, where the relative phase of all the dipoles is nearly equal to zero.

Between $t = 0$ and $t = T_{21}$, this occurs only near $t = 0$, where an FPD signal is emitted. However, following the second pulse, the state amplitudes in figure 10.4 intersect, and the dipoles are rephased at $t = 2T_{21}$, *independent of the value of* $\mathbf{v}$ or Δ. The pulse sequence has resulted in the dephasing–rephasing process represented schematically in figure 10.3. The echo signal is radiated at time $t \approx 2T_{21}$.

Analytical calculations of the signal intensity can be carried out using equations (10.7), (10.11), and (10.13)—namely,

$$\boldsymbol{\varrho}^I(t) = \mathbf{F}(t - T_{21})\mathbf{U}(T_{21}, \mathbf{k}_1)\mathbf{F}(T_{21})\mathbf{U}(0, \mathbf{k}_1)\boldsymbol{\varrho}^I(0), \tag{10.31}$$

where the explicit dependence of U on $\mathbf{k}_j$ has been indicated. The resulting expressions are rather complicated, in general. However, if $\sigma_w T_{21} \gg 1$ (solid) or $k_1 u T_{21} \gg 1$ (vapor), only those terms in the density matrix sequence indicated schematically in figure 10.4 survive the average over the inhomogeneous frequency distribution in the vicinity of the echo at $t \approx 2T_{21}$. All other terms are smaller by at least a factor of order $\exp[-(k_1 u T_{21})^2/2]$ or $\exp[-(\sigma_w T_{21})^2/2]$. You can convince yourself of this by working out all the terms and carrying out the average over Δ or **v**. *The survival of only a limited number of terms when there is inhomogeneous broadening is a key feature of coherent transients.* One draws phase diagrams to isolate the terms that contribute.

From figure 10.4, we see that to arrive at $\varrho_{21}(t)$, only the chain of interactions

$$\varrho_{11} \to \varrho_{12} \to \varrho_{21} \tag{10.32}$$

contributes significantly. Thus, using equations (10.7) and (10.11), we find

$$\varrho_{12}^I(0^+) = \frac{i}{2} \sin A_1 e^{-i\mathbf{k}_1\cdot[\mathbf{R}-\mathbf{v}t]}, \tag{10.33a}$$

$$\varrho_{12}^I\left(T_{21}^-\right) = \varrho_{12}^I(0^+)e^{-\gamma T_{21}}, \tag{10.33b}$$

$$\varrho_{21}^I\left(T_{21}^+\right) = \sin^2\left(\frac{A_2}{2}\right) e^{2i\mathbf{k}_1\cdot[\mathbf{R}-\mathbf{v}(t-T_{21})]} e^{2i(\delta_0+\Delta)T_{21}} \varrho_{12}^I\left(T_{21}^-\right), \tag{10.33c}$$

$$\varrho_{21}^I(t) = \varrho_{21}^I\left(T_{21}^+\right) e^{-\gamma(t-T_{21})}. \tag{10.33d}$$

At time $t > T_{21}$, we find an averaged density matrix element,

$$\begin{aligned}\langle\varrho_{21}(\mathbf{R},t)\rangle &= \left\langle \varrho_{21}^I(t)e^{-i(\delta_0+\Delta)t}\right\rangle e^{-i\omega t} \\ &= \frac{i}{2}e^{i\mathbf{k}_1\cdot\mathbf{R}} \sin A_1 \sin^2\left(\frac{A_2}{2}\right) e^{-i\omega t} \\ &\quad \times e^{-\gamma t} e^{-i\delta_0(t-2T_{21})} e^{-\sigma_w^2(t-2T_{21})^2/4} e^{-k_1^2u^2(t-2T_{21})^2/4}.\end{aligned} \tag{10.34}$$

The signal pulse propagates in the $\mathbf{k}_1$ direction. The corresponding power exiting the sample is

$$\begin{aligned}\mathcal{P}(L,t) &= \mathcal{P}_0 \left|\left\langle \varrho_{21}^I(t)e^{-i(\delta_0+\Delta)t}\right\rangle\right|^2 \\ &= \mathcal{P}_0 \frac{\sin^2 A_1 \sin^4(A_2/2)}{4} e^{-2\gamma t} e^{-\sigma_w^2(t-2T_{21})^2/2} e^{-k_1^2u^2(t-2T_{21})^2/2},\end{aligned} \tag{10.35}$$

where $\mathcal{P}_0$ is given by equation (10.28).

The echo signal is radiated for times $t \approx 2T_{21}$. The echo intensity near $t = 2T_{21}$ mirrors the FPD intensity immediately following the first pulse. Experimentally, one monitors the maximum echo intensity at $t = 2T_{21}$ or the integrated echo intensity about $t \approx 2T_{21}$ as a function of the pulse separation T_{21}. Both these signals vary as $e^{-4\gamma T_{21}}$, allowing one to extract the transverse relaxation time $T_2 = 1/\gamma$ from such experimental data.

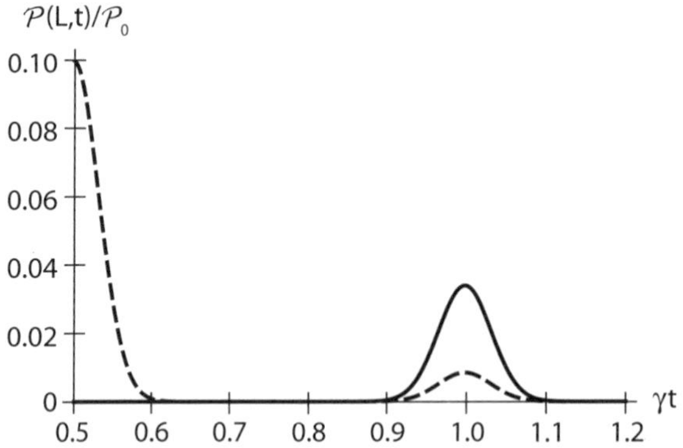

Figure 10.5. Free polarization decay and photon echo signal power following the second pulse in the photon echo pulse sequence for a vapor ($\sigma_w = 0$) having $k_1u/\gamma = 30$, $\gamma_2 = \gamma_{2,1} = 2\gamma$, and $\gamma T_{21} = 0.5$. The dashed curve corresponds to pulse areas ($A_1 = \pi/2$, $A_2 = \pi/2$), while the solid curve corresponds to pulse areas ($A_1 = \pi/2$, $A_2 = \pi$).

To prove that equation (10.34) corresponds to the main contribution to the echo signal at $t \approx 2T_{21}$ when $k_1uT_{21} \gg 1$, we can evaluate the exact expression (10.31) for $t > T_{21}$ using equations (10.8) and (10.12). For $\gamma_2 = \gamma_{2,1}$ and $\delta_0 = \sigma_w = 0$, we find

$$\begin{aligned}\langle \varrho_{21}(\mathbf{R}, t)\rangle = \frac{i}{2} e^{i\mathbf{k}_1\cdot\mathbf{R}} e^{-i\omega t} \Bigg\{ &\sin A_1 \sin^2\left(\frac{A_2}{2}\right) e^{-\gamma t} e^{-k_1^2u^2(t-2T_{21})^2/4} \\ &- \sin A_1 \cos^2\left(\frac{A_2}{2}\right) e^{-\gamma t} e^{-k_1^2u^2t^2/4} \\ &+ \sin A_2 \left[-1 + 2\sin^2\left(\frac{A_1}{2}\right) e^{-\gamma_2 T_{21}}\right] e^{-\gamma(t-T_{21})} e^{-k_1^2u^2(t-T_{21})^2/4} \Bigg\}.\end{aligned} \tag{10.36}$$

The first term corresponds to equation (10.34), the second term is the remnant of the free polarization decay signal following the first pulse (modified somewhat by the second pulse), and the third term is the FPD signal following the second pulse [proportional to the population difference, $\varrho_{22}(T_{21}) - \varrho_{11}(T_{21})$, at the time of the second pulse]. For $k_1uT_{21} \gg 1$, the second and third terms are negligibly small in the vicinity of the echo. In figure 10.5, the signal $\mathcal{P}(L, t)/\mathcal{P}_0$ is plotted as a function of γt for pulse areas ($A_1 = \pi/2$, $A_2 = \pi/2$) and ($A_1 = \pi/2$, $A_2 = \pi$) for $t \geq T_{21}$ and $k_1u/\gamma = 30$. The second term in equation (10.36) contributes negligibly, while the third term gives rise to the FPD signal following the second pulse when $A_2 = \pi/2$, but does not contribute when $A_2 = \pi$. The first term gives rise to the echo signal, with the optimal echo intensity occurring for ($A_1 = \pi/2$, $A_2 = \pi$).

For experimental reasons, it is often convenient to use a different propagation vector for the second pulse. Let $\mathbf{k}_1$ and $\mathbf{k}_2$ be the propagation vectors of the first and second pulses, which have identical carrier frequencies ω. In that case, the only difference is that equation (10.34) is replaced by

$$\begin{aligned}\langle \varrho_{21}(\mathbf{R},t)\rangle &= \left\langle \varrho_{21}^{I}(t)e^{-i(\delta_0+\Delta)t}\right\rangle e^{-i\omega t} \\ &= \frac{i}{2}e^{i(2\mathbf{k}_2-\mathbf{k}_1)\cdot\mathbf{R}}e^{-i\omega t}\sin A_1 \sin^2(A_2/2)e^{-\gamma t} \\ &\quad\times e^{-i\delta_0(t-2T_{21})}e^{-\sigma_w^2(t-2T_{21})^2/4}\int d\mathbf{v}\, W_0(\mathbf{v})e^{i[\mathbf{k}_1\cdot\mathbf{v}T_{21}-\mathbf{k}_2\cdot\mathbf{v}(t-T_{21})]}, \qquad (10.37)\end{aligned}$$

implying that signal propagates in the $\mathbf{k}_s = 2\mathbf{k}_2 - \mathbf{k}_1$ direction, enabling one to separate it from the incident field directions. Recall that $\sigma_w = 0$ for a vapor and $W_0(\mathbf{v}) = \delta_D(\mathbf{v})$ for a solid, where δ_D is the Dirac delta function.

In equation (10.37), there are two things to note. First, the phase-matching condition $k_s = \omega_s/c$ necessary for pulse propagation is no longer satisfied exactly, since $k_s = |2\mathbf{k}_2 - \mathbf{k}_1|$, $k_1 = k_2 = \omega/c$, and $\omega_s = \bar{\omega}_0$; however, for $\omega \approx \bar{\omega}_0$ and for fields that are nearly collinear, the effects of phase mismatch are negligible as long as $\left(k_s^2 - \bar{\omega}_0^2/c^2\right) L^2 \ll 1$. Second, there is now a qualitative difference between the solid and vapor case. Owing to the fact that the detuning depends on the propagation vectors for the vapor, it is not possible to exactly rephase all the dipoles in the vapor when $\mathbf{k}_1 \neq \mathbf{k}_2$. If $|\mathbf{k}_2 - \mathbf{k}_1| uT_{21} \ll 1$, however, nearly complete rephasing of the dipoles occurs for $t \approx 2T_{21}$.

As can be deduced from figure 10.4, the signal is sensitive only to off-diagonal density matrix elements in the entire time interval of interest. Thus, any disturbance of the off-diagonal density matrix elements or optical coherence is reflected as a decrease in the echo intensity. As such, echo signals can serve as a probe of all contributions to transverse relaxation. Transverse relaxation generally falls into two broad categories that can lead to qualitatively different modifications of the coherent transient signals. First, there are dephasing processes that produce an exponential damping of the coherence and contribute to γ. Second, there are *spectral diffusion* (solid) [10–13] or velocity-changing collisions (vapor) [12–15] in which the frequency associated with the optical coherence undergoes changes. Such terms enter the optical Bloch equations as *integral* terms, transforming the equations into differentio-integral equations. In a solid, the change in frequency can be produced by fluctuating fields acting at each atomic site. The situation in vapors is a bit more subtle. The phase-changing and velocity-changing aspects of collisions are entangled and cannot be separated, in general [14]. If the collision interaction is *state-independent*, however, as it is for some molecular transitions, then collisions are purely velocity changing in nature, leading to an echo that decays exponentially as T_{21}^3 for early times and T_{21} for later times [15]. For electronic transitions, collisions are mainly phase changing in nature, but there is a velocity-changing contribution that persists in the forward diffractive scattering cone. This diffractive scattering has been observed for Na- [16], Li- [17], and Yb- [18] rare gas collisions using photon echo techniques.

10.4 Stimulated Photon Echo

Up to this point, we have considered pulse sequences that are useful for measuring transverse relaxation times. Now, we examine *stimulated photon echoes,* which

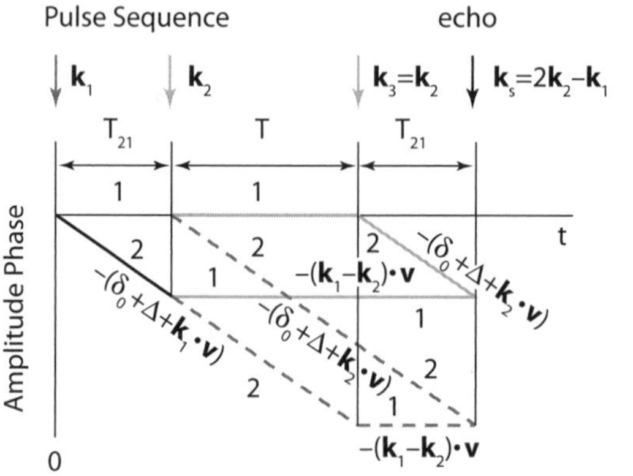

Figure 10.6. Phase diagram for a three-pulse stimulated photon echo. There are two sequences leading to this echo, one with ϱ_{22} and one with ϱ_{11} during the time interval T. The diagrams are drawn for $\mathbf{k}_1 \approx \mathbf{k}_2$.

can be used to *simultaneously* measure both transverse and longitudinal relaxation times. Stimulated photon echoes have become an important diagnostic probe of relaxation in condensed matter systems. The pulse sequence consists of three pulses, having areas A_1, A_2, A_3, and propagation vectors $\mathbf{k}_1, \mathbf{k}_2, \mathbf{k}_3$, with $k_i = \omega/c$. The time interval between the first two pulses is T_{21}, and pulse 3 occurs at time $T_3 = T_{21} + T$. The phase diagrams giving rise to a signal in the $(\mathbf{k}_2 + \mathbf{k}_3 - \mathbf{k}_1)$ direction are shown in figure 10.6 for $\mathbf{k}_3 = \mathbf{k}_2$. There are additional diagrams, giving rise to signals in other directions.

There are some new features in this phase diagram. First, there are *two* sets of paths that lead to an echo signal, one corresponding to population ϱ_{11} and the other to ϱ_{22} in the time interval T. Moreover, since the propagation vector changes from the first to the second pulse, there is no longer a single field interaction representation that can be applied to the entire chain of density matrix elements when one considers a vapor. As a consequence, it is convenient to view the state 1 amplitude as acquiring a phase $-(\mathbf{k}_1 - \mathbf{k}_2) \cdot \mathbf{v}(t - T_{21})$ following the second pulse interaction on the upper set of paths and phase $-(\mathbf{k}_1 - \mathbf{k}_2) \cdot \mathbf{v}(t - T_{21} - T)$ following the third pulse interaction on the lower set of paths. These phases are shown in the diagram and can also be calculated directly using the standard interaction representation, as we do later in this section. If $\mathbf{k}_1 \neq \mathbf{k}_2$, the two paths in each set of diagrams do not intersect identically at the echo time for a vapor and the signal is degraded.

It is assumed that T_{21} is greater than the inhomogeneous relaxation time T_2^*. The signal contains contributions from the optical coherence (off-diagonal density matrix elements) in the time interval $0 \leq t \leq T_{21}$ and for times $t > T_{21} + T$, and contributions from atomic state populations (diagonal density matrix elements) in the time interval T between the second and the third pulses. The echo appears when $t - (T_{21} + T) \approx T_{21}$.

To evaluate the echo signal, we need the sequence

$$\varrho_{11} \to \varrho_{12} \to (\varrho_{22} - \varrho_{11}) \to \varrho_{21}. \tag{10.38}$$

The calculation proceeds much as in the photon echo case. The only new feature is the free evolution of $(\varrho_{22} - \varrho_{11})$ between T_{21} and $T_{21} + T$. Following the second pulse, it follows from equations (10.7) to (10.10) that

$$\varrho_{22}(T_{21}^{+}) = -i \sin A_2 e^{i\Phi(T_2)} \varrho_{12}^{I}\left(T_{21}^{-}\right)/2\,, \tag{10.39a}$$

$$\varrho_{11}(T_{21}^{+}) = i \sin A_2 e^{i\Phi(T_2)} \varrho_{12}^{I}\left(T_{21}^{-}\right)/2. \tag{10.39b}$$

Note that $\varrho_{11}(T_{21}^{+})+\varrho_{22}(T_{21}^{+}) = 0$; these are contributions to the total population that are proportional to $\varrho_{12}^{I}(T_{21}^{-})$ and represent *changes* to and *not* the total population. From equations (10.39) and (10.12), we obtain

$$\varrho_{22}\left[(T_{21}+T)^{-}\right] - \varrho_{11}\left[(T_{21}+T)^{-}\right] = -i \sin A_2 e^{i\Phi(T_2)} G(T) \varrho_{12}^{I}(T_{21}^{-}), \tag{10.40}$$

where

$$G(T) = \frac{e^{-\gamma_2 T}}{2}\left(1 + \frac{\gamma_{2,1}}{\gamma_2}\right) + \frac{1}{2}\left(1 - \frac{\gamma_{2,1}}{\gamma_2}\right). \tag{10.41}$$

With this step established, it is easy to piece together the stimulated photon echo chain (10.38) as

$$\begin{aligned}\langle\varrho_{21}(\mathbf{R},t)\rangle &= (i/4)e^{i(2\mathbf{k}_2-\mathbf{k}_1)\cdot\mathbf{R}} e^{-i\omega t} \sin A_1 \sin A_2 \sin A_3 e^{-\gamma(t-2T_{21}-T)} \\ &\times G(T) e^{-2\gamma T_{21}} e^{-i\delta_0(t-2T_{21}-T)} e^{-\sigma_w^2(t-2T_{21}-T)^2/4} \\ &\times \int d\mathbf{v}\, W_0(\mathbf{v}) e^{i[\mathbf{k}_1\cdot\mathbf{v}T_{21}+(\mathbf{k}_1-\mathbf{k}_2)\cdot\mathbf{v}T-(2\mathbf{k}_2-\mathbf{k}_1)\cdot\mathbf{v}(t-T_{21}-T)]},\end{aligned} \tag{10.42}$$

and the output power is proportional to $|\langle\varrho_{21}(\mathbf{R},t)\rangle|^2$. The optimal pulse sequence consists of three $\pi/2$ pulses. Phase matching can be achieved only if $\left|\mathbf{k}_1 - \mathbf{k}_2\right| L \ll 1$. In a solid, the integral in equation (10.42) is equal to unity, and the echo signal is maximal for $t = T + 2T_{21}$. In a vapor, the echo signal is degraded if $\mathbf{k}_1 \neq \mathbf{k}_2$; however, if $\mathbf{k}_1 \approx \mathbf{k}_2$, then at $t = T + 2T_{21}$, the echo amplitude varies as

$$e^{-\gamma_2 T} e^{-2\gamma T_{21}} e^{-\left|\mathbf{k}_1-\mathbf{k}_2\right|^2 u^2 (T+2T_{21})^2/4}. \tag{10.43}$$

By varying the angle between $\mathbf{k}_1$ and $\mathbf{k}_2$, one can determine the Doppler width $k_1 u$. By monitoring the echo signal at $t = T + 2T_{21}$ as a function of T_{21} (T), one obtains information about the transverse (longitudinal) relaxation.

For $\mathbf{k}_1 \approx \mathbf{k}_2$, an echo signal is generated when the relative inhomogeneous phase, $(\delta_0+\Delta+\mathbf{k}_1\cdot\mathbf{v})T_{21}$, acquired in the time interval T_{21}, is canceled by the relative phase,

$$[(\delta_0 + \Delta + \mathbf{k}_2\cdot\mathbf{v}) - (\mathbf{k}_1 - \mathbf{k}_2)\cdot\mathbf{v}]\,(t - T_{21}) \approx (\delta_0 + \Delta + \mathbf{k}_1\cdot\mathbf{v})\,(t - T_{21}),$$

acquired in the interval T_{21} following the third pulse. If, between the second and third pulses, the frequency (solid) or velocity (vapor) has changed owing to spectral diffusion (solid) or velocity-changing collisions (vapor), the phase cancellation is not complete. Thus, the echo signal as a function of T provides information about these relaxation processes. For a closed two-level system ($\gamma_{2,1} = \gamma_2$), $G(T) = e^{-\gamma_2 T}$; as a consequence, the rate of spectral diffusion or velocity-changing collisions must be of order or greater than γ_2 to be observable. On the other hand, for *open* systems, $G(T)$ is *constant* for $\gamma_2 T \gg 1$ (see figure 10.7). As a result, one has a much longer time to observe such effects, since $G(T)$ is nonvanishing over some effective

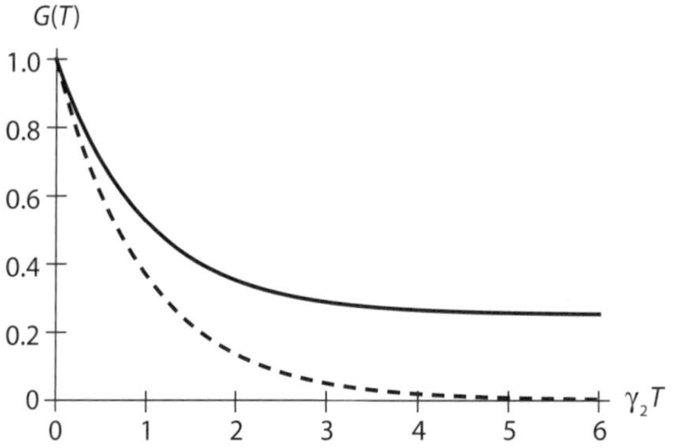

Figure 10.7. Graph of the function $G(T)$ as a function of $\gamma_2 T$. The stimulated echo signal is proportional to $G(T)^2$. For a closed system ($\gamma_{2,1} = \gamma_2$, dashed curve), $G(T)$ decays to zero, but for an open system ($\gamma_{2,1} = \gamma_2/2$, solid curve), $G(T)$ assumes a constant value for arbitrarily long T.

ground-state lifetime (taken here to be infinite). One can exploit this feature of open systems to study spectral diffusion or velocity-changing collisions with very high sensitivity since T can be chosen to be much larger than γ_2^{-1} [19, 20].

Open systems also offer interesting possibilities as storage devices. Since the effective ground-state lifetime can be as long as days in certain solids, one can write interferometric information into the sample by replacing one of the first two pulses by a signal pulse and reading it out at a later time with the third pulse. In the case of vapors, it is also possible to replace some of the incident pulses by standing-wave fields. In this manner, modulated *ground-state* populations with associated Doppler phases of order kuT can be created and rephased, providing sensitivity to velocity-changing collisions as small as a few cm/s.

If a Λ system is substituted for the two-level atom, then information can be stored in the coherence between levels 1 and 3 in the time interval T. Since these levels can be chosen as two sublevels of the ground-state manifold, this type of scheme offers an attractive way of storing information in a long-lived internal state coherence. An example of this nature is given in the problems.

10.5 Optical Ramsey Fringes

We have seen that coherent transients can be used to measure relaxation rates. With a slight modification of the stimulated photon echo geometry, it is also possible to use coherent transient signals to measure transition *frequencies* as well! By choosing $\mathbf{k}_3 = -\mathbf{k}_1$, we can generate a phase-matched signal in the

$$\mathbf{k}_s = \mathbf{k}_1 - \mathbf{k}_2 + \mathbf{k}_3 = -\mathbf{k}_2$$

direction. Moreover, in weak fields, the amplitude of the signal field is proportional to the conjugate of input field 2. As a consequence, the signal is referred to as a *phase conjugate* signal. To simplify matters, we set $\mathbf{k}_1 \approx \mathbf{k}_2$ and neglect terms of order $|\mathbf{k}_1 - \mathbf{k}_2| u(T + T_{21})$.

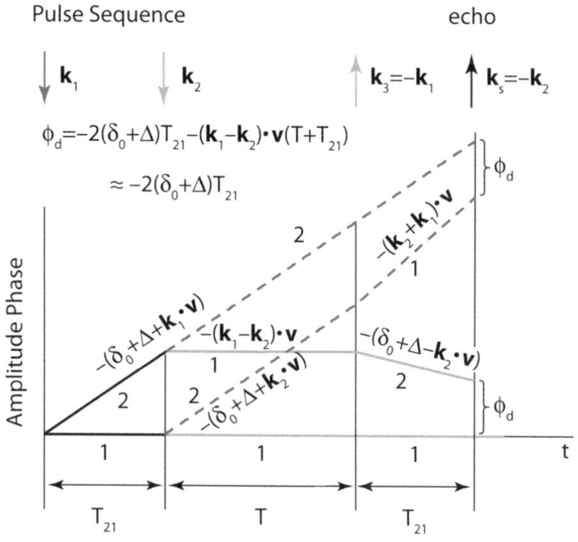

Figure 10.8. Phase diagram for a stimulated pulse echo sequence giving rise to optical Ramsey fringes in a vapor, but to no signal in an inhomogeneously broadened solid.

The echo signals produced in this manner are the optical analogues [21, 22] of Ramsey fringes [23], a method proposed by Norman Ramsey for measuring molecular transition frequencies using *spatially separated* oscillatory fields. As proposed by Ramsey, a molecular beam passes through two field interaction zones separated by a distance L. In effect, the first interaction starts a clock, and the second interaction stops the clock. In the interim, one measures the relative phase of the oscillatory field with that of the molecular transition as a function of the frequency of the oscillatory field. The resultant signal allows one to achieve a frequency resolution of order v_ℓ/L, where v_ℓ is the longitudinal speed of the beam. A method for producing Ramsey fringes using an atomic beam passing through several, spatially separated *optical* fields is discussed in appendix B.

To explore the possibility of obtaining optical Ramsey fringes in vapors and solids, we consider the phase diagram shown in figure 10.8. The density matrix chain leading to the optical Ramsey fringe signal is

$$\varrho_{11} \to \varrho_{21} \to (\varrho_{22} - \varrho_{11}) \to \varrho_{21}. \tag{10.44}$$

There is a qualitative difference between the phase diagrams of figure 10.8 and figure 10.6. At time $t = 2T_{21} + T$, the lines representing the state amplitudes *do not cross* in figure 10.8, even for $\mathbf{k}_1 = \mathbf{k}_2$. Rather, they are separated by a phase difference of $\phi_d = -2(\delta_0 + \Delta)T_{21}$. The phase shift resulting from *Doppler shifts* cancels at $t = 2T_{21}+T$ (for $\mathbf{k}_1 \approx \mathbf{k}_2$), but *not* the phase shift resulting from the atom–field detuning. The significance of these results will become apparent immediately. The averaged density matrix element in the vicinity of the echo can be calculated using the chain (10.44) as

$$\langle \varrho_{21}(\mathbf{R}, t)\rangle = (i/4)e^{-i\mathbf{k}_2\cdot\mathbf{R}}e^{-i\omega t}\sin A_1 \sin A_2 \sin A_3 G(T)e^{-\gamma(t-2T_{21}-T)}$$
$$\times e^{-2\gamma T_{21}}e^{-i\delta_0(t-T)}e^{-\sigma_w^2(t-T)^2/4}e^{-k^2u^2(t-2T_{21}-T)^2/4}. \tag{10.45}$$

In a solid, the signal is negligibly small near $t = 2T_{21} + T$ since $\sigma_w T_{21} \gg 1$. (There are no optical Ramsey fringes in an inhomogeneously broadened solid owing to the average over Δ.)

In a vapor, however, an echo is formed at time $t \approx T + 2T_{21}$. Although the lines corresponding to state amplitudes 1 and 2 do not cross at this time, the *relative* phase is the same for *all* the atoms. (Recall that $\Delta = 0$ in a vapor, so there is no average over Δ to kill the signal, as in solids.) At $t = T + 2T_{21}$, the averaged density matrix element $\langle \tilde{\varrho}_{21} \rangle$ varies as $e^{-2i\delta_0 T_{21}}$, a factor that was absent when all fields are nearly copropagating. This phase factor is the optical analogue of the phase factor that is responsible for the generation of Ramsey fringes. One can measure the phase factor directly by heterodyning the signal field with a reference field, or by converting the off-diagonal density matrix element to a population by the addition of a *fourth* pulse in the $-\mathbf{k}_2$ direction at time $t = T + 2T_{21}$. The population can be detected, for example, by the spontaneous emission signal emitted by the excited atoms. In either case, the signal varies as $\cos(2\delta_0 T_{21})$. In itself, this dependence is useless for determining the optical frequency, since one cannot identify the fringe corresponding to $\delta_0 = 0$. It is necessary to take data as a function of δ_0 for several values of T_{21}, and then average the data over T_{21}; in this manner, the central fringe can be identified [24]. Using a four-pulse sequence of this type on an ensemble of ultracold, laser-cooled atoms, Degenhardt *et al.* achieved a relative uncertainty of one part in 10^{14} on the Ca intercombination line [25].

A typical Ramsey fringe pattern, proportional to

$$H(\delta_0) = \int_0^\infty dT_{21} \cos(2\delta_0 T_{21}) e^{-2\gamma T_{21}} W(T_{21}) \tag{10.46}$$

for fixed T, is shown in figure 10.9 as a function of $a = \delta_0 \bar{T}_{21}$ for $\gamma \bar{T}_{21} = 0.2$ and $\Delta T_{21}/\bar{T}_{21} = 0.3$, assuming a distribution function

$$W(T_{21}) = \begin{cases} \frac{1}{\Delta T_{21}} & -\frac{\Delta T_{21}}{2} \leq T_{21} - \bar{T}_{21} \leq \frac{\Delta T_{21}}{2} \\ 0 & \text{otherwise} \end{cases}. \tag{10.47}$$

For $\delta_0 = 0$, all phases cancel, and the signal is a maximum. As $\delta_0 \bar{T}_{21}$ increases, the averaged signal begins to wash out. Experiments of this type allow one to measure optical frequencies with accuracy of order $\bar{T}_{21}^{-1}$. While this would appear to be "subnatural" resolution for $\bar{T}_{21} > \gamma_2^{-1}$, such a designation is somewhat misleading. The Ramsey fringe signal in this case diminishes as $e^{-2\gamma_2 \bar{T}_{21}}$, so the ability to measure a signal always boils down to a question of signal to noise.

The optical Ramsey fringe geometry has been reinterpreted as an atom interferometer [26, 27]. It is not difficult to see how such an interpretation is possible for any coherent transient signal that leads to echo formation. As is evident from figures 10.4 to 10.8, there are two paths that combine to produce the echo signal. One can interpret each path as representing one arm of an interferometer, although the analogue with optical interferometers is somewhat misleading. Optical interferometers rely on the wave nature of light for their operation, but as discussed in this chapter, the wave nature of the atoms' center-of-mass motion plays no role in echo formation. On the other hand, the interpretation becomes cogent when the atomic motion is quantized, since there are frequency shifts associated with

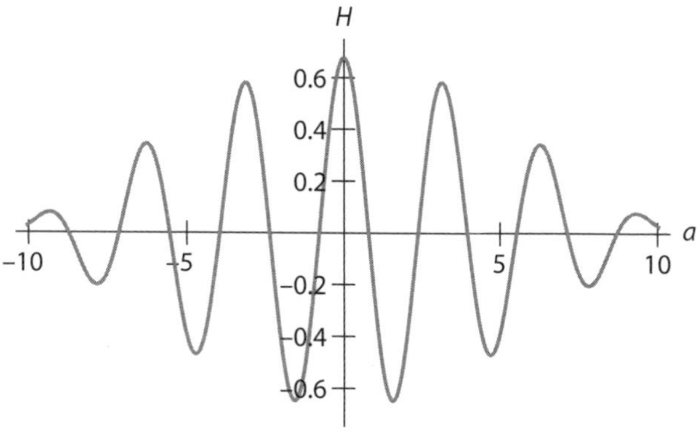

Figure 10.9. Ramsey fringe signal H as a function of $a = \delta_0 \bar{T}_{21}$ for $\gamma \bar{T}_{21} = 0.2$ and $\Delta T_{21}/\bar{T}_{21} = 0.3$.

the recoil atoms undergo on emitting or absorbing radiation that are different for the upper and lower pairs of paths shown in figure 10.8 (see appendix B). Atom interferometers are discussed in chapter 11, where the importance of quantum aspects of atomic motion are explored in more detail.

10.6 Frequency Combs

There has been a mini-revolution in metrology owing to the development of *frequency combs*. Frequency combs refer to the spectral distribution of the output of a *mode-locked* laser. By completely specifying and stabilizing the frequency of each "tooth" in the comb, one can use the beat frequency between the comb teeth and optical frequency standards to actually "count" optical cycles. Moreover, the combs allow one to compare frequency standards differing by frequencies on the order of optical frequencies. John Hall and Theodor Hänsch shared the Noble Prize (with Roy Glauber) in 2005 for their work on laser-based precision spectroscopy, involving, but not limited to, the development of frequency comb spectroscopy. We give a very brief introduction to the principles involved in the generation of frequency combs. More detailed accounts can be found in references [28–31].

In a laser cavity, the mode structure is determined by the length of the cavity. In other words, for a cavity of length L, possible mode frequencies are

$$\omega_n = n\omega_r, \tag{10.48}$$

where

$$\omega_r = \pi c/L = 2\pi/\tau_r, \tag{10.49}$$

and $\tau_r = 2L/c$ is the *round-trip time* of light in the cavity. To construct a laser using such a cavity, one must put an active medium inside the cavity. An active medium consists of atoms that have been incoherently excited to produce a population inversion of some of the atomic levels. In other words, the levels involved in the laser transition must be maintained with more population in the higher energy state. Depending on the degree of population inversion, the closeness of the atomic transition frequency to a cavity frequency, and the atomic density, one or more modes may be above *threshold* for lasing. The actual frequency of the modes will be changed somewhat from the cavity mode frequencies owing to the presence of the medium, but the spacing between modes remains an integer multiple of ω_r.

When a large number of modes are above threshold, they normally have random phases, resulting in a laser output that is quasi-continuous. To produce the short pulses needed to generate frequency combs, it is necessary to *phase lock* the various modes [32]. Mode locking can be accomplished *actively* by placing an acousto-optic or electro-optic modulator in the laser cavity that modulates the light in the cavity with a frequency ω_r. Such active mode locking can produce pulse widths on the order of picoseconds. To achieve the femtosecond pulses used in frequency combs, *passive mode locking* is employed. In one approach to passive mode locking, a saturable absorber is placed inside the laser cavity. As was shown in chapter 5, once a transition is saturated or bleached, the relative absorption of an incident field increases very slowly with increasing field strength. As such, a saturable absorber favors the passage of intense fields. When placed in a cavity, a saturable absorber can result in a nonlinear absorption coefficient that is modulated at frequency ω_r. The mode coupling in the presence of a saturable absorber is modified in such a manner that results in phase locking of the modes. The field inside the cavity then consists of a narrow, intense pulse that propagates between the cavity mirrors, a small part of which is coupled out of the cavity, resulting in a train of equally spaced pulses. An alternative method for achieving passive mode locking that is used in the titanium-sapphire laser is based on lensing produced by the *Kerr effect* [33]. The Kerr effect is the change in the index of refraction of a medium produced by intense fields (see problem 1 in chapter 7).

Let us imagine that N+1 modes are above threshold. We set the central frequency equal to

$$\omega_{central} = \omega_c + \omega_{ph}, \tag{10.50}$$

where $\omega_c = n_0\omega_r$ corresponds to one of the cavity modes in the absence of the medium, and ω_{ph} is the offset resulting from the presence of the medium. Thus, the output frequencies are given by

$$\omega_n = \omega_c + n\omega_r + \omega_{ph}, \tag{10.51}$$

and n runs from $-N/2$ to $N/2$ (assuming N to be even).

The Fourier transform of the positive frequency component of the field is assumed to be of the form of a *comb* of frequencies,

$$\tilde{E}_+(\omega) = g\left(\omega - \omega_c\right) \sum_{n=-N/2}^{N/2} \delta\left(\omega - \omega_c - n\omega_r - \omega_{ph}\right), \tag{10.52}$$

where $g(\bar{\omega})$ is a smooth function (think Gaussian) that is centered at $\bar{\omega} = 0$ and has a width that corresponds to anywhere between several thousand and a million modes. Since this spectral function cuts off the modes that are below threshold, we can extend the sum as

$$\tilde{E}_+(\omega) = g(\omega - \omega_c) \sum_{n=-\infty}^{\infty} \delta(\omega - \omega_c - n\omega_r - \omega_{ph}) \tag{10.53}$$

without making any significant errors. What we achieve here is that the sum is now a periodic function of $(\omega - \omega_c - \omega_{ph})$ and can be expressed as a Fourier series,

$$\sum_{n=-\infty}^{\infty} \delta\left(\omega - \omega_c - n\omega_r - \omega_{ph}\right) = \frac{1}{\omega_r} \sum_{n=-\infty}^{\infty} e^{2\pi i n\left(\omega - \omega_c - \omega_{ph}\right)/\omega_r}. \tag{10.54}$$

Combining equations (10.53) and (10.54), and taking the Fourier transform, we obtain

$$\begin{aligned} E_+(t) &= \frac{1}{\sqrt{2\pi}\,\omega_r} \int_{-\infty}^{\infty} e^{-i\omega t} g(\omega - \omega_c) \sum_{n=-\infty}^{\infty} e^{in\left(\omega - \omega_c - \omega_{ph}\right)\tau_r} d\omega \\ &= \frac{e^{-i\omega_c t}}{\omega_r} \sum_{n=-\infty}^{\infty} f\left(t - n\tau_r\right) e^{-in\omega_{ph}\tau_r}, \end{aligned} \tag{10.55}$$

where $f(t)$ is the Fourier transform of $g(\omega)$. Thus, if $g(\omega)$ is a Gaussian having a width on the order of $\Delta\omega$, then $f(t)$ is a pulse envelope having a temporal width of order $\tau_p = 2\pi/\Delta\omega$. We see that a frequency comb corresponds to an infinite number of pulses in the time domain, with the time between pulses equal to τ_r.

From equation (10.55), it follows that the maximum of the pulse *envelope* does not, in general, coincide with a maximum of the pulse *amplitude*. The phase difference between successive amplitude maxima relative to the pulse envelope maxima is given by

$$\Delta\phi_{ph} = \omega_c\tau_r + \omega_{ph}\tau_r = \omega_{ph}\tau_r \quad (\text{modulo } 2\pi), \tag{10.56}$$

since $\omega_c\tau_r = n_0\omega_r\tau_r = 2\pi n_0 = 0$ (modulo 2π). Thus, the phase shift from pulse to pulse is a linear function of pulse number. The phase slippage results from the difference between the phase velocity and the group velocity of the pulses in the cavity, which, in turn, is caused by the presence of the medium in the cavity. Some of these results are depicted in figure 10.10.

We can now characterize the laser output. In the frequency domain, $\omega_r/2\pi$ typically ranges from MHz to GHz, $\Delta\omega/2\pi$ (number of excited modes times $\omega_r/2\pi$) is typically 1 to 100 THz, and $\omega_{ph}/2\pi$ is in the MHz range. In principle, the frequency width of each tooth in the frequency comb can be sub-mHz, determined by the quality factor of the cavity (the inverse of the fraction of energy lost from the cavity each round-trip), the total power output of the laser pulses, and spontaneous emission processes, but in practice, the width is several orders of magnitude larger owing to technical noise. (The *relative* frequency stability of different comb teeth can be as small as a few μHz [34].) In the time domain, the width $\tau_p = 2\pi/\Delta\omega$ can be as small as several fs, and the time between pulses τ_r is typically in the 10-ns range.

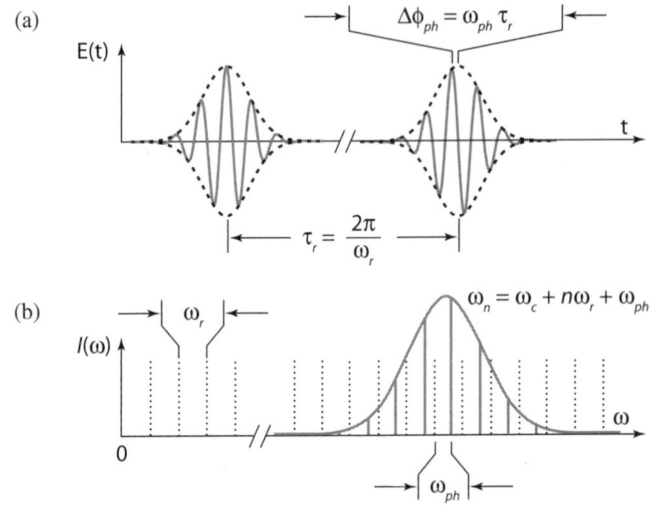

Figure 10.10. (a) Output of a mode-locked laser in the time domain. The duration of each pulse is of order $\tau_p = 2\pi/\Delta\omega$, where $\Delta\omega = N\omega_r$, and N is the number of modes above threshold. (b) Frequency spectrum of the output. The frequency range of excited modes is of order $\Delta\omega = N\omega_r = 2\pi N/\tau_r$. (Adapted from reference [31].)

The main point is that in order to maintain the integrity of the comb, one need only stabilize two radio frequencies, ω_r and ω_{ph}. Methods for doing this can be found in reference [31]. Stabilizing the comb to radio frequencies in effect connects radio frequency standards with optical frequencies. The only thing needed is a method to measure ω_{ph} such that the frequencies can be determined in an absolute sense. This has become possible recently with the development of optical fibers that can broaden the comb spectrum to the point where more than an octave of frequencies is contained in the comb, without introducing any additional distortion. In other words, the comb contains both frequencies ω and 2ω. By comparing

$$\begin{aligned} \omega_{2n} &= 2\left(\omega_c - m\omega_r\right) + \omega_{ph} \\ &= 2\left(n_0 - m\right)\omega_r + \omega_{ph} \equiv 2n\omega_r + \omega_{ph} \end{aligned} \tag{10.57}$$

$(n = n_0 - m)$ with the frequency

$$2\omega_n = 2\left(n\omega_r + \omega_{ph}\right) \tag{10.58}$$

obtained using a frequency-doubling crystal, one can subtract the two frequencies to obtain the value of ω_{ph} and then stabilize this frequency relative to a standard. In this manner, both the frequency separation and the absolute frequency in each tooth of the comb can be stabilized.

10.7 Summary

In this chapter, we have given a very brief introduction to optical coherent transients. Several examples were given to illustrate the manner in which coherent transients can be created and the manner in which they can be used to measure atomic frequencies and relaxation rates. The field of coherent transients continues to evolve at a rapid rate. In chemistry and biology, the use of "multidimensional" coherent transients (coherent transients observed as a function of time delays and carrier frequencies of the applied pulses) is developing as a analytical tool for unraveling relaxation dynamics in complex molecules [35]. We have alluded to the possibility of an interpretation of optical Ramsey fringes in terms of atom interferometry. We now turn our attention to atom optics and atom interferometry, where the importance of quantum aspects of atomic motion are explored in more detail.

10.8 Appendix A: Transfer Matrices in Coherent Transients

In this appendix, we give the evolution matrices in the field interaction representation and Bloch pictures. For a field interaction zone,

$$\begin{pmatrix} u \\ v \\ w \\ m \end{pmatrix}^{+} = \begin{pmatrix} 1 & 0 & 0 & 0 \\ 0 & \cos A & -\sin A & 0 \\ 0 & \sin A & \cos A & 0 \\ 0 & 0 & 0 & 1 \end{pmatrix} \begin{pmatrix} u \\ v \\ w \\ m \end{pmatrix}^{-}, \tag{10.59}$$

or

$$\begin{pmatrix} \varrho_{11} \\ \varrho_{22} \\ \tilde{\varrho}_{12} \\ \tilde{\varrho}_{21} \end{pmatrix}^{+} = \frac{1}{2} \begin{pmatrix} 1+\cos A & 1-\cos A & i\sin A & -i\sin A \\ 1-\cos A & 1+\cos A & -i\sin A & i\sin A \\ i\sin A & -i\sin A & 1+\cos A & 1-\cos A \\ -i\sin A & i\sin A & 1-\cos A & 1+\cos A \end{pmatrix} \begin{pmatrix} \varrho_{11} \\ \varrho_{22} \\ \tilde{\varrho}_{12} \\ \tilde{\varrho}_{21} \end{pmatrix}^{-}. \tag{10.60}$$

Note that the results are simpler in the (u, v, w, m) basis [equation (10.59)].

For free evolution between the interaction zones, the field is off, and the equations are simpler in the density matrix basis [equation (10.60)]. To be a little more general, we allow for an open system in which $\gamma_{2,1} \neq \gamma_2$. Moreover, we include an overall identical decay of both levels 1 and 2 with rate $\gamma_1 \ll \gamma_2$, intended to simulate transit time effects in vapors. (If atoms move out of the beam, they are lost.) In the field-free region, we need to consider the equations of motion for the atomic state populations:

$$\dot{m} = -\frac{\gamma_1 + \gamma_2^t - \gamma_{2,1}}{2} m - \frac{\gamma_2^t - \gamma_{2,1} - \gamma_1}{2} w, \tag{10.61a}$$

$$\dot{w} = -\frac{\gamma_2^t + \gamma_{2,1} - \gamma_1}{2} m - \frac{\gamma_2^t + \gamma_{2,1} + \gamma_1}{2} w, \tag{10.61b}$$

and

$$\dot{\varrho}_{11} = \gamma_{2,1}\varrho_{22} - \gamma_1\varrho_{11}\,, \tag{10.62a}$$
$$\dot{\varrho}_{22} = -\gamma_2^t\varrho_{22}\,, \tag{10.62b}$$
$$\dot{\tilde{\varrho}}_{12} = -(\gamma - i\tilde{\delta})\tilde{\varrho}_{12}\,, \tag{10.62c}$$
$$\dot{\tilde{\varrho}}_{21} = -(\gamma + i\tilde{\delta})\tilde{\varrho}_{21}\,, \tag{10.62d}$$

where $\tilde{\delta} = \delta_0 + \mathbf{k}\cdot\mathbf{v} + \Delta$ and

$$\gamma_2^t = \gamma_2 + \gamma_1. \tag{10.63}$$

The solutions of these equations are

$$\begin{pmatrix} u(t_0+T) \\ v(t_0+T) \\ w(t_0+T) \\ m(t_0+T) \end{pmatrix} = \begin{pmatrix} e^{-\gamma T}\cos\tilde{\delta}T & -e^{-\gamma T}\sin\tilde{\delta}T & 0 & 0 \\ e^{-\gamma T}\sin\tilde{\delta}T & e^{-\gamma T}\cos\tilde{\delta}T & 0 & 0 \\ 0 & 0 & F_{33} & F_{34} \\ 0 & 0 & F_{43} & F_{44} \end{pmatrix} \begin{pmatrix} u(t_0) \\ v(t_0) \\ w(t_0) \\ m(t_0) \end{pmatrix}, \tag{10.64}$$

where

$$F_{33} = \frac{1}{2}\left[e^{-\gamma_2^t T}\left(1 + \frac{\gamma_{2,1}}{\gamma_2^t - \gamma_1}\right) - e^{-\gamma_1 T}\left(1 - \frac{\gamma_{2,1}}{\gamma_2^t - \gamma_1}\right)\right], \tag{10.65a}$$

$$F_{34} = \frac{1}{2}\left(1 + \frac{\gamma_{2,1}}{\gamma_2^t - \gamma_1}\right)\left(e^{-\gamma_2^t T} - e^{-\gamma_1 T}\right), \tag{10.65b}$$

$$F_{43} = \frac{1}{2}\left(1 - \frac{\gamma_{2,1}}{\gamma_2^t - \gamma_1}\right)\left(e^{-\gamma_2^t T} - e^{-\gamma_1 T}\right), \tag{10.65c}$$

$$F_{44} = \frac{1}{2}\left[e^{-\gamma_2^t T}\left(1 - \frac{\gamma_{2,1}}{\gamma_2^t - \gamma_1}\right) + e^{-\gamma_1 T}\left(1 + \frac{\gamma_{2,1}}{\gamma_2^t - \gamma_1}\right)\right], \tag{10.65d}$$

and

$$\begin{pmatrix} \varrho_{11}(t_0+T) \\ \varrho_{22}(t_0+T) \\ \tilde{\varrho}_{12}(t_0+T) \\ \tilde{\varrho}_{21}(t_0+T) \end{pmatrix} = \frac{1}{2}\begin{pmatrix} e^{-\gamma_1 T} & \frac{\gamma_{2,1}}{\gamma_2^t-\gamma_1}\left(e^{-\gamma_1 T} - e^{-\gamma_2^t T}\right) & 0 & 0 \\ 0 & e^{-\gamma_2^t T} & 0 & 0 \\ 0 & 0 & e^{-(\gamma - i\tilde{\delta})T} & 0 \\ 0 & 0 & 0 & e^{-(\gamma + i\tilde{\delta})T} \end{pmatrix} \times \begin{pmatrix} \varrho_{11}(t_0) \\ \varrho_{22}(t_0) \\ \tilde{\varrho}_{12}(t_0) \\ \tilde{\varrho}_{21}(t_0) \end{pmatrix}. \tag{10.66}$$

We have left the factor $\gamma_2^t - \gamma_1 = \gamma_2$ explicitly in the denominators to show that an *overall* decay rate alone does not lead to a long-lived signal. For a closed system, except for overall decay, i.e. $\gamma_{2,1} = \gamma_2^t - \gamma_1$, $F_{33} \sim e^{-\gamma_2^t T}$, $F_{34} \sim e^{-\gamma_2^t T} - e^{-\gamma_1 T}$, $F_{43} \sim 0$, $F_{44} \sim e^{-\gamma_1 T}$—that is, F_{33} has no slowly decaying part, and as a consequence, there is no long-lived signal in the stimulated photon echo.

To calculate a coherent transient signal, one multiplies the appropriate transfer matrices together; however, it is efficient to use the diagrammatic technique to isolate only those terms that are needed.

10.9 Appendix B: Optical Ramsey Fringes in Spatially Separated Fields

An alternative means for observing coherent transient signals is to use an atomic beam that traverses a field interaction zone. A standard geometry involves a well-collimated atomic beam propagating in the Z direction that passes through one or more field interaction zones. The fields propagate in the $\pm X$ direction and have beam width equal to w (see figure 11.4 in chapter 11) in the Z direction. As such, if an atom is moving with longitudinal velocity v_ℓ, it "sees" a radiation pulse having duration w/v_ℓ in its rest frame. The velocity density distribution of an effusive thermal beam can be approximated as [36, 37]

$$W(\mathbf{v}_T, v_\ell) = \frac{4}{\pi^{1/2} u_\ell^3} v_\ell^2 e^{-v_\ell^2/u_\ell^2} W_T(\mathbf{v}_T), \tag{10.67}$$

where θ_b is the beam divergence, $W_T(\mathbf{v}_T)$ is the transverse velocity distribution (which can also depend on v_ℓ) having width of order $u_T \approx u_\ell \theta_b$, and u_ℓ is the most probable longitudinal speed of atoms in the beam. For a typical thermal, collimated effusive beam having $\theta_b \approx 1$ mrad, u_ℓ is of order 600 m/s, and u_T is of order 6.0 m/s. As such, the transverse Doppler width is of order $ku_T/2\pi \approx 1 - 10$ MHz. Supersonic beams have narrower longitudinal velocity spreads. Although ku_T is comparable to the excited-state decay rates considered throughout this text, optical Ramsey fringe experiments are often carried out on "forbidden" (e.g., intercombination) transitions or two-photon transitions between ground-state hyperfine levels (such as the Cs "clock" transition). In both these metrological applications, the decay times of the relevant levels is all but negligible, with line widths limited by other factors. In any event, the transverse Doppler width is much larger than all relaxation rates in such situations, so we can apply the same formalism that we have been using. Also important for a beam is the beam intensity or current density distribution given by

$$W_J(\mathbf{v}_T, v_\ell) = \frac{2}{u_\ell^4} v_\ell^3 e^{-v_\ell^2/u_\ell^2} W_T(\mathbf{v}_T), \tag{10.68}$$

for which

$$\bar{v}_\ell = 3\sqrt{\pi} u_\ell / 4 . \tag{10.69}$$

There are several differences between coherent transients involving pulsed fields acting on atoms in a vapor cell and those with atoms moving through spatially separated fields. Consider first the FPD signal radiated by a monoenergetic atomic beam a distance L from the scattering zone. If the atoms have longitudinal velocity v_ℓ, then the FPD signal, measured at a distance L from the field interaction zone, arises from atoms that were excited at time $t_e = t - L/v_\ell$. This implies that $\rho_{21}^I(t_e)$ is proportional to

$$\rho_{21}^I(t_e) = -i\mathcal{N}\frac{\sin A}{2}e^{i\delta t_e} \propto e^{i\delta(t-L/c)}. \tag{10.70}$$

As a consequence, $\rho_{21}(t, L)$ varies as

$$e^{i\delta(t-L/c)}e^{-i\omega_0 t}e^{-\gamma(t-t_e)} = e^{-i\omega t}e^{-\gamma L/c}.$$

The emitted field is radiated at the *incident* field frequency ω rather than the atomic frequency ω_0. Emission at the laser frequency rather than the atomic frequency is characteristic of the radiated signals associated with experiments involving spatially separated fields.

A second difference is that owing to a distribution of longitudinal velocities in the beam, the pulse area is different for different atoms, since the pulse duration in an atom's rest frame varies inversely with its longitudinal velocity. Moreover, for slowly moving atoms in the beam, the approximation of negligibly short interaction times [conditions (10.1)] is bound to be violated. In what follows, however, we will assume that conditions (10.1) hold for most of the atoms in the beam.

Without a doubt, the most important experiments involving atoms moving through spatially separated fields are those in which optical Ramsey fringes are produced [21, 22, 26, 38] (at least before the development of frequency combs). We can analyze this case using the results of section 10.2.2, with the time intervals between pulses replaced by the spatial separation of the field zones, divided by the longitudinal velocity of the atoms. The geometry we choose has three field zones, located at $Z = 0$, $Z = L_{21}$, and $Z = L_{21} + L$, in which the fields have propagation vectors $\mathbf{k}_1 = \mathbf{k}$, $\mathbf{k}_2 = \mathbf{k}$, and $\mathbf{k}_3 = -\mathbf{k}$, respectively. As was mentioned in section 10.2.2, one has the option of transforming the optical coherence to a population using a fourth pulse at position $Z = L + 2L_{21}$ ($t = T + 2T_{21}$) and measuring the population ϱ_{22} immediately following this fourth pulse, or one can monitor the field radiated at that position. In the case of atoms moving through spatially separated fields with different longitudinal velocities, one must first average equation (10.45) over *longitudinal* velocities before taking the absolute square to calculate the radiated field. As a result of this averaging, the radiated signal intensity is maximum for $\delta_0 = 0$. Consequently, for spatially separated fields, heterodyne detection or a fourth field is not necessary, since the radiated field intensity, as a function of δ_0, allows us to determine the line center.

Let us now calculate the Ramsey fringe signal for the case when a *fourth* field interaction zone, located at $Z = L + 2L_{21}$ is used to convert coherence to population. The field propagation vector for this fourth field is $\mathbf{k}_4 = -\mathbf{k}$. Thus, we need to evaluate the density matrix chain

$$\varrho_{11} \to \varrho_{21} \to (\varrho_{22} - \varrho_{11}) \to \varrho_{21} \to \varrho_{22} + \text{c.c.} \tag{10.71}$$

We have already calculated the first three steps of this chain to arrive at equation (10.45), which can be written in the interaction representation before the average over velocity is taken as

$$\varrho_{21}^{I}(\mathbf{R},\mathbf{v},t) = -(1/4)e^{-i\mathbf{k}\cdot\mathbf{R}}e^{i\delta_0 t}\sin A_1 \sin A_2 \sin A_3 G(T)e^{-\gamma(t-2T_{21}-T)} \times e^{-2\gamma T_{21}}e^{-i\delta_0(t-T)}e^{ikv_x(t-2T_{21}-T)}, \tag{10.72}$$

where

$$T_{21} = L_{21}/v_\ell, \quad T = L/v_\ell. \tag{10.73}$$

Using equation (10.8) to calculate the transformation of $\varrho_{21} \rightarrow \varrho_{22}$ produced by the fourth pulse, we find

$$\begin{aligned}\varrho_{22}(\mathbf{R},v_\ell,t) &= -(1/4)\sin A_1 \sin A_2 \sin A_3 \sin A_4 \\ &\quad \times G(T)e^{-\gamma_2(t-2T_{21}-T)}e^{-2\gamma T_{21}}e^{-2i\delta_0 T_{21}} + \text{c.c.} \\ &= -(1/2)\sin A_1 \sin A_2 \sin A_3 \sin A_4 \\ &\quad \times G(T)e^{-\gamma_2(t-2T_{21}-T)}e^{-2\gamma T_{21}}\cos(2\delta_0 T_{21}),\end{aligned} \tag{10.74}$$

exhibiting Ramsey fringe structure; there is no longer any dependence on the transverse velocity, since the fourth pulse is applied at the echo time. If the population is measured shortly after the fourth interaction zone, we find

$$\begin{aligned}\varrho_{22}(v_\ell, t \approx 2T_{21}+T) &= -(1/2)\prod_{j=1}^{4}\sin\left(A_j\bar{v}_\ell/v_\ell\right) \\ &\quad \times G(L/v_\ell)e^{-2\gamma L_{21}/v_\ell}\cos(2\delta_0 L_{21}/v_\ell),\end{aligned} \tag{10.75}$$

where equation (10.73) was used and we have assumed that pulse j has area A_j if $v_\ell = \bar{v}_\ell$. The fact that ϱ_{22} is negative should not be of concern, since this is only that *part* of ϱ_{22} exhibiting Ramsey fringe structure. There is an additional background term that ensures that the total excited-state population is positive, or else we would be in serious trouble. The background term, which does not exhibit Ramsey fringe structure, consists of contributions to ϱ_{22} that add to those given in the chain (10.71). They contain terms that represent the contributions to ϱ_{22} from each field acting separately, as well as the non-Ramsey-like population produced by multiple fields.

For a closed system with $\gamma_2 = 2\gamma$, $G(L/v_\ell) = e^{-2\gamma L/v_\ell}$, and the Ramsey fringe signal is determined by the function

$$H(\delta_0) = \int_0^\infty dv_\ell \cos(2\delta_0 L_{21}/v_\ell)e^{-2\gamma L/v_\ell}e^{-2\gamma L_{21}/v_\ell} W_J(v_\ell)\prod_{j=1}^{4}\sin\left(A_j\bar{v}_\ell/v_\ell\right), \tag{10.76}$$

where W_J is given by equation (10.68). For $\delta_0 = 0$, all phases cancel and the signal is a maximum. As $\delta_0 L_{21}$ increases, small changes in v_ℓ produce larger phase changes and the signal diminishes. You are asked to plot this result in the problems.

Often, instead of using three traveling wave fields, standing-wave fields are used for one or more of the interactions regions [26, 39–41]. For example, if $\mathbf{k}_1 = \mathbf{k}_2 = -\mathbf{k}_3 = \mathbf{k}$, one can collapse the second and third interaction zones ($L \sim 0$). In this manner, one has effectively replaced the second and third fields by a single, standing-wave field. In fact, in many of the original proposals for obtaining optical

Ramsey fringes, standing-wave fields were used for *each* field interaction zone. While this complicates the analysis somewhat, since all spatial harmonics of the optical coherence are created by standing-wave fields, the underlying physics remains the same.

The accuracy and precision of optical Ramsey fringe experiments is extraordinary. The recoil that atoms acquire when they absorb or emit radiation from oppositely propagating optical fields can be resolved using optical Ramsey fringes [42]. (The recoil shift, of order 23 kHz for Ca, results from recoil shifts of opposite sign for the two pairs of interferometric paths shown in figure 10.8.) It is not difficult to understand the origin of the recoil splitting by incorporating recoil into the phase diagram. We have seen that the excited-state amplitude for a vapor acquires a phase factor $\exp[-i\tilde{\delta}(t - T_j)]$ following absorption, and the ground-state amplitude acquires a phase factor $\exp[i\tilde{\delta}(t - T_j)]$ following emission, where

$$\tilde{\delta} = \delta_0 + \mathbf{k}_j \cdot \mathbf{v} = \frac{E_2 - E_1}{\hbar} - \omega_L + \mathbf{k}_j \cdot \mathbf{v}, \tag{10.77}$$

where E_j is the energy of state j. To include both the Doppler phase factor and the recoil energy, we replace this equation by

$$\begin{aligned}\tilde{\delta} &= \frac{E_2(\mathbf{P} + \hbar\mathbf{k}_j) - E_1(\mathbf{P})}{\hbar} - \omega_L \\ &= \frac{E_2 - E_1}{\hbar} + \frac{(\mathbf{P} + \hbar\mathbf{k}_j)^2}{2M} - \frac{P^2}{2M} - \omega_L \\ &= \delta_0 + \mathbf{k}_j \cdot \mathbf{v} + \omega_{k_j},\end{aligned} \tag{10.78}$$

where $\omega_{k_j} = \hbar k_j^2/2M$ is a recoil frequency, and $\mathbf{P} = M\mathbf{v}$ is the center-of-mass momentum before pulse j is applied.

In the interval between $t = 0$ and $t = T + 2T_{21}$ in figure 10.8, $\tilde{\varrho}_{21}$ for the lower pair of paths acquires a phase

$$\begin{aligned}&-\left[\delta_0 + \frac{(\mathbf{P} + \hbar\mathbf{k}_1)^2}{2M} - \frac{P^2}{2M}\right] T_{21} - \left[\frac{(\mathbf{P} + \hbar\mathbf{k}_1 - \hbar\mathbf{k}_2)^2}{2M} - \frac{P^2}{2M}\right] T \\ &-\left[\delta_0 + \frac{(\mathbf{P} + \hbar\mathbf{k}_3)^2}{2M} - \frac{P^2}{2M}\right] [t - (T - 2T_{21})] \\ &= -2\left(\delta_0 + \omega_k\right) T_{21},\end{aligned}$$

since $\mathbf{k}_1 = \mathbf{k}_2 = -\mathbf{k}_3 = \mathbf{k}$. However, $\tilde{\varrho}_{21}$ for the upper pair of paths acquires a phase

$$\begin{aligned}&-\left[\delta_0 + \frac{(\mathbf{P} + \hbar\mathbf{k}_1)^2}{2M} - \frac{P^2}{2M}\right] T_{21} - \left[\frac{(\mathbf{P} + \hbar\mathbf{k}_1)^2}{2M} - \frac{(\mathbf{P} + \hbar\mathbf{k}_2)^2}{2M}\right] T \\ &-\left\{\left[\delta_0 + \frac{(\mathbf{P} + \hbar\mathbf{k}_1)^2}{2M}\right] - \left[\frac{(\mathbf{P} + \hbar\mathbf{k}_2 - \hbar\mathbf{k}_3)^2}{2M}\right]\right\} [t - (T - 2T_{21})] \\ &= -2\left(\delta_0 - \omega_k\right) T_{21}.\end{aligned}$$

Thus, the net Ramsey signal is a maximum for $\delta_0 = -\omega_k$ for the lower pair of paths and $\delta_0 = \omega_k$ for the upper pair of paths, resulting in a recoil splitting ω_{split}

given by

$$\omega_{split} = 2\omega_k = \frac{\hbar k^2}{M}. \qquad (10.79)$$

The sensitivity of optical Ramsey fringe experiments is also sufficient to resolve the second-order Doppler shift associated with the motion of the atoms relative to the sources of the optical fields [43]. By detecting optical Ramsey fringes using thermal atomic beams passing through spatially separated fields, precisions have been achieved of 2.5×10^{-12} for the 657-nm Ca transition [44] and 2.5×10^{-12} for the 457-nm Mg transition [45], both of which are $^1S_0$ to $^3P_1$ intercombination transitions.

Problems

1. Derive equations (10.6) and (10.8).
2. For the two-pulse echo with copropagating pulses having areas $\pi/2$ and π, estimate the peak power exiting a sample of length L. Assume that the medium consists of an optically thin vapor having $k_1 u/2\pi = 1.0\,\text{GHz}$, $\gamma_2/2\pi = 10\,\text{MHz}$, $L = 1$ cm, $\omega_0 = 3 \times 10^{15}\ \text{s}^{-1}$, $T_{21} = 10$ ns, σ (pulse cross-sectional area) $= 10\,\text{mm}^2$, and $\mathcal{N} = 10^8$ atoms/cm^3. Estimate the number of photons in the echo signal.
3. Consider a "vapor" of stationary atoms that undergoes FPD following the application of a short $\pi/2$ pulse. Take the sample to be confined in a cylindrical volume of cross-sectional area A and length L. Calculate the power exiting the sample, assuming that the sample is optically thin. Show that for a sufficiently high density, the power exiting the sample exceeds the maximum energy that the initial pulse could have deposited in the sample. What is wrong here?
4. Consider FPD for an ensemble of two-level atoms having a Maxwellian distribution of velocities with most probable speed u. The atoms are subjected to a monochromatic field having constant amplitude for the time period $(-\infty, 0)$. At $t = 0$, the field is suddenly turned off. Calculate the FPD signal emitted for times $t \gg (1/ku)$, assuming that $|\delta| < ku$, $\gamma \ll ku$. Calculate the signal only to lowest *nonvanishing* order in the preparation field, and show that the lead term in $\langle\tilde{\varrho}_{12}\rangle$ varies as the field amplitude cubed. On physical grounds, explain why the signal, which decays in a time of order $(1/\gamma)$, has this field dependence.
5. Use the U and F matrices to calculate $\varrho^I_{21}(t)$ for $t > T$ for the two-pulse echo, including all contributions. Show that in the region $t \approx 2T$, the only term that contributes significantly to the signal is given by equation (10.34), assuming $kuT \gg 1$.
6. Go through the details of the stimulated photon echo calculation of section 10.2.1 for $\mathbf{k}_1 = \mathbf{k}_2 = \mathbf{k}_3 = \mathbf{k}$ and the following decay scheme:

$$\dot{\varrho}_{11} = -\Gamma_1\varrho_{11} + \gamma_2\varrho_{22}, \quad \dot{\varrho}_{22} = -(\Gamma_2 + \gamma_2)\varrho_{22}, \quad \dot{\varrho}_{12} = -\gamma\varrho_{12}.$$

Draw a Doppler phase diagram to show that an echo is produced at time T_{21} following the third pulse. Calculate $\varrho_{21}(\mathbf{R}, t)$ in the vicinity of the echo time.

Show that if $\Gamma_1 \ll \gamma_2$ and $\Gamma_1 \neq \Gamma_2$, one can have a photon echo for very long $T = T_{32}$ (times as long as one hour have been observed), but if $\Gamma_1 = \Gamma_2$, there is no long-lived echo. How do you explain this?

7. Integrate equation (10.76) numerically in the limit that $\gamma L/\bar{v}_\ell \ll 1$ and neglecting the dependence of pulse area on v_ℓ, that is, numerically integrate,

$$H(\delta_0) = \int_0^\infty dv_\ell \cos(2\delta_0 L_{21}/v_\ell) e^{-2\gamma L_{21}/v_\ell} W_J(v_\ell),$$

and plot the signal as a function of $a = \delta_0 L_{21}/\bar{v}_\ell$ for $b = \gamma L_{21}/\bar{v}_\ell = 0$, 0.5, 1.5 to display a typical Ramsey fringe pattern. The distribution $W_J(v_\ell)$ is given by equation (10.68). Does the decay affect the resolution of the signal significantly? See how the dependence of pulse area on v_ℓ modifies the signal by plotting

$$H'(\delta_0) = \int_0^\infty dv_\ell \cos(2\delta_0 L_{21}/v_\ell) e^{-2\gamma L_{21}/v_\ell} W_J(v_\ell) \sin^4\left(\frac{\pi}{2}\frac{\bar{v}_\ell}{v_\ell}\right)$$

for the same parameters.

8. Imagine a collimated atomic beam passing through two field zones that drive radio-frequency transitions between two ground-state hyperfine levels. Consider the **k** vector of the field to be in the X direction and the atomic beam to propagate in the Z direction. If the atoms are prepared in one of the hyperfine levels, show that the population of the other level, considered as a function of the detuning between the applied fields and the hyperfine transition frequency, exhibits Ramsey fringes. Why does this two-zone method not work for optical fields, in which an atomic beam propagating in the Z direction passes through two field regions in which the fields propagate in the X direction? Since the hyperfine levels have essentially infinite lifetimes, what will ultimately determine the resolution of the Ramsey fringes?

9. Derive equation (10.45).

10. Consider a three-pulse "echo" in an *homogeneously* broadened systems. Draw a phase diagram in which *field 1 acts first* that leads to a signal in the $\mathbf{k} = \mathbf{k}_1 - \mathbf{k}_2 + \mathbf{k}_3 = 2\mathbf{k}_1 - \mathbf{k}_2$ direction (with $\mathbf{k}_1 = \mathbf{k}_3$). We have not considered this contribution for inhomogeneously broadened systems, since such a diagram leads to an overall phase of $\phi_d = 2(\delta_0 + \Delta - 2\mathbf{k}_1 \cdot \mathbf{v})T_{21}$ at time $t = T + 2T_{21}$. On averaging over either Δ or $\mathbf{v}$ in an inhomogeneously broadened sample, this contribution would vanish. In a homogeneously broadened sample, however, $\Delta = 0$ and $\mathbf{v} = 0$, giving an identical relative phase ϕ_d to all the atoms. For $t > T + T_{21}$, calculate $\langle \tilde{\varrho}_{21}(\mathbf{R}, t) \rangle$. Show that the signal does not constitute an "echo" in the usual sense, since there is no dephasing–rephasing cycle. The signal appears promptly (it is actually an FPD signal) following the third pulse. If one measures the *time-integrated* intensity in the signal following the third pulse, however (as is often the case with ultrafast pulses in which time resolution of the echo is not possible), it is impossible to tell directly whether an "echo" has occurred. For such measurements, a signal emitted in the $\mathbf{k} = 2\mathbf{k}_1 - \mathbf{k}_2$ direction when pulse 1 acts first is a clear signature of an homogeneously broadened system, since such a signal vanishes for inhomogeneously broadened samples.

11. Calculate the overall phase of $\tilde{\varrho}_{21}$ using figure 10.6, and also calculate the phase of ϱ_{21}^I using equation (10.7) for the three field zones. Show that they are in agreement, given the fact that $\tilde{\varrho}_{21} = \varrho_{21}^I e^{-i(\delta_0+\Delta)t}$.
12. Consider a three-pulse echo in which the three fields have wave vectors $\mathbf{k}_1$, $\mathbf{k}_2$, $-\mathbf{k}_2$, with $\mathbf{k}_1 \approx \mathbf{k}_2$. Show that there is a Ramsey-fringe-type signal for vapors in the $\mathbf{k}_s = \mathbf{k}_1 - 2\mathbf{k}_2$ direction and a stimulated echo signal in inhomogeneously broadened solids in the $\mathbf{k}_s = -\mathbf{k}_1$ direction. To do this, consider the two sequences

$$\varrho_{11} \rightarrow \varrho_{21} \rightarrow (\varrho_{22} - \varrho_{11}) \rightarrow \varrho_{21}$$

and

$$\varrho_{11} \rightarrow \varrho_{12} \rightarrow (\varrho_{22} - \varrho_{11}) \rightarrow \varrho_{21}.$$

You need not calculate the signal explicitly. Just find the overall phase factor resulting from these two sequences.
13. Consider a *tri-level echo* for the Λ level scheme of chapter 9. This is a stimulated photon echo in which the first and third pulses have propagation vectors $\mathbf{k} = k\hat{\mathbf{z}}$ and are resonant with the 1–2 transition, while the second pulse has $\mathbf{k}' = \eta k'\hat{\mathbf{z}}$ ($\eta = \pm 1$) and is resonant with the 2–3 transition. The first two pulses are separated by T_{21}, and the second and third pulses by T. For an inhomogeneously broadened vapor, use a Doppler phase diagram to show that, for $\eta = 1$, there is a contribution to an echo on the 2–3 transition involving ϱ_{13} as an intermediate state. Show that there is no such contribution for $\eta = -1$, which is analogous to the result in the steady-state three-level system result (although you can convince yourself that there are no step-wise contributions to the echo in this scheme). Give an interpretation of this result in terms of Doppler phase cancellation. Use the phase diagram to calculate the time at which the echo occurs. This allows one to store information in the 1–3 coherence for a relatively long time. It differs from the slow light case, however, in that only a small portion of the input fields are "absorbed" in the media. Note that no density matrix calculations are needed in this problem; all results can be read off the phase diagrams (see reference [46]).
14. Prove equation (10.54).

References

[1] H. Metcalf and W. D. Phillips, *Time-resolved subnatural-width spectroscopy*, Optics Letters **5**, 540–542 (1980); W. D. Phillips and H. J. Metcalf, *Time resolved subnatural width spectroscopy*, in *Precision Measurements and Fundamental Constants II*, edited by B. N. Taylor and W. D. Phillips, National Bureau of Standards Special Publication 617 (U.S. Government Printing Office, Washington, DC, 1984) pp. 177–180.

[2] A. I. Lvovsky and S. R. Hartmann, *Superradiant self-diffraction*, Physical Review A **59**, 4052–4057 (1999).

[3] R. H. Dicke, *Coherence in spontaneous radiation processes*, Physical Review **93**, 99–110 (1954).

[4] See, for example, I. P. Herman, J. C. MacGillivray, N. Skribanowitz, and M. S. Feld, *Self-induced emission in optically pumped HF gas: the rise and fall of the superradiant state*, in *Laser Spectroscopy*, edited by R. G. Brewer and A. Mooradian (Plenum, New York, 1974), 379–412.

[5] D. Polder, M. F. H. Schuurmans, and Q. H. F. Vrehen, *Superfluorescence: quantum-mechanical derivation of Maxwell-Bloch description with fluctuating field source*, Physical Review A **19**, 1192–1203 (1979).

[6] N. A. Kurnit, I. D. Abella, and S. R. Hartmann, *Observation of a photon echo*, Physical Review Letters **13**, 567–568 (1964).

[7] E. L. Hahn, *Spin echoes*, Physical Review **80**, 580–594 (1950).

[8] P. R. Berman, *Theory of collision effects on line shapes using a quantum-mechanical description of the atomic center-of-mass motion—Application to lasers*, Physical Review A **2**, 2435–2454 (1970).

[9] R. Beach, S. R. Hartmann, and R. Friedberg, *Billiard ball echo model*, Physical Review A **25**, 2658–2666 (1982); R. Friedberg and S. R. Hartmann, *Billiard balls and matter-wave interferometry*, Physical Review A **48**, 1446–1472 (1993); R. Friedberg and S. Hartmann, *Echoes and billiard balls*, Laser Physics **3**, 1128–1137 (1993).

[10] P. W. Anderson, B. I. Halperin, and C. M. Varma, *Anomalous low-temperature thermal properties of glasses and spin glasses*, Philosophical Magazine **25**, 1–9 (1972).

[11] P. M. Selzer, *General techniques and experimental methods in laser spectroscopy of solids*, in *Laser Spectroscopy of Solids*, edited by W. M. Yen and P. M. Silzer, Springer Series in Topics in Applied Physics, vol. 49 (Springer-Verlag, Berlin, 1986), pp. 115–140.

[12] P. R. Berman, *Validity conditions for the optical Bloch equations*, Journal of the Optical Society of America B **3**, 564–571 (1986), and references therein.

[13] P. R. Berman, *Markovian relaxation processes for atoms in vapors and in solids: calculation of free-induction decay in the weak external-field limit*, Journal of the Optical Society of America B **3**, 572–586 (1986), and references therein.

[14] P. R. Berman, *Collisional effects in laser spectroscopy*, in *New Trends in Atomic Physics*, edited by G. Grynberg and R. Stora Les Houches Session XXXVIII, vol. 1 (North-Holland, Amsterdam, 1984), pp. 451–514.

[15] P. R. Berman, J. M. Levy, and R. G. Brewer, *Coherent optical transient study of molecular collisions: theory and observations*, Physical Review A **11**, 1668–1688 (1975).

[16] T. W. Mossberg, R. Kachru, and S. R. Hartmann, *Observation of collisional velocity changes associated with atoms in a superposition of dissimilar electronic states*, Physical Review Letters **44**, 73–77 (1980).

[17] R. Kachru, T. J. Chen, S. R. Hartmann, T. W. Mossberg, and P. R. Berman, *Measurement of a total atomic-radiator-perturber scattering cross section*, Physical Review Letters **47**, 902–905 (1981).

[18] R. A. Forber, L. Spinelli, J. E. Thomas, and M. S. Feld, *Observation of quantum diffractive velocity-changing collisions by use of two-level heavy optical radiators*, Physical Review Letters **50**, 331–335 (1982).

[19] R. Kachru, T. W. Mossberg, and S. R. Hartmann, *Stimulated photon echo study of* $Na(3^2S_{1/2})$*-CO velocity-changing collisions*, Optics Communications **30**, 57–62 (1979).

[20] T. Mossberg, A. Flusberg, R. Kachru, and S. R. Hartmann, *Total scattering cross section for Na on He measured by stimulated photon echoes*, Physical Review Letters **42**, 1665–1669 (1979).

[21] J. C. Bergquist, S. A. Lee, and J. L. Hall, *Saturated absorption with spatially separated laser fields: observation of optical Ramsey fringes*, Physical Review Letters **38**, 159–162 (1977).

[22] Ch. J. Bordé, Ch. Salomon, S. Avrillier, A. Van Lerberghe, Ch. Bréant, D. Bassi, and G. Scoles, *Optical Ramsey fringes with traveling waves*, Physical Review A **30**, 1836–1848 (1984).

[23] N. Ramsey, *A molecular beam resonance method with separated oscillating fields*, Physical Review **78**, 695–699 (1950).

[24] See, for example, L. S. Vasilenko, I. D. Matveyenko, and N. N. Rubtsova, *Study of narrow resonances of coherent radiation in time separated fields in* SF_6, Optics Communications **53**, 371–374 (1985).

[25] C. Degenhardt, H. Stoehr, C. Lisdat, G. Wilpers, H. Schnatz, B. Lipphardt, T. Nazarova, P.-E. Pottie, U. Sterr, J. Helmcke, and F. Riehle, *Calcium optical frequency standard with ultracold atoms: approaching* 10^{-15} *relative uncertainty*, Physical Review A **72**, 062111, 1–17 (2005).

[26] B. Ya. Dubetsky, A. P. Kazantsev, V. P. Chebotaev, and V. P. Yakolev, *Interference of atoms in separated optical fields*, Soviet Physics **62**, 685–693 (1985); V. P. Chebotaev, B. Ya. Dubetsky, A. P. Kazantsev, and V. P. Yakolev, *Interference of atoms in separated optical fields*, Journal of the Optical Society of America B **2**, 1791–1798 (1985).

[27] Ch. J. Bordé, *Atomic interferometry with internal state labeling*, Physics Letters A **140**, 10–12 (1989).

[28] Th. Udem, J. Reichert, R. Holzwarth, and T. W. Hänsch, *Absolute optical frequency measurement of the cesium* D_1 *line with a mode-locked laser*, Physical Review Letters **82**, 3568–3571 (1999).

[29] D. J. Jones, S. A. Diddams, J. K. Ranka, A. Stentz, R. S. Windeler, J. L. Hall, and S. T. Cundiff, *Carrier-envelope phase control of femtosecond mode-locked lasers and direct optical frequency synthesis*, Science **288**, 635–639 (2000).

[30] S. A. Diddams, D. J. Jones, J. Ye, S. T. Cundiff, J. L. Hall, J. K. Ranka, R. S. Windeler, R. Holzwarth, Th. Udem, and T. W. Hänsch, *Direct link between microwave and optical frequencies with a 300-THz femtosecond laser comb*, Physical Review Letters **84**, 5102–5105 (2000).

[31] For a review with many references, see M. C. Stowe, M. J. Thorpe, A. Pe'er, J. Ye, J. E. Stalnaker, V. Gerginov, and S. A. Diddams, *Direct frequency comb spectroscopy*, in *Advances in Atomic, Molecular, and Optical Physics*, edited by E. Arimondo, P. R. Berman, and C. C. Lin, vol. 55 (Elsevier, Amsterdam, 2008), pp. 1–60. Also, see S. T. Cundiff and J. Ye, *Colloquium: femtosecond optical frequency combs*, Reviews of Modern Physics **75**, 325–342 (2003).

[32] A. E. Siegman, *Lasers* (University Science Books, Mill Valley, CA, 1986), chap. 28; P. W. Milonni and J. H. Eberly, *Lasers* (Wiley, New York, 2009), chap. 12.

[33] U. Siegner and U. Keller, *Nonlinear optical processes for ultrashort pulse generation*, in *Handbook of Optics*, edited by M. Bass, J. M. Enoch, E. Van Stryland, and W. L. Wolf, Vol. IV (McGraw-Hill, New York, 2001), chap. 25, and references therein.

[34] J. K. Wahlstrand, J. T. Willis, C. R. Menyuk, and S. T. Cundiff, *The quantum-limited comb lineshape of a mode-locked laser: fundamental limits on frequency uncertainty*,

Optics Express **16**, 18624–18630 (2008); M. J. Martin, S. M. Foreman, T. R. Schibli, and J. Ye, *Testing ultrafast mode-locking at microhertz relative optical linewidth*, Optics Express **17**, 558–568 (2009).

[35] For a review with many references, see J. P. Ogilvie and K. J. Kubarych, *Multidimensional electronic and vibrational spectroscopy: an ultrafast probe of molecular relaxation and reaction dynamics*, in *Advances in Atomic, Molecular, and Optical Physics*, edited by E. Arimondo, P. R. Berman, and C. C. Lin, vol. 57 (Elsevier, Amsterdam, 2009), pp. 249–321.

[36] N. F. Ramsey, *Molecular Beams* (Oxford University Press, London, 1956) chap. II; E. A. Kennard, *Kinetic Theory of Gases* (McGraw-Hill, New York, 1938), section 37.

[37] P. T. Greenland, M. A. Lauder, and D. J. H. Wort, *Atomic beam velocity distributions*, Journal of Physics D **18**, 1223–1232 (1985).

[38] For a review with many references, see U. Sterr, K. Sengstock, W. Ertmer, F. Riehle, and W. Ertmer, *Atom interferometry based on separated light fields*, in *Atom Interferometry*, edited by P. R. Berman (Academic Press, San Diego, CA, 1997), pp. 293–362.

[39] Ye. V. Baklanov, B. Ya. Dubetsky, and V. P. Chebotaev, *Nonlinear Ramsey resonance in the optical region*, Applied Physics **11**, 171–173 (1976).

[40] V. P. Chebotaev, *Method of separated fields in optics*, Soviet Journal of Quantum Electronics **8**, 935–941 (1978).

[41] N. Hata and K. Shimoda, *Theory of optical Ramsey resonance in three separated fields produced by a corner reflector*, Applied Physics **22**, 1–9 (1980).

[42] R. L. Barger, J. C. Bergquist, T. C. English, and D. J. Glaze, *Resolution of photon-recoil structure of the 6573-Å calcium line in an atomic beam with optical Ramsey fringes*, Applied Physics Letters **34**, 850–852 (1979).

[43] R. L. Barger, *Influence of second-order Doppler effect on optical Ramsey fringe profiles*, Optics Letters **6**, 145–147 (1981).

[44] A. Morinaga, F. Riehle, J. Ishikawa, and J. Helmcke, *A Ca optical frequency standard: frequency stabilization by means of nonlinear Ramsey resonances*, Applied Physics B **48**, 165–171 (1989).

[45] J. Friebe, A. Pape, M. Riedmann, K. Moldenhauer, T. Mehlstäubler, N. Rehbein, C. Lisdat, E. M. Rasel, W. Ertmer, H. Schnatz, B. Lipphardt, and G. Grosche, *Absolute frequency measurement of the magnesium intercombination transition* $^1S_0 \rightarrow {}^3P_1$, Physical Review A **78**, 033830-1–7 (2008).

[46] T. W. Mossberg, R. Kachru, S. R. Hartmann, and A. M. Flusberg, *Echoes in gaseous media: a generalized theory of rephasing phenomena*, Physical Review A **20**, 1976–1996 (1979).

Bibliography

Many of the general references listed in chapter 1 have one or more chapters on coherent transients. There are thousands of articles in this field. Some review articles and books are listed here:

A. Abragam, *The Principles of Nuclear Magnetism* (Oxford University Press, New York, 1961).

P. R. Berman and R. G. Brewer, *Coherent transient spectroscopy in atomic and molecular vapors*, in *Encyclopedia of Modern Optics*, edited by B. D. Guenther and D. G. Steel (Elsevier, Amsterdam, 2004) pp. 154–163.

P. R. Berman and D. G. Steel, *Coherent optical transients*, in *Handbook of Optics*, edited by M. Bass, J. M. Enoch, E. Van Stryland, and W. L. Wolf, Vol. IV, (McGraw-Hill, New York, 2001), chap. 24.

R. G. Brewer, *Coherent optical spectroscopy*, in *Frontiers in Laser Spectroscopy*, Les Houches Session XXVII, edited by R. Balian, S. Haroche, and S. Liberman, vol. 1 (North-Holland, Amsterdam, 1977), pp. 341–396.

———, *Coherent optical spectroscopy*, in *Nonlinear Spectroscopy*, Proceedings of the International School of Physics, Enrico Fermi, Course 64, edited by N. Bloembergen (North-Holland, Amsterdam, 1977) pp. 87–137.

R. L. Shoemaker, *Coherent transient infrared spectroscopy*, in *Laser and Coherence Spectroscopy*, edited by J. L. Steinfeld (Plenum, New York, 1978), pp. 197–371.

C. P. Slichter, *Principles of Magnetic Resonance* (Harper and Row, New York, 1963).

W. Zinth and W. Kaiser, *Ultrafast coherent spectroscopy*, in *Ultrashort Laser Pulses and Applications*, edited by W. Kaiser, vol. 60 (Springer-Verlag, Berlin, 1978), pp. 235–277.

11

Atom Optics and Atom Interferometry

In chapter 5, we introduced a number of concepts related to atom optics. In particular, we examined the manner in which electromagnetic fields can exert forces on atoms. Moreover, in the previous chapter, we indicated that some coherent transient signals could be interpreted within the context of atom interferometry. In this chapter, we examine both atom optics and atom interferometry in more detail. Before doing so, however, it will prove useful to review some aspects of classical diffraction theory.

11.1 Review of Kirchhoff-Fresnel Diffraction

Diffraction plays an important role in both electromagnetism and quantum mechanics, since both are wave theories. Diffraction in quantum mechanics is actually a much easier problem than diffraction in electromagnetism, due to the fact that light is a vector field, whereas the quantum-mechanical wave function is a scalar. In reality, diffraction in electromagnetism is basically an unsolved problem, because of the difficulty of specifying the exact boundary conditions for the fields. Nevertheless, it is possible to get a good idea of diffraction phenomena that agrees well with experiment by using a theory in which the electric field is considered as a *scalar* field. This can be a good approximation if one limits the discussion to distances from the scatterer that are large compared with a wavelength, since in that region, the fields can be approximated as plane-wave fields. We present an approach to scalar diffraction based on wave-packet propagation that works for both electromagnetic and matter waves. In the end, we arrive at equations that coincide with those of Kirchhoff-Fresnel diffraction theory.

11.1.1 Electromagnetic Diffraction

Let us first consider the propagation of a wave in free space. Imagine that a pulse propagating in the Z direction strikes a diffracting screen at $Z = 0$. The screen is assumed to have an *amplitude* transmission function $T(X, Y)$. We approximate the positive frequency component of the field (considered as a scalar) at $t = 0$ as

$$E^{+}(\mathbf{R}, 0) = T(X, Y)F(Z, 0)e^{ik_0 Z}, \tag{11.1}$$

where $F(Z, 0)$ corresponds to a pulse function that is fairly well located in space. In other words, the spatial extent in the Z direction is much greater than a wavelength, but much less than the distance to the screen where the diffraction pattern is measured. It should be noted that equation (11.1) is not a solution of the scalar Helmholtz equation and already involves a type of paraxial approximation in which it is assumed that the maximum transverse propagation vector components associated with the Fourier transform of $T(X, Y)$ are much less than the average propagation constant k_0, such that $\mathbf{k} \approx k_0 \hat{\mathbf{z}}$.

At any time t,

$$E^{+}(\mathbf{R}, t) = \frac{1}{(2\pi)^{3/2}} \int d\mathbf{k}\, A(k_X, k_Y) B(k_Z) e^{i(\mathbf{k}\cdot\mathbf{R} - \omega t)}, \tag{11.2}$$

where $\omega = kc$, and the Fourier amplitudes are given by

$$A(k_X, k_Y) = \frac{1}{2\pi} \int T(X, Y) e^{-i(k_X X + k_Y Y)} dX dY, \tag{11.3a}$$

$$B(k_Z) = \frac{1}{(2\pi)^{1/2}} \int F(Z, 0) e^{-i(k_Z - k_0)Z} dZ. \tag{11.3b}$$

It is assumed that $B(k_Z)$ is a function that is sharply peaked about $k_Z = k_0$. Substituting the $A(k_X, k_Y)$ amplitudes back into equation (11.2), we have

$$E^{+}(\mathbf{R}, t) = \frac{1}{(2\pi)^2 (2\pi)^{1/2}} \int d\mathbf{k}_T dk_Z d\mathbf{R}'_T\, T(X', Y') B(k_Z) \times e^{i(\mathbf{k}_T\cdot\mathbf{R}_T - \omega t)} e^{-i\mathbf{k}_T\cdot\mathbf{R}'_T} e^{ik_Z Z}, \tag{11.4}$$

where

$$\mathbf{k}_T = k_X \hat{\mathbf{i}} + k_Y \hat{\mathbf{j}}, \tag{11.5}$$

$$\mathbf{R}_T = X\hat{\mathbf{i}} + Y\hat{\mathbf{j}}. \tag{11.6}$$

Since $B(k_Z)$ is a sharply peaked function centered at $k_Z = k_0$, we can expand $\omega(k)$ as

$$\begin{aligned}
\omega(k) &= c\sqrt{k_X^2 + k_Y^2 + k_Z^2} \\
&\approx \omega(k_Z = k_0) + \frac{d\omega}{dk_0}(k_Z - k_0) \\
&= c\sqrt{k_X^2 + k_Y^2 + k_0^2} + ck_0 \frac{(k_Z - k_0)}{\sqrt{k_X^2 + k_Y^2 + k_0^2}} \\
&\approx ck_0 + \frac{c}{2k_0}\left(k_X^2 + k_Y^2\right) + c(k_Z - k_0),
\end{aligned} \tag{11.7}$$

where we have used the fact that the maximum value of $(k_X^2 + k_Y^2)$ is much less than k_0^2, by assumption. Substituting the expression for $\omega(k)$ into equation (11.4), we find

$$\begin{aligned}
E^+(\mathbf{R}, t) &= \frac{e^{i(k_0 Z - \omega_0 t)}}{(2\pi)^2 (2\pi)^{1/2}} \int d\mathbf{k}_T dk_Z d\mathbf{R}'_T\, T(X', Y') B(k_Z) e^{i\mathbf{k}_T \cdot \mathbf{R}} \\
&\quad \times e^{-i(k_X X' + k_Y Y')} e^{i(k_Z - k_0)Z} e^{-i\frac{c}{2k_0}(k_X^2 + k_Y^2)t} e^{-i(k_Z - k_0)ct} \\
&= \frac{e^{i(k_0 Z - \omega_0 t)}}{(2\pi)^2} F(Z - ct, 0) \\
&\quad \times \int d\mathbf{k}_T d\mathbf{R}'_T\, T(X', Y') e^{i\mathbf{k}_T \cdot (\mathbf{R}_T - \mathbf{R}'_T)} e^{-i\frac{c}{2k_0}(k_X^2 + k_Y^2)t}. \qquad (11.8)
\end{aligned}$$

Thus, the pulse envelope is simply translated in the Z direction by ct, and the diffraction pattern is superimposed on the pulse in the transverse direction.

The integrals over k_X and k_Y are tabulated functions

$$\frac{1}{(2\pi)^{1/2}} \int_{-\infty}^{\infty} dk e^{ika} e^{-ik^2 b} = \frac{1}{\sqrt{2ib}} e^{i\frac{a^2}{4b}}, \qquad (11.9)$$

and we find

$$E^+(\mathbf{R}, t) = F(Z - ct, 0) e^{i(k_0 Z - \omega_0 t)} \Psi(X, Y), \qquad (11.10)$$

where the diffraction amplitude is given by

$$\begin{aligned}
\Psi(X, Y) &= \frac{1}{(2\pi)^2} \int d\mathbf{k}_T d\mathbf{R}'_T\, T(X', Y') e^{i\mathbf{k}_T \cdot (\mathbf{R}_T - \mathbf{R}'_T)} e^{-i\frac{c}{2k_0}(k_X^2 + k_Y^2)t} \\
&= \frac{-i}{(2\pi)} \frac{k_0}{ct} \int d\mathbf{R}'_T\, T(X', Y') e^{i\left(\frac{2k_0}{ct}\right)|\mathbf{R}_T - \mathbf{R}'_T|^2/4}. \qquad (11.11)
\end{aligned}$$

Since the pulse envelope function is sharply peaked at $Z = ct$, we can replace ct by Z in this expression to obtain

$$\Psi(X, Y) = \frac{-i}{\lambda_0 Z} \int d\mathbf{R}'_T\, T(X', Y') e^{ik_0|\mathbf{R}_T - \mathbf{R}'_T|^2/2Z}, \qquad (11.12)$$

where $\lambda_0 = 2\pi/k_0$. Equation (11.12) is essentially the same result obtained from the Kirchhoff-Fresnel theory [1].

11.1.1.1 Shadow, Fresnel, and Fraunhofer diffraction

The scattering can be classified as shadow, Fresnel, or Fraunhofer diffraction, according to

$$k_0 r_0^2/2Z \gg 1 \quad \text{shadow region}, \qquad (11.13a)$$

$$k_0 r_0^2/2Z \approx 1 \quad \text{Fresnel diffraction}, \qquad (11.13b)$$

$$k_0 r_0^2/2Z \ll 1 \quad \text{Fraunhofer diffraction}, \qquad (11.13c)$$

where r_0 is a characteristic dimension of the apertures in the diffracting screen. In all cases, it is assumed that

$$k_0 r_0 \gg 1, \tag{11.14}$$

$$k_0 Z \gg 1, \tag{11.15}$$

so the Kirchhoff-Fresnel theory is valid. We illustrate the idea of three zones using the one-dimensional problem of diffraction by a slit, but the results are quite general.

We refer to a one-dimensional aperture as one in which the transmission function is independent of Y. For such an aperture, equation (11.12) can be integrated over Y to yield the one-dimensional diffraction amplitude

$$\Psi(X) = \sqrt{\frac{-i}{\lambda_0 Z}} e^{ik_0 X^2/2Z} \int dX' T(X') e^{-ik_0 XX'/Z} e^{ik_0 X'^2/2Z}, \tag{11.16}$$

where $T(X)$ is the one-dimensional amplitude transmission function. Since the signal intensity is proportional to $|\Psi(X)|^2$, we can take as our signal intensity the quantity

$$I(X) = |D(X)|^2, \tag{11.17}$$

where

$$D(X) = \sqrt{\frac{-i}{\lambda_0 Z}} \int dX' T(X') e^{-ik_0 XX'/Z} e^{ik_0 X'^2/2Z}. \tag{11.18}$$

For a slit having width a, $T(X) = 1$ for $-a/2 \leq X \leq a/2$ and is zero otherwise. In that case,

$$|D(X)| = \sqrt{\frac{1}{\lambda_0 Z}} \left| \int_{-a/2}^{a/2} dX' e^{-ik_0 XX'/Z} e^{ik_0 X'^2/2Z} \right| \tag{11.19a}$$

$$= \frac{1}{2} \left| \begin{array}{l} \Phi\left[\sqrt{\frac{k_0 a^2}{Z}} \left(\frac{1-i}{4}\right) \left(1 - \frac{X}{a/2}\right)\right] \\ +\Phi\left[\sqrt{\frac{k_0 a^2}{Z}} \left(\frac{1-i}{4}\right) \left(1 + \frac{X}{a/2}\right)\right] \end{array} \right|, \tag{11.19b}$$

where

$$\Phi(x) = \frac{2}{\sqrt{\pi}} \int_0^x e^{-t^2} dt \tag{11.20}$$

is the error function. Let us look at the signal intensity as we go from the shadow to the Fraunhofer regions, using $r_0 = a/2$ in equations (11.13).

In the shadow zone, $k_0 a^2/8Z \gg 1$, $|D(X)| \approx 1$ for $|X| < a/2$ and quickly falls to zero for $|X| > a/2$—in other words, the signal exists in the *shadow* of the slit only. In the immediate vicinity of $|X| = a/2$, there are oscillations in the scattered intensity owing to diffraction from the sharp edges of the slit.

As we increase Z to the Fresnel zone, $k_0 a^2/8Z \approx 1$, most of the signal intensity is still confined spatially to dimensions on order of the aperture, but with interesting interference phenomena giving rise to a complex signal.

For still larger Z, $k_0 a^2/8Z \ll 1$, we enter the Fraunhofer region, where the results depend only on the ratio $X/Z = \sin\theta$, where θ is the diffraction angle. In the

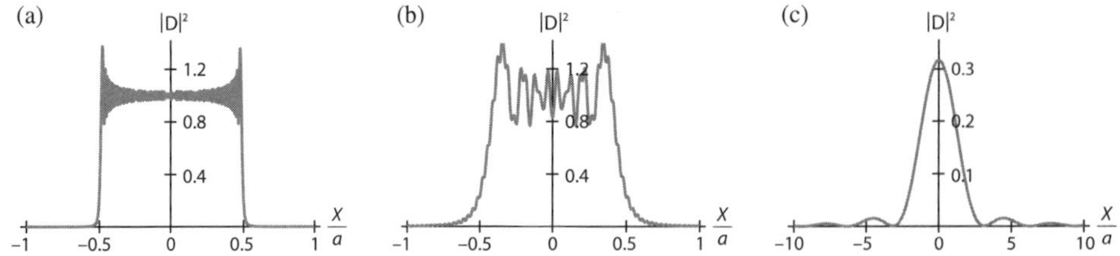

Figure 11.1. Graphs of the intensity of the signal diffracted from a slit having length a as a function of X/a for different values of $Q = k_0a^2/8Z$: (a) shadow region, $Q = 250$; (b) Fresnel diffraction, $Q = 25$; (c) Fraunhofer diffraction, $Q = 0.25$.

Fraunhofer region, we can neglect the second exponential in equation (11.19a) and obtain

$$|D(\theta)| = \sqrt{\frac{1}{\lambda_0 Z}} \int_{-a/2}^{a/2} dX' e^{-ik_0 X' \sin\theta} = \sqrt{\frac{a^2}{\lambda_0 Z}} \frac{\sin[k_0 a \sin(\theta)/2]}{k_0 a \sin(\theta)/2}, \tag{11.21}$$

a result from elementary physics. The first zero occurs when

$$\sin\theta = \frac{2\pi}{k_0 a} = \frac{\lambda_0}{a}. \tag{11.22}$$

This is a general result; an aperture of size a lets you resolve a signal to within an angle of order λ/a. Note that

$$\begin{aligned}\int_{-\pi/2}^{\pi/2} |D(\theta)|^2 Z d\theta &= \frac{a^2}{\lambda_0} \int_{-\infty}^{\infty} \left\{ \frac{\sin[k_0 a \sin(\theta)/2]}{k_0 a \sin(\theta)/2} \right\}^2 d\theta \\ &\approx \frac{2a}{\lambda_0 k_0} \int_{-\infty}^{\infty} \frac{\sin^2 x}{x^2} dx = a \\ &= a \int_{-a/2}^{a/2} \left| T(X') \right|^2 dX',\end{aligned} \tag{11.23}$$

which is a statement of conservation of energy, and the diffracted field flux is equal to the flux passing through the aperture. In deriving this equation, we used the fact that only values of $\theta \ll 1$ make a considerable contribution to the integral.

These diffraction features are illustrated in the graphs shown in figure 11.1, where $|D|^2$ is plotted as a function of X/a for several values of $Q = k_0a^2/2Z$. You can see easily the transition from the shadow to the Fraunhofer region. Note the change in horizontal scale as one goes into the Fraunhofer region, since the results now depend only on $\sin\theta = X/Z$.

11.1.2 Quantum-Mechanical Diffraction

In the case of quantum-mechanical diffraction, the calculation follows very closely that of the electromagnetic case, but the dispersion relation is different. For a free particle having mass M, the momentum is $\mathbf{P} = \hbar\mathbf{K}$, the energy is

$$E_k = \frac{(\hbar K)^2}{2M}, \tag{11.24}$$

and

$$\omega_K = \frac{E_K}{\hbar} = \frac{\hbar K^2}{2M}. \tag{11.25}$$

Owing to the dispersion relation, there is wave packet spreading that is absent in the electromagnetic case, but we can choose our initial wave packet sufficiently broad to ignore this spreading.

Since most of the examples we discuss involve one-dimensional scattering, we adopt this limit from the outset. At $t = 0$, the wave function is assumed to equal

$$\psi(X, Z; t) = T(X)F(Z, 0)e^{iK_0 Z}, \tag{11.26}$$

where the amplitude transmission function $T(X)$ has been imprinted on the wave function by the diffracting screen, and $F(Z, 0)$ corresponds to a pulse function that is fairly well located in space, but sufficiently broad to avoid wave-packet spreading on the time scale of the experiment. In other words, the spatial extent in the Z direction is much greater than a (de Broglie) wavelength, but much less than the distance to the screen where the diffraction pattern is measured. It is assumed that the transmission function has structure that leads to diffraction whose maximum transverse propagation vector components are much less than K_0.

The calculation now proceeds as in the electromagnetic case, except that the expansion needed for $\omega_K = \hbar K^2/2M$ is

$$\begin{aligned}\omega_K &= \frac{\hbar K^2}{2M} = \frac{\hbar\left(K_X^2 + K_Z^2\right)}{2M} \\ &= \frac{\hbar(K_Z - K_0 + K_0)^2}{2M} + \frac{\hbar K_X^2}{2M} \\ &\approx \frac{\hbar K_0^2}{2M} + \frac{\hbar K_0(K_Z - K_0)}{M} + \frac{\hbar K_X^2}{2M},\end{aligned} \tag{11.27}$$

leading to

$$\psi(X, Z; t) = F(Z - v_0 t, 0)e^{i(K_0 Z - \omega_0 t)}e^{i\left(\frac{M}{\hbar t}\right)X^2/2}D(X), \tag{11.28}$$

where the diffraction pattern is given by

$$D(X) \approx \sqrt{\frac{-iM}{2\pi\hbar t}} \int dX' T(X') e^{-i\left(\frac{M}{\hbar t}\right)XX'} e^{i\left(\frac{M}{\hbar t}\right)X'^2/2}, \tag{11.29}$$

$$v_0 = \frac{\hbar K_0}{M} \tag{11.30}$$

is the average velocity of the wave packet, and

$$\omega_0 = \frac{\hbar K_0^2}{2M}. \tag{11.31}$$

Thus, the pulse envelope is simply translated in the Z direction by $v_0 t$, and the diffraction pattern is superimposed on the pulse in the transverse direction.

Since the pulse envelope function is sharply peaked at $Z = v_0 t$, we can replace t by Z/v_0 in this expression; moreover, we can use the fact that

$$\frac{M}{\hbar t} = \frac{2\pi M v_0}{hZ} = \frac{2\pi}{\lambda_B Z} = \frac{K_0}{Z}, \tag{11.32}$$

where

$$\lambda_B = \frac{h}{M v_0} = \frac{2\pi}{K_0} \tag{11.33}$$

is the average de Broglie wavelength, to obtain

$$D(X) = \sqrt{\frac{-i}{Z\lambda_B}} \int dX' T(X') e^{-iK_0 XX'/Z} e^{iK_0 X'^2/2Z}, \tag{11.34}$$

which is essentially the same result (11.18) of the Kirchhoff-Fresnel theory but with λ replaced by λ_B and k_0 by K_0.

11.2 Atom Optics

Things can get confusing when one discusses atom optics and atom interferometry because it is possible to have an atom interferometer in which the wave aspects of the center-of-mass motion of the atoms play no role at all. For example, the photon echo illustrated in figure 10.4, in effect, can be considered an atom interferometer capable of detecting changes in atomic velocity, even though the motion of the atoms is classical. Thus, it is useful to distinguish between situations in which the wave nature of the center-of-mass motion is critical and situations where it is relatively unimportant.

In chapter 5, we looked at several applications involving the forces exerted on atoms by optical fields. In this section and the remainder of this chapter, we examine examples in which atoms' center-of-mass motion is modified via an interaction with a periodic potential or *grating*. The grating can be a microfabricated, material grating, or it can be formed using oppositely propagating optical fields. In both cases, as a result of interaction with the grating, the atoms acquire transverse momentum components. We consider one-dimensional gratings only.

A grating can be characterized by its period d. Moreover, for microfabricated gratings, the slit width a provides an additional length parameter. Although the total size of the grating can be very large, the *effective* size of the grating depends on the width D_b of the atomic beam that traverses the grating. In other words, if D_b is less than the size of the grating, the number of grating periods intersected by the beam is given by

$$N = D_b/d. \tag{11.35}$$

It is assumed that $N \gg 1$ (see figure 11.2).

Quantization of the center-of-mass motion becomes important once the atoms undergo scattering that takes them outside the shadow region behind the grating. In

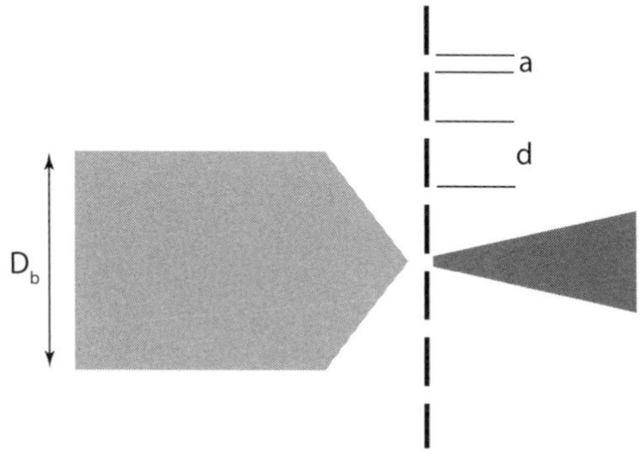

Figure 11.2. Schematic representation of scattering of an atomic beam by a material grating. Matter-wave effects for the grating become important when the transverse spreading of the beam resulting from diffraction at the slits is of order d.

passing through a slit having width a, a particle acquires a transverse momentum

$$\Delta P_X \approx h/a. \tag{11.36}$$

The particle leaves the shadow region behind the slit once [see equation (11.13a), with $r_0 = a/2$]

$$\frac{\pi a^2}{4\lambda_B Z} = \frac{Ma^2}{8\hbar}\frac{v_0}{Z} = \left(\frac{8\hbar}{Ma^2}t\right)^{-1} \approx \left[\frac{4}{\pi^2}\frac{(\Delta P_X)^2}{2M\hbar}t\right]^{-1} \lesssim 1. \tag{11.37}$$

In other words, quantization of the center-of-mass motion is necessary once

$$\omega_r t \gtrsim 1, \tag{11.38}$$

where

$$\omega_r = \frac{(\Delta P_X)^2}{2\hbar M} \tag{11.39}$$

is referred to as a *recoil frequency*. We return to this condition often in our analysis of atom optics. In this section, we assume that condition (11.38) holds, unless indicated otherwise.

Condition (11.38) alone is not sufficient to ensure that matter-wave interference between *different* slits occurs. For the grating to act truly as a grating, one must have interference between waves transmitted from different slits. Such interference occurs at distances when the transverse displacement resulting from diffraction at the slits is of order d—namely,

$$\frac{\Delta P_X}{M}\frac{Z}{v_0} \gtrsim d, \tag{11.40}$$

or

$$\frac{ad}{\lambda_B Z} \lesssim 1. \tag{11.41}$$

Last, by imposing the condition

$$\frac{a D_b}{\lambda_B Z} \gg 1, \tag{11.42}$$

we guarantee that diffraction by a single slit does not lead to transverse displacements that are larger than D_b, a limit we want to avoid in this discussion. In practice, we take a and d to be comparable.

Rather than deal with the conditions for shadow, Fresnel, and Fraunhofer diffraction in the abstract, we consider scattering by an amplitude grating to see how the various length scales determine the nature of the scattering.

11.2.1 Scattering by an Amplitude Grating

We consider first scattering by a microfabricated grating having period d and slit width a. The transmission function for a beam having diameter D_b is

$$T(X') = F(X')\Theta(|X'| - D_b/2), \tag{11.43}$$

where Θ is the Heaviside step function,

$$F(X') = \begin{cases} 1 & jd - \alpha d/2 \le X' \le jd + \alpha d/2 \\ 0 & jd + \alpha d/2 \le X' \le (j+1)d - \alpha d/2, \end{cases} \tag{11.44}$$

$$\alpha = a/d \tag{11.45}$$

is the ratio of the slit width to the period (*duty cycle*), and j is an integer that varies from $-\infty$ to ∞. In other words, the transmission function is that of an infinite grating multiplied by the beam width D_b (see figure 11.2).

Since $F(X')$ is a periodic function with period d, we can expand it as the Fourier series

$$F(X') = \sum_{n=-\infty}^{\infty} a_n e^{2\pi i n X'/d}, \tag{11.46}$$

where

$$\begin{aligned} a_n &= \frac{1}{d}\int_{-d/2}^{d/2} F(X')e^{-2\pi i n X'/d} dX' \\ &= \frac{1}{d}\int_{-\alpha d/2}^{\alpha d/2} e^{-2\pi i n X'/d} dX' = \frac{\sin(n\pi\alpha)}{n\pi}. \end{aligned} \tag{11.47}$$

From equations (11.34) and (11.43) to (11.46), one finds the diffraction amplitude

$$D(X) = \sqrt{\frac{-i}{Z\lambda_B}} \sum_{n=-\infty}^{\infty} a_n \int_{-D_b/2}^{D_b/2} dX' e^{2\pi i n X'/d} e^{-iK_0 XX'/Z} e^{iK_0 X'^2/2Z} \tag{11.48}$$

$$\begin{aligned} &= \frac{e^{-iK_0X^2/2Z}}{2} \sum_{n=-\infty}^{\infty} \frac{\sin(n\pi\alpha)}{n\pi} e^{2\pi i n X/d} e^{-2\pi i n^2 s} \\ &\times \left\{ \Phi\left[e^{-i\pi/4} \sqrt{\frac{2\pi}{s}} \left(ns - \frac{X}{2d} + \frac{N}{4} \right) \right] \right. \\ &\left. - \Phi\left[e^{-i\pi/4} \sqrt{\frac{2\pi}{s}} \left(ns - \frac{X}{2d} - \frac{N}{4} \right) \right] \right\}, \end{aligned} \tag{11.49}$$

where $N = D_b/d$ is the number of slits covered by the beam, Φ is the error function (11.20),

$$s = \frac{Z}{L_T}, \tag{11.50}$$

and

$$L_T = \frac{2d^2}{\lambda_B} = \frac{K_0 d^2}{\pi} \tag{11.51}$$

is referred to as the *Talbot length*, whose significance is discussed in section 11.2.2. The parameter s determines the scattering zone, classical, Fresnel, or Fraunhofer.

Owing to the $\sin(n\pi\alpha)/n\pi$ factor in equation (11.49), the maximum n that contributes in the summation is of order

$$n_{\max} \approx 1/\alpha. \tag{11.52}$$

If

$$2\pi n_{\max}^2 s = \pi \frac{Z\lambda_B}{a^2} \ll 1, \tag{11.53}$$

equation (11.49) reduces to

$$\begin{aligned} |D(X)| &\sim \frac{1}{2} \left| \sum_{n=-\infty}^{\infty} a_n e^{2\pi i n X/d} \left\{ \begin{array}{l} \Phi\left[e^{-i\pi/4} \sqrt{\frac{2\pi}{s}} \left(\frac{N}{4} - \frac{X}{2d} \right) \right] \\ +\Phi\left[e^{-i\pi/4} \sqrt{\frac{2\pi}{s}} \left(\frac{N}{4} + \frac{X}{2d} \right) \right] \end{array} \right\} \right| \\ &\approx F(X)\Theta(|X| - D_b/2) = T(X), \end{aligned} \tag{11.54}$$

since the sum of the error functions is approximately equal to zero for $|X| > D_b/2$ and approximately equal to 2 for $|X| < D_b/2$. The diffracted intensity is the same as the intensity transmission function—this is the shadow regime. We can write condition (11.53) in the suggestive form

$$v_0 t = Z \ll \frac{a^2}{\pi \lambda_B} = \frac{a^2 M v_0}{\pi h}, \tag{11.55}$$

or

$$\omega_r t \ll 1, \tag{11.56}$$

where ω_r is the recoil frequency, equation (11.39), associated with the transverse momentum $\Delta P_X \approx h/a$ imparted to the atoms by the grating apertures. If $\omega_r t \ll 1$, the scattering can be approximated as classical in nature. We are led to the important conclusion that *Fresnel or Fraunhofer diffraction for matter waves corresponds to times for which the recoil phase acquired by the atoms from apertures is greater than or of order unity.*

If $2\pi n_{\text{max}}^2 s \approx 1$ (Fresnel diffraction limit), the sum in equation (11.49) must be evaluated numerically. Since $n_{\text{max}}^2 \approx 1/\alpha^2$, it follows that $s \approx \alpha^2/2\pi < N$ in the Fresnel scattering zone.

For values of $2\pi n_{\text{max}}^2 s \gg 1$, we actually encounter *two* types of Fraunhofer diffraction limits for the grating. The "traditional" Fraunhofer limit is one in which

$$\frac{K_0 D_b^2}{2Z} < 1 \quad \text{or} \quad s > N^2, \tag{11.57}$$

allowing us to neglect the last exponential in equation (11.48). The diffraction pattern depends only on $X/Z = \sin\theta$ in this limit, and we recover the conventional Fraunhofer result for an N-slit grating. That is, maxima occur when $d \sin\theta = m\lambda_B$ for integral m, which is equivalent to the condition

$$\frac{X}{d} = 2ms\,. \tag{11.58}$$

For $\theta \ll 1$, the angular width of these principal maxima is of order $4\pi\lambda_B/Nd$, implying that the spatial width of these maxima in the X direction is of order

$$\Delta X \approx \frac{2\lambda_B Z}{Nd} = \frac{4sd}{N} = \frac{4s D_b}{N^2} > D_b. \tag{11.59}$$

On the other hand, under the somewhat less restrictive condition

$$\frac{K_0 d D_b}{2Z} < 1 \quad \text{or} \quad s > N, \tag{11.60}$$

we find a diffraction pattern in which the *entire beam* has diffraction maxima given by equation (11.58). In other words, the spatial width of the diffraction maxima is of order D_b, and diffraction separates the incident beam into a number of *nonoverlapping* beams, each having width of order D_b.

All these features are seen in the plots of $|D(X)|$ as a function of X/d given in figure 11.3, which are drawn for $N = 50$ and $\alpha = 0.2$. In figure 11.3(a), $s = 0.0004/2\pi$ and $2\pi n_{\text{max}}^2 s = 0.01$, corresponding to the shadow region. In figure 11.3(b), $s = 0.8/2\pi$ and $2\pi n_{\text{max}}^2 s = 20$, corresponding to the Fresnel zone. In figure 11.3(c), $s = 100$ and $2\pi n_{\text{max}}^2 s \approx 15{,}700$. This corresponds to the Fraunhofer zone, in which the entire beam has diffraction maxima, since $N < s < N^2$. It is easy to verify that the width of the maxima are of order $\Delta X/d = D_b/d = N = 50$. Last,

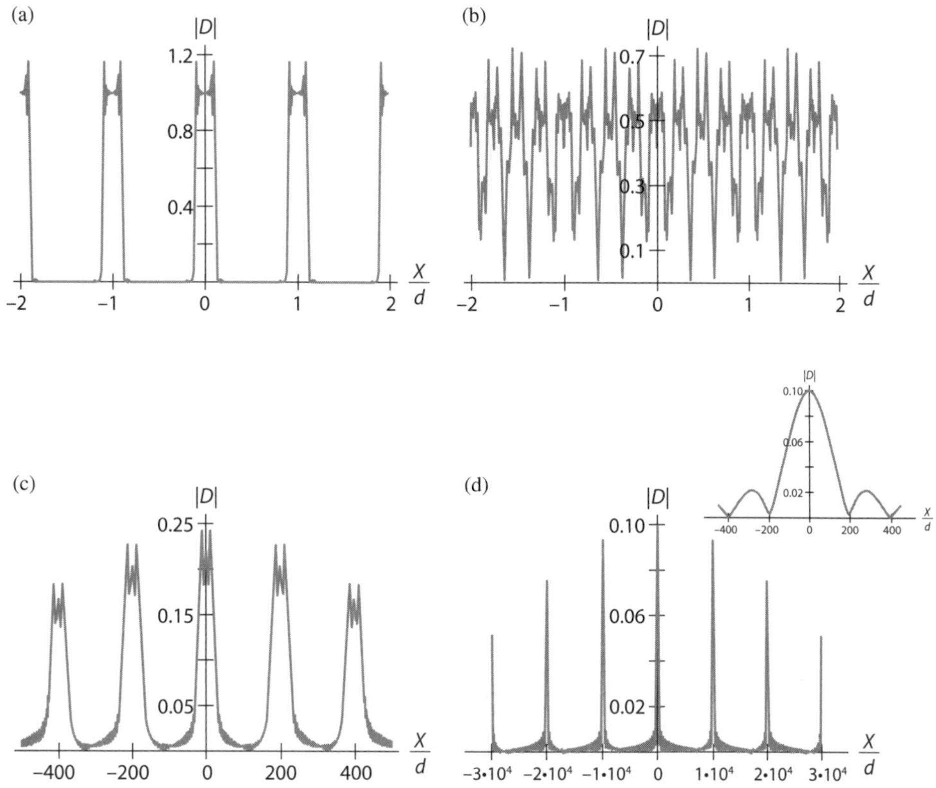

Figure 11.3. Diffraction by an amplitude grating having $N = 50$ and $\alpha = 0.2$. (a) $s = 0.0004/2\pi$—shadow region; (b) $s = 0.8/2\pi$—Fresnel zone; (c) $s = 100$—Fraunhofer zone in which the entire beam has diffraction maxima; (d) $s = 5000$—"traditional" Fraunhofer diffraction (the inset shows a principal maximum).

in figure 11.3(d), $s = 5000$ and $2\pi n_{\text{max}}^2 s \approx 785{,}000$, corresponding to "traditional" Fraunhofer diffraction. The width of the principal maximum shown in the inset is of order $400 = 4s/N$, in agreement with equation (11.59). Note the scale change as we go from the Fresnel to the Fraunhofer zones.

11.2.2 Scattering by Periodic Structures—Talbot Effect

In the limit of scattering by an *infinite* periodic structure

$$N = D_b/d \sim \infty, \tag{11.61}$$

equation (11.49) reduces to

$$|D(X)| = \left| \sum_{n=-\infty}^{\infty} a_n e^{2\pi i n X/d} e^{-2\pi i n^2 Z/L_T} \right|. \tag{11.62}$$

It is no longer possible to satisfy the conditions for Fraunhofer diffraction [equations (11.57) and (11.60)] in this limit; scattering is in either the classical or Fresnel zones. Equation (11.62) is quite general, with

$$a_n = \frac{1}{d}\int_{-d/2}^{d/2} T(X')e^{-2\pi inX'/d}. \tag{11.63}$$

Note that the Fourier transform $A(K)$ of $T(X)$ is

$$\begin{aligned} A(K) &= \frac{1}{\sqrt{2\pi}}\sum_{n=-\infty}^{\infty} a_n \int_{-\infty}^{\infty} dX' e^{2\pi inX'/d} e^{-iKX'} \\ &= \sqrt{2\pi}\sum_{n=-\infty}^{\infty} a_n \delta(K - 2\pi n/d). \end{aligned} \tag{11.64}$$

As such, the grating imparts integral multiples of momentum

$$\hbar K = h/d \tag{11.65}$$

to the atoms. The Fourier coefficients a_n determine the weight of the various momentum components.

Equation (11.62) contains a very interesting feature. For distances

$$Z = mL_T, \tag{11.66}$$

where m is a positive integer,

$$|D(X)| = \left|\sum_{n=-\infty}^{\infty} a_n e^{2\pi inX/d}\right| = |T(X)|\,. \tag{11.67}$$

An integral multiples of the Talbot distance, the grating structure is reproduced! This *self-imaging* effect was discovered by Talbot [2], and has been observed using both optical fields [3–6] and matter waves [7–9]. It is also possible to show that, at half-integral multiples of the Talbot length, the grating structure is reproduced with a half-period displacement. Moreover, at fractional Talbot distances, L_T/r, for integral r, it is possible to obtain grating structures having reduced period [10].

11.2.3 Scattering by Phase Gratings—Atom Focusing

The formalism given in the previous sections works equally well for phase gratings. It is a relatively easy matter to produce a phase grating using a standing-wave optical field instead of a microfabricated structure (see figure 11.4). Diffraction from a standing-wave field is characterized by a set of parameters that differs somewhat from those we encountered in scattering from microfabricated structures. For the optical field, the parameter set includes the period $d = \lambda/2$, field strength, and beam width w. For the atomic beam, the parameters are the beam width D_b, the longitudinal velocity v_0, and the angular divergence of the beam. The width D_b of the incident atomic beam is taken to be infinite—in other words, the number of lobes $N = D_b/d$ of the standing-wave field that is intersected by the atomic

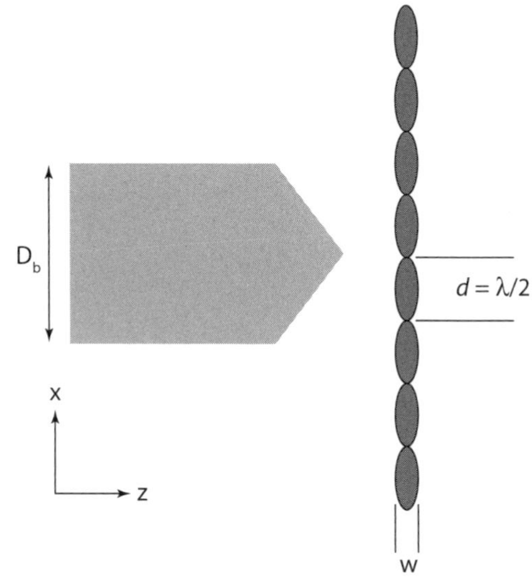

Figure 11.4. Diffraction by a standing-wave optical field.

beam is assumed to be much greater than unity. The atomic beam is taken to be a quasi-monochromatic pulse with average momentum $\mathbf{P}_0 = M\mathbf{v}_0$ in the Z direction. As a consequence, the transit time of an atom through the field is $\tau = w/v_0$. The atom–field interaction is a function of the detuning $\delta = \omega_0 - \omega$ (field frequency ω and atomic transition frequency ω_0), and the Rabi frequency (taken to be real and positive) is $\Omega_0 = 2\chi$. Although it might seem strange, the Rabi frequency actually plays the role of the slit width. In scattering by a microfabricated structure, the slit width determines the maximum transverse momentum imparted to the atoms. You will see that the field intensity serves this role in scattering by a standing-wave field.

To simplify this discussion, we assume that the longitudinal momentum P_0 is much greater than the transverse momentum acquired by the atoms in interacting with the fields. This assumption allows us to treat the scattering as a one-dimensional problem in the X direction. Moreover, we assume that $|\delta|\tau \gg 1$, $\chi/|\delta| \ll 1$ and that $\gamma_2\tau(\chi/|\delta|)^2 \ll 1$ (γ_2 is the excited state decay rate), ensuring that (1) the atom stays in a dressed state that returns to the ground state following the interaction with the field and (2) spontaneous decay during the atom–field interaction is negligible. In these limits, the net effect of the field is to provide a spatially dependent light shift or optical potential for the atomic ground state.

The transverse momentum acquired by the atoms is determined by the time the atom spends in the field and the strength of the optical potential. Each elementary scattering process in which momentum is transferred between counterpropagating standing-wave field components results in a transfer of $2\hbar k$ ($k = \omega/c$) momentum to an atom. Since the light shift (which serves as an effective two-photon Rabi frequency for momentum transfer) is of order $\chi^2/|\delta|$, the maximum transverse momentum $P_\perp$ imparted to an atom is of order

$$P_\perp = (2\hbar k)\left(\frac{\chi^2\tau}{|\delta|}\right) \ll P_0. \tag{11.68}$$

It is in this sense that a strong field corresponds to a narrow slit, since equation (11.68) can be written as $P_{\perp} = h/a_{eff}$, with

$$a_{eff} = \frac{\pi|\delta|}{k\chi^2\tau}. \tag{11.69}$$

A dimensionless parameter that helps to characterize the atom–field interaction is the transverse Doppler phase ϕ_D acquired during the atom–field interaction. This phase is of order

$$\phi_D = kv_{\perp}\tau = k(P_{\perp}/M)\tau = \left(\frac{\chi^2\tau}{|\delta|}\right)\omega_{2k}\tau, \tag{11.70}$$

where

$$\omega_q = \frac{\hbar q^2}{2M} \tag{11.71}$$

is the recoil frequency associated with a scattering process in which $\hbar q$ of momentum is transferred to the atom. In the examples considered in this chapter,

$$q = 2k. \tag{11.72}$$

For arbitrary values of ϕ_D, the atom–field dynamics can be solved numerically. There are two limits, however, for which analytic solutions can be obtained. The limit $\phi_D \ll 1$ corresponds to the *Raman-Nath approximation* [11, 12], in which all transverse motion of the atoms can be neglected *during* the atom–field interaction. The limit in which $\omega_q\tau \gg 1$ is referred to as the *Bragg limit*, for reasons that will become apparent in the following section. In this section, we discuss the Raman-Nath limit. Scattering of atoms from a standing-wave optical field is often referred to as *Kapitza-Dirac scattering* [13], since they proposed that standing-wave optical fields could be used to produce Bragg scattering of electrons.

11.2.3.1 Raman-Nath limit

In the Raman-Nath approximation, and with $|\delta|\tau \gg 1$, $\chi/|\delta| \ll 1$, and $\gamma_2\tau(\chi/|\delta|)^2 \ll 1$, the only effect of the atom–field interaction on the atoms is to modify the phase of the ground-state amplitude. In other words, the field results simply in a spatially modulated light shift of the ground-state energy. The positive frequency component of the field amplitude is given by

$$E^+(X, Z, t) = \frac{1}{2}E_0(Z)e^{-i\omega t}(e^{ikX} + e^{-ikX}), \tag{11.73}$$

where $E_0(Z)$ is the field envelope in the Z direction, having characteristic width w (see figure 11.4). After eliminating the excited-state amplitude, we find that the ground-state amplitude evolves as

$$i\hbar\dot{a}_1 = -\frac{4\hbar\chi^2(t)\cos^2(kX)}{\delta}a_1, \tag{11.74}$$

where

$$\chi(t) = -\frac{\mu_{21}E_0(t)}{2\hbar} \tag{11.75}$$

is one-half the Rabi frequency associated with each traveling-wave component of the field as seen in an atom's rest frame, and μ_{21} is the matrix element of the dipole moment operator along the direction of the field polarization. The Z dependence of the field envelope function in the laboratory frame is transformed into a time dependence in an atom's rest frame. Aside from an overall constant phase factor, it follows that the amplitude transmission function in the Raman-Nath approximation is simply

$$T(X) = e^{iA\cos(2kX)}, \tag{11.76}$$

where

$$A = 2\int_{-\infty}^{\infty} \frac{\chi^2(t)}{\delta} dt \tag{11.77}$$

is an effective pulse area. From the form of $T(X)$, it is clear that this off-resonant standing-wave field acts as a phase grating. We take $\delta > 0$ (red detuning).

It is instructive to calculate the Fourier transform of the transmission function,

$$\begin{aligned} \tilde{\Phi}(P_X) &= \frac{1}{\sqrt{2\pi\hbar}} \int_{-\infty}^{\infty} dX e^{iA\cos(2kX)} e^{-iP_X X/\hbar} \\ &= \sqrt{2\pi\hbar} \sum_{n=-\infty}^{\infty} (i)^n J_n(A)\delta(P_X - 2n\hbar k), \end{aligned} \tag{11.78}$$

where J_n is a Bessel function. To arrive at this equation, we used the expansion

$$e^{iA\cos(2kX)} = \sum_{n=-\infty}^{\infty} (i)^n J_n(A) e^{2inkX}. \tag{11.79}$$

The grating imparts integral multiples of momentum $2\hbar k$ to the atoms, with Bessel function weighting factors. The larger the pulse area, the larger the amount of transverse momentum given to the atoms. The different momentum components can be detected experimentally [14].

We now want to calculate the diffraction pattern. The transmission function is a periodic function of X having period $d = \pi/k$, and the Talbot length is

$$L_T = \frac{2d^2}{\lambda_B} = \frac{\pi M v_0}{\hbar k^2}. \tag{11.80}$$

As a consequence,

$$2\pi Z/L_T = 2\pi v_0 t/L_T = \omega_q t, \tag{11.81}$$

where the recoil frequency ω_q, with $q = 2k$, is defined by equation (11.71). With these results, we find that the diffraction amplitude, as given by equation (11.62), is

$$|D(X)| = \left| \sum_{n=-\infty}^{\infty} a_n e^{inqX} e^{-in^2\omega_q t} \right|, \tag{11.82}$$

with

$$a_n = \frac{k}{\pi}\int_{-\pi/2k}^{\pi/2k} dX e^{iA\cos(2kX)} e^{-2iknX}$$
$$= (i)^n J_n(A). \tag{11.83}$$

The diffraction intensity is proportional to

$$|D(X)|^2 = \sum_{n,n'=-\infty}^{\infty} (i)^n (-i)^{n'} J_n(A) J_{n'}(A) e^{i(n-n')qX} e^{-i(n^2-n'^2)\omega_q t}$$
$$= \sum_{m=-\infty}^{\infty} (i)^m e^{-im^2\omega_q t} e^{imqX}$$
$$\times \sum_{n'=-\infty}^{\infty} J_{n'+m}(A) J_{n'}(A) e^{-2imn'\omega_q t}. \tag{11.84}$$

In this form, we can use an addition theorem for Bessel functions [15],

$$\sum_{k=-\infty}^{\infty} J_{k+m}(A) J_k(A) e^{-ik\alpha} = J_m[2A\sin(\alpha/2)]\, e^{-im(\frac{\pi}{2}-\frac{\alpha}{2})}, \tag{11.85}$$

to rewrite equation (11.84) as

$$I(X) = |D(X)|^2 = 1 + 2\sum_{n=1}^{\infty} J_n[2A\sin(\omega_q t)]\cos(nqX) \tag{11.86}$$

$$= 1 + 2\sum_{n=1}^{\infty} J_n[2A\sin(2\pi Z/L_T)]\cos(2knX). \tag{11.87}$$

Although written in compact form, equation (11.87) displays a rich structure. At both integral and half-integral multiples of the Talbot distance, $I(X) = 1$, reproducing the original density pattern. At other distances, however, the phase grating is converted to a spatially varying atomic density containing all even spatial harmonics of the field. Although not apparent in equation (11.87), the standing-wave lobes act as lenses to focus the atoms at a distance $Z_f = L_T/(4\pi A)$ from the field for $A \gg 1$. This focal length can be obtained by a simple argument.

We approximate the optical potential near a lobe maximum as

$$U(X,t) = -\hbar\frac{2\chi^2(t)[1+\cos(2kX)]}{\delta} \approx -\hbar\frac{4\chi^2(t)(1-k^2X^2)}{\delta}, \tag{11.88}$$

which implies that the momentum impulse imparted to the atoms by the field as a function of X is

$$\Delta P_X = -\int dt \frac{\partial U(X,t)}{\partial X} = -4\hbar k^2 XA. \tag{11.89}$$

As a consequence, the atoms will be brought to a focus at

$$Z_f = v_0 t = \frac{v_0 MX}{|\Delta P_X|} = \frac{v_0 M}{4\hbar k^2 A} = \frac{L_T}{4\pi A}. \tag{11.90}$$

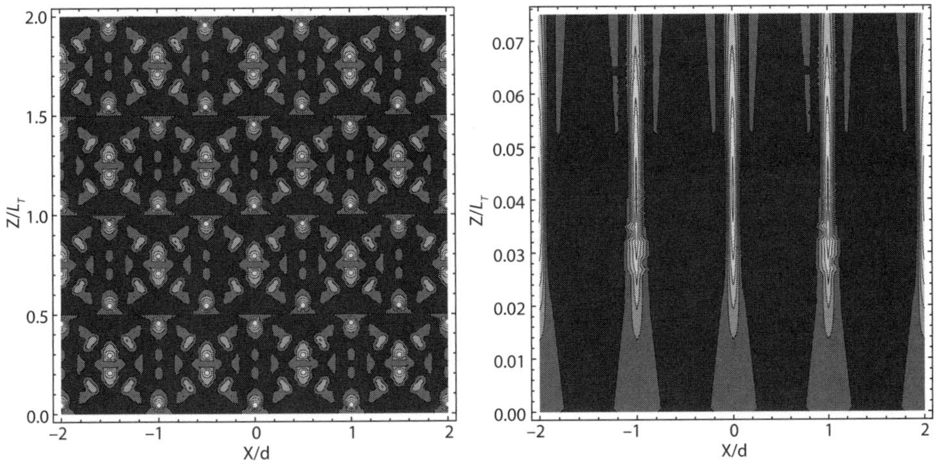

Figure 11.5. Contour plot of the atom density following passage through a standing-wave optical field. For a standing-wave field whose intensity varies as $\cos^2(kX)$, the period is $d = \lambda/2 = \pi/k$, implying that the centers of the intensity lobe maxima of the field are located at $X/d = m$, for integral m.

A somewhat better estimate,

$$Z_f = \frac{L_T}{4\pi A}\left(1 + \frac{1.27}{\sqrt{2A}}\right), \tag{11.91}$$

along with many more details, is given in the article by Cohen et al. [16]. A contour plot of the field intensity (11.87) is shown in figure 11.5 for $A = 3$. Focusing near $Z_f/L_T = [1 + 1.27/(2A)^{1/2}]/(4\pi A) = 0.040$ is clearly visible, as is the reimaging of the constant atom density for integral and half-integral multiples of the Talbot length.

Since we calculated the focal position using classical considerations, it is tempting to view the focusing as a classical effect. This is misleading, however, since the focal length is proportional to the Talbot length, which, in turn, is inversely proportional to the de Broglie wavelength of the atoms. As $\hbar \sim 0$, the de Broglie wavelength also goes to zero, giving an infinite focal length. In other words, a phase grating for matter is, by its very nature, of quantum-mechanical origin.

11.2.3.2 Bragg scattering

We now consider the limit opposite to that of the Raman-Nath approximation—that is, we assume that $\omega_q \tau \gg 1$. This limit will be seen to correspond to Bragg scattering. In conventional Bragg scattering, there is constructive interference in scattering of radiation from planes in a crystal for certain angles of the incident field propagation vector relative to the crystal planes. The same type of effect can occur in scattering of an atomic beam from a standing-wave field.

The difference between the Raman-Nath and Bragg regimes can be understood by making reference to the quantized transverse momentum states of the atoms. For a well-collimated beam, the atoms start in the atomic ground state with

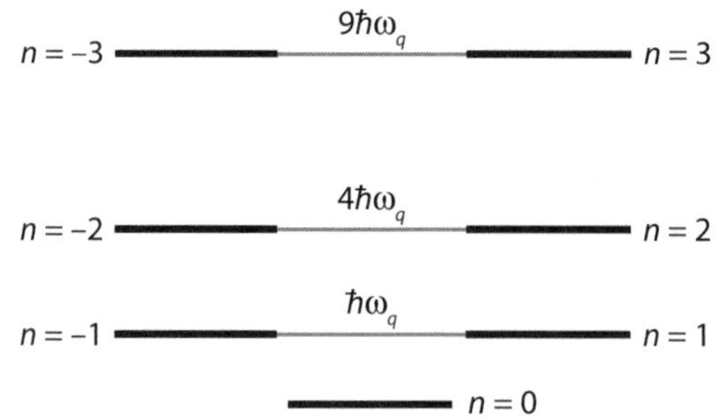

Figure 11.6. Bragg energy levels for atoms having $n\hbar q$ of momentum. The energies vary as $n^2\hbar\omega_q$.

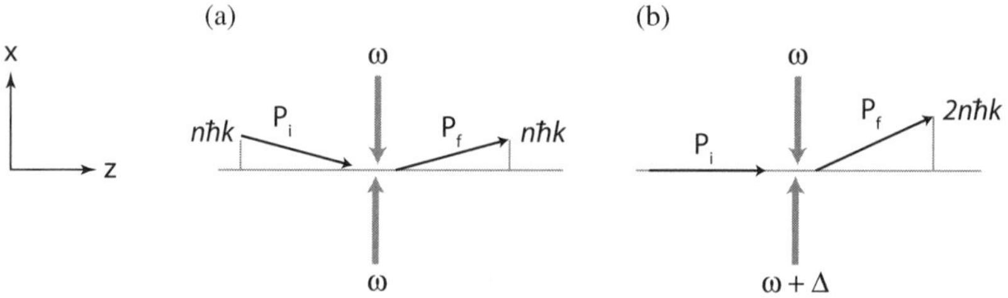

Figure 11.7. (a) Traditional geometry for Bragg scattering. (b) Alternative geometry for Bragg spectroscopy, in which Δ can be tuned to the Bragg resonances.

transverse momentum $P_{i_x} \approx 0$. The standing-wave field couples center-of-mass ground states differing in momentum by $2n\hbar k = n\hbar q$ or, in energy, by $\approx n^2\hbar\omega_q$ (see figure 11.6). In order to drive such transitions, the field intensity must contain frequency components equal to $n^2\omega_q$. The short duration pulses of the Raman-Nath regime have the frequency components necessary to drive all such transitions. Bragg scattering, however, corresponds to a "monochromatic" field limit, and the field intensity is, in effect, a static field. If the atomic beam is incident perpendicular to the field direction, no transitions to different momentum states can occur.

In order to conserve both momentum and energy when $\omega_q\tau \gg 1$, one must angle the incident atomic beam relative to the standing-wave optical field. Consider the geometry shown in figure 11.7(a), with $k = q/2$. A collimated atomic beam having momentum $\mathbf{P}_i = P_{i_x}\hat{\mathbf{x}} + P_{i_z}\hat{\mathbf{z}}$ is scattered by the standing-wave field into a final direction $\mathbf{P}_f = P_{f_x}\hat{\mathbf{x}} + P_{f_z}\hat{\mathbf{z}}$. For off-resonant scattering, we have seen that the initial and final components of momentum in the X direction differ by integral multiples n of $\hbar q$—that is, $P_{f_x} - P_{i_x} = n\hbar q$. In the Raman-Nath approximation, all components could be produced; however, if $\omega_q\tau \gg 1$, the energy-time uncertainty principle is respected, and there is destructive interference unless energy is conserved as well,

$$P_{i_z}^2 + \left(P_{i_x} + n\hbar q\right)^2 = P_{i_z}^2 + P_{i_x}^2, \tag{11.92}$$

or

$$P_{i_x} = -n\hbar q/2. \tag{11.93}$$

As a result, constructive interference occurs when

$$\tan\theta_n = \frac{P_{i_x}}{P_{i_z}} = -\frac{n\hbar q}{2P_{i_z}}, \tag{11.94}$$

where n is a positive or negative integer. Equation (11.94) can be rewritten as

$$2d\sin\theta_n = n\lambda_{dB}, \tag{11.95}$$

where $d = \lambda/2 = 2\pi/q$ is the grating period, and $\lambda_B = h/P = h/(P_{i_z}^2 + P_{i_x}^2)^{1/2}$ is the atomic de Broglie wavelength. Thus, in analogy with Bragg scattering, constructive interference occurs only for certain incident directions of the field relative to the "crystal planes." The resonance that occurs for fixed n is referred to as an *nth-order Bragg resonance*. Assuming that $\theta_n \ll 1$ and setting

$$v_0 = P_{i_z}/M \approx P_i/M, \tag{11.96}$$

we can rewrite the Bragg condition (11.94) as

$$\theta_n = \frac{n\omega_q}{2kv_0}. \tag{11.97}$$

Instead of changing the angle of the atomic beam relative to the field direction, it is also possible to drive Bragg resonances by using two traveling wave field components differing in frequency by Δ [figure 11.7(b)] and by choosing Δ equal to

$$\Delta_n = n\omega_q. \tag{11.98}$$

In terms of figure 11.6, resonances can occur for n positive and Δ_n positive by "absorbing" n field photons of frequency $(\omega + \Delta_n)$ and "emitting" n field photons having frequency ω, since

$$n(\omega + \Delta_n) - n\omega = n\Delta_n = n^2\omega_q. \tag{11.99}$$

For n negative and Δ_n negative, resonances occur by "absorbing" n field photons of frequency ω and "emitting" n field photons having frequency $(\omega + \Delta_n)$. By varying Δ and looking for scattering resonances, one is, in effect, carrying out a form of Bragg spectroscopy [17] that has proved to be very useful in analyzing scattering from cold gases and Bose condensates.

We present a brief derivation of the equations that describe Bragg spectroscopy. If the positive frequency component of the field amplitude is given by

$$E^+(X, Z, t) = \frac{1}{2}E_0(Z)e^{-i\omega t}\left[e^{i(kX-\Delta t)} + e^{-ikX}\right], \tag{11.100}$$

where $E_0(Z)$ is the field profile, then the effective Hamiltonian for the atom–field system is

$$H = \frac{P_X^2}{2M} - \frac{2\hbar\chi^2(t)}{\delta}\cos(2kX - \Delta t). \tag{11.101}$$

To arrive at the Hamiltonian (11.101), we adiabatically eliminated the excited state and neglected a light shift, $-2\hbar\chi^2(t)/\delta$, that is not spatially modulated. The Hamiltonian is written in the atomic rest frame with $t = Z/v_0$, and the motion of the atoms in the Z direction is treated classically. One sees that the effective

Hamiltonian contains contributions from the center-of-mass energy and the spatially and temporally modulated ground-state light shift produced by the combined action of the counterpropagating fields. We want to calculate the effect on the atoms as they pass thorugh the field region. As viewed from the atomic rest frame, this corresponds to a calculation of the atomic response to a pulse of duration w/v_0.

The wave function for the atom is expanded in an interaction representation as

$$\psi(X,t) = \frac{1}{\sqrt{2\pi\hbar}} \int_{-\infty}^{\infty} dp\, c(p,t) e^{-i\left(p^2/2M\right)t/\hbar} e^{ipX/\hbar}, \tag{11.102}$$

with $p \equiv P_X$. From the Schrödinger equation, with the Hamiltonian (11.101), one finds that the state amplitudes evolve as

$$\begin{aligned}\dot{c}(p,t) &= ir(t)e^{i\left(qpt/M-\omega_q-\Delta\right)t}c(p-\hbar q,t)\\ &\quad + ir(t)e^{i\left(-qpt/M-\omega_q+\Delta\right)t}c(p+\hbar q,t),\end{aligned} \tag{11.103}$$

where

$$r(t) = \frac{\chi^2(t)}{\delta} \tag{11.104}$$

is one-half of an effective Rabi frequency that couples different momentum states. Equation (11.103) represents a set of equations that must be solved numerically, in general, for a given $r(t)$ determined by the field's spatial profile in the Z direction.

For a well-collimated beam, we can take as our initial condition

$$c(p,0) = \sqrt{\delta(p)}, \tag{11.105}$$

where the square root of the Dirac delta function is defined as

$$\sqrt{\delta(p)} = \lim_{p_0\to 0} \frac{e^{-p^2/2p_0^2}}{\left(\pi p_0^2\right)^{1/4}}. \tag{11.106}$$

It is clear from equation (11.103) that with the initial condition (11.105), a trial solution

$$c(p,t) = \sum_{n=-\infty}^{\infty} c_n(t)\sqrt{\delta(p-n\hbar q)} \tag{11.107}$$

in equation (11.103) works, provided that the $c_n(t)$ satisfy

$$\begin{aligned}\dot{c}_n(t) &= ir(t)e^{i\left[(2n-1)\omega_q-\Delta\right]t}c_{n-1}(t)\\ &\quad + ir(t)e^{i\left[-(2n+1)\omega_q+\Delta\right]t}c_{n+1}(t),\end{aligned} \tag{11.108}$$

subject to the initial condition

$$c_n(0) = \delta_{n,0}. \tag{11.109}$$

Equations (11.108) can be solved numerically, for arbitrary values of $\omega_q\tau$, where $\tau = w/v_0$, and w is the field beam diameter through which the atoms pass. The number of terms needed will depend on the field strength, the recoil frequency, and the detuning Δ. In the Bragg regime, with $\omega_q\tau \gg 1$, nth-order Bragg resonances occur whenever condition (11.98) holds. One can see this by adiabatically eliminating all intermediate states between $n = 0$ and $n = \Delta/\omega_q$. It is safest, however, to solve the full set of coupled equations (11.108), as the adiabatic elimination is suspect for typical values of parameters encountered in experiments.

In the case of a first-order Bragg resonance with $\Delta \approx \omega_q$, equations (11.108) reduce to

$$\dot{c}_0(t) = ir(t)e^{i(\Delta-\omega_q)t}c_1(t), \tag{11.110a}$$

$$\dot{c}_1(t) = ir(t)e^{-i(\Delta-\omega_q)t}c_0(t), \tag{11.110b}$$

provided all other nonresonant amplitudes are neglected. Equations (11.110) are simply the equations for a two-level atom driven by a detuned field, having a pulse envelope proportional to $r(t)$. For a square pulse having duration τ and constant r, the solution for $|c_1(\tau)|^2$, which corresponds to the probability for the atom to have a transverse momentum $P_X = 2\hbar k$, is

$$|c_1(\tau)|^2 = \frac{4r^2}{4r^2 + (\Delta - \omega_q)^2}\sin^2\left[\frac{\sqrt{4r^2 + (\Delta - \omega_q)^2}}{2}\tau\right]. \tag{11.111}$$

The oscillations of $|c_1(\tau)|^2$ as a function of interaction time are referred to as *Pendellösung oscillations*. With an appropriate choice of Δ, one can excite higher-order Pendellösung oscillations corresponding to higher order Bragg resonances, but this requires additional field strength.

11.3 Atom Interferometry

Optical interferometers can serve as sensitive detectors of small changes in optical paths. The resolution of the measurements is usually limited by the wavelength of the optical fields. Typically, one measures a fringe shift in the interferometer as the length of one of the arms of the interferometer is varied. For example, consider a *Sagnac interferometer* [18] in which the phase difference between fields propagating in opposite directions around a ring cavity is measured (see figure 11.8). This type of device is also referred to as a *ring laser gyroscope*.

First, imagine that the cavity in figure 11.8 is actually in the form of a ring having radius R (an old physics joke—consider the horse as a sphere) and that the ring cavity is rotating clockwise. Owing to this rotation, in order for the field entering the interferometer at the beam splitter to return back to the beam splitter, the clockwise propagating field component must travel a distance $2\Omega R\tau$ longer than the counterpropagating field component, where Ω is the rotation rate, and $\tau = 2\pi R/c$ is the round-trip time for the light. This leads to a phase difference

$$\phi_L = \left(\frac{2\pi}{\lambda}\right)(2\Omega R\tau) = \frac{4A\omega\Omega}{c^2}, \tag{11.112}$$

where $\lambda = 2\pi c/\omega$ is the field wavelength, and A is the ring area. Another way to arrive at the same result is to note that the difference in Doppler shifts for the two waves in the rotating frame is

$$\Delta\omega = 2\omega\frac{v}{c} = 2\omega\frac{\Omega R}{c}, \tag{11.113}$$

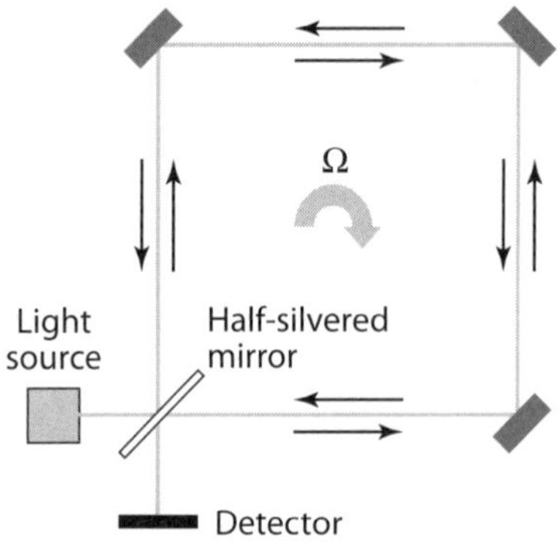

Figure 11.8. Sagnac interferometer.

giving rise to a fringe shift

$$\phi_L = \Delta\omega\tau = \frac{4A\omega\Omega}{c^2}, \tag{11.114}$$

as before.

Returning to figure 11.8, we use the Doppler shift method for a square cavity having side a to calculate the difference in Doppler shifts as

$$\Delta\omega = \frac{2\omega}{c}\frac{\int_0^{a/\sqrt{2}} \Omega r' dr'}{a/2} = \omega\frac{\Omega a}{c}, \tag{11.115}$$

where we averaged the linear velocity over the arms of the interferometer. As a consequence, we again find a fringe shift

$$\phi_L = \Delta\omega\tau = \frac{4\Delta\omega a}{c} = 4\omega\frac{\Omega a^2}{c^2} = \frac{4A\omega\Omega}{c^2}, \tag{11.116}$$

where A is the area of the cavity.

Using similar methods, these results can be generalized to [19]

$$\phi_L = \frac{4\alpha\omega\mathbf{A}\cdot\boldsymbol{\Omega}}{c^2}, \tag{11.117}$$

where $\mathbf{A} = A\hat{\mathbf{n}}$ and $\hat{\mathbf{n}}$ is the normal to the plane of the interferometer (obtained using a right-hand rule), and $\alpha = 1$ if the signal is measured at the input port (full round-trip), as in figure 11.8, or $\alpha = 1/2$ if the signal is measured at the midpoint of the interferometer where the beams are recombined, as in a Mach-Zehnder interferometer. Rotation rates on the order of 2×10^{-9} rad/s/$\sqrt{\text{Hz}}$ can be measured using a ring laser gyroscope [20]. This is considerably smaller than the Earth's rotation rate of 7.3×10^{-5} rad/s. The $\sqrt{\text{Hz}}$ in the sensitivity reflects the fact that the precision of the measurement increases with the integration time; thus, for a one second integration time, the sensitivity is 2×10^{-9} rad/s.

It is intriguing to consider the possibility of replacing the optical waves by matter waves in this device [21–23]. Then, equation (11.112) is replaced by

$$\phi_M = \left(\frac{2\pi}{\lambda_B}\right)(2\alpha\Omega R\tau) = \left(\frac{2\pi}{\lambda_B}\right)\left(2\alpha\Omega R\frac{2\pi R}{v_0}\right) = \frac{4\alpha M\Omega A}{\hbar}, \tag{11.118}$$

where v_0 is the average speed of the matter wave. For all things equal, the ratio

$$\frac{\phi_M}{\phi_L} = \frac{Mc^2}{\hbar\omega}, \tag{11.119}$$

which can represent a 10^{10} increase in sensitivity! Of course, the big catch here is "for all things equal." While it is possible to get large-angle beam splitters for optical fields, the beam splitters for matter waves usually produce relatively small momentum kicks (typically, several $\hbar k = \pi\hbar/d$, where d is the grating period). Moreover, laser field intensities are considerably larger than the equivalent atom fluxes. As a consequence, the factor of 10^{10} is reduced dramatically. Still, rotation rates with sensitivities approaching 6×10^{-10} rad/s/$\sqrt{\text{Hz}}$ have been achieved using matter-wave interferometers [24].

There is another advantage of matter wave over optical interferometers. Since matter waves carry mass, an atom interferometer is sensitive to inertial effects such as gravity, to the recoil that atoms undergo in scattering radiation, and to the changes in velocity or phase produced by collisions with atoms in the paths of the interferometers. The measurement of the recoil frequency, along with a sensitive measurement of the atom's mass, has the potential for yielding the most sensitive measurement of the fine structure constant. Thus, there are many reasons to pursue the concept of atom interferometry.

A generic interferometer contains a *beam splitter* that separates an incoming beam along different paths (which may be overlapping). A second element of the interferometer *recombines* the separated beams so that they can interfere at a *detector*. In the Michelson interferometer, for example, a half-reflecting, half-transmitting mirror serves as both the beam splitter and recombiner. The atom interferometers we discuss are assumed to work in the Raman-Nath regime—that is, the interactions of the atoms with the elements of the interferometer (microfabricated slits, standing-wave optical fields) occur sufficiently fast that any motion of the atoms *during* the interaction can be neglected.

In a typical atom interferometer, a collimated atomic beam passes through two sets of slits (or two standing-wave optical fields) separated in distance by L. The first set of slits serves as a beam splitter, and the second as a beam recombiner. Following passage through both sets of slits, the transverse atomic density is detected at a distance Z from the second slits. (In practice, one often uses a third set of slits as the detector, observing the atom flux as the slits are moved transverse to the atomic beam.) We discussed the first stage of this process in the section 11.2, where it was assumed that the incoming beam was mono-energetic.

Actually, for a perfectly mono-energetic beam, the recombiner is not necessary if the atom paths overlap. In fact, the diffraction pattern itself arises from atoms moving along different paths. However, the collimation requirements (transverse momentum $\ll \hbar k = \pi\hbar/d$, where d is the grating period) are severe in this case (typically, the beam angular divergence must be less than 10^{-4}), and it useful to use a scheme, similar to that of a photon echo, where the destructive interference resulting from the transverse momentum distribution can be canceled in certain echo planes. A device operating in this fashion is referred to as a *Talbot-Lau interferometer* [2, 25–27].

In a Talbot-Lau interferometer, the angular divergence of the incoming beam satisfies

$$\theta_b \gg d/L, \tag{11.120}$$

ensuring that any structure in the transverse density following the interaction with the beam splitter washes out in a distance $Z \ll L$. For a longitudinal velocity v_0 and a transverse velocity $u_\perp = v_0\theta_b$, this condition may be rewritten as

$$\theta_b = \frac{v_0\theta_b}{d}\frac{L}{v_0} = \frac{u_\perp}{d}T = \frac{ku_\perp T}{\pi} \gg 1 \tag{11.121}$$

and is reminiscent of that needed for a photon echo, $kuT \gg 1$, to ensure that the optical coherence is washed out before the second pulse. However, just as in a photon echo, the second grating can restore the signal at certain echo times or planes. Also, as in a photon echo, such points are determined by the condition that a Doppler phase cancels for *all* atoms, regardless of their transverse velocity. Since the Doppler phase does not contain $\hbar$, the location of the echo points can be determined from *classical* considerations, even if matter-wave effects are important in determining the actual density distribution at such points. The angular divergence of the beam cannot be arbitrarily large; we must require that $\theta_b \ll d/w$, where w is the grating thickness or beam diameter of the standing-wave optical field, in order to ensure that the atoms interact effectively with the beam splitter. This requirement is analogous to the one we used in coherent transients, where the condition $ku\tau \ll 1$ was needed to guarantee that all atoms were in the bandwith of the pulse.

As we mentioned earlier, there are so many different types of atom interferometers that things can get confusing. Let us give a brief classification. In all cases, we restrict the discussion to Talbot-Lau–type interferometers—that is, those in which there is a transverse velocity distribution satisfying condition (11.120).

11.3.1 Microfabricated Elements

If the beam splitters and recombiners consist of microfabricated slits, the interferometers operate in basically two regimes, the *moiré* or classical limit, and the *matter-wave* limit. A moiré interferometer relies on the shadow effect for its operation—the motion of the particles is treated classically. *Moiré* is a French word referring to the patterns one sees in insect wings. In a moiré interferometer, particles pass through two sets of slits separated by some distance $L \ll L_T$. By tracing the classical paths in such a geometry, one arrives at patterns following the second set of slits characteristic of a Talbot-Lau interferometer. This interferometer is sensitive to any changes in the atomic velocity, since they destroy the moiré pattern.

The matter-wave limit, on the other hand, corresponds to separation of the slits $L \gtrsim L_T$, where wave features can be seen. A detailed discussion of the theory of atom interferometers using microfabricated gratings can be found in the article by Dubetsky and Berman [10]. Here, we consider only atom interferometers in which

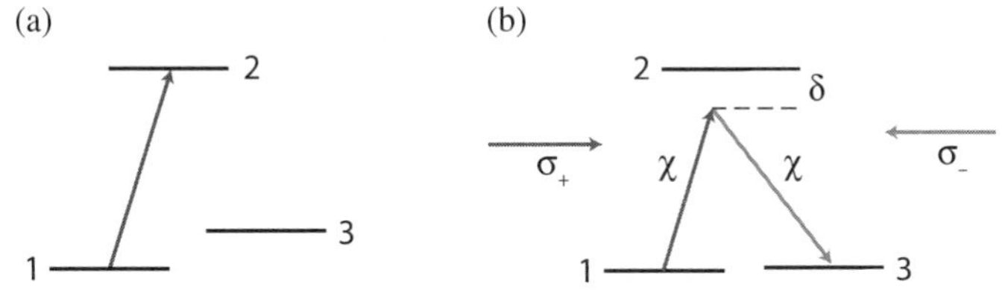

Figure 11.9. (a) Level scheme for an atom interferometer base on stimulated photon echoes. (b) Atom–field diagram for a Raman echo interferometer.

the beam splitters and combiners are constructed using counterpropagating optical fields.

11.3.2 Counterpropagating Optical Field Elements

Counterpropagating wave optical field atom interferometers break down into two general categories, classical motion and matter-wave interferometers. Moreover, such interferometers can operate in either *spatial* or *time* domains. A spatial interferometer is similar to one using microfabricated gratings—the counterpropagating optical fields replace the gratings. In such an interferometer, an atomic beam is incident perpendicular to two field regions separated by a distance L. In a time-domain interferometer, on the other hand, atoms in a magneto-optic trap (MOT) are subjected to two field pulses separated in time by T. If $L \gg L_T$ or $\omega_q T \gg 1$, the atomic center-of-mass motion must be treated quantum-mechanically, even if certain resonance positions can be determined classically.

11.3.2.1 Classical motion

The motion can be treated classically if $L \ll L_T$ or $\omega_q T \ll 1$. Since recoil frequencies $(\omega_q/2\pi)$ are of the order of 10 to 100 KHz or so, experiments having $T \lesssim 10\,\mu$s are in the classical regime. In the classical regime, there are several ways to design an atom interferometer. We are interested here only in atoms that spend most of their life in the interferometer in their ground-state sublevels, since decoherence in these states resulting from vacuum field interactions is at a minimum. Since the *total* ground-state population is equal to unity and is unmodulated, the key point is to construct a signal that depends on the population in a *single* ground-state sublevel or the *coherence* between ground-state sublevels.

For example, consider a stimulated photon echo in the three-level Λ scheme shown in figure 11.9(a). Levels 1 and 3 are two ground-state hyperfine levels, and level 2 is an excited state. The interaction pulses drive optical transitions between levels 1 and 2 *only*, while level 3 serves as a sink for *part* of the excited state decay. The spatial modulation created in the excited-state population by the first two pulses of the stimulated echo pulse sequence decays to *both* levels 1 and 3. Since the decay is to both levels, it does not totally refill the modulation "hole" left in level 1 by the field pulses. As a result, there is spatial modulation in *each* of levels 1 and 3 following spontaneous decay, even if the total ground-state population, $\varrho_{11} + \varrho_{33} = 1$, is unmodulated. The third pulse probes only level 1, such that the final signal is sensitive to changes in atomic velocity between the second and third pulses.

An atom interferometer based on a Raman photon echo, which relies on a modulated ground-state coherence for its operation, is discussed in detail in section 11.3.2.3.

11.3.2.2 Quantized center-of-mass motion

In "true" matter-wave interferometers, the recoil an atom undergoes on interacting with a field plays a critical role. For example, if a ground-state atom scatters radiation from one mode of a field having propagation vector $\mathbf{k}$ into another field mode having propagation vector $-\mathbf{k}$, the associated recoil is $\hbar q^2/2M$, where M is the atomic mass and $q = 2k$.[1] Clearly, this is a signature of a quantum effect, since $\hbar$ appears. Matter-wave effects become important once $L \gg L_T$ or $\omega_q T \gg 1$. In two-level atoms where the ground-state population is *constant* without any recoil, the matter-wave effects are the *only* ones that contribute. If internal state coherence is used as well, there can be matter-wave contributions to the signals, even if the echo times are determined by classical considerations [27].

Since we assume that the Raman-Nath approximation is valid, the theory of any interferometer can be formulated in terms of transfer or transmission matrices that determine how the fields act on the atoms and free-evolution matrices that characterize the atomic state evolution between the pulses. Next, we give one example of a time-domain internal state interferometer and one of a time-domain matter-wave interferometer.

11.3.2.3 Internal state interferometer

As an example of a time-domain internal state interferometer, we consider a Raman photon echo in a three-level Λ scheme. Levels 1 and 3 are the *degenerate* $m = \mp 1$ ground-state sublevels of a $J = 1$ ground state, and level 2 is a $J = 0$ excited state. The atoms are prepared in level 1 and driven by off-resonant fields having the same atom–field detuning δ and Rabi frequencies $\chi(t)$ [see figure 11.9(b)]. One field propagates in the Z direction and is σ_+ polarized (it drives only the 1–2 transition), while the other field propagates in the $-Z$ direction and is σ_- polarized (it drives only the 3–2 transition). By adiabatically eliminating the excited state and neglecting a light shift that is common to both levels, we find that the effective Hamiltonian for this atom field system is

$$H = -\frac{\hbar\chi^2(t)}{\delta}\left(e^{2ikZ}\left|3\right\rangle\left\langle 1\right| + e^{-2ikZ}\left|1\right\rangle\left\langle 3\right|\right). \tag{11.122}$$

This Hamiltonian is *identical* to the field interaction Hamiltonian for a two-level atom driven by a resonant pulse propagating in the Z direction with propagation

[1] There is also recoil associated with the Doppler effect. That is, an atom moving with velocity $\mathbf{v}_0$ changes its velocity to $\mathbf{v}_0 - \hbar\mathbf{q}/M$, implying that the frequency change (i.e., energy change divided by $\hbar$) is $-\mathbf{v}_0 \cdot \mathbf{q}/M + \omega_q$. The first term, which is a Doppler shift, is classical in origin, since it is independent of $\hbar$.

vector $\mathbf{q} = 2k\hat{\mathbf{z}}$ and two-photon Rabi frequency

$$\tilde{\Omega}(t) = -\frac{2\chi^2(t)}{\delta}. \tag{11.123}$$

Pairs of field pulses having two-photon Rabi frequencies $\tilde{\Omega}_1(t)$ and $\tilde{\Omega}_2(t)$ are applied at times $t = 0$ and $t = T$, respectively, to an ensemble of atoms having a velocity distribution

$$W(\mathbf{v}_0) = \left(\pi u^2\right)^{-3/2} e^{-v_0^2/u^2}. \tag{11.124}$$

We can take over the result of equation (10.34) for the two-pulse echo to write

$$\varrho_{31}(\mathbf{R}, \mathbf{v}, t) = -\frac{i}{2} e^{i\mathbf{q}\cdot\mathbf{R}} e^{-i\mathbf{q}\cdot\mathbf{v}(t-2T)} \sin A_1 \sin^2(A_2/2), \tag{11.125}$$

where $A_j = \int_{-\infty}^{\infty} \tilde{\Omega}_j(t)dt$, and $\mathbf{v}$ is an atomic velocity. The field pulse durations τ_j are assumed to satisfy $2ku\tau_j \ll 1$, so that they excite the entire velocity distribution, and $|\delta|\tau \gg 1$ to ensure that the *excited*-state population adiabatically follows the field pulses and vanishes following each pulse pair. Spontaneous emission during the pulse is assumed to be negligible.

From equation (11.125), we see that a spatially modulated echo in the Raman coherence $\varrho_{31}(\mathbf{R}, \mathbf{v}, t)$ is produced when $t = 2T$. This echo cannot be observed directly, since the Raman coherence does not radiate. To observe the echo, one applies a σ_+ polarized readout pulse around $t \approx 2T$ propagating in the Z direction. From equation (9.15f) for three-level dynamics, one sees that this field converts $\varrho_{31}(\mathbf{R}, \mathbf{v}, t)$ to a dipole density $\varrho_{32}(\mathbf{R}, \mathbf{v}, t)$, with

$$\varrho_{32}(\mathbf{R}, \mathbf{v}, t) \sim e^{i(\mathbf{q}-\mathbf{k})\cdot\mathbf{R}+i\omega t} e^{-i\mathbf{q}\cdot\mathbf{v}(t-2T)} \sin A_1 \sin^2(A_2/2), \tag{11.126}$$

where we have omitted some constants. On averaging this over velocity and using the fact that $\mathbf{q} = 2\mathbf{k} = 2k\hat{\mathbf{z}}$, we find

$$\varrho_{32}(Z, t) \sim e^{ikZ+i\omega t} e^{-k^2u^2(t-2T)^2} \sin A_1 \sin^2(A_2/2), \tag{11.127}$$

which corresponds to a signal propagating in the $-Z$ direction. In this manner, one can read out the Raman echo.

You might ask what this has to do with atom interferometry. Is this not just a coherent transient calculation? You would not be off-base to ask such a question. Although not essential, coherent transient experiments of this nature *can* be considered to be interferometric, however, since they involve an interference from two paths corresponding to different ways of achieving the final internal state coherence. This is seen clearly in all the Doppler phase diagrams of chapter 10. As such, the resultant signal is sensitive to velocity changes that occur in the arms of the interferometer.

As an example, consider the effect of gravity on the Raman echo signal (11.126). The signal depends on $\langle e^{-i\mathbf{q}\cdot\mathbf{R}(0)} e^{2i\mathbf{q}\cdot\mathbf{R}(T)}\rangle$, as in the two-photon echo. Assuming classical motion in a gravitational field with acceleration $\mathbf{g}$, we have

$$\mathbf{R}(0) = \mathbf{R} - \mathbf{v}_0 t - (1/2)\mathbf{g}t^2, \tag{11.128}$$

$$\mathbf{R}(T) = \mathbf{R} - (\mathbf{v}_0 + \mathbf{g}T)(t - T) - (1/2)\mathbf{g}(t - T)^2, \tag{11.129}$$

where $\mathbf{v}_0$ is the velocity of an atom before the pulses are applied. It then follows that for the velocity distribution (11.124),

$$\left\langle e^{-i\mathbf{q}\cdot\mathbf{R}(0)}e^{2i\mathbf{q}\cdot\mathbf{R}(T)}\right\rangle \sim e^{i\mathbf{q}\cdot\mathbf{R}}e^{-q^2u^2(t-2T)^2/4}e^{-i\phi_g}, \tag{11.130}$$

with

$$\phi_g = \mathbf{q}\cdot\mathbf{g}T^2. \tag{11.131}$$

This phase can be measured by heterodyning the signal field with a reference field. If the phase is measured as a function of T, one can obtain a value for $\mathbf{g}$. As we have already noted, experiments of this nature are also very sensitive to velocity-changing collisions and could be used to obtain the diffusion coefficient for ground-state scattering of the atoms with some perturber bath. Moreover, for a spatial interferometer undergoing rotation, one can measure the rotation rate using this interferometric scheme.

It might be noted that matter-wave effects can be incorporated into the result (11.126) without too much difficulty if one modifies the Doppler phase diagrams in chapter 10 to include recoil. For example, in the time interval T following the first pulse, ϱ_{13} acquires a relative phase

$$-E_{P,|\mathbf{P}+\hbar\mathbf{q}|}T/\hbar = -\left(\frac{P^2}{2M} - \frac{|\mathbf{P}+\hbar\mathbf{q}|^2}{2M}\right)T/\hbar = \left(\mathbf{q}\cdot\mathbf{v}+\omega_q\right)T, \tag{11.132}$$

containing both a Doppler and recoil shift. Thus, the recoil shift acts simply as a *detuning* in the Doppler phase diagrams. As such, for a Ramsey-fringe-type geometry, the echo signal contains an additional phase shift proportional to ω_q, allowing one to measure the recoil frequency.

11.3.2.4 Matter-wave interferometer

An example of a time domain, matter-wave interferometer is one that employs off-resonant, standing-wave optical field pulses as its "optical" elements and a cold vapor of two-level atoms as the "matter wave" [28]. Since the ground-state population remains equal to unity following the off-resonant pulses, the only way such an atom–field geometry can lead to interferometric fringes is if the wave nature of the center-of-mass motion plays a critical role.

Atoms in a magneto-optical trap (MOT) are subjected to two off-resonant standing-wave pulses at times $t = 0$ and $t = T$, respectively. The calculation is a relatively simple extension of the one given earlier for diffraction by a standing-wave field. One simply calculates the wave function following the first pulse, imposes the transmission function provided by the second field, and calculates the density following the second pulse. For an x-polarized standing wave field having positive frequency component

$$E^+(Z,t) = \frac{1}{2}E_0(t)e^{-i\omega t}\left(e^{ikZ}+e^{-ikZ}\right), \tag{11.133}$$

where $E_0(t)$ is the pulse envelope, the effective Hamiltonian for the atom–field system in the Raman-Nath approximation is

$$H = \frac{P^2}{2M} - \sum_{j=1}^{2} \hbar A_j \cos(\mathbf{q} \cdot \mathbf{R})\delta[t - (j-1)T], \tag{11.134}$$

where A_j $(j = 1, 2)$ is the area of pulse j [equation (11.77)] whose envelope is approximated by a delta function, and

$$\mathbf{q} = 2k\hat{\mathbf{z}}. \tag{11.135}$$

It simplifies matters if we take as our initial state wave function

$$\psi(\mathbf{R}, 0^-) = V^{-1/2} e^{i\mathbf{P}_0\cdot\mathbf{R}/\hbar}, \tag{11.136}$$

where 0^- is the time just before the application of the first pulse, and V is the interaction volume. As we have noted previously, any choice of wave function consistent with the initial state density matrix can be used for the calculation. The choice of plane-wave states, with an ultimate average over a Gaussian distribution of $\mathbf{v}_0 = \mathbf{P}_0/M$, is consistent with atoms having a homogenous spatial distribution and a Maxwellian distribution of velocities.

Following the first pulse,

$$\begin{aligned}\psi(\mathbf{R}, 0^+) &= V^{-1/2} e^{i\mathbf{P}_0\cdot\mathbf{R}/\hbar} e^{iA_1\cos(\mathbf{q}\cdot\mathbf{R})} \\ &= V^{-1/2} e^{i\mathbf{P}_0\cdot\mathbf{R}/\hbar} \sum_{n=-\infty}^{\infty} (i)^n J_n(A_1) e^{in\mathbf{q}\cdot\mathbf{R}},\end{aligned} \tag{11.137}$$

which implies that the momentum distribution is

$$\tilde{\Phi}(\mathbf{P}, 0^+) = \frac{(2\pi\hbar)^{3/2}}{V^{1/2}} \sum_{n=-\infty}^{\infty} (i)^n J_n(A_1)\delta(\mathbf{P} - \mathbf{P}_0 - n\hbar\mathbf{q}). \tag{11.138}$$

As a consequence, one finds that the wave function for $0 < t < T$ is given by

$$\begin{aligned}\psi(\mathbf{R}, t) &= (2\pi\hbar)^{-3/2} \int d\mathbf{P}\, \tilde{\Phi}(\mathbf{P}, 0^+) e^{-i(P^2/2M)t/\hbar} e^{i\mathbf{P}\cdot\mathbf{R}/\hbar} \\ &= V^{-1/2} e^{i(\mathbf{P}_0\cdot\mathbf{R} - E_{P_0}t)/\hbar} \\ &\quad \times \sum_{n=-\infty}^{\infty} (i)^n J_n(A_1) e^{-in^2\omega_q t} e^{in\mathbf{q}\cdot(\mathbf{R} - \mathbf{v}_0 t)}.\end{aligned} \tag{11.139}$$

The transverse Doppler phase factor $e^{-in\mathbf{q}\cdot\mathbf{v}_0 t}$ is contained in this result and leads to a dephasing in the density similar to the one in free polarization decay.

Following the second pulse, the wave function is

$$\begin{aligned}\psi(\mathbf{R}, T^+) &= e^{iA_2\cos(\mathbf{q}\cdot\mathbf{R})} \psi(\mathbf{R}, T^-) \\ &= V^{-1/2} e^{i(\mathbf{P}_0\cdot\mathbf{R} - E_{P_0}T)/\hbar} \sum_{n,m=-\infty}^{\infty} (i)^{n+m} J_n(A_1) J_m(A_2) \\ &\quad \times e^{im\mathbf{q}\cdot\mathbf{R}} e^{in\mathbf{q}\cdot(\mathbf{R} - \mathbf{v}_0 T)} e^{-in^2\omega_q T}.\end{aligned} \tag{11.140}$$

One can recalculate the momentum distribution as

$$\tilde{\Phi}(\mathbf{P},T^+) = (2\pi\hbar)^{-3/2}\int d\mathbf{R}\,\psi(\mathbf{R},T^+)e^{i(P^2/2M)T/\hbar}e^{i\mathbf{P}\cdot\mathbf{R}/\hbar}$$

$$= \frac{(2\pi\hbar)^{3/2}}{V^{1/2}}\sum_{n,m=-\infty}^{\infty}(i)^{n+m}J_n(A_1)J_m(A_2)e^{i(m+n)\mathbf{q}\cdot\mathbf{v}_0 T}e^{i(n+m)^2\omega_q T}$$

$$\times e^{-in\mathbf{q}\cdot\mathbf{v}_0 T}e^{-in^2\omega_q T}\delta[\mathbf{P}-\mathbf{P}_0-(n+m)\,\hbar\mathbf{q}]. \tag{11.141}$$

As a consequence, the wave function for $t > T$ is

$$\psi(\mathbf{R},t) = (2\pi\hbar)^{-3/2}\int d\mathbf{P}\,\tilde{\Phi}(\mathbf{P},T^+)e^{-i(P^2/2M)t/\hbar}e^{i\mathbf{P}\cdot\mathbf{R}/\hbar}$$

$$= e^{i\left(\mathbf{P}_0\cdot\mathbf{R}-E_{P_0}t\right)/\hbar}$$

$$\times\sum_{n,m=-\infty}^{\infty}(i)^{n+m}J_n(A_1)J_m(A_2)e^{-in\mathbf{q}\cdot\mathbf{v}_0 T}e^{-in^2\omega_q T}$$

$$\times e^{i(n+m)\mathbf{q}\cdot[\mathbf{R}-\mathbf{v}_0(t-T)]}e^{-i(n+m)^2\omega_q(t-T)}. \tag{11.142}$$

We are interested in the atomic density $\rho(\mathbf{R},t) = |\psi(\mathbf{R},t)|^2$ following the second pulse, averaged over the velocity distribution. Using equations (11.142) and (11.124), we find that the velocity averaged density is

$$\langle\rho(\mathbf{R},t)\rangle = V^{-1}\sum_{n,m,n'm'=-\infty}^{\infty}(i)^{n+m-n'-m'}J_n(A_1)J_m(A_2)J_{n'}(A_1)J_{m'}(A_2)$$

$$\times e^{-i\left(n^2-n'^2\right)\omega_q T}e^{-i\left[(n+m)^2-(n'+m')^2\right]\omega_q(t-T)}e^{i(n+m-n'-m')qZ}$$

$$\times e^{-[(n+m-n'-m')(t-T)+(n-n')T]^2q^2u^2/4}. \tag{11.143}$$

If we set $\bar{n} = n - n'$, $\bar{m} = m - m'$, and use the sum rule (11.85) in the form

$$i^{\bar{n}}\sum_{n'}J_{\bar{n}+n'}(A_1)J_{n'}(A_1)e^{-i\alpha(\bar{n}+2n')} = J_{\bar{n}}\left[2A_1\sin\alpha\right], \tag{11.144}$$

we can transform equation (11.143) into the simpler form

$$\langle\rho(\mathbf{R},t)\rangle = V^{-1}\sum_{\bar{n},\bar{m}=-\infty}^{\infty}e^{i(\bar{n}+\bar{m})qZ}e^{-[(\bar{n}+\bar{m})(t-T)+\bar{n}T]^2q^2u^2/4}$$

$$\times J_{\bar{n}}\left\{2A_1\sin[(\bar{n}+\bar{m})\omega_q(t-T)+\bar{n}\omega_q T]\right\}$$

$$\times J_{\bar{m}}\left\{2A_2\sin[(\bar{n}+\bar{m})\omega_q(t-T)]\right\}. \tag{11.145}$$

Although somewhat complicated, equation (11.145) displays an extraordinarily rich structure. If $\omega_q t \ll 1$ (classical motion limit), $\rho(\mathbf{R},t) \sim V^{-1}$; there is no structure in the classical motion limit. In other words, the diffraction pattern must arise from effects related to the quantized nature of the center-of-mass motion. If we assume that

$$quT = 2\omega_q T(Mu/\hbar q) \gg 1, \tag{11.146}$$

which is typical for atoms in a MOT (momentum greater than recoil momentum), then the interference pattern washes out except in the vicinity of *echo times*

$$t_e = \left(1 - \frac{\bar{n}}{\bar{n} + \bar{m}}\right) T > T. \tag{11.147}$$

At such times, the spatial density is constant, but there is spatial modulation having period $\lambda/[2|\bar{n} + \bar{m}|]$ in the transverse direction that occurs during a time interval $\Delta t \approx 1/qu$ about these points. This result shows that while the structures observed are products of matter-wave interferometry, the *positions* of the matter-wave gratings are determined by *classical* echo considerations, i.e., where a Doppler phase cancels, as in any Talbot-Lau interferometer.

Near the echo times $t = t_e + \Delta t = [1 - \bar{n}/(\bar{n} + \bar{m})]T + \Delta t$, equation (11.145) reduces to

$$\begin{aligned}\langle \rho(\mathbf{R}, t)\rangle = V^{-1} \sum_{\bar{n},\bar{m}=-\infty}^{\infty} & e^{i(\bar{n}+\bar{m})qZ} e^{-[(\bar{n}+\bar{m})qu\Delta t]^2/4} J_{\bar{n}} \left\{2A_1 \sin[\omega_q \Delta t(\bar{n} + \bar{m})]\right\} \\ & \times J_{\bar{m}} \left\{2A_2 \sin[-\bar{n}\omega_q T + \omega_q \Delta t(\bar{n} + \bar{m})]\right\},\end{aligned} \tag{11.148}$$

provided that $|\Delta t| \lesssim 1/qu$. For pulse areas of order unity and for $(Mu/\hbar q) \gg 1$, equation (11.148) becomes

$$\begin{aligned}\langle \rho(\mathbf{R}, t)\rangle \sim V^{-1} \sum_{\bar{n},\bar{m}=-\infty}^{\infty} & (-1)^{\bar{m}} (-1)^{(|\bar{n}|-\bar{n})/2} (|\bar{n}|!)^{-1} e^{i(\bar{n}+\bar{m})qZ} e^{-[(\bar{n}+\bar{m})qu\Delta t]^2/4} \\ & \times \left[A_1 \omega_q \Delta t(\bar{n} + \bar{m})\right]^{|\bar{n}|} J_{\bar{m}} \left[2A_2 \sin(\bar{n}\omega_q T)\right].\end{aligned} \tag{11.149}$$

All harmonics in the signal vanish at the echo times, but they exist in the vicinity of these points.

A contour plot of the signal is shown in figure 11.10, illustrating these features. The large amplitude near $t = T$ ($\bar{n} = 0$, $\bar{m} = \pm 1$) corresponds to a free polarization decay signal following the second pulse, with the density [obtained from equation (11.148)] varying as

$$V\langle \rho(\mathbf{R}, t)\rangle \sim 1 + 2(A_2 \omega_q \Delta t) e^{-(qu\Delta t)^2/4} \cos(qZ) \tag{11.150}$$

for $t \gtrsim T$. Near the echo plane at $t = 2T$ ($\bar{n} = \mp 1$, $\bar{m} = \pm 2$), the density varies as

$$V\langle \rho(\mathbf{R}, t)\rangle \sim 1 - 2(A_1 \omega_q \Delta t) J_2[2A_2 \sin(\omega_q T)] e^{-(qu\Delta t)^2/4} \cos(qZ), \tag{11.151}$$

with modulation period $2\pi/q$, and near the echo plane at $t = 3T/2$ ($\bar{n} = \mp 1$, $\bar{m} = \pm 3$), as

$$V\langle \rho(\mathbf{R}, t)\rangle \sim 1 - 2(A_1 \omega_q \Delta t) J_3[2A_2 \sin(\omega_q T)] e^{-(qu\Delta t)^2/4} \cos(2qZ), \tag{11.152}$$

with modulation period π/q. You can also see the modulation with period $2\pi/3q$ near the echo plane at $t = 4T/3$ ($\bar{n} = \mp 1$, $\bar{m} = \pm 4$), and the modulation with period $2\pi/q$ near the echo plane at $t = 3T$ ($\bar{n} = \mp 2$, $\bar{m} = \pm 3$), where the density varies as

$$V\langle \rho(\mathbf{R}, t)\rangle \sim 1 + \left(A_1 \omega_q \Delta t\right)^2 J_2[2A_2 \sin(2\omega_q T)] e^{-(qu\Delta t)^2/4} \cos(qZ). \tag{11.153}$$

The density in the echo planes is constant.

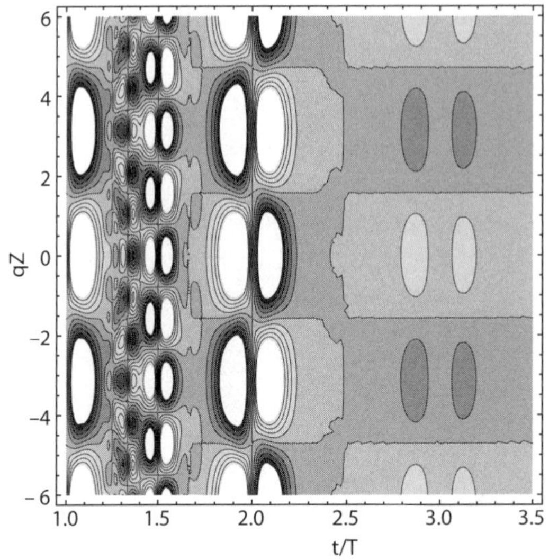

Figure 11.10. Contour plot of the density for a matter-wave interferometer, with $A_1 = 4$, $A_2 = 2$, $quT = 15$, and $\omega_q T = 1$. The darker colors correspond to lower densities.

Experimentally, it is easiest to observe the signal when $(\bar{n} + \bar{m}) = \pm 1$, since this value corresponds to a grating having period $\lambda/2$, from which one can back-scatter a traveling wave field having the same frequency. With $\bar{n} = -N$ and $\bar{m} = N + 1$,

$$t_e = (1 + N)T + \Delta t, \tag{11.154}$$

and the component of the grating varying as e^{iqZ} has an associated density equal to

$$\begin{aligned}\langle \rho(\mathbf{R}, t)\rangle_q &= -V^{-1}\,(N!)^{-1}\cos(2kZ)e^{-(qu\Delta t)^2/4} \\ &\quad \times \left(A_1\omega_q \Delta T\right)^N J_{N+1}[2A_2\sin(N\omega_q T)].\end{aligned} \tag{11.155}$$

Since the recoil frequency appears in this expression, one can measure ω_q by varying the interval between the pulses. Moreover, if we had included gravity, there would be an additional phase factor

$$\phi_g = \mathbf{k}\cdot\mathbf{g}T^2 N(N+1) \tag{11.156}$$

appearing in the argument of the cos function, allowing for a measure of the acceleration of gravity $\mathbf{g}$ using this matter-wave interferometer [28].

11.4 Summary

We have described scattering of radiation or matter waves in terms of shadow, Fresnel, and Fraunhofer regions. The shadow region corresponds to classical or straight line paths for the light or atoms, while the Fresnel and Fraunhofer regions correspond to wave diffraction. We have concentrated on the Fresnel region for matter-wave scattering and have seen how microfabricated gratings and standing-wave optical fields can be used as "optical elements" to diffract atoms. The

Talbot effect, or self-imaging of a periodic structure, was derived, as was the diffraction pattern for amplitude and phase gratings. The Raman-Nath and Bragg scattering limits were distinguished. Following a discussion of atom optics, we showed how atoms can be used in atom interferometric schemes to provide sensitive measurements of inertial effects and the recoil frequency. Interferometers in which the atomic motion could be taken to be classical were distinguished from those in which the center-of-mass motion had to be quantized. All calculations for atom interferometry were carried out in the Talbot-Lau regime, analogous to photon echo schemes in which there is a rephasing for different velocity groups at certain echo points or planes. We now switch gears and move on to the second part of this book, involving quantization of the radiation field.

Problems

1. Estimate the spreading of a Gaussian laser beam by using a transmission function

$$T(X, Y) = \frac{1}{\sqrt{\pi a^2}} e^{-(X^2+Y^2)/2a^2}$$

and calculating

$$\Psi(X, Y) \sim \frac{1}{\lambda_0 Z} \int d\mathbf{R}'_T T(X', Y') e^{-ik_0 \mathbf{R}_T \cdot \mathbf{R}'_T / Z},$$

valid in the Fraunhofer regime.

2. A lens can be modeled as having a transmission function

$$T(X, Y) = e^{ik_\alpha a\left[1-\left(\frac{R_T}{a}\right)^2\right]}$$

across the area of the lens, where a is the lens radius and k_α is a constant that is proportional to $n-1$, where n is the index of refraction of the lens. Neglecting the transmission from beyond the lens border, show that focusing occurs at a distance

$$Z = \frac{k_0 a}{2k_\alpha}$$

behind the lens. Moreover, prove that in the focal plane, the Fresnel (quadratic) contribution to the diffraction amplitude vanishes and that one is left with a Fraunhofer scattering result, even though the signal appears in the Fresnel zone. Estimate the spot size.

3. Assume that there is an angular divergence θ_b in the incident atomic beam resulting from a distribution of transverse velocities in the beam. Determine the maximum value of θ_b for which the Talbot effect is not "washed out" as a result of the beam divergence.

4. Use equation (11.48) or (11.49) to estimate the positions and widths of the intensity maxima for an N-slit grating in the traditional Fraunhofer limit, equation (11.57), when α is sufficiently small to neglect the diffraction

envelope of the slit. Plot $|D(X)|$ as a function of X/d for $N = 10$, $s = 10^4$, and $\alpha = 0.0001$.

5. Use equation (11.49) to estimate the positions and widths of the intensity maxima for an N-slit grating in the "separated beam" Fraunhofer limit, equation (11.60). Plot $|D(X)|$ as a function of X/d for $N = 10$, $s = 10$, and $\alpha = 0.2$. Show that under these conditions, the beam is separated into a number of beams, each having spatial width of order D.
6. Plot $|D(X)|$ as a function of X/d for $N = 40$, $\alpha = 0.2$, and $s = 0.5, 1, 2.3$. Show that the scattering for $s = 0.5$ and 1 is in agreement with the Talbot effect. Also show that for $s = 1/3$, there is evidence for the fractional Talbot effect.
7. Prove that at half-integral multiples of the Talbot length, the grating structure is reproduced with a half-period displacement.
8. Give a geometrical argument based on Huygen's principle to show that there can be a fractional Talbot effect. Assuming that the ratio of the slit size to the period is small, show, for a grating having period d, that there is a fractional Talbot effect having period $d/2m$ at distances equal to $L_T/4m$ for integer m, that there is a fractional Talbot effect having period $d/(2m+1)$ at distances equal to $L_T/(2m+1)$ for integer m, and that there is a displaced {shifted by $d/[2(2m+1)]$} fractional Talbot effect having period $d/(2m+1)$ at distances equal to $L_T/[2(2m+1)]$ for integer m.
9. Derive equations (11.103) and (11.108).
10. By adiabatically eliminating intermediate states assuming that $|\Delta - n\omega_q| \ll \omega_q$ and $n\omega_q\tau \gg 1$, show that the effective two-level equations for an nth-order Bragg resonance ($\Delta \approx n\omega_q > 0$) can be written as

$$\dot{c}_0(t) = i\chi_n(t)e^{i(n\Delta - n^2\omega_q)t}c_1(t),$$

$$\dot{c}_n(t) = i\chi_n(t)e^{-i(n\Delta - n^2\omega_q)t}c_0(t),$$

where

$$\chi_n(t) = \frac{r^n(t)}{[(n-1)!]^2\,\omega_q^{n-1}}.$$

11. Write equations (11.108) including $n = -1, 0, 1, 2$, assuming that Δ is tuned to the first-order Bragg resonance, $\Delta = \omega_q$. With $r(t) = (A/2)\,\text{sech}(\pi t/2)$ and $\omega_q = 5$, solve these equations numerically for $A = \pi/2, 3\pi/2, 5\pi/2$. Show that while the state amplitudes c_{-1} and c_2 fall to zero following the pulse, they still influence the final values of c_0 and c_1. Compare the values obtained from solving the four state equations with those obtained from the two-state equations. This should be a warning that even on-resonance, nonresonant states can modify the results by providing the equivalent of light shifts to the levels.
12. Consider a classical internal state interferometer in which the atoms have a three-level Λ scheme in which levels 1 and 3 are two ground-state hyperfine levels and level 2 is an excited state, as in figure 11.9. The interaction pulses drive optical transitions between levels 1 and 2 *only*. Consider a pulse sequence in which there are four pulses: pulse 1 at $t = 0$ with propagation vector $\mathbf{k}$; pulse 2 at time $T_{21} < \gamma_2^{-1}$ with propagation vector $-\mathbf{k}$; pulse 3 a standing-wave field

in the $\pm\mathbf{k}$ direction at time $T_{21}+T$; and pulse 4 at time $T_{21}+2T$ in direction $-\mathbf{k}$. Draw a Doppler phase diagram for this pulse sequence, and show that an echo can be produced at $t_e = 2T_{21}+2T$ in the $\mathbf{k}$ direction. What is the advantage of this grating-stimulated echo [29] over the conventional stimulated echo with regard to sensitivity to velocity-changing collisions?

13. Derive equation (11.145) starting from equation (11.143).
14. Draw a Doppler phase diagram analogous to that of figure 10.8 for resonant Raman transitions rather than optical transitions, including the effects of recoil. That is, take the difference of the propagation vectors equal to $\mathbf{q}$, $\mathbf{q}$, $-\mathbf{q}$ for the three field interaction zones. This can be done using the method outlined at the end of appendix B in chapter 10. Show that at the echo point $t = 2T_{21} + T$, $\tilde{\varrho}_{31}$ varies as $\cos(2\omega_q T_{21})$.
15. Draw Doppler phase diagrams to show how the phase shifts (11.131) and (11.156) arise.
16. Using arguments similar to those in arriving at equation (11.131), derive an expression for the fringe shift that occurs in the Raman echo for an atomic beam having velocity $\mathbf{v}_0 = v_0\hat{\mathbf{z}} + \mathbf{v}_T$ passing through two field zones separated by L as a result of the fact that the entire experiment is being carried out in a rotating frame (such as the Earth). To do this, assume that the acceleration of the atom in the interferometer is simply $2v_0\hat{\mathbf{z}} \times \boldsymbol{\Omega}$, the "fictitious" Coriolis acceleration. Show that your result agrees with equation (11.118). Thus, even if the entire apparatus is rotating, including the source, one can use an interferometer to measure the rotation rate. Assume that the transverse velocity $v_T \ll v_0$, but that $2kv_TL/v_0 = qv_TL/v_0 \gg 1$.
17. Estimate the fringe shift per radian of rotation rate when a supersonic sodium atomic beam passes though two microfabricated gratings separated by 0.66 m [22]. Assume that the beam velocity is 1.03×10^3 m/s and that the grating period is 200 nm. Consider only the matter-wave grating having the lowest order (that is, the largest) period.

References

[1] J. D. Jackson, *Classical Electrodynamics*, 3rd ed. (Wiley, New York, 1999), sec. 10.5.

[2] H. F. Talbot, *Facts relating to optical science, No. IV*, Philosophical Magazine **9**, 401–407 (1836).

[3] J. M. Cowley and A. F. Moodie, *Fourier images I. the point source; II. the out-of-focus patterns; III. finite sources*, Proceedings of the Physical Society of London B **70**, 486–496; 497–504; 505–513 (1957).

[4] G. L. Rogers, *Diffraction theory of insect vision II. theory and experiments with a simple model eye*, Proceedings of the Royal Society of London B **157**, 83–98 (1962).

[5] J. T. Winthrop and C. R. Worthington, *Theory of Fresnel images I. plane periodic objects in monochromatic light*, Journal of the Optical Society of America **55**, 373–381 (1965).

[6] For a review, see K. Patorski, *The self-imaging phenomenon and its application*, in *Progress in Optics XXVII*, edited by E. Wolf (North-Holland, Amsterdam, 1989), pp. 1–108.

[7] M. S. Chapman, C. R. Ekstrom, T. D. Hammond, J. Schmiedmayer, B. E. Tannian, S. Wehinger, and D. E. Pritchard, *Near-field imaging of atomic diffraction gratings: the atomic Talbot effect*, Physical Review A **51**, R14–R17 (1995).

[8] S. Nowak, Ch. Kurtsiefer, T. Pfau, and C. David, *High-order Talbot fringes for atomic matter waves*, Optics Letters **22**, 1430–1432 (1997).

[9] L. Deng, E. W. Hagley, J. Denschlag, J. E. Simsarian, M. Edwards, C. W. Clark, K. Helmerson, S. L. Rolston, and W. D. Phillips, *Temporal, matter-wave-dispersion Talbot effect*, Physical Review Letters **83**, 5407–5411 (1999).

[10] B. Dubetsky and P. R. Berman, *Atom interference using microfabricated structures*, in *Atom Interferometry*, edited by P. R. Berman (Academic Press, San Diego, CA, 1997), pp. 407–468.

[11] C. V. Raman and N. S. N. Nath, *Diffraction of light by H. F. sound waves. parts I and II*, Proceedings of the Indian Academy of Science, Section A **2**, 406–420 (1935).

[12] R. J. Cook and A. F. Bernhardt, *Deflection of atoms by a resonant standing wave field*, Physical Review A **18**, 2533–2537 (1978).

[13] P. L. Kapitza and P. A. M. Dirac, *The reflection of electrons from standing light waves*, Mathematical Proceedings of the Cambridge Philosophical Society **29**, 297–300 (1933).

[14] P. E. Moskowitz, P. L. Gould, S. R. Atlas, and D. E. Pritchard, *Diffraction of an atomic beam by standing wave radiation*, Physical Review Letters **51**, 370–373 (1983).

[15] M. Abramowitz and I. A. Stegun, *Handbook of Mathematical Functions with Formulas, Graphs, and Mathematical Tables* (U.S. Department of Commerce, Washington, DC, 1965), chap. 9.

[16] J. L. Cohen, B. Dubetsky, and P. R. Berman, *Atom focusing by far-detuned and resonant standing wave fields: thin-lens regime*, Physical Review A **60**, 4886–4901 (1999).

[17] P. R. Berman and B. Bian, *Pump-probe spectroscopy approach to Bragg scattering*, Physical Review A **55**, 4382–4385 (1997).

[18] G. Sagnac, *Luminous ether demonstrated by the effect of relative wind of ether in a uniform rotation of an interferometer*, Comptes Rendus Hebdomadaires des Seances de l'Academie des Sciences **157**, 708–710 (1913).

[19] For a review, see G. Rizzi and M. L. Ruggiero, *The relativistic Sagnac effect: two derivations*, in *Relativity in Rotating Frames*, edited by G. Rizzi and M. L. Ruggiero (Kluwer Academic Publishers, Dordrecht, The Netherlands, 2003), pp. 179–220, plus other chapters in this book.

[20] C. H. Rowe, U. K. Schreiber, S. J. Cooper, B. T. King, M. Poulton, and G. E. Stedman, *Design and operation of a very large ring laser gyroscope*, Applied Optics **38**, 2516–2523 (1999).

[21] F. Riehle, Th. Kisters, A. Witte, J. Helmcke, and Ch. J. Bordé, *Optical Ramsey spectroscopy in a rotating frame: Sagnac effect in a matter wave interferometer*, Physical Review Letters **67**, 177–180 (1991).

[22] A. Lenef, T. D. Hammond, E. T. Smith, M. S. Chapman, R. A. Rubenstein, and D. E. Pritchard, *Rotation sensing with an atom interferometer*, Physical Review Letters **78**, 760–763 (1997).

[23] T. L. Gustavson, P. Bouyer, and M. A. Kasevich, *Precision rotation measurements with an atom interferometer gyroscope*, Physical Review Letters **78**, 2046–2049 (1997).

[24] T. L. Gustavson, A. Landragin, and M. A. Kasevich, *Rotation sensing with a dual atom-interferometer Sagnac gyroscope*, Classical and Quantum Gravity **17**, 2385–2397 (2000).

[25] E. Lau, *Beugungserscheinungen an Doppelrastern (Diffraction phenomena from two parallel gratings)*, Annalen der Physik **6**, 417–423 (1948).

[26] J. L. Clauser and S. Li, *Talbot-vonLau atom interferometer with cold potassium*, Physical Review A **49**, R2213–R2216 (1994).

[27] A. V. Turlapov, A. Tonyushkin, and T. Sleator, *Talbot-Lau effect for atomic de Broglie waves manipulated with light*, Physical Review A **71**, 043612, 1–5 (2005).

[28] S. B. Cahn, A. Kumarakrishnan, U. Shim, T. Sleator, P. R. Berman, and B. Dubetsky, *Time domain de Broglie wave interferometer*, Physical Review Letters **79**, 784–787 (1997).

[29] B. Dubetsky, P. R. Berman, and T. Sleator, *Grating stimulated echo*, Physical Review A **46**, R2213–R2216 (1992).

Bibliography

There are more than one thousand references appropriate to this chapter. The following three reviews contain extensive lists of references:

C. S. Adams, M. Sigel, and J. Mlynek, *Atom Optics*, Physics Reports **240**, 143–210 (1994).

P. R. Berman, Ed. *Atom Interferometry* (Academic Press, San Diego,CA, 1997).

A. D. Cronin, J. Schmiedmayer, and D. E. Pritchard, *Optics and interferometry with atoms and molecules*, Reviews of Modern Physics **81**, 1051–1129 (2009).

12

The Quantized, Free Radiation Field

We now move on to problems in which it is necessary to quantize the electromagnetic field. As was mentioned in chapter 1, such problems fall into a number of categories, including cavity fields, spontaneous emission, and light scattering. There are many ways to quantize the radiation field. In the appendix, one such approach is given, following the treatment in Cohen-Tannoudji et al. [1]. More generally, however, the most foolproof way to quantize the field is to calculate the classical field modes associated with a given boundary value problem and then assign creation and annihilation operators that give rise to each of these distinct field modes.

There are many excellent texts on quantum optics that offer detailed discussions on quantum aspects of the radiation field. Some of these texts are listed in the bibliography of chapter 1. We provide a very limited introduction to quantum properties of the radiation field and photodetection in this and the following two chapters. In subsequent chapters, we concentrate on applications involving the interaction of the quantized radiation field with atoms.

12.1 Free-Field Quantization

We first consider fields in free space with no sources. Of course, there have to be sources present *somewhere* to produce these fields. To quantize the free field using a simplified approach, one can start with the spatial modes of a free, classical field satisfying periodic boundary conditions. Such spatial modes are plane waves, varying as $\exp(i\mathbf{k}\cdot\mathbf{R})$, where

$$\mathbf{k} = k_x\hat{\mathbf{x}} + k_y\hat{\mathbf{y}} + k_z\hat{\mathbf{z}}, \tag{12.1}$$

$$k_x = \frac{2\pi n_x}{L}, \quad k_y = \frac{2\pi n_y}{L}, \quad k_z = \frac{2\pi n_z}{L}, \tag{12.2}$$

and $L^3 = \mathcal{V}$ is the quantization volume. The quantities n_x, n_y, and n_z can take on all integer values.

The quantized electric field operator is written as

$$\mathbf{E}(\mathbf{R}) = i\sum_j E_j[a_j\boldsymbol{\epsilon}_j e^{i\mathbf{k}_j\cdot\mathbf{R}} - a_j^\dagger\boldsymbol{\epsilon}_j e^{-i\mathbf{k}_j\cdot\mathbf{R}}]$$
$$\equiv \mathbf{E}^+(\mathbf{R}) + \mathbf{E}^-(\mathbf{R}), \tag{12.3}$$

where $\boldsymbol{\epsilon}_j$ is the polarization of mode j,[1] E_j is a (real) normalization constant, a_j ($a_j^\dagger$) is the destruction (creation) operator for a mode having propagation vector $\mathbf{k}_j$ and polarization $\boldsymbol{\epsilon}_j$, and

$$\mathbf{E}^+(\mathbf{R}) = \left[\mathbf{E}^-(\mathbf{R})\right]^\dagger = i\sum_j E_j a_j\boldsymbol{\epsilon}_j e^{i\mathbf{k}_j\cdot\mathbf{R}} \tag{12.4}$$

is the positive frequency component of the field. The free-field modes are transverse, with $\mathbf{k}_j\cdot\boldsymbol{\epsilon}_j = 0$. There are two independent polarization vectors and creation and destruction operators associated with a given $\mathbf{k}_j$, and the subscript j is a shorthand notation for a set of quantum numbers (n_x, n_y, n_z). The destruction and creation operators satisfy the commutation relations

$$[a_i, a_j] = [a_i^\dagger, a_j^\dagger] = 0, \qquad [a_i, a_j^\dagger] = \delta_{i,j}. \tag{12.5}$$

In calculating any physically relevant quantities, the final answers must be independent of the choice of quantization volume $\mathcal{V}$.

Although a_i and $a_i^\dagger$ are operators, we do not write them as $\hat{a}_i$ and $\hat{a}_i^\dagger$, nor do we include a "hat" (ˆ) on the field operator. In fact, from this point onward, except where there is cause for confusion, we suppress the hat symbol on all operators.

For the free field, the total energy in the field is twice the integral of the energy density associated with the electric field integrated over the volume. As a consequence, the energy operator corresponding to the field (12.3) is

$$H = \epsilon_0\sum_{j,j'}\int d^3R\, E_j E_{j'}(a_j e^{i\mathbf{k}_j\cdot\mathbf{R}} - a_j^\dagger e^{-i\mathbf{k}_j\cdot\mathbf{R}})(a_{j'}^\dagger e^{-i\mathbf{k}_{j'}\cdot\mathbf{R}} - a_{j'} e^{i\mathbf{k}_{j'}\cdot\mathbf{R}})$$
$$= \epsilon_0\mathcal{V}\sum_j E_j^2(a_j a_j^\dagger + a_j^\dagger a_j) = 2\epsilon_0\mathcal{V}\sum_j E_j^2\left(a_j^\dagger a_j + \frac{1}{2}\right), \tag{12.6}$$

where ϵ_0 is the vacuum permittivity. If we choose

$$E_j = \sqrt{\frac{\hbar\omega_j}{2\epsilon_0\mathcal{V}}}, \tag{12.7}$$

where $\omega_j = k_j c$, then

$$H = \sum_j \hbar\omega_j\left(\hat{n}_j + \frac{1}{2}\right), \tag{12.8}$$

where the *number operator* $\hat{n}_j$ is defined by

$$\hat{n}_j = a_j^\dagger a_j. \tag{12.9}$$

[1] We adopt a notation in which the polarization vector for the quantized field is written without a ˆ ("hat"), although we continue to include the ˆ in the polarization vectors of classical fields.

The Hamiltonian is the same as that for an ensemble of harmonic oscillators having frequency ω_j. We will assume that equation (12.8) is the free-field Hamiltonian. For reference purposes, we list the momentum of the field

$$\mathbf{P} = \sum_j \hbar \mathbf{k}_j \left(a_j^\dagger a_j + \frac{1}{2} \right) = \sum_j \hbar \mathbf{k}_j a_j^\dagger a_j , \tag{12.10}$$

the vector potential

$$\mathbf{A}(\mathbf{R}) = \sum_j A_j (a_j \boldsymbol{\epsilon}_j e^{i\mathbf{k}_j \cdot \mathbf{R}} + a_j^\dagger \boldsymbol{\epsilon}_j e^{-i\mathbf{k}_j \cdot \mathbf{R}}), \tag{12.11}$$

the electric field

$$\mathbf{E}(\mathbf{R}) = i \sum_j E_j (a_j \boldsymbol{\epsilon}_j e^{i\mathbf{k}_j \cdot \mathbf{R}} - a_j^\dagger \boldsymbol{\epsilon}_j e^{-i\mathbf{k}_j \cdot \mathbf{R}}), \tag{12.12}$$

and the magnetic field

$$\mathbf{B}(\mathbf{R}) = i \sum_j B_j (a_j \boldsymbol{\kappa}_j \times \boldsymbol{\epsilon}_j e^{i\mathbf{k}_j \cdot \mathbf{R}} - a_j^\dagger \boldsymbol{\kappa}_j \times \boldsymbol{\epsilon}_j e^{-i\mathbf{k}_j \cdot \mathbf{R}}). \tag{12.13}$$

In these equations,

$$E_j = \sqrt{\frac{\hbar \omega_j}{2\epsilon_0 \mathcal{V}}}, \quad B_j = \frac{E_j}{c}, \quad A_j = \frac{E_j}{\omega_j}, \tag{12.14}$$

and

$$\boldsymbol{\kappa}_j = \frac{\mathbf{k}_j}{k_j}. \tag{12.15}$$

is a unit vector. For a single-mode field with $\omega_j = \omega$, the Hamiltonian is

$$H = \hbar\omega \left(a^\dagger a + \frac{1}{2} \right) = \hbar\omega \left(\hat{n} + \frac{1}{2} \right), \tag{12.16}$$

where $\hat{n} = a^\dagger a$. Since the Hamiltonian is the same as that of a harmonic oscillator, the eigenenergies are given by

$$E_n = \hbar\omega \left(n + \frac{1}{2} \right), \tag{12.17}$$

where $n = 0, 1, \ldots$. The eigenkets are denoted by $|n\rangle$, which are called *photon states* of the field. They are also referred to as *number states* or *Fock states*. They are *not* localized states, extending over the quantization volume. As in the case of the harmonic oscillator, the destruction operator a is a lowering operator,

$$a|n\rangle = \sqrt{n}|n-1\rangle, \tag{12.18}$$

while the creation operator $a^\dagger$ is a raising operator,

$$a^\dagger|n\rangle = \sqrt{n+1}|n+1\rangle, \tag{12.19}$$

and the number operator leaves the ket unchanged,

$$\hat{n}\,|n\rangle = a^{\dagger}a\,|n\rangle = n|n\rangle. \tag{12.20}$$

It follows from equation (12.19) that the ket $|n\rangle$ can be written as

$$|n\rangle = \frac{\left(a^{\dagger}\right)^{n}}{\sqrt{n!}}|0\rangle, \tag{12.21}$$

where the state $|0\rangle$ is referred to as the *vacuum state* of the field. For a multimode field,

$$|n_1 n_2 \ldots n_i\rangle = \frac{\left(a_1^{\dagger}\right)^{n_1}}{\sqrt{n_1!}}\frac{\left(a_2^{\dagger}\right)^{n_2}}{\sqrt{n_2!}}\cdots\frac{\left(a_i^{\dagger}\right)^{n_i}}{\sqrt{n_i!}}|0\rangle\,. \tag{12.22}$$

We have not said anything about how one can construct such number states. In general, this is a difficult task. The most common way to create a number state is to send a single atom in an excited state into a cavity and allow the atom to transfer its excitation to a single-cavity mode.

12.2 Properties of the Vacuum Field

The expectation value of the electric field in the vacuum state vanishes, since

$$\langle 0|\mathbf{E}|0\rangle = i\sum_j E_j\langle 0|(a_j\boldsymbol{\epsilon}_j e^{i\mathbf{k}_j\cdot\mathbf{R}} - a_j^{\dagger}\boldsymbol{\epsilon}_j e^{-i\mathbf{k}_j\cdot\mathbf{R}})|0\rangle = 0. \tag{12.23}$$

This is the type of result one would expect for the vacuum field. However, the fact that the average field vanishes does not imply that the energy density in the field, proportional to $\langle 0|\mathbf{E}^2|0\rangle$, also vanishes. In fact, one finds that the energy density

$$\begin{aligned}\epsilon_0\langle 0|\mathbf{E}^2|0\rangle &= -\epsilon_0\sum_{j,j'}\frac{\hbar\sqrt{\omega_j\omega_{j'}}}{2\epsilon_0\mathcal{V}}\boldsymbol{\epsilon}_j\cdot\boldsymbol{\epsilon}_{j'}[\langle 0|a_j^{\dagger}a_{j'}^{\dagger}|0\rangle e^{-i(\mathbf{k}_j+\mathbf{k}_{j'})\cdot\mathbf{R}}\\ &\quad+\langle 0|a_j a_{j'}|0\rangle e^{i(\mathbf{k}_j+\mathbf{k}_{j'})\cdot\mathbf{R}} - \langle 0|a_j a_{j'}^{\dagger}|0\rangle e^{i(\mathbf{k}_j-\mathbf{k}_{j'})\cdot\mathbf{R}} - \langle 0|a_j^{\dagger}a_{j'}|0\rangle e^{i(\mathbf{k}_j+\mathbf{k}_{j'})\cdot\mathbf{R}}]\\ &= \sum_j\frac{\hbar\omega_j}{2\mathcal{V}} = \sum_j\frac{\hbar k_j c}{2\mathcal{V}}.\end{aligned} \tag{12.24}$$

Going from a discrete sum to a continuum using the prescription

$$\sum_j = \frac{\mathcal{V}}{(2\pi)^3}\int d\mathbf{k}, \tag{12.25}$$

one obtains

$$\epsilon_0\langle 0|\mathbf{E}^2|0\rangle = \frac{\hbar c}{2}\frac{4\pi}{(2\pi)^3}\int_0^{\infty}dk\,k^3, \tag{12.26}$$

which diverges. The energy density of the vacuum field at *each* point in space is infinite. Depending on your point of view, this can be viewed as either an embarrassment or a simple inconvenience. The fact that the vacuum energy is infinite does not seem to have any physical consequence (apart from the cosmological constant). Nevertheless, *vacuum fluctuations* (e.g., the possibility to emit a photon while exciting an atom to a higher energy state) have measurable consequences such as the Lamb shift of atomic levels and the Casimir effect (e.g., attraction of two metal plates in vacuum). The field of quantum electrodynamics was developed in large part to deal with infinities associated with these vacuum fluctuations. When point charges and masses are renormalized to their measured values, vacuum fluctuations are found to lead to finite rather than infinite corrections to energy levels and interactions.

In expanding any state in the Schrödinger representation, it is useful to use a representation in which the (infinite) energy of the vacuum field is eliminated. This is the convention that we adopt such that, for the most part, we take as our effective Hamiltonian

$$H = \sum_j \hbar\omega_j a_j^\dagger a_j \,. \tag{12.27}$$

Moreover, any Lamb shifts of atomic energy levels resulting from the interaction with the vacuum field are assumed to be incorporated into the energies of the atomic states.

12.2.1 Single-Photon State

A multimode, *one-photon state* for the free field can be defined by

$$|\Psi(t)\rangle = \sum_j c_j(t) a_j^\dagger |0\rangle \,. \tag{12.28}$$

In effect, this state represents a superposition of single-photon states of the type $|1_j\rangle$—that is, a state where there is a single photon in mode j. From the Schrödinger equation with the Hamiltonian (12.27), it follows that the state amplitudes $c_j(t)$ evolve according to

$$\dot{c}_j(t) = -i\omega_j c_j(t) \,. \tag{12.29}$$

Therefore, we find

$$|\Psi(t)\rangle = \sum_j c_j(0) a_j^\dagger |0\rangle e^{-i\omega_j t} = \sum_j e^{-i\omega_j t} c_j(0) |1_j\rangle \,. \tag{12.30}$$

For this one-photon state, the expectation value of the positive frequency component of the field vanishes, since

$$\langle \mathbf{E}^+ \rangle = i \sum_j \left(\frac{\hbar\omega_j}{2\epsilon_0 \mathcal{V}} \right)^{1/2} e^{i\mathbf{k}_j \cdot \mathbf{R}} \boldsymbol{\epsilon}_j \langle \Psi(t) | a_j | \Psi(t) \rangle = 0. \tag{12.31}$$

This is not a surprising result—a one-photon state of the field cannot possess a well-defined phase, implying that a quantum-mechanical average of the field operator must vanish. On the other hand, the average value of $\langle \mathbf{E}^- \cdot \mathbf{E}^+ \rangle$, which is proportional to the average field intensity measured by a detector (since $\mathbf{E}^+$ "destroys" a photon at the detector), is

$$\begin{aligned}\langle \mathbf{E}^- \cdot \mathbf{E}^+ \rangle &= \langle \Psi | \sum_{j,j'} \frac{\hbar}{2\epsilon_0 \mathcal{V}} \boldsymbol{\epsilon}_j \cdot \boldsymbol{\epsilon}_{j'} \sqrt{\omega_j \omega_{j'}} a_j^\dagger a_{j'} e^{-i(\mathbf{k}_j - \mathbf{k}_{j'})\cdot \mathbf{R}} | \Psi \rangle \\ &= \sum_{j,j',l,l'} \langle 1_l | a_j^\dagger \frac{\hbar}{2\epsilon_0 \mathcal{V}} \boldsymbol{\epsilon}_j \cdot \boldsymbol{\epsilon}_{j'} \sqrt{\omega_j \omega_{j'}} e^{-i(\mathbf{k}_j - \mathbf{k}_{j'})\cdot \mathbf{R}} e^{-i\omega_{ll'} t} a_{j'} | 1_{l'} \rangle c_l^*(0) c_{l'}(0) .\end{aligned} \tag{12.32}$$

In the sum in equation (12.32), only terms with $j = l$ and $j' = l'$ contribute, leading to

$$\langle \mathbf{E}^- \cdot \mathbf{E}^+ \rangle = \frac{\hbar}{2\epsilon_0 \mathcal{V}} \left| \sum_j \sqrt{\omega_j} e^{i(\mathbf{k}_j \cdot \mathbf{R} - \omega_j t)} \boldsymbol{\epsilon}_j c_j(0) \right|^2 . \tag{12.33}$$

This expression is similar to that for the propagation of a classical electric field pulse. If all $\mathbf{k}_j$ are along the $\hat{\mathbf{x}}$ direction (plane wave), then the solution is a function of $(X - ct)$ only.

The result (12.33) is somewhat paradoxical. Recall that although the energy density at each point in space for the vacuum field is infinite, the average vacuum field intensity measured by a detector, proportional to $\langle 0 | \mathbf{E}^- \cdot \mathbf{E}^+ | 0 \rangle$, vanishes. On the other hand, the average value of $\mathbf{E}^- \cdot \mathbf{E}^+$ given in equation (12.33) for a single-photon state is proportional to the quantization volume when the sums over j are converted to integrals using the prescription (12.25). Since the quantization volume goes to infinity in free space, it would appear that the energy in a single-photon pulse measured by a detector is infinite. Clearly, something is wrong with this result.

Where did we go wrong? Well, we really didn't go wrong *yet*. In converting the sums to integrals in equation (12.33), we must also convert the state amplitudes $c_j(0)$ corresponding to discrete field modes to continuum-mode variables $c(\mathbf{k}, t = 0)$. To accomplish this conversion, we set

$$c_j(t) \to \sqrt{\frac{(2\pi)^3}{\mathcal{V}}} c(\mathbf{k}, t),$$

with the normalization such that $\int d\mathbf{k} \left| c(\mathbf{k}, 0) \right|^2 = 1$. With this substitution, equation (12.33) goes over into

$$\langle \mathbf{E}^- \cdot \mathbf{E}^+ \rangle = \frac{\hbar}{2\epsilon_0} \left| \int d\mathbf{k} \sqrt{kc} e^{i(\mathbf{k}\cdot \mathbf{R} - kct)} \boldsymbol{\epsilon}_{\mathbf{k}} c(\mathbf{k}, 0) \right|^2 , \tag{12.34}$$

which is independent of the quantization volume.

12.2.2 Single-Mode Number State

For a single mode of the field in a pure number or Fock state $|n\rangle$,

$$\Delta n^2 = \langle \hat{n}^2 \rangle - \langle \hat{n} \rangle^2 = 0, \tag{12.35}$$

and

$$\langle \mathbf{E}^{+} \rangle = 0, \tag{12.36a}$$

$$\langle \mathbf{E}^{-} \cdot \mathbf{E}^{+} \rangle = \frac{\hbar\omega n}{2\epsilon_0 \mathcal{V}}, \tag{12.36b}$$

$$\langle (\mathbf{E}^{-} \cdot \mathbf{E}^{+})^2 \rangle = \left(\frac{\hbar\omega}{2\epsilon_0 \mathcal{V}}\right)^2 \langle a^{\dagger} a a^{\dagger} a \rangle = \left(\frac{\hbar\omega}{2\epsilon_0 \mathcal{V}}\right)^2 \left(n^2 + n\right), \tag{12.36c}$$

$$\Delta \left(\mathbf{E}^{-} \cdot \mathbf{E}^{+}\right) = \left(\frac{\hbar\omega}{2\epsilon_0 \mathcal{V}}\right) \sqrt{n}, \tag{12.36d}$$

where Δ represents the standard deviation. Provided that $n \gg 1$, there is a relatively small uncertainty in the field intensity, since $\Delta \left(\mathbf{E}^{-} \cdot \mathbf{E}^{+}\right) / \langle \mathbf{E}^{-} \cdot \mathbf{E}^{+} \rangle = 1/\sqrt{n}$. On the other hand, a pure n state is *not* an eigenstate of field intensity. The fact that $\langle \mathbf{E}^{+} \rangle = 0$ is related to an uncertainty in the phase of the field. Phase is discussed in section 12.5, but for now, we note that a pure n state has a phase that can be considered to be totally random. In some sense, a pure n state is similar to a field having amplitude $E_0 = \sqrt{2}[\hbar\omega/(\epsilon_0 \mathcal{V})]^{1/2}\sqrt{n}$ and random phase ϕ such that $E = E_0 \cos(kz - \Omega t + \phi)$, $\langle E^{+} \rangle_{\phi} = 0$, and $\langle E^{-} E^{+} \rangle_{\phi} = E_0^2/4 = \hbar\omega n / 2\epsilon_0 \mathcal{V}$. The analogy is not complete, however, since there are no fluctuations in the intensity of this classical field.

12.2.3 Quasiclassical or Coherent States

In the early 1960s, the advent of the laser led Roy Glauber to consider a special class of states of the quantized radiation field that are known as *quasiclassical* or *coherent* states [2]. These states are chosen in such a fashion that their properties mirror those of classical radiation fields. More precisely, the coherent states are defined in a way that they reproduce the corresponding classical field results if quantum fluctuations of the field can be neglected. For his work, Glauber shared the Nobel Prize in 2005. Glauber has stated that his work aroused a great deal of skepticism at first, owing to some unusual properties of the coherent states.

In order to discuss coherent states, it will prove useful to introduce the Heisenberg representation. To this point, we have used the Schrödinger representation in which operators are time-independent and the state vector is a function of time. We will give a somewhat more detailed description of the Heisenberg representation in chapters 15 and 19. At this time, you need remember only that an operator $\hat{\mathrm{O}}^{\mathrm{H}}(t)$ in the Heisenberg representation is a function of time, obeying an evolution equation

$$i\hbar \frac{d\hat{\mathrm{O}}^{\mathrm{H}}(t)}{dt} = [\hat{\mathrm{O}}^{\mathrm{H}}(t), \hat{H}], \tag{12.37}$$

and that the state vector for the system in the Heisenberg representation, $|\Psi\rangle^{\mathrm{H}}$, is *constant* in time, equal to the state vector for the system in the Schrödinger representation at time $t = 0$. The Heisenberg representation is especially useful when one is interested in obtaining equations for operators rather than the state vector of a system, since the Heisenberg operators usually obey time-evolution equations that closely resemble those of the corresponding classical system. In the remainder of this chapter, Heisenberg operators will always be written with an explicit time dependence. The Heisenberg and Schrödinger operators are defined in such a manner that they coincide at $t = 0$. If the Hamiltonian is time-independent in the Schrödinger representation, it is also time-independent in the Heisenberg representation; moreover, the commutation relations of operators is the same in the Heisenberg and Schrödinger representation.

From equations (12.37), (12.27), and (12.5), it follows that, for free fields, the evolution equation for the annihilation operator $a_j(t)$ is

$$\dot{a}_j(t) = -i\omega_j a_j(t), \tag{12.38}$$

with solution

$$a_j(t) = e^{-i\omega_j t} a_j, \tag{12.39}$$

where $a_j = a_j(0)$ is a Schrödinger operator. Similarly,

$$a_j^\dagger(t) = e^{i\omega_j t} a_j^\dagger. \tag{12.40}$$

As a consequence, the electric field operator $\mathbf{E}(\mathbf{R}, t)$ (now written as a Heisenberg operator) for the free field is

$$\mathbf{E}(\mathbf{R}, t) = i \sum_j E_j (a_j \boldsymbol{\epsilon}_j e^{i\mathbf{k}_j\cdot\mathbf{R} - i\omega_j t} - a_j^\dagger \boldsymbol{\epsilon}_j e^{-i\mathbf{k}_j\cdot\mathbf{R} + i\omega_j t}). \tag{12.41}$$

Assume for the moment that there exists a single-mode quantum state of the field $|\alpha\rangle$ that is an eigenket of the annihilation operator a—that is,

$$\langle\alpha|a|\alpha\rangle = \alpha, \tag{12.42a}$$

$$\langle\alpha|a^\dagger|\alpha\rangle = \alpha^*, \tag{12.42b}$$

where α is a complex number. If such a state existed, then the expectation value of the electric field operator $\mathbf{E}(\mathbf{R}, t)$ for this single-mode field having frequency $\omega = kc$ and polarization $\boldsymbol{\epsilon}$ is

$$\langle\mathbf{E}(\mathbf{R}, t)\rangle = i E_\omega \boldsymbol{\epsilon} (\alpha e^{i\mathbf{k}\cdot\mathbf{R} - i\omega t} - \alpha^* e^{i\mathbf{k}\cdot\mathbf{R} - i\omega t}), \tag{12.43}$$

where $E_\omega = (\hbar\omega/2\epsilon_0\mathcal{V})^{1/2}$. In other words, the expectation value of the field is *identical* to that of a monochromatic plane-wave field having (complex) amplitude $E_\omega\alpha$. Moreover, the expectation values of the field energy and momentum equal those of the corresponding classical fields.

Thus, as long as quantum fluctuations of the fields are unimportant, the states $|\alpha\rangle$ can be associated with classical fields. Fluctuations tend to be unimportant if $\langle\hat{n}(t)\rangle = \langle a^\dagger(t)a(t)\rangle \gg 1$, although this proves to be a necessary, but not sufficient, condition. We have not yet indicated the manner in which one can create such quantum states of the field, but it is not unreasonable to believe that most of the radiation fields we encounter can be represented in first approximation by coherent states of the field. In fact, we will show in chapter 15 that a *classical* current distribution gives rise to a quantized coherent state of the radiation field.

In general, several coherent-state field modes can be excited. A multimode coherent state can be written as $|\alpha_1, \alpha_2, \ldots \alpha_j \ldots\rangle$, for which

$$\langle \mathbf{E}(\mathbf{R}, t)\rangle = i \sum_j E_j[\alpha_j \boldsymbol{\epsilon}_j e^{i\mathbf{k}_j\cdot\mathbf{R}-i\omega_j t} - \alpha_j^*(t)\boldsymbol{\epsilon}_j e^{-i\mathbf{k}_j\cdot\mathbf{R}+i\omega_j t}]. \tag{12.44}$$

For the remainder of this chapter, we deal mainly with single-mode fields.

The equation

$$a|\alpha\rangle = \alpha|\alpha\rangle \tag{12.45}$$

defines a coherent state. It is a fairly amazing equation, since the application of the *destruction* operator results in a state with the *same* value of $\langle\hat{n}\rangle$. It is precisely this property of coherent states that Glauber claims was disturbing to some people who read his articles. The fact that equation (12.45) leads to coherent states of the field does not guarantee the existence of such states, since the non-Hermitian operator a may not possess a set of normalizable eigenkets. However, we can try to construct such a state by writing

$$|\alpha\rangle = \sum_{n=0}^{\infty} \langle n|\alpha\rangle|n\rangle \tag{12.46}$$

and see whether we can find the expansion coefficients $\langle n|\alpha\rangle$.

We start by using equations (12.45), (12.46), and (12.18) to write

$$a|\alpha\rangle = \sum_n \sqrt{n}|n-1\rangle\langle n|\alpha\rangle = \alpha|\alpha\rangle. \tag{12.47}$$

Multiplying by $\langle m-1|$, we obtain

$$\sqrt{m}\langle m|\alpha\rangle = \alpha\langle m-1|\alpha\rangle. \tag{12.48}$$

Therefore,

$$\langle n|\alpha\rangle = \frac{\alpha}{\sqrt{n}}\langle n-1|\alpha\rangle = \frac{\alpha^2}{\sqrt{n(n-1)}}\langle n-2|\alpha\rangle = \frac{\alpha^n}{\sqrt{n!}}\langle 0|\alpha\rangle, \tag{12.49}$$

and it follows from equation (12.46) that

$$|\alpha\rangle = \sum_n \frac{\alpha^n}{\sqrt{n!}}\langle 0|\alpha\rangle|n\rangle. \tag{12.50}$$

To fix the value of $\langle 0|\alpha\rangle$, we assume that the $|\alpha\rangle$ states are normalized, such that

$$\langle\alpha|\alpha\rangle = \sum_n \frac{|\alpha|^{2n}}{n!}|\langle 0|\alpha\rangle|^2 = |\langle 0|\alpha\rangle|^2 e^{|\alpha|^2} = 1. \tag{12.51}$$

Taking $\langle 0|\alpha\rangle$ to be real and positive, we obtain

$$\langle 0|\alpha\rangle = e^{-|\alpha|^2/2} \tag{12.52}$$

and

$$|\alpha\rangle = \sum_n \frac{\alpha^n}{\sqrt{n!}} \langle 0|\alpha\rangle |n\rangle = \sum_n \frac{\alpha^n}{\sqrt{n!}} e^{-|\alpha|^2/2} |n\rangle$$

$$= \sum_n \frac{(\alpha a^\dagger)^n}{n!} e^{-|\alpha|^2/2} |0\rangle = e^{\alpha a^\dagger} e^{-|\alpha|^2/2} |0\rangle. \tag{12.53}$$

Although there was no guarantee that eigenkets of a exist, it has been possible to construct them. The coherent state with $\alpha = 0$ is the *vacuum state* of the field.

12.2.3.1 Properties of coherent states

The probability distribution of photon number n in a coherent state obeys a Poisson distribution

$$|\langle n|\alpha\rangle|^2 = \frac{|\alpha|^{2n}}{n!} e^{-|\alpha|^2}. \tag{12.54}$$

The mean, mean square, and variance of the photon number are

$$\langle \hat{n} \rangle = \langle a^\dagger a \rangle = \langle \alpha | a^\dagger a | \alpha \rangle = |\alpha|^2, \tag{12.55a}$$

$$\langle \hat{n}^2 \rangle = \langle a^\dagger a a^\dagger a \rangle = \langle \alpha | a^\dagger (a^\dagger a + 1) a | \alpha \rangle = |\alpha|^2 (|\alpha|^2 + 1), \tag{12.55b}$$

$$\langle \Delta \hat{n}^2 \rangle = \langle \hat{n}^2 \rangle - \langle \hat{n} \rangle^2 = |\alpha|^2, \tag{12.55c}$$

such that

$$\frac{\langle \Delta \hat{n} \rangle}{\langle \hat{n} \rangle} = \frac{1}{\sqrt{\langle \hat{n} \rangle}} = \frac{1}{|\alpha|}. \tag{12.56}$$

For large values of $|\alpha|$, the relative fluctuations of the intensity of a coherent state are minimal. These equations are valid only for $\alpha \neq 0$. The state with $\alpha = 0$ corresponds to the vacuum state $|0\rangle$, which is an eigenstate of the number operator having eigenvalue zero.

Since a is not a Hermitian operator, there is no guarantee that the eigenkets of a are orthogonal and complete. In fact, the coherent states are *not* orthogonal, since

$$\langle \alpha | \beta \rangle = \sum_{n,m} \frac{(\alpha^*)^n}{\sqrt{n!}} \frac{\beta^m}{\sqrt{m!}} e^{-(|\alpha|^2+|\beta|^2)/2} \langle n|m\rangle = e^{-(|\alpha|^2+|\beta|^2)/2} \sum_n \frac{(\alpha^* \beta)^n}{n!}$$

$$= e^{-\frac{1}{2}(|\alpha|^2+|\beta|^2)+\alpha^*\beta} = e^{-\frac{1}{2}|\alpha-\beta|^2} e^{\frac{1}{2}(\alpha^*\beta - \alpha\beta^*)}, \tag{12.57a}$$

$$|\langle \alpha|\beta\rangle|^2 = e^{-|\alpha-\beta|^2}. \tag{12.57b}$$

On the other hand, the coherent-state eigenkets form an *overcomplete* set. That is, although they obey a completeness-type relationship

$$\frac{1}{\pi} \int d^2\alpha |\alpha\rangle\langle\alpha| = 1, \tag{12.58}$$

owing to the lack of orthogonality, any coherent state can be expressed as a linear combination of *other* coherent states,

$$|\alpha\rangle = \frac{1}{\pi}\int d^2\beta|\beta\rangle\langle\beta|\alpha\rangle \,. \tag{12.59}$$

Since α and β are complex, the integrals in equation (12.58) and (12.59) are over the complex plane, $d^2\alpha = d\text{Re}(\alpha)d\text{Im}(\alpha)$, $d^2\beta = d\text{Re}(\beta)d\text{Im}(\beta)$. The overcompleteness of the eigenkets will be exploited when we consider the density matrix for the field.

12.2.3.2 Representation of a coherent state as a translation of the vacuum state

It turns out to be very useful to seek an alternative representation for coherent states. In particular, we want to show that a coherent state can be represented as a translation of the vacuum state. Recall that the vacuum state is a coherent state with $\alpha = 0$, for which $\langle\alpha|a|\alpha\rangle = \langle\alpha|a^\dagger|\alpha\rangle = 0$.

We seek unitary operators $D(\beta,\beta^*)$ and $D^\dagger(\beta,\beta^*)$ that act as a translation operators for the field creation and annihilation operators—namely,

$$D^\dagger(\beta,\beta^*)a\,D(\beta,\beta^*) = a + \beta, \tag{12.60a}$$

$$D^\dagger(\beta,\beta^*)a^\dagger D(\beta,\beta^*) = a^\dagger + \beta^*. \tag{12.60b}$$

Since β is complex, it is completely determined if we specify its real and imaginary parts or, alternatively, if we consider β and β^* as independent variables, which proves more convenient at the moment. Using equations (12.60a) and (12.47), we find

$$\begin{aligned} a\left[D^\dagger(\beta,\beta^*)|\alpha\rangle\right] &= \left[D^\dagger(\beta,\beta^*)a\,D(\beta,\beta^*) - \beta\right] D^\dagger(\beta,\beta^*)|\alpha\rangle \\ &= (\alpha-\beta)D^\dagger(\beta,\beta^*)|\alpha\rangle. \end{aligned} \tag{12.61}$$

In other words, $D^\dagger(\beta,\beta^*)|\alpha\rangle$ is an eigenket of a with eigenvalue $(\alpha-\beta)$. It follows that $D^\dagger(\alpha,\alpha^*)|\alpha\rangle = c|0\rangle$, where c is a constant that must have a magnitude of unity if $D(\beta,\beta^*)$ is unitary. Taking $c = 1$ leads to

$$|\alpha\rangle = D(\alpha,\alpha^*)|0\rangle, \tag{12.62}$$

which implies that $D(\alpha,\alpha^*)$ generates coherent states from the vacuum state.

To arrive at a specific form for $D(\alpha,\alpha^*)$, we consider first infinitesimal translations ϵ in the complex plane given by

$$D^\dagger(\epsilon,\epsilon^*)a\,D(\epsilon,\epsilon^*) = a + \epsilon. \tag{12.63}$$

To first order in ϵ and ϵ^*, we can write

$$D(\epsilon,\epsilon^*) = 1 + \epsilon A + \epsilon^* B, \tag{12.64}$$

which implies that, to this order,

$$D^\dagger(\epsilon,\epsilon^*) = 1 - \epsilon A - \epsilon^* B, \tag{12.65}$$

and

$$(1 - \epsilon A - \epsilon^* B) a (1 + \epsilon A + \epsilon^* B) = a + \epsilon \, [a, A] + \epsilon^* \, [a, B] \,. \tag{12.66}$$

If we compare equations (12.63) and (12.66), we find that $A = a^\dagger$ and that B is a function of a only. Using a similar procedure starting with

$$D^\dagger(\epsilon, \epsilon^*) a^\dagger D(\epsilon, \epsilon^*) = a^\dagger + \epsilon^*, \tag{12.67}$$

we obtain $B = -a$. As a consequence, equation (12.64) becomes

$$D(\epsilon, \epsilon^*) = 1 + \epsilon a^\dagger - \epsilon^* a. \tag{12.68}$$

To get $D(\alpha, \alpha^*)$ for finite translations, we use a standard "trick" and write

$$\begin{aligned} |\alpha(\lambda + d\lambda)\rangle &= D\left[\alpha(\lambda + d\lambda), \alpha^*(\lambda + d\lambda)\right] |0\rangle \\ &= D\left(\alpha d\lambda, \alpha^* d\lambda\right) D\left(\alpha\lambda, \alpha^*\lambda\right) |0\rangle, \end{aligned} \tag{12.69}$$

where λ is real. It then follows that

$$\frac{\partial D(\alpha\lambda, \alpha^*\lambda)}{\partial\lambda} d\lambda = (a^\dagger \alpha d\lambda - a\alpha^* d\lambda) D(\alpha\lambda, \alpha^*\lambda). \tag{12.70}$$

Integrating this equation from $\lambda = 0$ to $\lambda = 1$, using the fact that $D(0, 0) = 1$, we find

$$D\left(\alpha, \alpha^*\right) = e^{\alpha a^\dagger - a\alpha^*} = e^{-\frac{1}{2}|\alpha|^2} e^{\alpha a^\dagger} e^{-\alpha^* a}. \tag{12.71}$$

The coherent state (12.62) can be written as

$$|\alpha\rangle = D(\alpha, \alpha^*)|0\rangle = e^{-\frac{1}{2}|\alpha|^2} e^{\alpha a^\dagger} |0\rangle, \tag{12.72}$$

since $e^{-\alpha^* a}|0\rangle = |0\rangle$.

We have accomplished our goal of obtaining a displacement operator that acts on the vacuum state $|0\rangle$ to produce the coherent state $|\alpha\rangle$. Expressing a coherent state in terms of the displacement operator often simplifies calculations involving coherent states. Using the fact that $e^{\hat{A}+\hat{B}} = e^{\hat{A}} e^{\hat{B}} e^{-[\hat{A}, \hat{B}]/2}$, provided that both $\hat{A}$ and $\hat{B}$ commute with $[\hat{A}, \hat{B}]$, you can easily prove the "addition theorem" for translations,

$$D\left(\alpha_1 + \alpha_2, \alpha_1^* + \alpha_2^*\right) = D\left(\alpha_1, \alpha_1^*\right) D\left(\alpha_2, \alpha_2^*\right) e^{(\alpha_1 \alpha_2^* - \alpha_1^* \alpha_2)}. \tag{12.73}$$

12.3 Quadrature Operators for the Field

The operators a and $a^\dagger$ are not Hermitian and do not correspond to physical observables. So far, the only operators that we have discussed that could correspond to physical observables are $a^\dagger a$ and the field operators. Actually, in most experiments, it is a field intensity that is measured, so it is important to connect that intensity measurement with quantum-mechanical operators that characterize the field.

For a single-mode field,

$$\mathbf{E}(\mathbf{R}, t) = i E (a e^{i\mathbf{k}\cdot\mathbf{R} - i\omega t} - a^\dagger e^{-i\mathbf{k}\cdot\mathbf{R} + i\omega t}) \boldsymbol{\epsilon}. \tag{12.74}$$

If we define

$$a = a_1 + i a_2, \tag{12.75a}$$

$$a^\dagger = a_1 - i a_2, \tag{12.75b}$$

where $a_1 = a_1^\dagger$, $a_2 = a_2^\dagger$, are Hermitian operators, then

$$\mathbf{E}(\mathbf{R}, t) = -2E\boldsymbol{\epsilon}\left[a_1 \sin(\mathbf{k}\cdot\mathbf{R} - \omega t) + i a_2 \cos(\mathbf{k}\cdot\mathbf{R} - \omega t)\right]. \tag{12.76}$$

A classical field of the form

$$\mathbf{E}(\mathbf{R}, t) = iE\boldsymbol{\epsilon}(e^{i\mathbf{k}\cdot\mathbf{R} - i\omega t} - e^{-i\mathbf{k}\cdot\mathbf{R} + i\omega t}) \tag{12.77}$$

can be written as

$$\mathbf{E}(\mathbf{R}, t) = -2E\boldsymbol{\epsilon}\sin(\mathbf{k}\cdot\mathbf{R} - \omega t). \tag{12.78}$$

Thus, the term proportional to a_1 in equation (12.76) is referred to as the *in-phase* component of the field and that proportional to a_2 as the *out-of-phase* component of the field. As such, the operators a_1 and a_2 correspond to observables that can be measured in experiments by heterodyning the field with a reference field.

The Hermitian operators

$$a_1 = \frac{a + a^\dagger}{2}, \tag{12.79a}$$

$$a_2 = \frac{a - a^\dagger}{2i}, \tag{12.79b}$$

satisfy the commutation relations

$$[a_1, a_2] = \frac{1}{4i}\left[a + a^\dagger, a - a^\dagger\right] = -\frac{1}{2i} = \frac{i}{2}. \tag{12.80}$$

The operators do not commute, and the product of the variance of the operators satisfies

$$(\Delta a_1)^2 (\Delta a_2)^2 \geq 1/4.$$

Since these operators correspond to physical observables, it is useful to calculate their average values and variances for different quantum states of the field.

12.3.1 Pure *n* State

In a Fock or number state of the field

$$\langle n| a_1 |n\rangle = \langle n| a_2 |n\rangle = 0, \tag{12.81}$$

$$\langle n| a_1^2 |n\rangle = \frac{1}{4}\langle n| (a + a^\dagger)(a + a^\dagger) |n\rangle = \frac{1}{2}\left(n + \frac{1}{2}\right), \tag{12.82a}$$

$$\langle n| a_2^2 |n\rangle = -\frac{1}{4}\langle n| (a - a^\dagger)(a - a^\dagger) |n\rangle = \frac{1}{2}\left(n + \frac{1}{2}\right), \tag{12.82b}$$

$$\Delta a_1 = \Delta a_2 = \sqrt{n + 1/2}/\sqrt{2}, \tag{12.83}$$

and

$$\Delta a_1 \Delta a_2 = \frac{1}{2}\left(n + \frac{1}{2}\right). \tag{12.84}$$

The Fock state is not a minimum uncertainty state except for $n = 0$. The vacuum state *is* a minimum uncertainty state for the operators a_1, a_2.

12.3.2 Coherent State

In analogy with the operators a_1, a_2, we define real variables

$$\alpha_1 = \frac{\alpha + \alpha^*}{2}, \tag{12.85}$$

$$\alpha_2 = \frac{\alpha - \alpha^*}{2i}. \tag{12.86}$$

In a coherent state of the field, we then find

$$\begin{aligned} \langle \alpha | a_1 | \alpha \rangle &= \alpha_1, \\ \langle \alpha | a_2 | \alpha \rangle &= \alpha_2, \end{aligned} \tag{12.87}$$

and

$$\begin{aligned} \langle a_1^2 \rangle &= \langle \alpha | \frac{(a + a^\dagger)^2}{4} | \alpha \rangle = \frac{1}{4} \langle \alpha | a^2 + a^{\dagger 2} + a a^\dagger + a^\dagger a | \alpha \rangle \\ &= \frac{1}{4}(\alpha^2 + \alpha^{*2} + 1 + |\alpha|^2 + |\alpha|^2) = \alpha_1^2 + \frac{1}{4}. \end{aligned} \tag{12.88}$$

In the same way, we obtain

$$\langle \alpha | a_2^2 | \alpha \rangle = \alpha_2^2 + \frac{1}{4}, \tag{12.89}$$

such that

$$\Delta a_1 = \Delta a_2 = 1/2, \tag{12.90}$$

$$\Delta a_1 \Delta a_2 = 1/4. \tag{12.91}$$

The coherent state is a minimum uncertainty state. Note that $\Delta a_1 / \langle a_1 \rangle = 1/(2\alpha_1)$, which is of order $1/\left(\sqrt{2}\,\langle \hat{n} \rangle\right)$ (for $\alpha_1 = \alpha_2$), similar to the result for the uncertainty in photon number in a coherent state.

12.4 Two-Photon Coherent States or Squeezed States

In quantum mechanics, $\hat{x}$ and $\hat{p}$ are noncommuting operators. One can draw a phase-space diagram illustrating the relative uncertainty of these variables in a given quantum state. For example, in the ground state of an oscillator, the uncertainty diagram would be a circle centered at the origin. One can construct an analogous

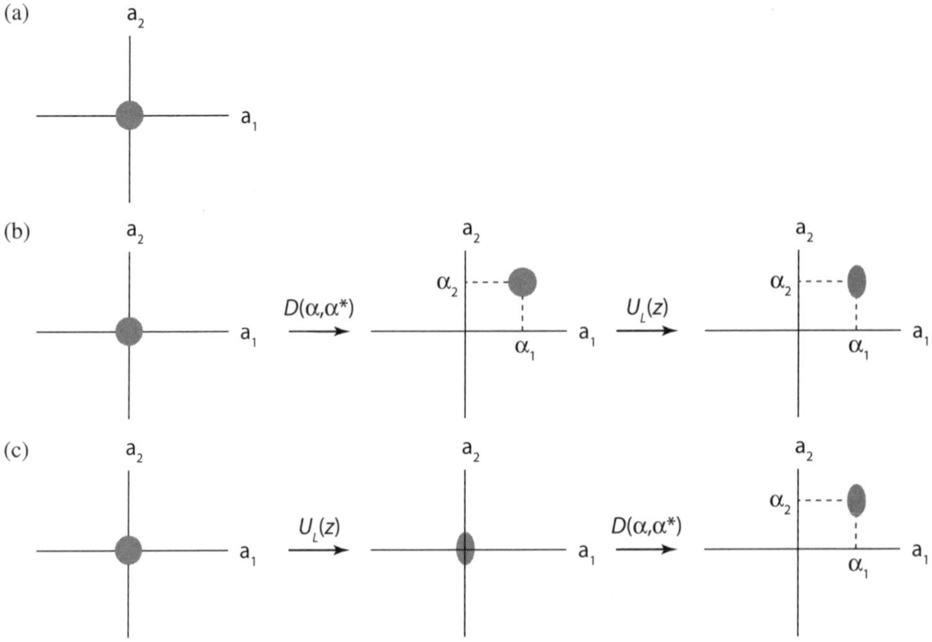

Figure 12.1. Representation of the uncertainties in the quadrature components of the field: (a) vacuum state (b) two-photon coherent state (c) squeezed state.

diagram for the operators a_1 and a_2. Since

$$\Delta a_1^2 + \Delta a_2^2 = 1/2 \tag{12.92}$$

in a coherent state of the field, the uncertainty diagram in the $a_1 - a_2$ plane is a circle of radius $1/\sqrt{2}$ centered at (α_1, α_2). For the vacuum state, the circle is centered at the origin, as shown in figure 12.1(a). For a coherent state, the uncertainties in a_1 and a_2 are always equal, $\Delta a_1 = \Delta a_2 = 1/2$, and this is referred to as the standard quantum limit (SQL).

Can we find new minimum uncertainty states where $\Delta a_1 \Delta a_2 = 1/4$ but $\Delta a_1 \ll 1/2$ *or* $\Delta a_2 \ll 1/2$? If so, it would be possible to measure one of the operators a_1, a_2 with an uncertainty below that of a coherent state and to beat the SQL for that operator (of course, the uncertainty of the other operator would be higher than that for a coherent state). Such a state is referred to as a *squeezed state* [3–7]. In the mid-1970s, squeezed states were being advertised as the future of reduced noise measurements in quantum optics. Although there has been some progress along these lines, it is probably safe to say that squeezed states of this nature have yet to live up to their potential [8].

To look for operators that can have squeezed uncertainties, one can define new operators obtained from a and $a^\dagger$ by a canonical transformation,

$$b = \mu a + \nu a^\dagger, \tag{12.93a}$$

$$b^\dagger = \mu^* a^\dagger + \nu^* a, \tag{12.93b}$$

in which

$$[b, b^{\dagger}] = |\mu|^2 - |\nu|^2 = 1. \tag{12.94}$$

The canonical transformation can be written in terms of a unitary operator U_L as

$$b = U_L a U_L^{\dagger}. \tag{12.95}$$

It is conventional to take

$$\mu = \cosh r \ , \tag{12.96a}$$

$$\nu = e^{i\theta} \sinh r, \tag{12.96b}$$

where r, θ are real. The complex number

$$z = re^{i\theta} \tag{12.97}$$

is referred to as a *squeezing parameter*. Since

$$b\left[U_L(z)|\alpha\rangle\right] = U_L(z)a|\alpha\rangle = \alpha\left[U_L(z)|\alpha\rangle\right], \tag{12.98}$$

$U_L(z)|\alpha\rangle$ is an eigenket of b with eigenvalue α. We denote this eigenket as

$$|z, \alpha\rangle \equiv U_L(z)|\alpha\rangle, \tag{12.99}$$

such that

$$b|z, \alpha\rangle = \alpha|z, \alpha\rangle. \tag{12.100}$$

The state $|z, \alpha\rangle$ is referred to as a *two-photon coherent state* (TPCS). This nomenclature is used since such states are generated from the vacuum by the operator $a^{\dagger 2}$, as is shown in the following section.

We want to obtain the expectation values and variances of the operators a_1 and a_2 in the state $|z, \alpha\rangle$. To do so, we need the inverse transform of equation (12.93), which is

$$\begin{aligned} a &= \mu^* b - \nu b^{\dagger}, \\ a^{\dagger} &= \mu b^{\dagger} - \nu^* b. \end{aligned} \tag{12.101}$$

Then, using equation (12.79), we find

$$\begin{aligned} \langle z, \alpha|a_1|z, \alpha\rangle &= \langle z, \alpha|\frac{a + a^{\dagger}}{2}|z, \alpha\rangle \\ &= \langle z, \alpha|\frac{\mu^* b - \nu b^{\dagger} + \mu b^{\dagger} - \nu^* b}{2}|z, \alpha\rangle = \frac{1}{2}\left[(\mu^* - \nu^*)\alpha + (\mu - \nu)\alpha^*\right] \end{aligned} \tag{12.102}$$

and

$$\langle z, \alpha|a_1^2|z, \alpha\rangle = \frac{1}{4}\langle z, \alpha|(\mu^* b + \mu b^{\dagger} - \nu b^{\dagger} - \nu^* b)(\mu^* b + \mu b^{\dagger} - \nu b^{\dagger} - \nu^* b)|z, \alpha\rangle. \tag{12.103}$$

To evaluate equation (12.103), it is easiest to use the commutator (12.94) to move all the b's to the right so that equation (12.100) and its adjoint can be used. In this manner, we obtain

$$\begin{aligned}\langle z,\alpha|a_1^2|z,\alpha\rangle &= \langle z,\alpha|a_1|z,\alpha\rangle^2 + (|\mu|^2 - \mu^*\nu - \nu^*\mu + |\nu|^2)/4 \\ &= \langle z,\alpha|a_1|z,\alpha\rangle^2 + |\mu-\nu|^2/4. \end{aligned} \tag{12.104}$$

In an analogous fashion, we find

$$\langle z,\alpha|a_2^2|z,\alpha\rangle = \langle z,\alpha|a_2|z,\alpha\rangle^2 + |\mu+\nu|^2/4. \tag{12.105}$$

From these results, it follows that

$$\Delta a_1^2 = |\mu-\nu|^2/4, \qquad \Delta a_2^2 = |\mu+\nu|^2/4, \tag{12.106a}$$

$$\Delta a_1 \Delta a_2 = |\mu^2 - \nu^2|/4 = |\cosh^2 r - \sinh^2 r e^{2i\theta}|/4. \tag{12.106b}$$

The state $|z,\alpha\rangle$ is not a minimum uncertainty state unless $\theta = 0$. If $\theta = 0$,

$$\Delta a_1 = e^{-r}/2, \qquad \Delta a_2 = e^{r}/2, \qquad \Delta a_1 \Delta a_2 = 1/4. \tag{12.107}$$

Thus, for large r, the variance of one of the quadrature operators can be well below the SQL.

It is possible to generalize the discussion and define Hermitian operators

$$A_1 = a_1 \cos(\theta/2) + a_2 \sin(\theta/2), \tag{12.108a}$$

$$A_2 = a_2 \cos(\theta/2) - a_1 \sin(\theta/2). \tag{12.108b}$$

For these operators, $\Delta A_1 = e^{-r}/2$ and $\Delta A_1 = e^{r}/2$ in the state $|z,\alpha\rangle$ with $z = re^{i\theta}$, such that the TPCS is always a minimum unceraintty state for these "rotated" operators. Thus, it is always possible to "squeeze" one of the variances below the SQL at the expense of the other.

The average value of the number operator in a TPCS is given by

$$\begin{aligned}\langle z,\alpha|\hat{n}|z,\alpha\rangle &= \langle z,\alpha|a^\dagger a|z,\alpha\rangle \\ &= \langle z,\alpha|(\mu b^\dagger - \nu^* b)(\mu^* b - \nu b^\dagger)|z,\alpha\rangle \\ &= -\mu\nu(\alpha^*)^2 + |\mu|^2|\alpha|^2 + |\nu|^2(|\alpha|^2+1) - \nu^*\mu^*\alpha^2. \end{aligned} \tag{12.109}$$

If $\alpha = 0$, then

$$\langle z,0|\hat{n}|z,0\rangle = |\nu|^2 = \sinh^2 r. \tag{12.110}$$

The state $|z,0\rangle$ is referred to as a *squeezed vacuum state*, although it is *not* a vacuum state of the field.

12.4.1 Calculation of $U_L(z)$

We can calculate $U_L(z)$ following a procedure that is similar to the one we used to arrive at the displacement operator. From equations (12.93a) and (12.95), we have

$$b = U_L(z) a U_L^\dagger(z) = \mu a + \nu a^\dagger. \tag{12.111}$$

Assume that the magnitude of the squeezing parameter $r \ll 1$, corresponding to an infinitesimal transformation with $\mu \approx \cosh r \approx 1$ and $\nu = e^{i\theta}\sinh r \approx re^{i\theta}$. Setting

$\epsilon = re^{i\theta}$, with $|\epsilon| \ll 1$, we can expand

$$U_L(\epsilon, \epsilon^*) = 1 + \epsilon A + \epsilon^* B, \tag{12.112a}$$

$$U_L^\dagger(\epsilon, \epsilon^*) = 1 - \epsilon A - \epsilon^* B. \tag{12.112b}$$

To order ϵ, equation (12.111) becomes

$$(1 + \epsilon A + \epsilon^* B)a(1 - \epsilon A - \epsilon^* B) = a + \epsilon a^\dagger, \tag{12.113}$$

which implies that

$$[A, a] = a^\dagger, \quad [B, a] = 0. \tag{12.114}$$

As a consequence, we can conclude that

$$A = -(a^\dagger)^2/2 \tag{12.115}$$

and that B is a function of a only.

Using the equation adjoint to equation (12.111) and following a similar procedure, we obtain

$$B = a^2/2. \tag{12.116}$$

By combining equations (12.112a), (12.115), and (12.116), we arrive at

$$U_L(\epsilon, \epsilon^*) = 1 - \epsilon \frac{(a^\dagger)^2}{2} + \epsilon^* \frac{a^2}{2}. \tag{12.117}$$

As in the case of the displacement operator, we can generalize this result to a finite transformation. The resulting squeeze operator is

$$U_L(z) = e^{[-z(a^\dagger)^2 + z^* a^2]/2} = U_L^\dagger(-z), \tag{12.118}$$

where $z = re^{i\theta}$.

We can now understand the origin of the nomenclature TPCS. Suppose that there is an effective Hamiltonian for the field given by

$$H = c_1(a^\dagger)^2 + c_1^* a^2. \tag{12.119}$$

The state evolution operator $e^{-iHt/\hbar}$ associated with this Hamiltonian is of the form (12.118), with $z = ic_1 t/\hbar$. Since

$$|z, \alpha\rangle = U_L(z)|\alpha\rangle, \tag{12.120}$$

this form of $U_L(z)$ implies that if the field is prepared in a coherent state, a Hamiltonian involving two-photon operators, a^2, $(a^\dagger)^2$, can lead to a TPCS.

Although they exhibit squeezing, the TPCS

$$|z, \alpha\rangle = U_L(z)|\alpha\rangle = U_L(z)\mathcal{D}(\alpha)|0\rangle \tag{12.121}$$

is usually not referred to per se as a squeezed state. As shown in equation (12.121) and represented in figure 12.1(b), a TPCS is generated via a translation of the vacuum state and a "squeezing" of this translated state. It can be shown using properties of $U_L(z)$ and $\mathcal{D}(\alpha)$ that

$$|\alpha, z\rangle = |z, \alpha'\rangle = \mathcal{D}(\alpha)U_L(z)|0\rangle \tag{12.122}$$

is also a two-photon coherent state, provided that

$$\alpha' = \alpha \cosh r + \alpha^* \sinh r e^{i\theta} \tag{12.123}$$

and $z = re^{i\theta}$.

The nomenclature *squeezed state* is usually reserved for the state $|\alpha, z\rangle$, which results from a "squeezing" of the vacuum state, followed by a translation of the resultant state; see figure 12.1(c). For the vacuum ($\alpha = 0$), the TPCS and squeezed states are identical. The average value of the number operator in the squeezed state (12.122) is

$$\begin{aligned}
\langle \alpha, z|\hat{n}|\alpha, z\rangle &= \langle 0|U_L^\dagger(z)\mathcal{D}^\dagger(\alpha)a^\dagger a\mathcal{D}(\alpha)U_L(z)|0\rangle \\
&= \langle 0|U_L^\dagger(z)\mathcal{D}^\dagger(\alpha)a^\dagger\mathcal{D}(\alpha)\mathcal{D}^\dagger(\alpha)a\mathcal{D}(\alpha)U_L(z)|0\rangle \\
&= \langle 0|U_L^\dagger(z)(a^\dagger + \alpha^*)(a + \alpha)U_L(z)|0\rangle \\
&= \langle z, 0|a^\dagger a + a^\dagger\alpha + \alpha^* a + |\alpha|^2|z, 0\rangle \\
&= \sinh^2 r + |\alpha|^2,
\end{aligned} \tag{12.124}$$

where equation (12.110) and the fact that $\langle z, 0|a|z, 0\rangle = \langle z, 0|a^\dagger|z, 0\rangle = 0$ have been used. One can also show that

$$\langle \Delta n^2\rangle = |\alpha \cosh r - \alpha^* \sinh r e^{i\theta}|^2 + 2\sinh^2 r \cosh^2 r \tag{12.125}$$

for these states. Moreover, for the quadrature operators, we find

$$\langle \alpha, z|a_1|\alpha, z\rangle = a_1, \tag{12.126a}$$

$$\langle \alpha, z|a_2|\alpha, z\rangle = a_2, \tag{12.126b}$$

$$\Delta a_1^2 = \frac{1}{4}|\mu - \nu|^2, \tag{12.126c}$$

$$\Delta a_2^2 = \frac{1}{4}|\mu + \nu|^2. \tag{12.126d}$$

Thus, the variances in a_1 and a_2 are the same as for a TPCS. We show in chapter 14 how it is possible to use squeezed states to reduce quantum noise in measurements.

12.5 Phase Operator

Measuring the in-phase and out-of-phase components of a classical monochromatic field is equivalent to measuring the amplitude and phase of the field. However, since the operators a_1 and a_2 do not commute, they cannot be measured simultaneously for a quantized state of the field. One might ask, therefore, whether it is possible to construct operators for the amplitude and phase of the field. The simple answer to this question is "no," since one cannot define *measurable* Hermitian operators that correspond to these quantities. Instead of measuring the amplitude of the field, one is often content to measure the field intensity. For the phase operator, it is probably best to follow Mandel and Wolf and use operational definitions of operators that correspond to the sine and cosine of the phase difference between two fields [9]. In other words, one relates the measured field intensities in a given experiment to these phase operators.

The definition of a phase operator for the quantized field is not unique and remains a topic of current research [10]. It is possible to define an Hermitian phase operator following the suggestion of Pegg and Barnett [11], but it does not appear that this operator corresponds to a physical observable. Alternatively, one can use non-Hermitian operators that still correspond, for the most part, to what one would expect from a phase operator. In what follows, we use the Susskind and Glogower definition of phase operators [12], also discussed in the texts of Loudon [13] and Mandel and Wolf [14].

Since the operator a in equation (12.74) has the appearance of a complex state amplitude, it is not unreasonable to write $a = \hat{A}\widehat{\exp(i\phi)}$, where $\hat{A}$ and $\widehat{\exp(i\phi)}$ are amplitude and phase operators, respectively. The Susskind and Glogower phase operators have this general structure and are defined via

$$a = \sqrt{\hat{n}+1}\,\widehat{\exp(i\phi)}, \tag{12.127a}$$

$$a^{\dagger} = \widehat{\exp(-i\phi)}\sqrt{\hat{n}+1}. \tag{12.127b}$$

With this definition, the phase operators can be written in functional form as

$$\widehat{\exp(i\phi)} = (\hat{n}+1)^{-1/2}a, \tag{12.128a}$$

$$\widehat{\exp(-i\phi)} = a^{\dagger}(\hat{n}+1)^{-1/2}. \tag{12.128b}$$

It then follows that

$$\widehat{\exp(-i\phi)} = [\widehat{\exp(i\phi)}]^{\dagger} \tag{12.129}$$

and

$$\widehat{\exp(i\phi)}\,\widehat{\exp(-i\phi)} = (\hat{n}+1)^{-1/2}aa^{\dagger}(\hat{n}+1)^{-1/2} = 1, \tag{12.130}$$

but

$$\widehat{\exp(-i\phi)}\,\widehat{\exp(i\phi)} = a^{\dagger}(\hat{n}+1)^{-1}a \neq 1.$$

The fact that this product of operators is not equal to unity is shown easily by taking its expectation value in the vacuum state of the field. Suppose that we had defined the phase operator as $e^{i\hat{\phi}}$. Since the operator $\widehat{\exp(i\phi)}$ is not unitary, $\hat{\phi}$ would not be Hermitian with this definition. To arrive at Hermitian operators, we can define cosine and sine phase operators by

$$\widehat{\cos\phi} = \frac{1}{2}[\widehat{\exp(i\phi)} + \widehat{\exp(-i\phi)}], \tag{12.131a}$$

$$\widehat{\sin\phi} = \frac{1}{2i}[\widehat{\exp(i\phi)} - \widehat{\exp(-i\phi)}] \tag{12.131b}$$

and hope that they correspond to a physical observable. The matrix elements of the phase operator between number states are given by

$$\langle n-1|\widehat{\cos\phi}|n\rangle = \langle n|\widehat{\cos\phi}|n-1\rangle = \frac{1}{2} \quad \text{for} \quad n \neq 0, \tag{12.132}$$

$$\langle n-1|\widehat{\sin\phi}|n\rangle = -\langle n|\widehat{\sin\phi}|n-1\rangle = \frac{1}{2i} \quad \text{for} \quad n \neq 0, \tag{12.133}$$

$$\langle 1|\widehat{\cos\phi}|0\rangle = \frac{1}{2}, \qquad \langle 1|\widehat{\sin\phi}|0\rangle = -\frac{1}{2i}. \tag{12.134}$$

All other matrix elements vanish. We also find

$$\langle n|\widehat{\cos\phi}|n\rangle = \langle n|\widehat{\sin\phi}|n\rangle = 0, \tag{12.135a}$$

$$\langle n|\widehat{\cos^2\phi}|n\rangle = \langle n|\widehat{\sin^2\phi}|n\rangle = \frac{1}{2} \text{ for } n \neq 0, \tag{12.135b}$$

$$\langle n|\widehat{\cos^2\phi}|n\rangle = \langle n|\widehat{\sin^2\phi}|n\rangle = \frac{1}{4} \text{ for } n = 0. \tag{12.135c}$$

Other properties of the phase operators are left to the problems.

We would like to think of a number state as one for which the phase is random. The expectation values given in equations (12.135a) and (12.135b) are consistent with this idea. The fact that the average value of $\widehat{\cos^2\phi}$ or $\widehat{\sin^2\phi}$ equals 1/4 in the vacuum state is somewhat troubling and throws into doubt the reliability of these operators.

For a coherent state with $\alpha = |\alpha|e^{i\theta}$, one would expect to find $\langle\widehat{\cos\phi}\rangle_\alpha \approx \cos\theta$ and a minimum value of $\Delta n \Delta(\cos\phi)$. In other words, a coherent state is one in which both the amplitude and phase of the field are fairly well-defined. Calculations of these expectation values are left to the problems.

12.6 Summary

We have given a brief introduction to properties of the free, quantized radiation field. Various possible states of the field were discussed, including number states, coherent states, and squeezed states. Methods for *generating* such states is still an active research area. Generally speaking, one generates number states in a cavity by passing an excited-state atom through the cavity and verifying that the atom left the cavity in its ground state, transferring one photon to the cavity field [15]. Coherent states can be generated from classical current sources, as is discussed in chapter 15. Moreover, the output of laser resonators, while not a coherent state of the field, closely resembles one. Squeezed states are generated in two-photon processes, such as those encountered in parametric down conversion of a single "source" photon into a "signal" and "idler" photon, four-wave mixing, and resonance fluorescence. The quantum properties of the field are invariably linked to the quantum properties of the oscillating charges that produce the fields. We now proceed to discuss and compare coherence properties of classical and quantized fields.

12.7 Appendix: Field Quantization

In this appendix, we indicate a possible path for field quantization, following the approach given in reference [1]. In the presence of charge density ρ and current density **J**, Maxwell's equations are

$$\nabla \cdot \mathbf{E} = \rho/\varepsilon_0, \tag{12.136a}$$

$$\nabla \times \mathbf{E} = -\frac{\partial \mathbf{B}}{\partial t}, \tag{12.136b}$$

$$\nabla \times \mathbf{B} = \mu_0 \left(\mathbf{J} + \varepsilon_0 \frac{\partial \mathbf{E}}{\partial t}\right), \tag{12.136c}$$

$$\nabla \cdot \mathbf{B} = 0, \tag{12.136d}$$

and the Lorentz force on a charge q_α having mass m_α located at position $\mathbf{R}_\alpha$ is

$$m_\alpha \ddot{\mathbf{R}}_\alpha = q_\alpha \left(\mathbf{E} + \mathbf{v}_\alpha \times \mathbf{B}\right), \tag{12.137}$$

where $\mathbf{v}_\alpha = \dot{\mathbf{R}}_\alpha$.

The equation of continuity is

$$\frac{\partial \rho}{\partial t} + \nabla \cdot \mathbf{J} = 0, \tag{12.138}$$

where

$$\rho = \sum_\alpha q_\alpha \delta \left[\mathbf{R} - \mathbf{R}_\alpha(t)\right], \tag{12.139}$$

$$\mathbf{J} = \sum_\alpha q_\alpha \mathbf{v}_\alpha(t) \delta \left[\mathbf{R} - \mathbf{R}_\alpha(t)\right], \tag{12.140}$$

and the sum is over all charges. The Hamiltonian or total energy for the system of charges and fields is

$$H = \sum_\alpha \frac{1}{2} m_\alpha v_\alpha^2 + \frac{\epsilon_0}{2} \int d^3R \left(\mathbf{E}^2 + c^2 \mathbf{B}^2\right), \tag{12.141}$$

while the total momentum is

$$\mathbf{P} = \sum_\alpha m_\alpha \mathbf{v}_\alpha + \epsilon_0 \int d^3R \, \mathbf{E} \times \mathbf{B}. \tag{12.142}$$

The electric $\mathbf{E}$ and magnetic $\mathbf{B}$ field are related to the vector potential $\mathbf{A}$ and scalar potential ϕ by

$$\mathbf{B} = \nabla \times \mathbf{A}, \tag{12.143a}$$

$$\mathbf{E} = -\frac{\partial \mathbf{A}}{\partial t} - \nabla \phi. \tag{12.143b}$$

Substituting equation (12.143) into Maxwell equations, we obtain

$$\nabla^2 \phi = -\frac{\partial \left(\nabla \cdot \mathbf{A}\right)}{\partial t} - \rho / \epsilon_0, \tag{12.144a}$$

$$\left(\frac{1}{c^2} \frac{\partial^2}{\partial t} - \nabla^2\right) \mathbf{A} = \mu_0 \mathbf{J} - \nabla \left(\nabla \cdot \mathbf{A} + \frac{1}{c^2} \frac{\partial \phi}{\partial t}\right). \tag{12.144b}$$

The electric and magnetic fields, equation (12.143), are invariant with respect to the gauge transformation

$$\mathbf{A} \to \mathbf{A} - \nabla \Xi, \tag{12.145a}$$

$$\phi \to \phi + \frac{\partial \Xi}{\partial t}, \tag{12.145b}$$

where the gauge function Ξ is an arbitrary function of position and time.

One can impose various conditions on $\mathbf{A}$ and ϕ. In *Lorentz gauge*, we choose Ξ such that

$$\nabla \cdot \mathbf{A} + \frac{1}{c} \frac{\partial \phi}{\partial t} = 0. \tag{12.146}$$

This condition simplifies the field equations, equation (12.144), which become

$$\Box\phi = \rho/\epsilon_0, \tag{12.147a}$$

$$\Box\mathbf{A} = \mu_0\mathbf{J}, \tag{12.147b}$$

where

$$\Box = \frac{1}{c^2}\frac{\partial^2}{\partial t^2} - \nabla^2. \tag{12.148}$$

In *Coulomb gauge*, we choose

$$\nabla\cdot\mathbf{A} = 0, \tag{12.149}$$

so that equations (12.144) become

$$\nabla^2\phi = -\rho/\epsilon_0, \tag{12.150a}$$

$$\Box\mathbf{A} = \mu_0\mathbf{J} + \frac{1}{c^2}\nabla\frac{\partial\phi}{\partial t}. \tag{12.150b}$$

12.7.1 Reciprocal Space

It proves convenient to introduce reciprocal space in which

$$\mathbf{E}(\mathbf{k}, t) = \frac{1}{(2\pi)^{3/2}}\int d^3 R\,\mathbf{E}(\mathbf{R}, t)e^{-i\mathbf{k}\cdot\mathbf{R}} \equiv \tilde{\mathbf{E}}. \tag{12.151}$$

It is easy to see that

$$\mathbf{E}(-\mathbf{k}, t) = \mathbf{E}^*(\mathbf{k}, t). \tag{12.152}$$

In reciprocal space, Maxwell equations become

$$i\mathbf{k}\cdot\tilde{\mathbf{E}} = \tilde{\rho}/\epsilon_0, \tag{12.153a}$$

$$i\mathbf{k}\cdot\tilde{\mathbf{B}} = 0, \tag{12.153b}$$

$$i\mathbf{k}\times\tilde{\mathbf{E}} = -\frac{\partial\tilde{\mathbf{B}}}{\partial t}, \tag{12.153c}$$

$$i\mathbf{k}\times\tilde{\mathbf{B}} = \frac{1}{c^2}\frac{\partial\tilde{\mathbf{E}}}{\partial t} + \mu_0\tilde{\mathbf{J}}, \tag{12.153d}$$

where

$$\tilde{f} \equiv \tilde{f}(\mathbf{k}, t) = \frac{1}{(2\pi)^{3/2}}\int d^3\mathbf{R}\, f(\mathbf{R}, t)e^{-i\mathbf{k}\cdot\mathbf{R}}. \tag{12.154}$$

The equation of continuity is

$$\frac{\partial\tilde{\rho}}{\partial t} + i\mathbf{k}\cdot\tilde{\mathbf{J}} = 0, \tag{12.155}$$

and the scalar and vector potentials are described via

$$\tilde{\mathbf{B}} = i\mathbf{k}\times\tilde{\mathbf{A}}, \tag{12.156}$$

$$\tilde{\mathbf{E}} = -\frac{\partial\tilde{\mathbf{A}}}{\partial t} - i\mathbf{k}\tilde{\phi}. \tag{12.157}$$

Gauge transformations are defined by

$$\tilde{\mathbf{A}} \to \tilde{\mathbf{A}} + i\mathbf{k}\tilde{\Xi}, \tag{12.158a}$$

$$\tilde{\phi} \to \tilde{\phi} - \frac{\partial \tilde{\Xi}}{\partial t}, \tag{12.158b}$$

leading to the equations

$$\frac{1}{c^2}\tilde{\mathbf{A}} + \mathbf{k}^2\tilde{\mathbf{A}} = \mu_0\tilde{\mathbf{J}} - i\mathbf{k}\left(i\mathbf{k}\cdot\tilde{\mathbf{A}} + \frac{1}{c^2}\frac{\partial\tilde{\phi}}{\partial t}\right), \tag{12.159a}$$

$$\mathbf{k}^2\tilde{\phi} = \tilde{\rho}/\epsilon_0 + i\mathbf{k}\cdot\frac{\partial\tilde{\mathbf{A}}}{\partial t}. \tag{12.159b}$$

The Lorentz gauge condition is

$$i\mathbf{k}\cdot\tilde{\mathbf{A}} + \frac{1}{c^2}\frac{\partial\tilde{\phi}}{\partial t} = 0, \tag{12.160}$$

and Coulomb gauge is defined by

$$\mathbf{k}\cdot\tilde{\mathbf{A}} = 0\,. \tag{12.161}$$

12.7.2 Longitudinal and Transverse Vector Fields

It is always possible to write a vector field $\mathbf{F}$ as

$$\mathbf{F} = \mathbf{F}_L + \mathbf{F}_T,$$

where

$$\nabla\times\mathbf{F}_L = 0;\;\; \nabla\cdot\mathbf{F}_T = 0\,. \tag{12.162}$$

In reciprocal space, the corresponding equations are

$$\tilde{\mathbf{F}} = \tilde{\mathbf{F}}_T + \tilde{\mathbf{F}}_L$$

and

$$i\mathbf{k}\times\tilde{\mathbf{F}}_L = 0,\;\; i\mathbf{k}\cdot\tilde{\mathbf{F}}_T = 0, \tag{12.163}$$

where

$$\tilde{\mathbf{F}}_L = \boldsymbol{\kappa}(\boldsymbol{\kappa}\cdot\tilde{\mathbf{F}}),\;\; (\tilde{\mathbf{F}}_T)_i = \sum_j\left(\delta_{i,j} - \frac{k_i k_j}{k^2}\right)(\tilde{\mathbf{F}})_j, \tag{12.164}$$

with

$$\boldsymbol{\kappa} = \mathbf{k}/|\mathbf{k}|\,. \tag{12.165}$$

Taking a Fourier transform of the transverse field and using the convolution theorem, we obtain

$$(\mathbf{F}_T)_i = \sum_j\int d^3\mathbf{R}'\delta^T_{ij}(\mathbf{R}-\mathbf{R}')\left[\mathbf{F}(\mathbf{R}')\right]_j, \tag{12.166}$$

where

$$\delta_{ij}^{T}(\mathbf{R}) = \frac{1}{(2\pi)^{3/2}} \int d^3\mathbf{k}\, e^{i\mathbf{k}\cdot\mathbf{R}} \left(\delta_{i,j} - \frac{k_i k_j}{k^2} \right) = \delta_{i,j}\delta(\mathbf{R}) - \frac{\partial^2}{\partial \mathbf{R}_i \partial \mathbf{R}_j} \frac{1}{\mathbf{R}} \tag{12.167}$$

is a *transverse Dirac delta function*. Although this decomposition is not relativistically invariant, it is still useful for quantization.

12.7.3 Transverse Electromagnetic Field

Since $\nabla \cdot \mathbf{B} = 0$, the magnetic field is purely transverse

$$\mathbf{B}_L = 0, \;\; \mathbf{B} = \mathbf{B}_T . \tag{12.168}$$

In decomposing the electric field, one associates the longitudinal component of the field with the free charges,

$$\mathbf{E}_L(\mathbf{R}) = \frac{1}{4\pi\epsilon_0} \int d^3R' \rho(\mathbf{R}', t) \frac{\mathbf{R} - \mathbf{R}'}{|\mathbf{R} - \mathbf{R}'|^3} \sim \frac{1}{4\pi\epsilon_0} \sum_{\alpha} q_\alpha \frac{\mathbf{R} - \mathbf{R}_\alpha}{|\mathbf{R} - \mathbf{R}_\alpha|^3}, \tag{12.169}$$

which, in reciprocal space, is

$$\tilde{\mathbf{E}}_L = \boldsymbol{\kappa}(\boldsymbol{\kappa} \cdot \tilde{\mathbf{E}}) = -i \frac{\tilde{\rho}}{\epsilon_0 k^2} \mathbf{k} . \tag{12.170}$$

In terms of the scalar and vector potentials, the electric field components are

$$\mathbf{E}_T = -\frac{\partial \mathbf{A}_T}{\partial t}, \tag{12.171}$$

$$\mathbf{E}_L = -\frac{\partial \mathbf{A}_L}{\partial t} - \nabla \phi, \tag{12.172}$$

since $\nabla\phi$ is purely longitudinal.

From this point onward, it will be convenient to work in Coulomb gauge, $\nabla \cdot \mathbf{A} = 0$. In Coulomb gauge, $\mathbf{A} = \mathbf{A}_T$ is purely transverse, and

$$\mathbf{E}_L = -\nabla \phi, \tag{12.173}$$

$$\phi(\mathbf{R}, t) = \frac{1}{4\pi\epsilon_0} \int d^3R' \frac{\rho(\mathbf{R}', t)}{|\mathbf{R} - \mathbf{R}'|} \sim \frac{1}{4\pi\epsilon_0} \sum_{\alpha} \frac{q_\alpha}{|\mathbf{R} - \mathbf{R}_\alpha|} . \tag{12.174}$$

In terms of these quantities, Maxwell equations become

$$\nabla \cdot \mathbf{E}_L = \rho/\epsilon_0, \tag{12.175a}$$

$$\nabla \times \mathbf{E}_T = -\frac{\partial \mathbf{B}}{\partial t}, \tag{12.175b}$$

$$\nabla \cdot \mathbf{B} = 0, \tag{12.175c}$$

$$\nabla \times \mathbf{B} = \frac{1}{c^2} \frac{\partial}{\partial t} (\mathbf{E}_L + \mathbf{E}_T) + \frac{1}{\epsilon_0 c^2} (\mathbf{J}_L + \mathbf{J}_T), \tag{12.175d}$$

$$i\mathbf{k}\cdot\tilde{\mathbf{E}}_L = \tilde{\rho}/\epsilon_0, \tag{12.176a}$$

$$i\mathbf{k}\times\tilde{\mathbf{E}}_T = -\frac{\partial\tilde{\mathbf{B}}}{\partial t}, \tag{12.176b}$$

$$\tilde{\mathbf{B}}_L = 0, \tag{12.176c}$$

$$i\mathbf{k}\times\tilde{\mathbf{B}} = \frac{1}{c^2}\frac{\partial\tilde{\mathbf{E}}_T}{\partial t} + \frac{1}{\epsilon_0 c^2}\tilde{\mathbf{J}}_T, \tag{12.176d}$$

and the energy is

$$H_{field} = H_L + H_T, \tag{12.177}$$

where

$$H_L = \frac{1}{2}\epsilon_0\int d^3k|\tilde{\mathbf{E}}_L|^2 = \frac{1}{2}\epsilon_0\int d^3R|\mathbf{E}_L|^2 = V_{coul}, \tag{12.178}$$

and

$$H_T = \frac{1}{2}\epsilon_0\int d^3k\left(|\tilde{\mathbf{E}}_T|^2 + c^2\tilde{\mathbf{B}}|^2\right) = \frac{1}{2}\epsilon_0\int d^3R\left(|\mathbf{E}_T|^2 + c^2|\mathbf{B}|^2\right). \tag{12.179}$$

[The longitudinal part of equation (12.175d) simply reproduces the equation of continuity.] The total energy of the charges plus fields is

$$H_{tot} = \frac{1}{2}\sum_\alpha m_\alpha\dot{R}_\alpha^2 + V_C + H_T, \tag{12.180}$$

where the Coulomb potential for an ensemble of point charges is given by

$$V_C = \frac{1}{8\pi\epsilon_0}\sum_{\alpha,\beta\neq\alpha}\frac{q_\alpha q_\beta}{|\mathbf{R}_\alpha - \mathbf{R}_\beta|}, \tag{12.181}$$

neglecting the self-energy of each charge.

Similarly, for the momentum,

$$\mathbf{P}_L = \epsilon_0\int d^3R\,\mathbf{E}_L\times\mathbf{B} = \epsilon_0\int d^3k\,\tilde{\mathbf{E}}_L^*\times\tilde{\mathbf{B}}. \tag{12.182}$$

Taking into account equation (12.156) and equation (12.170), we obtain

$$\mathbf{P}_L = \int d^3R\,\rho\mathbf{A} = \sum_\alpha q_\alpha\mathbf{A}(\mathbf{R}_\alpha), \tag{12.183}$$

$$\mathbf{P}_T = \epsilon_0\int d^3R\,\mathbf{E}_T\times\mathbf{B}, \tag{12.184}$$

$$\mathbf{P}_{tot} = \sum_\alpha m_\alpha\dot{\mathbf{R}}_\alpha + q_\alpha\mathbf{A}(\mathbf{R}_\alpha) + \mathbf{P}_T, \tag{12.185}$$

$$H_{tot} = \sum_\alpha\frac{1}{2m_\alpha}\left[\mathbf{p}_\alpha - q_\alpha\mathbf{A}(\mathbf{R}_\alpha)\right]^2 + V_C + H_T. \tag{12.186}$$

We now define

$$\boldsymbol{\alpha}(\mathbf{k}, t) = -\frac{i}{2N(\mathbf{k})}\left(\tilde{\mathbf{E}}_T - c\boldsymbol{\kappa} \times \tilde{\mathbf{B}}\right) \tag{12.187}$$

and

$$\boldsymbol{\beta}(\mathbf{k}, t) = -\frac{i}{2N(\mathbf{k})}\left(\tilde{\mathbf{E}}_T + c\boldsymbol{\kappa} \times \tilde{\mathbf{B}}\right) = -\boldsymbol{\alpha}^*(-\mathbf{k}, t), \tag{12.188}$$

where $N(\mathbf{k})$ is a normalization factor that will be fixed at a later time. Using equations (12.187) and (12.188), we find

$$\tilde{\mathbf{E}}_T = i N(\mathbf{k}) \left[\boldsymbol{\alpha}(\mathbf{k}, t) - \boldsymbol{\alpha}^*(-\mathbf{k}, t)\right], \tag{12.189}$$

$$\tilde{\mathbf{B}} = \frac{i}{c} N(\mathbf{k}) \left[\boldsymbol{\kappa} \times \boldsymbol{\alpha}(\mathbf{k}, t) + \boldsymbol{\kappa} \times \boldsymbol{\alpha}^*(-\mathbf{k}, t)\right], \tag{12.190}$$

$$\frac{\partial \boldsymbol{\alpha}(\mathbf{k}, t)}{\partial t} + i\omega\boldsymbol{\alpha}(\mathbf{k}, t) = \frac{i}{2\epsilon_0 N(\mathbf{k})}\tilde{\mathbf{J}}_T, \tag{12.191}$$

where $\omega = ck$.

Including two polarizations $\boldsymbol{\epsilon}_{\mathbf{k}}$ for each $\mathbf{k}$, we write

$$\boldsymbol{\alpha}(\mathbf{k}, t) = \sum_{\epsilon_{\mathbf{k}}} \boldsymbol{\epsilon}_{\mathbf{k}} \alpha_\epsilon(\mathbf{k}, t), \tag{12.192}$$

$$\frac{\partial \alpha_\epsilon(\mathbf{k}, t)}{\partial t} + i\omega\alpha_\epsilon(\mathbf{k}, t) = \frac{i}{2\epsilon_0 N(\mathbf{k})}\boldsymbol{\epsilon}_{\mathbf{k}} \cdot \tilde{\mathbf{J}}. \tag{12.193}$$

If we introduce

$$\begin{aligned} \boldsymbol{\alpha} &= \boldsymbol{\alpha}(\mathbf{k}, t), \\ \boldsymbol{\alpha}_- &= \boldsymbol{\alpha}(-\mathbf{k}, t), \end{aligned} \tag{12.194}$$

then the transverse field energy can be written as

$$\begin{aligned} H_T &= \frac{1}{2}\epsilon_0 \int d^3\mathbf{k} \left(|\mathbf{E}_T|^2 + c^2\mathbf{B}|^2\right) \\ &= \epsilon_0 \int d^3\mathbf{k}\, N^2(\mathbf{k}) [\alpha^*_{\epsilon_{\mathbf{k}}} \cdot \alpha_{\epsilon_{\mathbf{k}}} + (\alpha_-)_{\epsilon_{-\mathbf{k}}} (\alpha_-)^*_{\epsilon_{-\mathbf{k}}}] \\ &= \epsilon_0 \sum_{\epsilon_{\mathbf{k}}} \int d^3\mathbf{k}\, N^2(\mathbf{k}) \left(\alpha^*_{\epsilon_{\mathbf{k}}} \alpha_{\epsilon_{\mathbf{k}}} + \alpha_{\epsilon_{\mathbf{k}}} \alpha^*_{\epsilon_{\mathbf{k}}}\right). \end{aligned} \tag{12.195}$$

Even though everything is classical to this point, we now choose the normalization constant in such a way that it leads to an energy of $\hbar\omega$ in each mode of the field (you can think of $\hbar$ as some undefined constant)—that is, we set

$$N(\mathbf{k}) = \left(\frac{\hbar\omega}{2\epsilon_0}\right)^{1/2} \left(\frac{L}{2\pi}\right)^{3/2}, \tag{12.196}$$

where $L^3 = \mathcal{V}$ is the quantization volume in space. Then,

$$H_T = \left(\frac{L}{2\pi}\right)^3 \int d^3\mathbf{k} \sum_{\epsilon_k} \frac{1}{2}\hbar\omega \left(\alpha^*_{\epsilon_k}\alpha_{\epsilon_k} + \alpha_{\epsilon_k}\alpha^*_{\epsilon_k}\right), \tag{12.197}$$

$$\mathbf{P}_T = \left(\frac{L}{2\pi}\right)^3 \int d^3\mathbf{k} \sum_{\epsilon_k} \frac{1}{2}\hbar\mathbf{k} \left(\alpha^*_{\epsilon_k}\alpha_{\epsilon_k} + \alpha_{\epsilon_k}\alpha^*_{\epsilon_k}\right), \tag{12.198}$$

$$\mathbf{E}_T = i \int d^3\mathbf{k} \sum_{\epsilon_k} E_\omega (\alpha_{\epsilon_k}\boldsymbol{\epsilon}_k e^{i\mathbf{k}\cdot\mathbf{R}} - \alpha^*_{\epsilon_k}\boldsymbol{\epsilon}_k e^{-i\mathbf{k}\cdot\mathbf{R}}), \tag{12.199}$$

$$\mathbf{B} = i \int d^3\mathbf{k} \sum_{\epsilon_k} B_\omega (\alpha_{\epsilon_k}\boldsymbol{\kappa} \times \boldsymbol{\epsilon}_k e^{i\mathbf{k}\cdot\mathbf{R}} - \alpha^*_{\epsilon_k}\boldsymbol{\kappa} \times \boldsymbol{\epsilon}_k e^{-i\mathbf{k}\cdot\mathbf{R}}); \tag{12.200}$$

$$\mathbf{A} = i \int d^3\mathbf{k} \sum_{\epsilon_k} A_\omega (\alpha_{\epsilon_k}\boldsymbol{\epsilon}_k e^{i\mathbf{k}\cdot\mathbf{R}} + \alpha^*_{\epsilon_k}\boldsymbol{\epsilon}_k e^{-i\mathbf{k}\cdot\mathbf{R}}), \tag{12.201}$$

where

$$E_\omega = \sqrt{\frac{\hbar\omega\mathcal{V}}{2\epsilon_0(2\pi)^3}}, \qquad B_\omega = \frac{E_\omega}{c}, \qquad A_\omega = \frac{E_\omega}{\omega}. \tag{12.202}$$

We can go over to discrete variables by using the prescription (12.25) with $\mathbf{k}_j$ defined by equations (12.1) and, (12.2). In this manner, we obtain

$$H_T = \sum_j \frac{1}{2}\hbar\omega_j \left(\alpha^*_j\alpha_j + \alpha_j\alpha^*_j\right), \tag{12.203}$$

$$\mathbf{P}_T = \sum_j \frac{1}{2}\hbar\mathbf{k}_j \left(\alpha^*_j\alpha_j + \alpha_j\alpha^*_j\right), \tag{12.204}$$

$$\mathbf{A} = \sum_j A_j (\alpha_j\boldsymbol{\epsilon}_j e^{i\mathbf{k}_j\cdot\mathbf{R}} + \alpha^*_j\boldsymbol{\epsilon}_j e^{-i\mathbf{k}_j\cdot\mathbf{R}}), \tag{12.205}$$

$$\mathbf{E}_T = i \sum_j E_j (\alpha_j\boldsymbol{\epsilon}_j e^{i\mathbf{k}_j\mathbf{R}} - \alpha^*_j\boldsymbol{\epsilon}_j e^{-i\mathbf{k}_j\mathbf{R}}), \tag{12.206}$$

$$\mathbf{B} = i \sum_j B_j (\alpha_j\boldsymbol{\kappa}_j \times \boldsymbol{\epsilon}_j e^{i\mathbf{k}_j\cdot\mathbf{R}} - \alpha^*_j\boldsymbol{\kappa}_j \times \boldsymbol{\epsilon}_j e^{-i\mathbf{k}_j\cdot\mathbf{R}}), \tag{12.207}$$

where $\alpha_j \equiv \alpha_{\mathbf{k}_j,\epsilon_{k_j}}$, and

$$E_j = \sqrt{\frac{\hbar\omega_j}{2\epsilon_0\mathcal{V}}}, \quad B_j = \frac{E_j}{c}, \quad A_j = \frac{E_j}{\omega_j}, \quad \boldsymbol{\kappa}_j = \mathbf{k}_j/|\mathbf{k}_j|. \tag{12.208}$$

The α_j obey the equation of motion

$$\dot{\alpha}_j + i\omega_j\alpha_j = \frac{i}{\sqrt{2\epsilon_0\hbar\omega_j}}\tilde{J}_j, \tag{12.209}$$

with

$$\tilde{J}_j = \frac{1}{\sqrt{\mathcal{V}}}\int d\mathbf{R}\, e^{-i\mathbf{k}_j\cdot\mathbf{R}}\boldsymbol{\epsilon}_j\cdot\mathbf{J}(\mathbf{R}, \mathbf{t}). \tag{12.210}$$

12.7.4 Free Field

In the absence of sources, equation (12.209) becomes

$$\dot{\alpha}_j + i\omega_j\alpha_j = 0. \tag{12.211}$$

Therefore,

$$\alpha_j = \alpha_j(0)e^{-i\omega_j t}, \tag{12.212a}$$

$$\alpha_j^* = \alpha_j^*(0)e^{i\omega_j t}. \tag{12.212b}$$

Using equation (12.203), we then find

$$\frac{\partial H_T}{\partial\alpha_j} = \hbar\omega_j\alpha_j^* = -i\hbar\dot{\alpha}_j^*, \tag{12.213a}$$

$$\frac{1}{i\hbar}\frac{\partial H_T}{\partial\alpha_j^*} = -\frac{i}{\hbar}\frac{\partial H_T}{\partial\alpha_j^*} = -i\omega_j\alpha_j = \dot{\alpha}_j. \tag{12.213b}$$

Therefore, α_j and $i\hbar\alpha_j^*$ are canonical variables. Quantization of the free field is then carried out using the prescription $\alpha_i \to a_i$, $\alpha_i^* \to a_i^\dagger$, with

$$[a_i, a_j] = \left[a_i^\dagger, a_j^\dagger\right] = 0,\ \left[a_i, a_j^\dagger\right] = \delta_{i,j}. \tag{12.214}$$

In this manner, we arrive at equations (12.8) to (12.13) of the text. The same quantization procedure in the presence of sources leads to the correct equations of motion for both the charges and the fields.

Problems

1. Given the Hamiltonian $H = \sum_i(1/2)\hbar\omega_i(\alpha_i^*\alpha_i + \alpha_i\alpha_i^*)$, where α_i satisfies $\dot{\alpha}_i + i\omega_i\alpha_i = 0$, show that two sets of canonical variables are (α_i and $i\hbar\alpha_i^*$) and [$q_i = (\hbar/2\omega_i)^{1/2}(\alpha_i + \alpha_i^*)$ and $p_i = i(\hbar\omega_i/2)^{1/2}(\alpha_i^* - \alpha_i)$]. Use these results to write H in two distinct quantized forms.
2. Prove that $[E_x(\mathbf{R}), B_y(\mathbf{R}')] = -(i\hbar/\epsilon_0)\partial/\partial z[\delta(\mathbf{R}-\mathbf{R}')]$—that is, the fields do not commute.
3. Evaluate $\langle\mathbf{E}(\mathbf{R}, t)\rangle$ and $\langle\mathbf{E}(\mathbf{R}, t)^2\rangle$ for a single-mode coherent state and a pure n state, using the Heisenberg operators for the fields. Also calculate ΔE for these states.

4. Prove that for the operators

$$A_1 = a_1 \cos(\theta/2) + a_2 \sin(\theta/2),$$
$$A_2 = a_2 \cos(\theta/2) - a_1 \sin(\theta/2),$$

$\Delta A_1 = e^{-r}/2$, $\Delta A_1 = e^{r}/2$ in a single-mode TPCS.
5. Derive equations (12.126).
6. Prove that $a^\dagger$ has no normalizable eigenstates.
7. Prove that $[\hat{n}, \widehat{\exp(i\phi)}] = -\widehat{\exp(i\phi)}$; $[\hat{n}, \widehat{\exp(-i\phi)}] = \widehat{\exp(-i\phi)}$; $[\hat{n}, \widehat{\cos(\phi)}] = -i\widehat{\sin\phi}$; $[\hat{n}, \widehat{\sin\phi}] = i\widehat{\cos(\phi)}$, and that

$$\widehat{\exp(i\phi)}|n\rangle = \frac{\sqrt{n}}{(n+1)^{1/2}}|n-1\rangle, \qquad \widehat{\exp(-i\phi)}|n\rangle = |n+1\rangle.$$

8. Prove that, in a single-mode coherent state with $|\alpha| \gg 1$,

$$\langle\widehat{\cos\phi}\rangle_\alpha = \cos\theta\left(1 - \frac{1}{8|\alpha|^2}\right),$$
$$\langle\widehat{\cos^2\phi}\rangle_\alpha = \cos^2\theta - \frac{\cos 2\theta}{4|\alpha|^2},$$
$$\langle\Delta(\cos\phi)\,\Delta n\rangle \approx \frac{1}{2}\sin\theta \approx \frac{1}{2}\langle\widehat{\sin\phi}\rangle,$$

where $\alpha = |\alpha|e^{i\theta}$. In other words, show that both the intensity and phase are fairly well-defined in a coherent state whose average energy is much greater than $\hbar\omega$. You will need to use the following asymptotic expansions:

$$\sum_{n=0}^{\infty} \frac{|\alpha|^{2n}}{\sqrt{n+1}n!} = \frac{e^{|\alpha|^2}}{|\alpha|}\left(1 - \frac{1}{8|\alpha|^2}\right),$$
$$\sum_{n=0}^{\infty} \frac{|\alpha|^{2n}}{\sqrt{(n+1)(n+2)}n!} = \frac{e^{|\alpha|^2}}{|\alpha|^2}\left(1 - \frac{1}{2|\alpha|^2}\right),$$

valid for $|\alpha| \gg 1$.
9. Prove that

$$|n\rangle\langle n| = \pi^{-2}\int d^2\alpha\, d^2\beta\, e^{-[|\alpha|^2+|\beta|^2]/2}(n!)^{-1}(\alpha^*\beta)^n|\alpha\rangle\langle\beta|$$

and

$$|n\rangle\langle n| = \int P(\alpha)|\alpha\rangle\langle\alpha|d^2\alpha, \qquad \text{with} \qquad P(\alpha) = \frac{n!e^{r^2}}{2\pi r(2n)!}\left(-\frac{\partial}{\partial r}\right)^{2n}\delta(r),$$

where $\alpha = |r|e^{i\theta}$. How can this operator be either diagonal or nondiagonal in the α basis?

A two-photon coherent state (TPCS) is defined by

$$b|z,\beta\rangle = \beta|z,\beta\rangle,$$

where

$$b = U(z)a[U(z)]^\dagger = a\cosh(r) + a^\dagger e^{i\theta}\sinh(r),$$
$$U(z) = e^{(z^* a^2 - z a^{\dagger 2})/2}$$

and $z = re^{i\theta}$.

In the text, it is shown that $|z,\beta\rangle = U(z)|\beta\rangle$, where $|\beta\rangle$ is now the normal coherent state defined through $a|\beta\rangle = \beta|\beta\rangle$, $|\beta\rangle = D(\beta)|0\rangle$, $D(\beta) = e^{\beta a^\dagger - \beta^* a}$, and $[D(\beta)]^\dagger a D(\beta) = a + \beta$.

10. Using the facts that

$$D(\beta)D(\alpha) = e^{(\beta\alpha^* - \alpha\beta^*)/2} D(\alpha+\beta),$$

$$[U_L(z)]^\dagger D(\beta) U_L(z) = D(\beta'),$$

where

$$\beta' = \beta\cosh(r) + \beta^*\sinh(r)e^{i\theta},$$

prove that

$$\left|z, \alpha + \beta'\right\rangle = e^{-(\beta'\alpha^* - \beta'^*\alpha)} D(\beta)|z,\alpha\rangle.$$

In other words, $D(\beta)\,|z,\alpha\rangle$ is also a TPCS. In particular, show that

$$|\alpha, z\rangle \equiv |z,\alpha'\rangle = D(\alpha)|z,0\rangle = D(\alpha)U_L(z)|0\rangle,$$

where

$$\alpha' = \alpha\cosh(r) + \alpha^*\sinh(r)e^{i\theta}.$$

11. For the states $|z,\alpha'\rangle$ defined in problem 10, prove that

$$\Delta n^2 = |\alpha''|^2 + 2\sinh^2(r)\cosh^2(r), \qquad \alpha'' = \alpha\cosh(r) - \alpha^*\sinh(r)e^{i\theta}.$$

{Hint: Use of the relationship $[D(\alpha)]^\dagger a D(\alpha) = a + \alpha$ and the fact that $U_L(-z) a U_L(z) = b' = b(-z) = \mu a - \nu a^\dagger$ may save you some time.}

12. Calculate $\langle \mathbf{E}(\mathbf{R},t)\rangle$, $\langle \mathbf{E}(\mathbf{R},t)^2\rangle$, and ΔE for the single-mode states defined in problem 10, using the Heisenberg operators for the fields. Compare the results with those of problem 3.

References

[1] C. Cohen-Tannoudji, J. Dupont-Roc, and G. Grynberg, *Photons and Atoms—Introduction to Quantum Electrodynamics* (Wiley-Interscience, New York, 1989), chaps. 1 and 2.

[2] R. G. Glauber, *The Quantum Theory of Optical Coherence*, Physical Review **130**, 2529–2539 (1963); R. G. Glauber *Coherent and incoherent states of the radiation field*, Physical Review **131**, 2766–2788 (1963).

[3] D. Stoler, *Equivalence classes of minimum uncertainty packets*, Physical Review D **1**, 3217–3219 (1970); D **4**, 1925–1926 (1971).

[4] H. P. Yuen, *Two-photon coherent state of the radiation field*, Physical Review A **13**, 2226–2243 (1976).

[5] D. F. Walls, *Squeezed states of light*, Nature **206**, 141–146 (1983).

[6] H. J. Kimble and D. F. Walls, Eds., *Squeezed States of the Electromagnetic Field*, Special Issue of Journal of the Optical Society of America B **4**, 1449–1741 (1987); R. Loudon and P. L. Knight, Eds., Special Issue of the Journal of Modern Optics on Squeezed Light, **34**, 707–1020 (1987).

[7] S. Reynaud, A Heidmann, E. Giacobino, and C. Fabre, *Quantum fluctuations in optical systems*, in *Progress in Optics*, edited by E. Wolf vol. XXX, (North-Holland, Amsterdam, 1992) 3–85.

[8] For a more recent article with promising applications to detection of gravitational waves, see H. Vahlbruch, M. Mehmet, S. Chelkowski, B. Hage, A. Franzen, N. Lastzka, S. Goßler, K. Danzmann, and R. Schnabel, *Observation of squeezed light with 10-dB quantum-noise reduction*, Physical Review Letters **100**, 033602, 1–4 (2008).

[9] L. Mandel and E. Wolf, *Optical Coherence and Quantum Optics* (Cambridge University Press, Cambridge, UK, 1995), sec. 10.7.4.

[10] S. M. Barnett and J. A. Vaccarro Eds., *The Quantum Phase Operator: A Review* (CRC Press, Taylor and Francis Group, Boca Raton, FL, 2007).

[11] D. T. Pegg and S. M. Barnett, *Phase properties of the quantized single-mode electromagnetic field*, Physical Review A **39**, 1665–1675 (1989).

[12] L. Susskind and J. Glogower, *Quantum mechanical phase and time operator*, Physics (Long Island City, NY) **1**, 49–61 (1964).

[13] R. Loudon, *The Quantum Theory of Light*, 2nd ed. (Oxford University Press, Oxford, UK, 1983), sec. 4.8.

[14] L. Mandel and E. Wolf, *Optical Coherence and Quantum Optics* (Cambridge University Press, Cambridge, UK, 1995), sec. 10.7.

[15] See, for example, S. Haroche and J. M. Raimond, *Manipulation of nonclassical field states in a cavity by atom interferometry*, in *Advances in Atomic, Molecular, and Optical Physics, Supplement I*, edited by P. Berman (Academic Press, New York, 1994), pp. 123–170.

Bibliography

See the books on quantum optics listed in chapter 1.

13

Coherence Properties of the Electric Field

13.1 Coherence: Some General Concepts

Now that we have discussed quantized states of the radiation field, it is probably a good time to consider ways in which it is possible to characterize optical fields. The coherence properties of fields are discussed extensively in Mandel and Wolf [1]. Here, we adopt a simplified approach but still attempt to touch upon many of the essential components needed to piece together a theory of coherence.

13.1.1 Time versus Ensemble Averages

The coherence properties of any time-dependent process, denoted here in some generic fashion by the function $f(t)$, are often described in terms of a quantity $F(t, \tau)$ defined by

$$F(t, \tau) = \langle f(t) f(t + \tau) \rangle / \left\langle f^2(t) \right\rangle . \tag{13.1}$$

This type of definition of coherence is appealing, since it compares the function with itself at different times; clearly, if the function $f(t)$ is constant, then $F(t, \tau) = 1$, and the function is perfectly coherent according to this definition.

The confusion associated with equation (13.1) is hidden in the brackets $\langle \ldots \rangle$. Exactly what do these brackets mean? For the time being, let us restrict the discussion to classical rather than quantum-mechanical systems. Most likely, you have heard it told that a monkey typing at random would eventually produce all the works of literature, given an infinite time. (Alas, if only we all had an infinite amount of time.) On the other hand, you could also achieve the same result with an infinite number of monkeys typing for a much shorter time. The two processes illustrate the difference between a *time average* and an *ensemble average*. Often, the *ergodic theorem* is invoked to guarantee the equivalence of both averages. If a system has enough time to experience all possible random configurations of its components, there is no difference between a time and an ensemble average. For finite times, and experiments are usually carried out over a finite time, there can be a fundamental difference between time and ensemble averages. We illustrate this concept by considering a simple physical system.

Imagine that a two-level atom, whose transition frequency is ω_0 in the absence of interactions, is placed in a bath of stationary perturber atoms located at random positions in space. The net effect of each perturber atom is to alter the transition frequency of the two-level atom by an amount that depends on its separation from the atom (e.g., shifts resulting from van der Waals interactions). The atom is prepared in a superposition of its ground and excited states at $t = 0$ and allowed to radiate. To simplify matters, we neglect any effects of spontaneous decay—the atom acts as an undamped oscillator. The question is, "What is the line width associated with the oscillator?"

Since the perturber atoms are stationary, the only effect of the bath, in this limit, is to change the transition frequency of the atom. It still radiates as an undamped oscillator—in other words, the radiation is perfectly coherent with "zero" line width. On the other hand, if we place a large number of these oscillators in the sample and excite them all at $t = 0$, the situation is similar to the one we encountered in free polarization decay—owing to the inhomogeneous broadening of the sample (different atoms have their frequencies shifted by a different amount by the bath), the coherent signal decays very rapidly, and the line width is the inhomogeneous line width of the sample. In this instance, the difference between a time average for a single atom and an ensemble average is dramatic. The research field of single-molecule spectroscopy [2] depends critically on the fact that the response of a single molecule is totally different from the response of an ensemble of such atoms embedded in a host material.

The situation changes somewhat if we allow the bath atoms to move. Now if we wait for a sufficiently long time, all possible bath configurations will have been experienced by the (stationary) single two-level atom. In this case, a time average taken over a sufficiently long time reproduces the same result as an ensemble average of a large number of these atoms placed in the bath at random.

The output field of a laser cavity is another case in point. The laser field is close to that of a coherent state with a fairly well-defined phase. But this phase must be a *global* phase, since the laser starts from spontaneous emission. In other words, if one turns on the laser at different times, the global phase of the output would be different. An ensemble average over different realizations of the laser output yields an average field amplitude at any time that vanishes, whereas the output field amplitude for a laser can be considered almost as a classical field.

It should be clear from this discussion that the results of a given experiment must be examined on a case-by-case basis to see what type of average enters. In most cases, such as atoms in a vapor cell, one encounters ensemble rather than time averages; however, if one is truly following a single realization of a stochastic process, such as in single-molecule spectroscopy or the output of a laser cavity, it is a time average that is needed.

The lesson to be learned here is that some care must be taken in evaluating the $\langle \ldots \rangle$ in equation (13.1). Even in quantum mechanics, where you might think that the $\langle \ldots \rangle$ refers to a quantum-mechanical average in the usual sense, the situation is not so simple. The reason for this is that it is usually all but impossible to carry out the average over a complicated atom–bath system, and some types of approximations, corresponding to ensemble averages, are often invoked. With this warning in mind, let us now turn our attention to optical fields. We are interested in determining what parameters are needed to characterize these fields.

13.1.2 Classical Fields

For monochromatic fields in vacuum, it is necessary to specify only the frequency and amplitude of the field. One could also specify a phase, but this assignment is equivalent to choosing a time origin. It is natural to think of a monochromatic field as a coherent field, since its amplitude and phase are well-defined. But what happens if the field consists of a sum of two monochromatic fields? Of course, we must now specify not only the amplitude and frequency of each field, but also the relative phase of the two fields. Is this superposition state of two fields coherent? The answer to this question depends on whether we consider an ensemble or a time average. In this case, as for any complex wave form, an ensemble average has no meaning, since we are not considering fluctuations in the *sources* of the field. In this sense, the field is still a coherent field, since it is totally deterministic. On the other hand, you will see that the field is *not* coherent, according to the standard definition (13.1), when the brackets in that equation represent a time average.

The situation can change if we take into account any fluctuations in the *sources* giving rise to the field. Imagine that the field is produced by an ensemble of oscillators having the *same* natural frequency but random, *constant* phases. Since the phases are constant, if one uses the time-averaged definition, the field radiated by these oscillators would be purely coherent. That is, for a *specific* realization of the phases, the output is perfectly coherent. If one takes an ensemble average, however, corresponding to repeating the experiment many times with *different* sets of phases, the coherence properties of the radiation field are altered. Examples are given in the following section. For phases that are fluctuating functions of time, the time- and ensemble-averaged results are identical for sufficiently long sampling times.

13.1.3 Quantized Fields

In some sense, the description of quantized fields is more straightforward than classical fields. As in any quantum system, the state vector provides a complete description of the system. However, the quantum state of a field that fluctuates as a result of its interactions with its environment is an extremely complex entity. The radiation field can become entangled with the environment owing to these interactions. If we do not observe the environment, we are, in effect, tracing over the environment states. In this limit, it is the reduced density matrix of the field that provides a complete description of our knowledge of the field. Thus, in discussing quantized fields, we concentrate on the density matrix or *photon statistics* (density matrix elements of the field in the number representation) of the field. You will see that there are properties of the quantized field that can distinguish it in an unambiguous way from classical fields. You have already seen this for atomic states, where negative values of the density matrix elements in the Wigner representation are a signature of quantum effects.

13.2 Classical Fields: Correlation Functions

13.2.1 First-Order Correlation Function

We limit our discussion to one polarization component of an electric field. For the moment, we neglect the position dependence and look at the field as a function of time at $\mathbf{R} = 0$, setting $\mathbf{E}(\mathbf{R}, t) = \mathbf{E}(0, t)$. It is convenient to expand $\mathbf{E}(0, t)$ in terms of positive and negative frequency components as

$$\mathbf{E}(0, t) = \boldsymbol{\epsilon}\left[E^+(t) + E^-(t)\right], \tag{13.2}$$

where $\boldsymbol{\epsilon}$ is the field polarization

$$E^+(t) = \frac{1}{\sqrt{2\pi}} \int_{-\infty}^{\infty} d\omega \tilde{E}(\omega) e^{-i\omega t}, \tag{13.3}$$

and $E^- = (E^+)^*$. The separation into positive and negative frequency components makes sense only for $\omega > 0$, but we can formally allow for negative values of ω by setting $\tilde{E}(\omega) = 0$ for $\omega < 0$. The quantity $\tilde{E}(\omega)$ is the Fourier transform of the positive frequency component of the field.

The *power spectrum* of the field is proportional to the absolute square of this quantity,

$$\begin{aligned} |\tilde{E}(\omega)|^2 &= \frac{1}{2\pi} \int_{-\infty}^{\infty} dt' E^+(t') e^{i\omega t'} \int_{-\infty}^{\infty} dt E^-(t) e^{-i\omega t} \\ &= \frac{1}{2\pi} \int_{-\infty}^{\infty} d\tau \int_{-\infty}^{\infty} dt E^-(t) E^+(t+\tau) e^{i\omega\tau}. \end{aligned} \tag{13.4}$$

If we define

$$h(\tau) \equiv \langle E^-(t) E^+(t+\tau) \rangle = \lim_{T\to\infty} \frac{1}{T} \int_{-T/2}^{T/2} dt E^-(t) E^+(t+\tau), \tag{13.5}$$

then

$$|\tilde{E}(\omega)|^2 = \frac{T}{2\pi} \int_{-\infty}^{\infty} d\tau h(\tau) e^{i\omega\tau}. \tag{13.6}$$

Note that, at this point, $h(\tau)$ represents a time average.

The power spectrum $F(\omega)$ is defined as

$$F(\omega) \equiv \frac{|\tilde{E}(\omega)|^2}{\displaystyle\int_{-\infty}^{\infty} d\omega |\tilde{E}(\omega)|^2} = \frac{|\tilde{E}(\omega)|^2}{\displaystyle\int_{-\infty}^{\infty} dt |E^+(t)|^2} = \frac{|\tilde{E}(\omega)|^2}{T\langle E^-(t) E^+(t) \rangle}, \tag{13.7}$$

such that

$$F(\omega) = \frac{1}{2\pi} \int_{-\infty}^{\infty} d\tau g^{(1)}(\tau) e^{i\omega\tau}, \tag{13.8}$$

where

$$g^{(1)}(\tau) = \frac{\langle E^-(t) E^+(t+\tau) \rangle}{\langle E^-(t) E^+(t) \rangle} \tag{13.9}$$

is the *degree of first-order temporal coherence* or the *first-order correlation function.* If $\left|g^{(1)}(\tau)\right| = 1$, the field is said to be *first-order coherent.* The power spectrum and first-order correlation functions are Fourier transforms of one another.

From equation (13.5), it follows that

$$\langle E^{-}(t)E^{+}(t-\tau)\rangle = \langle E^{-}(t+\tau)E^{+}(t)\rangle = \langle E^{-}(t)E^{+}(t+\tau)\rangle^{*} . \tag{13.10}$$

Using equations (13.9) and (13.10), we can show that

$$g^{(1)}(-\tau) = \left[g^{(1)}(\tau)\right]^{*} \tag{13.11}$$

and

$$|g^{(1)}(\tau)| \leq 1 . \tag{13.12}$$

Moreover, from equation (13.11), we can derive

$$\begin{aligned}\int_{-\infty}^{\infty} d\tau g^{(1)}(\tau)e^{i\omega\tau} &= \int_{-\infty}^{0} d\tau g^{(1)}(\tau)e^{i\omega\tau} + \int_{0}^{\infty} d\tau g^{(1)}(\tau)e^{i\omega\tau} \\ &= \int_{0}^{\infty} d\tau\{[g^{(1)}(\tau)]^{*} + g^{(1)}(\tau)\}e^{i\omega\tau} \\ &= 2\mathrm{Re}\int_{0}^{\infty} d\tau g^{(1)}(\tau)e^{i\omega\tau} ,\end{aligned} \tag{13.13}$$

and

$$F(\omega) = \frac{1}{\pi}\mathrm{Re}\int_{0}^{\infty} d\tau g^{(1)}(\tau)e^{i\omega\tau} . \tag{13.14}$$

The first-order correlation function is a measure of the coherence of the field amplitudes—in other words, it is proportional to $\langle E^{-}(t)E^{+}(t+\tau)\rangle$. Clearly, if $E^{-}(t)$ and $E^{+}(t+\tau)$ are uncorrelated, then $g^{(1)}(\tau) \sim 0$. One might think that $g^{(1)}(\infty) \sim 0$, and this is generally the case. However, if $E^{+}(t)$ is a periodic function of time (e.g., a sum of a *finite* number of monochromatic fields, or an infinite train of pulses), then $g^{(1)}(\tau)$ need not vanish as $t \sim 0$. One can say that the *correlation time* τ_c of such fields is infinite, even if the magnitude of the first-order correlation function is not equal to unity. Let us look at some examples, in which we take $\tau > 0$.

Monochromatic field. A monochromatic field is characterized by

$$E^{+}(t) = \frac{1}{2}E_0 e^{-i\omega_L t} . \tag{13.15}$$

As a consequence, one finds from equations (13.9) and (13.14) that

$$g^{(1)}(\tau) = e^{-i\omega_L\tau} , \tag{13.16}$$

$$F(\omega) = \delta(\omega - \omega_L) . \tag{13.17}$$

A monochromatic field is first-order coherent.

Two-and multifrequency fields. Now consider a field consisting of the sum of two monochromatic fields,

$$E^{+}(t) = \frac{1}{2}\left(E_1 e^{-i\omega_1 t} + E_2 e^{-i\omega_2 t}\right) . \tag{13.18}$$

You might think that this sum of monochromatic fields is first-order coherent, but

$$g^{(1)}(\tau) = \frac{\langle E^-(t)E^+(t+\tau)\rangle}{\langle E^-(t)E^+(t)\rangle} = \frac{|E_1|^2 e^{-i\omega_1\tau} + |E_2|^2 e^{-i\omega_2\tau}}{|E_1|^2 + |E_2|^2}. \tag{13.19}$$

Clearly, $\left|g^{(1)}(\tau)\right| \neq 1$, even for equal field amplitudes. Although the field has no fluctuations, we find that the field is not first-order coherent, according to the time-averaged definition. On the other hand, since $\left|g^{(1)}(\tau)\right|$ does not tend to zero for τ greater than some correlation time τ_c, the correlation *time* of this field is infinite.

If there is a *continuous* distribution of frequencies $I(\omega) = \left|\tilde{E}(\omega)\right|^2$, then equation (13.19) yields

$$g^{(1)}(\tau) = \frac{\int d\omega I(\omega)\, e^{-i\omega\tau}}{\int d\omega I(\omega)}. \tag{13.20}$$

For example, if $\left|\tilde{E}(\omega)\right|^2$ is a Gaussian centered at ω_L with some characteristic width $\Delta\omega$, the $\left|g^{(1)}(\tau)\right|$ is Gaussian function of τ^2 having a width of order $\Delta\omega^{-1}$. Inhomogeneous broadening leads to a decay of the first-order correlation function. In this case, the time average is equivalent to an ensemble average over sources having a distribution of frequencies.

Field produced by oscillators having the same frequency and constant random phases. In this case,

$$E^+(t) = \frac{1}{2}\sum_j E_0 e^{-i\omega_L t + i\phi_j} = \frac{1}{2} A E_0 e^{-i\omega_L t}, \tag{13.21}$$

where $A = \sum_j e^{i\phi_j}$. Since the phases are time-independent, A is also time-independent, and one finds that the first-order correlation function is equal to $e^{-i\omega_L\tau}$, implying that $\left|g^{(1)}(\tau)\right| = 1$; the field is first-order coherent. Nevertheless, there *is* a difference between time and ensemble averages in this case. There is no time variation of A, implying that $\langle A\rangle = A$ if a time average is taken; on the other hand, an ensemble average for random phases gives $\langle A\rangle = 0$. Moreover, the average value of the intensity $\left\langle E^-(t)E^+(t)\right\rangle = |AE_0|^2/4$ if the average corresponds to a time average, but if an ensemble average is taken,

$$\begin{aligned}\left\langle E^-(t)E^+(t)\right\rangle &= \frac{1}{4}|E_0|^2 \left\langle \sum_{j,j'=1}^{N} e^{i(\phi_j - \phi_{j'})}\right\rangle \\ &= \frac{1}{4}|E_0|^2 \sum_{j,j'=1}^{N} \delta_{jj'} = \frac{N}{4}|E_0|^2,\end{aligned} \tag{13.22}$$

where N is the number of oscillators. Thus, although both averaging procedures result in $\left|g^{(1)}(\tau)\right| = 1$, they lead to values of $\langle A\rangle$ and $\left\langle E^-(t)E^+(t)\right\rangle$ that differ.

Field produced by oscillators having random phases that fluctuate in time (collision model). We have already discussed a model in which collisions of a two-level atom with perturber atoms produce phase changes in the atoms' off-diagonal density matrix elements. The phase changes occur "instantaneously" at random collision times. To model this process, we assume that $E^+(t)$ is given by

$$E^+(t) = \frac{1}{2}\sum_{j=1}^{N} E_0 e^{-i\omega_L t + i\phi_j(t)}, \tag{13.23}$$

where the sum is over the N two-level atoms in the sample.

To calculate the first-order correlation function (13.9), we need

$$\langle E^-(t)E^+(t+\tau)\rangle = \frac{|E_0|^2}{4T}\sum_{j,j'=1}^{N}\int_{-T/2}^{T/2} dt e^{i\omega_L t} e^{-i\omega_L(t+\tau)} e^{-i\phi_j(t)} e^{i\phi_{j'}(t+\tau)}. \tag{13.24}$$

This is a very complicated expression, since the phase jumps at each atom depend on the specific collision history. To make some progress in evaluating this quantity, we carry out an *ensemble* average and assume that (1) the phases of different atoms are uncorrelated $\langle e^{-i\phi_j(t)} e^{i\phi_{j'}(t+\tau)}\rangle = \langle e^{i[\phi_j(t+\tau)-\phi_j(t)]}\rangle \delta_{j,j'}$; (2) the overall collision process is *stationary*, that is, $\langle e^{i[\phi_j(t+\tau)-\phi_j(t)]}\rangle$ is independent of t; and (3) each atom, on average, sees the same collision history. With these three assumptions, equation (13.24) is transformed into

$$\langle E^-(t)E^+(t+\tau)\rangle = \frac{1}{4}N|E_0|^2 e^{-i\omega_L\tau} G(\tau), \tag{13.25}$$

where

$$G(\tau) = \langle e^{i[\phi_j(t+\tau)-\phi_j(t)]}\rangle = \langle e^{i[\phi_j(\tau)-\phi_j(0)]}\rangle. \tag{13.26}$$

The first-order correlation function (13.9) is given by

$$g^{(1)}(\tau) = e^{-i\omega_L\tau} G(\tau). \tag{13.27}$$

To calculate $G(\tau)$, we employ the method used in appendix B in chapter 3. We assume that an *impact* approximation is valid, allowing us to calculate the change in $G(\tau)$ in a time interval $d\tau$ that contains at most one collision, but contains the entire collision (collision duration much less than the time between collisions). In this manner, we find

$$\begin{aligned}\langle \delta G(\tau)\rangle &= \langle G(\tau+d\tau) - G(\tau)\rangle = \langle [e^{i\phi(b,v)} - 1]\mathcal{P}(b,v)\, d\tau\, G(\tau)\rangle \\ &\approx -(\Gamma + iS)\langle G(\tau)\rangle\, d\tau,\end{aligned} \tag{13.28}$$

where $\mathcal{P}(b,v)$ is the probability density per unit time for a collision having impact parameter b and relative speed v,

$$\Gamma + iS = \int W_r(v)dv \int 2\pi b db \mathcal{P}(b,v)[1 - e^{i\phi(b,v)}], \tag{13.29}$$

and $W_r(v)$ is the relative speed distribution. In going to the second line in equation (13.28), we made a *Markov approximation*, assuming that each collision is independent of all past collisions.

If equation (13.28) is converted to a differential equation, the solution is

$$G(\tau) = e^{-(\Gamma+iS)\tau}, \tag{13.30}$$

which, when combined with equations (13.9) and (13.25), yields the correlation function

$$g^{(1)}(\tau) = e^{-i(\omega_L+S)\tau}e^{-\Gamma\tau}. \tag{13.31}$$

As a result of collisions, the correlation function decays with a *coherence time* of order Γ^{-1} and *coherence length* of order c/Γ. Although we formulated this theory in terms of collisions, it is applicable to any type of phase diffusion process. The power spectrum associated with this exponential decay of the correlation function is a Lorentzian, since

$$F(\omega) = \frac{1}{\pi}\text{Re}\int_0^\infty d\tau e^{-i(\omega_L+S-\omega)\tau}e^{-\Gamma\tau} = \frac{1}{\pi}\frac{\Gamma}{(\omega-\omega_L-S)^2+\Gamma^2}. \tag{13.32}$$

The fluctuating phases lead to a broadening and a shift in the power spectrum.

13.2.2 Young's Fringes

One immediate application of the first-order correlation function is the interference of two fields that start from a common source point and travel to a field point along different paths. A Young's interferometer, such as the one shown in figure 13.1, can be used to illustrate this type of interference. A field passes through a pinhole that acts as a spatial filter for the field and then through two pinholes in a screen on its way to the observation plane. Other types of interferometers, such as the Mach-Zehnder and Michelson interferometers, in which an incident plane wave is split along two paths of unequal length and then recombined at a detector, are discussed in sections 13.2.4 and 14.2. In this section, we are interested in determining how the fringe contrast of an interferometer depends on the noise properties of the field.

As in chapter 11, we use a scalar theory of diffraction in which the polarization properties of the field are neglected. Moreover, we assume (1) that the distances between the screens shown in figure 13.1, as well as the distance from the second screen to the observation plane, are much greater than a wavelength; (2) that the aperture sizes are much greater than a wavelength (justifying the use of Kirchhoff-Fresnel diffraction theory); and (3) that the scattering angles are much less than unity. In these limits, the field following each circular aperture takes the form of an outgoing spherical wave. The positive frequency component of the field at the observation point $\mathbf{R}$ at time t is given by [3, 4]

$$\begin{aligned} E^+(\mathbf{R},t) &= \frac{k_L\sigma}{2\pi i}\left[\frac{e^{ik_Ls_1}}{s_1}e^{-i\omega_L(t-t_1)}E^+(\mathbf{R}_1,t_1) + \frac{e^{ik_Ls_2}}{s_2}e^{-i\omega_L(t-t_2)}E^+(\mathbf{R}_2,t_2)\right] \\ &= \frac{k_L\sigma}{2\pi i}\left[\left(\frac{1}{s_1}\right)E^+(\mathbf{R}_1,t_1) + \left(\frac{1}{s_2}\right)E^+(\mathbf{R}_2,t_2)\right], \end{aligned} \tag{13.33}$$

where $k_L = \omega_L/c$, σ is a constant having units of area (in effect, a quantity proportional to the area of each pinhole), and

$$t_1 = t - s_1/c, \quad t_2 = t - s_2/c. \tag{13.34}$$

The field has been expressed in terms of the field amplitudes at the positions of the two diffracting apertures located at positions $\mathbf{R}_1$ and $\mathbf{R}_2$, taking into account the propagation of the fields to the observation point.

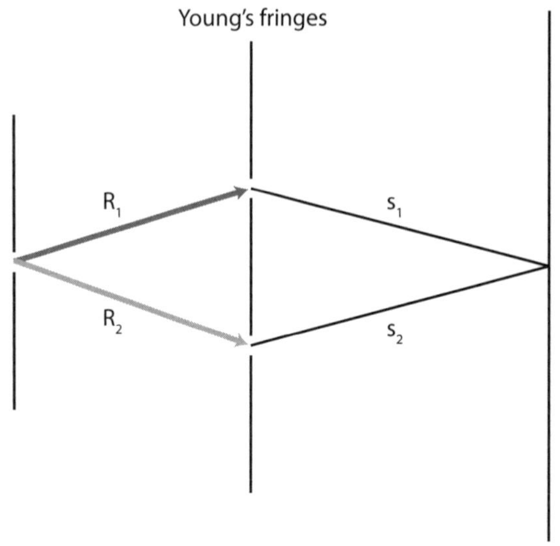

Figure 13.1. Schematic representation of a Young's interferometer. Incident radiation passes through a pinhole and then a pair of pinholes on its way to a screen.

The field intensity at the observation point is

$$\bar{I}(\mathbf{R}, t) = \langle E^-(\mathbf{R}, t)E^+(\mathbf{R}, t)\rangle = u_1^2\bar{I}_1(\mathbf{R}_1, t_1) + u_2^2\bar{I}_2(\mathbf{R}_2, t_2) \\ +2u_1u_2\,\mathrm{Re}\langle E^-(\mathbf{R}_1, t_1)E^+(\mathbf{R}_2, t_2)\rangle, \tag{13.35}$$

where

$$\bar{I}_j = \langle E^-(\mathbf{R}_j, t_j)E^+(\mathbf{R}_j, t_j)\rangle, \tag{13.36a}$$

$$u_j = k\sigma/2\pi s_j, \tag{13.36b}$$

for $j = 1, 2$.

If we define

$$\langle E^-(\mathbf{R}_1, t_1)E^+(\mathbf{R}_2, t_2)\rangle \equiv G^{(1)}(\mathbf{R}_1, t_1; \mathbf{R}_2, t_2) \\ = |G^{(1)}(\mathbf{R}_1, t_1; \mathbf{R}_2, t_2)|e^{i\phi(\mathbf{R}_1, t_1; \mathbf{R}_2, t_2)}, \tag{13.37}$$

then

$$\bar{I}(\mathbf{R}, t) = u_1^2\bar{I}_1 + u_2^2\bar{I}_2 + 2u_1u_2|G^{(1)}(\mathbf{R}_1, t_1; \mathbf{R}_2, t_2)|\cos[\phi(\mathbf{R}_1, t_1; \mathbf{R}_2, t_2)]\,. \tag{13.38}$$

As the observation point on the screen is varied, interference fringes can be observed, since the phase $\phi(\mathbf{R}_1, t_1; \mathbf{R}_2, t_2)$ depends on the difference in distance from each of the apertures to the observation point. The fringe contrast is defined as

$$\xi = \frac{\bar{I}_{max} - \bar{I}_{min}}{\bar{I}_{max} + \bar{I}_{min}} = \frac{2u_1u_2|G^{(1)}(\mathbf{R}_1, t_1; \mathbf{R}_2, t_2)|}{u_1^2\bar{I}_1 + u_2^2\bar{I}_2}\,. \tag{13.39}$$

This expression can be recast in a suggestive form if the definition of the first-order correlation function is generalized as

$$g^{(1)}(\mathbf{R}_1, t_1; \mathbf{R}_2, t_2) = \frac{G^{(1)}(\mathbf{R}_1, t_1; \mathbf{R}_2, t_2)}{\sqrt{\bar{I}_1 \bar{I}_2}} = \frac{\langle E^-(\mathbf{R}_1, t_1) E^+(\mathbf{R}_2, t_2)\rangle}{\sqrt{\bar{I}_1 \bar{I}_2}}. \tag{13.40}$$

In terms of $g^{(1)}(\mathbf{R}_1, t_1; \mathbf{R}_2, t_2)$, equation (13.39) becomes

$$\xi = \frac{2u_1 u_2 \sqrt{\bar{I}_1 \bar{I}_2}}{u_1^2 \bar{I}_1 + u_2^2 \bar{I}_2} |g^{(1)}(\mathbf{R}_1, t_1; \mathbf{R}_2, t_2)| . \tag{13.41}$$

The maximum fringe contrast occurs if $|g^{(1)}(\mathbf{R}_1, t_1; \mathbf{R}_2, t_2)| = 1$, that is, if the fields are first-order coherent. In what follows, we assume that $R_1 = R_2$, such that $\bar{I}_1 = \bar{I}_2 = I$, and also assume that $u_1 \approx u_2 \equiv u$. As a consequence, the field intensity (13.38) and fringe contrast (13.41) are given by

$$\bar{I}(\mathbf{R}, t) = 2uI \left\{1 + |g^{(1)}(\mathbf{R}_1, t_1; \mathbf{R}_2, t_2)| \cos\left[\phi(\mathbf{R}_1, t_1; \mathbf{R}_2, t_2)\right]\right\} \tag{13.42}$$

and

$$\xi = |g^{(1)}(\mathbf{R}_1, t_1; \mathbf{R}_2, t_2)|, \tag{13.43}$$

respectively.

For coherent fields with

$$E^+(\mathbf{R}, t) = \frac{1}{2} E_0 e^{i(\mathbf{k}\cdot\mathbf{R} - \omega_L t)}, \tag{13.44}$$

($\mathbf{k}$ is the field propagation vector before the first screen), the first-order correlation function is

$$g^{(1)}(\mathbf{R}_1, t_1; \mathbf{R}_2, t_2) = \exp\left\{i\left[\mathbf{k}\cdot(\mathbf{R}_2 - \mathbf{R}_1) + k(s_2 - s_1)\right]\right\}, \tag{13.45}$$

where the fact that

$$t_2 - t_1 = (s_1 - s_2)/c \tag{13.46}$$

has been used. The field intensity (13.42) is

$$\bar{I}(\mathbf{R}, t) = 2uI\left\{1 + \cos\left[\mathbf{k}\cdot(\mathbf{R}_2 - \mathbf{R}_1) + k(s_2 - s_1)\right]\right\}, \tag{13.47}$$

and fringes are observed as $(s_2 - s_1)$ is varied. The fringe contrast (13.43) is unity. Since the correlation length for a coherent field is infinite, the fringe contrast does not change as $(s_2 - s_1)$ is varied (neglecting changes in the u_j's).

In the phase diffusion or collision model, it follows from equations (13.40) and (13.31) that

$$g^{(1)}(\mathbf{R}_1, t_1; \mathbf{R}_2, t_2) = \exp\left\{i\left[\mathbf{k}\cdot(\mathbf{R}_2 - \mathbf{R}_1) + k(s_2 - s_1) - S\tau\right]\right\} e^{-\Gamma\tau}, \tag{13.48}$$

where $\tau = |s_2 - s_1|/c$. In this case, the fringe contrast (13.43) is

$$\xi = e^{-\Gamma|s_2 - s_1|/c}. \tag{13.49}$$

Owing to the fact that the field is not first-order coherent, the fringes wash out if $\Gamma|s_2 - s_1|/c > 1$ or $|s_2 - s_1| > \lambda_{coh}$, where $\lambda_{coh} = c/\Gamma$ is the coherence length.

Thus, when phase noise is present in the fields, the difference in arm lengths of an interferometer is limited by the coherence length of the source. For stabilized laser fields, the coherence length can be on the order of kilometers or more.

13.2.3 Intensity Correlations—Second-Order Correlation Function

The first-order correlation function provides us with some information about the statistical properties of the field, but as with any statistical distribution, higher order moments are needed to fully characterize the distribution. We are content to limit ourselves to the second-order correlation function, but note that higher order correlation functions have been studied experimentally [5].

Our discussion of the second-order correlation function begins with the definition

$$I(t) \equiv E^{-}(t)E^{+}(t). \tag{13.50}$$

The quantity $I(t)$ is proportional to the field intensity measured at a detector. There can be fluctuations in the field intensity, so it is appropriate to introduce the various moments:

$$\bar{I} = \langle I(t) \rangle, \; \overline{I^2} = \langle I^2(t) \rangle, \; \Delta I = \sqrt{\overline{I^2} - \bar{I}^2}. \tag{13.51}$$

As was the case for the field amplitudes, we can compare the field intensity at different times. This leads to a definition of the *second-order correlation function*:

$$g^{(2)}(\tau) \equiv \frac{\langle I(t)I(t+\tau)\rangle}{\bar{I}^2} = \frac{\langle E^{-}(t)E^{-}(t+\tau)E^{+}(t+\tau)E^{+}(t)\rangle}{\langle E^{-}(t)E^{+}(t)\rangle^2}. \tag{13.52}$$

The quantities $E^{\pm}(t)$ are classical functions of time, but we write the expression in a way that will allow us to make a connection with the analogous equation for quantized fields. It is relatively easy to prove that $g^{(2)}(t)$ satisfies

$$g^{(2)}(-\tau) = g^{(2)}(\tau), \tag{13.53a}$$

$$g^{(2)}(0) \geq 1, \tag{13.53b}$$

$$g^{(2)}(\tau) \leq g^{(2)}(0). \tag{13.53c}$$

If

$$g^{(2)}(\tau) = 1, \tag{13.54}$$

the field is said to be second-order coherent. Whereas the time-average definition of $g^{(1)}(\tau)$ yields consistent results even for pulses of finite temporal extent, it leads to $g^{(2)}(0) = \infty$ and ill-defined values of $g^{(2)}(\tau)$ for pulses of finite extent (see problem 5).

The second-order correlation function is a measure of the coherence of the field intensity—that is, it represents a comparison of the field intensity at two times separated by τ. Clearly, if $I(t)$ and $I(t+\tau)$ are uncorrelated, then $g^{(2)}(\tau) \sim 1$. In general, $g^{(2)}(\tau)$ decays in a smooth fashion from its maximum value at $\tau = 0$ to a value of unity for $\tau \gg \tau_c$, unless $I(t)$ is a periodic function of time. As with the first-order correlation function, a periodic field can have an infinite correlation *time*, even if the field is not second-order coherent.

Monochromatic field. With $E^{+}(t) = \frac{1}{2}E_0 e^{-i\omega_L t}$, $I(t) = \bar{I} = \frac{1}{4}|E_0|^2$, $\langle I(t)I(t+\tau)\rangle = \frac{1}{16}|E_0|^4$, and $g^{(2)}(\tau) = 1$. A monochromatic field is second-order coherent.
Two- and multifrequency fields. The field amplitude is given by equation (13.18), implying that

$$I(t) = \frac{1}{4}\left[|E_1|^2 + |E_2|^2 + E_1 E_2^* e^{-i(\omega_1-\omega_2)t} + E_2 E_1^* e^{i(\omega_1-\omega_2)t}\right], \tag{13.55}$$

$$\bar{I} = \frac{1}{4}\left(|E_1|^2 + |E_2|^2\right), \tag{13.56}$$

$$\langle I(t)I(t+\tau)\rangle = \bar{I}^2 + \frac{1}{8}|E_1|^2|E_2|^2 \cos\left[(\omega_2-\omega_1)\tau\right], \tag{13.57}$$

and

$$g^{(2)}(\tau) = 1 + \frac{2|E_1|^2|E_2|^2 \cos\left[(\omega_2-\omega_1)\tau\right]}{\left(|E_1|^2 + |E_2|^2\right)^2}. \tag{13.58}$$

The field is not second-order coherent, but the correlation time is infinite, since $g^{(2)}(\tau)$ does not tend to unity as $\tau \sim \infty$. For equal field intensities, $g^{(2)}(\tau)$ oscillates between values of 3/2 and 1/2. If there is a *continuous* distribution of frequencies having characteristic width $\Delta\omega$, $g^{(2)}(\tau)$ decays in a smooth fashion from its maximum value at $\tau = 0$ to a value of unity for $\tau \gg \tau_c \approx \Delta\omega^{-1}$.
Field produced by oscillators having constant random phases. For the field amplitude given in equation (13.21),

$$I(t) = I = \frac{1}{4}|E_0|^2 \sum_{j,j'=1}^{N} e^{i(\phi_j-\phi_{j'})}, \tag{13.59}$$

where N is the number of oscillators. Clearly, since there is no time dependence in I, $g^{(2)} = 1$, if we use the time-average definition for the correlation function.

On the other hand, if we use the ensemble-average definition, then

$$g^{(2)} = \frac{\left\langle \sum_{j,j',k,k'=1}^{N} e^{i(\phi_j+\phi_{j'}-\phi_k-\phi_{k'})} \right\rangle}{\left\langle \sum_{j,j'=1}^{N} e^{i(\phi_j-\phi_{j'})} \right\rangle^2}, \tag{13.60}$$

which is also time-independent. The phases ϕ_j are random numbers between 0 and 2π. As a consequence,

$$\left\langle \sum_{j,j'=1}^{N} e^{i(\phi_j-\phi_{j'})} \right\rangle = \left\langle \sum_{j,j'=1}^{N} e^{i(\phi_j-\phi_{j'})}\delta_{j,j'} \right\rangle = \sum_{j=1}^{N} 1 = N. \tag{13.61}$$

The average of the sum in the numerator in equation (13.60) is nonvanishing only if ($j = k$ and $j' = k'$) or ($j = k'$ and $j' = k$). We consider first the term with $j = k = j' = k'$,

$$\left\langle \sum_{j}^{N} 1 \right\rangle = N. \tag{13.62}$$

Next, we look at the terms with ($j = k$ and $j' = k'$). We can choose j in any of N ways; however, once we choose j, there are only $(N-1)$ choices left for j' since we must exclude terms with $j = j'$. As a result, this contribution, as well as the remaining contribution from terms with ($j = k'$ and $j' = k$), is

$$\left(\sum_{j=1}^{N} 1 \sum_{j'=1}^{N-1} 1 \right) = N^2 - N, \tag{13.63}$$

such that the numerator in equation (13.60) is given by

$$\left\langle \sum_{j,j',k,k'=1}^{N} e^{i(\phi_j + \phi_{j'} - \phi_k - \phi_{k'})} \right\rangle = N + 2(N^2 - N) = 2N^2 - N, \tag{13.64}$$

implying that

$$g^{(2)} = \frac{2N^2 - N}{N^2}. \tag{13.65}$$

The field is *not* second-order coherent if one employs the ensemble-average definition, with $g^{(2)} \approx 2$ for $N \gg 1$. For a *single* realization of the phases, the output is that of a coherent source; however, the output averaged over ensembles of oscillators having different sets of initial phases results in different statistical properties of the field. One would expect the output of a laser cavity or a field that has been sent through a narrow-band spectral filter, such as a Fabry-Pérot cavity, to correspond to the time-average definition.

Field produced by oscillators having random phases that fluctuate in time (collision model). The field amplitude is given by equation (13.23), from which one obtains

$$\bar{I} = \frac{1}{4} |E_0|^2 \sum_{j,j'=1}^{N} e^{i[\phi_j(t) - \phi_{j'}(t)]} = \frac{1}{4} N |E_0|^2 \tag{13.66}$$

and

$$\begin{aligned} &\langle E^-(t) E^-(t+\tau) E^+(t+\tau) E^+(t) \rangle \\ &= \frac{|E_0|^4}{16} \left\langle \sum_{j,j',k,k'=1}^{N} e^{i[\phi_j(t) + \phi_{j'}(t+\tau) - \phi_k(t+\tau) - \phi_{k'}(t)]} \right\rangle. \end{aligned} \tag{13.67}$$

As in the case of constant phases, the sum is nonvanishing only if ($j = k$ and $j' = k'$) or ($j = k'$ and $j' = k$), and using the same method that we used to arrive at

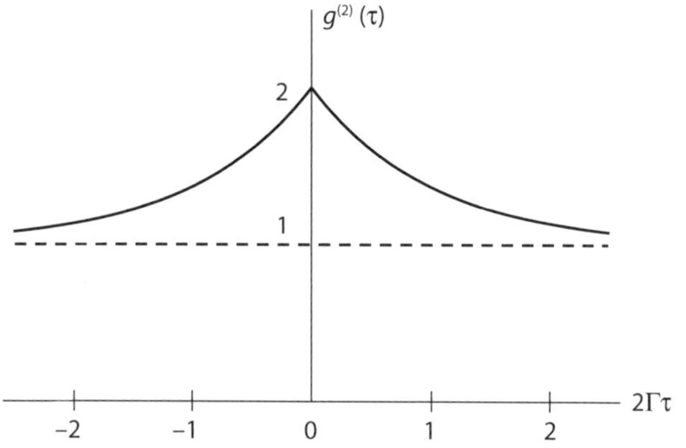

Figure 13.2. Second-order correlation function for the collision or phase diffusion model of the field sources.

equation (13.64) and the fact that $|\langle e^{i(\phi_j(t)-\phi_j(t+\tau))}\rangle|^2$ is independent of j, we find

$$\left\langle \sum_{j,j',k,k'=1}^{N} e^{i[\phi_j(t)+\phi_{j'}(t+\tau)-\phi_k(t+\tau)-\phi_{k'}(t)]} \right\rangle$$
$$= N + N(N-1) + N(N-1)|\langle e^{i[\phi_j(t)-\phi_j(t+\tau)]}\rangle|^2. \tag{13.68}$$

As a consequence, the second-order correlation function is

$$g^{(2)}(\tau) = \{N + N(N-1)[1 + |\langle e^{i[\phi_j(t)-\phi_j(t+\tau)]}\rangle|^2]\}/N^2$$
$$= \frac{1}{N} + \left(1 - \frac{1}{N}\right)[1 + |g^{(1)}(\tau)|^2], \tag{13.69}$$

where the fact that $|\langle e^{i[\phi_j(t)-\phi_j(t+\tau)]}\rangle| = |g^{(1)}(\tau)|$ [see equations (13.26) and (13.27)] has been used. In the limit that $N \gg 1$ and with $|g^{(1)}(\tau)| = e^{-\Gamma\tau}$, one finds that $g^{(2)}(\tau) \sim 1 + e^{-2\Gamma\tau}$ (see figure 13.2). Equation (13.53a) is used to extend the results to negative τ.

Often, the fact that the correlation function is a maximum at $\tau = 0$ and falls off with increasing τ is associated with *photon bunching*, a process that is connected with the fact that photons are bosons. Clearly, our discussion has made no mention of the quantum properties of the field. The maximum in the second-order correlation function at $\tau = 0$ is simply a consequence of the fact that $\langle I^2(t)\rangle \geq \langle I(t)\rangle^2$.

13.2.4 Hanbury Brown and Twiss Experiment

As in the case of the first-order correlation function, the definition of the second-order correlation function associated with light at space-time points $(\mathbf{R}_1, t_1)$ and $(\mathbf{R}_2, t_2)$ can be generalized as

$$g^{(2)}(\mathbf{R}_1, t_1; \mathbf{R}_2, t_2) = \frac{\langle E^-(\mathbf{R}_1, t_1)E^-(\mathbf{R}_2, t_2)E^+(\mathbf{R}_2, t_2)E^+(\mathbf{R}_1, t_1)\rangle}{\bar{I}_1 \bar{I}_2}, \tag{13.70}$$

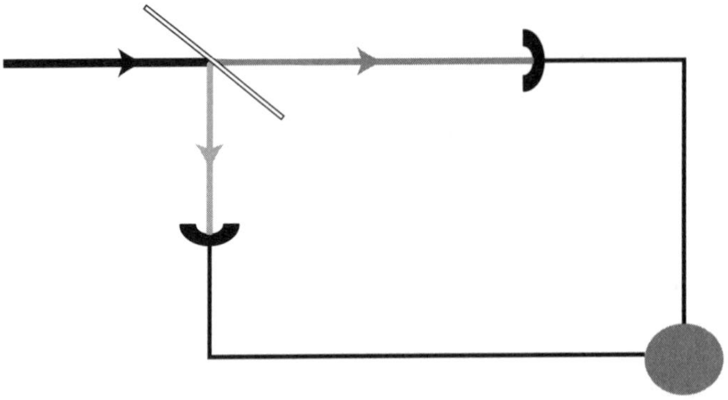

Figure 13.3. Mach-Zehnder interferometer.

where $\bar{I}_j$ is defined in equation (13.36a). In 1956, Hanbury Brown and Twiss used a Mach-Zehnder-type interferometer (figure 13.3) to measure the second-order correlation function of an optical field [6]. For an incident field intensity I, two identical detectors, and a 50/50 beam splitter,

$$\bar{I}_1 = \langle I_1(\mathbf{R}, t_1)\rangle = \bar{I}_2 = \langle I_2(\mathbf{R}, t_2)\rangle = \frac{1}{2}\bar{I}\,. \tag{13.71}$$

In their experiment, Hanbury Brown and Twiss measured the product of the fluctuations at the two detectors,

$$\begin{aligned}\left\langle\left[I_1(\mathbf{R}, t_1) - \bar{I}_1\right]\left[I_2(\mathbf{R}, t_2) - \bar{I}_2\right]\right\rangle &= \langle I_1(\mathbf{R}, t_1)I_2(\mathbf{R}, t_2)\rangle - \bar{I}_1\bar{I}_2 \\ &= \frac{1}{4}[\langle I(\mathbf{R}, t_1)I(\mathbf{R}, t_2)\rangle - \bar{I}^2] \\ &= \frac{1}{4}\bar{I}^2\left[g^{(2)}(\tau) - 1\right],\end{aligned} \tag{13.72}$$

where $\tau = \Delta s/c$ and Δs is the magnitude of the difference in path lengths for the two arms of the interferometer. They found that there was a correlation for $\tau = 0$ that disappeared as Δs was increased.

The motivation of Hanbury Brown and Twiss to carry out this experiment was based on a proposal they had for measuring stellar diameters with an interferometer of this type. In the experiment, they actually measured the spatial rather than temporal coherence of the source [7]. For separations of the detectors $\Delta s \gtrsim \lambda/\Delta\theta$ (λ is the wave length, $\Delta\theta$ is the angular diameter of the source), the correlation begins to decay. They showed that their measurements on Sirius were consistent with an angular diameter of $0.0063''$ for this star.

13.3 Quantized Fields: Density Matrix for the Field and Photon Optics

In those cases where a quantized description of the radiation field is required, a quantum-mechanical approach is needed to discuss statistical properties of the field.

Just as for atomic states, we can define a density matrix for field states. That is, we can write

$$\varrho = \sum \varrho_{nm} |n\rangle \langle m|, \tag{13.73}$$

in the n basis. Let us consider some possible quantized field states.

13.3.1 Coherent State

If the field is in a single-mode coherent state

$$|\alpha\rangle = \sum_{n=0}^{\infty} \frac{\alpha^n}{\sqrt{n!}} e^{-\frac{1}{2}|\alpha|^2} |n\rangle, \tag{13.74}$$

density matrix elements are given by

$$\varrho_{nm} = \frac{\alpha^n \alpha^{*m}}{\sqrt{n!m!}} e^{-|\alpha|^2}, \tag{13.75}$$

and the density matrix is *not* diagonal in the number basis. Of course, we could equally well express the density matrix as a diagonal matrix in the coherent-state basis,

$$\varrho = |\alpha\rangle \langle \alpha|. \tag{13.76}$$

Note that in the number-state basis,

$$P_n = \varrho_{nn} = \frac{|\alpha|^{2n}}{n!} e^{-|\alpha|^2} \tag{13.77}$$

represents a Poisson distribution, with

$$\langle \hat{n} \rangle = \sum_{n=0}^{\infty} n P_n = |\alpha|^2 \tag{13.78}$$

and

$$\langle \hat{n}^2 \rangle = \sum_{n=0}^{\infty} n^2 P_n = |\alpha|^4 + |\alpha|^2 . \tag{13.79}$$

For a multimode coherent state,

$$\varrho = |\alpha_1 \alpha_2 \ldots\rangle \langle \alpha_1 \alpha_2 \ldots |, \tag{13.80}$$

or, in the number-state basis,

$$\varrho = \prod_{j=1}^{N} \sum_{n_j, m_j = 0}^{\infty} \frac{\alpha_j^{n_j} \left(\alpha_j^{m_j}\right)^*}{\sqrt{n_j! m_j!}} e^{-|\alpha_j|^2} |n_j\rangle \langle m_j|, \tag{13.81}$$

where N is the number of modes.

13.3.2 Thermal State

For a field in thermal equilibrium with some reservoir at temperature T,

$$\varrho = \frac{e^{-\beta H}}{\mathrm{Tr}(e^{-\beta H})}, \tag{13.82}$$

where $\beta = 1/(k_B T)$, and k_B is Boltzmann's constant. For a single-mode free field, $H = \hbar\omega_L a^\dagger a$, and

$$\begin{aligned}\varrho_{nm} &= \frac{\langle n|e^{-qa^\dagger a}|m\rangle}{\sum_{n=0}^{\infty}\langle n|e^{-qa^\dagger a}|n\rangle} \\ &= \frac{e^{-nq}}{\sum_{n=0}^{\infty} e^{-nq}}\delta_{n,m} = (1-e^{-q})e^{-nq}\delta_{n,m},\end{aligned} \tag{13.83}$$

where

$$q = \frac{\hbar\omega_L}{k_B T}. \tag{13.84}$$

The density matrix of a thermal state is diagonal in the number-state basis and can be written as

$$\varrho = \sum P_n |n\rangle\langle n|, \tag{13.85}$$

where

$$P_n = (1-e^{-q})e^{-nq} \tag{13.86}$$

is the probability for being in a state of n photons.

The average number of photons in the thermal state is

$$\begin{aligned}\bar{n} = \langle\hat{n}\rangle &= \sum_n nP_n = \sum_n n(1-e^{-q})e^{-nq} = -(1-e^{-q})\frac{d}{dq}\sum_n e^{-nq} \\ &= -(1-e^{-q})\frac{d}{dq}\frac{1}{1-e^{-q}} = \frac{e^{-q}}{1-e^{-q}}.\end{aligned} \tag{13.87}$$

Therefore,

$$e^{-q} = \frac{\bar{n}}{1+\bar{n}}, \tag{13.88}$$

$$1-e^{-q} = \frac{1}{1+\bar{n}}, \tag{13.89}$$

and

$$P_n = \varrho_{nn} = \frac{1}{1+\bar{n}}\left(\frac{\bar{n}}{1+\bar{n}}\right)^n. \tag{13.90}$$

For a multimode thermal field,

$$\begin{aligned} \varrho &= \left(\sum_{n_1} P_{n_1}|n_1\rangle\langle n_1|\right)\left(\sum_{n_2} P_{n_2}|n_2\rangle\langle n_2|\right)\cdots\left(\sum_{n_k} P_{n_k}|n_k\rangle\langle n_k|\right) \\ &= \sum_{n_1 n_2 \ldots n_k} |n_1 n_2 \ldots\rangle\langle n_1 n_2 \ldots| \prod_k P_{n_k}. \end{aligned} \tag{13.91}$$

13.3.3 P(α) Distribution

Sometimes, it is convenient to expand the field density matrix in the coherent-state basis rather than the number-state basis. The density matrix can be expanded in terms of coherent states as

$$\varrho = \int \langle\alpha|\varrho|\beta\rangle|\alpha\rangle\langle\beta| d^2\alpha d^2\beta, \tag{13.92}$$

where $d^2\alpha = d\,\mathrm{Re}\,(\alpha)\, d\,\mathrm{Im}\,(\alpha)$. Since the $|\alpha\rangle$ are overcomplete, this expansion is not unique. On the other hand, owing to this property, it may be possible to find a *diagonal* representation in which

$$\varrho = \int P(\alpha')|\alpha'\rangle\langle\alpha'| d^2\alpha', \tag{13.93}$$

where $P(\alpha') \equiv P(\alpha', \alpha'^*)$. The distribution $P(\alpha)$ is not necessarily positive definite; in general, it cannot be interpreted as the probability of being in state α. The usefulness of this representation is that it becomes trivial to calculate the expectation value of any operator that can be written in "normal" order with all a's to right and $a^\dagger$'s to left.

As a first step to obtain $P(\alpha)$, we write

$$\varrho(\alpha) \equiv \langle\alpha|\varrho|\alpha\rangle = \int d^2\alpha' P(\alpha')|\langle\alpha'|\alpha\rangle|^2 = \int d^2\alpha' P(\alpha') e^{-|\alpha-\alpha'|^2}, \tag{13.94}$$

where ϱ is the density matrix in some arbitrary representation, and equation (12.57) was used. The problem is to find $P(\alpha)$, given ϱ and $\langle\alpha|\varrho|\alpha\rangle$. With the substitutions

$$\alpha = \frac{q + ip}{\sqrt{2}}, \tag{13.95a}$$

$$\alpha' = \frac{q' + ip'}{\sqrt{2}}, \tag{13.95b}$$

we can rewrite equation (13.94) as the integral equation

$$\varrho(q, p) = \int dq' dp' P(q', p') K(q - q', p - p'), \tag{13.96}$$

where the kernel $K(q, p)$ is given by

$$K(q, p) = \frac{1}{2} e^{-\frac{1}{2}q^2} e^{-\frac{1}{2}p^2}. \tag{13.97}$$

Since the kernel of this integral equation is a function of $(q - q')$ and $(p - p')$ only, it can be solved by integral transform techniques. Taking Fourier transforms of each side, we obtain a solution

$$\tilde{\varrho}(x,k) = 2\pi \tilde{K}(x,k)\tilde{P}(x,k), \tag{13.98}$$

where the Fourier transforms are defined by

$$\tilde{A}(x,k) = \frac{1}{2\pi}\int e^{i(xp-kq)}A(q,p)dqdp. \tag{13.99}$$

Since the Fourier transform of the Gaussian function is also a Gaussian, we find

$$\tilde{\varrho}(x,k) = \pi e^{-\frac{1}{2}(x^2+k^2)}\tilde{P}(x,k), \tag{13.100}$$

or

$$\tilde{P}(x,k) = \frac{1}{\pi}e^{\frac{1}{2}(x^2+k^2)}\tilde{\varrho}(x,k). \tag{13.101}$$

We can get $P(\alpha,\alpha^*)$ by taking the inverse transform, *if it exists* [8]. Using this method, we can show that, for a thermal state,

$$P(\alpha,\alpha^*) = \frac{1}{\pi\bar{n}}e^{-\frac{|\alpha|^2}{\bar{n}}}. \tag{13.102}$$

13.3.4 Correlation Functions for the Field

We are now in a position to calculate the correlation functions analogous to those calculated for classical fields. It is convenient to use the Heisenberg representation for operators.

13.3.4.1 First-order correlation function

The first-order correlation function is defined by

$$g^{(1)}(\mathbf{R}_1,t_1;\mathbf{R}_2,t_2) \equiv \frac{\langle E_1^- E_2^+\rangle}{\sqrt{\langle E_1^- E_1^+\rangle\langle E_2^- E_2^+\rangle}}, \tag{13.103}$$

where $\langle A\rangle \equiv \mathrm{Tr}\,(\varrho A)$, and

$$E_1^\pm \equiv E^\pm(\mathbf{R}_1,t_1), \qquad E_2^\pm \equiv E^\pm(\mathbf{R}_2,t_2). \tag{13.104}$$

13.3.4.1.1 Single-mode field: For a single-mode field having frequency ω_L and propagation vector **k**, the positive frequency component of the field operator is

$$E^+(\mathbf{R},t) = i\sqrt{C}e^{i\sigma(\mathbf{R},t)}a, \tag{13.105}$$

where

$$\sigma(\mathbf{R},t) = \mathbf{k}\cdot\mathbf{R} - \omega_L t \tag{13.106}$$

and

$$C = \hbar\omega_L/(2\epsilon_0\mathcal{V}). \tag{13.107}$$

It follows that

$$\langle E_1^- E_1^+ \rangle = C\mathrm{Tr}\left(\varrho a^\dagger a\right) = C\sum_n n\varrho_{nn} \tag{13.108}$$

and

$$\langle E_1^- E_2^+ \rangle = Ce^{i\sigma_{21}}\mathrm{Tr}\left(\varrho a^\dagger a\right) = Ce^{i\sigma_{21}}\sum_n n\varrho_{nn}, \tag{13.109}$$

where

$$\sigma_{21} = \mathbf{k}\cdot(\mathbf{R}_2 - \mathbf{R}_1) - \omega_L(t_2 - t_1). \tag{13.110}$$

From the definition (13.103), we then find

$$g^{(1)}(\mathbf{R}_1, t_1; \mathbf{R}_2, t_2) = e^{i\sigma_{21}}, \tag{13.111}$$

$$|g^{(1)}(\mathbf{R}_1, t_1; \mathbf{R}_2, t_2)| = 1, \tag{13.112}$$

and

$$g^{(1)}(\tau) \equiv g^{(1)}(\mathbf{R}_1, t_1; \mathbf{R}_1, t_1 + \tau) = e^{-i\omega_L\tau}. \tag{13.113}$$

Any single-mode field is first-order coherent since only a single frequency is present. That is, even a *thermal* single-mode field does not correspond to a field whose phases fluctuate in time. It is closer to a filtered field or a field produced by oscillators with *constant* random phases.

13.3.4.1.2 Multimode fields: In all cases, we take the modes to have the same polarization and to propagate in the $\hat{\mathbf{z}}$ direction.

13.3.4.1.2.1 Two-Photon State: The state vector is denoted by $|\Psi\rangle = |11\rangle$, a single photon in each of two modes. The positive frequency component of the two-mode field is

$$E^+(Z, t) = i\sqrt{C_1}a_1e^{i\sigma_1(Z,t)} + i\sqrt{C_2}a_2e^{i\sigma_2(Z,t)}, \tag{13.114}$$

where

$$\sigma_j(Z, t) = k_jZ - \omega_jt, \qquad C_j = \hbar\omega_j/(2\epsilon_0\mathcal{V}) \tag{13.115}$$

($j = 1, 2$). Reminiscent of equations (13.108) and (13.109), we find

$$\langle E_1^- E_1^+ \rangle = C_1 + C_2 \tag{13.116}$$

and

$$\langle E_1^- E_2^+ \rangle = C_1e^{-i\omega_1\tau} + C_1e^{-i\omega_2\tau}, \tag{13.117}$$

with $\tau = t_2 - t_1 - (Z_2 - Z_1)/c$. Therefore,

$$g^{(1)}(\tau) = \frac{C_1e^{-i\omega_1\tau} + C_1e^{-i\omega_2\tau}}{C_1 + C_2}, \qquad |g^{(1)}(\tau)| \neq 1. \tag{13.118}$$

As was the case for a superposition of two classical monochromatic fields, the total field is not first-order coherent. In the quantum case considered, however, the *reason* that $|g^{(1)}(\tau)| \neq 1$ is different than in the classical case. In the classical case, the fact that $|g^{(1)}(\tau)| \neq 1$ is a consequence of the definition of $g^{(1)}(\tau)$ in terms of a time average. In the quantum case, $g^{(1)}(\tau)$ is defined in terms of a quantum-mechanical average, not a time average. The fact that $|g^{(1)}(\tau)| \neq 1$ in the quantum case can be traced to a state of the field in which the two modes are in pure number states, such that the relative phase of the two field modes is not well-defined, e.g., $\langle 11| a_2^\dagger a_1 |11\rangle = 0$.

13.3.4.1.2.2 Multimode coherent state: The positive frequency component of the field is

$$E^+(Z,t) = i \sum_j \sqrt{C_j} a_j e^{i\sigma_j(Z,t)}, \tag{13.119}$$

and the state vector for the field is

$$|\Psi\rangle = |\{\alpha_k\}\rangle = |\alpha_1 \alpha_2 \ldots\rangle. \tag{13.120}$$

we find that

$$\langle E_1^- E_1^+ \rangle = \sum_{jj'} \sqrt{C_j C_{j'}} e^{i[\sigma_j(1) - \sigma_{j'}(1)]} \langle a_{j'}^\dagger a_j \rangle = |A_1|^2, \tag{13.121}$$

where

$$A_1 = \sum_j \sqrt{C_j} \alpha_j e^{i\sigma_j(1)}, \tag{13.122}$$

and

$$\sigma_j(n) = k_j Z_n - \omega_j t_n, \tag{13.123}$$

having used the fact that $\langle a_{j'}^\dagger a_j \rangle = \alpha_j \alpha_{j'}^*$. Moreover, it is a simple matter to show that

$$\langle E_1^- E_2^+ \rangle = A_1^* A_2, \tag{13.124}$$

and therefore,

$$g^{(1)}(\tau) = \frac{A_1^* A_2}{|A_1 A_2|}, \qquad |g^{(1)}(\tau)| = 1 \tag{13.125}$$

and the field is first-order coherent. This result differs from that for a classical multifrequency field. As noted earlier, the difference can be traced to the time average taken in the classical case. In the quantum case, we are taking an ensemble average. Note that $\langle E_1^- E_1^+ \rangle = |A_1|^2$ is a function of time for a multifrequency quantized field.

13.3.4.1.2.3 Multimode thermal state: The density matrix for a multimode thermal state is

$$\varrho = \sum_{n_1, n_2, \ldots} |n_1, n_2, \ldots\rangle\langle n_1, n_2, \ldots| \prod_k P(n_k), \tag{13.126}$$

where

$$P(n_k) = \frac{(\bar{n}_k)^{n_k}}{(1+\bar{n}_k)^{1+n_k}}. \tag{13.127}$$

The average field intensity $\langle E_1^- E_1^+ \rangle = \mathrm{Tr}\left(\varrho E_1^- E_1^+\right)$ can be calculated as

$$\langle E_1^- E_1^+ \rangle = \sum_{n_1,n_2,\ldots} \langle n_1, n_2, \ldots | \prod_k P(n_k) \sqrt{C_j C_{j'}} e^{i[\sigma_j(1)-\sigma_{j'}(1)]} \times a_{j'}^\dagger a_j | n_1, n_2, \ldots \rangle, \tag{13.128}$$

where $\sigma_j(1)$ is defined by equation (13.123). The terms in equation (13.128) contribute only if $j = j'$; therefore,

$$\begin{aligned}\langle E_1^- E_1^+ \rangle &= \sum_{n_1,n_2,\ldots} \prod_k P(n_k) \sum_j C_j n_j \\ &= \sum_{n_1,n_2,\ldots} P(n_1)P(n_2)\ldots(C_1 n_1 + C_2 n_2 + \cdots) \\ &= \sum_j C_j \bar{n}_j .\end{aligned} \tag{13.129}$$

In a similar manner, we obtain

$$\langle E_1^- E_2^+ \rangle = \sum_j C_j \bar{n}_j e^{i[\sigma_j(2)-\sigma_j(1)]}, \tag{13.130}$$

such that the first-order correlation function is

$$g^{(1)}(\tau) = \frac{\sum_j C_j \bar{n}_j e^{-i\omega_j \tau}}{\sum_j C_j \bar{n}_j} = \frac{\sum_j \omega_j \bar{n}_j e^{-i\omega_j \tau}}{\sum_j \omega_j \bar{n}_j}, \tag{13.131}$$

where $\tau = t_2 - t_1 - (Z_2 - Z_1)/c$.

In essence, $\omega_j \bar{n}_j$, properly normalized, gives the frequency distribution of the modes. For example, if we set

$$\omega_j \bar{n}_j = \frac{N(\omega)}{\pi} \frac{\Gamma}{(\omega - \omega_L)^2 + \Gamma^2}, \tag{13.132}$$

in which $N(\omega)$ is a slowly varying normalization factor that can be evaluated at $\omega = \omega_L$, treat $\omega_j \to \omega$ as a continuous variable, and replace $\sum_j \Rightarrow 2\pi L/c \int d\omega$ for our one-dimensional modes, we obtain (for $\tau > 0$)

$$g^{(1)}(\tau) = e^{-i\omega_L \tau} e^{-\Gamma \tau}, \tag{13.133}$$

in agreement with the classical phase diffusion or collision model in which the shift S is neglected.

13.3.4.2 Second-order correlation function

In analogy with the classical second-order correlation function (13.52), we define $g^{(2)}$ for quantized fields as

$$g^{(2)}(\mathbf{R}_1, t_1; \mathbf{R}_2, t_2) \equiv \frac{\langle E_1^- E_2^- E_2^+ E_1^+ \rangle}{\langle E_1^- E_1^+ \rangle \langle E_2^- E_2^+ \rangle}\,. \tag{13.134}$$

Since $E_1^- E_2^- E_2^+ E_1^+ = \left(E_2^+ E_1^+\right)^\dagger E_2^+ E_1^+$, and since E^+ is a superposition of destruction operators, the second-order correlation function is related to the joint probability of destroying a photon at position $\mathbf{R}_1$ at time t_1 *and* destroying a photon at position $\mathbf{R}_2$ at time t_2. In contrast to the result for classical fields, it is possible to have $g^{(2)}(\mathbf{R}, t; \mathbf{R}, t) < 1$. If this is the case, one has a clear signature of the quantized nature of the field.

13.3.4.2.1 Single-mode field: For a single-mode field,

$$\langle E_1^- E_1^+ \rangle = C \sum n \varrho_{nn} = C\bar{n} \tag{13.135}$$

and

$$\langle E_1^- E_2^- E_2^+ E_1^+ \rangle = C^2 \langle a^\dagger a^\dagger a a \rangle = C^2 \langle a^\dagger a a^\dagger a - a^\dagger a \rangle = C^2 \left(\overline{n^2} - \bar{n}\right), \tag{13.136}$$

implying that

$$g^{(2)}(\tau) = \frac{\overline{n^2} - \bar{n}}{\bar{n}^2} \tag{13.137}$$

is independent of τ.

13.3.4.2.1.1 Number state: For a number state of the field $|\psi\rangle = |n\rangle$,

$$\langle n^2 \rangle = \langle n \rangle^2 = n^2 \tag{13.138}$$

and

$$g^{(2)} = 1 - \frac{1}{n} < 1\,. \tag{13.139}$$

A pure number state has no classical analogue since $g^{(2)} < 1$. If $n = 1$, $g^{(2)} = 0$, reflecting the fact that if we destroy (measure) a photon and there is only one photon present, we cannot destroy (measure) a second photon.

13.3.4.2.1.2 Coherent state: For a coherent state of the field $|\psi\rangle = |\alpha\rangle$,

$$\bar{n} = |\alpha|^2, \tag{13.140}$$

$$\langle n^2 \rangle = |\alpha|^4 + |\alpha|^2, \tag{13.141}$$

and

$$g^{(2)} = 1\,. \tag{13.142}$$

The coherent state is second-order coherent. This is not surprising, since the coherent states were constructed in a fashion so that they would be n-order coherent.

13.3.4.2.1.3 Thermal state: For a thermal state of the field with $\varrho = \sum_n P(n)|n\rangle\langle n|$ and

$$P_n = \frac{\bar{n}^n}{(1+\bar{n})^{n+1}}, \tag{13.143}$$

we find

$$\langle n^2 \rangle = \frac{1}{1+\bar{n}} \sum n^2 x^n, \tag{13.144}$$

where

$$x = \bar{n}/(1+\bar{n}). \tag{13.145}$$

To evaluate the sum, we note that

$$\begin{aligned}\langle n(n-1)\rangle = \langle n^2\rangle - \langle n\rangle &= \frac{x^2}{1+\bar{n}} \frac{d^2}{dx^2} \sum x^n = \frac{x^2}{1+\bar{n}} \frac{d^2}{dx^2} \frac{1}{1-x} \\ &= \frac{2x^2}{(1+\bar{n})(1-x)^3} = 2x^2(1+\bar{n})^2 = 2\bar{n}^2.\end{aligned} \tag{13.146}$$

Therefore,

$$\langle n^2 \rangle = 2\bar{n}^2 + \bar{n}, \quad g^{(2)} = 2. \tag{13.147}$$

A thermal state corresponds to the *ensemble*-averaged result of a field produced by oscillators having constant random phases. This is a reasonable result, since the relative phases of the various n states in a thermal state are completely uncertain.

13.3.4.2.1.4 Two-photon coherent state: We know that a TPCS represents a quantized state of the field since it is created from the vacuum by the operator $a^{\dagger 2}$, which has no classical analogue. For the squeezed state $|\alpha, z\rangle = |z, \alpha'\rangle$, it follows from equations (13.137), (12.124), and (12.125) that

$$g^{(2)} = \frac{2s^4 + 2s^2 |\alpha|^2 + \left|\alpha^2 - cse^{i\theta}\right|^2}{\left(s^2 + |\alpha|^2\right)^2}, \tag{13.148}$$

where $s = \sinh r$, $c = \cosh r$. For the TPCS $|z, \alpha\rangle$, the result for $g^{(2)}$ is obtained most easily by substituting the inverse of equation (12.123)—namely,

$$\alpha = c\alpha' - se^{i\theta}\alpha'^*, \tag{13.149}$$

into equation (13.148). In general, $g^{(2)}$ can be greater or less than unity, depending on the values of α, r, and θ. For a squeezed vacuum ($\alpha = 0$), $g^{(2)} > 1$.

13.3.4.2.2 Multimode fields: In all cases, we take the modes to have the same polarization and to propagate in the $\hat{\mathbf{z}}$ direction. We state the results and leave details of the calculations to the problems.

13.3.4.2.2.1 Multimode coherent state:

$$g^{(2)}(\tau) = 1. \tag{13.150}$$

13.3.4.2.2.2 Multimode thermal state:

$$g^{(2)}(\tau) = 1 + |g^{(1)}(\tau)|^2 . \tag{13.151}$$

This is the same result as for the classical case of a field produced by N oscillators having random, fluctuating phases, if the limit $N \to \infty$ is taken.

13.4 Summary

In this chapter, we presented an elementary treatment of the coherence properties of classical and quantized radiation fields. The difference between a time average and ensemble average was discussed in some detail, along with several examples illustrating how these two averages can lead to different results. First-order and second-order correlation functions were defined for classical and quantized fields, and these functions were evaluated for a variety of field states. Two types of interferometers were described that could be used to measure these correlation functions. In the next chapter, we explore the theory of interferometric detection in more detail and show how the use of squeezed states can reduce quantum noise to below the standard quantum limit. To prepare the grounds for that discussion, an elementary introduction to photodetection is also presented.

Problems

1. Prove that $|g^{(1)}(\tau)| \leq 1$.
2. Consider the quantity $S(N) = |A(N,t)|^2$, where $A(N,t) = \sum_{j=1}^{N} e^{-i\omega t + i\phi_j}$, which is supposed to simulate the output field of a number of oscillators with random phases ϕ_j. Calculate $\langle A \rangle$, $\langle S \rangle$, $\langle S^2 \rangle$, and $\Delta S^2 = \langle S^2 \rangle - \langle S \rangle^2$.
3. Consider the field intensity produced by an ensemble of 100 oscillators. That is, evaluate
$$S = \left| \sum_{j=1}^{100} e^{-i\omega t + i\phi_j} \right|^2 ,$$
where the ϕ_j are random phases. Use a random number generator to choose these phases, and evaluate S for three specific realizations of the random phases. Then repeat the calculation a thousand times and take the average, and see how close you get to the average value calculated in problem 1.
4. Prove, for classical fields, that $g^{(2)}(\tau) \leq g^{(2)}(0)$ and that $g^{(2)}(0) \geq 1$.
5. Calculate and graph $g^{(2)}(\tau)$ when
$$I(t) = \sum_{n=-\infty}^{\infty} \Theta(t + 0.1 - n) - \Theta(t - 0.1 - n) .$$
Why doesn't $g^{(2)}(\tau) \sim 1$ as $\tau \sim \infty$? Note that this intensity pattern corresponds to a frequency comb.

Now consider a pulse having positive frequency component

$$E^{+}(t)=\frac{e^{-i\omega_L t}}{\pi^{1/4}}e^{-t^2/2}.$$

Show that $g^{(1)}(\tau)$ can be calculated, but that $g^{(2)}(\tau)$ is ill-defined and that $g^{(2)}(0)=\infty$.

6. Calculate the fringe pattern in a Michelson interferometer as a function of the difference in arm lengths assuming a chaotic source having coherence time τ_c. In order to be able to see fringes, what is the maximum arm length separation for (a) light from a discharge lamp and (b) laser light? For visible light, what is the maximum number of fringes that can be seen as the relative arm length is varied?
7. Calculate the fringe pattern for the interferometer of problem 6 when a sodium discharge lamp containing the Na D1 and D2 lines is used to illuminate the interferometer. Can the interferometer be used to measure the frequency difference between the D1 and D2 lines? Explain.
8. Derive equation (13.98).
9. Use the method outlined in the text to obtain the diagonal coherent-state representation for the thermal state having density matrix

$$\varrho=\sum_n(1+\bar{n})^{-1}[\bar{n}/(1+\bar{n})]^n\,|n\rangle\,\langle n|\,,$$

that is, derive equation (13.102).

10. Prove that $g^{(2)}(\tau)=1$ for a multimode coherent state $|\{\alpha_k\}\rangle$.
11. Derive equation (13.151) for a quantized multimode thermal state.
12. Calculate $g^{(2)}(\tau)$ for the squeezed state $|\alpha,z\rangle=\left|z,\alpha'\right\rangle$, and show that $g^{(2)}(0)$ is greater than 1 for $\alpha=0$ but may be less than 1 for certain values of α.
13. To model the stellar interferometer of Hanbury Brown and Twiss, consider two detectors located at $\mathbf{R}_1=-(d/2)\hat{\mathbf{i}}$ and $\mathbf{R}_2=(d/2)\hat{\mathbf{i}}$ that receive signals from a star whose center is at $D\hat{\mathbf{z}}$. Consider the star to be composed of a line of N oscillators located at positions $\mathbf{R}_j=D\hat{\mathbf{z}}+X_j\hat{\mathbf{i}}$, with the oscillators distributed between $-a/2$ and $a/2$. Assuming that each oscillator has a random phase, calculate the classical second-order correlation function $g^{(2)}(\mathbf{R}_1,t;\mathbf{R}_2,t)$ and show that it decays for separations $d\gtrsim\lambda/\Delta\theta$ (λ is the wavelength, $\Delta\theta=a/D$ is the angular diameter of the source). To do this, assume that the field at the detector position $\mathbf{R}_\alpha$ ($\alpha=1,2$) is given by $E^{+}(\mathbf{R}_\alpha,t)=\frac{1}{2}\sum_j E_0 e^{-i\omega t_j+i\phi_j(t_j)}\approx\frac{1}{2}\sum_j E_0 e^{-i\omega t_j+i\phi_j(t-D/c)}$, where $t_j=t-|\mathbf{R}_j-\mathbf{R}_\alpha|/c$ is the retarded time.

References

[1] L. Mandel and E. Wolf, *Optical Coherence and Quantum Optics* (Cambridge University Press, Cambridge, UK, 1995), chaps. 1–8.

[2] For a review, see W. E. Moerner, *A dozen years of single-molecule spectroscopy in physics, chemistry, and biophysics*, Journal of Physical Chemistry B **106**, 910–927 (2002).

[3] J. D. Jackson, *Classical Electrodynamics*, 3rd ed. (Wiley, New York, 1999), sec. 10.5.
[4] A. Sommerfeld, *Optics* (Academic Press, New York, 1964), chaps. V and VI.
[5] See, for example, S. Chopra and L. Mandel, *Higher-order correlation properties of a laser beam*, Physical Review Lettters **30**, 60–63 (1973).
[6] R. Hanbury Brown and R. Q. Twiss, *Correlation between photons in two coherent beams of light*, Nature **177**, 27–29 (1956).
[7] R. Hanbury Brown and R. Q. Twiss, *A test of a new type of stellar interferometer*, Nature **178**, 1046–1048 (1956).
[8] H. Nussenzveig, *Introduction to Quantum Optics* (Gordon and Breach, London, 1973), secs. 3.3 and 3.4.

Bibliography

See the books on quantum optics listed in chapter 1.

14

Photon Counting and Interferometry

14.1 Photodetection

In quantum mechanics, there is no way to avoid the measurement problem. That is, although quantum mechanics represents an incredibly successful theory of matter and of matter–field interactions, it does not provide a prescription for coupling the quantum-mechanical system of interest to a classical measuring apparatus. Quantum mechanics enables one to make predictions for the outcome of measurements but has very little to say about the measurements themselves. In this chapter, we give a very elementary discussion of photodetection and discuss some applications to interferometry. There are chapters in most quantum optics textbooks on this topic, and you are urged to consult these texts for more details. Suffice it to say, however, that the theory of photodetection and photodetectors is far from complete.

One simple model for a photodetector is an optically thick medium consisting of inhomogeneously broadened two-level atoms. If we consider a time that is sufficiently short, any radiation incident on such a medium is absorbed by the atoms and is not scattered into other modes of the vacuum field. In contrast to the scattering process that we have been discussing throughout this text, this is true absorption. If the inhomogeneous width is sufficiently large, the medium acts as a broadband detector. The excited-state population can be used as a measure of the "number of photons" absorbed in the medium. Of course, we need to specify some way in which this excited-state population is measured. Let us not worry about this. In most theories of photodetection, the "excited state" of the atoms is actually the continuum, and the ejected electron is assumed to produce a measurable signal.

You can appreciate why the theory of photodetection is complicated. Imagine that we have some initial field pulse that is heading toward the detector. If we know the quantum state of this field, we have a well-defined quantum-mechanical problem in which the atoms in the detector are all in their ground states and the field state is specified. In principle, we can calculate the state of the system at any future time and obtain the probability to have n atoms excited in the detector. The problem is that, in an ideal photodetector, each time an atom is excited, it leads to a measurable signal, which implies that the quantum state of the system is modified. Somehow, this measurement must be fed back into the system and the quantum state of the system updated. Although there are theories of continuous measurement [1] that address questions of this nature, quantum mechanics runs into problems once the measuring apparatus is included in the system being studied.

14.1.1 Photodetection of Classical Fields

Let us assume that the probability of photo-emission in the detector is proportional to the intensity $I(t) = |E^+(t)|^2$ of an incident classical field, integrated over the surface of the detector. Since a photodetector is a quantum device, there are always quantum fluctuations in the number of "counts" registered by a photodetector. That is, if we prepare the same classical pulse and send it into a detector several times, the number of photo-counts recorded each time will differ. For a given incident field intensity, we can define $P_m(t, \tau)$ as the probability of registering m counts in a time between t and $(t + \tau)$. We assume that $\tau > 0$.

We can calculate $P_m(t, \tau)$ in much the same way that we calculated the correlation function for collisions—that is, we assume that we can choose a $\delta\tau$ sufficiently short to guarantee that at most one photo-count occurs in $\delta\tau$, but sufficiently long to ensure that a "complete" photo-count occurs (in other words, $\delta\tau$ is much greater than the response time of the photodetector). This is reminiscent of the impact approximation used to examine the role of collisions on line shapes.

With these approximations, we can write

$$P_m(t, \tau + \delta\tau) = P_0(t + \tau, \delta\tau)P_m(t, \tau) + P_1(t, \delta\tau)P_{m-1}(t, \tau), \tag{14.1}$$

where $P_0(t, \delta\tau)$ is the probability for no counts in a time between t and $(t + \delta\tau)$, and $P_1(t, \delta\tau)$ is the probability of one count in that interval. Let $p(t)$ be the probability per unit time that a photo-count occurs at time t. We assume that $p(t)$ is proportional to the field intensity,

$$p(t) = \mu I(t), \tag{14.2}$$

where μ is related to the detector efficiency. As a consequence,

$$P_1(t, \delta\tau) = p(t)\delta\tau = \mu I(t)\delta\tau. \tag{14.3}$$

Equation (14.1) can be rewritten as

$$P_m(t, \tau) + \frac{dP_m(t, \tau)}{d\tau}d\tau = [1 - p(t + \tau)d\tau]\, P_m(t, \tau) + p(t + \tau)d\tau\, P_{m-1}(t, \tau), \tag{14.4}$$

or

$$\frac{dP_m(t, \tau)}{d\tau} = -p(t + \tau)P_m(t, \tau) + p(t + \tau)P_{m-1}(t, \tau). \tag{14.5}$$

For $m = 0$,

$$\frac{dP_0(t, \tau)}{d\tau} = -p(t + \tau)P_0(t, \tau), \tag{14.6}$$

and $P_0(t, 0) = 1$. Therefore,

$$P_0(t, \tau) = e^{-\int_0^\tau p(t+\tau')d\tau'} = e^{-\mu \int_t^{t+\tau} I(t')dt'} \equiv e^{-\mu \bar{I}(t,T)T}, \tag{14.7}$$

where

$$I(t, \tau) \equiv \frac{1}{\tau} \int_t^{t+\tau} I(t')dt'. \tag{14.8}$$

If we try a solution to equation (14.5) of the form

$$P_m(t, \tau) = \tilde{P}_m(t, \tau) e^{-\int_0^\tau p(t+\tau')d\tau'}, \tag{14.9}$$

with $\tilde{P}_m(t, 0) = 1$, then [see equation (14.7)]

$$\tilde{P}_0(t, \tau) = 1, \tag{14.10}$$

and, for $m > 0$, $\tilde{P}_m(t, \tau)$, satisfies the differential equation

$$\frac{d\tilde{P}_m(t, \tau)}{d\tau} = p(t+\tau)\tilde{P}_{m-1}(t, \tau). \tag{14.11}$$

Therefore,

$$\tilde{P}_1(t, \tau) = \int_0^\tau p(t+\tau')d\tau' = \int_t^{t+\tau} p(t')dt', \tag{14.12}$$

$$\tilde{P}_2(t, \tau) = \int_t^{t+\tau} dt' \int_t^{t+t'} dt'' p(t')p(t''), \tag{14.13}$$

and so on.

Recall that, for *functions* $f(t)$, the time-ordered exponential $\mathcal{T}e^{\int_0^t f(t')dt'}$ and the exponential $e^{\int_0^t f(t')dt'}$ are identical, as we have already noted in chapter 2. Since each of the $\tilde{P}_m(t, \tau)$ represent terms in the expansion of the time-ordered exponential $\mathcal{T}e^{\int_t^{t+\tau} p(t')dt''} = e^{\int_t^{t+\tau} p(t')dt''}$, we conclude that

$$\tilde{P}_m(t, \tau) = \frac{1}{m!}\left[\int_t^{t+\tau} p(t')dt'\right]^m = \frac{1}{m!}\left[\mu I(t, \tau)\tau\right]^m, \tag{14.14}$$

where $I(t, \tau)$ is defined by equation (14.8). Combining this result with equation (14.9), we obtain

$$P_m(t, \tau) = \frac{1}{m!}\left[\mu I(t, \tau)\tau\right]^m e^{-\mu I(t,\tau)\tau}. \tag{14.15}$$

If a radiation pulse is incident on the detector, then it is clear that $P_m(t, \tau)$ depends on t, insofar as the pulse envelope is a function of t. For continuous fields that can be considered to be stationary (that is, their noise properties do not change in time), it is customary to define a photo-count distribution that is averaged over time—namely,

$$P_m(\tau) = \langle P_m(t, \tau)\rangle_t. \tag{14.16}$$

Unless noted otherwise, we limit our discussions to stationary fields.

If $I(t, \tau) = I$ is *constant*, as would be the case for a monochromatic field, then

$$P_m(\tau) = \frac{X^m}{m!}e^{-X}, \tag{14.17}$$

where

$$X = \mu I \tau. \tag{14.18}$$

Even for a constant field intensity, there is a distribution of photo-counts, owing to the statistical nature of the quantum-mechanical detection mechanism. The distribution (14.17) is a Poisson distribution, with $\bar{m} = X$, $\overline{m^2} = \bar{m}^2 + \bar{m}$, $\Delta m^2 = \bar{m}$, and $\Delta m/\bar{m} \sim 1/\sqrt{\bar{m}}$. The unavoidable noise that is associated with the photodetection process is referred to as *shot noise*.

When the field itself has noise properties, the signal recorded by a photodetector reflects both the noise properties of the field and that of the detector. If the response time of the detector is longer than the correlation time of the field, the detection process tends to average out the noise properties of the field. We consider only the opposite limit, in which the integral in equation (14.8) is carried out for a time τ that is short compared with the correlation time of the field intensity, but sufficiently long to contain a large number of counts. In this limit, $I(t, \tau)$ given by equation (14.8), is equal to $I(t)$.

The noise properties of the field can be attributed to the noise properties of the sources of the field. If a sufficiently large number of successive measurements is taken at the photodetector, the time average of these measurements is equivalent to an ensemble average over the sources, such that the photo-count distribution, given by equations (14.15) and (14.16), is

$$P_m(\tau) = \frac{1}{m!} \left\langle (\mu I \tau)^m e^{-\mu I \tau} \right\rangle, \tag{14.19}$$

where the average now corresponds to an ensemble average over the sources of the field.

We can characterize the signal by the number of photo-counts $\bar{m}(\tau)$ in a time interval τ and the variance of this quantity. The average number of counts is given by

$$\bar{m}(\tau) = \left\langle \sum_m m P_m(t, \tau) \right\rangle = \mu \langle I \rangle \tau \tag{14.20}$$

and the second moment by

$$\overline{m^2}(\tau) = \mu^2 \left\langle I^2 \right\rangle \tau^2 + \mu \langle I \rangle \tau, \tag{14.21}$$

such that

$$\Delta m^2(\tau) = \mu \langle I \rangle \tau + \mu^2 \tau^2 \left[\langle I^2 \rangle - \langle I \rangle^2 \right]. \tag{14.22}$$

Thus, fluctuations in number count have a contribution from shot noise, $\mu \langle I \rangle \tau$, and one from field fluctuations, $\mu^2\tau^2[\langle I^2 \rangle - \langle I \rangle^2]$. The shot noise can be reduced by increasing the counting statistics.

For constant field intensity,

$$\Delta m^2(\tau) = \mu \langle I \rangle \tau = \bar{m}(\tau), \tag{14.23}$$

as was found earlier. For a field produced by sources having fluctuating phases,

$$E^{+}(t) = \frac{1}{2}|E_0| \sum_{j=1}^{N} e^{-i\omega_L t + i\phi_j(t)}, \tag{14.24}$$

$$\langle I \rangle = \frac{1}{4} N |E_0|^2, \tag{14.25}$$

$$\langle I^2 \rangle = \frac{1}{16}\left(2N^2 - N\right)|E_0|^4, \tag{14.26}$$

such that

$$\Delta m^2(\tau) = \left[4\mu N\tau\,|E_0|^2 + \mu^2\tau^2\left(N^2 - N\right)|E_0|^4\right]/16. \tag{14.27}$$

The shot noise and source noise fluctuations can be distinguished by their τ dependence.

In the limit of large N, the moments for the phase diffusion model are approximately given by $\langle I^n \rangle = n!(N|E_0|^2/4)^n$, which is consistent with a Poisson intensity distribution

$$P(I) = \frac{1}{\bar{I}} e^{-I/\bar{I}}, \tag{14.28}$$

where $\bar{I} = N|E_0|^2/4$. In this approximation, the photo-count distribution (14.19) can be evaluated as

$$\begin{aligned} P_m(\tau) &= \int_0^\infty dI \frac{1}{m!} P(I)\,(\mu I\tau)^m e^{-\mu I\tau} \\ &= \frac{[\bar{m}(\tau)]^m}{\{[\bar{m}(\tau)] + 1\}^{m+1}}, \end{aligned} \tag{14.29}$$

where $\bar{m}(\tau) = \mu\bar{I}\tau$. This is a scaled version of the photon *number* distribution for a thermal source. You will see that this result is typical when the sampling time is much shorter than the correlation time of the field—the photon counting statistics is a scaled version of the photon number distribution, the scaling factor depending on the detector efficiency μ and counting time τ.

14.1.2 Photodetection of Quantized Fields

As we mentioned, the problem of photodetection of quantized fields is a complex one, since the measurement process itself modifies the field. We shall neglect such complications for the most part and assume that the signal at the detectors is proportional to

$$\left\langle E^{-}(\mathbf{R}_d, t)E^{+}(\mathbf{R}_d, t)\right\rangle = \mathrm{Tr}[\varrho\overline{E^{-}(\mathbf{R}_d, t)E^{+}(\mathbf{R}_d, t)}], \tag{14.30}$$

where ϱ is the density matrix for the field, and the bar indicates an average over the surface $\mathbf{R}_d$ of the detector. In other words, the signal is assumed to be proportional to the average value of the intensity *operator*, averaged over the surface of the detector. Moreover, we assume that the field is uniform over the detector and suppress the field label $\mathbf{R}_d$. Of course, this procedure is oversimplified, since no detector has an instantaneous response. In actual cases involving rapid changes in $E^-(\mathbf{R}_d, t)E^+(\mathbf{R}_d, t)$, the correlation time of the detector must be taken into account.

For a quantized radiation field, one must destroy m photons in the field to get m photo-counts. The probability for m photo-counts depends on averages of the form

$$\left\langle |E^+(t_m)E^+(t_{m-1})\ldots|^2\right\rangle = \left\langle E^-(t_1)E^-(t_2)\ldots E^-(t_m)E^+(t_m)\ldots E^+(t_2)E^+(t_1)\right\rangle .$$

That is, all E^+'s appear to the right, in what is called *normal order*. The calculation proceeds much as in the classical field case, except that one must now account for the fact that the intensity is an operator and that the photo-count distribution must involve the field operators taken in normal order.

A formal expression for the photo-count probability distribution can be taken as [2]

$$P_m(t,\tau) = \text{Tr}\left\{\varrho\mathfrak{N}\frac{[\mu\hat{I}(t,\tau)\tau]^m}{m!}e^{-\mu\hat{I}(t,\tau)\tau}\right\}, \tag{14.31}$$

where

$$\hat{I}(t,\tau) \equiv \frac{1}{\tau}\int_t^{t+\tau} E^-(t')E^+(t')dt' , \tag{14.32}$$

and $\mathfrak{N}$ is a "normal order" operator that moves all E^+'s in the expansion to the right. As in any operator equation, the exponential function is defined by its series expansion. The fields are to be evaluated at the detector surface. At this point, we consider single-mode fields only, for which

$$E^+(\mathbf{R},t) \sim i\left(\frac{\hbar\omega}{2\epsilon_0\mathcal{V}}\right)^{1/2} a e^{i(\mathbf{k}\cdot\mathbf{R}-\omega t)} \tag{14.33}$$

and

$$P_m(\tau) = \text{Tr}\left[\varrho\mathfrak{N}\frac{(\xi a^\dagger a)^m}{m!}e^{-\xi a^\dagger a}\right] , \tag{14.34}$$

where

$$\xi = \mu\left(\frac{\hbar\omega}{2\epsilon_0\mathcal{V}}\right)\tau \tag{14.35}$$

is known as the *quantum efficiency* of the detector.

If we expand the exponential and orders the resulting operators in normal order, we find

$$P_m(\tau) = \text{Tr}\left[\varrho\sum_{q=0}^{\infty}\frac{(-\xi)^q\xi^m(a^\dagger)^{m+q}a^{m+q}}{m!q!}\right] . \tag{14.36}$$

In the n-representation, where

$$\varrho = \sum_{n,n'}\varrho_{nn'}\,|n\rangle\,\langle n'|, \tag{14.37}$$

it is easy to show that only diagonal terms enter into the expression (14.36) for $P_m(\tau)$, since matrix elements of the form $\langle n'|(a^\dagger)^{m+q}a^{m+q}\,|n\rangle$ are needed. Moreover, it is obvious that one must have $m+q<n$, which is satisfied for $n>m$, $q<n-m$. Since

$$a^{m+q}|n\rangle = \sqrt{n(n-1)\dots[n-(m+q)]}|0\rangle, \tag{14.38}$$

it follows from equations (14.36) to (14.38) that

$$P_m(\tau) = \sum_{n=m}^{\infty}\sum_{q=0}^{n-m} P_n \frac{(-\xi)^q \xi^m}{m!q!} \frac{n!}{(n-m-q)!} = \sum_{n=m}^{\infty} P_n \binom{n}{m} \xi^m (1-\xi)^{n-m}, \tag{14.39}$$

where

$$P_n = \varrho_{nn}, \tag{14.40}$$

$$\binom{n}{m} = \frac{n!}{m!\,(n-m)!} \tag{14.41}$$

is a binomial coefficient, and we used the fact that

$$\sum_{q=0}^{p} \frac{(-a)^q}{q!(p-q)!} = \sum_{q=0}^{p} \binom{p}{q} \frac{(-a)^q}{p!} = \frac{(1-a)^p}{p!}. \tag{14.42}$$

We can now evaluate $P_m(\tau)$ for various quantized states of the field.

14.1.2.1 *Coherent state*

For a coherent state $|\alpha\rangle$, ϱ is not diagonal, but only diagonal elements,

$$P_n = \frac{\bar{n}^n}{n!} e^{-\bar{n}} = \frac{|\alpha|^{2n}}{n!} e^{-|\alpha|^2}, \tag{14.43}$$

are of importance in the photo-count distribution. From equation (14.39), we find

$$P_m(\tau) = \frac{(\xi\bar{n})^m}{m!} e^{-\xi m}. \tag{14.44}$$

The photo-count distribution is just a scaled version of the photon number distribution.

14.1.2.2 *Thermal state*

For a thermal state with

$$P_n = \frac{\bar{n}^n}{(1+\bar{n})^{1+n}}, \tag{14.45}$$

we find

$$P_m(\tau) = \frac{(\xi\bar{n})^m}{(1+\xi\bar{n})^{1+m}}. \tag{14.46}$$

Again, the photo-count distribution is just a scaled version of the photon number distribution.

14.1.2.3 Number state

For the number state $|q\rangle$,

$$P_n = \delta_{q,n}, \tag{14.47}$$

and

$$P_m(\tau) = \begin{cases} \binom{p}{q}\xi^m(1-\xi)^{q-m}, & \text{if } m \leq q, \\ 0, & \text{if } m > q. \end{cases} \tag{14.48}$$

In this case, the photo-count distribution, which is simply a binomial distribution for $m \leq q$, is no longer a scaled version of the photon number distribution. Of course, it is clear that our assumption that the measurement does not affect the quantum state of the field cannot remain valid here; if q photons are detected, there cannot be any field left.

One can also relate the photo-count distribution to the second-order correlation function for a single-mode field. (The magnitude of the first-order correlation function is unity.) From equations (13.137) and (14.39), we find

$$\sum_m \frac{m(m-1)P_m(\tau)}{\bar{m}^2} = \sum_n \frac{n(n-1)P_n}{\bar{n}^2} = g^{(2)}(\tau). \tag{14.49}$$

Moreover, for the photo-count distribution (14.39), it is straightforward to calculate

$$\bar{m} = \xi\bar{n}, \tag{14.50a}$$

$$\overline{m^2} = \xi\bar{n} + \xi^2\left(\overline{n^2} - \bar{n}\right), \tag{14.50b}$$

$$\Delta m^2 = \xi^2\Delta n^2 + \xi\bar{n}(1-\xi) = \left(\frac{\bar{m}}{\bar{n}}\right)^2 \Delta n^2 + \bar{m}\left(1 - \frac{\bar{m}}{\bar{n}}\right). \tag{14.50c}$$

For an ideal detector, $\xi = 1$ and $\Delta m^2 = \Delta n^2$.

14.2 Michelson Interferometer

Photodetectors are often used as the measuring devices in optical interferometers. The general configuration of an interferometer involves a beam splitter that divides an incoming field along two paths. After traversing different paths, the beams are recombined on a detector. The interference pattern is measured as the optical path length between the two arms of the interferometer is varied. Phase changes between the two arms translate into fringe shifts at the detector. The optical path length between the two arms can result in a number of ways. One can physically change the path length by using a micrometer screw or a piezoelectric transducer. More often, one uses the interferometer to monitor changes in optical paths resulting from some transitory effect such as a gravitational wave (not so easy to measure), the introduction of a transparent test body into one of the arms, or a difference in the velocity of light in the two arms owing to an "ether drift" (as in the Michelson-Morley experiment).

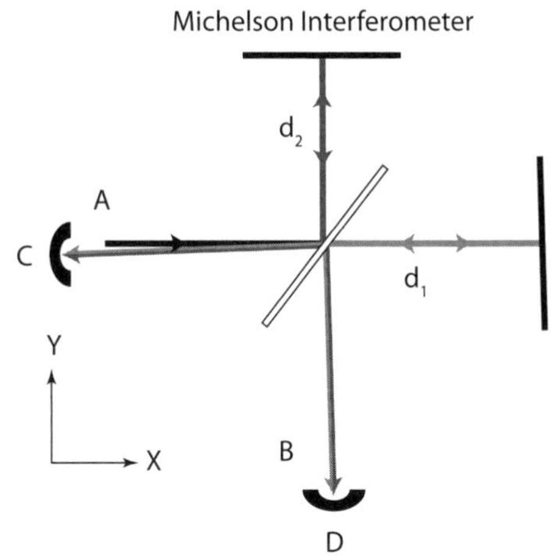

Figure 14.1. Michelson interferometer. There are two input ports A and B, but only radiation entering port A is shown.

14.2.1 Classical Fields

One of the most famous of all interferometers, the Michelson interferometer, is illustrated in figure 14.1. If light enters the interferometer from the left, the recombined light can be detected at either detector C or D, or both. Let us first consider what happens when classical, monochromatic light enters this interferometer. In anticipation of quantizing the field modes, we consider a classical field mode corresponding to a wave incident from the left,

$$\Psi_A(\mathbf{R}) = \begin{cases} e^{ikX} - Te^{-ikX}e^{ikd_1} - Re^{-ikX}e^{ikd_2} & \text{for } X < 0, \\ -\sqrt{RT}(e^{ikd_1} + e^{ikd_2})e^{-ikY} & \text{for } Y < 0, \\ \sqrt{T}(e^{ikX} - e^{-ik(X-d_1)}) & \text{for } X > 0, \\ \sqrt{R}(e^{ikY} - e^{-ik(Y-d_2)}) & \text{for } Y > 0, \end{cases} \tag{14.51}$$

where the beam splitter is located at the origin, mirror 1 at $\mathbf{R} = d_1\hat{\mathbf{x}}$, and mirror 2 at $\mathbf{R} = d_2\hat{\mathbf{y}}$. Note that the mode function vanishes on the mirrors. The parameters $\sqrt{R}$ and $\sqrt{T}$ are the amplitude reflection and transmission coefficients for the beam splitter and can be complex numbers. We shall assume a lossless beam splitter, such that

$$|R| + |T| = 1. \tag{14.52}$$

There is a phase shift of π on reflection from the mirrors. The relative phases of $\sqrt{R}$ and $\sqrt{T}$ will be fixed by conservation of energy.

To understand the consequences of conservation of energy, we must calculate the signal at each of the detectors and equate the total flux at the detectors to the incoming flux. The signal measured at detector D is proportional to

$$|\Psi_D|^2 = 2\,|R|\,|T|\,[1 + \cos(k\Delta d)]\,, \tag{14.53}$$

where

$$\Delta d = d_2 - d_1, \tag{14.54}$$

and the signal at detector C is proportional to

$$|\Psi_C|^2 = |R|^2 + |T|^2 + RT^* e^{ik\Delta d} + R^* T e^{-ik\Delta d}. \tag{14.55}$$

Since $(|R| + |T|)^2 = 1$, this equation can be transformed into

$$|\Psi_C|^2 = 1 - 2\,|R|\,|T| + RT^* e^{ik\Delta d} + R^* T e^{-ik\Delta d}. \tag{14.56}$$

By equating the total flux striking the detectors to the incident flux, $|\Psi_D|^2 + |\Psi_C|^2 = 1$, we obtain the condition

$$e^{ik\Delta d}(|R|\,|T| + RT^*) + e^{-ik\Delta d}(|R|\,|T| + R^* T) = 0. \tag{14.57}$$

If we set

$$R = |R|\, e^{2i\phi_R}, \tag{14.58a}$$

$$T = |T|\, e^{2i\phi_T}, \tag{14.58b}$$

condition (14.57) implies that

$$\phi_T - \phi_R = \pm\pi/2\,. \tag{14.59}$$

This is a general result for *any* lossless beam splitter; the relative phase of the (amplitude) transmission and reflection coefficients is $\pi/2$. Without loss of generality, we take $\phi_T = 0$, $\phi_R = \pi/2$, such that

$$\sqrt{R} = i\sqrt{|R|}, \qquad \sqrt{T} = \sqrt{|T|}. \tag{14.60}$$

If

$$\phi \equiv 2k\Delta d \tag{14.61}$$

is defined as the optical path difference for the two arms of the interferometer and if equation (14.60) is used, it follows that equations (14.53) and (14.56) can be rewritten in the form

$$|\Psi_D|^2 = 4\,|R|\,|T| \cos^2\left(\frac{\phi}{4}\right), \tag{14.62a}$$

$$|\Psi_C|^2 = 1 - 4\,|R|\,|T| \cos^2\left(\frac{\phi}{4}\right). \tag{14.62b}$$

The asymmetry in the signals at the two detectors arises from the fact that the signal reaching detector C is the sum of two field components, one of which is reflected twice at the beam splitter and the other is transmitted twice at the beam splitter, while the signal at detector D is the sum of two field components, both of which are reflected once at the beam splitter and transmitted once at the beam splitter.

Fringe maxima and minima occur when $\phi = 2n\pi$, for integer n. To achieve maximum sensitivity to small phase shifts, one needs to choose a region where the slope of the signal is a maximum with respect to changes in ϕ. Moreover, to help reduce the effects of any fluctuations in the incident fields on the signals, one can *subtract* the signals at the two detectors. The difference signal is proportional to a quantity S defined by

$$\begin{aligned} S &= |E_A^+|^2 \left(|\Psi_D|^2 - |\Psi_C|^2\right) \\ &= |E_A^+|^2 \,[4\,|R|\,|T| \cos(\phi/2) + 4\,|R|\,|T| - 1], \end{aligned} \tag{14.63}$$

where $|E_A^+|^2$ is proportional to the incident field intensity. In the examples to follow, we will always assume a 50/50 beam splitter with $|R| = |T| = 1/2$; in this limit,

$$S = |E_A^+|^2 \cos(\phi/2)\,. \tag{14.64}$$

The signal vanishes when $\phi/2 = (n + 1/2)\pi$, which also corresponds to the maximum slope of the signal as a function of ϕ. If one adjusts the arm lengths of the interferometer such that $\phi/2 = (n + 1/2)\pi + \delta\phi$ with $|\delta\phi| \ll 1$,

$$S \approx \pm |E_A^+|^2 \delta\phi, \tag{14.65}$$

the signal is linearly proportional to $\delta\phi$. For a classical coherent field, the sensitivity is limited only by shot noise.

14.2.2 Quantized Fields

It is not difficult to go over to the quantized field case if we use the prescription mentioned in chapter 12—that is, we write the Hamiltonian for a single-mode field as

$$H = \hbar\omega |\Psi_A(\mathbf{R})|^2 a_A^\dagger a_A\,, \tag{14.66}$$

where a_A is a destruction operator for this mode. This method is especially good for considering an infinite number of modes (as is needed for the vacuum field), but we limit the discussion to a single mode.

Several new features enter when we consider quantized rather than classical fields. First, there is the annoying but persistent problem that measurement of the field invariably modifies the quantum state of the field. Thus, if one wants to start with quantum states of the field corresponding to single- or two-photon pulses, some care in modeling the photodetectors is needed. Second, there are invariably number-state fluctuations for a quantum field, except for those prepared in a pure number state. This can translate into scaled photo-count distributions, as was discussed earlier. However, there can be an *additional* contribution to the signal noise resulting from fluctuations in the difference in "photon number" in each arm of the interferometer—this contribution is absent for classical fields. We concentrate on formulating a theory that can be used to illustrate the manner in which squeezed light can be used to reduce signal noise, but the formalism is also useful for understanding the contributions to the signal at a given detector.

The starting point of the calculation is the expression for the positive frequency component of the electric field amplitude,

$$E^+ = i \left(\frac{\hbar\omega}{2\epsilon_0 \mathcal{V}}\right)^{\frac{1}{2}} \Psi_A(\mathbf{R}) a_A\,. \tag{14.67}$$

Most of the results for the classical field can be taken over directly to the quantized field case. In all cases, we assume a 50/50 beam splitter.

14.2.2.1 *Signal at detector D*

The generalization of equation (14.62a) for the field intensity operator at detector D is

$$I_D = E^-(\mathbf{R}_D, t)E^+(\mathbf{R}_D, t)\cos^2(\phi/4). \tag{14.68}$$

Assuming stationary fields, we can calculate the photo-count distribution in the same manner as we did in going from equations (14.31) to (14.50). The only difference is that the detector efficiency ξ is replaced by $\xi\cos^2(\phi/4)$. That is, equations (14.50) remain valid provided that ξ is replaced by $\xi\cos^2(\phi/4)$.

As a consequence,

$$\Delta m^2 = \left(\frac{\bar{m}}{\bar{n}}\right)^2 \Delta n^2 + \bar{m}\left(1 - \frac{\bar{m}}{\bar{n}}\right), \tag{14.69}$$

with

$$\bar{m} = \xi\bar{n}\cos^2(\phi/4). \tag{14.70}$$

For different incident field states, we find

$$\Delta m^2 = \begin{cases} \bar{m}\left(1 - \frac{\bar{m}}{\bar{n}}\right) & \text{number state,} \\ \bar{m} & \text{coherent state,} \\ \bar{m}(1 + \bar{m}) & \text{thermal state.} \end{cases} \tag{14.71}$$

The fluctuations for a pure number state are lower than that for a coherent state; this is sometimes referred to as *number-state squeezing*.

Notice that both the signal and the fluctuations vanish at the minima for a 50/50 beam splitter, since there is total interference at the detector. (Of course, when the signal vanishes at detector D, it is a maximum at detector C.) The absence of any fluctuations at these points can be traced to the fact that we are dealing with a single-mode field. For a multimode field, there is a distribution of k's, and it is impossible to satisfy the condition for signal maxima or minima *simultaneously* for all modes. The first term in equation (14.69) represents scaled fluctuations. For an ideal detector, $\bar{m} = \bar{n}\cos^2(\phi/4)$, and the second term can be interpreted as a fluctuation due to a difference in "photon number" in each arm of the interferometer [2]; this term vanishes at *both* signal maxima and minima (recall that a maximum at one detector is a minimum at the other) since there is total interference at such points.

14.2.2.2 *Correlated signals at both detectors: difference photo-count signal*

In this section, we want to illustrate the concept of correlated measurements and to calculate the probability of observing a photo-count at *both* detectors C and D. Moreover, to prepare for calculations involving more than one input field, we obtain expressions for the average number of photo-counts and their fluctuations in terms of the field operators at each detector. To simplify matters, we assume that the detector efficiency is equal to unity. For an ideal detector ($\xi = 1$), the average number of counts at detector D and its variance can be obtained from equations (14.70) and (14.69) as

$$\bar{m}_D = \cos^2(\phi/4)\bar{n} = \cos^2(\phi/4)\left\langle a^\dagger a\right\rangle, \tag{14.72a}$$

$$\begin{aligned}\overline{m_D^2} &= \cos^2(\phi/4)\bar{n} + \cos^4(\phi/4)\left(\overline{n^2} - \bar{n}\right)\\ &= \cos^2(\phi/4)\left\langle a^\dagger a\right\rangle + \cos^4(\phi/4)\left\langle a^\dagger a^\dagger a a\right\rangle.\end{aligned} \tag{14.72b}$$

We can re-express these quantities in terms of the field operators acting at detector D by

$$\bar{m}_D = \frac{1}{K}\left\langle E_D^- E_D^+\right\rangle, \tag{14.73a}$$

$$\overline{m_D^2} = \frac{1}{K}\left\langle E_D^- E_D^+\right\rangle + \frac{1}{K^2}\left\langle E_D^- E_D^- E_D^+ E_D^+\right\rangle, \tag{14.73b}$$

where $K = \hbar\omega/(2\epsilon_0\mathcal{V})$. In an analogous manner, the corresponding expressions for detector C are

$$\bar{m}_C = \frac{1}{K}\left\langle E_C^- E_C^+\right\rangle, \tag{14.74a}$$

$$\overline{m_C^2} = \frac{1}{K}\left\langle E_C^- E_C^+\right\rangle + \frac{1}{K^2}\left\langle E_C^- E_C^- E_C^+ E_C^+\right\rangle. \tag{14.74b}$$

There is one additional quantity we need to calculate, the *joint* probability of measuring m_C counts at detector C *and* m_D counts at detector D—that is, $\overline{m_C m_D}$ [3]. Since this corresponds to the *simultaneous* destruction of m_C photons at detector C and m_D photons at detector D, the joint probability distribution must correspond to a time-ordered expression involving $\left(E_C^+ E_D^+\right)^\dagger E_C^+ E_D^+$. We will assume that [3]

$$\overline{m_C m_D} = \overline{m_D m_C} = \sin^2(\phi/4)\cos^2(\phi/4)\left\langle a^\dagger a^\dagger a a\right\rangle = \frac{1}{K^2}\left\langle E_D^- E_C^- E_C^+ E_D^+\right\rangle. \tag{14.75}$$

To show that these definitions are reasonable, let us calculate the expectation values of these operators for a single-photon state of the field, having the spatial mode defined by equation (14.51), such that

$$E_C^+ = -\sqrt{K}e^{2i\Phi}e^{-ikX}\sin(\phi/4)\,a, \tag{14.76a}$$

$$E_D^+ = \sqrt{K}e^{2i\Phi}e^{-ikY}\cos(\phi/4)\,a, \tag{14.76b}$$

where

$$\Phi = k(d_1 + d_2)/4. \tag{14.77}$$

Although there are problems with measurement of a one-photon state since the measurement destroys the state, the probabilities for photodetection of such states can be calculated using standard procedures. In practice, one requires a multimode field to get a one-photon *pulse*, but we neglect such complications.

For a one-photon state $|1\rangle$ having field mode (14.51), we can use equations (14.74a), (14.73a), (14.75), and (14.76) to obtain

$$\bar{m}_C = \sin^2(\phi/4), \qquad \bar{m}_D = \cos^2(\phi/4), \tag{14.78}$$

and

$$\overline{m_C m_D} = \sin^2(\phi/4)\cos^2(\phi/4)\left\langle a^\dagger a^\dagger a a\right\rangle = 0. \tag{14.79}$$

The one photon can be detected in either detector, but it is (obviously) impossible to measure a count at *both* detectors C and D for a one-photon field.

As in the case of classical fields, one can maximize the detection sensitivity to small phase changes by measuring the difference signal from the two detectors. We will assume that the photo-current distributions are proportional to the photo-count distributions. Let $\hat{S} = \hat{M}_D - \hat{M}_C$, where $\hat{M}_\alpha$ is the photo-count operator at detector α, defined in such a manner that $\langle \hat{M}_\alpha\rangle = \bar{m}_\alpha$ and $\langle \hat{M}_C \hat{M}_D\rangle = \overline{m_C m_D}$. Then, the average difference signal and its variance are given by

$$\langle \hat{S}\rangle = \bar{m}_D - \bar{m}_C = \frac{1}{K}\left(\langle E_D^- E_D^+\rangle - \langle E_C^- E_C^+\rangle\right), \tag{14.80a}$$

$$\begin{aligned}\langle \hat{S}^2\rangle &= \overline{m_D^2} + \overline{m_C^2} - 2\overline{m_C m_D}\\ &= \frac{1}{K}\left(\langle E_D^- E_D^+\rangle + \langle E_C^- E_C^+\rangle\right) + \frac{1}{K^2}\left(\langle E_D^- E_D^- E_D^+ E_D^+\rangle + \langle E_C^- E_C^- E_C^+ E_C^+\rangle\right)\\ &\quad - \frac{2}{K^2}\left\langle E_D^- E_C^- E_C^+ E_D^+\right\rangle.\end{aligned} \tag{14.80b}$$

14.2.2.3 *Coherent state*

We now want to study the difference signal between detectors D and C, and the fluctuations in this signal when the incident field is in a coherent state $|\alpha\rangle$, assuming ideal detectors and a 50/50 beam splitter. The average signal and its variance, obtained using equations (14.80) and (14.76), are

$$\langle \hat{S}\rangle = \cos(\phi/2)\,|\alpha|^2, \tag{14.81a}$$

$$\langle \hat{S}^2\rangle = |\alpha|^2 + \cos^2(\phi/2)\,|\alpha|^4. \tag{14.81b}$$

Note that

$$\Delta S^2 = \langle \hat{S}^2\rangle - \langle \hat{S}\rangle^2 = |\alpha|^2 \tag{14.82}$$

independent of ϕ.

Suppose that the interferometer is adjusted such that

$$\phi = \pi - 2\epsilon, \tag{14.83}$$

where $|\epsilon| \ll 1$. In this limit,

$$\langle \hat{S}\rangle = \epsilon\,|\alpha|^2, \tag{14.84}$$

and the signal is a linear function of ϵ. Since $\langle \hat{S}\rangle = \epsilon\,|\alpha|^2$ and $\Delta S = |\alpha|$, we are sensitive to phase shifts

$$\epsilon \approx |\alpha|^{-1} = \langle \hat{n}_A\rangle^{-1/2}, \tag{14.85}$$

where $\hat{n}_A$ is the number operator for the incoming field mode. For smaller values of ϵ, the noise begins to dominate the signal. The value $\epsilon = |\alpha|^{-1}$ corresponds to the standard quantum limit (SQL) for this interferometer.

14.2.2.4 *Entrance port B: reduction of noise below the SQL*

The Michelson interferometer shown in figure 14.1 actually has *two* input ports. The field mode we have discussed up to now is one associated with a field entering from $X < 0$ into port A. There is, in addition, a field mode entering from $Y < 0$ into port B. The mode function for this field mode is

$$\Psi_B(\mathbf{R}) = \begin{cases} e^{ikY} - Te^{-ikY}e^{ikd_2} - Re^{-ikY}e^{ikd_1} & \text{for } Y < 0, \\ -\sqrt{RT}(e^{ikd_1} + e^{ikd_2})e^{-ikX} & \text{for } X < 0, \\ \sqrt{R}[e^{ikX} - e^{-ik(X-d_1)}] & \text{for } X > 0, \\ \sqrt{T}[e^{ikY} - e^{-ik(Y-d_2)}] & \text{for } Y > 0. \end{cases} \tag{14.86}$$

The question arises as to whether we can reduce the noise below the SQL by replacing the vacuum field, which has been assumed implicitly to enter port B, with a field of our choosing.

Although one should allow for an infinite number of modes into port B in considering the vacuum field, we consider only the limit in which single-mode fields having the same frequency and polarization enter ports A and B. In this limit, the Hamiltonian is

$$H = \hbar\omega\left[|\Psi_A(\mathbf{R})|^2\, a_A^\dagger a_A + |\Psi_B(\mathbf{R})|^2\, a_B^\dagger a_B\right], \tag{14.87}$$

where a_α is an annihilation operator for mode α. For a 50/50 beam splitter with $R = -1/2$, $T = 1/2$, it follows from equations (14.51) and (14.86) that the positive frequency components of the fields at detectors C and D are given by

$$\begin{aligned} E_C^+ &= i(\sqrt{K}/2)e^{-ikX}\left[\left(e^{ikd_2} - e^{ikd_1}\right)a_A - i(e^{ikd_2} + e^{ikd_1})a_B\right] \\ &= \sqrt{K}e^{2i\Phi}e^{-ikX}\left[\cos(\phi/4)a_B - \sin(\phi/4)a_A\right] \end{aligned} \tag{14.88}$$

and

$$\begin{aligned} E_D^+ &= i(\sqrt{K}/2)e^{-ikY}\left[-i\left(e^{ikd_2} + e^{ikd_1}\right)a_A - \left(e^{ikd_2} - e^{ikd_1}\right)a_B\right] \\ &= \sqrt{K}e^{2i\Phi}e^{-ikY}\left[\cos(\phi/4)\,a_A + \sin(\phi/4)\,a_B\right], \end{aligned} \tag{14.89}$$

respectively.

One can use equations (14.74a), (14.73a), (14.88), and (14.89) to show that

$$\bar{m}_C + \bar{m}_D = \langle a_A^\dagger a_A\rangle + \langle a_B^\dagger a_B\rangle, \tag{14.90}$$

which is consistent with the conservation of probability. We are interested in calculating $\langle\hat{S}\rangle$ and $\langle\hat{S}^2\rangle$ given in equations (14.80). From equations (14.80a), (14.88), and (14.89), we find

$$\langle\hat{S}\rangle = \left[\cos(\phi/2)\langle a_A^\dagger a_A - a_B^\dagger a_B\rangle - \sin(\phi/2)\langle a_A^\dagger a_B - a_B^\dagger a_A\rangle\right]. \tag{14.91}$$

In what follows, let us assume that a coherent state $|\alpha\rangle$ enters port A and that α is real. Then,

$$\langle \hat{S} \rangle = \cos(\phi/2)(\alpha^2 - \bar{n}_B) + \alpha \sin(\phi/2) \langle a_B^\dagger + a_B \rangle, \tag{14.92}$$

where $\bar{n}_B = \langle a_B^\dagger a_B \rangle$. If the vacuum field enters port B, $\bar{n}_B = 0$, $\langle a_B^\dagger + a_B \rangle = 0$, and equation (14.92) reduces to (14.81a).

There is now a lot of algebra involved in calculating $\langle \hat{S}^2 \rangle$ and ΔS. Using equations (14.80b), (14.88), and (14.89), one can obtain a very complicated expression for $\langle \hat{S}^2 \rangle$. However, we are interested only in fluctuations about the point $\phi = \pi - 2\epsilon$ where the signal is most sensitive to changes in ϕ. Since we know from equation (14.82) that ΔS has a contribution of order ϵ^0 at $\phi = \pi$, to zeroth order in ϵ, we need only calculate the fluctuations at the point $\phi = \pi$. Using equations (14.80b), (14.88), and (14.89) with $\phi = \pi$ and a coherent state in port A, we find [using the fact that all operators in equation (14.80b) are normal-ordered, allowing us to replace a by α and $a^\dagger$ by $\alpha^* = \alpha$ in that equation]

$$\begin{aligned}
\langle \hat{S}^2 \rangle &= \bar{n}_A + \bar{n}_B + \frac{1}{4} \langle (\alpha + a_B^\dagger)(\alpha + a_B^\dagger)(\alpha + a_B)(\alpha + a_B) \rangle \\
&\quad + \frac{1}{4} \langle (a_B^\dagger - \alpha)(a_B^\dagger - \alpha)(a_B - \alpha)(a_B - \alpha) \rangle \\
&\quad - \frac{1}{2} \langle (\alpha + a_B^\dagger)(a_B^\dagger - \alpha)(a_B - \alpha)(\alpha + a_B) \rangle \\
&= \bar{n}_A + \bar{n}_B + \frac{1}{2} \langle \{\alpha^4 + \alpha^2 [a_B^2 + (a_B^\dagger)^2 + 4 a_B^\dagger a_B] + (a_B^\dagger a_B)^2 \} \rangle \\
&\quad - \frac{1}{2} \langle \{\alpha^4 - \alpha^2 [a_B^2 + (a_B^\dagger)^2] + (a_B^\dagger a_B)^2 \} \rangle \\
&= \bar{n}_A + \bar{n}_B + \alpha^2 \langle (a_B^\dagger)^2 + a_B^2 + 2 a_B^\dagger a_B \rangle \\
&= \bar{n}_A + \bar{n}_B + \alpha^2 \langle (a_B^\dagger + a_B)^2 + (a_B^\dagger a_B - a_B a_B^\dagger) \rangle \\
&= \alpha^2 + \bar{n}_B + \alpha^2 \langle (a_B^\dagger + a_B)^2 \rangle - \alpha^2 = \bar{n}_B + \alpha^2 \langle (a_B + a_B^\dagger)^2 \rangle \\
&= \bar{n}_B + 4\alpha^2 \langle a_{B_1}^2 \rangle ,
\end{aligned} \tag{14.93}$$

where $a_{B_1} = 2(a_B + a_B^\dagger)$ is one of the field quadrature operators for the field entering port B.

Therefore, at $\phi = \pi$,

$$\Delta S^2 = \langle \hat{S}^2 \rangle - \langle \hat{S} \rangle^2 = 4\alpha^2 \Delta a_{B_1}^2 + \bar{n}_B. \tag{14.94}$$

For the vacuum field entering port B, $\Delta a_{B_1}^2 = 1/4$, $\bar{n}_B = 0$, and we recover equation (14.82). However, for a squeezed vacuum entering port B, $\Delta a_{B_1}^2 \sim \frac{1}{4} e^{-r}$ and $\bar{n}_B = \sinh^2 r$, where r is the squeezing parameter. Thus, if

$$\sqrt{4\alpha^2 \Delta a_{B_1}^2 + \bar{n}_B} < \alpha, \tag{14.95}$$

the noise is reduced below the SQL. This will be the case if $\alpha^2 > e^r \sinh^2 r$. Thus, by replacing the vacuum field with a squeezed field, one can reduce the sensitivity of the measurement to below the SQL.

14.2.2.4.1 Two-photon state: Last, we would like to see what happens when we send a one-photon state into each port—that is, the incident state is $|11\rangle$, with one photon in mode A and one in mode B. We assume that the modes have the same frequency and polarization. From equations (14.75), (14.88), and (14.89), we find

$$\begin{aligned}\overline{m_C m_D} &= \left\langle \left[\cos(\phi/4)a_A^\dagger + \sin(\phi/4)a_B^\dagger\right] \left[\cos(\phi/4)a_B^\dagger - \sin(\phi/4)a_A^\dagger\right] \right.\\ &\quad \left. \times \left[\cos(\phi/4)a_B - \sin(\phi/4)a_A\right] \left[\cos(\phi/4)a_A + \sin(\phi/4)a_B\right] \right\rangle \\ &= \cos^4(\phi/4) + \sin^4(\phi/4) - 2\sin^2(\phi/4)\cos^2(\phi/4) \\ &= \cos^2(\phi/2). \end{aligned} \tag{14.96}$$

For $\phi = \pi$, the average signal produced at detectors C and D is *identical* for each input field acting separately; however, as a result of interference, $\overline{m_C m_D} = 0$ when both fields are present. This is impossible classically, since the interferometer is set at a position where the average intensity at each detector does not vanish. The same type of result is obtained for a beam splitter alone. That is, if one replaces the mirrors in figure 14.1 by detectors, there can be a detected signal at either of these detectors, but not both. As a result of interference, the "photons" follow one path or the other. A better treatment would include more modes to allow for "pulsed" one-photon states to enter each port.

14.3 Summary

We have presented a very elementary theory of photodetection for both classical and quantized fields. You are urged to consult the bibliography for more advanced treatments. We have seen that there is an inherent "shot noise limit" that is associated with photodetection, resulting from the quantum nature of the detection process. Moreover, we have related the photo-count distribution to the noise properties of the fields that are incident on the detectors. It was shown that the use of squeezed states in one port of an interferometer can reduce the noise below the standard quantum limit. We now return to problems involving the interaction of radiation with matter.

Problems

1. In the limit of large N, show that the moments for the phase diffusion model are given approximately by $\langle I^n\rangle = n!\left(N|E_0|^2/4\right)^n$.
2. Derive equations (14.50). It may help to use the auxiliary function

$$f(a) = [ax + (1-x)]^n = \sum_{m=0}^{n} \binom{n}{m} (ax)^m (1-x)^{n-m}$$

   and express the moments in terms of f and its derivatives.
3. Prove equation (14.49).
4. A dielectric slab can serve as a beam splitter. Look up the amplitude transmission and reflection coefficients for a dielectric slab (you can derive

them if you want), and show that they are consistent with a relative phase shift of $\pi/2$.

5. We have assumed that the photo-count signal is proportional to the intensity at the detector. However, if we assume that the signal actually results from exciting atoms, we might expect that the signal for a counting time τ would be proportional to

$$S_1 = \int_t^{t+\tau} dt' \int_t^{t+\tau} dt'' \tilde{E}^-(t')\tilde{E}^+(t'')e^{-i\delta(t'-t'')},$$

where $E^+(t) = \tilde{E}^+(t)e^{-i\omega_L t}$, since the amplitude to excite a state is proportional to $\int_t^{t+\tau} dt' E^+(t')e^{-i\delta t}$. In this expression, δ is the detuning of the incident radiation from the transition frequency in the detector. Show that, in the limit of a *broadband detector*, the signal is proportional to

$$S_2 = \int_t^{t+\tau} dt' \tilde{E}^-(t')\tilde{E}^+(t') = \int_t^{t+\tau} dt' I(t').$$

To do this, assume that the incident field has a frequency spectrum centered at ω_L with width $\Delta\omega_L$. For simplicity, assume that the distribution of detector transition frequencies is given by

$$W(\omega_0) = \frac{1}{\sqrt{\pi\delta_d^2}} e^{-(\omega_0-\omega_L)^2/\delta_d^2}.$$

Average S_1 over ω_0 and show that it leads to S_2, provided that $\delta_d \gg \Delta\omega_L$—that is, for a broadband detector. See Cohen-Tannoudji et al. [3] for more details.

6. For the Michelson interferometer, we found that the average number of counts at detector D is proportional to $\cos^2(\phi/4)$. For a classical field in the phase diffusion model, we showed in chapter 13 that the fringes should wash out when the difference in arm lengths is greater than the coherence length of the field. How can these results be consistent? [Hint: The field mode (14.51) corresponds to a single mode only.]

7. You might think that, since the detectors C and D in the Michelson interferometer are independent, $\overline{m_C m_D} = \bar{m}_C \bar{m}_D$. Give a simple example to show that this cannot be true, in general. For a single-mode field, ideal detectors, and a 50/50 beam splitter, evaluate and compare $\overline{m_C m_D}$ and $\bar{m}_C \bar{m}_D$ for a coherent state, a pure number state, and a thermal state of the field.

8–9. A beam splitter is shown in figure 14.2 in which the difference in photo-counts between the two detectors is measured.

The "local oscillator" field A can be considered as a quantum coherent state of the field while some arbitrary quantum field S enters the second port. Obtain expressions for the field modes and the fields incident on each detector. Prove that by choosing S as a squeezed state, the noise in the signal can be reduced below the level it would have if the "vacuum" entered the port. Consider noise due only to fluctuations in the fields.

10. Consider the beam splitter and detectors shown in figure 14.2. Show that if a single photon enters each of the ports (that is, a two-photon state), with each photon having the same frequency and polarization, then the joint probability

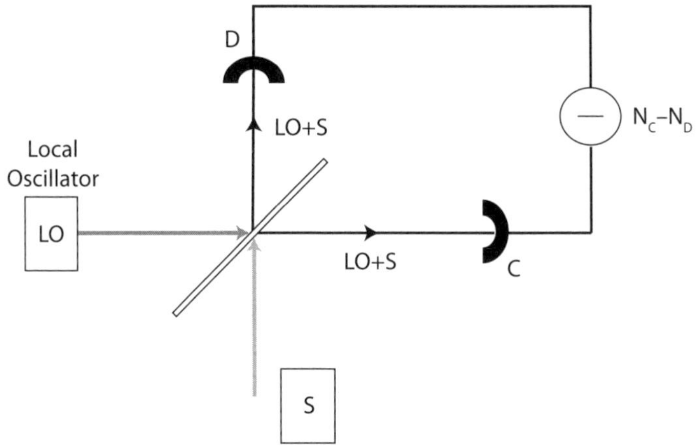

Figure 14.2. Schematic of a beam splitter.

of having a photo-count in each detector vanishes, assuming a lossless, 50/50 beam splitter. To do this, write expressions for the field modes, obtain the fields incident on each detector, and use equation (14.75). Also calculate the average number of photo-counts at each detector.

References

[1] A. Barchielli and V. P. Belavkin, *Measurements continuous in time and a posteriori states in quantum mechanics*, Journal of Physics A **24**, 1495–1514 (1991), and references therein; M. Ueda, N. Imoto, and T. Ogawa, *Quantum theory for continuous photodetection processes*, Physical Review A **41**, 3891–3904 (1990), and references therein; P. Goetsch and R. Graham, *Linear stochastic wave equations for continuously measured quantum systems*, Physical Review A **50**, 5242–5255 (1994), and references therein.

[2] R. Loudon, *The Quantum Theory of Light*, 2nd ed. (Oxford University Press, Oxford, UK, 1983), chap. 6.

[3] C. Cohen-Tannoudji, J. Dupont-Roc, and G. Grynberg, *Atom-Photon Interactions* (Wiley-Interscience, New York, 1992), complement A_{II}.

Bibliography

Most of the quantum optics books listed in chapter 1 have treatments of photodetection. Two early papers on quantum noise reduction using squeezed light are:

C. M. Caves, *Quantum-mechanical noise in an interferometer*, Physical Review D **23**, 1693–1708 (1981).

S. Reynaud and A. Heidmann, *Can photon noise be reduced?* Annales de Physique **10**, 227–239 (1985).

15

Atom–Quantized Field Interactions

In the previous chapters, we considered quantized radiation fields, but not their interaction with matter. In this chapter, we introduce a Hamiltonian that can be used to model the interaction of quantized radiation fields with atoms. We first give a brief review of the Heisenberg picture and then obtain an expression for the atom–field Hamiltonian in dipole approximation. To explore a relatively simple application of the formalism, we look at the interaction of a monochromatic cavity field with a single two-level atom that is stationary within the cavity—this problem is referred to as the Jaynes-Cummings problem [1]. As a second application of the formalism, we discuss dressed states of the atom–field system for a monochromatic field interacting with a two-level atom. These quantized dressed states differ in a fundamental manner from the semiclassical dressed states that we have been using up to this point. The quantized dressed states are true stationary states of the atom–field Hamiltonian in the rotating-wave approximation (RWA). Last, we indicate a method for creating coherent and squeezed states of the radiation field.

15.1 Interaction Hamiltonian and Equations of Motion

We have seen that there are times that the Heisenberg representation offers advantages over the Schrödinger representation. The Heisenberg state vector $|\psi\rangle^{\rm H}$ is related to the Schrödinger state vector $|\psi(t)\rangle$ via

$$|\psi\rangle^{\rm H} = e^{\frac{i}{\hbar}Ht}|\psi(t)\rangle, \tag{15.1}$$

where H is the *total* Hamiltonian for the system.

15.1.1 Schrödinger Representation

In the Schrödinger representation, the state vector, $|\psi(t)\rangle$, is time-dependent, as is the density matrix

$$\varrho^S(t) = |\psi(t)\rangle\langle\psi(t)|. \tag{15.2}$$

Operators are time-independent, but their expectation values are time-dependent, owing to the time dependence of the state vector. The equation of motion for the expectation value of an operator $\mathcal{O}$ is

$$i\hbar\frac{d}{dt}\langle\psi(t)|\mathcal{O}|\psi(t)\rangle = \langle\psi(t)|[\mathcal{O}, H]|\psi(t)\rangle, \tag{15.3}$$

where

$$\begin{aligned}\langle\mathcal{O}\rangle = \langle\psi(t)|\mathcal{O}|\psi(t)\rangle &= \sum_{E,E'}\langle\psi(t)|E\rangle\langle E|\mathcal{O}|E'\rangle\langle E'|\psi(t)\rangle \\ &= \sum_{E,E'}\langle E'|\psi(t)\rangle\langle\psi(t)|E\rangle\langle E|\mathcal{O}|E'\rangle \\ &= \sum_{E,E'}\langle E'|\varrho^S(t)|E\rangle\langle E|\mathcal{O}|E'\rangle = \mathrm{Tr}[\varrho^S(t)\mathcal{O}]\,. \end{aligned}\tag{15.4}$$

15.1.2 Heisenberg Representation

In the Heisenberg representation, the state vector is time-independent, as is the density matrix of the system, but the operators themselves are generally functions of time. We have

$$|\psi\rangle^{\mathrm{H}} = e^{\frac{i}{\hbar}Ht}|\psi(t)\rangle = e^{\frac{i}{\hbar}Ht}e^{-\frac{i}{\hbar}Ht}|\psi(0)\rangle = |\psi(0)\rangle, \tag{15.5}$$

$$\varrho^{\mathrm{H}} = e^{\frac{i}{\hbar}Ht}\varrho^S(t)e^{-\frac{i}{\hbar}Ht} = \varrho^S(0), \tag{15.6}$$

and

$$\mathcal{O}^{\mathrm{H}}(t) = e^{\frac{i}{\hbar}Ht}\mathcal{O}e^{-\frac{i}{\hbar}Ht}. \tag{15.7}$$

The expectation values of operators are still time time-dependent, but the time dependence originates from the *operator* rather than the state vector—namely,

$$\langle\mathcal{O}\rangle = \mathrm{Tr}[\varrho^{\mathrm{H}}\mathcal{O}^{\mathrm{H}}(t)]. \tag{15.8}$$

Using the definition (15.7), it is easy to show that $\mathcal{O}^{\mathrm{H}}(t)$ satisfies the equation of motion

$$i\hbar\frac{d\mathcal{O}^{\mathrm{H}}(t)}{dt} = [\mathcal{O}^{\mathrm{H}}(t), H]. \tag{15.9}$$

The advantage of the Heisenberg representation is that the operators often obey equations of motion that are identical to their classical counterparts. It is important to note that commutation laws such as $[x(t), p(t)] = i\hbar$ remain valid in the Heisenberg representation, but that $[x(t), p(t')] \neq i\hbar$ in general, if $t \neq t'$.

Things can get a bit confusing. The operator $|E\rangle\langle E|$ is constant in the Heisenberg representation since $e^{\frac{i}{\hbar}Ht}|E\rangle\langle E|e^{-\frac{i}{\hbar}Ht} = |E\rangle\langle E|$. On the other hand, if atoms interact with a field via a Hamiltonian $H = H_A + H_F + V_{AF}$, where $H_A|A\rangle = E_A|A\rangle$, $H_F|F\rangle = E_F|F\rangle$, then $\hat{A} \equiv |A\rangle\langle A|$ is a time-independent operator, but $A^{\mathrm{H}}(t) = e^{\frac{i}{\hbar}Ht}|A\rangle\langle A|e^{-\frac{i}{\hbar}Ht}$ is a function of time. In other words, if one considers a two-level atom interacting with a quantized radiation field, the raising operator

$\sigma_+ = |2\rangle\langle 1|$ is time-independent, but the corresponding Heisenberg operator $\sigma_+(t) = e^{\frac{i}{\hbar}Ht}|2\rangle\langle 1|e^{-\frac{i}{\hbar}Ht}$ is a function of time.

15.1.3 Hamiltonian

The Hamiltonian in Coulomb gauge derived in the appendix in chapter 12, appropriate to an N-electron atom interacting with a quantized radiation field, is given by

$$H = \sum_{\alpha=1}^{N} \frac{1}{2m} [\mathbf{p}_\alpha + e\mathbf{A}(\mathbf{r}_\alpha)]^2 + H_T + V_C\,, \tag{15.10}$$

where m is the electron mass,

$$V_C = -\frac{1}{4\pi\epsilon_0}\sum_{\alpha=1}^{N}\frac{Ne^2}{r_\alpha} + \frac{1}{8\pi\epsilon_0}\sum_{\alpha,\beta=1;\alpha\neq\beta}^{N}\frac{e^2}{|\mathbf{r}_\alpha - \mathbf{r}_\beta|} \tag{15.11}$$

is the electrostatic Coulomb potential, assuming that the nucleus is fixed at $\mathbf{R} = 0$ and has charge Ne, $\mathbf{r}_\alpha$ is the coordinate of electron α,

$$H_T = \sum_{\mathbf{k},\epsilon_k} \hbar\omega_k a^\dagger_{\mathbf{k},\epsilon_\mathbf{k}} a_{\mathbf{k},\epsilon_\mathbf{k}} \tag{15.12}$$

is the Hamiltonian associated with the transverse field modes,

$$\mathbf{A}(\mathbf{R}) = \sum_{\mathbf{k},\epsilon_\mathbf{k}} \left(\frac{\hbar}{2\epsilon_0\omega_\mathbf{k}\mathcal{V}}\right)^{1/2} \left(a_{\mathbf{k},\epsilon_\mathbf{k}}\boldsymbol{\epsilon}_\mathbf{k}e^{i\mathbf{k}\cdot\mathbf{R}} + a^\dagger_{\mathbf{k},\epsilon_\mathbf{k}}\boldsymbol{\epsilon}_\mathbf{k}e^{-i\mathbf{k}\cdot\mathbf{R}}\right) \tag{15.13}$$

is the vector potential (transverse in Coulomb gauge),

$$\boldsymbol{\epsilon}_\mathbf{k} \cdot \mathbf{k} = 0, \tag{15.14}$$

and the sum over $\hat{\boldsymbol{\epsilon}}_\mathbf{k}$ is over two independent free-field polarizations for each field mode having propagation vector $\mathbf{k}$. The quantity $\mathbf{p}_\alpha$ is the momentum canonical to $\mathbf{r}_\alpha$ for particle α having charge $-e$.

If we quantize as in the free atom and free-field case using

$$\left[(\mathbf{r}_\alpha)_i\,, (\mathbf{p}_{\alpha'})_j\right] = i\hbar\delta_{\alpha,\alpha'}\delta_{i,j}, \tag{15.15}$$

$$\left[a_i, a_j^\dagger\right] = i\hbar\delta_{i,j}, \tag{15.16}$$

and work in the Heisenberg representation, we regain Newton's equations for the atomic operators (with the Lorentz force law) and Maxwell's equations for the fields, i.e.,

$$\dot{a}_j + i\omega_j a_j = \left(\frac{i}{2\epsilon_0\hbar\omega_j}\right)\tilde{J}_{T_j}, \tag{15.17}$$

where $\tilde{J}_{T_j}$ is the transverse current defined by equation (12.210). Note that $a_j(t) \neq a_j(0)e^{-i\omega_j t}$ if $\tilde{J}_{T_j} \neq 0$ at any time.

When one deals with atom–classical field interactions, one often makes the dipole approximation, assuming that $ka \ll 1$, where k is the propagation constant for the field, and a is a characteristic atomic dimension on the order of the Bohr radius a_0. For a quantized field, however, the approximation $ka \ll 1$ cannot be true for *all* the field modes. Thus, if we run into any infinities involving summations over quantized field modes, *some* of those infinities can be removed by imposing a cutoff at $k \approx a_0^{-1}$. For the time being, let us suppose that the dipole approximation is valid for the field modes of interest.

In the dipole approximation, it is useful to carry out a unitary transformation on the state vector of the form

$$|\tilde{\psi}(t)\rangle = T|\psi(t)\rangle, \tag{15.18}$$

where

$$T = e^{-\frac{i}{\hbar}\boldsymbol{\mu}\cdot\mathbf{A}(\mathbf{R}=0)}, \tag{15.19}$$

and $\boldsymbol{\mu} = -\sum_{\alpha=1}^{N} e\mathbf{r}_\alpha$ is the electric dipole moment operator. Under this unitary transformation (15.19), the Hamiltonian (15.10) becomes

$$\tilde{H} = THT^\dagger = \sum_{\alpha=1}^{N} \frac{\left(T\mathbf{p}_\alpha T^\dagger\right)^2}{2m} + V_C + T\sum_j \hbar\omega_j a_j^\dagger a_j T^\dagger, \tag{15.20}$$

where the sum over j is a sum over all field modes.

Using the fact that

$$e^A B e^{-A} = B + [A, B] + \frac{1}{2!}[A, [A, B]] + \cdots, \tag{15.21}$$

we find that

$$T\mathbf{p}_\alpha T^\dagger = \mathbf{p}_\alpha - e\mathbf{A}(0). \tag{15.22}$$

Moreover, since

$$T = e^{\sum_j \lambda_j^* a_j - \lambda_j a_j^\dagger} = \prod_j \mathcal{D}^\dagger(\lambda_j, \lambda_j^*) \tag{15.23}$$

is just a product of field displacement operators (12.71), with

$$\lambda_j = \frac{i\hat{\boldsymbol{\epsilon}}_j \cdot \boldsymbol{\mu}}{\sqrt{2\epsilon_0 \mathcal{V}\hbar\omega_j}}, \tag{15.24}$$

it follows that

$$Ta_jT^\dagger = \mathcal{D}^\dagger(\lambda_j, \lambda_j^*)a_j\mathcal{D}(\lambda_j, \lambda_j^*) = a_j + \lambda_j \tag{15.25}$$

and

$$Ta_j^\dagger T^\dagger = a_j^\dagger + \lambda_j^*. \tag{15.26}$$

As a result, $\tilde{H}$ can be written as

$$\tilde{H}_T = \sum_{\alpha=1}^{N} \frac{p_\alpha^2}{2m} + V_C + \sum_j \hbar\omega_j a_j^\dagger a_j + \sum_j \hbar\omega_j(\lambda_j^* a_j + \lambda_j a_j^\dagger) + \sum_j \hbar\omega_j |\lambda_j|^2. \tag{15.27}$$

The $\sum_j \hbar\omega_j(\lambda_j^* a_j + \lambda_j a_j^\dagger)$ term is simply equal to $-\boldsymbol{\mu}\cdot\mathbf{E}_T(0)$, where

$$\mathbf{E}_T(\mathbf{R}) = i\sum_{\mathbf{k},\boldsymbol{\epsilon}_\mathbf{k}} \left(\frac{\hbar\omega_\mathbf{k}}{2\epsilon_0\mathcal{V}}\right)^{1/2} \left(a_{\mathbf{k},\boldsymbol{\epsilon}_\mathbf{k}}\boldsymbol{\epsilon}_\mathbf{k} e^{i\mathbf{k}\cdot\mathbf{R}} - a_{\mathbf{k},\boldsymbol{\epsilon}_\mathbf{k}}^\dagger \boldsymbol{\epsilon}_\mathbf{k} e^{-i\mathbf{k}\cdot\mathbf{R}}\right) \tag{15.28}$$

is the quantum operator for the transverse electric field. The $\sum_j \hbar\omega_j|\lambda_j|^2$ term is infinite, but independent of the field—it represents a self-energy term that can be taken into account by a renormalization of the atomic state energies. Consequently, we subtract out this term (the zero point field energy has already been excluded) and take as our Hamiltonian in dipole approximation

$$H = \sum_{\alpha=1}^{N} \frac{p_\alpha^2}{2m} + V_C(\mathbf{r}_\alpha) - \boldsymbol{\mu}\cdot\mathbf{E}_T(0) + \sum_j \hbar\omega_j a_j^\dagger a_j. \tag{15.29}$$

Any expectation values of operators calculated with the new Hamiltonian are the same as those for the original one, although technically speaking, in a dielectric medium the correct operator in this equation should be $\mathbf{D}_T(0)$ rather than $\mathbf{E}_T(0)$, if we use our original basis states to take expectation values [2, 3].

This Hamiltonian can be generalized to include the center-of-mass motion of the atom. In that case, we take

$$H = \frac{P^2}{2M} + H_{\text{Atom}} + H_{\text{Field}} + V_{AF}, \tag{15.30}$$

$$H_{\text{Atom}} = \sum_{\alpha=1}^{N} \frac{p_\alpha^2}{2m} + V_C(\mathbf{r}_\alpha), \tag{15.31}$$

$$H_{\text{Field}} = \sum_j \hbar\omega_j a_j^\dagger a_j, \tag{15.32}$$

$$V_{AF} = -\boldsymbol{\mu}\cdot\mathbf{E}_T(\mathbf{R}), \tag{15.33}$$

where $\mathbf{R}$ is the center-of-mass coordinate of the atom, $\mathbf{P}$ is the center-of-mass momentum of the atom, M is the atomic mass, m is the reduced mass of an electron located at position $\mathbf{r}_\alpha$ relative to the nucleus having canonical momentum $\mathbf{p}_\alpha$, and $\mathbf{E}_T(\mathbf{R})$ is given by equation (15.28). We will usually use the Schrödinger representation when working with probability amplitudes and the Heisenberg representation when working with density matrix and operator equations. The Hamiltonian (15.30) is the starting point for many of the calculations in the remainder of this book.

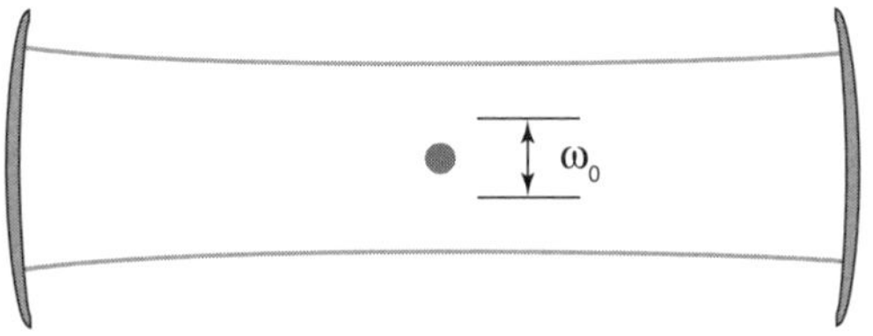

Figure 15.1. Jaynes-Cummings model. A single, stationary two-level atom interacts with a single-mode cavity field.

15.1.4 Jaynes-Cummings Model

Let us consider first one of the most fundamental problems involving the interaction of a quantized field with a stationary two-level atom—the so-called Jaynes-Cummings model [1]. Even though it is difficult to achieve conditions where the Jaynes-Cummings model is applicable in experimental situations, it is instructive nevertheless to study this model in some detail. Moreover, there has been an increased interest in using high-finesse optical cavities as a springboard for many applications in quantum information. In these applications, one encounters problems that involve the interaction of one or more atoms with a cavity field. The Jaynes-Cummings model is the simplest possible realization of *cavity quantum electrodynamic* problems of this nature [4].

The Jaynes-Cummings Hamiltonian can be taken as the limit of equation (15.30) when a single-mode cavity field interacts with a stationary two-level atom in the RWA (figure 15.1). The Hamiltonian in this case is

$$\mathrm{H} = \mathrm{H}_A + \mathrm{H}_F + \mathrm{V}_{AF} = \frac{\hbar\omega_0}{2}\boldsymbol{\sigma}_z + \hbar\omega a^\dagger a + \hbar(g\boldsymbol{\sigma}_+ a + g^* a^\dagger \boldsymbol{\sigma}_-), \qquad (15.34)$$

where ω_0 is the atomic transition frequency, ω is the single-mode field frequency,

$$g = -i\left(\frac{\omega}{2\hbar\epsilon_0\mathcal{V}}\right)^{1/2}\mu_{21}, \qquad (15.35)$$

μ_{21} is the projection of the dipole moment matrix element along the field polarization direction, and the $\boldsymbol{\sigma}$ operators have their conventional meaning. The spatial mode function for the cavity field has been set equal to unity for simplicity. Since the atom is assumed to be stationary, the actual form of the mode function is not important. For moving atoms, the spatial field pattern becomes important since the field amplitude changes as the atom moves through the field. In the following section, we obtain the eigenvalues and eigenkets of the Hamiltonian (15.34). Here, we concentrate on the time evolution of the state amplitudes of the atom–field system.

We write the state vector for the atom–field system in a field interaction representation as

$$\begin{aligned}|\psi(t)\rangle &= \tilde{c}_{1,0}(t)e^{i\omega t/2}\,|1;0\rangle \\ &\quad + \sum_{n=1}^{\infty}\tilde{c}_{1,n}(t)e^{-in\omega t}e^{i\omega t/2}\,|1;n\rangle \\ &\quad + \sum_{n=1}^{\infty}\tilde{c}_{2,n-1}(t)e^{-in\omega t}e^{i\omega t/2}|2;n-1\rangle, \end{aligned} \qquad (15.36)$$

where the first index refers to the atomic state and the second to the field state. We have separated off the state $|1;0\rangle$, since this is an eigenstate of the system in the RWA, having energy $-\hbar\omega_0/2$. In the field interaction representation, the ground-state amplitude evolves according to

$$\tilde{c}_{1,0}(t) = \tilde{c}_{1,0}(0)e^{i\delta t/2}, \tag{15.37}$$

where $\delta = \omega_0 - \omega$. Using the Hamiltonian (15.34) and Schrödinger equation, we can show easily that the evolution equations for the remaining state amplitudes are

$$d\tilde{c}_{1,n}/dt = i\frac{\delta}{2}\tilde{c}_{1,n} - ig_n^*\tilde{c}_{2,n-1}, \tag{15.38a}$$

$$d\tilde{c}_{2,n-1}/dt = -i\frac{\delta}{2}\tilde{c}_{2,n-1} - ig_n\tilde{c}_{1,n}, \tag{15.38b}$$

where

$$g_n = \sqrt{n}g. \tag{15.39}$$

These equations are identical to those solved earlier in the field interaction representation for a classical field interacting with a two-level atom. Thus, we can use the result (2.116) to write the solution as

$$\begin{pmatrix} \tilde{c}_{1,n} \\ \tilde{c}_{2,n-1} \end{pmatrix} = \begin{pmatrix} \cos\left(\frac{\Omega_n t}{2}\right) + \frac{i\delta}{\Omega_n}\sin\left(\frac{\Omega_n t}{2}\right) & -\frac{2ig_n^*}{\Omega_n}\sin\left(\frac{\Omega_n t}{2}\right) \\ -\frac{2ig_n}{\Omega_n}\sin\left(\frac{\Omega_n t}{2}\right) & \cos\left(\frac{\Omega_n t}{2}\right) - \frac{i\delta}{\Omega_n}\sin\left(\frac{\Omega_n t}{2}\right) \end{pmatrix}$$

$$\times \begin{pmatrix} \tilde{c}_{1,n}(0) \\ \tilde{c}_{2,n-1}(0) \end{pmatrix}, \tag{15.40}$$

where

$$\Omega_n = \sqrt{\delta^2 + 4\,|g_n|^2}. \tag{15.41}$$

Equations (15.37) and (15.40) provide a complete solution to the problem. The field connects states $|1;n\rangle$ and $|2;n-1\rangle$ $(n > 0)$ with a characteristic Rabi frequency Ω_n that depends on n through the coupling constant g_n.

Recall that the state amplitudes for a two-level atom interacting with a classical monochromatic field oscillate in time with a frequency equal to one-half the Rabi frequency. We want to see when this is also true for a quantized coherent state of the field. Assume that at $t = 0$, the atom is in its ground state and the field is in a coherent state

$$|\psi(0)\rangle = |1\rangle\,|\alpha\rangle, \tag{15.42}$$

with α real, which implies that

$$\tilde{c}_{1,n}(0) = \frac{\alpha^n e^{-\alpha^2/2}}{\sqrt{n!}}, \quad \tilde{c}_{2,n-1}(0) = 0. \tag{15.43}$$

From equation (15.40), we find that the probability that the atom is in its excited state at time t is

$$
\begin{aligned}
P_2 &= \sum_{n=1}^{\infty} \left|\tilde{c}_{2,n-1}(t)\right|^2 \\
&= \sum_{n=1}^{\infty} \frac{4\left|g_n\right|^2}{\Omega_n^2} \sin^2\left(\frac{\Omega_n t}{2}\right) \left|\tilde{c}_{1,n}(0)\right|^2 \\
&= \sum_{n=1}^{\infty} \frac{4\left|g_n\right|^2}{\delta^2 + 4\left|g_n\right|^2} \frac{\alpha^{2n} e^{-\alpha^2}}{n!} \sin^2\left(\frac{\sqrt{\delta^2 + 4\left|g_n\right|^2}}{2} t\right).
\end{aligned}
\tag{15.44}
$$

Note that if we take g_n to be constant, we recover the results for a classical field. However, g_n is *not* constant, since it varies as $\sqrt{n}$. You might think that for large α, the result reduces to that for a classical field, but you would be only partially correct with this conclusion.

If $\alpha \gg 1$, the distribution $(\alpha^{2n} e^{-\alpha^2})/n!$ is sharply peaked about $n = \alpha^2 = \bar{n}$ and, to first approximation, we can take $g_n = g\alpha^2$. In this limit, we recover the classical result with $\chi = 2\alpha g$. To see the role that fluctuations about this average value produce, we write

$$
g_n = \sqrt{n} g = \sqrt{\bar{n} + (n - \bar{n})} g \approx \sqrt{\bar{n}} g + \frac{1}{2} \frac{(n - \bar{n})}{\sqrt{\bar{n}}} g,
\tag{15.45}
$$

such that the *phase* appearing in equation (15.40) is

$$
\begin{aligned}
\Phi &= \frac{\sqrt{\delta^2 + 4\left|g_n\right|^2}}{2} t \approx \frac{\sqrt{\delta^2 + 4\bar{n}\left|g\right|^2 + 4(n - \bar{n})\left|g\right|^2}}{2} t \\
&\approx \frac{\sqrt{\delta^2 + 4\bar{n}\left|g\right|^2}}{2} t + \frac{(n - \bar{n})\left|g\right|^2}{\sqrt{\delta^2 + 4\bar{n}\left|g\right|^2}} t.
\end{aligned}
\tag{15.46}
$$

You may now remember the admonition about a small dog given earlier in this book. To neglect a term in a *phase*, that term must have magnitude much less than unity, regardless of its value relative to other phase terms. Thus, as soon as the second term in equation (15.46) becomes of order unity, the various n components get out of phase, giving rise to a "collapse" of the Rabi oscillations. This occurs for times of order

$$
t_c \approx \frac{\sqrt{\delta^2 + 4\alpha^2\left|g\right|^2}}{\alpha\left|g\right|^2},
\tag{15.47}
$$

where we set $(n - \bar{n}) \approx \sqrt{\bar{n}} = \alpha$ in the numerator of equation (15.46) to evaluate t_c. If $\delta = 0$, the collapse time is given by $|g|t_c \approx 2$, as is seen in figure 15.2.

This is an interesting example where a "strong" quantum mechanical coherent state of the field is not equivalent to a classical field. The field fluctuations of the coherent state eventually are felt in the phase of the atomic response [5]. Note that as $g \sim 0$ (as it would in free space), but $\alpha g \sim$ constant, the collapse time is infinite.

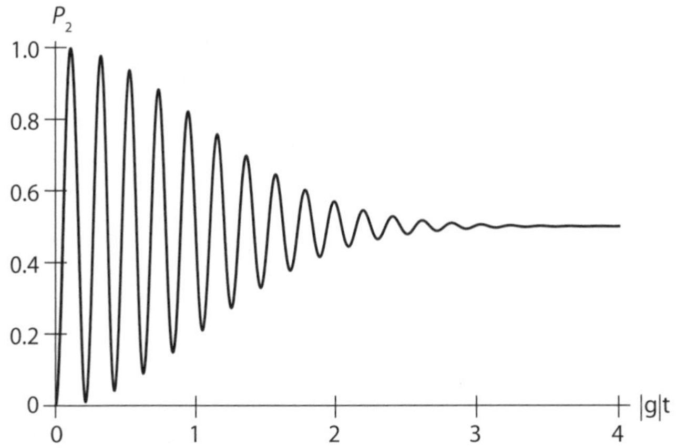

Figure 15.2. Collapse of Rabi oscillations in the Jaynes-Cummings model, with $\delta = 0$, and $\alpha = 15$.

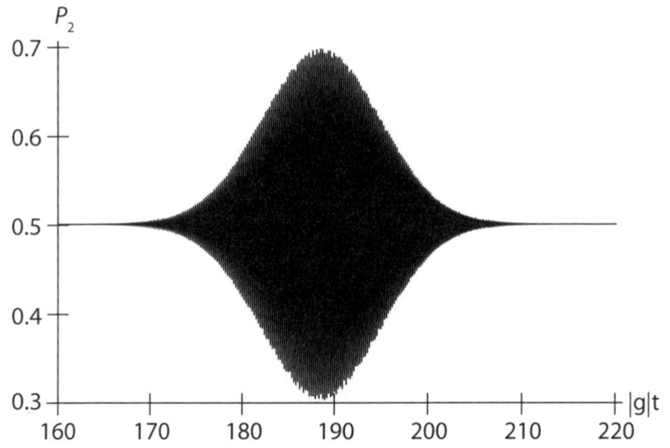

Figure 15.3. Revival of Rabi oscillations in the Jaynes-Cummings model.

The collapse is not the end of the story. To illustrate another phenomenon that occurs, we set $\delta = 0$ in equation (15.46), leading to

$$\sin\Phi \approx \sin\left[\sqrt{\bar{n}}\,|g|\,t + \frac{(n-\bar{n})\,|g|}{2\sqrt{\bar{n}}}t\right]. \tag{15.48}$$

We see that $\sin\Phi$ is independent of n for times when

$$t \approx t_r = 2(2m\pi)\sqrt{\bar{n}}/|g| = 4m\pi\alpha/|g|, \tag{15.49}$$

for integer m. This leads to *quantum revivals* (not religious) at such times. Figure 15.3 shows the revival for $m = 1$ and $\alpha = 15$, centered at $|g|t = 4\pi\alpha \approx 188$. Both collapses and revivals are intrinsically quantum phenomena.

Last, consider an initial state in which the atom is in its excited state and the field is in the vacuum state

$$|\psi(0)\rangle = |2\rangle\,|0\rangle. \tag{15.50}$$

Now, $\tilde{c}_{2,0}(0) = 1$, and the excited state probability is

$$P_2 = \left|\tilde{c}_{2,0}(t)\right|^2 = \left[\cos^2\left(\frac{\sqrt{\delta^2 + 4|g|^2}}{2}t\right) + \frac{\delta^2}{\delta^2 + 4|g|^2}\sin\left(\frac{\sqrt{\delta^2 + 4|g|^2}}{2}t\right)\right]. \tag{15.51}$$

There is no spontaneous decay because there is only a single mode and the system is totally reversible. The Rabi oscillations in this case are known as *vacuum Rabi oscillations* [6].

15.2 Dressed States

To introduce quantized dressed states [7], we start with the Jaynes-Cummings Hamiltonian (15.34). We have seen already that the cavity field couples only pairs of states such as $|1, n\rangle$ and $|2, n-1\rangle$. In other words, for any state of the field, the interaction breaks down into a number of effective two-level problems. This is the reason that we were able to solve the Jaynes-Cummings model in analytic form. The fact that the problem is exactly soluble suggests that the Hamiltonian has a relatively simple structure. Moreover, it should be possible to diagonalize **H** exactly.

In the $|\alpha, n\rangle$ basis, where $\alpha = 1, 2$ labels the atomic states and $n = 0, 1, 2, \ldots$ the field states, matrix elements of **H** [equation (15.34)] are given by

$$\langle 1, n|\mathbf{H}|1, m\rangle = (-\hbar\omega_0/2 + n\hbar\omega)\,\delta_{n,m}, \tag{15.52a}$$

$$\langle 2, n|\mathbf{H}|2, m\rangle = (\hbar\omega_0/2 + n\hbar\omega)\,\delta_{n,m}, \tag{15.52b}$$

$$\langle 2, n-1|\mathbf{H}|1, n\rangle = \hbar g\langle 2, n-1|\boldsymbol{\sigma}_+ a|1, n\rangle = \sqrt{n}\hbar g, \tag{15.52c}$$

$$\langle 1, n|\mathbf{H}|2, n-1\rangle = \hbar g^*\langle 1, n|a^\dagger\boldsymbol{\sigma}_-|2, n-1\rangle = \sqrt{n}\hbar g^*, \tag{15.52d}$$

and all other matrix elements vanish. There is a nondegenerate ground state $|1, 0\rangle$ having energy

$$E_0 = -\hbar\omega_0/2 = -\hbar\omega/2 - \hbar\delta/2. \tag{15.52e}$$

The remaining energy levels of the unperturbed Hamiltonian $\mathbf{H}_A + \mathbf{H}_F$ are given by a ladder of nearly degenerate doublets shown in figure 15.4, where the frequency separation between two levels *within* a doublet is

$$(E_{2,n-1} - E_{1,n})/\hbar = \omega_0 - \omega = \delta. \tag{15.53}$$

The separation between the corresponding levels in *successive* doublets is the field frequency ω.

For example, if we isolate four states of the Hamiltonian, $|2, n-1\rangle$, $|1, n\rangle$, $|2, n-2\rangle$, $|1, n-1\rangle$ with $n \geq 2$, the corresponding submatrix of the Hamiltonian would be

$$\mathbf{H} = \hbar\begin{pmatrix} \frac{\omega_0}{2} + (n-1)\omega & g_n & 0 & 0 \\ g_n^* & -\frac{\omega_0}{2} + n\omega & 0 & 0 \\ 0 & 0 & \frac{\omega_0}{2} + (n-2)\omega & g_{n-1} \\ 0 & 0 & g_{n-1}^* & -\frac{\omega_0}{2} + (n-1)\omega \end{pmatrix}, \tag{15.54}$$

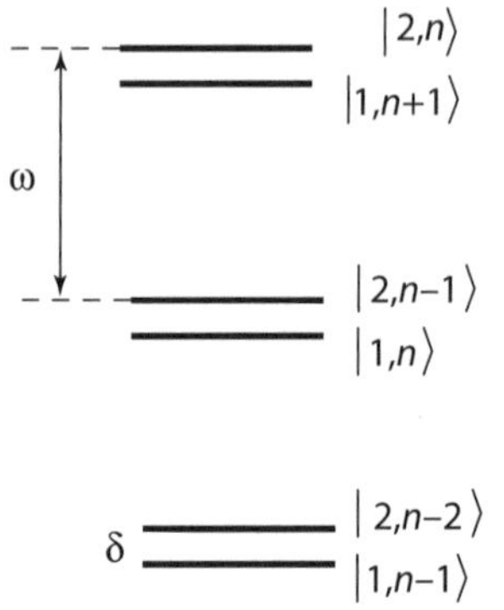

Figure 15.4. Quantum states of the atom plus field in the absence of the atom–field interaction.

where $g_n = \sqrt{n}g$. Since **H** breaks down into a series of 2×2 submatrices, each submatrix can be diagonalized separately. The diagonalization of each submatrix is the same as that we encountered using semiclassical dressed states.

Let us isolate a single 2×2 matrix

$$\mathbf{H}_n = \hbar \begin{pmatrix} \frac{\omega_0}{2} + (n-1)\omega & g_n \\ g_n^* & -\frac{\omega_0}{2} + n\omega \end{pmatrix}$$

$$= \left(n - \frac{1}{2}\right)\hbar\omega\hat{\mathbf{I}} + \hbar \begin{pmatrix} \frac{\delta}{2} & g_n \\ g_n^* & -\frac{\delta}{2} \end{pmatrix} \tag{15.55}$$

for the states $|2, n-1\rangle$, $|1, n\rangle$, with $n \geq 1$. (Note that the order of the states differs from the $|1\rangle$, $|2\rangle$ order we have taken normally.) If we relabel the states as

$$\begin{array}{cc} & |1n\rangle \; |2n-1\rangle \\ \begin{array}{c} |1n\rangle \\ |2n-1\rangle \end{array} & \begin{pmatrix} \dots & \dots \\ \dots & \dots \end{pmatrix} \end{array}, \tag{15.56}$$

then

$$\mathbf{H}_n = \left(n - \frac{1}{2}\right)\hbar\omega\hat{\mathbf{I}} + \frac{\hbar}{2}\begin{pmatrix} -\delta & \Omega_n^{0\,*} \\ \Omega_n^0 & \delta \end{pmatrix}, \tag{15.57}$$

where

$$\Omega_n^0 = 2g_n. \tag{15.58}$$

Aside from the term proportional to the unit matrix, this is identical to the matrix we diagonalized for the semiclassical dressed states. Therefore, the eigenvalues associated with H_n are

$$E_{n_\pm} = \left(n - \frac{1}{2}\right)\hbar\omega \pm \frac{\hbar\Omega_n}{2}, \tag{15.59}$$

with

$$\Omega_n = \sqrt{|\Omega_n^0|^2 + \delta^2}, \tag{15.60}$$

and the eigenkets are

$$|I_n\rangle = \cos\theta_n|1, n\rangle - \sin\theta_n|2, n-1\rangle, \tag{15.61a}$$

$$|II_n\rangle = \sin\theta_n|1, n\rangle + \cos\theta_n|2, n-1\rangle, \tag{15.61b}$$

where $0 \leq \theta_n \leq \pi/2$,

$$\tan 2\theta_n = \Omega_n^0/\delta, \tag{15.62a}$$

$$\sin\theta_n = \left[\frac{1-\delta/\Omega_n}{2}\right]^{1/2}, \tag{15.62b}$$

$$\cos\theta_n = \left[\frac{1+\delta/\Omega_n}{2}\right]^{1/2}, \tag{15.62c}$$

and the coupling constant $g = \Omega_n^0/2$ has been taken as real. As was the case for the semiclassical dressed states, state $|II_n\rangle$ is always higher in energy than state $|I_n\rangle$. These states have different asymptotic forms as Ω_n^0 goes to zero depending on the sign of δ. Specifically,

$$|I_n\rangle \sim \begin{cases} |1, n\rangle & \text{for } \delta > 0, \\ -|2, n-1\rangle & \text{for } \delta < 0, \end{cases} \tag{15.63a}$$

$$|II_n\rangle \sim \begin{cases} |2, n-1\rangle & \text{for } \delta > 0, \\ |1, n\rangle & \text{for } \delta < 0, \end{cases} \tag{15.63b}$$

as $\Omega_n^0 \sim 0$. State $|1, n\rangle$ is higher in energy than $|2, n-1\rangle$ if $\delta < 0$ and lower if $\delta > 0$. In the new basis, **H** is diagonal with eigenvalues and eigenvectors defined by equation (15.59) and equation (15.61), respectively. The new eigenstates are referred to as (quantized) dressed states of the atom–field system [7]. The first few eigenstates are shown schematically in figure 15.5.

We use the dressed states in chapter 20 to gain insight into light scattering in intense fields. The strength of the quantized dressed states, aside from the fact that they are true stationary states of a time-independent Hamiltonian, is that they often can be used to provide a "picture" that helps one to understand atom–field interactions. For example, we will see in chapter 20 that the arrows shown in figure 15.6 between eigenkets $\{|II_n\rangle, |I_n\rangle\}$ and $\{|II_{n-1}\rangle, |I_{n-1}\rangle\}$ represent possible paths for spontaneous emission. As such, it is obvious immediately from the figure that the radiation scattered by an intense field consists of three lines centered at the transition frequencies shown in the figure. On the other hand, in almost all cases where a strong field drives a transition, calculations are most easily performed using a semiclassical dressed-state approach rather than a quantized dressed-state approach, since the former involves a single classical monochromatic field, while the latter requires a ladder of quantized field states.

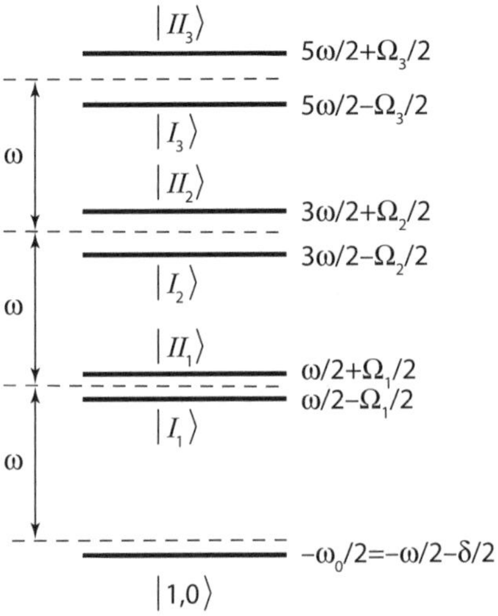

Figure 15.5. The first few eigenkets of the Jaynes-Cummings Hamiltonian (frequency separations not drawn to scale). The separation of each doublet increases with increasing n.

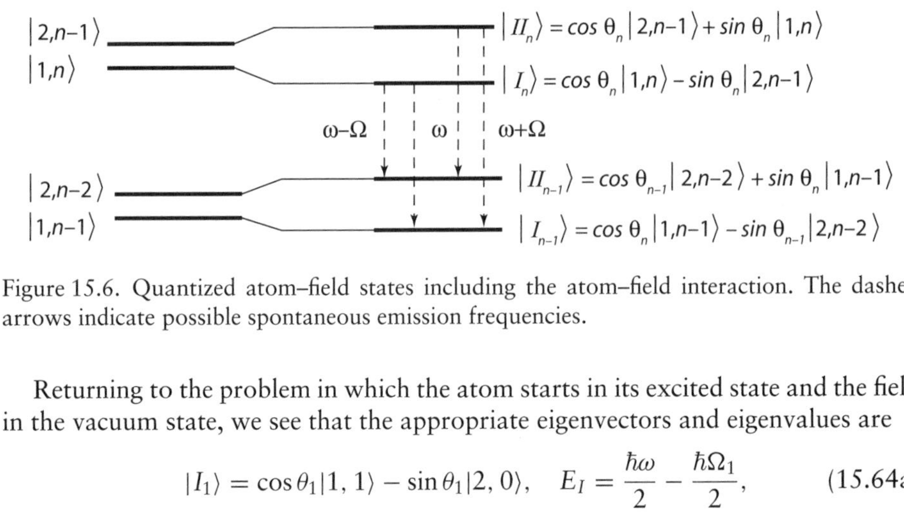

Figure 15.6. Quantized atom–field states including the atom–field interaction. The dashed arrows indicate possible spontaneous emission frequencies.

Returning to the problem in which the atom starts in its excited state and the field in the vacuum state, we see that the appropriate eigenvectors and eigenvalues are

$$|I_1\rangle = \cos\theta_1|1,1\rangle - \sin\theta_1|2,0\rangle, \quad E_I = \frac{\hbar\omega}{2} - \frac{\hbar\Omega_1}{2}, \tag{15.64a}$$

$$|II_1\rangle = \sin\theta_1|1,1\rangle + \cos\theta_1|2,0\rangle, \quad E_{II} = \frac{\hbar\omega}{2} + \frac{\hbar\Omega_1}{2}. \tag{15.64b}$$

The initial state in which the field is in the vacuum state is *not* an energy eigenstate of the system. As a consequence, the state vector for the system is

$$|\psi(t)\rangle = e^{-i\omega t/2}\left(\cos\theta_1 e^{-i\Omega_1 t/2}|II_1\rangle - \sin\theta_1 e^{i\Omega_1 t/2}|I_1\rangle\right), \tag{15.65}$$

that is, a superposition of energy eigenstates that oscillates in time. Although the oscillations result from the interaction of the vacuum field with the excited atom, the nomenclature *vacuum Rabi oscillations* is somewhat misleading. It is simply the fact that the system is not prepared in an energy eigenstate that leads to the oscillations. This is similar to the scenario that gives rise to neutrino oscillations, since the neutrino is not produced in a mass eigenstate.

15.3 Generation of Coherent and Squeezed States

Last, let us look at how to generate coherent and squeezed states of the radiation field.

15.3.1 Coherent States

First, we show that a classical current distribution gives rise to a coherent state. The atom–field interaction is taken as

$$V_{AF} = -\boldsymbol{\mu}(t) \cdot \mathbf{E}(\mathbf{R}), \tag{15.66}$$

where $\boldsymbol{\mu}(t)$ is the dipole moment associated with a *classical* source [$\boldsymbol{\mu}(t)$ is *not* an operator, simply a vector function of time], and $\mathbf{E}(\mathbf{R})$ is the quantized field operator. The total Hamiltonian for the free field and the atoms is

$$H(t) = \sum_{\mathbf{k}} \hbar\omega_k a_{\mathbf{k}}^{\dagger} a_{\mathbf{k}} - \boldsymbol{\mu}(t) \cdot \left[i \sum_{\mathbf{k}} \left(\frac{\hbar\omega_k}{2\epsilon_0 \mathcal{V}} \right)^{1/2} \boldsymbol{\epsilon}_{\mathbf{k}} a_{\mathbf{k}} e^{i\mathbf{k}\cdot\mathbf{R}} + \text{adj.} \right], \tag{15.67}$$

where the sum over polarization modes has been suppressed. We transform this Hamiltonian into a field interaction representation via

$$\begin{aligned}\tilde{H}(t) &= e^{i\omega_k a_{\mathbf{k}}^{\dagger} a_{\mathbf{k}} t} H(t) e^{-i\omega_k a_{\mathbf{k}}^{\dagger} a_{\mathbf{k}} t} \\ &= -\boldsymbol{\mu}(t) \cdot \left[i \sum_{\mathbf{k}} \left(\frac{\hbar\omega_k}{2\epsilon_0 \mathcal{V}} \right)^{1/2} \boldsymbol{\epsilon}_{\mathbf{k}} a_{\mathbf{k}} e^{i(\mathbf{k}\cdot\mathbf{R} - \omega_k t)} + \text{adj.} \right],\end{aligned} \tag{15.68}$$

where we used the fact that $e^{i\omega_k a_{\mathbf{k}}^{\dagger} a_{\mathbf{k}} t} a_{\mathbf{k}} e^{-i\omega_k a_{\mathbf{k}}^{\dagger} a_{\mathbf{k}} t} = a_{\mathbf{k}} e^{-i\omega_k t}$.

Since $[\tilde{H}(t), \tilde{H}(t')] \neq 0$, one cannot simply exponentiate $\tilde{H}(t)$ to get a solution. On the other hand, the solution of the Schrödinger equation

$$i\hbar \frac{\partial |\tilde{\psi}(t)\rangle}{\partial t} = \tilde{H}(t) |\tilde{\psi}(t)\rangle, \tag{15.69}$$

with $|\tilde{\psi}(0)\rangle = |0\rangle$, is

$$|\tilde{\psi}(t)\rangle = \lim_{\Delta t \to 0} e^{-\frac{i}{\hbar}\tilde{H}(t_n)\Delta t} e^{-\frac{i}{\hbar}\tilde{H}(t_{n-1})\Delta t} \ldots e^{-\frac{i}{\hbar}\tilde{H}(t_2)\Delta t} e^{-\frac{i}{\hbar}\tilde{H}(t_1)\Delta t} |0\rangle, \tag{15.70}$$

which can be verified by expanding the exponentials. If $[\tilde{H}(t_1), \tilde{H}(t_2)]$ is a c-number, as it is, then

$$e^{-\frac{i}{\hbar}\tilde{H}(t_2)\Delta t} e^{-\frac{i}{\hbar}\tilde{H}(t_1)\Delta t} \approx e^{-\frac{i}{\hbar}\left[\tilde{H}(t_2) + \tilde{H}(t_1)\right]\Delta t} e^{\frac{i}{\hbar}\left[\tilde{H}(t_1), \tilde{H}(t_2)\right]\Delta t}, \tag{15.71}$$

which is the product of $e^{-\frac{i}{\hbar}[\tilde{H}(t_2) + \tilde{H}(t_1)]\Delta t}$ and a c-number phase factor.

As a consequence, the solution (15.70) for $|\tilde{\psi}(t)\rangle$ is

$$|\tilde{\psi}(t)\rangle = e^{-\frac{i}{\hbar}\int_0^t \tilde{H}(t')dt'} |0\rangle \times \text{phase factor}. \tag{15.72}$$

Using equation (15.68), we see that

$$-\frac{i}{\hbar} \int_0^t \tilde{H}(t')dt' = \sum_{\mathbf{k}} \left[-\alpha_{\mathbf{k}}^*(t) a_{\mathbf{k}} + \alpha_{\mathbf{k}}(t) a_{\mathbf{k}}^{\dagger} \right], \tag{15.73}$$

where

$$\alpha_{\mathbf{k}}^*(t) = \frac{1}{\hbar}\left(\frac{\hbar\omega_k}{2\epsilon_0\mathcal{V}}\right)^{1/2}\int_0^t dt'\boldsymbol{\mu}(t')\cdot\boldsymbol{\epsilon}_{\mathbf{k}}e^{i(\mathbf{k}\cdot\mathbf{R}-\omega_k t')}. \tag{15.74}$$

Thus, to within a constant phase factor,

$$|\tilde{\psi}(t)\rangle = e^{\sum_{\mathbf{k}}\left(-\alpha_{\mathbf{k}}^* a_{\mathbf{k}}+\alpha_{\mathbf{k}}a_{\mathbf{k}}^\dagger\right)}|0\rangle = \prod_{\mathbf{k}} e^{\left(-\alpha_{\mathbf{k}}^* a_{\mathbf{k}}+\alpha_{\mathbf{k}}a_{\mathbf{k}}^\dagger\right)}|0\rangle. \tag{15.75}$$

In other words, a classical current source creates a multimode coherent state of the field.

15.3.2 Squeezed States

To generate a squeezed state, it is necessary to produce two photons that are correlated. This can be done in four-wave mixing by converting two pump photons to two correlated signal photons or in parametric down conversion, where a single photon is converted to two photons. A prototype Hamiltonian in a field interaction representation is taken to be of the form

$$\tilde{H} = \hbar\Lambda a^2 + \hbar\Lambda^* a^{\dagger 2}, \tag{15.76}$$

where Λ can represent either the second-order susceptibility (parametric down conversion) or third-order susceptibility (four-wave mixing). There are very few treatments, if any, in which Λ is calculated from first principles. The solution of the Schrödinger equation for this Hamiltonian is

$$|\tilde{\psi}(t)\rangle = e^{\xi a^2 - \xi^*\left(a^\dagger\right)^2}|0\rangle, \tag{15.77}$$

where

$$\xi = -i\Lambda t. \tag{15.78}$$

We recognize that the state (15.77) is just the squeezed vacuum.

15.4 Summary

The atom–field Hamiltonian has been obtained in dipole approximation. This is the Hamiltonian we use for the rest of this book. The simple, but important, Jaynes-Cummings model was studied. It was found that a coherent, quantized state of the field can lead to atom–field dynamics that differs from that of a classical field, even if the field strength is large. The Jaynes-Cummings model was also the starting point for a discussion of dressed states of a quantized field interacting with a two-level atom. In contrast to semiclassical dressed states, these quantized field dressed states are stationary states of the atom–field Hamiltonian. Last, we showed how the specific atom–field Hamiltonians can give rise to coherent and squeezed states of the radiation field. Now, we turn our attention to multimode quantized field states to see how the vacuum field can lead to spontaneous decay.

Problems

1–2. For the Hamiltonian

$$H = \sum_{\alpha} [\mathbf{p}_\alpha - q_\alpha \mathbf{A}(\mathbf{r}_\alpha)]^2/(2m_\alpha) + V_C + H_T,$$

prove that

$$m\ddot{\mathbf{r}}_\alpha = q_\alpha \mathbf{E}(\mathbf{r}_\alpha) + q_\alpha [\dot{\mathbf{r}}_\alpha \times \mathbf{B}(\mathbf{r}_\alpha) - \mathbf{B}(\mathbf{r}_\alpha) \times \dot{\mathbf{r}}_\alpha]/2$$

and that

$$\dot{a}_j + i\omega_j a_j = i(2\epsilon_0 \hbar\omega_j)^{-1/2} \tilde{J}_j,$$

where the operators are written in the Heisenberg representation, V_C is the electrostatic Coulomb interaction potential, and

$$H_T = \sum_j \hbar\omega_j \left(a_j^\dagger a_j + 1/2\right),$$

$$\tilde{\mathbf{J}}_j = (1/\sqrt{\mathcal{V}}) \int d\mathbf{r}\, e^{-i\mathbf{k}_j\cdot\mathbf{r}} \boldsymbol{\epsilon}_j \cdot \mathbf{J}(\mathbf{r}),$$

$$\mathbf{J}(\mathbf{r}) = (1/2) \sum_\alpha q_\alpha [\dot{\mathbf{r}}_\alpha \delta(\mathbf{r} - \mathbf{r}_\alpha) + \delta(\mathbf{r} - \mathbf{r}_\alpha)\dot{\mathbf{r}}_\alpha],$$

$$\mathbf{A}(\mathbf{r}) = \sum_j \left(\frac{\hbar}{2\epsilon_0 \omega_j \mathcal{V}}\right)^{1/2} \left(a_j \boldsymbol{\epsilon}_j e^{i\mathbf{k}_j\cdot\mathbf{r}} + a_j^\dagger \boldsymbol{\epsilon}_j e^{-i\mathbf{k}_j\cdot\mathbf{r}}\right).$$

In other words, the operators satisfy the Lorentz force equation for the charges and Maxwell's equations for the fields, properly symmetrized.

3. For the Hamiltonian given by equation (15.30), introduce the unitary transformation $H' = T(t)H[T(t)]^\dagger$, with

$$T(t) = \Pi_i \exp(\lambda_i^* a_i - \lambda_i a_i^\dagger), \qquad \lambda_i = \alpha_i e^{-i\Omega_i t}.$$

Show that, under this transformation, the electric field operator transforms as

$$\mathbf{E}'(\mathbf{R}, t) = T\mathbf{E}T^\dagger = \mathbf{E}_{cl}(\{\alpha_i\}, \mathbf{R}, t) + \mathbf{E}(\mathbf{R}),$$

and if the initial state function was the coherent state $|\{\alpha_i\}\rangle$, the new transformed state function is just the vacuum. Consequently, the true quantized coherent state field can be approximated by a classical field, provided that the effects of vacuum fluctuations are not important relative to those of the classical field.

4. Consider a three-level cascade atom interacting with two single-mode quantized radiation fields. The first field drives the 1–2 transition, and the second field the 2–3 transition. The second field is weak and can be treated in perturbation theory, while the first field is of arbitrary intensity and is best treated using a dressed-atom basis. Using a dressed-atom energy-level diagram, explain why one would expect two resonance peaks (Autler-Townes doublet)

in the absorption spectrum for the second field, and calculate the positions of these resonances.

5. Draw an energy-level diagram for the first four dressed-state manifolds.
6. Imagine that the initial state of the system in the Jaynes-Cummings problem is $|\psi(0)\rangle = \frac{1}{\sqrt{2}}|1\rangle(|1\rangle + |2\rangle)$—that is, the atom in state 1 and the field in a superposition of one- and two-photon states. Assume that $\delta = 0$. Find and plot the probability that the atom is in its excited state as a function of time. Show that there can never be a complete revival of the system, nor can the excited-state population be exactly equal to unity.
7. Choose your own set of parameters to illustrate collapse and revival in the Jaynes-Cummings model.
8. Prove that $[\tilde{H}(t_1), \tilde{H}(t_2)]$ is a c-number, for $\tilde{H}(t)$ given in equation (15.68).

References

[1] E. T. Jaynes and F. W. Cummings, *Comparison of quantum and semiclassical radiation theories with application to the beam maser*, Proceedings of the IEEE **51**, 89–109 (1963).

[2] J. R. Ackerhalt and P. Milonni, *Interaction Hamiltonian of quantum optics*, Journal of the Optical Society of America B **1**, 116–120 (1984).

[3] C. Cohen-Tannoudji, J. Dupont-Roc, and G. Grynberg, *Atom-Photon Interactions* (Wiley-Interscience, New York, 1992), app. 5.

[4] P. R. Berman, Ed., *Cavity Quantum Electrodynamics* (Academic Press, New York, 1994).

[5] T. von Foerster, *A comparison of quantum and semiclassical theories of the interaction between a two-level atom and the radiation field*, Journal of Physics A **8**, 95–103 (1975).

[6] M. Brune, F. Schmidt-Kaler, A. Maali, J. Dreyer, E. Hagley, J. M. Raimond, and S. Haroche, *Quantum Rabi oscillation: a direct test of field quantization in a cavity*, Physical Review Letters **76**, 1800–1803 (1996).

[7] C. Cohen-Tannoudji, *Atoms in Strong Resonant Fields*, in *Frontiers in Laser Spectroscopy*, Les Houches Session XXVII, vol. 1, edited by R. Balian, S. Haroche, and S. Liberman (North Holland, Amsterdam, 1975), pp. 3–103; C. Cohen-Tannoudji, J. Dupont-Roc, and G. Grynberg, *Atom-Photon Interactions* (Wiley-Interscience, New York, 1992), chap. VI.

Bibliography

See the references in chapter 1. In particular, additional coverage and references for the material in this chapter is given in

B. W. Shore, *The Theory of Coherent Excitation*, vols. 1 and 2 (Wiley-Interscience, New York, 1990), secs. 9.15–10.6.

16

Spontaneous Decay

When you first learn quantum mechanics, you spend a lot of time studying stationary states of the hydrogen atom, as you should. However, a little later in life, you find out that these states are not stationary at all. In fact, some of the low-lying states decay back to the ground state in times of the order of nanoseconds. Clearly, something was left out in the treatment of the stationary states of the hydrogen atom. That "something" was the vacuum field. Even though the ether has disappeared, the vacuum field has remained an essential component of modern physics.

In this chapter, we look at the interaction of the vacuum field with an atom prepared in an excited state. This is one of the most basic atom–field interactions, and one of practical importance. We calculate the rate at which the excited state decays and the properties of the field emitted by the atom.

Suffice it to say that spontaneous emission leads to some interesting problems that have yet to be resolved. One starts with a perfectly unitary Hamiltonian and ends up with irreversible behavior. How can this happen? The irreversibility is linked to the assumption that any radiation emitted by an atom as a result of the atom–vacuum field interaction cannot come back and excite the atom at some later time. In terms of field quantization, the equivalent assumption is that the spacing of the field modes approaches zero or the quantization volume approaches infinity. Even with this approximation, there are problems with unitarity. Some *additional* assumptions, sometimes not rigorously justified, are needed to save the day.

16.1 Spontaneous Decay Rate

Despite the apparent simplicity of spontaneous emission—put an atom in an excited state and watch it decay—spontaneous emission poses all sorts of mathematical difficulties. Before considering spontaneous emission, let us first look at the problem of an atom in its ground state and *no* photons present in the field. It is apparent from the Hamiltonian (15.29) that this state is *not* an eigenstate of the system. The atom can "emit" a photon while going to an excited state. Of course, such a process does not conserve energy, so the atom must then very quickly "reabsorb" the photon and return to its ground state. The overall reaction is reminiscent of a two-photon transition that begins and ends in the same level, only in this case, the non-energy-conserving scattering is provided by the vacuum field. The net effect is something like a light shift, and the energy of the ground state is lowered. Unfortunately, it is lowered by an infinite amount, as you will now see.

We write the state vector in the interaction representation as

$$|\Psi(t)\rangle = \sum_{\alpha;\mathbf{k},\epsilon_\mathbf{k}} c_{\alpha;n_{\mathbf{k},\epsilon_\mathbf{k}}}(t) e^{-in_{\mathbf{k},\epsilon_\mathbf{k}}\omega_k t} |\alpha; n_{\mathbf{k},\epsilon_\mathbf{k}}\rangle e^{-i\omega_\alpha t}, \tag{16.1}$$

where α labels the atomic state, $\hbar\omega_\alpha$ is the energy of state α, and $n_{\mathbf{k},\epsilon_\mathbf{k}}$ is the number of photons in the field mode having propagation vector $\mathbf{k}$, frequency ω_k, and polarization $\epsilon_\mathbf{k}$. From the Schrödinger equation with the Hamiltonian (15.29), we find

$$\dot{c}_{1;0}(t) = -\frac{i}{\hbar} \sum_{\alpha,\mathbf{k},\epsilon_\mathbf{k}} e^{-i(\omega_\alpha+\omega_k)t} \langle 1;0| V_{AF} |\alpha; 1_{\mathbf{k},\epsilon_\mathbf{k}}\rangle c_{\alpha;1_{\mathbf{k},\epsilon_\mathbf{k}}}(t), \tag{16.2}$$

where $\alpha = 1$ corresponds to the ground state. The state $|1_{\mathbf{k},\epsilon_\mathbf{k}}\rangle$ corresponds to a one-photon state of the field. To get a slowly varying component for $c_{1;0}(t)$, we solve adiabatically for the excited-state amplitude $c_{\alpha;n_{\mathbf{k},\epsilon_\mathbf{k}}}(t)$, as we did for the Bloch-Siegert and light shifts. Formally, this can be done using integration by parts on the evolution equation

$$\dot{c}_{\alpha;1_{\mathbf{k},\epsilon_\mathbf{k}}}(t) = -\frac{i}{\hbar} \langle \alpha; 1_{\mathbf{k},\epsilon_\mathbf{k}} | V_{AF} | 1;0\rangle \int_0^t dt' e^{i(\omega_\alpha+\omega_k)t'} c_{1;0}(t'), \tag{16.3}$$

yielding

$$c_{\alpha;1_{\mathbf{k},\epsilon_\mathbf{k}}}(t) \sim -\frac{1}{\hbar} \frac{e^{i(\omega_\alpha+\omega_k)t}}{\omega_\alpha + \omega_k} \langle \alpha; 1_{\mathbf{k},\epsilon_\mathbf{k}} | V_{AF} | 1;0\rangle c_{1;0}(t), \tag{16.4}$$

where V_{AF}, as given by equations (15.33) and (15.28), is

$$V_{AF} = -i\hat{\boldsymbol{\mu}}\cdot \sum_{\mathbf{k},\epsilon_\mathbf{k}} \left(\frac{\hbar\omega_\mathbf{k}}{2\epsilon_0 \mathcal{V}}\right)^{1/2} \left(a_{\mathbf{k},\epsilon_\mathbf{k}} \boldsymbol{\epsilon}_\mathbf{k} e^{i\mathbf{k}\cdot\mathbf{R}} - a^\dagger_{\mathbf{k},\epsilon_\mathbf{k}} \boldsymbol{\epsilon}_\mathbf{k} e^{-i\mathbf{k}\cdot\mathbf{R}}\right), \tag{16.5}$$

and $\hat{\boldsymbol{\mu}}$ is the dipole moment operator. We have included only those contributions to $c_{\alpha;1_{\mathbf{k},\epsilon_\mathbf{k}}}(t)$ from $c_{1;0}(t)$, since only these terms lead to a slowly varying amplitude for $c_{1;0}(t)$.

Substituting this back into the equation for $\dot{c}_{1;0}(t)$, we find

$$\dot{c}_{1;0}(t) = \frac{i}{\hbar^2} \sum_{\alpha;\mathbf{k},\epsilon_\mathbf{k}} \frac{|\langle 1;0| V_{AF} |\alpha; 1_{\mathbf{k},\epsilon_\mathbf{k}}\rangle|^2}{\omega_\alpha + \omega_k} c_{1;0}(t), \tag{16.6}$$

and using equation (16.5), we obtain

$$i\hbar\dot{c}_{1;0}(t) = -\hbar\Delta_1 c_{1;0}(t) \tag{16.7a}$$

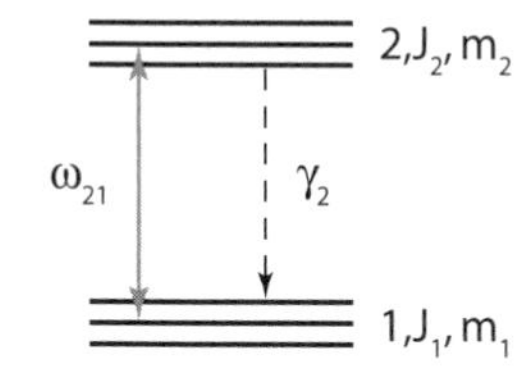

Figure 16.1. Spontaneous emission between two manifolds of levels.

where

$$\Delta_1 = \frac{1}{\hbar} \sum_{\alpha;\mathbf{k},\boldsymbol{\epsilon}_\mathbf{k}} \left(\frac{\hbar\omega_k}{2\epsilon_0 \mathcal{V}}\right) \frac{|\langle 1|\hat{\boldsymbol{\mu}} \cdot \boldsymbol{\epsilon}_\mathbf{k}|\alpha\rangle|^2}{\omega_\alpha + \omega_k} \tag{16.7b}$$

is the ground-state vacuum field shift or *ground-state Lamb shift*. When this is converted to an integral using the prescription

$$\sum_{\mathbf{k},\boldsymbol{\epsilon}_\mathbf{k}} = \frac{\mathcal{V}}{(2\pi)^3} \sum_{\boldsymbol{\epsilon}_\mathbf{k}} \int d^3k, \tag{16.8}$$

the resultant expression diverges as $\sim \int_0^\infty \omega_k^3 d\omega_k$. One can get a finite result by taking into account mass renormalization and by imposing either a relativistic ($\hbar k_{max} = mc$) or dipole approximation ($k_{max}a_0 \sim 1$) cutoff, but a proper calculation of the Lamb shift involves the use of quantum electrodynamic approach. We neglect all Lamb shifts and assume that they are incorporated into the energies of the states.

We now return to the decay problem and assume that an atom has been excited to a state $|J_2, m_2\rangle$ at time $t = 0$. The notation here is that J_2 is the total angular momentum of state 2 and m_2 a magnetic quantum number associated with J_2. Of course, it is impossible to *instantaneously* excite the atom at a given time, so what we really mean is that the atom is excited in a time τ that is much less than the decay rate of the system, but much longer than the inverse of the transition frequency. The interaction of the atom with the vacuum radiation field then drives the atom back to the manifold of ground states denoted by $|J_1, m_1\rangle$ (see figure 16.1). It is important not to forget that it is simply a vacuum field–induced decay process with which we are dealing, since the algebra can get a little messy.

Initially, the system is in the state $|J_2, m_2; 0\rangle$, with zero photons in the field. The excited-state amplitude evolves as

$$\dot{c}_{2,J_2,m_2;0}(t) = \frac{1}{i\hbar} \sum_{\mathbf{k},\boldsymbol{\epsilon}_\mathbf{k}} \sum_{m_1} e^{-i(\omega_k - \omega_{21})t} \times \langle 2, J_2, m_2; 0|V_{AF}|1, J_1, m_1; \mathbf{k}, \boldsymbol{\epsilon}_\mathbf{k}\rangle c_{1,J_1,m_1;\mathbf{k},\boldsymbol{\epsilon}_\mathbf{k}}(t), \tag{16.9}$$

where ω_{21} is the transition frequency between levels $(2, J_2)$ and $(1, J_1)$, and sums over m_J go from $-J$ to J. We have neglected transitions to states other than state 1, since they are virtual transitions that lead only to contributions to the Lamb shift of state 2. A formal solution for the ground-state amplitude is

$$c_{1,J_1,m_1;\mathbf{k},\epsilon_\mathbf{k}}(t) = \frac{1}{i\hbar}\int_0^t \sum_{J_2,m_2} dt'\, e^{i(\omega_k-\omega_{21})t'}$$
$$\times\langle 1, J_1, m_1; \mathbf{k}, \epsilon_\mathbf{k}| V_{AF} |2, J_2, m_2; 0\rangle c_{2,J_2,m_2;0}(t'). \qquad (16.10)$$

When equation (16.10) is substituted into equation (16.9), we find

$$\dot{c}_{2,J_2,m_2;0}(t) = -\frac{e^{-i(\omega_k-\omega_{21})t}}{\hbar^2}\sum_{\mathbf{k},\epsilon_\mathbf{k}}\sum_{J_2',m_2',m_1}\int_0^t dt'\, e^{i(\omega_k-\omega_{2'1})t'} c_{2,J_2',m_2';0}(t')$$
$$\times\langle 2, J_2, m_2; 0| V_{AF} |1, J_1, m_1; \mathbf{k}, \epsilon_\mathbf{k}\rangle$$
$$\times\langle 1, J_1, m_1; \mathbf{k}, \epsilon_\mathbf{k}| V_{AF} |2, J_2', m_2'; 0\rangle, \qquad (16.11)$$

where we have allowed for transitions to another level J_2' in the same state 2 electronic state manifold, and $\omega_{2'1}$ is a shorthand notation for $\omega_{J_2'J_1}$.

One can try to find a self-consistent solution in which $c_{2,J_2',m_2';0}(t')$ is assumed to vary slowly with respect to $e^{i\omega_k t'}$ so that it can be removed from the integrand and evaluated at $t' = t$. Alternatively, one can guess a solution of the form

$$c_{2,J_2,m_2;0}(t) = e^{-\gamma t} c_{2,J_2,m_2;0}(0), \qquad (16.12)$$

where γ is to be determined. Substituting the trial solution (16.12) into equation (16.11), we find

$$\gamma e^{-\gamma t} c_{2,J_2,m_2;0}(0)$$
$$= \sum_{\mathbf{k},\epsilon_\mathbf{k}}\sum_{J_2',m_2',m_1} \frac{\langle 2, J_2, m_2; 0| V_{AF} |1, J_1, m_1; \mathbf{k}, \epsilon_\mathbf{k}\rangle\langle 1, J_1, m_1; \mathbf{k}, \epsilon_\mathbf{k}| V_{AF} |2, J_2', m_2'; 0\rangle}{\hbar^2\left[\gamma - i(\omega_k - \omega_{2'1})\right]}$$
$$\times\left[e^{-i(\omega_k-\omega_{21})t} - e^{-\gamma t + i\omega_{22'}t}\right] c_{2,J_2',m_2';0}(0), \qquad (16.13)$$

where $\omega_{22'}$ is the transition frequency between levels $(2, J_2)$ and $(2, J_2')$. We now use equation (16.8) to convert the sum over $\mathbf{k}$ to an integral and evaluate the matrix elements explicitly using equation (16.5). In this manner, we obtain

$$\gamma e^{-\gamma t} c_{2,J_2,m_2;0}(0) = \frac{1}{\hbar^2}\frac{\mathcal{V}}{(2\pi)^3}\sum_{J_2',m_2',m_1}\int d^3k\left(\frac{\hbar\omega_k}{2\epsilon_0\mathcal{V}}\right)$$
$$\times\sum_{\epsilon_\mathbf{k}}\frac{\langle 2, J_2, m_2|\hat{\mu}\cdot\epsilon_\mathbf{k}|1, J_1, m_1\rangle\langle 1, J_1, m_1|\hat{\mu}\cdot\epsilon_\mathbf{k}|2, J_2', m_2'\rangle}{\gamma - i(\omega_k - \omega_{2'1})}$$
$$\times\left[e^{-i(\omega_k-\omega_{21})t} - e^{-\gamma t + i\omega_{22'}t}\right] c_{2,J_2',m_2';0}(0). \qquad (16.14)$$

The matrix elements depend only on the angular components of $\mathbf{k}$, but not on its magnitude, so it is possible to first carry out the integral over $\omega_k = kc$. The integrand varies as

$$\frac{\omega_k^3}{\gamma - i(\omega_k - \omega_{2'1})}\left[e^{-i(\omega_k-\omega_{21})t} - e^{-\gamma t + i\omega_{22'}t}\right]. \qquad (16.15)$$

In principle, the integral over ω_k diverges for large ω_k. However, if we impose a cutoff on the ω_k integral based on the range of validity of the dipole approximation, then the major contribution to the integral comes from a region $|\omega_k - \omega_{2'1}| \sim \gamma \ll \omega_{2'1}$ for optical transitions. (The imaginary part diverges, but this is similar to the infinities encountered in the Lamb shift.) Thus, we can set the ω_k^3 equal to $\omega_{2'1}^3$, remove it from the integral, and extend the ω_k integral from (0 to ∞) to ($-\infty$ to ∞). These two approximations comprise what is known as the *Weisskopf–Wigner approximation* [1]. Setting $d^3k = \omega_k^2 d\Omega_k d\omega_k/c^3$ in equation (16.14), and carrying out the integration over ω_k in the Weisskopf-Wigner approximation, we find

$$\begin{aligned}\gamma c_{2,J_2,m_2;0}(0) = \frac{1}{\hbar}\frac{\pi}{2\epsilon_0 c^3(2\pi)^3} \sum_{J_2',m_2',m_1} \int d\Omega_k \omega_{2'1}^3 e^{i\omega_{22'}t} \\ \times \sum_{\epsilon_\mathbf{k}} \langle 2, J_2, m_2|\hat{\mu}\cdot\epsilon_\mathbf{k}|1, J_1, m_1\rangle \\ \times \langle 1, J_1, m_1|\hat{\mu}\cdot\epsilon_\mathbf{k}|2, J_2', m_2'\rangle c_{2,J_2',m_2';0}(0)\,,\end{aligned} \tag{16.16}$$

where we have used

$$\int_{-\infty}^{\infty} d\omega_k \frac{\left[e^{-i(\omega_k-\omega_{21})t} - e^{-\gamma t + i\omega_{22'}t}\right]}{\gamma - i(\omega_k - \omega_{2'1})} = \pi e^{-\gamma t} e^{i\omega_{22'}t}\,, \tag{16.17}$$

and the integral in equation (16.16) is over solid angle $d\Omega_k$ in k space. The solution (16.16) can be correct only if the term multiplying $c_{2,J_2',m_2';0}(0)e^{i\omega_{22'}t}$ is proportional to $\delta_{J_2,J_2'}\delta_{m_2,m_2'}$. We now check this.

To carry out the angular integration, we must write explicit expressions for the unit polarization vectors. The unit vectors $\epsilon_\mathbf{k}^{(1)}$, $\epsilon_\mathbf{k}^{(2)}$, and $\hat{\mathbf{k}}$ make up a right-handed system, with

$$\hat{\mathbf{k}} = \sin\theta_k \cos\phi_k \hat{\mathbf{x}} + \sin\theta_k \sin\phi_k \hat{\mathbf{y}} + \cos\theta_k \hat{\mathbf{z}}, \tag{16.18a}$$

$$\epsilon_\mathbf{k}^{(1)} = \hat{\boldsymbol{\theta}}_k = \cos\theta_k \cos\phi_k \hat{\mathbf{x}} + \cos\theta_k \sin\phi_k \hat{\mathbf{y}} - \sin\theta_k \hat{\mathbf{z}}, \tag{16.18b}$$

$$\epsilon_\mathbf{k}^{(2)} = \hat{\boldsymbol{\phi}}_k = -\sin\phi_k \hat{\mathbf{x}} + \cos\phi_k \hat{\mathbf{y}}\,. \tag{16.18c}$$

It is then straightforward to carry out the summations and integrations in equation (16.16) to obtain

$$\begin{aligned}\sum_{\lambda=1}^{2}\sum_{i,j=1}^{3} d_i \bar{d}_j \int d\Omega_k \left[\epsilon_\mathbf{k}^{(\lambda)}\right]_i \left[\epsilon_\mathbf{k}^{(\lambda)}\right]_j &= \frac{8\pi}{3}\sum_{i,j}\delta_{i,j} d_i \bar{d}_j \\ &= \frac{8\pi}{3}\mathbf{d}\cdot\bar{\mathbf{d}},\end{aligned} \tag{16.19}$$

where we have denoted first and second matrix elements in equation (16.16) by $\mathbf{d}$ and $\bar{\mathbf{d}}$, respectively.

To proceed further, we write the components of the matrix elements $\mathbf{d}$ and $\bar{\mathbf{d}}$, as well as those of the dipole moment operator, in spherical form as

$$\hat{\mu}_1 = -\frac{\hat{\mu}_x + i\hat{\mu}_y}{\sqrt{2}}, \quad \hat{\mu}_{-1} = \frac{\hat{\mu}_x - i\hat{\mu}_y}{\sqrt{2}}, \quad \hat{\mu}_0 = \hat{\mu}_z, \tag{16.20a}$$

$$d_1 = -\frac{d_x + id_y}{\sqrt{2}}, \quad d_{-1} = \frac{d_x - id_y}{\sqrt{2}}, \quad d_0 = d_z, \tag{16.20b}$$

such that

$$\mathbf{d} \cdot \bar{\mathbf{d}} = \sum_q (-1)^q d_q \bar{d}_{-q}\,. \tag{16.21}$$

We can now use the Wigner-Eckart theorem [2] to evaluate

$$\begin{aligned} d_q &= \langle 2, J_2, m_2 | \hat{\mu}_q | 1, J_1, m_1 \rangle \\ &= \frac{1}{\sqrt{2J_2+1}} \begin{bmatrix} J_1 & 1 & J_2 \\ m_1 & q & m_2 \end{bmatrix} \langle 2, J_2 \| \mu^{(1)} \| 1, J_1 \rangle\,, \end{aligned} \tag{16.22}$$

$$\begin{aligned} \bar{d}_{-q} &= \langle 2, J_2', m_2' | \hat{\mu}^{\dagger}_{-q} | 1, J_1, m_1 \rangle^* \\ &= (-1)^q \langle 2, J_2' m_2' | \hat{\mu}_q | 1, J_1, m_1 \rangle^* \\ &= \frac{1}{\sqrt{2J_2+1}} (-1)^q \begin{bmatrix} J_1 & 1 & J_2 \\ m_1 & q & m_2' \end{bmatrix} \langle 2, J_2 \| \mu^{(1)} \| 1, J_1 \rangle^*\,, \end{aligned} \tag{16.23}$$

where the quantities in square brackets are *Clebsch-Gordan coefficients* and $\langle 2, J_2 \| \mu^{(1)} \| 1, J_1 \rangle$ is the *reduced matrix element* of the dipole moment operator between states $|2, J_2\rangle$ and $|1, J_1\rangle$ [2]. It then follows that

$$\begin{aligned} \sum_{m_1} \sum_q (-1)^q d_q d_{-q} &= \frac{1}{2J_2+1} |\langle 2, J_2 \| \mu^{(1)} \| 1, J_1 \rangle|^2 \sum_{q,m_1} \begin{bmatrix} J_1 & 1 & J_2 \\ m_1 & q & m_2 \end{bmatrix} \begin{bmatrix} J_1 & 1 & J_2' \\ m_1 & q & m_2' \end{bmatrix} \\ &= \frac{1}{2J_2+1} |\langle 2, J_2 \| \mu^{(1)} \| 1, J_1 \rangle|^2 \delta_{J_2, J_2'} \delta_{m_2, m_2'}\,, \end{aligned} \tag{16.24}$$

where the orthogonality property of the Clebsch-Gordan coefficients has been used. The sum is proportional to $\delta_{J_2, J_2'} \delta_{m_2, m_2'}$, giving a result that is consistent with our trial solution (16.12). Combining all the terms, we obtain finally

$$2\gamma = \gamma_{2,J_2;1,J_1} = \frac{4}{3} \frac{\alpha_{FS}}{2J_2+1} |\langle 2, J_2 \| r \| 1, J_1 \rangle|^2 \frac{\omega_{21}^3}{c^2}\,, \tag{16.25}$$

where $\alpha_{FS} = e^2/(4\pi\epsilon_0 \hbar c)$ is the fine-structure constant, $\langle 2, J_2 \| r \| 1, J_1 \rangle$ is the reduced matrix element for the position operator, and $\gamma_{2,J_2;1,J_1}$ is the decay rate from state $|2, J_2\rangle$ to $|1, J_1\rangle$.

The decay rate is proportional to the cube of the transition frequency and to the absolute square of the reduced matrix element. The reduced matrix element is a measure of the overlap of the initial and final state wave functions. For example, the absolute square of the reduced matrix element for a harmonic oscillator having natural frequency ω_0 varies as n/ω_0 for an n to $(n-1)$ transition in the oscillator. Thus, the overall spontaneous decay rate for such a transition varies as $n\omega_0^2$. On the other hand, the $nP - 1S$ spontaneous decay rate in hydrogen *decreases* with increasing n owing to rapid decrease in the overlap of the wave functions and the reduced matrix element.

Using the quantized vacuum field, we have derived a fundamental and important equation for the spontaneous emission rate. The fact that $\dot{c}_{2,J_2,m_2;0}$ is not coupled to $c_{2,J_2',m_2';0}$ for $J_2' \neq J_2$ or $m_2' \neq m_2$ can be understood simply in terms of conservation of angular momentum. States in the excited-state manifold differing in total angular momentum or the z-component of angular momentum cannot be coupled by "emitting" and "absorbing" the *same* photon. We had to invoke the Weisskopf-Wigner approximation to carry out the calculation. The Weisskopf-Wigner approximation is incredible in that there is no formal justification for its use (even if it seems reasonable), yet when used consistently, this approximation results in overall conservation of probability for the atom–field system.

Up to this point, J is some generic angular momentum in the state 2 manifold of levels. In general, an electronic state manifold of levels can contain both fine and hyperfine structure. The frequency separation of hyperfine levels of low-lying states is often in the 10s of MHz to GHz range, while fine-structure separations of the first excited manifolds of the alkali metal atoms are on the order of 1 to 10 THz. In an LS coupling scheme, a state can be labeled by $|n, L_n, S, J_n, I, F\rangle$, where the quantum number n specifies the electronic state, L_n the total orbital angular momentum, S the total spin angular momentum, J_n the combined orbital and spin angular momentum that characterizes a fine structure manifold, I the total nuclear spin angular momentum, and F the total angular momentum that characterizes a given hyperfine state. Let H label the total angular momentum of an excited hyperfine state and G the total angular momentum of a ground hyperfine state. In that case, equation (16.25) is transformed into

$$\gamma_{H;G} = \frac{4}{3}\alpha_{FS}\frac{1}{2H+1}|\langle 2, J_2, I, H\|r\|1, J_1, I, G\rangle|^2\frac{\omega_{HG}^3}{c^2}, \tag{16.26}$$

where the L and S indices have been suppressed for the moment.

The quantity $\gamma_{H;G}$ is now the *partial decay rate* from state H to state G. To get the total decay rate out of H, one must sum over G. In doing so, one can usually neglect variations in ω_{HG} and replace ω_{HG} by $\omega_{J_2J_1}$, since this leads to corrections that are of the order of the hyperfine splitting over the electronic state splitting (of order 10^{-6}). The reduced matrix element can be expressed as [2]

$$\begin{aligned}\langle 2, J_2, I, H\|r\|1, J_1, I, G\rangle &= \langle 2, J_2\|r\|1, J_1\rangle(-1)^{G+I+J_2+1}\\ &\quad\times\sqrt{(2H+1)(2G+1)}\begin{Bmatrix} J_2 & 1 & J_1 \\ G & I & H \end{Bmatrix},\end{aligned} \tag{16.27}$$

where {} is a six-J symbol. Therefore,

$$\gamma_H = \sum_G \gamma_{H;G} = \frac{4}{3}\alpha_{FS}\frac{1}{2H+1}|\langle 2, J_2\|r\|1, J_1\rangle|^2(2H+1)$$
$$\times \sum_G (2G+1)\begin{Bmatrix} J_1 & 1 & J_2 \\ H & I & G \end{Bmatrix}\begin{Bmatrix} J_1 & 1 & J_2 \\ H & I & G \end{Bmatrix}\frac{\omega_{J_2J_1}^3}{c^2}$$
$$= \frac{4}{3}\alpha_{FS}\frac{1}{2J_2+1}\frac{\omega_{J_2J_1}^3}{c^2}|\langle 2, J_2\|r\|1, J_1\rangle|^2 . \tag{16.28}$$

Each hyperfine state in a given fine-structure multiplet decays at the same rate, with corrections of order of the ground-or excited-state hyperfine splittings divided by the electronic state separation. Equation (16.28) reproduces equation (16.25) for a state of given J.

This process can be taken one step further to look at different fine structure states in a state of given L. If we neglect terms of order of the fine structure to electronic state splitting (typically, 10^{-2} to 10^{-3}), then, using the relationship

$$\langle 2, L_2, S, J_2\|r\|1, L_1, S, J_1\rangle = \langle 2, L_2\|r\|1, L_1\rangle(-1)^{J_1+S+L_2+1}$$
$$\times\sqrt{(2J_2+1)(2J_1+1)}\begin{Bmatrix} L_2 & 1 & L_1 \\ J_1 & S & J_2 \end{Bmatrix}, \tag{16.29}$$

and using the same reasoning that led to equation (16.28), we find

$$\gamma_J = \frac{4}{3}\alpha_{FS}\frac{1}{2L_2+1}\frac{\omega_{L_2L_1}^3}{c^2}|\langle 2, L_2\|r\|1, L_1\rangle|^2 . \tag{16.30}$$

Each fine-structure state in a given state of orbital angular momentum decays at the same rate (with corrections of order of the fine structure to electronic state splitting).

We can combine equations (16.26), (16.27), and (16.29) to write the decay rate $\gamma_{H;G}$ as

$$\gamma_{H;G} = \frac{4}{3}\alpha_{FS}(2G+1)(2J_2+1)(2J_1+1)|\langle 2, L_2\|r\|1, L_1\rangle|^2\frac{\omega_{HG}^3}{c^2}$$
$$\times\begin{Bmatrix} J_2 & 1 & J_1 \\ G & I & H \end{Bmatrix}^2\begin{Bmatrix} L_2 & 1 & L_1 \\ J_1 & S & J_2 \end{Bmatrix}^2 . \tag{16.31}$$

In this form, it is a simple matter to compare the partial decay rates for transitions from levels H in the (L_2, J_2) manifold to levels G in the (L_1, J_1) manifold.

16.2 Radiation Pattern and Repopulation of the Ground State

Having calculated the decay rate, we now turn our attention to the spectrum of the emitted radiation and to the rate at which the ground state is repopulated by spontaneous decay. Note that the atomic dipole moment, proportional to $\mathrm{Re}\,\varrho_{12}$, equals *zero* at all times in this problem, since the atom is in a superposition of states $|1, J_1, m_1; \mathbf{k}, \boldsymbol{\epsilon}_\mathbf{k}\rangle$ and $|2, J_2, m_2; 0\rangle$, for which $\varrho_{12} = 0$. The fact that the atomic dipole moment vanishes implies that the *average* field emitted by the atom also vanishes, where the average is a quantum-mechanical average over a number of identically prepared systems. Since we start in an excited state with no well-defined phase, it is not surprising that the average field vanishes. As you will see, however, even though the average field vanishes, the radiation pattern of the field intensity can be the same as that of a classical dipole oscillator, in certain limits.

16.2.1 Radiation Pattern

The calculation of the radiation spectrum begins with the expression for the amplitude to be in the ground state with a photon of type $\mathbf{k}, \boldsymbol{\epsilon}_\mathbf{k}$ in the field. This amplitude is equal to [see equations (16.10) and (16.12)]

$$c_{1,J_1,m_1;\mathbf{k},\boldsymbol{\epsilon}_\mathbf{k}}(t) = \frac{1}{i\hbar}\langle 1, J_1, m_1; \mathbf{k}, \boldsymbol{\epsilon}_\mathbf{k}|V_{AF}|2, J_2, m_2; 0\rangle \times \sum_{m_2}\int_0^t dt' e^{i(\omega_k-\omega_{21})t'} e^{-\gamma_2 t'/2} c_{2,J_2,m_2;0}(0). \tag{16.32}$$

In this expression, J can be considered to be a generic angular momentum; the calculation is valid equally for transitions between manifolds of hyperfine-or fine-structure states. However, we restrict the discussion to transitions between a single excited-state and a single ground-state manifold, with ω_{21} the transition frequency between these manifolds of levels.

The radiation emitted by the atom is characterized by the frequency spectrum, angular distribution, and polarization of the radiated field. Information about all these quantities is contained in the probability $|c_{1,J_1,m_1;\mathbf{k},\boldsymbol{\epsilon}_\mathbf{k}}(\infty)|^2$, since this quantity represents the probability that the radiation is found in mode $\mathbf{k}, \boldsymbol{\epsilon}_\mathbf{k}$ after the excited state has decayed. Thus, $|c_{1,J_1,m_1;\mathbf{k},\boldsymbol{\epsilon}_\mathbf{k}}(\infty)|^2 \left(\omega_k^2/c^3\right) d\omega_k d\Omega_k$ is the probability that the photon has frequency between ω_k and $\omega_k + d\omega_k$ and is emitted with polarization $\boldsymbol{\epsilon}_\mathbf{k}$ in an element of solid angle $d\Omega_k$ around the direction $\hat{\mathbf{k}}$. The normalization is such that

$$\frac{\mathcal{V}}{(2\pi)^3}\int_0^\infty \left(\omega_k^2/c^3\right) d\omega_k \int d\Omega_k \sum_{\boldsymbol{\epsilon}_\mathbf{k}}\sum_{m_1} |c_{1,J_1,m_1;\mathbf{k},\boldsymbol{\epsilon}_\mathbf{k}}(\infty)|^2 = 1. \tag{16.33}$$

We define the spectral density $dI_{\mathbf{k},\boldsymbol{\epsilon}_\mathbf{k}}(2, J_2; 1, J_1)/d\omega_k d\Omega_k$ for the $J_2; J_1$ transition as

$$dI_{\mathbf{k},\boldsymbol{\epsilon}_\mathbf{k}}(2, J_2; 1, J_1)/d\omega_k d\Omega_k = \left(\omega_k^2/c^3\right)\sum_{m_1}|c_{1,J_1,m_1;\mathbf{k},\boldsymbol{\epsilon}_\mathbf{k}}(\infty)|^2 = \frac{\mathcal{V}\omega_k^2}{8\pi^3\hbar^2 c^3}$$
$$\times \sum_{m_1,m_2,m_2'} \frac{\langle 2, J_2, m_2'; 0|V_{AF}|1, J_1, m_1; \mathbf{k}, \boldsymbol{\epsilon}_\mathbf{k}\rangle\langle 1, J_1, m_1; \mathbf{k}, \boldsymbol{\epsilon}_\mathbf{k}|V_{AF}|2, J_2, m_2; 0\rangle}{(\omega_k - \omega_{21})^2 + (\gamma_2/2)^2}$$
$$\times \varrho_{2,J_2,m_2;2,J_2,m_2'}(0). \tag{16.34}$$

The spectrum is Lorentzian if the ω_k's appearing in the matrix elements and the ω_k^2 multiplicative factor are evaluated at $\omega_k = \omega_{21}$. In fact, the normalization given by equation (16.33) is valid only if all multiplicative values of ω_k appearing in equation (16.33) are evaluated at $\omega_k = \omega_{21}$ *and* if the integral is extended to $-\infty$. In some sense, one maintains unitarity only if a Weisskopf-Wigner-like approximation is also used for the radiated field intensity.

The product of matrix elements and the initial density matrix determines the angular distribution of radiation, on which we now concentrate. We evaluate the matrix elements explicitly using equation (16.5), integrate over ω_k, and use equation (16.25) to define the angular distribution as

$$\begin{aligned} dI_{\mathbf{k},\boldsymbol{\epsilon}_\mathbf{k}}(2, J_2; 1, J_1)/d\Omega_k &= \int_0^\infty d\omega_k \, dI_{\mathbf{k},\boldsymbol{\epsilon}_\mathbf{k}}(2, J_2; 1, J_1)/d\omega_k d\Omega_k \\ &= \frac{3}{8\pi}(2J_2+1)\sum_{m_2,m_2'} A_{m_2,m_2'}\varrho_{J_2,m_2;J_2,m_2'}(0), \end{aligned} \tag{16.35}$$

where

$$A_{m_2,m_2'} = \frac{\sum_{m_1}\left(\boldsymbol{\epsilon}_\mathbf{k}\cdot\langle J_1,m_1|\hat{\boldsymbol{\mu}}|J_2,m_2\rangle\right)\left(\boldsymbol{\epsilon}_\mathbf{k}\cdot\langle J_1,m_1|\hat{\boldsymbol{\mu}}|J_2,m_2'\rangle\right)^*}{|\langle J_1\|\hat{\boldsymbol{\mu}}\|J_2\rangle|^2}, \tag{16.36}$$

and some indices have been suppressed. This definition is consistent with a normalization in which

$$\sum_{\boldsymbol{\epsilon}_\mathbf{k}}\int d\Omega_k\left[dI_{\mathbf{k},\boldsymbol{\epsilon}_\mathbf{k}}(2, J_2; 1, J_1)/d\Omega_k\right] = 1. \tag{16.37}$$

To evaluate the matrix elements, we write

$$\boldsymbol{\epsilon}\cdot\hat{\boldsymbol{\mu}} = \sum_{q=-1}^{1}(-1)^q\epsilon_{-q}\hat{\mu}_q\,, \tag{16.38}$$

where

$$\epsilon_1 = -\frac{\epsilon_x + i\epsilon_y}{\sqrt{2}}, \tag{16.39a}$$

$$\epsilon_{-1} = \frac{\epsilon_x - i\epsilon_y}{\sqrt{2}}, \tag{16.39b}$$

$$\epsilon_0 = \epsilon_z\,, \tag{16.39c}$$

with similar equations for the spherical tensor operators $\hat{\mu}_q$.

Writing $\boldsymbol{\epsilon}\cdot\hat{\boldsymbol{\mu}}$ in this form allows us to use the Wigner-Eckart theorem to obtain

$$\begin{aligned} A_{m_2,m_2'} &= \sum_{q,q',m_1}\frac{(-1)^{q+q'}\epsilon_{-q}\left(\epsilon_{q'}\right)^*}{2J_1+1}\begin{bmatrix} J_2 & 1 & J_1 \\ m_2 & q & m_1 \end{bmatrix}\begin{bmatrix} J_2 & 1 & J_1 \\ m_2' & q' & m_1 \end{bmatrix} \\ &= \sum_{q,q',m_1}\frac{\epsilon_q\left(\epsilon_{q'}\right)^*}{2J_2+1}\begin{bmatrix} J_1 & 1 & J_2 \\ m_1 & q & m_2 \end{bmatrix}\begin{bmatrix} J_1 & 1 & J_2 \\ m_1 & q' & m_2' \end{bmatrix}. \end{aligned} \tag{16.40}$$

By combining this equation with equations (16.35), we find

$$dI_{\mathbf{k},\epsilon_\mathbf{k}}(2,J_2;1,J_1)/d\Omega_k = \frac{3}{8\pi}\sum_{m_1,q,q',m_2,m_2'} \epsilon_q\epsilon^*_{q'}\begin{bmatrix} J_1 & 1 & J_2 \\ m_1 & q & m_2 \end{bmatrix}$$
$$\times \begin{bmatrix} J_1 & 1 & J_2 \\ m_1 & q' & m_2' \end{bmatrix} \varrho_{J_2,m_2;J_2,m_2'}(0)\,. \tag{16.41}$$

In effect, this equation determines how the initial excited-state polarization, represented by density matrix elements $\varrho_{J_2,m_2;J_2,m_2'}(0)$, is transferred to the radiated field. Consider first the limit where the initial state is unpolarized

$$\varrho_{J_2,m_2;J_2,m_2'}(0) = \frac{1}{2J_2+1}\delta_{m_2,m_2'}\,. \tag{16.42}$$

In this case,

$$\begin{aligned} dI_{\mathbf{k},\epsilon_\mathbf{k}}(2,J_2;1,J_1)/d\Omega_k &= \sum_{m_1,q,q',m_2} \frac{3}{8\pi}\frac{\epsilon_q\left(\epsilon_{q'}\right)^*}{(2J_2+1)}\begin{bmatrix} J_1 & 1 & J_2 \\ m_1 & q & m_2 \end{bmatrix}\begin{bmatrix} J_1 & 1 & J_2 \\ m_1 & q' & m_2 \end{bmatrix} \\ &= \frac{1}{8\pi}\sum_{m_1,q,q',m_2} \epsilon_q\left(\epsilon_{q'}\right)^*\begin{bmatrix} J_1 & J_2 & 1 \\ -m_1 & m_2 & q \end{bmatrix}\begin{bmatrix} J_1 & J_2 & 1 \\ -m_1 & m_2 & q' \end{bmatrix} \\ &= \frac{1}{8\pi}\sum_{q,q'}\epsilon_q(\epsilon_{q'})^*\delta_{q,q'} = \frac{1}{8\pi}\sum_q |\epsilon_q|^2 = \frac{1}{8\pi}. \end{aligned} \tag{16.43}$$

As expected, the emitted radiation is unpolarized and isotropic in this limit, since the initial atomic state is unpolarized. The integral of $(1/8\pi)$ over solid angle, summed over the two independent polarizations, equals unity.

Next, consider a polarized initial atomic state and a $J_1 = 0$, $J_2 = 1$ transition. From equation (16.41), we find

$$dI_{\mathbf{k},\epsilon_\mathbf{k}}(2,1;1,0)/d\Omega_k = \epsilon_{m_2}\epsilon^*_{m_2'}\varrho_{J_2,m_2;J_2,m_2'}(0). \tag{16.44}$$

Not surprisingly, the angular distribution and polarization of the emitted radiation depends on how the initial state was prepared. If a z-polarized field is used to prepare the atom in the $m_2 = 0$ sublevel, then

$$\varrho_{J_2,m_2;J_2,m_2'}(0) = \delta_{m_2,0}\delta_{m_2',0}, \tag{16.45a}$$

and

$$dI_{\mathbf{k},\epsilon_\mathbf{k}}(2,1;1,0)/d\Omega_k = \frac{3}{8\pi}\left[(\epsilon_\mathbf{k})_0\right]^2 = \frac{3}{8\pi}\left[(\epsilon_\mathbf{k})_z\right]^2\,. \tag{16.45b}$$

For a given direction $\hat{\mathbf{k}}$, there are two independent polarization directions given by equations (16.18b) and (16.18c). For these polarization components, we find

$$(\epsilon^{(1)}_\mathbf{k})_z = (\hat{\boldsymbol{\theta}}_\mathbf{k})_z = -\sin\theta_k\,, \tag{16.46a}$$
$$(\epsilon^{(2)}_\mathbf{k})_z = (\hat{\boldsymbol{\phi}}_\mathbf{k})_z = 0\,. \tag{16.46b}$$

Therefore,

$$dI_{\mathbf{k},\hat{\boldsymbol{\theta}}_\mathbf{k}}(2,1;1,0)/d\Omega_k = \frac{3}{8\pi}\sin^2\theta_k \tag{16.47}$$

polarized in the $\boldsymbol{\epsilon}_\mathbf{k}^{(1)}$ or $\hat{\boldsymbol{\theta}}_\mathbf{k}$ direction. The angular distribution and polarization is identical to that of a classical radiating dipole aligned along the z axis, *even though the dipole moment of the atom vanishes identically.* It turns out that this result is unique to a $J = 0 - 1$ transition.

Suppose, instead, that the atom is prepared in the $m_2 = 1$ state using a circularly polarized field, $\varrho_{J_2,m_2;J_2,m_2'}(0) = \delta_{m_2,1}\delta_{m_2',1}$. The radiation pattern is given by

$$dI_{\mathbf{k},\boldsymbol{\epsilon}_\mathbf{k}}(2,1;1,0)/d\Omega_k = \frac{3}{8\pi}\left|(\boldsymbol{\epsilon}_\mathbf{k})_1\right|^2 = \frac{3}{8\pi}\frac{[(\boldsymbol{\epsilon}_\mathbf{k})_x]^2 + \left[(\boldsymbol{\epsilon}_\mathbf{k})_y\right]^2}{2}. \tag{16.48}$$

Therefore, the intensity of the $\hat{\boldsymbol{\theta}}_\mathbf{k}$ polarization component is

$$dI_{\mathbf{k},\hat{\boldsymbol{\theta}}_\mathbf{k}}(2,1;1,0)/d\Omega_k = \frac{\left[\epsilon_x^{(1)}\right]^2 + \left[\epsilon_y^{(1)}\right]^2}{2} = \frac{\cos^2\theta_k}{2}, \tag{16.49a}$$

and that of the $\hat{\boldsymbol{\phi}}_\mathbf{k}$ component is

$$dI_{\mathbf{k},\hat{\boldsymbol{\phi}}_\mathbf{k}}(2,1;1,0)/d\Omega_k = \frac{\left[\epsilon_x^{(2)}\right]^2 + \left[\epsilon_y^{(2)}\right]^2}{2} = \frac{1}{2}. \tag{16.49b}$$

For $\theta_k = 0$ or π, the intensity in both components is the same, a result consistent with circularly polarized radiation.

To prove that the field is circularly polarized, we need an alternative expansion for the vacuum field. In the decomposition of the free quantized electric field, we took $\boldsymbol{\epsilon}_\mathbf{k}^{(1)} = \hat{\boldsymbol{\theta}}_\mathbf{k}$ and $\boldsymbol{\epsilon}_\mathbf{k}^{(2)} = \hat{\boldsymbol{\phi}}_\mathbf{k}$ to be real, corresponding to two independent linear polarizations ($\hat{\boldsymbol{\theta}}_\mathbf{k}\times\hat{\boldsymbol{\phi}}_\mathbf{k} = \hat{\mathbf{k}}$). We could just as well have used complex $\boldsymbol{\epsilon}$'s defined by

$$\boldsymbol{\epsilon}_\mathbf{k}^{\pm} = \frac{\boldsymbol{\epsilon}_\mathbf{k}^{(1)} \pm i\boldsymbol{\epsilon}_\mathbf{k}^{(2)}}{\sqrt{2}} = \frac{\hat{\boldsymbol{\theta}}_\mathbf{k} \pm i\hat{\boldsymbol{\phi}}_\mathbf{k}}{\sqrt{2}}. \tag{16.50}$$

The field is then written as

$$\mathbf{E}_T(\mathbf{R}) = i\sum_{\mathbf{k},\lambda=\pm}\left(\frac{\hbar\omega_\mathbf{k}}{2\epsilon_0\mathcal{V}}\right)^{1/2}\left[a_{\mathbf{k}_\lambda}\boldsymbol{\epsilon}_\mathbf{k}^{(\lambda)}e^{i\mathbf{k}\cdot\mathbf{R}} - a_{\mathbf{k}_\lambda}^{\dagger}\left(\boldsymbol{\epsilon}_\mathbf{k}^{(\lambda)}\right)^* e^{-i\mathbf{k}\cdot\mathbf{R}}\right], \tag{16.51}$$

where $a_{\mathbf{k}_\pm}$ is a destruction operator for photons having polarization $\boldsymbol{\epsilon}_\mathbf{k}^{\pm}$. As is shown in appendix A, for field propagation in the $\hat{\mathbf{k}}$ direction, $\boldsymbol{\epsilon}_\mathbf{k}^{\pm}$ correspond to the polarization of left and right circularly polarized photons, respectively. In calculating the emitted field (as we do in chapter 19), one finds that, for radiation emitted parallel to the z axis, one circular polarization component only of the field is created if the atom is prepared in either sublevel $m = 1$ or $m = -1$.

For a somewhat more detailed discussion of the manner in which the radiation pattern is linked to the fields that create the initial density matrix and for transitions other than $J_1 = 0$, $J_2 = 1$, see the discussion in appendix B.

16.2.2 Repopulation of the Ground State

Although the probability that the atom radiates into one *specific* mode of the field is negligibly small, it is a sure thing that the atom returns to the ground state as a result of spontaneous emission when we sum over all possible modes of emission. The repopulation of the ground state is of critical importance in the processes of optical pumping and sub-Doppler laser cooling to be considered in the next two chapters.

To calculate the rate at which the ground-state sublevels are repopulated, we sum over all modes of the emitted field—namely,

$$\dot{\varrho}_{1,J_1,m_1;1,J_1,m_1'}(t) = \frac{d}{dt}\sum_{\mathbf{k},\epsilon_\mathbf{k}} c_{1,J_1,m_1;\mathbf{k},\epsilon_\mathbf{k}}(t)c^*_{1,J_1,m_1';\mathbf{k},\epsilon_\mathbf{k}}(t)\,. \tag{16.52}$$

You might think that only diagonal elements of the ground state are repopulated, but, in general, *both* diagonal and nondiagonal ground-state density matrix elements are repopulated. Substituting equation (16.32) into equation (16.52), we find

$$\begin{aligned}\dot{\varrho}_{1,J_1,m_1;1,J_1,m_1'}(t) &= \frac{1}{\hbar^2}\frac{d}{dt}\sum_{\mathbf{k},\epsilon_\mathbf{k}}\sum_{m_2,m_2'}\int_0^t dt' e^{i(\omega_k-\omega_{21})t'}\int_0^t dt'' e^{-i(\omega_k-\omega_{21})t''}\\ &\times\langle 1,J_1,m_1;\mathbf{k},\epsilon_\mathbf{k}|V_{AF}|2,J_2,m_2;0\rangle\langle 1,J_1,m_1';\mathbf{k},\epsilon_\mathbf{k}|V_{AF}|2,J_2,m_2';0\rangle^*\\ &\times c_{2,J_2,m_2;0}(t')c^*_{2,J_2,m_2';0}(t'')\,.\end{aligned} \tag{16.53}$$

The sum over $\mathbf{k}$ is transformed into an integral using the prescription (16.8). As in the calculation of the spontaneous decay rate, the integral over k leads to a delta function, $\delta(t'-t'')$, in the Weisskopf-Wigner approximation. Evaluating the matrix elements and summing over angles and polarizations as we did for spontaneous decay yields

$$\dot{\varrho}_{1,J_1,m_1;1,J_1,m_1'}(t) = \sum_{m_2,m_2'}\gamma^{2,J_2,m_2;2,J_2,m_2'}_{1,J_1,m_1;1,J_1,m_1'}\varrho_{2,J_2,m_2;2,J_2,m_2'}(t)\,, \tag{16.54}$$

where the transfer rate $\gamma^{2,J_2,m_2;2,J_2,m_2'}_{1,J_1,m_1;1,J_1,m_1'}$ is given by

$$\gamma^{2,J_2,m_2;2,J_2,m_2'}_{1,J_1,m_1;1,J_1,m_1'} = \gamma_{2,J_2;1,J_1}\sum_q\begin{bmatrix}J_1 & 1 & J_2\\ m_1 & q & m_2\end{bmatrix}\begin{bmatrix}J_1 & 1 & J_2\\ m_1' & q & m_2'\end{bmatrix}. \tag{16.55}$$

Both population and coherence (off-diagonal density matrix elements) can be transferred to the ground state from the excited state via spontaneous decay. The selection rule $(m_2 - m_1) = (m_2' - m_1')$ is a statement of conservation of the z component of angular momentum.

If we look at decay from a specific excited-state sublevel m_2, the *partial decay rate* to a specific ground-state sublevel m_1 is given by

$$\gamma^{2,J_2,m_2;2,J_2,m_2}_{1,J_1,m_1;1,J_1,m_1} = \gamma_{2,J_2;1,J_1}\begin{bmatrix}J_1 & 1 & J_2\\ m_1 & m_2-m_1 & m_2\end{bmatrix}^2. \tag{16.56}$$

The partial decay rate divided by the total decay rate defines the *branching ratio* for the $(J_2,m_2;J_1,m_1)$ transition. The branching ratio is simply the value of the Clebsch-Gordan coefficient squared.

For an arbitrary excited-state density matrix, equation (16.54) is not all that transparent. Additional physical insight can be obtained with a change of representation. In some sense, the m representation is not a natural one for spontaneous decay. If we expand the density matrix in an *irreducible tensor basis* with matrix elements defined by [2]

$$\varrho_Q^K(J_2, J_2') = \sum_{K,Q} (-1)^{J_2'-m_2'} \begin{bmatrix} J_2 & J_2' & K \\ m_2 & -m_2' & Q \end{bmatrix} \varrho_{J_2,m_2;J_2',m_2'}, \tag{16.57}$$

we can use equation (16.80) in appendix B to show that

$$\dot{\varrho}_Q^K(J_1, J_1) = \gamma^{(K)}_{2,J_2;1,J_1} \varrho_Q^K(J_2, J_2), \tag{16.58}$$

where the multipole decay rate

$$\gamma^{(K)}_{2,J_2;1,J_1} = (-1)^{J_2+J_1+K+1}(2J_2+1) \begin{Bmatrix} J_2 & J_2 & K \\ J_1 & J_1 & 1 \end{Bmatrix} \gamma_{2,J_2;1,J_1} \tag{16.59}$$

is independent of Q. The quantity in curley braces is a 6-J symbol. In an irreducible tensor basis, each $\varrho_Q^K(J, J)$ has properties that are well defined under rotation. Under rotation, the total angular momentum K is unchanged, but the z component Q can change. However, on *average*, spontaneous emission acts as a scalar under rotation, so it changes neither K nor Q.

Although you may not have a lot of experience using the irreducible tensor basis, there is no need to fear it. It is a simple matter to write computer programs to convert from the m basis to the irreducible tensor basis. Moreover, you can see from the discussion of spontaneous decay that the irreducible tensor basis can offer distinct advantages. From equation (16.54), it would appear that one must specify $[(2J_1+1)(2J_2+1)]^2$ values of $\gamma^{2,J_2,m_2;2,J_2,m_2'}_{1,J_1,m_1;1,J_1,m_1'}$ to characterize the repopulation. The symmetry properties of the Clebsch-Gordan coefficients reduces this somewhat, but the number of independent decay parameters is not all that obvious. However, it is clear from equation (16.58) that the number of independent decay parameters is $\min[(2J_1+1), (2J_2+1)]$, quite a reduction from $[(2J_1+1)(2J_2+1)]^2$.

For reference purposes, the $\varrho_Q^K(J, J)$ are listed here for $J = 1/2$ and $J = 1$:

$J = 1/2$

$$\varrho_0^0 = \frac{1}{\sqrt{2}}(\varrho_{\frac{1}{2}\frac{1}{2}} + \varrho_{-\frac{1}{2}-\frac{1}{2}}), \tag{16.60a}$$

$$\varrho_0^1 = \frac{1}{\sqrt{2}}(\varrho_{\frac{1}{2}\frac{1}{2}} - \varrho_{-\frac{1}{2}-\frac{1}{2}}), \tag{16.60b}$$

$$\varrho_1^1 = -\varrho_{\frac{1}{2}-\frac{1}{2}}. \tag{16.60c}$$

$J = 1$

$$\varrho_0^0 = \frac{1}{\sqrt{3}}(\varrho_{11} + \varrho_{00} + \varrho_{-1-1}), \tag{16.61a}$$

$$\varrho_0^1 = \frac{1}{\sqrt{2}}(\varrho_{11} - \varrho_{-1-1}), \tag{16.61b}$$

$$\varrho_1^1 = -\frac{1}{\sqrt{2}}(\varrho_{10} + \varrho_{0-1}), \tag{16.61c}$$

$$\varrho_0^2 = \frac{1}{\sqrt{6}}(\varrho_{11} + \varrho_{-1-1} - 2\varrho_{00}), \tag{16.61d}$$

$$\varrho_1^2 = -\frac{1}{\sqrt{2}}(\varrho_{10} - \varrho_{0-1}), \tag{16.61e}$$

$$\varrho_2^2 = \varrho_{1-1}, \tag{16.61f}$$

with

$$\varrho_Q^K = (-1)^Q \left(\varrho_{-Q}^K\right)^*. \tag{16.62}$$

The density matrix element $\varrho_0^0(J,J) = 1/(\sqrt{2J+1})$ is proportional to the total population, $\varrho_0^K(J,J)$ depends on differences of populations, while $\varrho_Q^K(J,J)$, with $Q \neq 0$, depends on Zeeman-level coherence (off-diagonal density matrix elements in the m basis). Since the ϱ_Q^K have well-defined properties under rotation, they can be associated with the multipole moments [e.g., magnetic dipole ($K = 1$), electric quadrupole ($K = 2$)] of a given state.

16.3 Summary

We have considered the interaction of the vacuum field with an atom that has been prepared in an excited state. The rate of spontaneous emission, the radiation pattern, and the manner in which the ground state is repopulated have been derived from first principles. Aside from some slight embarrassments involving infinite-level shifts and the use of the Weisskopf-Wigner approximation without formal justification, we arrived at results that are consistent with conservation of energy and probability. In the next chapter, we will see how the combined action of spontaneous emission and a weak driving field modifies the ground-state density matrix. A method for including spontaneous emission in atom–field interactions using an *amplitide* rather than a density matrix approach is given in appendix C.

16.4 Appendix A: Circular Polarization

There can be some confusion when one considers the selection rules for circular polarization. The origin of the difficulty is that circular polarization is defined relative to the propagation vector of the field, while the magnetic quantum numbers are defined relative to the quantization axis. One can always choose the quantization axis along the propagation direction of a field, but this may not be particularly convenient, especially if there are several fields incident or if there is a constant external magnetic field present. In this appendix, we choose the z axis as the axis of quantization and consider fields propagating in either of the $\pm\hat{\mathbf{z}}$ directions.

For a classical monochromatic field, the field amplitude is

$$\mathbf{E}(Z,t) = \frac{1}{2}E_0\hat{\boldsymbol{\epsilon}}e^{i\beta} + \text{c.c.}\,, \tag{16.63}$$

where

$$\beta = \pm kZ - \omega t + \alpha, \tag{16.64}$$

α is a constant phase, E_0 is a real field amplitude, and $\hat{\boldsymbol{\epsilon}}$ lies in the XY plane and can be *complex*. Note that $\boldsymbol{\epsilon}$ real corresponds to linearly polarized light. For light propagating in the $+\hat{\mathbf{z}}\,(-\hat{\mathbf{z}})$ directions, one takes the $+\,(-)$ sign in equation (16.64). It is convenient to write the polarization in a spherical tensor basis as

$$\hat{\boldsymbol{\epsilon}} = \sum_{q=-1}^{1}(-1)^q\epsilon_{-q}\hat{\boldsymbol{\epsilon}}_q\,, \tag{16.65}$$

with

$$\epsilon_{\pm 1} = \mp\frac{\epsilon_x \pm i\epsilon_y}{\sqrt{2}}, \tag{16.66a}$$

$$\epsilon_0 = \epsilon_z, \tag{16.66b}$$

and

$$\hat{\boldsymbol{\epsilon}}_{\pm 1} = \mp\frac{\hat{\mathbf{x}} \pm i\hat{\mathbf{y}}}{\sqrt{2}}, \tag{16.67a}$$

$$\hat{\boldsymbol{\epsilon}}_0 = \hat{\mathbf{z}}\,. \tag{16.67b}$$

For $\mathbf{k}$ in the $\pm\hat{\mathbf{z}}$ directions, $\epsilon_0 = 0$. Let us look at some examples:
$\epsilon_x = \cos\phi$; $\epsilon_y = \sin\phi$. In this case,

$$\mathbf{E}(Z,t) = E_0\,(\hat{\mathbf{x}}\cos\phi + \hat{\mathbf{y}}\sin\phi)\cos\beta. \tag{16.68}$$

The radiation is linearly polarized at an angle ϕ relative to the x axis.

$\epsilon_x = 1/\sqrt{2}$; $\epsilon_y = i/\sqrt{2}$. From equations (16.63) to (16.67), we find

$$\begin{aligned}\epsilon_1 &= 0, \quad \epsilon_{-1} = 1,\\ \hat{\boldsymbol{\epsilon}} &= -\hat{\boldsymbol{\epsilon}}_1 = (\hat{\mathbf{x}} + i\hat{\mathbf{y}})/\sqrt{2},\end{aligned} \tag{16.69}$$

and

$$\begin{aligned}\mathbf{E}(Z,t) &= \frac{1}{2}E_0\left(\frac{\hat{\mathbf{x}} + i\hat{\mathbf{y}}}{\sqrt{2}}e^{i\beta} + \text{c.c.}\right)\\ &= \frac{1}{\sqrt{2}}E_0\,(\hat{\mathbf{x}}\cos\beta - \hat{\mathbf{y}}\sin\beta)\,.\end{aligned} \tag{16.70}$$

This field corresponds to *circular* polarization.

Left circular polarization (LCP) radiation has angular momentum directed *along* its propagation direction. When viewed head-on with the radiation approaching you, the polarization vector has constant amplitude and rotates in a counterclockwise direction. Right circular polarization (RCP) radiation has angular momentum directed *opposite* to its propagation direction. When viewed head-on with the radiation approaching you, the polarization has constant amplitude and rotates in a clockwise direction.

To determine the polarization of the field (16.70), we set $Z = 0$ to see how the polarization evolves in time. For $Z = 0$, $\beta = (-\omega t + \alpha)$, and equation (16.70) reduces to

$$\mathbf{E}(0, t) = \frac{1}{\sqrt{2}} E_0 \left[\hat{\mathbf{x}} \cos(\omega t - \alpha) + \hat{\mathbf{y}} \sin(\omega t - \alpha)\right]. \tag{16.71}$$

The field polarization vector rotates in a counterclockwise direction with constant amplitude in the xy plane. Such a field corresponds to LCP if $\mathbf{k} = k\hat{\mathbf{z}}$ and RCP if $\mathbf{k} = -k\hat{\mathbf{z}}$. In *both* cases, these fields induce $\Delta m = +1$ transitions on absorption and $\Delta m = -1$ transitions on emission, (as is discussed later in this section). Such fields are referred to as having σ_+ polarization or, simply, as σ_+ radiation. Thus, σ_+ radiation can be either LCP or RCP, depending on the direction of propagation.

$\epsilon_x = 1/\sqrt{2}$; $\epsilon_y = -i/\sqrt{2}$. From equations (16.63) to (16.67), we find

$$\begin{aligned} \epsilon_1 &= -1; \quad \epsilon_{-1} = 0, \\ \hat{\boldsymbol{\epsilon}} &= \hat{\boldsymbol{\epsilon}}_{-1} = (\hat{\mathbf{x}} - i\hat{\mathbf{y}})/\sqrt{2}, \end{aligned} \tag{16.72}$$

and

$$\begin{aligned} \mathbf{E} &= \frac{1}{2} E_0 \left(\frac{\hat{\mathbf{x}} - i\hat{\mathbf{y}}}{\sqrt{2}} e^{i\beta} + \text{c.c.} \right) \\ &= \frac{1}{\sqrt{2}} E_0 \left(\hat{\mathbf{x}} \cos\beta + \hat{\mathbf{y}} \sin\beta\right). \end{aligned} \tag{16.73}$$

The field polarization vector rotates in a clockwise direction with constant amplitude in the xy plane. Such a field corresponds to RCP if $\mathbf{k} = k\hat{\mathbf{z}}$ and LCP if $\mathbf{k} = -k\hat{\mathbf{z}}$. In *both* cases, these fields induce $\Delta m = -1$ transitions on absorption and $\Delta m = +1$ transitions on emission. Such fields are referred to as having σ_- polarization or, simply, as σ_- radiation.

To derive the Δm selection rules alluded to earlier, note that the interaction is of the form $V = -\hat{\boldsymbol{\mu}} \cdot \mathbf{E}$. On absorption from state 1 to 2, the amplitude equation in the interaction representation is of the form $\dot{c}_2 = V_{21} e^{i\omega_{21} t} c_1$, where V_{21} is a matrix element. Therefore, we must take the part of the field varying as $e^{-i\omega t}$ in the RWA, which is proportional to $\hat{\boldsymbol{\epsilon}}$. From equations (16.63) and (16.64), we deduce that the V_{21} coupling term varies as

$$\hat{\boldsymbol{\epsilon}} \cdot \hat{\boldsymbol{\mu}} = \sum_q (-1)^q \epsilon_{-q} \hat{\mu}_q, \tag{16.74}$$

where

$$\hat{\mu}_{\pm 1} = \mp(\hat{\mu}_x \pm i\hat{\mu}_y)/\sqrt{2}, \tag{16.75a}$$

$$\hat{\mu}_0 = \hat{\mu}_z. \tag{16.75b}$$

Thus, for absorption from state $|1, J_1, m_1\rangle$ to state $|2, J_2, m_2\rangle$, the transition amplitude is proportional to

$$\langle 2, J_2, m_2| \sum_q (-1)^q \epsilon_{-q} \hat{\mu}_q |1, J_1, m_1\rangle$$
$$= \sum_q (-1)^q \langle 2, J_2 || \mu^{(1)} || 1, J_1\rangle \frac{\epsilon_{-q}}{\sqrt{2J_2+1}} \begin{bmatrix} J_1 & 1 & J_2 \\ m_1 & q & m_2 \end{bmatrix}. \tag{16.76}$$

From the properties of the Clebsch-Gordan coefficients, it follows that q must be equal to $m_2 - m_1 \equiv \Delta m$, or else the Clebsch-Gordan coefficient vanishes. If $\Delta m = 1$ (a σ_+ transition), the coupling term vanishes unless $\epsilon_{-1} \neq 0$.

On emission, we must take the part of the field varying as $e^{i\omega t}$ in the RWA, which is proportional to $\hat{\boldsymbol{\epsilon}}^*$. This gives a term varying as $\hat{\boldsymbol{\epsilon}}^* \cdot \hat{\boldsymbol{\mu}}$; as a consequence, for emission from state $|2, J_2, m_2\rangle$ to a state $|1, J_1, m_2 - 1\rangle$, we require $(\epsilon_1)^* \neq 0$, which is equivalent to having $\epsilon_{-1} \neq 0$. Therefore σ_+ radiation has $|\epsilon_{-1}| = 1$, $\epsilon_1 = \epsilon_0 = 0$, $\hat{\boldsymbol{\epsilon}} = \pm(\hat{\mathbf{x}} + i\hat{\mathbf{y}})/\sqrt{2}$ and drives $\Delta m = 1$ transitions on absorption and $\Delta m = -1$ transitions on emission.

This is an expected result. Light propagating in the $\hat{\mathbf{z}}$ direction has angular momentum along $\hat{\mathbf{z}}$ if the polarization is LCP, $\hat{\boldsymbol{\epsilon}} = \pm(\hat{\mathbf{x}} + i\hat{\mathbf{y}})/\sqrt{2}$. If one photon is absorbed from this field, the angular momentum of the field decreases along $\hat{\mathbf{z}}$, so the angular momentum of the atom must increase.

Similar arguments can be used for σ_- radiation. To summarize the results, we find

- σ_+ radiation: $\hat{\boldsymbol{\epsilon}} = \pm(\hat{\mathbf{x}} + i\hat{\mathbf{y}})/\sqrt{2}$, $\epsilon_q = \pm\delta_{q,-1}$; LCP $\mathbf{k} = k\hat{\mathbf{z}}$; RCP $\mathbf{k} = -k\hat{\mathbf{z}}$.
- σ_- radiation: $\hat{\boldsymbol{\epsilon}} = \pm(\hat{\mathbf{x}} - i\hat{\mathbf{y}})/\sqrt{2}$, $\epsilon_q = \mp\delta_{q,-1}$; LCP $\mathbf{k} = -k\hat{\mathbf{z}}$; RCP $\mathbf{k} = k\hat{\mathbf{z}}$.

16.5 Appendix B: Radiation Pattern

We can generalize the results for the radiation pattern somewhat by allowing for several initial and final hyperfine states. We label the ground-state angular momenta by G and the excited-state angular momenta by H. If the *frequency* of the emitted radiation from the various excited to ground hyperfine states is resolved by the detector, we would have to consider each $H; G$ transition separately. If the frequencies are not resolved, we must *trace* over final ground hyperfine states and *average* over initial hyperfine states. For the moment, we assume that the emission frequencies are not resolved and allow for the possibility that the atom is prepared in a linear superposition of several excited-state hyperfine levels.

The generalization of equation (16.41) is

$$dI_{\mathbf{k},\epsilon_k}/d\Omega_k = \sum_G dI_{\mathbf{k},\epsilon_k}(H, H'; G)/d\Omega_k = \frac{3}{8\pi}\epsilon_q \left(\epsilon_{q'}\right)^* \begin{bmatrix} G & 1 & H \\ m_1 & q & m_2 \end{bmatrix}$$
$$\times \begin{bmatrix} G & 1 & H' \\ m_1 & q' & m_2' \end{bmatrix} \varrho_{H,m_2;H'm_2'}(0), \tag{16.77}$$

where the electronic state labels are suppressed, and we introduce a *summation convention* in which all repeated indices on the right-hand side of an equation are summed unless they also appear on the left-hand side of the equation.

A "natural" basis for the density matrix is the irreducible tensor basis defined by equation (16.57), whose inverse is

$$\varrho_{H,m_2;H'm'_2} = (-1)^{H'-m'_2} \begin{bmatrix} H & H' & K \\ m_2 & -m'_2 & Q \end{bmatrix} \varrho^K_Q(H, H'), \tag{16.78}$$

such that

$$\begin{aligned} dI_{\mathbf{k},\epsilon_\mathbf{k}}/d\Omega_k &= \frac{3}{8\pi}\epsilon_q \left(\epsilon_{q'}\right)^* (-1)^{H'-m'_2} \begin{bmatrix} G & 1 & H \\ m_1 & q & m_2 \end{bmatrix} \\ &\quad \times \begin{bmatrix} G & 1 & H' \\ m_1 & q' & m'_2 \end{bmatrix} \begin{bmatrix} H & H' & K \\ m_2 & -m'_2 & Q \end{bmatrix} \varrho^K_Q(H, H') \\ &= \sqrt{(2H+1)(2H'+1)}\epsilon_q \left(\epsilon_{q'}\right)^* (-1)^{H-m_2} \begin{bmatrix} G & H & 1 \\ m_1 & m_2 & q \end{bmatrix} \begin{bmatrix} H' & G & 1 \\ m'_2 & m_1 & q' \end{bmatrix} \\ &\quad \times \begin{bmatrix} H' & H & K \\ m'_2 & -m_2 & -Q \end{bmatrix} (-1)^{-G+H'+1+Q} \varrho^K_Q(H, H'). \end{aligned} \tag{16.79}$$

Using the fact that [2]

$$\begin{aligned} &\begin{bmatrix} G & H & j_3 \\ M_1 & -M_2 & m_3 \end{bmatrix} \begin{bmatrix} J_3 & G & j_2 \\ M_3 & M_1 & m_2 \end{bmatrix} \begin{bmatrix} J_3 & H & j_1 \\ M_3 & M_2 & m_1 \end{bmatrix} (-1)^{H+M_2} \\ &= [(2j_1+1)(2j_2+1)]^{1/2} (-1)^{m_2+m_3} (-1)^{2j_1+H+J_3} \\ &\quad \times \begin{bmatrix} j_1 & j_2 & j_3 \\ -m_1 & m_2 & m_3 \end{bmatrix} \begin{Bmatrix} j_1 & j_2 & j_3 \\ G & H & J_3 \end{Bmatrix}, \end{aligned} \tag{16.80}$$

we find

$$\begin{aligned} dI_{\mathbf{k},\epsilon_\mathbf{k}}/d\Omega_k &= (8\pi)^{-1} \sqrt{3(2K+1)(2H+1)(2H'+1)}(-1)^{G+H+1}\epsilon_q \left(\epsilon_{q'}\right)^* \\ &\quad \times \begin{Bmatrix} K & 1 & 1 \\ Q & q' & q \end{Bmatrix} \begin{Bmatrix} K & 1 & 1 \\ G & H & H' \end{Bmatrix} \varrho^K_Q(H, H') \\ &= (-1)^{G+H+K} \epsilon_q \left(\epsilon_{-q'}\right)^* \begin{bmatrix} 1 & 1 & K \\ q & q' & Q \end{bmatrix} \\ &\quad \times (-1)^{q'} \begin{Bmatrix} K & 1 & 1 \\ G & H & H' \end{Bmatrix} \varrho^K_Q(H, H'). \end{aligned} \tag{16.81}$$

As in many problems involving addition of angular momenta, it proves useful to define a *coupled tensor basis* by

$$\epsilon^K_Q = (-1)^{q'} \epsilon_q \left(\epsilon_{-q'}\right)^* \begin{bmatrix} 1 & 1 & K \\ q & q' & Q \end{bmatrix}, \tag{16.82}$$

that allows us to transform equation (16.81) into

$$\begin{aligned} dI_{\mathbf{k},\epsilon_\mathbf{k}}/d\Omega &= \frac{3}{8\pi} \frac{1}{2G+1} (-1)^{G+H+K} \sqrt{(2H+1)(2H'+1)} \\ &\quad \times \epsilon^K_Q \begin{Bmatrix} 1 & 1 & K \\ H & H' & G \end{Bmatrix} \varrho^K_Q(H, H'). \end{aligned} \tag{16.83}$$

Each spherical component of the initial state density matrix leads to a corresponding spherical component of the radiated field.

The coupled tensor components ϵ^K_Q are given by equation (16.82) in terms of the spherical components ϵ_q of the unit polarization vectors. For each propagation direction, there are two independent polarization directions, conventionally chosen in the $\hat{\boldsymbol{\theta}}_{\mathbf{k}}$ and $\hat{\boldsymbol{\phi}}_{\mathbf{k}}$ directions. Thus, we take (dropping the $\mathbf{k}$ subscripts)

$$\epsilon^K_Q(\hat{\boldsymbol{\theta}}) = (-1)^{q'} \epsilon^{(1)}_q \left(\epsilon^{(1)}_{-q'}\right)^* \begin{bmatrix} 1 & 1 & K \\ q & q' & Q \end{bmatrix}, \tag{16.84}$$

$$\epsilon^K_Q(\hat{\boldsymbol{\phi}}) = (-1)^{q'} \epsilon^{(2)}_q \left(\epsilon^{(2)}_{-q'}\right)^* \begin{bmatrix} 1 & 1 & K \\ q & q' & Q \end{bmatrix}, \tag{16.85}$$

where, using equations (16.39), (16.18b), and (16.18c), we have

$$\epsilon^{(1)}_{\pm 1} = \mp \frac{\cos\theta e^{\pm i\phi}}{\sqrt{2}}, \tag{16.86a}$$

$$\epsilon^{(1)}_{0} = -\sin\theta, \tag{16.86b}$$

$$\epsilon^{(2)}_{\pm 1} = -\frac{i e^{\pm i\phi}}{\sqrt{2}}, \tag{16.87a}$$

$$\epsilon^{(2)}_{0} = 0. \tag{16.87b}$$

As a consequence, equations (16.84) and (16.85) become

$$\begin{aligned} \epsilon^K_Q(\hat{\boldsymbol{\theta}}) &= (-1)^{q'} \epsilon^{(1)}_q \left(\epsilon^{(1)}_{-q'}\right)^* \begin{bmatrix} 1 & 1 & K \\ q & q' & Q \end{bmatrix} \\ &= -\frac{1}{\sqrt{3}}\delta_{K,0} \\ &+ \begin{bmatrix} \frac{1}{2}\cos^2\theta\, e^{-2i\phi}\delta_{Q,-2} - \frac{1}{2}\sin(2\theta)e^{-i\phi}\delta_{Q,-1} \\ + \left(\sqrt{\frac{2}{3}}\sin^2\theta - \sqrt{\frac{1}{6}}\cos^2\theta\right)\delta_{Q,0} \\ + \frac{1}{2}\sin(2\theta)e^{i\phi}\delta_{Q,1} + \frac{1}{2}\cos^2\theta e^{2i\phi}\delta_{Q,2} \end{bmatrix} \delta_{K,2} \end{aligned} \tag{16.88}$$

and

$$\begin{aligned} \epsilon^K_Q(\hat{\boldsymbol{\phi}}) &= (-1)^{q'} \epsilon^{(2)}_q \left(\epsilon^{(2)}_{-q'}\right)^* \begin{bmatrix} 1 & 1 & K \\ q & q' & Q \end{bmatrix} \\ &= -\frac{1}{\sqrt{3}}\delta_{K,0} - \left(\frac{1}{2}e^{-2i\phi}\delta_{Q,-2} + \sqrt{\frac{1}{6}}\delta_{Q,0} + \frac{1}{2}e^{2i\phi}\delta_{Q,2}\right)\delta_{K,2}. \end{aligned} \tag{16.89}$$

We now look at some examples in which we take $H = H'$—that is, we assume that the atom is prepared in a given hyperfine excited-state manifold. Moreover, we assume that there is only a single final state G.

16.5.1 Unpolarized Initial State

If the atom is prepared in an unpolarized initial state,

$$\varrho^K_Q(H, H) = \frac{1}{\sqrt{2H+1}}\delta_{K,0}\delta_{Q,0}\,, \tag{16.90}$$

one can use the fact that

$$\begin{Bmatrix} 1 & 1 & 0 \\ H & H & G \end{Bmatrix} = \frac{(-1)^{H+G}}{\sqrt{3\,(2H+1)}} \tag{16.91}$$

to obtain

$$\begin{aligned} dI_{\mathbf{k},\epsilon_\mathbf{k}}/d\Omega &= \frac{3}{8\pi}(-1)^{G+H}\sqrt{(2H+1)}\epsilon^0_0\frac{(-1)^{H+G}}{\sqrt{3\,(2H+1)}} \\ &= (-1)^{q'}\frac{\sqrt{3}}{8\pi}\epsilon_q\left(\epsilon_{-q'}\right)^*\begin{bmatrix} 1 & 1 & 0 \\ q & q' & 0 \end{bmatrix} \\ &= \frac{1}{8\pi}\sum_q |\epsilon_q|^2 = \frac{1}{8\pi}\,. \end{aligned} \tag{16.92}$$

The radiation is unpolarized and isotropic, as expected.

16.5.2 *z*-Polarized Excitation

To examine other initial conditions, we need to specify $\varrho^K_Q(H, H)$. The initial state density matrix elements $\varrho^K_Q(H, H)$ are created by the external excitation fields. In perturbation theory, one can show that the values of $\varrho^K_Q(H, H)$ are determined by the coupled tensor $\epsilon^K_Q(i)$ of the *excitation* field according to

$$\varrho^K_Q(H, H) \propto \epsilon^K_{-Q}(i)\begin{Bmatrix} 1 & 1 & K \\ H & H & G \end{Bmatrix}\,. \tag{16.93}$$

This result can be derived using the formalism presented in the following chapter. Combining equation (16.93) with equation (16.83), we find an (unnormalized) radiation pattern given by

$$dI_{\mathbf{k},\epsilon_\mathbf{k}}/d\Omega_k = (-1)^K\epsilon^K_Q\epsilon^K_{-Q}(i)\begin{Bmatrix} 1 & 1 & K \\ H & H & G \end{Bmatrix}^2\,. \tag{16.94}$$

For a *z*-polarized light field ($\epsilon_q = \delta_{q,0}$), we can use equation (16.82) to calculate

$$\epsilon^K_Q(i) \propto \left(-\frac{1}{\sqrt{3}}\delta_{K,0} + \frac{2}{\sqrt{6}}\delta_{K,2}\right)\delta_{Q,0}\,, \tag{16.95}$$

which implies that

$$dI_{\mathbf{k},\epsilon_\mathbf{k}}/d\Omega_k = \epsilon^K_0\left(-\frac{1}{\sqrt{3}}\delta_{K,0} + \frac{2}{\sqrt{6}}\delta_{K,2}\right)\begin{Bmatrix} 1 & 1 & K \\ H & H & G \end{Bmatrix}^2\,. \tag{16.96}$$

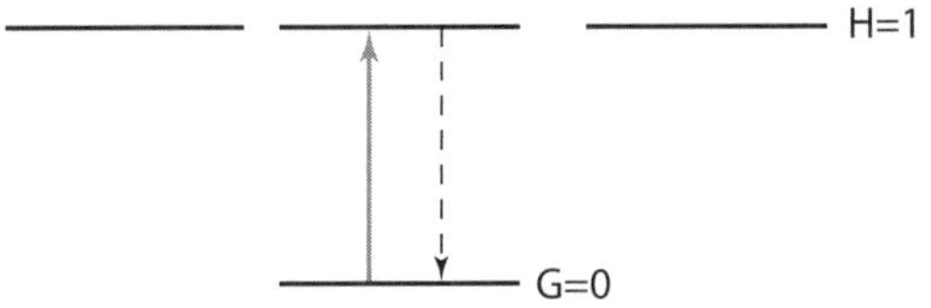

Figure 16.2. Decay from $H = 1$ to $G = 0$ for excitation with linearly polarized light.

Using equations (16.88) and (16.89), we then obtain for the two polarization components of the radiated intensity

$$dI_{\mathrm{k}}(\hat{\boldsymbol{\theta}})/d\Omega_k = \frac{1}{3}\begin{Bmatrix} 1 & 1 & 0 \\ H & H & G \end{Bmatrix}^2 + \frac{2}{3}\left(\sin^2\theta - \frac{1}{2}\cos^2\theta\right)\begin{Bmatrix} 1 & 1 & 2 \\ H & H & G \end{Bmatrix}^2 \tag{16.97}$$

and

$$dI_{\mathrm{k}}(\hat{\boldsymbol{\phi}})/d\Omega_k = \frac{1}{3}\left(\begin{Bmatrix} 1 & 1 & 0 \\ H & H & G \end{Bmatrix}^2 - \begin{Bmatrix} 1 & 1 & 2 \\ H & H & G \end{Bmatrix}^2\right). \tag{16.98}$$

These equations are the principal result of this appendix.

For $G = 0$, $H = 1$, equations (16.97) and (16.98) reduce to

$$\begin{aligned} dI_{\mathrm{k}}(\hat{\boldsymbol{\theta}})/d\Omega_k &= \frac{1}{3}\frac{1}{9} + \frac{2}{3}\left(\sin^2\theta - \frac{1}{2}\cos^2\theta\right)\frac{1}{9} \\ &= \frac{1}{27}(1 + 2\sin^2\theta - \cos^2\theta) = \frac{1}{9}\sin^2\theta \end{aligned} \tag{16.99}$$

and

$$dI_{\mathrm{k}}(\hat{\boldsymbol{\phi}})/d\Omega_k = \frac{1}{3}\left(\frac{1}{9} - \frac{1}{9}\right) = 0, \tag{16.100}$$

respectively. The field is polarized in the $\hat{\boldsymbol{\theta}}$ direction, as was found previously.

In general, however,

$$dI_{\mathrm{k}}(\hat{\boldsymbol{\theta}})/d\Omega_k = \frac{1}{3}\left(\begin{Bmatrix} 1 & 1 & 0 \\ H & H & G \end{Bmatrix}^2 - \begin{Bmatrix} 1 & 1 & 2 \\ H & H & G \end{Bmatrix}^2\right) + \sin^2\theta\begin{Bmatrix} 1 & 1 & 2 \\ H & H & G \end{Bmatrix}^2 \tag{16.101}$$

and

$$dI_{\mathrm{k}}(\hat{\boldsymbol{\phi}})/d\Omega_k = \frac{1}{3}\left(\begin{Bmatrix} 1 & 1 & 0 \\ H & H & G \end{Bmatrix}^2 - \begin{Bmatrix} 1 & 1 & 2 \\ H & H & G \end{Bmatrix}^2\right). \tag{16.102}$$

Since the lead term in equation (16.101) is identical to equation (16.102), there is, in general, an isotropic, unpolarized background term, in addition to the $\hat{\boldsymbol{\theta}}$-polarized term that has a $\sin^2\theta$ dependence.

For $G = 0$ and $H = 1$, the isotropic background term is absent owing to a cancellation of the 6-J symbols. In this case, all emission is back along the $m_1 = 0$ to $m_2 = 0$ excitation path (see figure 16.2). This may explain why we recover a simple dipole radiation pattern corresponding to a classical dipole oscillating in the $\hat{\mathbf{z}}$ direction.

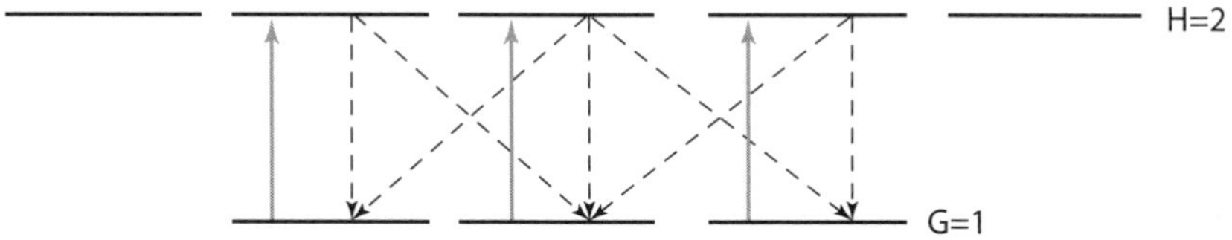

Figure 16.3. Decay from $H = 2$ to $G = 1$ for excitation with linearly polarized light.

On the other hand, for $G = 1$ and $H = 2$, spontaneous emission is no longer confined to $\Delta m = 0$ transitions, as it was on excitation (see figure 16.3). This explains why the simple linear dipole pattern is lost and there is an isotropic, unpolarized contribution to the radiation pattern. There is no ϕ dependence in the signal owing to the axially symmetric (z-polarized) nature of the excitation process.

16.5.3 Other Than *z*-Polarized Excitation

Other initial field polarizations can be considered in a similar manner using equation (16.94). For example, for σ_+ polarized light ($\epsilon_q = \delta_{q,-1}$),

$$\epsilon_Q^K(i) = -\left(\frac{1}{\sqrt{3}}\delta_{K,0} - \frac{1}{\sqrt{2}}\delta_{K,1} + \frac{1}{\sqrt{6}}\delta_{K,2}\right)\delta_{Q,0} \tag{16.103}$$

and

$$dI_{\mathrm{k},\epsilon_{\mathrm{k}}}/d\Omega_k = -\epsilon_0^K\left(\frac{1}{\sqrt{3}}\delta_{K,0} - \frac{1}{\sqrt{2}}\delta_{K,1} + \frac{1}{\sqrt{6}}\delta_{K,2}\right)\begin{Bmatrix} 1 & 1 & K \\ H & H & G \end{Bmatrix}^2, \tag{16.104}$$

such that

$$dI_{\mathrm{k}}(\hat{\boldsymbol{\theta}})/d\Omega_k = \frac{1}{3}\begin{Bmatrix} 1 & 1 & 0 \\ H & H & G \end{Bmatrix}^2 + \frac{1}{6}\begin{Bmatrix} 1 & 1 & 2 \\ H & H & G \end{Bmatrix}^2 - \frac{1}{2}\sin^2\theta\begin{Bmatrix} 1 & 1 & 2 \\ H & H & G \end{Bmatrix}^2 \tag{16.105}$$

and

$$dI_{\mathrm{k}}(\hat{\boldsymbol{\phi}})/d\Omega_k = \frac{1}{3}\begin{Bmatrix} 1 & 1 & 0 \\ H & H & G \end{Bmatrix}^2 + \frac{1}{6}\begin{Bmatrix} 1 & 1 & 2 \\ H & H & G \end{Bmatrix}^2. \tag{16.106}$$

16.6 Appendix C: Quantum Trajectory Approach to Spontaneous Decay

In chapter 3, we said that it was necessary to use a density matrix approach to account for the "in" term in spontaneous emission, $\dot{\varrho}_{11} = \gamma_2\varrho_{22}$. However, it *is* possible to use an amplitude approach if one uses a *quantum trajectory* or Monte Carlo technique [3]. Since this method has received increased attention in the last few years, we give a brief introduction to the ideas behind such an approach.

The method works because spontaneous emission can be considered as a *quantum jump* process owing to the fact that the correlation time of the vacuum field is so small ($\leq 10^{-16}$ s in the optical region). Thus, one can assume that quantum jumps occur according to some random algorithm and then average over a large number of possible histories.

The basic idea is to look at the wave function and density matrix at time intervals separated by δt. We consider only a two-level atom. At a given time, the wave function is of the form

$$|\Psi(t)\rangle = a_1(t)|1\rangle + a_2(t)|2\rangle \tag{16.107}$$

and is assumed to evolve under the *non-Hermitian* Hamiltonian

$$H = H_0 + \begin{pmatrix} 0 & 0 \\ 0 & -i\hbar\gamma \end{pmatrix}, \tag{16.108}$$

where $\gamma = \gamma_2/2$.

In a time interval δt small compared with (Rabi frequencies)$^{-1}$, (detunings)$^{-1}$, and γ_2^{-1}, the probability dP for spontaneous emission is

$$dP = \gamma_2 |a_2(t)|^2 \delta t\,. \tag{16.109}$$

We now choose a random variable ϵ uniformly distributed between 0 and 1. If $\epsilon < dP$, no emission occurs, and we set

$$|\Psi(t+\delta t)\rangle = \frac{\left(1 - \dfrac{i}{\hbar} H\delta t\right)|\Psi(t)\rangle}{\langle\Psi(t+\delta t)|\Psi(t+\delta t)\rangle^{1/2}}\,. \tag{16.110}$$

In other words, we allow the system to evolve according to the Hamiltonian H, but renormalize the state vector for the system. On the other hand, if $\epsilon > dP$, emission occurs, and we set

$$|\Psi(t+\delta t)\rangle = |1\rangle, \tag{16.111}$$

that is, we restart the atom in its ground state. Note that

$$\begin{aligned} \langle\Psi(t+\delta t)|\Psi(t+\delta t)\rangle &= \langle\Psi(t)|\left(1 + \frac{i}{\hbar}H^\dagger\delta t\right)\left(1 - \frac{i}{\hbar}H\delta t\right)|\Psi(t)\rangle \\ &\approx \langle\Psi(t)|\Psi(t)\rangle - \gamma_2\langle\Psi(t)|\begin{pmatrix} 0 & 0 \\ 0 & 1 \end{pmatrix}|\Psi(t)\rangle\delta t \\ &= 1 - \gamma_2\delta t|a_2(t)|^2 = 1 - dP\,. \end{aligned} \tag{16.112}$$

Now we construct the density matrix at time $(t+\delta t)$ using

$$\overline{\varrho(t+\delta t)} = P_0(\delta t)\overline{\frac{(1 - \frac{i}{\hbar}H\delta t)\,|\Psi(t)\rangle\langle\Psi(t)|\,(1 + \frac{i}{\hbar}H^\dagger\delta t)}{1 - dP}} + \overline{dP|1\rangle\langle 1|}\,, \tag{16.113}$$

where $P_0(\delta t) = (1 - dP)$ is the probability of no emission in time δt and the "line" denotes averaging over different histories. From equations (16.108) and (16.113), we obtain, to order δt,

$$\begin{aligned} \overline{\varrho(t+\delta t)} = \left\{\overline{\varrho(t)} + \frac{1}{i\hbar}\left[H_0, \overline{\varrho(t)}\right]\right\}\delta t - \frac{\gamma_2}{2}\left[\overline{\varrho(t)}\begin{pmatrix} 0 & 0 \\ 0 & 1 \end{pmatrix} + \begin{pmatrix} 0 & 0 \\ 0 & 1 \end{pmatrix}\overline{\varrho(t)}\right]\delta t \\ + \gamma_2\overline{|a_2(t)|^2}\delta t|1\rangle\langle 1|\,, \end{aligned} \tag{16.114}$$

or

$$\dot{\bar{\varrho}}(t) = \frac{1}{i\hbar}[H_0, \bar{\varrho}(t)] - \frac{\gamma_2}{2}[\bar{\varrho}_{12}(t)|1\rangle\langle 2| + \bar{\varrho}_{21}(t)|2\rangle\langle 1|]$$
$$-\gamma_2\bar{\varrho}_{22}(t)|2\rangle\langle 2| + \gamma_2\bar{\varrho}_{22}(t)|1\rangle\langle 1|, \qquad (16.115)$$

which is identical to the optical Bloch equations.

Thus, one can use amplitude equations for a number of histories and then form density matrix elements. The advantage is that for n coupled amplitude equations, one need solve only n equations (for a number of histories) rather than n^2 equations for a single history. This method is especially useful if one has to use a quantized description of the center-of-mass motion, since the number of quantized motion states that must be included can become very large.

Problems

1. Estimate the $2P$–$1S$ decay rate in hydrogen and also the $2S$–$2P$ decay rate. You will need to look up the wave functions and frequency spacings of these levels.
2. Use equations (16.28) and (16.29) to estimate the ratio of the decay rates for the D1 and D2 transitions in Na.
3. Consider spontaneous emission from state $|2, J_2\rangle$ to state $|1, J_1\rangle$ in an ensemble of independently decaying atoms. Calculate the radiation pattern as a function of $\mathbf{k}$ and the polarization defined by $\boldsymbol{\epsilon} = \cos(\psi)\hat{\boldsymbol{\theta}} + \sin(\psi)\hat{\boldsymbol{\phi}}$, assuming an initial density matrix $\langle J_2, m_2|\varrho|J_2, m_2'\rangle$ for the atoms. The unit vectors $\hat{\boldsymbol{\theta}}$ and $\hat{\boldsymbol{\phi}}$ are defined relative to the direction of emission $\mathbf{k}$. For an unpolarized initial state, $\langle J_2, m_2|\varrho|J_2, m_2'\rangle = (2J_2+1)^{-1}\delta_{m_2,m_2'}$, show that the radiation in any direction is unpolarized (that is, the radiated intensity is independent of ψ for any θ and ϕ). For $J_1 = 0$ and $J_2 = 1$, show that the radiation emitted in the z direction is consistent with circularly polarized radiation if $\langle J_2, m_2|\varrho|J_2, m_2'\rangle = \delta_{m_2,1}\delta_{m_2',1}$.
4. Prove that
$$\sum_{\lambda=1}^{2}\int d\Omega_k \sum_i \left[\boldsymbol{\epsilon}_{\mathbf{k}}^{(\lambda)}\right]_i d_i \sum_j \left[\boldsymbol{\epsilon}_{\mathbf{k}}^{(\lambda)}\right]_j \bar{d}_j = \frac{8\pi}{3}\sum_i d_i\bar{d}_i .$$
5. Prove that
$$\sum_{m_1}\langle 2, J_2, m_2|\mathbf{d}|1, J_1, m_1\rangle \cdot \langle 1, J_1, m_1|\mathbf{d}|2, J_2', m_2'\rangle$$
$$= \frac{|\langle 2, J_2\|d\|1, J_1\rangle|^2}{2J_2+1}\delta_{m_2,m_2'}\delta_{J_2,J_2'} .$$
6. Derive equation (16.54).
7. Use equation (16.80) to derive equation (16.57) from equation (16.54).
8. Use perturbation theory and equation (16.80) to derive equation (16.93).
9. Write a program to calculate density matrix elements in the irreducible tensor basis, given the density matrix elements in the m basis. Use your program to verify equations (16.60) and (16.61).

10. If we look at the spontaneous emission problem for a single excited state and a single ground state and neglect polarization effects, the equation for the excited-state amplitude for $t > 0$ can be written as

$$\dot{c}_2 = -\frac{1}{2\pi}\int_{-\infty}^{\infty} d\omega \int_0^t dt' \tilde{\gamma}_2(\omega) \exp[-i(\omega - \omega_0)(t - t')] c_2(t'),$$

where

$$\tilde{\gamma}_2(\omega) = \begin{cases} \gamma_2(\omega/\omega_0)^3 & \omega > 0\,, \\ 0 & \omega < 0\,. \end{cases}$$

Set

$$\gamma_2(\tau) = \frac{1}{2\pi}\int_0^{\infty} d\omega \tilde{\gamma}_2(\omega) \exp[-i(\omega - \omega_0)\tau]$$

such that

$$\dot{c}_2 = -\int_0^t d\tau \; \gamma_2(\tau) c_2(t - \tau).$$

Solve this equation by Laplace transform techniques to obtain an integral expression for $c_2(t)$. What assumptions are then needed to arrive at an exponential decay law?

11. Suppose that at $t = 0$, an atom is in the linear superposition of ground (G) and excited (H) states

$$|\psi(0)\rangle = \sum_{m_G} c_{m_G;0}(0) \, |m_G; 0\rangle + \sum_{m_H} c_{m_H;0}(0) \, |m_H; 0\rangle \, e^{-i\omega_H t}\,,$$

with no photons in the field. The system undergoes spontaneous emission such that the state vector at any time is

$$|\psi(t)\rangle = \sum_{m_G} c_{m_G;0}(t) \, |m_G; 0\rangle + \sum_{m_H} c_{m_H;0}(t) \, |m_H; 0\rangle \, e^{-i\omega_H t} + \sum_{m_G, \mathbf{k}_\lambda} c_{m_G;\mathbf{k}_\lambda}(t) \, \big|m_G; \mathbf{k}_\lambda\big\rangle \, e^{-i\omega_k t}\,.$$

Show that the density matrix element (in this interaction representation) decays as

$$\dot{\varrho}_{m_G m_H} = -\frac{\gamma_H}{2}\varrho_{m_G m_H}\,.$$

(Note: This is a problem with a two-line solution.)

12. An atom undergoes spontaneous decay with a central wavelength of 600 nm and a lifetime of 100 ns. Estimate the Rabi frequency if the field from this atom interacts with a similar atom 1 cm away. If, instead, the single-photon pulse can be focused to an area of λ^2 when it strikes the atom, show that the Rabi frequency is of the order of the decay rate and the pulse area is of order unity—that is, a single-photon pulse focused to a wavelength can fully excite an atom.

References

[1] V. Weisskopf and E. Wigner, *Berechnung der natürlichen Linienbreite auf Grund der Diracschen Lichttheorie*, Zeitschrift für Physik **63**, 54–73 (1930); a translation is available, *Calculation of the natural line width on the basis of Dirac's theory of light*, in W. R. Hindmarsh, *Atomic Spectra* (Pergamon Press, London, 1967), pp. 304–327.

[2] We follow the phase conventions given in vol. II, app. C, of A. Messiah, *Quantum Mechanics* (North Holland, Amsterdam, 1966). Properties of the Clebsch-Gordan, 3-J, 6-J, and 9-J symbols are discussed in this appendix, as well as irreducible tensor operators and rotation matrices.

[3] See, for example, J. Dalibard, Y. Castin, and K. Mølmer, *Wave-function approach to dissipative processes in quantum optics*, Physical Review Letters **68**, 580–583 (1992); R. Dum, A. S. Parkins, P. Zoller, and C. W. Gardiner, *Monte Carlo simulation of master equations in quantum optics for vacuum, thermal, and squeezed reservoirs,* Physical Review A **46**, 4382–4396 (1992); K. Mølmer, Y. Castin, and J. Dalibard, *Monte Carlo wave-function method in quantum optics*, Journal of the Optical Society of America B **10**, 524–538 (1993); H. J. Carmichael, *Quantum trajectory theory for cascaded open systems*, Physical Review Letters **70**, 2273–2276 (1993); Y. Castin and K. Mølmer, *Monte Carlo wave functions and nonlinear master equations*, Physical Review A **54**, 5275–5290 (1996).

Bibliography

W. Heitler, *The Quantum Theory of Radiation,* 3rd ed. (Oxford University Press, London,1954), chaps. IV and V.

Some books on angular momentum and irreducible tensor operators are listed here:

K. Blum, *Density Matrix Theory and Applications* (Plenum Press, New York, 1981).

D. M. Brink and G. R. Satchler, *Angular Momentum*, 3rd ed. (Clarendon Press, Oxford, UK, 1994).

A. R. Edmonds, *Angular Momentum in Quantum Mechanics* (Princeton University Press, Princeton, NJ, 1996).

U. Fano and G. Racah, *Irreducible Tensorial Sets* (Academic Press, New York, 1959).

M. E. Rose, *Elementary Theory of Angular Momentum* (Dover, New York, 1995).

W. J. Thompson, *Angular Momentum* (Wiley, New York, 1994).

For quantum trajectory theory, see

H. J. Carmichael, *An Open Systems Approach to Quantum Optics*, Lecture Notes in Physics, New Series M—Monographs, vol. M18 (Springer-Verlag, Berlin, 1993).

17

Optical Pumping and Optical Lattices

17.1 Optical Pumping

In 1950, Alfred Kastler proposed *optical pumping* [1] as a means for using optical fields to transfer internal state angular momentum to atoms. For his discoveries, Kastler was awarded the Nobel Prize in 1966. Optical pumping refers to a process in which the combined action of (1) stimulated absorption and emission produced by optical fields and (2) spontaneous decay results in a polarized ground state. [A polarized ground state is one in which $\varrho_{J,m;J,m'} \neq \delta_{m,m'}/(2J+1)$; in irreducible tensor language, $\varrho^K_Q \neq \delta_{K,0}\delta_{Q,0}/(2J+1)^{1/2}$.] Actually, as has been stressed throughout this book, one cannot really represent the scattering of a cw field by atoms as absorption followed by spontaneous emission. The entire process is one in which the incident field is scattered into previously unoccupied modes of the vacuum field, resulting also in a modification of the ground-state density matrix of the atoms. This should be kept in mind, even if we use the terms *absorption* and *spontaneous emission* in this chapter.

Although the density matrix equations can get algebraically complicated, the idea is simple. As an example, consider transitions between a $G=1$ ground state and an $H=0$ excited state driven by a z-polarized field (figure 17.1). (In all examples and calculations, we limit the discussion to a single ground-state hyperfine level having total angular momentum G and excited state having total angular momentum H.) The dipole selection rules are $\Delta m = \pm 1, 0$, with $\Delta m = 0$ forbidden if $G=H$ and G is integral. For the level scheme of figure 17.1, optical pumping results ultimately in steady-state populations

$$\varrho_{0,0} = 0, \quad \varrho_{-1-1} = \varrho_{11} = \frac{1}{2}, \tag{17.1}$$

and moreover, the atom is decoupled from the field. In this case, the ground state becomes polarized, but the expectation value of the z component of angular momentum is equal to zero. A linearly polarized field has no intrinsic or spin angular momentum associated with it [2]. It can transfer linear momentum to the atom's

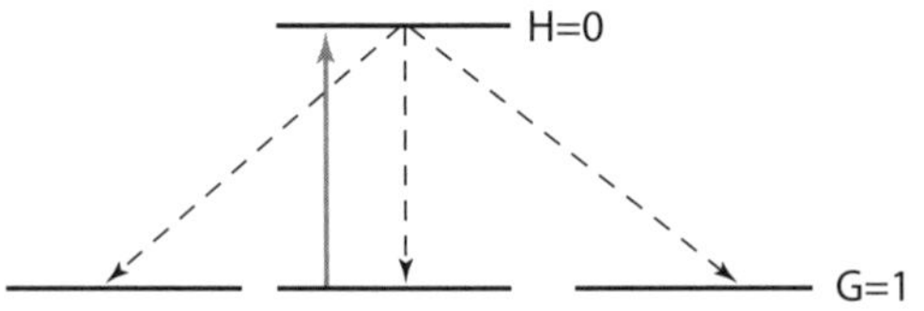

Figure 17.1. Optical pumping in a $G=1$ to $H=0$ transition with z-polarized excitation.

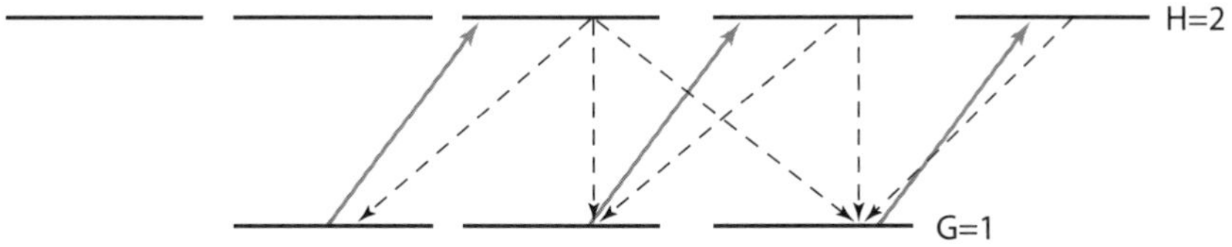

Figure 17.2. Optical pumping in a $G = 1$ to $H = 2$ transition with σ_+-polarized excitation.

center-of-mass motion, but this does not result in a net transfer of internal state angular momentum.

As another example, consider a $G=1$, $H=2$ transition driven by σ_+ circularly polarized radiation (figure 17.2). Optical pumping results in a final state in which ground-state density matrix elements $\varrho_{-1,-1}=\varrho_{0,0}=0$, and the field ultimately drives transitions in a "two-state" atom between levels $G=1, m_G=1$ and $H=2, m_H=2$. Thus, optical pumping constitutes an important tool for reducing a multilevel atomic scheme to that of an effective two-level atom. As such, optical pumping is often used in a preparatory stage in many experiments. In this example, optical pumping results in a transfer of the intrinsic or spin angular momentum of a circularly polarized field to the atom. It is also possible to construct fields that contain orbital angular momentum that can be transferred to the atom's center-of-mass motion [3]. In this chapter, we neglect any effects related to changes in the atom's center-of-mass motion.

To work out the theory of optical pumping, we must generalize the two-level problem to include magnetic degeneracy and include the "in" terms of spontaneous emission calculated previously. We consider first a traveling-wave field and, subsequently, a standing-wave field.

17.1.1 Traveling-Wave Fields

We treat the incident field classically, taking the electric field vector as

$$\mathbf{E}_c(\mathbf{R}, t) = \frac{1}{2}\mathcal{E}\hat{\boldsymbol{\epsilon}}e^{i(\mathbf{k}\cdot\mathbf{R}-\omega t)} + \text{c.c.}, \tag{17.2}$$

where *both* the field amplitude $\mathcal{E}$ *and* the polarization vector

$$\hat{\boldsymbol{\epsilon}} = \epsilon_x\hat{\mathbf{x}} + \epsilon_y\hat{\mathbf{y}} + \epsilon_z\hat{\mathbf{z}} \tag{17.3}$$

can be complex. (The subscript c stands for "classical.") For plane-wave fields, $(\mathbf{k}\cdot\hat{\boldsymbol{\epsilon}}) = 0$. As before, we define spherical polarization components as

$$\epsilon_{\pm 1} = \mp\frac{\epsilon_x \pm i\epsilon_y}{\sqrt{2}}, \quad \epsilon_0 = \epsilon_z. \tag{17.4}$$

In treating spontaneous decay, the polarization vectors for the radiated field were taken to be real [see equations (16.18b) and (16.18c)] such that $\boldsymbol{\epsilon}_{\mathbf{k}} = \boldsymbol{\epsilon}_{\mathbf{k}}^*$. Since we allow for complex polarization vectors for the classical field (17.2), this relation no longer need hold for the field polarization $\hat{\boldsymbol{\epsilon}}$. As a consequence, there can be a difference between $(\epsilon_q)^*$ and $(\epsilon^*)_q$, since

$$(\epsilon_q)^* = (-1)^q (\epsilon^*)_{-q}, \quad q = 0, \pm 1. \tag{17.5}$$

An atom placed in the field (17.2) experiences an atom–field interaction of the form

$$V_{AF}(\mathbf{R}, t) = -\hat{\boldsymbol{\mu}} \cdot \mathbf{E}_c(\mathbf{R}, t). \tag{17.6}$$

We obtain equations of motion for density matrix elements just as we did in chapter 3, except that we must include the magnetic state labels and account for repopulation of the ground-state sublevels via spontaneous decay. The matrix elements of $V_{AF}(\mathbf{R}, t)$ that enter are proportional to

$$\begin{aligned}
\langle Hm_H|\hat{\boldsymbol{\mu}} \cdot \hat{\boldsymbol{\epsilon}}|Gm_G\rangle &= \sum_q (-1)^q \epsilon_{-q} \langle H, m_H|\hat{\mu}_q|G, m_G\rangle \\
&= \langle H||\mu^{(1)}||G\rangle \sum_q (-1)^q \epsilon_{-q} \begin{bmatrix} G & 1 & H \\ m_G & q & m_H \end{bmatrix} \frac{1}{\sqrt{2H+1}} \\
&= \langle H||\mu^{(1)}||G\rangle \sum_q (-1)^q \epsilon_{-q} (-1)^{H-m_H} \begin{pmatrix} H & 1 & G \\ -m_H & q & m_G \end{pmatrix},
\end{aligned} \tag{17.7}$$

where $[\ldots]$ is a Clebsch-Gordan coefficient, $(\ldots)$ is a 3-J symbol, the $\hat{\mu}_q$ are given by equation (16.20a), and $\langle H||\mu^{(1)}||G\rangle$ is a reduced matrix element.

With a field interaction representation defined by

$$\varrho_{GH}(\mathbf{R}, \mathbf{v}, t) = \tilde{\varrho}_{GH}(\mathbf{v}, t) e^{-i(\mathbf{k}\cdot\mathbf{R} - \omega t)}, \tag{17.8}$$

We can use the Schrödinger equation and equation (17.7) to show that the density matrix elements obey the following equations in the RWA.

$$\begin{aligned}
\frac{\partial \varrho_{Gm_G;Gm'_G}}{\partial t} &= \sum_{m_H, m'_H} \gamma_{Gm_G;Gm'_G}^{Hm_H;Hm'_H} \varrho_{Hm_H;Hm'_H} \\
&\quad - i\chi_{HG}^* \sum_{q, m_H} (\epsilon_q)^* \begin{bmatrix} G & 1 & H \\ m_G & q & m_H \end{bmatrix} \tilde{\varrho}_{Hm_H;Gm'_G} \\
&\quad + i\chi_{HG} \sum_{q, m_H} (-1)^q \epsilon_{-q} \begin{bmatrix} G & 1 & H \\ m'_G & q & m_H \end{bmatrix} \tilde{\varrho}_{Gm_G;Hm_H},
\end{aligned} \tag{17.9a}$$

$$\begin{aligned}
\frac{\partial \varrho_{Hm_H;Hm'_H}}{\partial t} &= -\gamma_H \varrho_{Hm_H;Hm'_H} + i\chi_{HG}^* \sum_{q, m_G} (\epsilon_q)^* \begin{bmatrix} G & 1 & H \\ m_G & q & m'_H \end{bmatrix} \tilde{\varrho}_{Hm_H;Gm_G} \\
&\quad - i\chi_{HG} \sum_{q, m_G} (-1)^q \epsilon_{-q} \begin{bmatrix} G & 1 & H \\ m_G & q & m_H \end{bmatrix} \tilde{\varrho}_{Gm_G;Hm'_H},
\end{aligned} \tag{17.9b}$$

$$\frac{\partial \tilde{\varrho}_{Gm_G;Hm_H}}{\partial t} = -\frac{\gamma_H}{2}\tilde{\varrho}_{Gm_G;Hm_H} + i\delta(\mathbf{v})\tilde{\varrho}_{Gm_G;Hm_H}$$
$$-i\chi_{HG}^{*}\sum_{q,m'_H}(\epsilon_q)^{*}\begin{bmatrix} G & 1 & H \\ m_G & q & m'_H \end{bmatrix}\varrho_{Hm'_H;Hm_H}$$
$$+i\chi_{HG}^{*}\sum_{q,m'_G}(\epsilon_q)^{*}\begin{bmatrix} G & 1 & H \\ m'_G & q & m_H \end{bmatrix}\varrho_{Gm_G;Gm'_G}, \tag{17.9c}$$

$$\tilde{\varrho}_{Hm_H;Gm_G} = \left(\tilde{\varrho}_{Gm_G;Hm_H}\right)^{*}, \tag{17.9d}$$

where

$$\chi_{HG} = -\frac{\langle H||\mu^{(1)}||G\rangle\mathcal{E}}{2\hbar\sqrt{2H+1}}, \tag{17.10}$$

$$\delta(\mathbf{v}) = \omega_{HG} - \omega + \mathbf{k}\cdot\mathbf{v}, \tag{17.11}$$

and relaxation terms arising from spontaneous decay have been included. Equations (17.9) are a generalization of equations (5.9), with $\gamma_{Gm_G;Gm'_G}^{Hm_H;Hm'_H}$ given by equation (16.55). Although equations (17.9) are algebraically complicated, they can be solved in steady state (or as a function of time) numerically quite easily using a computer program, since they are linear equations.

In optical pumping, we usually are concerned only with the evolution of ground-state density matrix elements in the presence of weak fields. In weak fields, the time scale needed for ground-state density matrix elements to reach their steady-state values is much longer than it is for other density matrix elements. As a consequence, ground-state–excited-state and excited-state–excited-state density matrix elements can be solved for in a quasi-static limit, similar to the one we used for intermediate-state amplitudes in applications such as the Bloch-Siegert shift and two-photon transitions. If we restrict our discussion to fields having constant amplitude for $t > 0$ and consider only times for which $t > |\gamma_H/2 + i\delta(\mathbf{v})|^{-1}$ (assuming that the fields reach their constant amplitude at time $t = 0$), then ground-state–excited-state density matrix elements, as well as excited-state population and coherence, achieve their steady-state values. This is the limit we will consider.

For weak fields, $|\chi_{HG}|^2 \ll \left[(\gamma_H/2)^2 + \delta^2(\mathbf{v})\right]$, one can neglect the ϱ_{HH} terms in equation (17.9c) and solve in steady state to obtain the ground-state–excited-state coherence

$$\tilde{\varrho}_{Gm_G;Hm_H} \simeq i\frac{\chi_{HG}^{*}\sum_{q,m'_G}(\epsilon_q)^{*}\begin{bmatrix} G & 1 & H \\ m'_G & q & m_H \end{bmatrix}\varrho_{Gm_G;Gm'_G}}{\gamma_H/2 - i\delta(\mathbf{v})}. \tag{17.12}$$

The next step in the calculation is to see how this electronic state coherence is coupled back to atomic state population by the field. To order $|\chi_{HG}|^2$, this coupling occurs via two channels. First, the coherence couples directly to ground-state density matrix elements via the field [equation (17.9a)]. Second, the coherence couples to excited-state density matrix elements via the field [equation (17.9b)], which, in turn, are coupled back to the ground state as a result of spontaneous decay [first term in equation (17.9a)].

The excited-state density matrix elements are obtained by substituting equation (17.12) into equation (17.9b) and solving in steady state. In this manner, we find

$$\varrho_{Hm_H;Hm'_H} \simeq \frac{1}{\gamma_H} \sum_{q,q',m_G,m'_G} \left\{ \frac{|\chi_{HG}|^2 (-1)^q \epsilon_{-q} \left(\epsilon_{q'}\right)^*}{\gamma_H/2 - i\delta(\mathbf{v})} \right.$$
$$\times \begin{bmatrix} G & 1 & H \\ m_G & q & m_H \end{bmatrix} \begin{bmatrix} G & 1 & H \\ m'_G & q & m'_H \end{bmatrix} \varrho_{Gm_G;Gm'_G} + \frac{|\chi_{HG}|^2 \epsilon_{q'} \left(\epsilon_q\right)^*}{\gamma_H/2 + i\delta(\mathbf{v})}$$
$$\left. \times \begin{bmatrix} G & 1 & H \\ m_G & q & m'_H \end{bmatrix} \begin{bmatrix} G & 1 & H \\ m'_G & q' & m_H \end{bmatrix} \varrho_{Gm'_G;Gm_G} \right\}. \tag{17.13}$$

Last, by substituting equations (17.12) and (17.13) into equation (17.9a), and setting $\gamma_H/2 = \gamma_{GH}$, we obtain

$$\frac{\partial \varrho_{m_G m'_G}}{\partial t} = \sum_{\bar{m}_G, \bar{m}'_G} \mathcal{R}_{m_G m'_G}^{\bar{m}_G \bar{m}'_G} \varrho_{\bar{m}_G \bar{m}'_G}, \tag{17.14}$$

where

$$\mathcal{R}_{m_G m'_G}^{\bar{m}_G \bar{m}'_G} = \frac{|\chi_{HG}|^2}{\gamma_H} \sum_{q,q',m_H,m'_H} \gamma_{m_G m'_G}^{m_H m'_H} \left\{ \begin{bmatrix} G & 1 & H \\ \bar{m}_G & q & m_H \end{bmatrix} \begin{bmatrix} G & 1 & H \\ \bar{m}'_G & q' & m'_H \end{bmatrix} \right.$$
$$\left. \times \left[\frac{(-1)^q \epsilon_{-q} \left(\epsilon_{q'}\right)^*}{\gamma_{GH} - i\delta(\mathbf{v})} + \frac{\left(\epsilon_{q'}\right)^* \epsilon_q}{\gamma_{GH} + i\delta(\mathbf{v})} \right] \right\}$$
$$- |\chi_{HG}|^2 \sum_{q,q',m_H} \left\{ \frac{(-1)^q \epsilon_{-q} \left(\epsilon_{q'}\right)^*}{\gamma_{GH} - i\delta(\mathbf{v})} \begin{bmatrix} G & 1 & H \\ m'_G & q & m_H \end{bmatrix} \begin{bmatrix} G & 1 & H \\ \bar{m}'_G & q' & m_H \end{bmatrix} \delta_{m_G,\bar{m}_G} \right.$$
$$\left. + \frac{\left(\epsilon_q\right)^* \epsilon_{q'}}{\gamma_{GH} + i\delta(\mathbf{v})} \begin{bmatrix} G & 1 & H \\ m_G & q & m_H \end{bmatrix} \begin{bmatrix} G & 1 & H \\ \bar{m}_G & q' & m_H \end{bmatrix} \delta_{m'_G,\bar{m}'_G} \right\}, \tag{17.15}$$

and the labels G and H have been suppressed in equations (17.14) and (17.15). Although these coupled equations appear complicated, they are often relatively easy to solve. Equations (17.14) and (17.15) are the principal results of this section.

A cursory glance at equations (17.14) and (17.15) reveals that, in steady state, the $|\chi_{HG}|^2$ terms drop out! Even though the *rate* at which steady state is established depends on field strength, the steady-state values are *independent of field strength* (unless of course the atoms leave the interaction region). The steady-state result is one that is *zeroth order* in the applied field amplitude. This is the bottom-line result of optical pumping—no matter how weak the field, the same steady state is reached. In our weak field approximation, $\sum_{m_G} \varrho_{m_G m_G} = 1$, which is a necessary condition for a nontrivial steady-state solution of equations (17.9) for the ground-state density matrix elements. Terms of order $|\chi_{HG}|^2/\left[\gamma_{GH}^2 + \delta^2(\mathbf{v})\right]$ have been neglected, corresponding to excited-state populations.

17.1.2 z-Polarized Excitation

As an example, we calculate the optical pumping when the excitation radiation is z-polarized—that is, the field has polarization $\hat{\boldsymbol{\epsilon}} = \hat{\mathbf{z}}$, corresponding to spherical components

$$\epsilon_0 = 1, \epsilon_{\pm 1} = 0. \tag{17.16}$$

In this limit,

$$\begin{aligned}\mathcal{R}_{m_G m'_G}^{\bar{m}_G \bar{m}'_G} = -|\chi_{HG}|^2 &\left\{ \frac{1}{\gamma_{GH} - i\delta(\mathbf{v})} \begin{bmatrix} G & 1 & H \\ m'_G & 0 & m'_G \end{bmatrix}^2 \right. \\ &\left. + \frac{1}{\gamma_{GH} + i\delta(\mathbf{v})} \begin{bmatrix} G & 1 & H \\ m_G & 0 & m_G \end{bmatrix}^2 \right\} \delta_{m_G, \bar{m}_G} \delta_{m'_G, \bar{m}'_G} \\ &+ \frac{2\gamma_{GH}}{\gamma_H} \frac{|\chi_{HG}|^2}{\gamma_{GH}^2 + \delta^2(\mathbf{v})} \begin{bmatrix} G & 1 & H \\ \bar{m}_G & 0 & \bar{m}_G \end{bmatrix} \begin{bmatrix} G & 1 & H \\ \bar{m}'_G & 0 & \bar{m}'_G \end{bmatrix} \gamma_{m_G m'_G}^{\bar{m}_G \bar{m}'_G},\end{aligned} \tag{17.17}$$

with

$$\gamma_{m_G m'_G}^{\bar{m}_G \bar{m}'_G} = \gamma_H \sum_q \begin{bmatrix} G & 1 & H \\ m_G & q & \bar{m}_G \end{bmatrix} \begin{bmatrix} G & 1 & H \\ m'_G & q & \bar{m}'_G \end{bmatrix}. \tag{17.18}$$

The selection rule imposed by the Clebsch-Gordan coefficients in equation (17.18) is

$$m_G - m'_G = \bar{m}_G - \bar{m}'_G. \tag{17.19}$$

This selection rule, coupled with equations (17.17) and (17.14), implies that there are two qualitatively different types of evolution equations for ground-state density matrix elements when the incident field is linearly polarized. If one prepares the ground state with some coherence $\varrho_{m_G m'_G}$ ($m_G \neq m'_G$) *before* the application of the optical pumping field, then the optical pumping connects this coherence to other ground-state coherence having the same value of $(m_G - m'_G)$. In other words, for a $G = 1$ ground state, $\dot{\varrho}_{01}$ depends only on ϱ_{01} and ϱ_{-10}. The only *steady-state* solution for these off-diagonal ground-state density matrix elements is $\varrho_{m_G m'_G} = 0$ ($m_G \neq m'_G$). Any initial ground-state coherence decays away as a result of the combined action of the incident field and spontaneous decay (i.e., scattering).

On the other hand, the steady-state ground-state populations can differ from one another. That is, the atoms are left in a polarized state (one having different ground-state sublevel populations) as a result of optical pumping. To calculate the evolution of the ground-state populations, we combine equations (17.14), (17.17), and (17.18) to obtain

$$\begin{aligned}\dot{\varrho}_{m_G m_G} = &-\frac{2\gamma_{GH}|\chi_{HG}|^2}{\gamma_{GH}^2 + \delta^2(\mathbf{v})} \begin{bmatrix} G & 1 & H \\ m_G & 0 & m_G \end{bmatrix}^2 \varrho_{m_G m_G} \\ &+ \frac{2\gamma_{GH}|\chi_{HG}|^2}{\gamma_{GH}^2 + \delta^2(\mathbf{v})} \frac{1}{\gamma_H} \begin{bmatrix} G & 1 & H \\ m_G & 0 & m_G \end{bmatrix}^2 \sum_{m'_G} \gamma_{m_G m_G}^{m'_G m'_G} \varrho_{m'_G m'_G},\end{aligned} \tag{17.20}$$

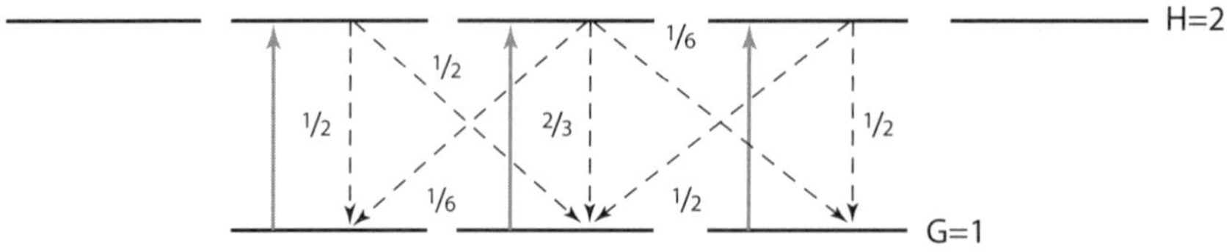

Figure 17.3. Optical pumping in a $G = 1$ to $H = 2$ transition with z-polarized excitation. The fractions are the square of the Clebsch-Gordan coefficients associated with each transition.

where

$$\gamma_{m_G m_G}^{m'_G m'_G} = \gamma_H \begin{bmatrix} G & 1 & H \\ m_G & m'_G - m_G & m'_G \end{bmatrix}^2 . \tag{17.21}$$

The first term represents pumping out of level m_G by the field, while the second term represents repopulation of level m_G resulting from the combined action of the field and spontaneous decay.

For a $G = 1 \leftrightarrow H = 2$ transition (see figure 17.3), the needed Clebsch-Gordan coefficients are

$$\begin{aligned} &\begin{bmatrix} 1 & 1 & 2 \\ \pm1 & 0 & \pm1 \end{bmatrix}^2 = \frac{1}{2}, \quad \begin{bmatrix} 1 & 1 & 2 \\ 0 & 0 & 0 \end{bmatrix}^2 = \frac{2}{3}, \\ &\begin{bmatrix} 1 & 1 & 2 \\ 1 & -1 & 0 \end{bmatrix}^2 = \begin{bmatrix} 1 & 1 & 2 \\ -1 & 1 & 0 \end{bmatrix}^2 = \frac{1}{6}, \\ &\begin{bmatrix} 1 & 1 & 2 \\ 0 & 1 & 1 \end{bmatrix}^2 = \begin{bmatrix} 1 & 1 & 2 \\ 0 & -1 & -1 \end{bmatrix}^2 = \frac{1}{2}. \end{aligned} \tag{17.22}$$

As a consequence, we find from equation (17.20) that the populations evolve as

$$\dot{\varrho}_{11} = \Gamma' \left[-\frac{1}{2}\varrho_{11} + \frac{1}{2}\left(\frac{1}{2}\right)\varrho_{11} + \frac{2}{3}\left(\frac{1}{6}\right)\varrho_{00} \right], \tag{17.23a}$$

$$\dot{\varrho}_{00} = \Gamma' \left[-\frac{2}{3}\varrho_{00} + \frac{2}{3}\left(\frac{2}{3}\right)\varrho_{00} + \frac{1}{2}\left(\frac{1}{6}\right)(\varrho_{11} + \varrho_{-1-1}) \right], \tag{17.23b}$$

$$\dot{\varrho}_{-1-1} = \Gamma' \left[-\frac{1}{2}\varrho_{-1-1} + \frac{1}{2}\left(\frac{1}{2}\right)\varrho_{-1-1} + \frac{2}{3}\left(\frac{1}{6}\right)\varrho_{00} \right], \tag{17.23c}$$

where

$$\Gamma' = \frac{2\gamma_{GH}|\chi_{HG}|^2}{\gamma_{GH}^2 + \delta^2(\mathbf{v})} \tag{17.24}$$

is an *optical pumping rate*.

The numerical coefficients appearing in equations (17.23) could be obtained by inspection from figure 17.3, using the square of the Clebsch-Gordan coefficients for excitation and decay. For example, the first term in equation (17.23b) represents the loss from sublevel 0 produced by the field, the second term represents the repopulation of sublevel 0 resulting from the field driving population out of ground-state sublevel 0 to excited-state sublevel 0 and spontaneous emission restoring some of the population to ground-state sublevel 0, while the last term represents the repopulation of sublevel 0 resulting from the field driving population out of ground-state sublevels ± 1 to excited-state sublevels ± 1 and spontaneous emission transferring some of the population back to ground-state sublevel 0. The steady-state solution of equations (17.23) is

$$\varrho_{11} = \varrho_{-1-1} = \frac{4}{17}, \tag{17.25a}$$

$$\varrho_{00} = \frac{9}{17}, \tag{17.25b}$$

with all other density matrix elements equal to zero. As discussed earlier, although the ground state is polarized, no net internal state angular momentum has been transferred to the atoms by the z-polarized optical field.

17.1.3 Irreducible Tensor Basis

For other than z-polarized fields, there can be off-diagonal density matrix elements in steady state. If we use linearly polarized light in the $\hat{\mathbf{x}}$ rather than the $\hat{\mathbf{z}}$ direction, for example, drawing simple diagrams is not too much help, since the field creates coherence in the excited state that is maintained in the decay. The final density matrix elements can be obtained by "rotating" the z-polarized results using angular momentum rotation matrices (see problems 1–2), giving steady-state values

$$\varrho_{11} = \varrho_{-1-1} = 13/34, \tag{17.26a}$$

$$\varrho_{00} = 4/17, \tag{17.26b}$$

$$\varrho_{1-1} = \varrho_{-11} = -5/34, \tag{17.26c}$$

with all other density matrix elements equal to zero. There is now a nonvanishing ground-state coherence (off-diagonal density matrix element), but the average value of the z component of angular momentum remains equal to zero, since the excitation is via a linearly polarized field. This result also indicates that what looks like unequal populations and zero coherence in one basis can appear as both unequal populations and *nonvanishing* coherence in a rotated basis. *The only ground-state density matrix that is invariant under rotation is an unpolarized state.*

In this and more complicated cases, it is easier to use an irreducible tensor basis since decay is diagonal in that basis. As was mentioned in the previous chapter, many people get very nervous when irreducible tensors are mentioned, but there is no need to panic. The irreducible tensor basis represents simply an alternative basis to the usual angular momentum basis. Moreover, it often leads to simplified equations when systems have simple properties under rotation.

To transform equation (17.14) into an irreducible tensor basis requires a lot of algebra, which we do not reproduce here [4]. We simply give the result; the ground-state density matrix elements evolve as

$$\dot{\varrho}^{K}_{Q}(G) = \sum_{K,K',\bar{K},Q'} A(K,K',\bar{K};Q,Q';G,H)P(G,H;K,K',\bar{K})\varrho^{K'}_{Q'}(G), \tag{17.27}$$

where

$$A(K,K',\bar{K};Q,Q';G,H) = (-1)^{Q'}\sum_{\bar{Q}}\begin{pmatrix} K' & K & \bar{K} \\ -Q' & Q & \bar{Q}\end{pmatrix}|\chi_{GH}|^2\epsilon_{\bar{Q}}^{\bar{K}} \times\left[\frac{1}{\gamma_{GH}-i\delta(\mathbf{v})}+\frac{(-1)^{K+K'+\bar{K}}}{\gamma_{GH}+i\delta(\mathbf{v})}\right], \quad (17.28)$$

$$P(G,H;K,K',\bar{K}) = -3(-1)^{K+K'+H-G}\sqrt{(2K+1)(2K'+1)(2\bar{K}+1)} \times\left[\begin{matrix}\begin{Bmatrix} K & K' & \bar{K} \\ G & G & G\end{Bmatrix}\begin{Bmatrix} 1 & 1 & \bar{K} \\ G & G & H\end{Bmatrix} \\ -(-1)^{2G}(2H+1)\begin{Bmatrix} H & H & K \\ G & G & 1\end{Bmatrix}\begin{Bmatrix} K & K' & \bar{K} \\ H & G & 1 \\ H & G & 1\end{Bmatrix}\end{matrix}\right], \quad (17.29)$$

$$\epsilon_Q^K = \sum_{q,q'}(-1)^{q'}\epsilon_q\left(\epsilon_{-q'}\right)^*\begin{bmatrix} 1 & 1 & K \\ q & q' & Q\end{bmatrix}, \quad (17.30)$$

$$\chi_{HG} = -\frac{\langle H||\mu^{(1)}||G\rangle\mathcal{E}}{2\hbar\sqrt{2H+1}}, \quad (17.31)$$

and the last {} in equation (17.29) is a 9-J symbol. It is a simple matter to program equation (17.27). Clebsch-Gordan, 3-J, 6-J, and 9-J symbols are standard subroutines in most symbolic math programs.

To get some idea of the power of the irreducible tensor basis, we consider optical pumping on a $G = 1$ to $H = 2$ transition by an $\hat{\mathbf{x}}$-polarized field, starting with an unpolarized initial state $\varrho_Q^K = (1/\sqrt{3})\delta_{K,0}\delta_{Q,0}$. For an $\hat{\mathbf{x}}$-polarized field (see the appendix),

$$\epsilon_q^{(x)} = \left(\delta_{q,-1} - \delta_{q,-1}\right)/\sqrt{2} \quad (17.32)$$

and

$$\epsilon_Q^K = -\frac{1}{\sqrt{3}}\delta_{K,0}\delta_{Q,0} + \delta_{K,2}\left[-\frac{1}{\sqrt{6}}\delta_{Q,0} + \frac{1}{2}\left(\delta_{Q,2} + \delta_{Q,-2}\right)\right]. \quad (17.33)$$

Substituting equation (17.33) into equation (17.27), setting $G = 1$ and $H = 2$, and using the fact that $\varrho_0^0 = 1/\sqrt{3}$, we obtain the evolution equations

$$\dot{\varrho}_0^2 = \frac{\Gamma'}{120}\left(-22\varrho_0^2 + 4\sqrt{6}\varrho_2^2 + 5\sqrt{\frac{2}{3}}\right), \quad (17.34a)$$

$$\dot{\varrho}_2^2 = \frac{\Gamma'}{120}\left(-30\varrho_2^2 + 2\sqrt{6}\varrho_0^2 - 5\right), \quad (17.34b)$$

where Γ' is given by equation (17.24), and $\varrho_{-2}^2 = \varrho_2^2$. The steady-state solution is

$$\varrho_0^2 = \frac{5}{17}\frac{1}{\sqrt{6}}, \quad \varrho_{\pm 2}^2 = -\frac{5}{34}, \quad (17.35)$$

which is consistent with equation (17.26). There are only two density matrix elements that need to be calculated using an irreducible tensor basis.

Equation (17.27) actually has a very simple structure, although you might be somewhat skeptical about this conclusion. The dynamics is governed by two factors, $A(K, K', \bar{K}; Q, Q'; G, H)$ and $P(G, H; K, K', \bar{K})$. The quantity $P(G, H; K, K', \bar{K})$ is a "geometric" factor that is independent of field polarization, while $A(K, K', \bar{K}; Q, Q'; G, H)$ reflects the polarization of the driving field, written in terms of the coupled polarization tensor ϵ^K_Q. The coupled tensor description is especially useful for determining the manner in which population and coherence is transferred from the ground-state to the excited-state and back to the ground-state manifolds. For reference purposes, values of ϵ^K_Q for various field polarizations are given in the appendix.

17.1.4 Standing-Wave and Multiple-Frequency Fields

The results can be generalized to any number of incident fields. This generalization is especially important since it will allow us to consider the standard geometries for sub-Doppler cooling that are discussed in the next chapter. The total electric field is taken to be of the form

$$\mathbf{E}_c(\mathbf{R}, t) = \frac{1}{2}\sum_j \mathcal{E}^{(j)}\hat{\boldsymbol{\epsilon}}^{(j)}e^{i(\mathbf{k}\cdot\mathbf{R}-\omega_j t)} + \text{c.c.}, \tag{17.36}$$

that is, a sum of a number of monochromatic fields. Density matrix equations are now written in the *standard* interaction representation, since the field frequencies and wave vectors of the various fields need not be equal. In general, the equations cannot be solved analytically, since there are several frequencies present in the atomic rest frame. Let us assume, however, that the adiabatic elimination of the ϱ_{GH} and ϱ_{HH} density matrix elements remains valid [4], allowing us to arrive at the same type of evolution equation for $\varrho^K_Q(G)$ that was found earlier—namely [4],

$$\begin{aligned}\frac{\partial \varrho^K_Q(G)}{\partial t} + \mathbf{v}\cdot\nabla\varrho^K_Q(G) = &\sum_{K,K',\bar{K},Q'} A(K, K', \bar{K}; Q, Q'; G, H; \mathbf{R}, t)\\ &\times P(G, H; K, K', \bar{K})\varrho^{K'}_{Q'}(G),\end{aligned} \tag{17.37}$$

where the factor $A(K, K', \bar{K}; Q, Q'; G, H; \mathbf{R}, t)$ now contains all combinations $e^{i\phi_{jj'}(\mathbf{R},t)}$ of spatial and temporal beats between the different fields with

$$\phi_{jj'}(\mathbf{R}, t) = (\mathbf{k}_j - \mathbf{k}_{j'})\cdot\mathbf{R} - (\omega_j - \omega_{j'})t. \tag{17.38}$$

Explicitly, we find [4]

$$\begin{aligned}A(K, K', \bar{K}; Q, Q'; G, H; \mathbf{R}, t) = (-1)^{Q'} &\sum_{j,j',\bar{Q}} \begin{pmatrix} K' & K & \bar{K} \\ -Q' & Q & \bar{Q}\end{pmatrix}\\ \times\bar{\chi}^{(j)}_{HG}\left(\bar{\chi}^{(j')}_{HG}\right)^* \epsilon^{\bar{K}}_{\bar{Q}}(j, j')e^{i\phi_{jj'}(\mathbf{R},t)} &\left[\frac{1}{\gamma_{GH} - i\delta^{(j')}(\mathbf{v})} + \frac{(-1)^{K+K'+\bar{K}}}{\gamma_{GH} + i\delta^{(j)}(\mathbf{v})}\right],\end{aligned} \tag{17.39}$$

where

$$\bar{\chi}_{HG}^{(j)} = -\frac{\langle H||\mu^{(1)}||G\rangle \mathcal{E}^{(j)}}{2\hbar\sqrt{3}}, \tag{17.40}$$

$$\epsilon_Q^K(j,j') = \sum_{q,q'} (-1)^{q'} \epsilon_q^{(j)} \left[\epsilon_{q'}^{(j')}\right]^* \begin{bmatrix} 1 & 1 & K \\ q & q' & Q \end{bmatrix}, \tag{17.41}$$

and

$$\delta^{(j)}(\mathbf{v}) = \omega_{HG} - \omega_j + \mathbf{k}_j \cdot \mathbf{v}. \tag{17.42}$$

Of course, these equations look complicated, but the basic physical content is simple. Each pair of fields i, j (including $i = j$) drives population and coherence from the ground-state manifold to the excited-state manifold of levels in a manner dependent on the spherical tensor components $\epsilon_Q^K(j, j')$. Spontaneous emission returns some of this population and coherence to the ground state. Since there is a spatial and temporal phase given by equation (17.38) that occurs for fields having different **k** vectors and frequencies, respectively, it is possible for the fields to create both spatial modulation and temporal beats in the ground-state density matrix elements. However, for our "closed" G–H transition, the *total ground-state population is constant*,

$$\frac{d}{dt}\varrho_0^0(G) = 0, \tag{17.43}$$

and

$$\varrho_0^0(G) = \frac{1}{\sqrt{2G+1}} \tag{17.44}$$

to zeroth order in the optical pumping rate Γ'. On the other hand, ground-state density matrix elements, $\varrho_Q^K(G)$ for $K \neq 0$, $Q \neq 0$, can be modulated in space and time. We refer to this as *modulated atomic state polarization*. As you will see, this feature is critical for sub-Doppler cooling.

17.1.4.1 Counterpropagating fields

Let us look at some limiting cases to see what happens to the ground-state sublevels for two counterpropagating fields. We set $\mathbf{k}_1 = -\mathbf{k}_2 = \mathbf{k}$, $\mathcal{E}_1 = \mathcal{E}_2$, and $\omega_1 = \omega_2 = \omega = kc$.

17.1.4.1.1 ∥ polarization: If both fields have the same polarization, $\epsilon_Q^K(j, j')$ is independent of j or j'; therefore,

$$A(K, K', \bar{K}; Q, Q'; G, H; \mathbf{R}, t) \propto \left(1 + e^{2i\mathbf{k}\cdot\mathbf{R}} + e^{-2i\mathbf{k}\cdot\mathbf{R}}\right), \tag{17.45}$$

and

$$\dot{\varrho}_Q^K = \frac{\partial \varrho_Q^K}{\partial t} + \mathbf{v}\cdot\nabla\varrho_Q^K = \sum_{K',Q'} T_{KQ}^{K'Q'} \varrho_{Q'}^{K'} \left(1 + e^{2i\mathbf{k}\cdot\mathbf{R}} + e^{-2i\mathbf{k}\cdot\mathbf{R}}\right), \tag{17.46}$$

where $T_{KQ}^{K'Q'}$ is some coupling coefficient. Let us assume a steady-state solution of the form

$$\varrho_Q^K(G) = \varrho_Q^K(G,0) + \varrho_Q^K(G,+)e^{2i\mathbf{k}\cdot\mathbf{R}} + \varrho_Q^K(G,-)e^{-2i\mathbf{k}\cdot\mathbf{R}}. \tag{17.47}$$

The second two terms constitute the modulated atomic state polarization.

Substituting this trial solution into equation (17.46), we find, in steady state,

$$\sum_{K',Q'} T_{KQ}^{K'Q'} \varrho_{Q'}^{K'}(G,0) = 0, \tag{17.48a}$$

$$\pm 2i\mathbf{k}\cdot\mathbf{v}\varrho_Q^K(G,\pm) = \sum_{K',Q'} T_{KQ}^{K'Q'} \varrho_{Q'}^{K'}(G,\pm). \tag{17.48b}$$

Equations (17.48a) and (17.48b) each correspond to a set of $(2G+1)^2$ linear equations. Equation (17.48a) has a nontrivial solution owing to the fact that $\varrho_0^0(G) = 1/(2G+1)^{1/2}$; that is, the $\varrho_{Q'}^{K'}(G,0)$ are not linearly independent. On the other hand, in equation (17.48b), the $\varrho_Q^K(G,\pm)$ *are* linearly independent, and the only solution of these equations is

$$\varrho_Q^K(G,\pm) = 0. \tag{17.49}$$

For parallel polarization, it then follows from equation (17.47) that *there is no modulated atomic state polarization.* It is easy to understand why this is the case. No matter where in the sample an atom finds itself, it experiences the same field polarization, implying that the steady state cannot be a function of position, provided that an atom does not remain in a node of the field. Each atom is optically pumped to the same steady state.

17.1.4.1.2 lin⊥lin polarization: For crossed, linearly polarized fields $\boldsymbol{\epsilon}_1 = \hat{\mathbf{x}}$; $\boldsymbol{\epsilon}_2 = \hat{\mathbf{y}}$, the spherical polarization tensor components that are needed are (see the appendix)

$$\begin{aligned}
\epsilon_Q^K(x,x) &= -\frac{1}{\sqrt{3}}\delta_{K,0}\delta_{Q,0} + \delta_{K,2}\left[-\frac{1}{\sqrt{6}}\delta_{Q,0} + \frac{1}{2}\left(\delta_{Q,2} + \delta_{Q,-2}\right)\right],\\
\epsilon_Q^K(y,y) &= -\frac{1}{\sqrt{3}}\delta_{K,0}\delta_{Q,0} - \delta_{K,2}\left[\frac{1}{\sqrt{6}}\delta_{Q,0} + \frac{1}{2}\left(\delta_{Q,2} + \delta_{Q,-2}\right)\right],\\
\epsilon_Q^K(x,y) &= \left[\epsilon_{-Q}^K(y,x)\right]^* = \frac{i}{\sqrt{2}}\left[\delta_{K,1}\delta_{Q,0} + \frac{1}{\sqrt{2}}\delta_{K,2}\left(\delta_{Q,2} - \delta_{Q,-2}\right)\right].
\end{aligned} \tag{17.50}$$

In contrast to the case of parallel polarization, the polarization of the total field (17.36) is now a function of position, implying that the atomic response can be modulated as well.

If the counterpropagating, crossed-polarized fields are incident on atoms that are initially in an unpolarized ground state, the resulting ground-state dynamics can be quite complicated. However, for a $G=1/2$ ground state, the only values of ϱ_Q^K that can enter are $\varrho_0^0 = 1/\sqrt{2}$ and ϱ_Q^1, since $K = 0, 1$ when one couples two spin 1/2 systems. Moreover, for the values of ϵ_Q^K given in equation (17.50), and for $K=0, 1$, it is clear that only $Q=0$ can enter. Thus, the only density matrix element that need be evaluated is ϱ_0^1 when $G=1/2$ for crossed-polarized fields.

Using equations (17.37) and (17.50) and the fact that $\varrho_0^0 = 1/\sqrt{2}$, we find that, for a $G = 1/2$ ground state and with $\bar{\chi}_{HG}^{(j)} \equiv \bar{\chi}_{HG}$,

$$\frac{\partial \varrho_0^1}{\partial t} + \mathbf{v} \cdot \nabla \varrho_0^1 = -\Gamma'(1/2, 1/2)\varrho_0^1 - \frac{1}{\sqrt{2}}\Gamma'(1/2, 1/2)\sin 2kZ \tag{17.51}$$

if $H = 1/2$ and

$$\frac{\partial \varrho_0^1}{\partial t} + \mathbf{v} \cdot \nabla \varrho_0^1 = -\Gamma'(1/2, 3/2)\varrho_0^1 - \frac{1}{\sqrt{2}}\Gamma'(1/2, 3/2)\sin 2kZ \tag{17.52}$$

if $H = 3/2$, where

$$\Gamma'(1/2, 1/2) = 2\Gamma'(3/2, 1/2) = \frac{2}{3}\frac{|\bar{\chi}_{HG}|^2 \gamma_H}{\gamma_{GH}^2 + \delta^2}, \tag{17.53}$$

the appropriate values of G and H are to be inserted in $\bar{\chi}_{HG}$, and $\gamma_H = 2\gamma_{GH}$. Since

$$\varrho_0^1 = \left(\varrho_{1/2,1/2} - \varrho_{-1/2,-1/2}\right)/\sqrt{2},$$

it is clear from these equations that the population difference, $\left(\varrho_{1/2,1/2} - \varrho_{-1/2,-1/2}\right)$, is spatially modulated. In the next chapter, we show that this leads to polarization gradient, sub-Doppler cooling.

17.1.4.1.3 $\sigma_+ - \sigma_-$ polarization: For σ_+–σ_- radiation, $\hat{\boldsymbol{\epsilon}}_1 = (\hat{\mathbf{x}}+i\hat{\mathbf{y}})/\sqrt{2}\ [\epsilon_q = \delta_{q,-1}]$, $\hat{\boldsymbol{\epsilon}}_2 = -(\hat{\mathbf{x}} - i\hat{\mathbf{y}})/\sqrt{2}\ [\epsilon_q = \delta_{q,1}]$, and (see the appendix)

$$\begin{aligned}
\epsilon_Q^K(\sigma_+, \sigma_+) &= -\left(\frac{1}{\sqrt{3}}\delta_{K,0} - \frac{1}{\sqrt{2}}\delta_{K,1} + \frac{1}{\sqrt{6}}\delta_{K,2}\right)\delta_{Q,0}, \\
\epsilon_Q^K(\sigma_-, \sigma_-) &= -\left(\frac{1}{\sqrt{3}}\delta_{K,0} + \frac{1}{\sqrt{2}}\delta_{K,1} + \frac{1}{\sqrt{6}}\delta_{K,2}\right)\delta_{Q,0}, \\
\epsilon_Q^K(\sigma_+, \sigma_-) &= -\delta_{K,2}\delta_{Q,-2}, \quad \epsilon_Q^K(\sigma_-, \sigma_+) = -\delta_{K,2}\delta_{Q,2}.
\end{aligned} \tag{17.54}$$

In the next chapter, we will see that, for a $G = 1$ ground state, $\varrho_2^2 = \varrho_{1-1}$ is modulated as e^{2ikz} and $\varrho_{-2}^2 = \varrho_{-11}$ as e^{-2ikz}, but the population differences are not spatially modulated. This leads to a "corkscrew" type of sub-Doppler cooling.

17.2 Optical Lattice Potentials

If the incident fields are detuned by a sufficiently large amount, one can in first approximation neglect the repopulation of the ground state as a result of spontaneous emission. In other words, one neglects the "in" terms of spontaneous emission and works with an effective Hamiltonian for the system that is obtained by adiabatically eliminating the excited-state *amplitudes*. An amplitude approach is possible, since the repopulation of the ground state is neglected. One can even retain the decay parameters in the equations for the ground-state–excited-state coherence. In doing so, one arrives at an effective non-Hermitian Hamiltonian for the ground-state sublevels. Since the ground-state polarization can be modulated, the Hamiltonian is also spatially modulated. In other words, the atoms find themselves in a modulated potential or *optical lattice* that can actually serve as a network of traps for the atoms. For a review of this subject area, see the article by Jessen and Deutsch [5]. Optical lattices provide a controllable environment for simulating condensed matter systems.

We use the same formalism as for optical pumping, but work with state amplitudes rather than density matrix elements. We assume that all the fields have the same frequency and neglect any Doppler shifts of the atoms, assuming that they have been cooled to very low temperatures, such that $|\mathbf{k}_j \cdot \mathbf{v}| \ll \gamma_H$. In a field interaction representation, the excited-state amplitudes evolve as

$$d\tilde{c}_{Hm_H}/dt = -\left(\frac{\gamma_H}{2} + i\delta\right)\tilde{c}_{Hm_H} + \frac{i}{2\hbar}\sum_{j,m_G}\langle H, m_H|\hat{\boldsymbol{\mu}} \cdot \hat{\boldsymbol{\epsilon}}^{(j)}|G, m_G\rangle e^{i\mathbf{k}_j\cdot\mathbf{R}}\mathcal{E}^{(j)}\tilde{c}_{Gm_G}. \tag{17.55}$$

Assuming that the optical pumping rate Γ' is much less than $|\gamma_H/2 + i\delta|$, the excited-state amplitude becomes approximately constant on a time scale much less than $1/\Gamma'$. Thus, we can adiabatically eliminate the upper-state amplitude by taking $d\tilde{c}_{H,m_H}/dt \approx 0$, or

$$\tilde{c}_{Hm_H} \approx \frac{i}{2\hbar(\gamma_H/2 + i\delta)}\sum_{j,m'_G}\langle H, m_H|\hat{\boldsymbol{\mu}} \cdot \hat{\boldsymbol{\epsilon}}^{(j)}|G, m'_G\rangle e^{i\mathbf{k}_j\cdot\mathbf{R}}\mathcal{E}^{(j)}\tilde{c}_{Gm'_G}. \tag{17.56}$$

The ground-state amplitudes evolve as

$$\begin{aligned} d\tilde{c}_{Gm_G}/dt &= \frac{i}{2\hbar}\sum_{j',m_H}\langle G, m_G|\hat{\boldsymbol{\mu}}\cdot\left[\hat{\boldsymbol{\epsilon}}^{(j')}\right]^*|H, m_H\rangle e^{-i\mathbf{k}_{j'}\cdot\mathbf{R}}\mathcal{E}^{(j')*}\tilde{c}_{Hm_H} \\ &= \frac{i}{2\hbar}\sum_{j',m_H}\left[\hat{\boldsymbol{\epsilon}}^{(j')}\right]^*\cdot\langle H, m_H|\hat{\boldsymbol{\mu}}|G, m_G\rangle^* e^{-i\mathbf{k}_{j'}\cdot\mathbf{R}}\mathcal{E}^{(j')*}\tilde{c}_{Hm_H}. \end{aligned} \tag{17.57}$$

Combining equations (17.56) and (17.57), and using

$$\hat{\boldsymbol{\mu}} \cdot \hat{\boldsymbol{\epsilon}} = \sum_{q=-1}^{1}(-1)^q\,\hat{\mu}_q\epsilon_{-q} \tag{17.58}$$

and the Wigner-Eckart theorem to evaluate matrix elements of $\hat{\boldsymbol{\mu}}$, we find

$$i\hbar d\tilde{c}_{Gm_G}/dt = \sum_{m'_G} H_{m_G m'_G}\tilde{c}_{Gm'_G},$$

where

$$\begin{aligned} H_{m_G m'_G} &= -\frac{1}{2H+1}\frac{3\hbar}{\delta - i\gamma_H/2}\sum_{q,q',m_H,j,j'}(-1)^{q+q'}\begin{bmatrix} G & 1 & H \\ m'_G & q & m_H\end{bmatrix} \\ &\times \begin{bmatrix} G & 1 & H \\ m_G & -q' & m_H\end{bmatrix}\bar{\chi}_{HG}^{(j)}\left[\bar{\chi}_{HG}^{(j')}\right]^*\epsilon_{-q}^{(j)}\left[\epsilon_{q'}^{(j')}\right]^* e^{i\mathbf{k}_{jj'}\cdot\mathbf{R}}. \end{aligned} \tag{17.59}$$

The matrix elements $H_{m_G m'_G}$ form a non-Hermitian, effective Hamiltonian for the ground-state manifold.

For any given field polarizations, we calculate these matrix elements and diagonalize $\mathbf{H}$ to get the optical lattice potentials. It is possible to get $\mathbf{H}$ into a simple form if the coupled tensors defined by equation (17.41) are used, as in the optical pumping equations. In terms of the coupled tensors defined by equation (17.30), equation (17.59) can be rewritten as

$$H_{m_G m'_G} = -\frac{3\hbar}{\delta - i\gamma_H/2} \sum_{q,q',j,j',m_H,K,Q} \frac{\bar{\chi}_{HG}^{(j)} \left[\bar{\chi}_{HG}^{(j')}\right]^*}{2H+1} (-1)^{Q+q'} \times \begin{bmatrix} G & 1 & H \\ m'_G & q & m_H \end{bmatrix} \begin{bmatrix} G & 1 & H \\ m_G & -q' & m_H \end{bmatrix} \begin{bmatrix} 1 & 1 & K \\ -q & -q' & Q \end{bmatrix} e^{i\mathbf{k}_{jj'}\cdot\mathbf{R}} \epsilon_Q^K(j, j'). \tag{17.60}$$

This can be summed over q, q', m_H to give

$$H_{m_G m'_G} = -\frac{3\hbar}{\delta - i\gamma_H/2} \sum_{j,j',K,Q} \bar{\chi}_{HG}^{(j)} \left[\bar{\chi}_{HG}^{(j')}\right]^* e^{i\mathbf{k}_{jj'}\cdot\mathbf{R}} (-1)^{K+H+G+Q} \times \epsilon_Q^K(j, j') (-1)^{G-m'_G} \begin{bmatrix} G & G & K \\ m_G & -m'_G & Q \end{bmatrix} \begin{Bmatrix} G & G & K \\ 1 & 1 & H \end{Bmatrix}, \tag{17.61}$$

where we used equation (16.80). This shows that $|\Delta m_G| = |m_G - m'_G| \leq 2$, since the maximum value of K is equal to 2.

For z-polarized light,

$$\epsilon_Q^K(z, z) = -\frac{1}{\sqrt{3}}\delta_{K,0}\delta_{Q,0} + \sqrt{\frac{2}{3}}\delta_{K,2}\delta_{Q,0}. \tag{17.62}$$

As a consequence, the Hamiltonian (17.61) is already in diagonal form, implying that the eigenstates are the normal m-basis states. For lin⊥lin polarization with the coupled tensor components given by equation (17.50), the Hamiltonian is not diagonal, except for $G = 1/2$.

Once the Hamiltonian is diagonalized, it is possible to analyze the motion of atoms in the effective potential. To do this, one must use either a classical, semiclassical, or quantized description of the motion in the potentials. For atoms having energy much less than the well depth of the potential, a quantized approach is often needed, although semiclassical approaches can sometimes give a good approximate solution.

As an example of optical lattice potentials, let us take the simple case of a $G = 1/2$ ground state and an $H = 1/2$ or $3/2$ excited state for the case of equal amplitude, crossed-polarized fields, with the coupled field polarization tensor given by equation (17.50). For this polarization, the matrix (17.61) is already diagonal, and it is a simple matter to calculate

$$\mathbf{H}(1/2, 3/2) = -\frac{|\bar{\chi}_{HG}|^2}{\delta - i\gamma_H/2} \begin{pmatrix} 1 - \frac{1}{2}\sin 2kZ & 0 \\ 0 & 1 + \frac{1}{2}\sin 2kZ \end{pmatrix} \tag{17.63}$$

and

$$\mathbf{H}(1/2, 1/2) = -\frac{|\bar{\chi}_{HG}|^2}{\delta - i\gamma_H/2} \begin{pmatrix} 1 + \sin 2kZ & 0 \\ 0 & 1 - \sin 2kZ \end{pmatrix}, \tag{17.64}$$

where the order of the matrix elements is $m = 1/2, -1/2$. It is clear that the energy levels of the atoms are spatially modulated.

17.3 Summary

In this chapter, we have seen how weak radiation fields can modify ground-state sublevel populations and coherence. In conventional optical pumping, a single traveling-wave field redistributes population among the ground-state sublevels and can also lead to ground-state coherence. The polarization properties of the field are imprinted on the atoms, regardless of the strength of the field. Optical pumping can be generalized to include an arbitrary number of incident fields. In this manner, one can also imprint spatial and temporal beats into ground-state density matrix elements. In the case of far-detuned fields having the same frequency, one arrives at optical lattice potentials. An immediate application of the formalism developed in this chapter is sub-Doppler laser cooling, to which we now turn our attention.

17.4 Appendix: Irreducible Tensor Formalism

17.4.1 Coupled Tensors

Many details of the calculations involving the irreducible components of the density matrix are included in the paper by Rogers et al. [4]. On the web site (http://press.princeton.edu/titles/9376.html) devoted to this book there is a link to examples of Mathematica programs that can be used to obtain some of the equations related to optical pumping. For reference purposes, the coupled tensor elements for the most common field polarizations are given in this appendix.

The polarization $\hat{\boldsymbol{\epsilon}}$ and spherical tensor components of the polarization are defined in equations (17.3) and (17.4), respectively. The coupled tensor components are given by

$$\epsilon_Q^K(j, j') = \sum_{q,q'} (-1)^{q'} \epsilon_q^{(j)} \left[\epsilon_{-q'}^{(j')}\right]^* \begin{bmatrix} 1 & 1 & K \\ q & q' & Q \end{bmatrix}, \tag{17.65}$$

such that

$$\epsilon_Q^K(j, j') = (-1)^Q \left[\epsilon_{-Q}^K(j', j)\right]^*. \tag{17.66}$$

For linearly polarized fields along x, y, and z,

$$\epsilon_q^{(x)} = \left(\delta_{q,-1} - \delta_{q,-1}\right)/\sqrt{2}, \tag{17.67a}$$

$$\epsilon_q^{(y)} = -i\left(\delta_{q,-1} + \delta_{q,-1}\right)/\sqrt{2}, \tag{17.67b}$$

$$\epsilon_q^{(z)} = \delta_{q,0}, \tag{17.67c}$$

and for σ_+ and σ_- circularly polarized fields,

$$\epsilon_q^{(+)} = \delta_{q,-1}, \tag{17.68a}$$

$$\epsilon_q^{(-)} = \delta_{q,1}. \tag{17.68b}$$

This choice corresponds to $\hat{\boldsymbol{\epsilon}}^{(\pm)} = \pm\left(\hat{\mathbf{x}} \pm i\hat{\mathbf{y}}\right)/\sqrt{2}$.

From these individual field polarizations, we use equation (17.65) to construct

$$\epsilon_Q^K(x,x) = -\frac{1}{\sqrt{3}}\delta_{K,0}\delta_{Q,0}+\delta_{K,2}\left[-\frac{1}{\sqrt{6}}\delta_{Q,0}+\frac{1}{2}\left(\delta_{Q,2}+\delta_{Q,-2}\right)\right], \quad (17.69a)$$

$$\epsilon_Q^K(y,y) = -\frac{1}{\sqrt{3}}\delta_{K,0}\delta_{Q,0}-\delta_{K,2}\left[\frac{1}{\sqrt{6}}\delta_{Q,0}+\frac{1}{2}\left(\delta_{Q,2}+\delta_{Q,-2}\right)\right], \quad (17.69b)$$

$$\epsilon_Q^K(z,z) = -\frac{1}{\sqrt{3}}\delta_{K,0}\delta_{Q,0}+\sqrt{\frac{2}{3}}\delta_{K,2}\delta_{Q,0}, \quad (17.69c)$$

$$\epsilon_Q^K(x,y) = \frac{i}{\sqrt{2}}\left[\delta_{K,1}\delta_{Q,0}+\frac{1}{\sqrt{2}}\delta_{K,2}\left(\delta_{Q,2}-\delta_{Q,-2}\right)\right], \quad (17.69d)$$

$$\epsilon_Q^K(x,z) = \frac{1}{2}\left[\delta_{K,1}\left(\delta_{Q,1}+\delta_{Q,-1}\right)-\delta_{K,2}\left(\delta_{Q,1}-\delta_{Q,-1}\right)\right], \quad (17.69e)$$

$$\epsilon_Q^K(y,z) = -\frac{i}{2}\left[\delta_{K,1}\left(\delta_{Q,1}-\delta_{Q,-1}\right)+\delta_{K,2}\left(\delta_{Q,1}+\delta_{Q,-1}\right)\right], \quad (17.69f)$$

$$\epsilon_Q^K(+,+) = -\left(\frac{1}{\sqrt{3}}\delta_{K,0}-\frac{1}{\sqrt{2}}\delta_{K,1}+\frac{1}{\sqrt{6}}\delta_{K,2}\right)\delta_{Q,0}, \quad (17.69g)$$

$$\epsilon_Q^K(-,-) = -\left(\frac{1}{\sqrt{3}}\delta_{K,0}+\frac{1}{\sqrt{2}}\delta_{K,1}+\frac{1}{\sqrt{6}}\delta_{K,2}\right)\delta_{Q,0}, \quad (17.69h)$$

$$\epsilon_Q^K(+,-) = -\delta_{K,2}\delta_{Q,-2}. \quad (17.69i)$$

Additional components are obtained using equation (17.66).

17.4.2 Density Matrix Equations

We can also write the density matrix equations in an irreducible tensor notation. The incident laser fields drive transitions between a ground-state manifold characterized by quantum numbers L_G (total orbital angular momentum), J_G (coupling of L_G and S_G), I (total nuclear-spin angular momentum), G (total angular momentum–coupling of J_G and I), and excited-state manifold characterized by quantum numbers L_H, S_H, J_H, I, and H. In the rotating-wave approximation (RWA) and for *equal field frequencies*, the appropriate equations in the field interaction representation defined by equation (3.46) are [4]

$$\begin{aligned}\dot{\tilde{\varrho}}_Q^K(G,G) &= \gamma^{(K)}(H;G)\tilde{\varrho}_Q^K(H,H) \\ &\quad -i\bar{\chi}_{HG}^{(j)}(-1)^{2G}e^{i\mathbf{k}_j\cdot\mathbf{R}}(-1)^q\epsilon_{-q}^{(j)}\Lambda_{Q'qQ}^{K'1K}(G,G,H)\tilde{\varrho}_{Q'}^{K'}(G,H) \\ &\quad +i\left[\bar{\chi}_{HG}^{(j)}\right]^*(-1)^{2H+1+K'-K+Q'}e^{-i\mathbf{k}_j\cdot\mathbf{R}} \\ &\quad \times\left[\epsilon_q^{(j)}\right]^*\Lambda_{Q'qQ}^{K'1K}(G,G,H)[\tilde{\varrho}_{-Q'}^{K'}(G,H)]^*, \end{aligned} \quad (17.70a)$$

$$\dot{\tilde{\varrho}}^K_Q(H,H) = -\Gamma_{HH'}\tilde{\varrho}^K_Q(H,H)$$
$$+i\bar{\chi}^{(j)}_{HG}(-1)^{2H+K'-K+1}e^{i\mathbf{k}_j\cdot\mathbf{R}}(-1)^q\epsilon^{(j)}_{-q}\Lambda^{K'1K}_{Q'qQ}(H,H,G)\tilde{\varrho}^{K'}_{Q'}(G,H)$$
$$-i\left[\bar{\chi}^{(j)}_{HG}\right]^*(-1)^{2G+Q'}e^{-i\mathbf{k}_j\cdot\mathbf{R}}(\epsilon^{(j)}_q)^*$$
$$\times\Lambda^{K'1K}_{Q'qQ}(H,H,G)[\tilde{\varrho}^{K'}_{-Q'}(G,H)]^*, \tag{17.70b}$$
$$\dot{\tilde{\varrho}}^K_Q(G,H) = -(\gamma_{GH}-i\delta)\,\tilde{\varrho}^K_Q(G,H)$$
$$-i\left[\bar{\chi}^{(j)}_{HG}\right]^*(-1)^{2G}e^{-i\mathbf{k}_j\cdot\mathbf{R}}\left[\epsilon^{(j)}_q\right]^*\Lambda^{K'1K}_{Q'qQ}(H,G,G)\tilde{\varrho}^{K'}_{Q'}(G,G)$$
$$+i\left[\bar{\chi}^{(j)}_{HG}\right]^*(-1)^{2G+Q'+K'-K+1}e^{-i\mathbf{k}_j\cdot\mathbf{R}}\left[\epsilon^{(j)}_q\right]^*$$
$$\times\Lambda^{K'1K}_{Q'qQ}(G,H,H')[\tilde{\varrho}^{K'}_{-Q'}(H,H)]^*, \tag{17.70c}$$
$$\tilde{\varrho}^K_Q(F',F) = (-1)^{F-F'+Q}[\tilde{\varrho}^K_{-Q}(F,F')]^*. \tag{17.70d}$$

It should be noted that the Rabi frequency, as defined in reference [4], is the negative of ours, as is the detuning δ.

Density matrix elements in the irreducible tensor basis are related to those in the m basis by

$$\varrho^K_Q(F,F') = (-1)^{F'-m'}\begin{bmatrix} F & F' & K \\ m & -m' & Q \end{bmatrix}\varrho(F,m;F',m'), \tag{17.71}$$

where the quantity in square brackets is a Clebsch-Gordan coefficient. All other symbols in equation (17.70) are defined in the text, except

$$\Lambda^{K'kK}_{Q'qQ}(A,B,C) = (-1)^{k+K}[(2k+1)(2K'+1)]^{1/2}$$
$$\times\begin{bmatrix} K' & k & K \\ Q' & q & Q \end{bmatrix}\begin{Bmatrix} K' & k & K \\ A & B & C \end{Bmatrix}. \tag{17.72}$$

The time derivatives in equations (17.70) are total time derivatives in the sense that

$$\frac{d}{dt} = \frac{\partial}{\partial t} + \mathbf{v}\cdot\nabla, \tag{17.73}$$

where $\mathbf{v}$ is the atomic velocity. There is a summation convention implicit in equations (17.70) to (17.72). Repeated indices appearing on the right-hand side of an equation are to be summed over, except if these indices also appear on the left-hand side of the equation. These equations have been written for a single ground-state and a single excited-state manifold of levels—more general equations, allowing for a number of nearly degenerate manifolds of levels and several field frequencies, are given in reference [4].

For specific experimental conditions, it might be necessary to add additional terms to equation (17.70), such as those arising from external magnetic fields, collisions, "source" terms that bring atoms into the interaction volume, or loss terms resulting from atoms leaving the interaction volume. In obtaining equation (17.70), it has been assumed implicitly that all the magnetic sublevels within a state of given F are degenerate.

Problems

1–2. Calculate the steady-state ground-state density matrix elements when x-polarized radiation drives a $G = 1 \rightarrow H = 2$ transition. Do this as a rotation of the z-polarized results given by equation (17.25) using both the irreducible and m-state bases.

In solving this problem, you can use the fact that, under a rotation, elements of an irreducible tensor operator transform as

$$\varrho_Q^{\prime K} = \sum_{Q'} D_{Q'Q}^{(K)}(\alpha, \beta, \gamma)\, \varrho_{Q'}^K,$$

where $D_{Q'Q}^{(K)}(\alpha, \beta, \gamma)$ are elements of the (active) angular momentum rotation matrices given by

$$D_{mm'}^{(K)}(\alpha, \beta, \gamma) = e^{-im\alpha} r_{mm'}^{(K)}(\beta) e^{-im'\gamma},$$
$$r_{mm'}^{(K)}(\beta) = \langle Km| e^{-i\beta J_y} |Km'\rangle,$$

where

$$D_{0m}^{(K)}(\alpha, \beta, \gamma) = (-1)^m \sqrt{\frac{4\pi}{2K+1}} \left[Y_Q^K(\beta, \gamma)\right]^*,$$

$Y_Q^K(\theta, \phi)$ is a spherical harmonic, and

$$r^{(1)}(\beta) = \begin{pmatrix} \frac{1}{2}(1+\cos\beta) & -\frac{1}{\sqrt{2}}\sin\beta & \frac{1}{2}(1-\cos\beta) \\ \frac{1}{\sqrt{2}}\sin\beta & \cos\beta & -\frac{1}{\sqrt{2}}\sin\beta \\ \frac{1}{2}(1-\cos\beta) & \frac{1}{\sqrt{2}}\sin\beta & \frac{1}{2}(1+\cos\beta) \end{pmatrix},$$

with the order $(1, 0, -1)$. In the m basis, the rotation is given by

$$\varrho'_{mm'} = \sum_{qq'} D_{mq}^{(G)}(\alpha, \beta, \gamma) \left[D_{m'q'}^{(G)}(\alpha, \beta, \gamma)\right]^* \varrho_{qq'}.$$

3–4. Write a program for the density matrix equations (17.27) in terms of the coupled tensor polarization components. Test the program for the $G = 1$ to $H=2$ transition and an x–polarized field to show that the evolution equations and steady-state values agree with equations (17.34) and (17.35), respectively.

5. Assume that a weak magnetic field $\mathbf{B}$ is applied to the atoms in an optical pumping experiment in some arbitrary direction. Calculate the additional terms that must be added to the right-hand side of equation (17.27) to account for this field. This can be done by using equation (17.70a) without the "in" term, setting $H = G$, and relating $\bar{\chi}_{HG}^{(j)}$ to the analogous term for the magnetic interaction. Assume that the magnetic field interaction can be written as $-\boldsymbol{\mu}_{\text{mag}} \cdot \mathbf{B} = \mu_B g_G \mathbf{G} \cdot \mathbf{B}$, where μ_B is the Bohr magneton, $\mathbf{G}$ is the angular momentum in units of $\hbar$, and g_G is the Landé g factor.

6–7. Write a program to obtain the lattice matrix elements given by equation (17.61) in terms of the coupled tensor components $\epsilon_Q^K(j, j')$. Use your program with $\gamma_H = 0$ to obtain the lattice potentials for a $G = 1$ ground state with $H = 2$ for both lin⊥lin and $\sigma_+ - \sigma_-$ field polarizations. Show that there is no modulation in the $\sigma_+ - \sigma_-$ case.

8. Derive any one of equations (17.9) without the decay terms for the interaction potential given by equation (17.9).

References

[1] A. Kastler, *Quelques suggestions concernant la production optique et la détection optique d'une inégalité de population des niveaux de quantification spatiale des atomes—application à l'experience de Stern et Gerlach et à resonance magnetique* (*Some ideas about the production and detection by optical means of an inequality in magnetic sublevels—application to the Stern-Gerlach experiment and magnetic resonance*), Journal de Physique et le Radium **11**, 255–265 (1950); J. Brossel, A. Kastler, and J. Winter, *Création optique d'une inégalité de population entre les sous-niveaux Zeeman de l'état fondamental des atoms (Creation of an inequality in ground-state magnetic sublevels)*, Journal de Physique et le Radium **13**, 668 (1952).

[2] L. Mandel and E. Wolf, *Optical Coherence and Quantum Optics* (Cambridge University Press, Cambridge, UK, 1995), sec. 10.6.

[3] L. Allen, M. J. Padgett, and M. Babiker, *The orbital angular momentum of light*, in *Progress in Optics*, edited by E. Wolf, vol. XXXIX (Elsevier Science, Amsterdam, 1999), pp. 291–372; M. F. Andersen, C. Ryu, P. Cladé, V. Natarajan, A. Vaziri, K. Helmerson, and W. D. Phillips, *Quantized rotation of atoms from photons with orbital angular momentum*, Physical Review Letters **97**, 170406, 1–4 (2006).

[4] G. Rogers, P. R. Berman, and B. Dubetsky, *Rate equations between electronic-state manifolds*, Physical Review A **48**, 1506–1513 (1993).

[5] P. S. Jessen and I. H. Deutsch, *Optical lattices*, in *Advances in Atomic, Molecular, and Optical Physics*, edited by B. Bederson and H. Walther, vol. 37 (Academic Press, New York, 1996), pp. 95–138.

Bibliography

W. Happer, *Optical pumping*, Reviews of Modern Physics **44**, 169–249 (1972). This review article contains a comprehensive list of references.

A. Omont, *Irreducible components of the density matrix. Application to optical pumping*, Progress in Quantum Electronics **5**, 69–138 (1977).

See also the references in chapter 16.

18

Sub-Doppler Laser Cooling

In chapter 5, we found a limit of $k_BT \approx \hbar\gamma_2/2$ for laser cooling. One of the most dramatic violations of Murphy's law (if something can go wrong, it will) occurred in the late 1980s, when several research groups tried to verify this limit of Doppler cooling. The temperature they obtained was *lower* than that predicted by theory. This discovery and the subsequent explanation of the *sub-Doppler cooling mechanism* was rewarded with a Nobel Prize in 1997 for Steven Chu, Claude Cohen-Tannoudji, and William Phillips. The work also paved the way for experiments that ultimately resulted in a Bose condensate of neutral atoms.

It was not that the theory of Doppler cooling was wrong, it was only that it did not quite go far enough. In theories of laser cooling, the atoms were modeled as two-level quantum systems. This is a useful model in many cases, but for laser cooling, it missed the boat. We have already seen in the previous chapter how optical pumping using standing-wave optical fields can result in a spatial modulation of the ground-state polarization. Moreover, this steady-state modulation is achieved on a time scale that is related to the optical pumping rate Γ' rather than the excited-state decay rate γ_H. Thus, optical pumping introduces a *new* timescale into the problem. Since, for sufficiently weak fields, $\Gamma' \ll \gamma_H$, it is possible that a relationship such as $k_BT = \hbar\gamma_2/2$ could be replaced by $k_BT = \hbar\Gamma'/2 \ll \hbar\Gamma/2$. Even if this expression does not provide the correct limit, it is clear that one must reevaluate the limits of laser cooling, taking into account the effects of optical pumping.

There are essentially three mechanisms by which optical pumping using counter-propagating fields can result in cooling:

(1) Optical pumping results in a spatially modulated ground-state polarization (that is, a spatially modulated population difference or coherence between different ground-state sublevels), allowing for an exchange of momentum between counterpropagating fields.
(2) Optical pumping results in an *unmodulated* imbalance in ground-state populations, leading to different scattering rates for each of the counter-propagating waves.

(3) Optical pumping may be velocity selective and pump atoms into a state with $\mathbf{v}=0$, which is a dark state that is decoupled from the fields (coherent population trapping).

For a review of (1) and (2), see the articles by Dalibard and Cohen-Tannoudji [1], Cohen-Tannoudji [2], and Finkelstein et al. [3]. For a review of (3), see the article by Aspect et al. [4]. In this chapter, we sketch some of the results.

18.1 Cooling via Field Momenta Exchange and Differential Scattering

Let us first discuss the mechanisms (1) and (2), which can be treated by a common formalism. A proper treatment of sub-Doppler cooling can necessitate the use of a quantized picture of atomic motion [5], but we consider the motion to be classical in this section. As long as the atoms are not trapped in the optical lattice potentials, the classical motion approximation is satisfactory. We calculate the spatially averaged friction force experienced by the atoms. Recall that the friction force was first calculated in chapter 5 for an ensemble of two-level atoms. Of course, in order to get the final energy distribution for the atoms, one has to also include diffusion. The appendix contains an outline of an approach based on the Wigner distribution in which the friction force and diffusion coefficient are calculated using a common formalism.

We have discussed ways in which different fields can exchange momenta using atoms to accomplish this exchange. In effect, we saw that some spatial pattern of the fields was transferred to the atoms. Provided that there was a phase shift between the field and matter spatial patterns, momentum could be exchanged by the fields. You will see that this mechanism is at the heart of one type of sub-Doppler cooling referred to as *polarization gradient cooling*. This cooling mechanism has also been explained in terms of a *Sisyphus effect*, which we also describe.

In addition to this polarization gradient cooling, there can be sub-Doppler cooling resulting from an imbalance in ground-state population caused by optical pumping. The fields then undergo differential scattering from these states, again leading to sub-Doppler cooling. Dalibard and Cohen-Tannoudji explain this process in terms of an effective magnetic field [1]. In reality, both types of cooling mechanisms are usually present for ground-state manifolds having angular momentum $G > 1/2$. The formalism we develop allows us to identify the contributions from each process.

The average friction force on an atom is given by

$$\mathbf{F}(\mathbf{R}, \mathbf{v}) = \mathrm{Tr}\left\{\varrho(\mathbf{R}, \mathbf{v}, t)\nabla\left[\hat{\boldsymbol{\mu}} \cdot \mathbf{E}_c(\mathbf{R}, t)\right]\right\}. \tag{18.1}$$

We assume that the incident field is the sum of equal frequency fields given by

$$\mathbf{E}_c(\mathbf{R}, t) = \frac{1}{2}\sum_j \mathcal{E}^{(j)}\hat{\boldsymbol{\epsilon}}^{(j)}e^{i(\mathbf{k}_j \cdot \mathbf{R}-\omega t)} + \text{c.c.} \tag{18.2}$$

and express $\hat{\boldsymbol{\mu}} \cdot \hat{\boldsymbol{\epsilon}}^{(j)}$ in the form

$$\hat{\boldsymbol{\mu}} \cdot \hat{\boldsymbol{\epsilon}}^{(j)} = \sum_{q=-1}^{1} (-1)^q \epsilon_{-q}^{(j)} \hat{\mu}_q. \tag{18.3}$$

In the RWA, we then find from equation (18.1) that

$$\mathbf{F}(\mathbf{R}, \mathbf{v}) = \frac{i}{2} \sum_{j,q,m,m'} \mathcal{E}^{(j)} \mathbf{k}_j (-1)^q \epsilon_{-q}^{(j)} e^{i(\mathbf{k}_j \cdot \mathbf{R} - \omega t)} \langle H, m' | \hat{\mu}_q | G, m \rangle \varrho_{Gm;Hm'}(\mathbf{R}, \mathbf{v}, t) + \text{c.c.} \tag{18.4}$$

Using the fact that

$$\varrho_{Gm;Hm'} = \sum_{K,Q} (-1)^{H-m'} \begin{bmatrix} G & H & K \\ m & -m' & Q \end{bmatrix} \varrho_Q^K(G, H), \tag{18.5}$$

evaluating matrix elements of $\hat{\mu}_q$ using the Wigner-Eckart theorem, and noting that

$$\begin{aligned} &\sum_{m,m',K,Q} (-1)^{H-m'} \begin{bmatrix} G & H & K \\ m & -m' & Q \end{bmatrix} \begin{bmatrix} G & 1 & H \\ m & q & m' \end{bmatrix} \frac{\varrho_Q^K(G, H)}{\sqrt{2H+1}} \langle H \| \mu^{(1)} \| G \rangle \\ &= \sum_{m,m',K,Q} \begin{bmatrix} G & H & K \\ m & -m' & Q \end{bmatrix} \begin{bmatrix} G & H & 1 \\ m & -m' & -q \end{bmatrix} \frac{(-1)^{H-G+q} \langle H \| \mu^{(1)} \| G \rangle}{\sqrt{3}} \varrho_Q^K(G, H) \\ &= \frac{(-1)^{H-G+q}}{\sqrt{3}} \langle H \| \mu^{(1)} \| G \rangle \varrho_{-q}^K(G, H), \end{aligned} \tag{18.6}$$

we can rewrite equation (18.1) as

$$\mathbf{F}(\mathbf{R}, \mathbf{v}) = -\sum_{j,q} i\hbar \mathbf{k}_j (-1)^{H-G} \bar{\chi}_{HG}^{(j)} \epsilon_q^{(j)} \tilde{\varrho}_q^1(G, H; \mathbf{R}, \mathbf{v}) e^{i\mathbf{k}_j \cdot \mathbf{R}} + \text{c.c.}, \tag{18.7}$$

where

$$\tilde{\varrho}_q^1(G, H; \mathbf{R}, \mathbf{v}) = \varrho_q^1(G, H; \mathbf{R}, \mathbf{v}, t) e^{-i\omega t}, \tag{18.8}$$

$$\delta = \omega_{HG} - \omega, \tag{18.9a}$$

and[1]

$$\bar{\chi}_{HG}^{(j)} = -\frac{\langle H || \mu^{(1)} || G \rangle \mathcal{E}^{(j)}}{2\hbar\sqrt{3}}. \tag{18.9b}$$

We have reached the desired result, an expression for the friction force in terms of atomic state density matrix elements. The density matrix elements, in turn, are obtained as steady-state solutions of equations (17.70) given in section 17.4.2.

Density matrix elements $\varrho(G, H; \mathbf{R}, \mathbf{v})$ depend on $\varrho(G, G; \mathbf{R}, \mathbf{v})$ and $\varrho(H, H; \mathbf{R}, \mathbf{v})$. As in the previous chapter, we assume that the incident fields are weak, allowing us to neglect the dependence of $\varrho(G, H; \mathbf{R}, \mathbf{v})$ on $\varrho(H, H; \mathbf{R}, \mathbf{v})$. The quasi-static solution

[1] To make a connection with the seminal article of Dalibard and Cohen-Tannoudji, one sets $\bar{\chi}_{HG} = \Omega/\sqrt{3}$ for lin⊥lin polarization and $\bar{\chi}_{HG} = \Omega/[2(3/5)^{1/2}]$ for $\sigma_+ - \sigma_-$ polarization, where Ω is the Rabi frequency given in their article [1].

for the optical coherence is given by an equation analogous to equation (17.12), obtained from equation (17.70c) as

$$\tilde{\varrho}^1_q(G, H; \mathbf{R}, \mathbf{v}) = -i \sum_{Q,j,j'} (-1)^{2G} [\bar{\chi}^{(j')}_{HG}]^* e^{-i\mathbf{k}\cdot\mathbf{R}} [\epsilon^{(j')}_{q'}]^* \sqrt{3(2K+1)}$$
$$\times \begin{bmatrix} K & 1 & 1 \\ Q & q' & q \end{bmatrix} \begin{Bmatrix} K & 1 & 1 \\ H & G & G \end{Bmatrix} \frac{1}{\gamma_{GH} - i\delta} \varrho^K_Q(G; \mathbf{R}, \mathbf{v}), \quad (18.10)$$

where $\varrho^K_Q(G; \mathbf{R}, \mathbf{v}) \equiv \varrho^K_Q(G, G; \mathbf{R}, \mathbf{v})$. When this expression is substituted into equation (18.7) and equation (17.41) is used, we find

$$\mathbf{F}(\mathbf{R}, \mathbf{v}) = 3 \sum_{K,Q,j,j'} (-1)^{K+H+G} \hbar \mathbf{k}_j e^{i\mathbf{k}_{jj'}\cdot\mathbf{R}} \frac{\bar{\chi}^{(j)}_{HG} [\bar{\chi}^{(j')}_{HG}]^*}{\gamma_{GH} - i\delta}$$
$$\times \begin{Bmatrix} K & 1 & 1 \\ H & G & G \end{Bmatrix} \epsilon^K_Q(j, j') \varrho^K_Q(G; \mathbf{R}, \mathbf{v}) + \text{c.c.}, \quad (18.11)$$

where

$$\mathbf{k}_{jj'} = \mathbf{k}_j - \mathbf{k}_{j'}. \quad (18.12)$$

We have arrived at another important step in the calculation. The friction force is expressed in terms of ground-state matrix elements $\varrho^K_Q(G)$ and the coupled spherical tensor components $\epsilon^K_Q(j, j')$ defined by equation (17.41). We derived the evolution equation for the irreducible tensor components of the ground-state density matrix elements, equation (17.37), in the previous chapter.

The velocity dependence in the detuning has been dropped in all the preceeding equations under the assumption that any Doppler shifts are negligible compared with the excited-state decay rates. As such, this expression for the force does not include conventional Doppler cooling. In some sense, it has been assumed that conventional Doppler cooling has been used in a first stage to "precool" the atoms to the sub-Doppler cooling stage.

18.1.1 Counterpropagating Fields

We now limit the discussion to a one-dimensional problem involving two counterpropagating fields ($\mathbf{k}_1 = -\mathbf{k}_2 = k\hat{\mathbf{z}}$) having equal amplitudes. It is clear from equations (17.37) to (17.39), as well as from the fact that the nonlinear interaction of the atoms with the fields leads to all even spatial harmonics in atomic state populations, that a steady-state solution for $\varrho^K_Q(G)$ can be written as a Fourier series

$$\varrho^K_Q(G; Z, v) = \sum_{n=-\infty}^{\infty} A^K_Q(n; v) e^{2inkZ}, \quad (18.13)$$

where $v = v_z$. When this result is substituted into equation (18.11), we find a *spatially averaged force* that depends only on $A^K_Q(0; v)$ and $A^K_Q(\pm 1; v)$—namely,

$$\langle F(v)\rangle = \frac{(-1)^{H+G} 3\hbar k |\bar{\chi}_{HG}|^2}{\gamma_{GH} - i\delta} \sum_{K,Q} (-1)^K \left\{ \left[\epsilon^K_Q(1,1) - \epsilon^K_Q(2,2)\right] A^K_Q(0; v) \right.$$
$$\left. + \epsilon^K_Q(1,2) A^K_Q(-1; v) - \epsilon^K_Q(2,1) A^K_Q(1; v) \right\} \begin{Bmatrix} K & 1 & 1 \\ H & G & G \end{Bmatrix} + \text{c.c.}, \tag{18.14}$$

where $F(v)$ is the z component of the force. The term proportional to $A^K_Q(0; v)$ gives the difference in scattering off the imbalanced ground-state sublevel populations produced by optical pumping. The remaining terms arise from transfer of momenta between the fields scattering from the atomic polarization gratings produced by the fields. In what follows, for the most part we suppress the explicit dependence of $\varrho^K_Q(G; Z, v)$ on Z and v, and that of $A^K_Q(n; v)$ on v.

We can see the relationship between the fields and the gratings a bit more clearly if we use the relationships

$$\epsilon^K_{-Q}(j, j') = (-1)^Q \left[\epsilon^K_Q(j', j)\right]^*, \tag{18.15}$$

$$\varrho^K_{-Q}(G) = (-1)^Q \left[\varrho^K_Q(G)\right]^*, \tag{18.16}$$

$$A^K_{-Q}(n) = (-1)^Q \left[A^K_Q(-n)\right]^*, \tag{18.17}$$

and set

$$\epsilon^K_Q(j, j') = |\epsilon^K_Q(j', j)| e^{i\varphi^K_Q(j,j')}, \tag{18.18}$$

$$A^K_Q(n) = |A^K_Q(n)| e^{i\varphi^K_Q(n)}, \tag{18.19}$$

where, in general, $\varphi^K_Q(n)$ is a function of v. The last two terms in equation (18.14) can then be written as

$$\sum_Q \left[\epsilon^K_Q(1,2) A^K_Q(-1) - \epsilon^K_Q(2,1) A^K_Q(1)\right]$$
$$= \sum_Q \left\{\epsilon^K_Q(1,2) A^K_Q(-1) - \left[\epsilon^K_{-Q}(1,2) A^K_{-Q}(-1)\right]^*\right\}$$
$$= \sum_Q \left\{\epsilon^K_Q(1,2) A^K_Q(-1) - \left[\epsilon^K_Q(1,2) A^K_Q(-1)\right]^*\right\}$$
$$= 2i\,\mathrm{Im}\left[\sum_Q \epsilon^K_Q(1,2) A^K_Q(-1)\right] = 2i\,\mathrm{Im}\left\{\sum_Q \epsilon^K_Q(1,2)(-1)^Q \left[A^K_{-Q}(1)\right]^*\right\}$$
$$= 2i\,\mathrm{Im}\left\{\sum_Q (-1)^Q \left|\epsilon^K_Q(1,2) A^K_{-Q}(1)\right| e^{-i\left[\varphi^K_{-Q}(1) - \varphi^K_Q(1,2)\right]}\right\}. \tag{18.20}$$

Therefore, if $\epsilon^K_Q(1,2)$ and $A^K_{-Q}(1)$ are "in phase"—that is, if $\varphi^K_{-Q}(1) = \varphi^K_Q(1,2)$—then there is no exchange of momentum between fields. This result is similar to the two-level case, where we found that there was no exchange of momenta between the counterpropagating fields if the phase difference between the matter gratings and field intensity gratings vanished.

Last, we write equation (18.14) for the spatially averaged force as

$$\langle F(v)\rangle = \frac{(-1)^{H+G}3\hbar k|\bar{\chi}_{HG}|^2}{\gamma_{GH}-i\delta}$$
$$\times\left(\sum_{K,Q}\left[\epsilon_Q^K(1,1)-\epsilon_Q^K(2,2)\right]A_Q^K(0)\begin{Bmatrix}K&1&1\\H&G&G\end{Bmatrix}(-1)^K\right.$$
$$\left.+\,2i\,\mathrm{Im}\sum_{K,Q}(-1)^K\epsilon_Q^K(1,2)A_Q^K(-1)\begin{Bmatrix}K&1&1\\H&G&G\end{Bmatrix}\right)+\text{c.c.} \qquad (18.21)$$

Recall that

$$\epsilon_Q^K(j,j') = \sum_{q,q'}(-1)^{q'}\epsilon_q^{(j)}[\epsilon_{-q'}^{(j')}]^*\begin{bmatrix}1&1&K\\q&q'&Q\end{bmatrix}, \qquad (18.22)$$

$$\varrho_Q^K(G) = \sum_{n=-\infty}^{\infty}A_Q^K(n)e^{2inkZ}, \qquad (18.23)$$

and, in steady state,

$$v\frac{\partial\varrho_Q^K(G)}{\partial Z} = \sum_{K,K,'\bar{K},Q'}A(K,K',\bar{K};Q,Q';G,H;Z,t)P(K,K',\bar{K};G,H)\varrho_{Q'}^{K'}(G), \qquad (18.24)$$

$$A(K,K',\bar{K};Q,Q';G,H;R,t) = \sum_{j,j',\bar{Q}}(-1)^{Q'}\begin{pmatrix}K'&K&\bar{K}\\-Q'&Q&\bar{Q}\end{pmatrix}$$
$$\times|\bar{\chi}_{HG}|^2\epsilon_{\bar{Q}}^{\bar{K}}(j,j')e^{ik_{jj'}Z}\left[\frac{1}{\gamma_{GH}-i\delta}+\frac{(-1)^{K+K'+\bar{K}}}{\gamma_{GH}+i\delta}\right], \qquad (18.25)$$

$$k_{11}=k_{22}=0,\quad k_{12}=-k_{21}=2k, \qquad (18.26)$$

and $P(K,K',\bar{K};G,H)$ is given by equation (17.29). As a consequence, the spatially averaged force is completely determined once we solve equation (18.24).

We now solve for the average force for three cases: $\|$ polarization, crossed polarization for a $G=1/2\leftrightarrow H=3/2$ transition, and $\sigma_\pm$ polarization for a $G=1\leftrightarrow H=2$ transition.

18.1.1.1 Parallel polarization

When the polarization of both fields is the same,

$$\epsilon_Q^K(1,1)=\epsilon_Q^K(1,2)=\epsilon_Q^K(2,1)=\epsilon_Q^K(2,2), \qquad (18.27)$$

$\epsilon_Q^K(1,1)-\epsilon_Q^K(2,2)=0$, implying that scattering term vanishes. On the other hand, you have seen in chapter 17 that there is no spatial modulation, $A_Q^K(n)=0$ for $n\neq 0$.

Therefore, the field momentum exchange term also vanishes, and there is no sub-Doppler cooling.

18.1.1.2 *Crossed polarization:* G=1/2 ↔ H=3/2 *transition*

We consider now the crossed or lin⊥lin geometry for which $\boldsymbol{\epsilon}^{(1)} = \hat{\mathbf{x}}$, $\boldsymbol{\epsilon}^{(2)} = \hat{\mathbf{y}}$. The values of the spherical polarization tensor components are given in equation (17.50). Moreover, we showed in the previous chapter that the only nonvanishing elements of ϱ_Q^K (and, consequently, A_Q^K) are ϱ_0^0 and ϱ_0^1 when $G = 1/2$. For $K < 2$, it follows from equations (17.50) that

$$\epsilon_Q^K(2,2) - \epsilon_Q^K(1,1) = \epsilon_Q^K(y,y) - \epsilon_Q^K(x,x) = 0. \tag{18.28}$$

As a consequence, the differential scattering term in equation (18.21) does not contribute. For higher angular momentum ground states, this is no longer true.

To evaluate the friction force using the momentum exchange term in equation (18.21), we must calculate $\varrho_0^0(1/2)$ and $\varrho_0^1(1/2)$. Recall that

$$\varrho_0^0(1/2) = \frac{1}{\sqrt{2}}\left(\varrho_{1/2,1/2} + \varrho_{-1/2,-1/2}\right) = \frac{1}{\sqrt{2}} \tag{18.29}$$

is proportional to the (constant) total ground-state population, while

$$\varrho_0^1(1/2) = \frac{1}{\sqrt{2}}\left(\varrho_{1/2,1/2} - \varrho_{-1/2,-1/2}\right) \tag{18.30}$$

is a measure of the population difference of the ground-state sublevels. The spatially averaged friction force for a $G = 1/2$ ground state and lin⊥lin polarization is obtained from equation (18.21), with $\epsilon_Q^K(1,2) = i\delta_{K,1}\delta_{Q,0}/\sqrt{2}$ as

$$\langle F(v)\rangle = -\frac{(-1)^{H+G}3\hbar k|\tilde{\chi}_{HG}|^2}{\gamma_{GH} - i\delta}\sqrt{2}i\,\mathrm{Im}\left\{i\left[A_0^1(1)\right]^*\right\}\begin{Bmatrix} 1 & 1 & 1 \\ H & 1/2 & 1/2 \end{Bmatrix} + \text{c.c.} \tag{18.31}$$

We see from this equation and equation (18.23) that the spatially averaged force depends only on the spatially modulated population difference of the ground-state sublevels through $A_0^1(\pm 1)$.

The equation of motion (18.24) for $\varrho_0^1(1/2)$ is

$$\begin{aligned} v\frac{\partial \varrho_0^1(1/2)}{\partial Z} &= \sum_{\bar{K}} A\left(1,1,\bar{K};0,0;\frac{1}{2},\frac{3}{2};Z,t\right) P\left(1,1,\bar{K},\frac{1}{2},\frac{3}{2}\right)\varrho_0^1(1/2) \\ &\quad + A\left(1,0,1;0,0;\frac{1}{2},\frac{3}{2};Z,t\right) P\left(1,0,1,\frac{1}{2},\frac{3}{2}\right)\varrho_0^0(1/2) \\ &= -\Gamma_{3/2}^{\perp}\varrho_0^1(1/2) - \frac{1}{\sqrt{2}}\Gamma_{3/2}^{\perp}\sin(2kZ), \end{aligned} \tag{18.32}$$

where

$$\Gamma_{3/2}^{\perp} = \Gamma'(1/2, 3/2) = \frac{1}{3}\frac{|\tilde{\chi}_{HG}|^2\gamma_H}{\gamma_{GH}^2 + \delta^2} \tag{18.33}$$

is an optical pumping rate [see equation (17.53)]. With a trial solution of the form given in equation (18.13):

$$\varrho_0^1(1/2) = A_0^1(1)e^{2ikZ} + \left[A_0^1(1)\right]^* e^{-2ikZ}, \tag{18.34}$$

we find the steady-state solution

$$A_0^1(1) = \frac{i\Gamma_{3/2}^{\perp}}{2\sqrt{2}} \frac{1}{\Gamma_{3/2}^{\perp} + 2ikv}. \tag{18.35}$$

Therefore, for the spatially averaged force (18.31), we have

$$\begin{aligned} \langle F(v)\rangle &= -\sqrt{2}i\frac{3\hbar k|\bar{\chi}_{HG}|^2}{\gamma_{GH} - i\delta}\mathrm{Im}\left[\frac{-\Gamma_{3/2}^{\perp}/(2\sqrt{2})}{\Gamma_{3/2}^{\perp} - 2ikv}\right]\begin{Bmatrix} 1 & 1 & 1 \\ 3/2 & 1/2 & 1/2 \end{Bmatrix} + \text{c.c.} \\ &= \frac{3i\hbar k|\bar{\chi}_{HG}|^2}{\gamma_{GH} - i\delta}\frac{kv\Gamma_{3/2}^{\perp}}{(\Gamma_{3/2}^{\perp})^2 + 4k^2v^2}\left(-\frac{1}{6}\right) + \text{c.c.} \\ &= -\frac{\hbar k|\bar{\chi}_{HG}|^2}{\gamma_{GH}^2 + \delta^2}\frac{\delta kv\Gamma_{3/2}^{\perp}}{(\Gamma_{3/2}^{\perp})^2 + 4k^2v^2}, \end{aligned} \tag{18.36}$$

and taking into account equation (18.33), we obtain

$$\langle F(v)\rangle = -\beta_f v = -3\hbar k^2 v\frac{\delta}{\gamma_H}\frac{\left(\Gamma_{3/2}^{\perp}\right)^2}{\left(\Gamma_{3/2}^{\perp}\right)^2 + 4k^2v^2}, \tag{18.37}$$

which is a friction force for red detuning ($\delta > 0$). For $H = 1/2$, the sign is changed (there is a friction force for blue detuning), and $\Gamma_{3/2}^{\perp}$ is replaced by

$$\Gamma_{1/2}^{\perp} = \Gamma'(1/2, 1/2) = \frac{2}{3}\frac{|\bar{\chi}_{HG}|^2\gamma_H}{\gamma_{GH}^2 + \delta^2}. \tag{18.38}$$

From our relatively simple optical pumping calculation, we have arrived at a very important result. Near $v = 0$, the friction coefficient,

$$\beta_f = 3\hbar k^2\frac{\delta}{\gamma_H} \tag{18.39}$$

is of order $(\gamma_{GH}^2 + \delta^2)/(\gamma_H\Gamma_{3/2}^{\perp}) \gg 1$ times the Doppler cooling friction coefficient (5.47) calculated in chapter 5. Of course, the *capture range* (range of velocities for which atoms experience the friction force) is smaller by a factor $\Gamma_{3/2}^{\perp}/\gamma_H$ than that for Doppler cooling, so atoms must be precooled to this capture range.

Moreover, we see that the origin of the friction force is an exchange of momenta between the counterpropagating fields. From equations (18.18) and (17.50), one calculates that the phase associated with the field polarization grating is $\varphi_0^1(1,2) = \pi/2$, while from equations (18.19) and (18.35), one finds that the phase associated with the *matter* polarization grating is $\varphi_0^1(1) = \pi/2 + \tan^{-1}(2kv/\Gamma_{3/2}^{\perp})$. Thus, the relative phase between the field and matter polarization gratings appearing in equation (18.20) is

$$\varphi_0^1(1,2) - \varphi_0^1(1) = \tan^{-1}\left(2kv/\Gamma_{3/2}^{\perp}\right). \tag{18.40}$$

For values of $v \neq 0$, the relative phase is nonvanishing, and the fields can exchange momenta, leading to the sub-Doppler friction force of laser cooling. To show the momentum exchange explicitly, one must write coupled Maxwell-Bloch equations for both fields [6].

To get the final energy and temperature, diffusion must be included. Details of such a calculation using the Wigner distribution are given in the appendix. The diffusion coefficient has two contributions. The first, D_0, results from scattering of the incident fields into previously unoccupied vacuum modes and is similar to that for the two-level problem. The second, D_1, results from fluctuations in the ground-state polarization that is created by the fields. We write the spatially averaged diffusion coefficient as

$$\langle D \rangle = D_0 + D_1, \tag{18.41}$$

with

$$D_0 = \frac{41\hbar^2 k^2}{20}\Gamma^{\perp}_{3/2}\left[1 - \frac{4}{41}\frac{4\left(\Gamma^{\perp}_{3/2}\right)^2}{\left(\Gamma^{\perp}_{3/2}\right)^2 + 4k^2v^2}\right], \tag{18.42}$$

and

$$D_1 = \frac{9}{2}\hbar^2 k^2 \Gamma^{\perp}_{3/2}\left(\frac{\delta}{\gamma_H}\right)^2 \frac{\left(\Gamma^{\perp}_{3/2}\right)^2}{\left(\Gamma^{\perp}_{3/2}\right)^2 + 4k^2v^2}. \tag{18.43}$$

If $|\delta/(\gamma_H)| \gg 1$, $D_1 \gg D_0$, and

$$k_B T \approx \frac{D_1}{\beta_f} = \frac{3}{2}\hbar\Gamma^{\perp}_{3/2}\left|\frac{\delta}{\gamma_H}\right|. \tag{18.44}$$

The temperature achieved in sub-Doppler cooling is much less than that of the Doppler limit, $k_B T = 7\hbar\gamma_H/20$, if the field strength is taken to be weak. It appears that $k_B T \sim 0$ if the field strength goes to zero, but this is not the case. In this limit, D_0 must also be included (see the appendix). The lowest energy achievable is of order of several recoil energies E_r defined by

$$E_r = \frac{\hbar^2 k^2}{2M} = \frac{1}{2}k_B T_r \tag{18.45}$$

In Cs, $T_r \approx 2.5\ \mu$K (corresponding to approximately 13 times the recoil energy) was obtained using this cooling method [7].

18.1.1.3 $\sigma_+ - \sigma_-$ *polarization:* $G = 1 \Rightarrow H = 2$ *transition*

We now assume that the counterpropagating fields have σ_+ and σ_- polarizations, respectively, such that

$$\begin{aligned} \mathbf{E}^+(Z) &= \frac{1}{2}\left(\hat{\boldsymbol{\epsilon}}_1 e^{ikZ} + \hat{\boldsymbol{\epsilon}}_2 e^{-ikZ}\right)\mathcal{E}, \\ &= \frac{1}{2}\left(\frac{\hat{\mathbf{x}} + i\hat{\mathbf{y}}}{\sqrt{2}} e^{ikZ} - \frac{\hat{\mathbf{x}} - i\hat{\mathbf{y}}}{\sqrt{2}} e^{-ikZ}\right)\mathcal{E} \\ &= \frac{i}{\sqrt{2}}\left(\hat{\mathbf{x}}\sin kZ + \hat{\mathbf{y}}\cos kZ\right)\mathcal{E}. \end{aligned} \tag{18.46}$$

The magnitude of the field is constant, but its direction rotates as a function of kZ; this is referred to as *corkscrew polarization*. As a consequence, for a $G = 1$ ground state, there is no spatial modulation of the force with this polarization—the spatial variation of $\epsilon_Q^K(j, j')\varrho_Q^K(G)$ in equation (18.11) exactly cancels that of the factor $e^{i\mathbf{k}_{jj'}\cdot\mathbf{R}}$.

The values of the spherical polarization tensor components are given in equation (17.54). Using equations (17.54) and (18.14) and the values

$$\begin{Bmatrix} 1 & 1 & 1 \\ 2 & 1 & 1 \end{Bmatrix} = \frac{1}{6}, \qquad \begin{Bmatrix} 2 & 1 & 1 \\ 2 & 1 & 1 \end{Bmatrix} = \frac{1}{30}, \tag{18.47}$$

we find that the force is

$$F(v) = \frac{3\hbar k|\bar{\chi}_{HG}|^2}{\gamma_{GH} - i\delta}\left[A_0^1(0)\frac{\sqrt{2}}{6} + \frac{1}{15}i\,\mathrm{Im}\,A_{-2}^2(-1)\right] + \text{c.c.} \tag{18.48}$$

Thus, to calculate $F(v)$, we need $A_0^1(0)$ and $A_{-2}^2(-1)$. Note that there are now contributions from both the differential scattering and momentum exchange terms.

For σ_+–σ_- polarization, the combined action of the counterpropagating fields results in $\Delta m = \pm 2$. That is, if one starts in an $m = 0$ sublevel, the combined action of the fields can result in transitions to $m = \pm 2, \pm 4, \pm 6$, etc. Each successive transition corresponds to an additional spatial harmonic in ground-state polarization that is produced. For a $G = 1$ ground state, the maximum Δm is 2, which means that only the second harmonic contributes. This is very different from the lin⊥lin geometry, in which *all* harmonics can be generated regardless of the value of G, owing to the spatial modulation of the field polarization.

Using equation (18.24) and the fact that $\varrho_0^0 = 1/\sqrt{3}$, we find that, in steady state, the equations for the relevant density matrix elements are

$$v\frac{\partial \varrho_2^2}{\partial Z} = -\Gamma_c'\varrho_2^2 - \Gamma_c' e^{2ikZ}\left(\frac{1}{5}\sqrt{\frac{2}{3}}\varrho_0^2 + \frac{1}{5\sqrt{2}}\frac{i\delta}{\gamma_{GH}}\varrho_0^1\right) + \frac{1}{6}\Gamma_c' e^{2ikZ}, \tag{18.49a}$$

$$v\frac{\partial \varrho_0^2}{\partial Z} = -\frac{11}{15}\Gamma_c'\varrho_0^2 - \frac{\Gamma_c'}{5}\sqrt{\frac{2}{3}}\left(\varrho_2^2 e^{-2ikZ} + \varrho_{-2}^2 e^{2ikZ}\right) + \frac{1}{3\sqrt{6}}\Gamma_c', \tag{18.49b}$$

$$v\frac{\partial \varrho_0^1}{\partial Z} = -\frac{1}{5}\Gamma_c'\varrho_0^1 - \frac{1}{5\sqrt{2}}\frac{i\Gamma_c'\delta}{\gamma_{GH}}\left(\varrho_2^2 e^{-2ikZ} - \varrho_{-2}^2 e^{2ikZ}\right), \tag{18.49c}$$

where $\varrho^2_{-2} = \left(\varrho^2_2\right)^*$, and

$$\Gamma'_c = \frac{\gamma_{GH}|\bar{\chi}_{GH}|^2}{\gamma^2_{GH} + \delta^2}. \tag{18.50}$$

A trial solution of the form

$$\varrho^2_2 = A^2_2 e^{2ikZ}, \tag{18.51a}$$
$$\varrho^2_{-2} = A^2_{-2} e^{-2ikZ}, \tag{18.51b}$$
$$\varrho^2_0 = A^2_0, \tag{18.51c}$$
$$\varrho^1_0 = A^1_0, \tag{18.51d}$$

leads to the equations

$$\left(\Gamma'_c + 2ikv\right) A^2_2 + \frac{1}{5}\sqrt{\frac{2}{3}}\Gamma'_c A^2_0 + \frac{1}{5\sqrt{2}}\frac{i\Gamma'_c\delta}{\gamma_{GH}} A^1_0 = \frac{1}{6}\Gamma'_c, \tag{18.52a}$$

$$\left(\Gamma'_c - 2ikv\right) A^2_{-2} + \frac{1}{5}\sqrt{\frac{2}{3}}\Gamma'_c A^2_0 - \frac{1}{5\sqrt{2}}\frac{i\Gamma'_c\delta}{\gamma_{GH}} A^1_0 = \frac{1}{6}\Gamma'_c, \tag{18.52b}$$

$$A^2_0 = \frac{15}{11}\left[\frac{1}{3\sqrt{6}} - \frac{1}{5}\sqrt{\frac{2}{3}}\left(A^2_2 + A^2_{-2}\right)\right], \tag{18.52c}$$

$$A^1_0 = \frac{1}{\sqrt{2}}\frac{i\delta}{\gamma_{GH}}\left(A^2_{-2} - A^2_2\right). \tag{18.52d}$$

These equations can be solved in closed form, with the solution given by

$$A^2_2 = \left(A^2_{-2}\right)^* = \frac{3}{22}\Gamma'_c \frac{\Gamma'_c\left(1 + \frac{1}{5}\frac{\delta^2}{\gamma^2_{GH}}\right) - 2ikv}{\Gamma'^2_c\left(\frac{51}{55} + \frac{51}{275}\frac{\delta^2}{\gamma^2_{GH}}\right) + 4k^2v^2}, \tag{18.53a}$$

$$A^1_0 = \sqrt{2}\frac{\delta}{\gamma_{GH}}\mathrm{Im}\,A^2_2, \tag{18.53b}$$

$$A^2_0 = \frac{15}{11}\left(\frac{1}{3\sqrt{6}} - \frac{2}{5}\sqrt{\frac{2}{3}}\mathrm{Re}\,A^2_2\right). \tag{18.53c}$$

Substituting the results for $A^2_{\pm 2}$ and A^1_0 into equation (18.48) for $F(v)$, we obtain

$$F(v) = -\frac{150}{17}\hbar k^2 v\left(\frac{5}{30} + \frac{1}{30}\right)\frac{\gamma_H\delta}{\delta^2 + 5\gamma^2_{GH} + \frac{275}{51}\left(\frac{kv\gamma_H}{\Gamma'_c}\right)^2}, \tag{18.54}$$

where the term with $5/30$ is related to differential scattering, and the term with $1/30$ is related to momentum exchange of the fields. Near $v = 0$ [8],

$$F(v) = -\frac{30}{17}\hbar k^2 v \frac{\gamma_H\delta}{\delta^2 + 5\gamma^2_{GH}}. \tag{18.55}$$

This term is smaller than the corresponding lin$\perp$lin cooling force $F(v)$ for $G = 1/2$. If $\delta/\gamma_{GH} > 1$, it is smaller by a factor of γ^2_{GH}/δ^2; at optimum detuning, it is four times smaller. The lowest temperature reached is comparable to that for crossed-polarized fields, however, since the diffusion coefficient is also smaller.

18.2 Sisyphus Picture of the Friction Force for a $G = 1/2$ Ground State and Crossed-Polarized Fields

For a $G = 1/2$ ground state, Dalibard and Cohen-Tannoudji [1] have given a picture of sub-Doppler cooling in terms of a *Sisyphus effect*. Strictly speaking, it works only for a $G = 1/2$ ground state, but the general idea is applicable to other lattice potentials. For crossed-polarized counter propagating fields, the electric field is

$$\mathrm{E}_c(Z,t) = \frac{1}{2}\left[\mathrm{E}(Z)e^{-i\omega t} + \mathrm{E}^*(Z)e^{i\omega t}\right], \tag{18.56}$$

with

$$\begin{aligned}\mathrm{E}(Z) &= \left(\hat{\mathbf{x}}\mathcal{E}e^{ikZ} + \hat{\mathbf{y}}\mathcal{E}e^{-ikZ}\right)/2 \\ &= \left[(\hat{\mathbf{x}}+\hat{\mathbf{y}})\cos kZ + i\,(\hat{\mathbf{x}}-\hat{\mathbf{y}})\sin kZ\right]\mathcal{E}/2.\end{aligned} \tag{18.57}$$

The polarization varies as a function of Z in the medium, going from linear to circular to linear, etc. Since

$$\boldsymbol{\epsilon}_1 \equiv -\frac{\hat{\mathbf{x}} + i\hat{\mathbf{y}}}{\sqrt{2}}, \tag{18.58a}$$

$$\boldsymbol{\epsilon}_{-1} \equiv \frac{\hat{\mathbf{x}} - i\hat{\mathbf{y}}}{\sqrt{2}}, \tag{18.58b}$$

we can write the field as

$$\begin{aligned}\mathrm{E}(Z) &= \frac{\mathcal{E}}{2}\left[\frac{1}{\sqrt{2}}\left(\boldsymbol{\epsilon}_{-1} - \boldsymbol{\epsilon}_1\right)e^{ikZ} + \frac{i}{\sqrt{2}}\left(\boldsymbol{\epsilon}_{-1} + \boldsymbol{\epsilon}_1\right)e^{-ikZ}\right] \\ &= \frac{\mathcal{E}}{2}\left[\frac{1}{\sqrt{2}}\left(\boldsymbol{\epsilon}_{-1} - \boldsymbol{\epsilon}_1\right)e^{ikZ} + \frac{1}{\sqrt{2}}\left(\boldsymbol{\epsilon}_{-1} + \boldsymbol{\epsilon}_1\right)e^{-ikZ+i\pi/2}\right] \\ &= \frac{\mathcal{E}}{2}\left[\frac{1}{\sqrt{2}}\left(\boldsymbol{\epsilon}_{-1} - \boldsymbol{\epsilon}_1\right)e^{i(kZ-\pi/4)} + \frac{i}{\sqrt{2}}\left(\boldsymbol{\epsilon}_{-1} + \boldsymbol{\epsilon}_1\right)e^{-i(kZ-\pi/4)}\right]e^{i\pi/4} \\ &= \sqrt{2}\mathcal{E}e^{i\pi/4}\left[\cos(kZ-\pi/4)\boldsymbol{\epsilon}_{-1} - i\sin(kZ-\pi/4)\boldsymbol{\epsilon}_1\right]/\sqrt{2} \\ &= \mathcal{E}e^{i\pi/4}\left[\cos(kZ)\,\boldsymbol{\epsilon}_{-1} - i\sin(kZ)\,\boldsymbol{\epsilon}_1\right]/\sqrt{2},\end{aligned} \tag{18.59}$$

where the origin has been shifted by $kZ = \pi/4$; that is, $kZ \to kZ + \pi/4$. The $\boldsymbol{\epsilon}_{-1}$ term leads to $\Delta m = -1$ transitions and the $\boldsymbol{\epsilon}_1$ term to $\Delta m = 1$ transitions on absorption. From equation (18.59), we see clearly that the field can be viewed as a superposition of two standing-wave σ_+ and σ_- fields. Since the field polarization is spatially modulated, the cooling resulting from this field polarization is referred to as *polarization gradient cooling*.

Let us first consider a $H = 3/2$ excited state. The ground-state light shifts can be obtained by inspection using the field amplitude $\mathrm{E}(Z)$ given by equation (18.59) and the fact that the atom–field interaction varies as $\sum_q(-1)^q\hat{\mu}_{-q}\epsilon_q$. One can use

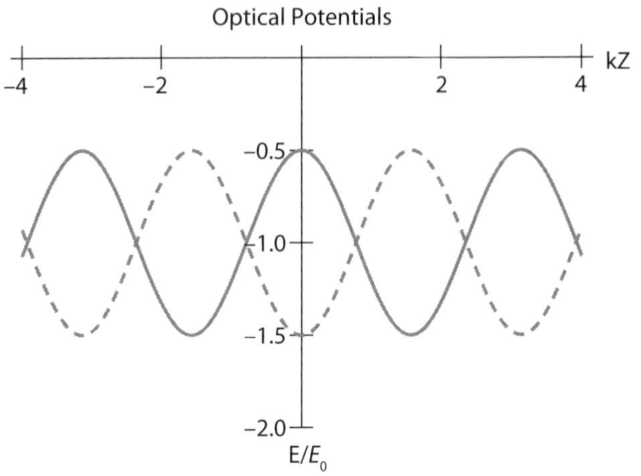

Figure 18.1. Optical potential in units of E_0. The solid curve is the optical potential for the $m = 1/2$ sublevel, and the dashed curve is the optical potential for the $m = -1/2$ sublevel.

an amplitude approach just as we did for the optical lattice potentials, adiabatically eliminate the excited state amplitudes, and obtain an equation for the time evolution of the ground-state amplitudes. In this manner, we find that the shift of the $m = 1/2$ sublevel in weak fields is simply

$$\Delta E_{1/2} = -\frac{\hbar\delta}{\gamma_{GH}^2 + \delta^2}\left(\frac{\mathcal{E}^2}{2\hbar^2}\right)\left[\begin{array}{l}\left\langle \frac{3}{2}, \frac{3}{2}\right| \hat{\mu}_1 \left|\frac{1}{2}, \frac{1}{2}\right\rangle^2 \sin^2(kZ) \\ +\left\langle \frac{3}{2}, -\frac{1}{2}\right| \hat{\mu}_{-1} \left|\frac{1}{2}, \frac{1}{2}\right\rangle^2 \cos^2(kZ)\end{array}\right]$$

$$= -\frac{\hbar\delta|\langle H||\mu^{(1)}||G\rangle|^2\mathcal{E}^2}{8\left(\gamma_{GH}^2 + \delta^2\right)}\left[\sin^2(kZ) + \frac{1}{3}\cos^2(kZ)\right]. \tag{18.60}$$

In other words, the $\boldsymbol{\epsilon}_1$ component of the polarization acts only on the $m_G = \frac{1}{2} \to m_H = \frac{3}{2}$ transition and the $\boldsymbol{\epsilon}_{-1}$ component only on the $m_G = \frac{1}{2} \to m_H = -\frac{1}{2}$ transition. There is an analogous result for $\Delta E_{-1/2}$. Assuming that $\delta \gg \gamma_{GH}$ and using equation (18.9b), we find that the ground-state light shifts are

$$\Delta E_{1/2} = -E_0 + \frac{\hbar|\bar{\chi}_{HG}|^2}{2\delta}\cos 2kZ, \tag{18.61a}$$

$$\Delta E_{-1/2} = -E_0 - \frac{\hbar|\bar{\chi}_{HG}|^2}{2\delta}\cos 2kZ, \tag{18.61b}$$

where

$$E_0 = \frac{\hbar|\bar{\chi}_{HG}|^2}{\delta}. \tag{18.62}$$

These results agree with equation (17.63) once the potentials (18.61) are shifted back by $\pi/4$, $kZ \to kZ - \pi/4$. The potentials are plotted in figure 18.1 for red detuning (δ positive).

From equation (18.32), for $v = 0$, we have

$$\varrho_{1/2,1/2} - \varrho_{-1/2,-1/2} = -\sin 2kZ \tag{18.63}$$

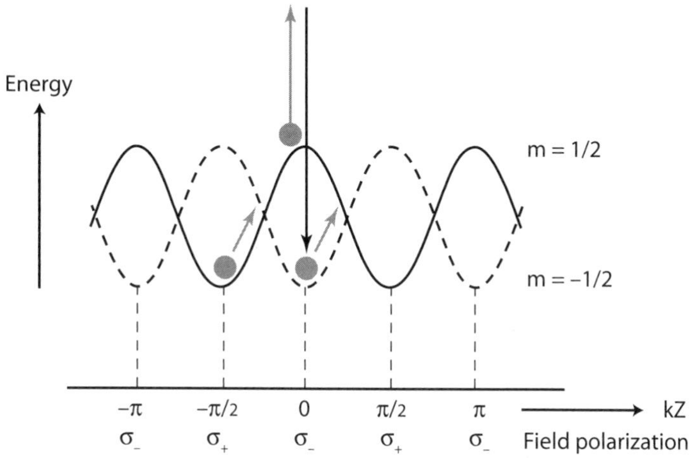

Figure 18.2. Schematic picture of Sisyphus cooling.

or

$$\varrho_{1/2,1/2} - \varrho_{-1/2,-1/2} = -\cos 2kZ, \tag{18.64}$$

with the origin shift of $\pi/4$ that we have introduced. Using equation (18.64) and the fact that $\varrho_{1/2,1/2} + \varrho_{-1/2,-1/2} = 1$, we obtain

$$\varrho_{1/2,1/2} = \sin^2 kZ, \qquad \varrho_{-1/2-1/2} = \cos^2 kZ. \tag{18.65}$$

Therefore, the maximum population is at the minimum of the potential wells, as expected. Moreover, the maximum sublevel populations occur at *antinodes* of the appropriate component of the circularly polarized field that drives those sublevels.

What happens as an atom moves in these potentials? As an atom in the $m = 1/2$ state moves to the right, it tries to reestablish equilibrium by being optically pumped into the $m = -1/2$ state, but this takes some time, of order $(\Gamma^{\perp}_{3/2})^{-1} = \tau_p$. If $kv\tau_p \leqslant 1$, the atom loses energy on climbing the potential and then is pumped into the $m = -1/2$ state, where it climbs and falls again. This is referred to as *Sisyphus cooling*, in analogy with the Greek myth where a poor lad rolled a stone up a hill only to have it roll down again (see figure 18.2).

Had we considered a $J = 1/2$ excited state and negative δ, the maxima and minima of the potentials are reversed. As a result, the maximum sublevel populations occur at *nodes* of the appropriate component of the circularly polarized field that drives those sublevels. This feature reduces the off-resonant scattering of the fields and lowers the achievable temperature. Optical lattices in which atoms are trapped near nodes of the field are referred to as *gray* or *dark* optical lattices [3], [9]–[11]. They have the advantage that they reduce heating produced by Rayleigh scattering of the fields by the atoms.

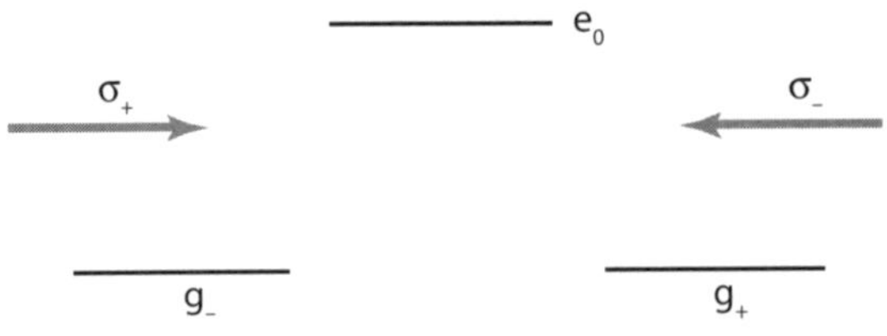

Figure 18.3. Level scheme for coherent population trapping.

18.3 Coherent Population Trapping

Sub-Doppler friction cooling results in energy widths of order of the recoil energy. It is possible to go below this limit using *velocity selective coherent population trapping* (VSCPT) [4]. In effect, one has to create a dark state for atoms corresponding to a very narrow velocity distribution. When atoms fall into this state, they are decoupled from the fields. We consider VSCPT for cooling in one dimension only—cooling in two and three dimensions using VSCPT is somewhat more problematic [12].

The basic idea is as follows: Circularly polarized fields having propagation constant k and polarizations $\sigma_\pm$ counterpropagate in the Z direction and drive transitions between a $G = 1$ ground state and an $H = 1$ excited state. The atoms are prepared in a superposition of ground-state sublevels in which there is no population in the $m_G = 0$ sublevel. As a consequence, the $m_G = 0$ sublevel is *never* populated, since the incident fields can excite only the $m_H = 0$ sublevel and the $m_H = 0$ sublevel is *not* coupled to the $m_G = 0$ sublevel by spontaneous emission (owing to selection rules). In effect, the atom–field geometry reduces to the Λ scheme shown in figure 18.3, in which $g_\pm$ are the $m_G = \pm 1$ sublevels and e_0 is the $m_H = 0$ sublevel. States are labled as $|\alpha, P\rangle$, where α is an internal state of the atom, and P is the Z component of the atom's (quantized) center-of-mass momentum. In the absence of spontaneous decay, states $|g_-, P - \hbar k\rangle$ and $|g_+, P + \hbar k\rangle$ are coupled by $\sigma_\pm$ counterpropagating fields to state $|e_0, P\rangle$. The center-of-mass energies associated with these states are $E_{P-\hbar k}$, $E_{P+\hbar k}$, and E_P, where $E_P = P^2/2M$.

We know from chapter 9 that there is a dark state if the detunings of the two fields are equal. For equal frequency fields, the detunings are given by

$$\delta_+ = \delta + E_{P,P-\hbar k} = \delta + \frac{kP}{M} - \frac{\hbar k^2}{2M}, \tag{18.66a}$$

$$\delta_- = \delta + E_{P,P+\hbar k} = \delta - \frac{kP}{M} - \frac{\hbar k^2}{2M}, \tag{18.66b}$$

where $E_{P,P'} = E_P - E_{P'}$. As a consequence, for equal-intensity fields, the state

$$|\Psi(P)\rangle = \frac{1}{\sqrt{2}}\left(|g_-, P - \hbar k\rangle - |g_+, P + \hbar k\rangle\right) \tag{18.67}$$

is a dark state, *provided* that $P = 0$. If scattering brings an atom into the state $|\Psi(0)\rangle$, it is decoupled from the fields. Atoms must have $P < 2\hbar k$ to be able to fall into the trapped state via scattering. Eventually, all atoms scatter into the $P = 0$ state. Note that the trapped state is not an eigenstate of momentum, so measurement yields two peaks, centered at $\pm\hbar k$, provided that $P = 0$. The widths of each of these peaks decreases the longer the field interacts with the atoms, a process that is governed by *Lévy statistics*. For some limits on this type of cooling, see Bardou et al. [13].

18.4 Summary

We have shown that optical pumping gives rise to two distinct sub-Doppler cooling mechanisms. The first involves an exchange of momentum between the fields resulting from scattering of the fields from a matter-state polarization grating that is created by the fields. The second mechanism involves a differential scattering of the fields from a ground-state sublevel imbalance produced by optical pumping. Both of these mechanisms require ground states having magnetically degenerate substates. A third type of sub-Doppler cooling, coherent population trapping, involves optical pumping of zero-velocity atoms into a dark state. In the next chapter, we return to our discussion of atom-quantized field interactions. Instead of using the Schrödinger representation, however, we obtain equations of motion for operators in the Heisenberg representation and derive a source-field expression for the electric field operator in terms of atomic state operators.

18.5 Appendix: Fokker-Planck Approach for Obtaining the Friction Force and Diffusion Coefficients

18.5.1 Fokker-Planck Equation

This appendix contains a calculation of the diffusion coefficient and the final energy distribution for sub-Doppler cooling. There is quite a bit of algebra needed for the derivation of these quantities, so the calculation may appear somewhat complicated. You should keep in mind that we are trying to find the Wigner function for the ground-state population density. To do this, we obtain a Fokker-Planck equation for the ground-state density matrix elements.

We have seen that the modifications of ground-state populations and coherence give rise to two contributions to the sub-Doppler friction force, one produced by differential scattering from any unmodulated population imbalance and one by momentum exchange of the fields via a spatially modulated ground-state polarization. Optical pumping also results in momentum *diffusion* for the atoms. This *spontaneous* contribution to diffusion results from the recoil an atom undergoes when a field is scattered into a previously unoccupied vacuum field mode. We have already encountered this contribution to diffusion in conventional Doppler cooling—the additional degenerate sublevels simply complicate the algebra. There is, in addition, a *stimulated* contribution to diffusion resulting from fluctuations in the spatially modulated ground-state polarization.

We assume that the atomic momentum is much greater than the recoil momentum and that the energy of the atoms is greater that the ground-state light shifts. If this is not the case, one must solve the full quantum dynamics for the atom–field system using the eigenstates of the atoms in the periodic light shift potentials. If the momentum is much less than the atomic momentum, changes in atomic velocity can be treated in a diffusion or Brownian motion approximation. In statistical physics, the *Fokker-Planck equation* for the phase-space distribution function describes such diffusive motion. The Fokker-Planck equation can be derived as an appropriate limit of the transport equation. A general form of the Fokker-Planck equation is [14]

$$\frac{\partial f(Z,P,t)}{\partial t}+v\frac{\partial f(Z,P,t)}{\partial Z}=\frac{\partial}{\partial P}\left[\beta P f(Z,P,t)\right]+D\frac{\partial^2 f(Z,P,t)}{\partial P^2}, \tag{18.68}$$

where $f(Z,P,t)$ is the distribution function, β is a friction coefficient, D is a diffusion constant, and $v=P/M$. We consider one-dimensional motion only. It is assumed that the transverse velocity distribution undergoes negligible cooling when the incident fields are directed along the Z axis.

The goal in this appendix is to derive the analogue of the Fokker-Planck equation for the Wigner function associated with the ground-state population. You may recall that the Wigner function was discussed in chapter 5 and is the quantum analogue of the distribution function. The starting point for the calculation is the generalization of equation (17.27) to include the center-of-mass motion of the atoms, starting from equation (5.73). There is considerable algebra needed to derive such an equation, and we do not give the details; however, the final equation simply reflects the recoil processes described earlier. Explicitly, we find [15]

$$\begin{aligned}&\frac{\partial \varrho_Q^K(P,P',t;G)}{\partial t}+i\omega_{P,P'}\varrho_Q^K(P,P';G)=(-1)^{Q'}\begin{pmatrix}K & K' & \tilde{K}\\ Q & -Q' & \tilde{Q}\end{pmatrix}\\ &\times\Big\{\epsilon_{\tilde{Q}}^{\tilde{K}}(j,j')S_+(K,K',\tilde{K},G,H,j,j')\varrho_{Q'}^{K'}\left(P,P'+\hbar k_{jj'},t;G\right)\\ &+\epsilon_{\tilde{Q}}^{\tilde{K}}(j,j')S_-(K,K',\tilde{K},G,H,j,j')\varrho_{Q'}^{K'}\left(P-\hbar k_{jj'},P',t;G\right)\\ &+T(K,K',\tilde{K},\bar{K},\bar{K}',G,H,j,j')\int d\Omega\,\lambda(\bar{K},\bar{K}',\tilde{K},\tilde{Q},j,j';\Omega)\\ &\times\varrho_{Q'}^{K'}\left[P+\hbar(q-k_j),P'+\hbar(q-k_{j'}),t;G)\right]\Big\},\end{aligned} \tag{18.69}$$

where

$$\omega_{P,P'}=\frac{\left(P^2-P'^2\right)}{2\hbar M}, \tag{18.70}$$

$$\begin{aligned}&T(K,K',\tilde{K},\bar{K},\bar{K}',G,H,j,j')=3(-1)^{1+\bar{K}+K'}(2H+1)\,B_{\bar{K}}\\ &\times[(2K+1)(2K'+1)(2\tilde{K}+1)(2\bar{K}+1)(2\bar{K}'+1)]^{1/2}\\ &\times|\bar{\chi}_{HG}|^2\left(\gamma_{GH}^2+\delta^2\right)^{-1}A(K,K',\tilde{K},\bar{K},\bar{K}',G,H),\end{aligned} \tag{18.71}$$

$$A(K,K',\tilde{K},\bar{K},\bar{K}',G,H)=(2X+1)\begin{Bmatrix}K & K' & \tilde{K}\\ \bar{K}' & \bar{K} & X\end{Bmatrix}\begin{Bmatrix}K & X & \bar{K}\\ G & H & 1\\ G & H & 1\end{Bmatrix}\begin{Bmatrix}X & K' & \bar{K}'\\ H & G & 1\\ H & G & 1\end{Bmatrix}, \tag{18.72}$$

$$S_+(K, K', \tilde{K}, G, H, j, j') = -3(-1)^{H-G+\tilde{K}} \left[(2K+1)\left(2K'+1\right)\right]^{1/2}$$
$$\times(2\tilde{K}+1)^{1/2}|\bar{\chi}_{HG}|^2 (\gamma_{GH} - i\delta)^{-1} \begin{Bmatrix} K & K' & \tilde{K} \\ G & G & G \end{Bmatrix} \begin{Bmatrix} 1 & 1 & \tilde{K} \\ G & G & H \end{Bmatrix}, \quad (18.73)$$

$$S_-(K, K', \tilde{K}, G, H, j, j') = (-1)^{K+K'-\tilde{K}} \left[S_+(K, K', \tilde{K}, G, H, j, j')\right]^*, \quad (18.74)$$

$$\lambda(\bar{K}, \bar{K}', \tilde{K}, \tilde{Q}, j, j'; \Omega) = \frac{3}{8\pi}(-1)^{\bar{K}+\bar{K}'+\bar{Q}}\sqrt{2\tilde{K}+1} \begin{pmatrix} \bar{K} & \bar{K}' & \tilde{K} \\ Q & -Q' & \bar{Q} \end{pmatrix} Y_{\bar{K}\bar{Q}}(\Omega)\epsilon^{\bar{K}'}_{\bar{Q}'}(j, j'), \quad (18.75)$$

$$B_{\bar{K}} = \sqrt{16\pi/3}\delta_{\bar{K},0} + \sqrt{8\pi/15}\delta_{\bar{K},2}, \quad (18.76)$$

$$q = k\cos\theta, \quad (18.77)$$

$$k_1 = -k_2 = k, \quad k_{jj'} = k_j - k_{j'}, \quad (18.78)$$

$Y_{\bar{K}\bar{Q}}(\Omega)$ is a spherical harmonic, and all other symbols have been defined in the text. A summation convention is used in equation (17.9), and all subsequent equations in this appendix in which any repeated indices appearing on the right-hand side of an equation are summed over, unless they also appear on the left-hand side of the equation. There is no sum over G or H, since only one value for each of these parameters is considered.

Admittedly, this is an algebraically complicated equation, containing an integral over the projection, $\hbar q \cos\theta$, onto the z axis of the recoil momentum associated with scattering of the field into the vacuum mode. Nevertheless, it is a relatively simple matter to program this equation; a link to a Mathematica notebook that accomplishes this task is posted at http://press.princeton.edu/titles/9376.html. The next step is to convert equation (18.69) to the Wigner function using equation (5.77c),

$$\varrho^K_Q(Z, P, t; G) = (2\pi\hbar)^{-1} \int_{-\infty}^{\infty} du\, \varrho^K_Q(P + u/2, P - u/2, t; G)e^{iuZ/\hbar}. \quad (18.79)$$

From equations (18.69) and (18.79), and using the fact that

$$(2\pi\hbar)^{-1} \int_{-\infty}^{\infty} du\, u\varrho^K_Q(P + u/2, P - u/2, t; G)e^{iuZ/\hbar} = (\hbar/i)\, \partial\varrho^K_Q(Z, P, t; G)/\partial Z, \quad (18.80)$$

we can obtain

$$\left(\frac{\partial}{\partial t} + v\frac{\partial}{\partial Z}\right) \varrho^K_Q(Z, P, t; G) = (-1)^{Q'} \begin{pmatrix} K & K' & \bar{K} \\ Q & -Q' & \bar{Q} \end{pmatrix} e^{ik_{jj'}Z}$$
$$\times \Big\{ T(K, K', \tilde{K}, \bar{K}, \bar{K}', G, H, j, j')$$
$$\times \int d\Omega\, \lambda(\bar{K}, \bar{K}', \tilde{K}, \tilde{Q}, j, j'; \Omega)\varrho^{K'}_{Q'} \left[Z, P + \hbar(q - k_j/2 - k_{j'}/2), t; G\right]$$
$$+\epsilon^{\bar{K}}_{\bar{Q}}(j, j')S_+(K, K', \tilde{K}, G, H, j, j')\varrho^{K'}_{Q'} \left(Z, P + \hbar k_{jj'}/2, t; G\right)$$
$$+ \epsilon^{\bar{K}}_{\bar{Q}}(j, j')S_-(K, K', \tilde{K}, G, H, j, j')\varrho^{K'}_{Q'} \left(Z, P - \hbar k_{jj'}/2, t; G\right) \Big\}. \quad (18.81)$$

The ground-state population Wigner distribution is equal to $(2G+1)^{1/2} \times \varrho_0^0(Z, P, t; G)$, so we need to get an equation for $\varrho_0^0(Z, P, t; G)$. To arrive at a Fokker-Planck equation for $\varrho_0^0(Z, P, t; G)$, we expand equation (18.81) to second order in $\hbar$ to arrive at

$$\begin{aligned}
\left(\frac{\partial}{\partial t} + v\frac{\partial}{\partial Z}\right)\varrho_0^0(Z, P, t; G) &= \frac{(-1)^{K'} e^{ik_{jj'}Z}}{\sqrt{2K'+1}} \\
&\times \left\{ T(0, K', K', \bar{K}, \bar{K}', G, H, j, j') \int d\Omega\, \lambda(\bar{K}, \bar{K}', K', Q', j, j'; \Omega) \right. \\
&\times \left[\hbar(k\cos\theta - k_j/2 - k_{j'}/2)\frac{\partial \varrho_{Q'}^{K'}(Z, P, t; G)}{\partial P} \right. \\
&\left. + \hbar^2(k\cos\theta - k_j/2 - k_{j'}/2)^2 \frac{\partial^2 \varrho_{Q'}^{K'}(Z, P, t; G)}{\partial P^2} \right] \\
&+ \epsilon_{Q'}^{K'}(j, j')\hbar k_{jj'} \begin{bmatrix} S_+(0, K', K', G, H, j, j') \\ -S_-(0, K', K', G, H, j, j') \end{bmatrix} \frac{\partial \varrho_{Q'}^{K'}(Z, P, t; G)}{\partial P} \\
&\left. + \epsilon_{Q'}^{K'}(j, j')\frac{\hbar^2 k_{jj'}^2}{2} \begin{bmatrix} S_+(0, K', K', G, H, j, j') \\ +S_-(0, K', K', G, H, j, j') \end{bmatrix} \frac{\partial^2 \varrho_{Q'}^{K'}(Z, P, t; G)}{\partial P^2} \right\},
\end{aligned} \tag{18.82}$$

where equations (18.71) and (18.72) were used. Note that there is no zeroth-order term in $\hbar$ on the right-hand side of the equation, since the total population is constant if recoil is neglected.

The next step is to find equations for the remaining elements $\varrho_Q^K(Z, P, t; G)$ for $K \neq 0$. We need keep terms only of order $\hbar$ in these equations, since higher order terms lead to terms of order $\hbar^3 k^3$ in the equation for $\varrho_0^0(Z, P, t; G)$. Moreover, we can solve for these terms in quasi-steady-state, since they adiabatically follow the total ground population, once any initial transients have died away. Thus, for $K \neq 0$, we expand equation (18.81) in steady state to first order in $\hbar$ to arrive at

$$\begin{aligned}
v\frac{\partial \varrho_Q^K(Z, P, t; G)}{\partial Z} &= (-1)^{Q'} \begin{pmatrix} K & K' & \bar{K} \\ Q & -Q' & \bar{Q} \end{pmatrix} e^{ik_{jj'}Z} \\
&\times \left\{ T(K, K', \tilde{K}, \bar{K}, \bar{K}', G, H, j, j') \int d\Omega\, \lambda(\bar{K}, \bar{K}', \tilde{K}, \tilde{Q}, j, j'; \Omega) \right. \\
&\times \left[\varrho_{Q'}^{K'}(Z, P, t; G) + \hbar(k\cos\theta - k_j/2 - k_{j'}/2)\partial \varrho_{Q'}^{K'}(Z, P, t; G)/\partial P \right] \\
&+ \epsilon_{\tilde{Q}}^{\tilde{K}}(j, j') \begin{bmatrix} S_+(K, K', \tilde{K}, G, H, j, j') \\ +S_-(K, K', \tilde{K}, G, H, j, j') \end{bmatrix} \varrho_{Q'}^{K'}(Z, P, t; G) \\
&\left. + \epsilon_{\tilde{Q}}^{\tilde{K}}(j, j')\hbar k_{jj'} \begin{bmatrix} S_+(K, K', \tilde{K}, G, H, j, j') \\ -S_-(K, K', \tilde{K}, G, H, j, j') \end{bmatrix} \partial \varrho_{Q'}^{K'}(Z, P, t; G)/\partial P \right\}.
\end{aligned} \tag{18.83}$$

We now have all the ingredients in place to calculate the diffusion coefficient.

18.5.2 $G = 1/2$; lin⊥lin Polarization

For $G = 1/2$ and lin⊥lin polarization (17.50), only ϱ_0^0 and ϱ_0^1 contribute. Equations (18.82) and (18.83) reduce to

$$\frac{\partial \varrho_0^0(Z,P,t)}{\partial t} + v\frac{\partial \varrho_0^0(Z,P,t)}{\partial Z} = \Gamma_H^{\perp}\left\{ -A_H \frac{\hbar k \delta}{\gamma_H}\cos\xi \frac{\partial \varrho_0^1(Z,P,t)}{\partial P} + \frac{\hbar^2 k^2}{2}\left[C_H \frac{\partial^2 \varrho_0^0(Z,P,t)}{\partial P^2} - D_H \sin\xi \frac{\partial^2 \varrho_0^1(Z,P,t)}{\partial P^2}\right]\right\} \tag{18.84}$$

and

$$v\frac{\partial \varrho_0^1(Z,P,t)}{\partial Z} = \Gamma_H^{\perp}\left[-\varrho_0^1(Z,P,t) - \sin\xi\, \varrho_0^0(Z,P,t) - E_H \frac{\hbar k \delta}{\gamma_H}\cos\xi \frac{\partial \varrho_0^0(Z,P,t)}{\partial P}\right], \tag{18.85}$$

where

$$\xi = 2kZ, \tag{18.86}$$

$$\begin{aligned} &A_{1/2} = 3,\ C_{1/2} = 2,\quad D_{1/2} = 1,\ E_{1/2} = 3, \\ &A_{3/2} = -3,\ C_{3/2} = 41/10,\quad D_{3/2} = -4/5,\ E_{3/2} = -3, \end{aligned} \tag{18.87}$$

and it is understood that $G = 1/2$. The first term in equation (18.84) gives rise to the friction force. It also gives rise to a contribution to the diffusion coefficient, D_1, that results from fluctuations in the spatially modulated ground state polarization. The last term in equation (18.84) gives rise to a contribution to the diffusion coefficient, D_0, that results from fluctuations in the way the incident fields are scattered from the (average) ground-state population and coherence. These fluctuations result from both the "out" terms, reflecting a difference in the scattering rates for the two fields and the "in" terms, reflecting the diffusion resulting from scattering into unoccupied vacuum field modes.

To solve these equations, we make an approximation akin to the rotating-wave approximation. We neglect any rapid spatial variations (e.g., $\sin 2\xi$ variations) in $\varrho_0^0(Z,P,t)$ and assume that $\varrho_0^0(Z,P,t) \approx \varrho_0^0(P,t)$, independent of Z. This approximation is based on the neglect of particle localization in the optical potentials. With this assumption, we find that the periodic solution of equation (18.85) is

$$\begin{aligned} \varrho_0^1(Z,P,t) &\approx -\frac{1}{\alpha_H}\int_{-\infty}^{\xi} d\xi' e^{-(\xi-\xi')/\alpha_H} \left[\begin{array}{c} \sin\xi'\, \varrho_0^0(P,t) \\ +E_H \frac{\hbar k\delta}{\gamma_H}\cos\xi' \frac{\partial \varrho_0^0(P,t)}{\partial P} \end{array}\right] \\ &= \frac{-1}{1+\alpha_H^2}\left[\begin{array}{c} (\sin\xi - \alpha_H\cos\xi)\, \varrho_0^0(P,t) \\ +E_H \frac{\hbar k\delta}{\gamma_H}(\cos\xi + \alpha_H \sin\xi)\frac{\partial \varrho_0^0(P,t)}{\partial P} \end{array}\right], \end{aligned} \tag{18.88}$$

where

$$\alpha_H = \frac{2kv}{\Gamma_H^\perp} = \frac{2kP}{M\Gamma_H^\perp}, \tag{18.89}$$

and the $\Gamma_H^\perp$ are given by equations (18.38) and (18.33).

We substitute equation (18.88) into equation (18.84), replace $\sin^2\xi$ and $\cos^2\xi$ by 1/2, and neglect rapidly varying terms—that is, terms varying as $\sin 2\xi$ or $\cos 2\xi$. In this way, we arrive at the spatially averaged Fokker-Planck equation

$$\frac{\partial \varrho_0^0(P,t)}{\partial t} = \frac{\partial}{\partial P}\left\{ -\langle F(H,P)\rangle\, \varrho_0^0(P,t) + \langle D_1(H,P)\rangle \frac{\partial \varrho_0^0(P,t)}{\partial P} + \frac{\partial\left[\langle D_0(H,P)\rangle\, \varrho_0^0(P,t)\right]}{\partial P} \right\}, \tag{18.90}$$

where the spatially averaged friction force is

$$\langle F(H,P)\rangle = A_H \frac{\hbar k^2 v\delta}{\gamma_H} \frac{1}{1+\alpha_H^2}, \tag{18.91}$$

and the spatially averaged diffusion coefficients are

$$\langle D_0(H,P)\rangle = \frac{\hbar^2 k^2 \Gamma_H^\perp}{2}\left(C_H + \frac{1}{2}\frac{D_H}{1+\alpha_H^2}\right) \tag{18.92}$$

and

$$\langle D_1(H,P)\rangle = \frac{A_H E_H}{2}\left(\frac{\hbar k\delta}{\gamma_H}\right)^2 \frac{\Gamma_H^\perp}{1+\alpha_H^2}. \tag{18.93}$$

For $H = 1/2$,

$$\langle F(1/2,P)\rangle = \frac{3\hbar k^2 v\delta}{\gamma_{1/2}} \frac{1}{1+\alpha_{1/2}^2}, \tag{18.94a}$$

$$\langle D_0(1/2,P)\rangle = \hbar^2 k^2 \Gamma_{1/2}^\perp \left(1 + \frac{1}{4}\frac{1}{1+\alpha_{1/2}^2}\right), \tag{18.94b}$$

$$\langle D_1(1/2,P)\rangle = \frac{9}{2}\left(\frac{\hbar k\delta}{\gamma_{1/2}}\right)^2 \frac{\Gamma_H^\perp}{1+\alpha_{1/2}^2}, \tag{18.94c}$$

while for $H = 3/2$,

$$\langle F(3/2,P)\rangle = -\frac{3\hbar k^2 v\delta}{\gamma_{3/2}} \frac{1}{1+\alpha_{3/2}^2}, \tag{18.95a}$$

$$\langle D_0(3/2,P)\rangle = \frac{41\hbar^2 k^2 \Gamma_{3/2}^\perp}{20}\left(1 - \frac{4}{41}\frac{1}{1+\alpha_{3/2}^2}\right), \tag{18.95b}$$

$$\langle D_1(3/2,P)\rangle = \frac{9}{2}\left(\frac{\hbar k\delta}{\gamma_{3/2}}\right)^2 \frac{\Gamma_{3/2}^\perp}{1+\alpha_{3/2}^2}. \tag{18.95c}$$

18.5.3 $G=1$ to $H=2$ Transition; $\sigma_+ - \sigma_-$ Polarization

For $G=1$, $H=2$, and $\sigma_+ - \sigma_-$ polarization, only ϱ_0^0 and

$$\varrho_0^1 = A_0^1, \quad \varrho_0^2 = A_0^2, \quad \varrho_{\pm 2}^2 = A_{\pm 2}^2 e^{\pm i\xi} \tag{18.96}$$

contribute. We have suppressed the explicit dependence on P and t in these variables. With $\gamma_{GH} \equiv \gamma = \gamma_H/2$, equations (18.82) and (18.83) reduce to

$$\begin{aligned}\frac{\partial \varrho_0^0}{\partial t} &= \frac{\Gamma_c'}{\sqrt{3}} \left\{ \hbar k \frac{\partial}{\partial P} [-\sqrt{2} A_0^1 + (i\delta/5\gamma)\left(A_2^2 - A_{-2}^2\right)] \right. \\ &\left. + \frac{\hbar^2 k^2}{450} \frac{\partial^2}{\partial P^2} [407\sqrt{3}\varrho_0^0 + 28\sqrt{6} A_0^2 - 33\left(A_2^2 + A_{-2}^2\right)] \right\}\end{aligned} \tag{18.97}$$

for the ground-state population density and to the quasi-steady-state *matrix* equations

$$\mathbf{M}_1 \mathbf{A} = \mathbf{M}_2 \partial \mathbf{A}/\partial P + \boldsymbol{\lambda}^{(0)} + \hbar k \boldsymbol{\lambda}^{(1)}, \tag{18.98}$$

for the other elements, where

$$\mathbf{M}_1 = \begin{pmatrix} \frac{1}{5} & 0 & \frac{i\delta}{5\sqrt{2}\gamma} & \frac{-i\delta}{5\sqrt{2}\gamma} \\ 0 & \frac{11}{15} & \frac{\sqrt{6}}{15} & \frac{\sqrt{6}}{15} \\ \frac{i\delta}{5\sqrt{2}\gamma} & \frac{\sqrt{6}}{15} & 1+iy & 0 \\ \frac{-i\delta}{5\sqrt{2}\gamma} & \frac{\sqrt{6}}{15} & 0 & 1-iy \end{pmatrix}, \tag{18.99}$$

$$\mathbf{M}_2 = \begin{pmatrix} 0 & -\frac{\sqrt{3}}{5} & \frac{1}{5\sqrt{2}} & \frac{1}{5\sqrt{2}} \\ \frac{-7}{5\sqrt{3}} & 0 & \frac{i\delta}{5\sqrt{6}\gamma} & \frac{-i\delta}{5\sqrt{6}\gamma} \\ \frac{-1}{5\sqrt{2}} & \frac{-i\delta}{5\sqrt{6}\gamma} & 0 & 0 \\ \frac{-1}{5\sqrt{2}} & \frac{i\delta}{5\sqrt{2}\gamma} & 0 & 0 \end{pmatrix}, \tag{18.100}$$

and

$$\mathbf{A} = \begin{pmatrix} A_0^1 \\ A_0^2 \\ A_2^2 \\ A_{-2}^2 \end{pmatrix}, \quad \boldsymbol{\lambda}^{(0)} = \varrho_0^0 \begin{pmatrix} 0 \\ \frac{1}{3\sqrt{2}} \\ \frac{1}{2\sqrt{3}} \\ \frac{1}{2\sqrt{3}} \end{pmatrix}, \quad \boldsymbol{\lambda}^{(1)} = \frac{\partial \varrho_0^0}{\partial P} \begin{pmatrix} -\sqrt{\frac{3}{2}} \\ 0 \\ -\frac{i\delta}{5\sqrt{3}\gamma} \\ \frac{i\delta}{5\sqrt{3}\gamma} \end{pmatrix}, \tag{18.101}$$

with

$$y = \frac{2kv}{\Gamma_c'} = \frac{2kP}{M\Gamma_c'} \tag{18.102}$$

and Γ_c' given by equation (18.50).

The solution of equation (18.98), written in the form

$$\mathbf{A} = \mathbf{A}^{(0)} + \hbar k \mathbf{A}^{(1)}, \tag{18.103}$$

is

$$\mathbf{A}^{(0)} \equiv \mathbf{A}[0] = \mathbf{M}_1^{-1}\boldsymbol{\lambda}^{(0)} \tag{18.104}$$

and

$$\mathbf{A}^{(1)} = \mathbf{M}_1^{-1}\boldsymbol{\lambda}^{(1)} + \mathbf{M}_1^{-1}\mathbf{M}_2\mathbf{M}_1^{-1}\frac{\partial\boldsymbol{\lambda}^{(0)}}{\partial P} + \mathbf{M}_1^{-1}\mathbf{M}_2\frac{\partial\mathbf{M}_1^{-1}}{\partial P}\boldsymbol{\lambda}^{(0)}. \tag{18.105}$$

The solution for $\mathbf{A}^{(0)}$ is given by equations (18.53), multiplied by $\sqrt{3}\varrho_0^0(Z, P, t)$. The three terms in equation (18.105) are designated as

$$\mathbf{A}[1] = \mathbf{M}_1^{-1}\boldsymbol{\lambda}^{(1)}, \tag{18.106a}$$

$$\mathbf{A}[2] = \mathbf{M}_1^{-1}\mathbf{M}_2\mathbf{M}_1^{-1}\frac{\partial\boldsymbol{\lambda}^{(0)}}{\partial P}, \tag{18.106b}$$

$$\mathbf{A}[3] = \mathbf{M}_1^{-1}\mathbf{M}_2\frac{\partial\mathbf{M}_1^{-1}}{\partial P}\boldsymbol{\lambda}^{(0)}. \tag{18.106c}$$

There is considerable algebra required to evaluate equations (18.104) and (18.105), but they are handled easily by a computer program. After solving for $\mathbf{A}^{(0)}$ and $\mathbf{A}^{(1)}$, we substitute these values into equation (18.97) and write that equation in the form (18.90). The friction force is given by

$$\begin{aligned} F(y) &= -\frac{\hbar k\Gamma_c'}{\sqrt{3}}[-\sqrt{2}A_0^1[0] + (i\delta/5\gamma)\left(A_2^2[0] - A_{-2}^2[0]\right)] \\ &= -\frac{30}{17}\hbar k^2 v\frac{\gamma_H\delta}{\delta^2 + 5\gamma^2 + c_1 y^2\gamma^2}, \end{aligned} \tag{18.107}$$

where y is defined in equation (18.102),

$$c_1 = 275/51, \tag{18.108}$$

and $\gamma = \gamma_{GH} = \gamma_H/2$. This result agrees with equation (18.54). There is an *additional* contribution to the force arising from the last term in equation (18.105) that can be included; however, this term is small for the range of parameters for which the theory is valid (see problem 10). The diffusion coefficients are given by

$$\begin{aligned} D_0(y) &= \frac{\hbar^2k^2\Gamma_c'}{450}\left[407\sqrt{3} + 28\sqrt{2}A_0^2[0]/\varrho_0^0 - 11\sqrt{3}\left(A_2^2[0] + A_{-2}^2[0]\right)/\varrho_0^0\right] \\ &= \frac{383}{425}\hbar^2k^2\Gamma_c'\frac{\delta^2 + 5\gamma^2 + c_2 y^2\gamma^2}{\delta^2 + 5\gamma^2 + c_1 y^2\gamma^2}, \end{aligned} \tag{18.109}$$

with

$$c_2 = 4275/766, \tag{18.110}$$

and

$$D_1(y) = D_a(y) + D_b(y), \tag{18.111}$$

with

$$D_a(y) = \frac{\hbar^2 k^2 \Gamma_c'}{\sqrt{3}}[-\sqrt{2}A_0^1[1] + (i\delta/5\gamma)\left(A_2^2[1] - A_{-2}^2[1]\right)]\left(\frac{\partial \varrho_0^0}{\partial P}\right)^{-1}$$
$$= \frac{\hbar^2 k^2 \Gamma_c'}{5}\frac{25\left(5\gamma^2 + c_1 y^2\gamma^2\right) - \delta^2}{\delta^2 + 5\gamma^2 + c_1 y^2\gamma^2}; \tag{18.112a}$$

$$D_b(y) = \frac{\hbar^2 k^2 \Gamma_c'}{\sqrt{3}}[-\sqrt{2}A_0^1[2] + (i\delta/5\gamma)\left\{A_2^2[2] - A_{-2}^2[2]\right\}]\left(\frac{\partial \varrho_0^0}{\partial P}\right)^{-1}$$
$$= \frac{\hbar^2 k^2 \Gamma_c'}{2601}\frac{34375 y^4\gamma^4 + 75 y^2\gamma^2\left(425\gamma^2 - 13\delta^2\right) + 306\delta^2\left(\delta^2 + 5\gamma^2\right)}{\left(\delta^2 + 5\gamma^2 + c_1 y^2\gamma^2\right)^2}. \tag{18.112b}$$

Near zero velocity ($y = 0$), we can write a single diffusion coefficient [1, 8],

$$D = D_0 + D_1 = \hbar^2 k^2 \Gamma_c'\left(\frac{348}{425} + \frac{432}{17}\frac{\gamma^2}{\delta^2 + 5\gamma^2}\right). \tag{18.113}$$

18.5.4 Equilibrium Energy

Having calculated the friction and diffusion coefficients, we can estimate the equilibrium energy starting from the Fokker-Planck equation (18.90). In steady state, the solution of that equation for the ground-state population density $\psi(P) = (2G+1)^{1/2}\varrho_0^0(P;G)$ is

$$-\langle F(P)\rangle\,\psi(P) + \langle D_1(P)\rangle\frac{\partial\psi(P)}{\partial P} + \frac{\partial\left[\langle D_0(P)\rangle\,\psi(P)\right]}{\partial P} = \text{constant} = 0, \tag{18.114}$$

where we take the constant equal to zero to insure that $\psi(0)$ is finite. Once we calculate $\psi(P)$, we can find the equilibrium energy by evaluating the average value of $P^2/2M$. In terms of dimensionless variables, equation (18.114) can be written as

$$\left[-\tilde{\delta}\beta\tilde{F}(\beta) + \frac{\tilde{\Gamma}'}{2}\frac{\partial\tilde{D}_0(\beta)}{\partial\beta}\right]\psi(\beta) + \frac{\tilde{\Gamma}'}{2}\tilde{D}(\beta)\frac{\partial\psi(\beta)}{\partial\beta} = 0, \tag{18.115}$$

where

$$\tilde{F}(\beta) = \frac{M\,\langle F(\beta)\rangle}{\hbar^2 k^3\tilde{\delta}\beta}, \tag{18.116a}$$

$$\tilde{D}_0(\beta) = \frac{\langle D_0(\beta)\rangle}{\hbar^2 k^2\Gamma'}, \tag{18.116b}$$

$$\tilde{D}(\beta) = \frac{\left[\langle D_1(\beta)\rangle + \langle D_0(\beta)\rangle\right]}{\hbar^2 k^2\Gamma'}, \tag{18.116c}$$

$$\beta = \frac{P}{\hbar k} = \frac{Mv}{\hbar k}, \qquad \beta^2 = \frac{Mv^2/2}{\hbar^2 k^2/2M} = \frac{E}{E_R}, \tag{18.117}$$

$$\tilde{\delta} = \delta/\gamma_H, \tag{18.118}$$

$$\tilde{\Gamma}' = \frac{\Gamma'}{\hbar k^2/2M}, \tag{18.119}$$

and Γ' is the optical pumping rate appropriate to a given level scheme ($\Gamma_H^{\perp}$ or Γ_c). The $(\tilde{\Gamma}'/2)(\partial \tilde{D}_0/\partial\beta)$ term in equation (18.115) acts as an effective force, but usually can be neglected (see problem 9). The solution of equation (18.115) with the neglect of this term is

$$\psi(\beta) = \psi(0)\exp\left\{2\tilde{\delta}/\tilde{\Gamma}' \int_0^{\beta} \beta' d\beta' \left[\tilde{F}(\beta')/\tilde{D}(\beta')\right]\right\}. \tag{18.120}$$

To proceed further, we must put in the specific forms for the friction and diffusion coefficients.

18.5.4.1 lin⊥lin polarization

In this case, from equations (18.116), (18.91), (18.93), and (18.92), we have

$$\tilde{F}(\beta) = \tilde{F}_H(\beta) = \frac{A_H}{1 + 16\beta^2/\tilde{\Gamma}_H^{\perp 2}}, \tag{18.121}$$

$$\tilde{D}(\beta) = \frac{C_H}{2} + \frac{D_H/2 + A_H E_H \tilde{\delta}^2}{2\left(1 + 16\beta^2/\tilde{\Gamma}_H^{\perp 2}\right)}. \tag{18.122}$$

Setting

$$x = 4\beta/\tilde{\Gamma}_H^{\perp}, \tag{18.123}$$

we can write equation (18.120) as

$$\psi(x) = \psi(0)\exp\left[-a_H \int_0^{x} x' dx' \frac{A_H}{J_H + (1 + x'^2)\, G_H}\right] \tag{18.124a}$$

$$= \psi(0)\exp\left[-\frac{a_H A_H}{2G_H}\ln\left(1 + \frac{G_H x^2}{G_H + J_H}\right)\right]$$

$$= \psi(0)\left(1 + \frac{G_H x^2}{G_H + J_H}\right)^{-\frac{a_H A_H}{2G_H}}, \tag{18.124b}$$

where

$$G_H = C_H/2, \tag{18.125}$$

$$J_H = \left(D_H/2 + A_H E_H \tilde{\delta}^2\right)/2, \tag{18.126}$$

$$a_H = -\tilde{\delta}\tilde{\Gamma}_H^{\perp}/8, \tag{18.127}$$

and A_H, C_H, D_H, and E_H are defined in equation (8.87). (Note that $a_H A_H > 0$. If $H = 1/2$, $A_H = 3$, but $\delta < 0$ for cooling; if $H = 3/2$, $A_H = -3$, and $\delta > 0$ for cooling.)

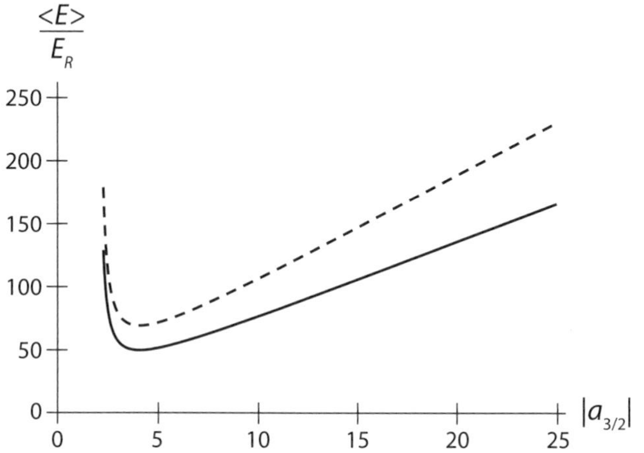

Figure 18.4. Graph of $\langle E\rangle/E_R$ as a function of $|a_{3/2}|$ when $H = 3/2$ and $\tilde{\delta} = 5$ (solid line) and $\tilde{\delta} = 1$ (dashed line).

The average equilibrium energy, in units of the recoil energy $E_R = \hbar^2k^2/2M$, is given by

$$\frac{\langle E\rangle}{E_R} = \frac{\int_{-\infty}^{\infty}\psi(\beta)\beta^2 d\beta}{\int_{-\infty}^{\infty}\psi(\beta)d\beta} = \frac{\tilde{\Gamma}_H^{\perp 2}}{16}\frac{\int_{-\infty}^{\infty}\psi(x)x^2 dx}{\int_{-\infty}^{\infty}\psi(x)dx}. \tag{18.128}$$

The integrals are tabulated, enabling us to obtain

$$\begin{aligned}\frac{\langle E\rangle}{E_R} &= \frac{\tilde{\Gamma}_H^{\perp 2}}{16}\frac{G_H + J_H}{A_H a_H - 3G_H} \\ &= \frac{4a_H^2}{\tilde{\delta}^2}\frac{G_H + J_H}{A_H a_H - 3G_H}.\end{aligned} \tag{18.129}$$

Since a_H is proportional to the field intensity, the average energy decreases linearly with decreasing intensity for $A_H a_H \gg 3G_H$, reaches a minimum, and then increases as the field intensity is lowered even more, owing to spontaneous contributions to the diffusion coefficient that dominate at these low field intensities. A graph of $\langle E\rangle/E_R$ as a function of $|a_{3/2}|$ is shown in figure 18.4 for $\tilde{\delta} = 1$ and 5.

The average energy actually diverges when $A_H a_H = 3G_H$, but the calculation is not valid for such field intensities. The minimum energy occurs for $a_H A_H = 6G_H$, such that

$$\begin{aligned}\left.\frac{\langle E\rangle}{E_R}\right)_{\min} &= 48G_H\frac{G_H + J_H}{A_H^2\tilde{\delta}^2} \\ &= 12C_H\frac{C_H + \left(D_H/2 + A_H E_H\tilde{\delta}^2\right)}{A_H^2\tilde{\delta}^2}.\end{aligned} \tag{18.130}$$

For $H = 1/2$,

$$\left.\frac{\langle E\rangle}{E_R}\right)_{\min} = 24\left[1 + 5/\left(18\tilde{\delta}^2\right)\right], \tag{18.131}$$

and for $H = 3/2$,

$$\left.\frac{\langle E\rangle}{E_R}\right)_{\min} = \frac{246}{5}\left[1 + 37/\left(90\tilde{\delta}^2\right)\right]. \tag{18.132}$$

The minimum average energy is roughly 20 to 50 recoil energies, justifying our approach in which it was assumed that $E/E_R \gg 1$. On the other hand, the ratio of the particle energy to the optical potential depth U,

$$\left[U(H=1/2) = 2\hbar|\bar{\chi}_{HG}|^2/\delta\,,\ U(H=3/2) = \hbar|\bar{\chi}_{HG}|^2/\delta\right],$$

is of order unity, implying that spatial modulation of the density may play a role. Nevertheless, the solution (18.129) is in good agreement with the fully quantum-mechanical result [5]. New features also enter when cooling in two and three dimensions is considered [16]–[18].

18.5.4.2 $\sigma_+ - \sigma_-$ polarization

In this case, the expressions for the diffusion coefficient are much more complicated. Following a procedure identical to crossed polarization and again neglecting the $\partial\tilde{D}_0/\partial\beta$ term, we arrive at

$$\begin{aligned}\frac{\langle E\rangle}{E_R} &= \frac{\int_{-\infty}^{\infty}\psi(\beta)\beta^2 d\beta}{\int_{-\infty}^{\infty}\psi(\beta)d\beta} = \frac{\tilde{\Gamma}_c'^2}{16c_1 g}\frac{\int_{-\infty}^{\infty}\psi(x)x^2dx}{\int_{-\infty}^{\infty}\psi(x)dx}\\ &= \frac{4c_1a_c^2 g}{\tilde{\delta}^2}\frac{\int_{-\infty}^{\infty}\psi(x)x^2dx}{\int_{-\infty}^{\infty}\psi(x)dx},\end{aligned} \tag{18.133}$$

where

$$\psi(x) = \psi(0)\exp\left[\frac{a_c}{2}\int_0^{x^2}dz'\frac{\tilde{F}(z')}{\tilde{D}(z')}\right], \tag{18.134}$$

$$\tilde{F}(z) = -\frac{120}{17}g\frac{1}{1+z}, \tag{18.135}$$

$$\begin{aligned}\tilde{D}(z) &= \frac{383}{425} + 5 + \frac{\frac{383}{425}\left(\frac{c_2}{c_1}-1\right)z}{1+z} - \frac{104}{5}\frac{g\tilde{\delta}^2}{1+z}\\ &+ \frac{\frac{8g\tilde{\delta}^2}{17} + \frac{g}{187}\left(425 - 52\tilde{\delta}^2\right)z + \frac{5}{11}z^2}{(1+z)^2},\end{aligned} \tag{18.136}$$

$$g = \frac{1}{5+4\tilde{\delta}^2}, \tag{18.137}$$

$$x^2 = z = \frac{16c_1 g\beta^2}{\tilde{\Gamma}_c'^2} = c_1 g y^2, \tag{18.138}$$

and

$$a_c = \frac{\tilde{\delta}\tilde{\Gamma}'_c}{8c_1 g}. \tag{18.139}$$

The integrals must now be done numerically. One arrives at a behavior qualitatively similar to the lin⊥lin case.

In both the lin⊥lin and corkscrew polarization cases considered, the velocity distribution is not Gaussian and the half width can be considerably smaller than $\langle v^2\rangle^{1/2}$. Examples are left to the problems.

Problems

1–2. Write a computer program for equation (18.24)—that is, write a program that gives these differential equations once the field polarizations are specified. Using your program, verify that equations (18.32) and (18.49) are correct.

3. Calculate the friction force for the lin⊥lin geometry and a $G = 1/2$, $H = 1/2$ transition. Show that the detuning must be negative (to the blue) for the force to be a friction force.

4. In the model of Sisyphus cooling, calculate the light shifts for a $G = 1/2$ ground state and both $H = 1/2, 3/2$ excited states.

5. In the model of Sisyphus cooling, show that the $v = 0$ atoms are localized near the nodes of the field for a $G = 1/2$ ground state and an $H = 1/2$ excited state.

6. Using the light shifts and steady-state ground-state populations in Sisyphus cooling for a $G = 1/2$ ground state and an $H = 1/2$ excited state, calculate the force on atoms having $v = 0$ as a function of Z and show that the average force vanishes.

7. Continue problem 6 by calculating the force for atoms moving with velocity v. To do this, use the steady-state values for the ground-state population difference (phase shifted by $\pi/4$) that can be obtained using equations (18.34) and (18.35), and show that the result agrees with equation (18.37).

8. Write equations for the state amplitudes for the three levels in coherent population trapping, using a Hamiltonian in which the center-of-mass motion of the atoms is quantized,

$$H = \frac{P^2}{2M} + \hbar\omega_0 \left|e_0\right\rangle\left\langle e_0\right| + \hbar\chi(\left|e_0\right\rangle\left\langle g_-\right| e^{i\mathbf{k}\cdot\mathbf{R}-i\omega_0 t} + \left|e_0\right\rangle\left\langle g_+\right| e^{-i\mathbf{k}\cdot\mathbf{R}-i\omega_0 t} + \text{adj.}),$$

where $\mathbf{k} = k\hat{\mathbf{z}}$. The energies of states $|g_\pm\rangle$ are taken to be zero and the fields are resonant with the transitions. Show that the families of states $|g_-, P - \hbar k\rangle$, $|g_+, P + \hbar k\rangle$, $|e_0, P\rangle$ are coupled by the fields and that $|\Psi\rangle = \frac{1}{\sqrt{2}}(|g_-, P - \hbar k\rangle - |g_+, P + \hbar k\rangle)$, while not coupled directly to state $|e_0, P\rangle$ by the fields, is a dark state only for $P = 0$.

9. Show that the $(\tilde{\Gamma}'/2)(\partial\tilde{D}_0/\partial\beta)$ term in equation (18.115) leads to additional force terms such that

$$\tilde{F} \rightarrow \tilde{F} - \frac{8}{\tilde{\delta}y\tilde{\Gamma}'}\frac{\partial\tilde{D}_0}{\partial y},$$

where $y = 2kv/\Gamma' = 4\beta/\tilde{\Gamma}'$. As a consequence, show that the exponentials in equations (18.124a) and (18.134) are replaced by

$$\exp\left\{-\int_0^x x'dx' \frac{a_H A_H + D_H/\left[2(1+x'^2)\right]}{J_H + \left(1+x'^2\right) G_H}\right\}$$

and

$$\exp\left[\frac{1}{2}\int_0^{x^2} dz' \frac{a_c \tilde{F}(z') - \frac{59}{935}\frac{1}{(1+z')^2}}{\tilde{D}(z')}\right],$$

respectively. Note that the new terms are down by order a_H or a_c from the leading terms and generally can be neglected for values of a_H or a_c where the theory is valid. Of course, since the corrections appear in *exponents*, to be neglected, the integrals of these terms must be much less than unity over the range of a_H or a_c for which the distribution function is nonvanishing.

10. Show that the third term in equation (18.105) leads to an additional force term of the form

$$F_1 = -\frac{\hbar k \Gamma'_c}{\sqrt{3}}\left[-\sqrt{2}A_0^1[3] + (i\delta/5\gamma)\left\{A_2^2[3] - A_{-2}^2[3]\right\}\right]\left(\varrho_0^0\right)^{-1}$$
$$= \frac{2\hbar^2 k^3}{M}\frac{250y}{44217}\frac{21675(1+c_1y^2/5)\gamma^6 + 5(663+1100y^2)\gamma^4\delta^2 - 204\gamma^2\delta^4}{\left(\delta^2+5\gamma^2+c_1y^2\gamma^2\right)^3}$$

and that this changes the exponential in equation (18.134) to

$$\exp\left[\frac{1}{2}\int_0^{x^2} dz' \frac{a_c \tilde{F}(z') - \tilde{T}(z')}{\tilde{D}(z')}\right],$$

where

$$\tilde{T}(z) = \frac{250}{44217}\frac{g^2}{c_1}\frac{21675\left(1+\frac{z}{5g}\right) + 20\left(663 + 1100\frac{z}{c_1 g}\right)\tilde{\delta}^2 - 3264\tilde{\delta}^4}{(1+z)^3}.$$

Again, this is down by order a_c from the leading term and generally can be neglected for values of a_c where the theory is valid.

11. Graph equation (18.129) as a function of a_H for $H = 1/2$, $\tilde{\delta} = -5$, and $\tilde{\delta} = -1$. Numerically obtain a graph of $\langle E\rangle/E_R$ as a function of a_c from equation (18.133) for $\tilde{\delta} = 5$ and $\tilde{\delta} = 1$.
12. Graph the distribution given by equation (18.124b) as a function of x for $H = 1/2$ and $\tilde{\delta} = -5$ and both $a_H = 4$ and $a_H = 1.5$. Write an expression for a Gaussian having the same value of $\langle x^2\rangle$, and graph both the actual and equivalent thermal (Gaussian) distributions. Compare the two graphs, and show that width of the ψ distribution is smaller than that of the Gaussian—in other words, mean energy is larger than the half width of the distribution.
13. Repeat problem 12 for corkscrew polarization and $\tilde{\delta} = 5$ for both $a_c = 800$, 400. Numerically find $\langle x^2\rangle$ that corresponds to these values, and graph the

actual distribution and the (unnormalized) Gaussian distribution having a mean square width equal to $\langle x^2 \rangle$. Compare the two graphs, and show that width of the ψ distribution is smaller than that of the Gaussian—in other words, mean energy is larger than the half width of the distribution.

References

[1] J. Dalibard and C. Cohen-Tannoudji, *Laser cooling below the Doppler limit by polarization gradients: simple theoretical models*, Journal of the Optical Society B **6**, 2023–2045 (1989).

[2] C. Cohen-Tannoudji, *New laser cooling mechanisms*, in *Laser Manipulation of Atoms and Ions*, Enrico Fermi Summer School Course CXVIII, edited by E. Arimondo, W. D. Phillips, and S. Strumia (North-Holland, Amsterdam, 1992), pp. 99–169.

[3] V. Finkelstein, P. R. Berman, and J. Guo, *One-dimensional laser cooling below the Doppler limit*, Physical Review A **45**, 1829–1842 (1992).

[4] A. Aspect, E. Arimondo, R. Kaiser, N. Vansteenkiste, and C. Cohen-Tannoudji, *Laser cooling below the one-photon recoil energy by velocity selective coherent population trapping: theoretical analysis*, Journal of the Optical Society B **6**, 2112–2124 (1989).

[5] Y. Castin, J. Dalibard, and C. Cohen-Tannoudji, *The limits of Sisyphus cooling*, in *Light-Induced Kinetic Effects on Atoms, Ions, and Molecules*, edited by L. Moi, S. Gozzini, C. Gabbanini, E. Arimondo, and F. Strumia (ETS Editrice, Pisa, 1991), pp. 5–24; J. Dalibard and Y. Castin, *Laser cooling from the semi-classical to the quantum regime*, in *Frontiers in Laser Spectroscopy*, Enrico Fermi Summer School Course CXX, edited by T. W. Hänsch and M. Inguscio (North-Holland, Amsterdam, 1994) pp. 445–476.

[6] P. R. Berman, *Nonlinear spectroscopy and laser cooling*, in *Laser Spectroscopy*, edited by L. Ducloy, E. Giacobino, and G. Gamy (World Scientific, Singapore, 1992), pp. 15–20.

[7] C. Salomon, J. Dalibard, W. D. Phillips, A. Clairon, and S. Guellati, *Laser cooling of cesium atoms below 3 microkelvins*, Europhysics Letters **12**, 683–688 (1990).

[8] Y. Castin and K. Mølmer, *Atomic momentum diffusion in a $\sigma_+ - \sigma_-$ laser configuration: influence of internal level structure*, Journal of Physics B **23**, 4101–4110 (1990).

[9] G. Grynberg and J.-Y. Courtois, *Proposal for a magneto-optic lattice for trapping atoms in nearly dark states*, Europhysics Letters **27**, 41–46 (1994).

[10] A. Hemmerich, M. Weidemüller, T. Esslinger, G. Zimmermann, and T. Hänsch, *Trapping atoms in dark optical lattices*, Physical Review Letters **75**, 37–40 (1995).

[11] K. I. Petsas, J.-Y. Courtois, and G. Grynberg, *Temperature and magnetism of gray optical lattices*, Physical Review A **53**, 2533–2538 (1996).

[12] F. Mauri and E. Arimondo, *Coherent trapping subrecoil cooling in two dimensions aided by a force*, Europhysics Letters **16**, 717–722 (1991); M. A. Ol'shanii and V. G. Minogin, *Three-dimensional velocity selective coherent population trapping in resonant light fields*, in *Light-Induced Kinetic Effects on Atoms, Ions, and Molecules*, edited by L. Moi, S. Gozzini, C. Gabbanini, E. Arimondo, and F. Strumia (ETS Editrice, Pisa, 1991), pp. 99–110; J. Lawall, S. Kulin, B. Saubamea, N. Bigelow, M. Leduc, and C. Cohen-Tannoudji, *Three-dimensional laser cooling of helium beyond the single photon recoil limit*, Physical Review Letters **75**, 4194–4197 (1995).

[13] F. Bardou, J.-P. Bouchaud, A. Aspect, and C. Cohen-Tannoudji, *Levy Statistics and Laser Cooling* (Cambridge University Press, Cambridge, UK, 2001).
[14] H. Risken, *The Fokker-Planck Equation: Methods of Solutions and Applications*, 2nd ed. (Springer-Verlag, Berlin, 1989).
[15] B. Dubetsky and P. R. Berman, *Atom-field interactions: density matrix equations including quantization of the center-of-mass motion*, Physical Review A **53**, 390–399 (1996).
[16] V. Finkelstein, J. Guo, and P. R. Berman, *Two-dimensional forces and atomic motion in a sub-Doppler limit*, Physical Review A **46**, 7108–7122 (1992).
[17] K. Mølmer, *Friction and diffusion coefficients for cooling of atoms in laser fields with multidimensional periodicity*, Physical Review A **44**, 5820–5832 (1991); Y. Castin and K. Mølmer, *Monte-Carlo wave function analysis of 3D optical molasses,* Physical Review Letters **74**, 3772–3775 (1995).
[18] J. Javanainen, *Numerical experiments in semi-classical laser cooling theory of multi-state atoms*, Physical Review A **46**, 5819–5835 (1992).

Bibliography

H. J. Metcalf and P. van der Straten, *Laser Cooling and Trapping*, (Springer-Verlag, New York, 1999). This book is an excellent reference, covering many topics on laser cooling not discussed in this chapter. In addition, it contains an extensive list of references.

See also:

C. Cohen-Tannoudji, *Laser Manipulation of Atoms and Ions*, in Enrico Fermi Summer School Course CXVIII, edited by E. Arimondo, W. D. Phillips, and S. Strumia (North-Holland, Amsterdam, 1992), pp. 99–169.

J. Javanainen, *Cooling and trapping*, in *Springer Handbook of Atomic, Molecular, and Optical Physics*, edited by G. Drake (Springer, Würzburg, Germany, 2006), pp. 1091–1106.

Journal of the Optical Society B **6** (1989). Special issue on laser cooling.

K. Mølmer, *The optimum Fokker-Planck equation for laser cooling*, Journal of Physics B **27**, 1889–1898 (1994).

S. Stenholm, *The semiclassical theory of laser cooling*, Reviews of Modern Physics **58**, 699–739 (1986).

19

Operator Approach to Atom–Field Interactions: Source-Field Equation

For the most part, we have used the Schrödinger representation to this point. In this chapter, we consider the equations of motion for atomic and field operators in the Heisenberg representation. The atomic and field operators already contain an implicit trace over other state variables. For example, the atomic population difference operator $\boldsymbol{\sigma}_z$ contains a trace over field variables, while the number operator $\hat{n} = a^\dagger a$ involves a trace over all atomic variables. Thus, use of the Heisenberg representation can greatly simplify calculations when we are interested in field or atomic variables *only*. On the other hand, if one is interested in correlations between field and atomic states, the Schrödinger representation is often more useful, but computationally more cumbersome. As an application of the Heisenberg operator approach, we derive a source-field expression for the electric field operator that relates the properties of the field to those of the atomic sources that create the field. In this fashion, one arrives at an equation that closely mirrors its classical counterpart.

The Heisenberg representation was discussed in chapter 15. Of particular relevance to this chapter are the equations for the time development for the expectation values of operators. Recall that in the Schrödinger representation, the density matrix $\varrho^S(t)$ is a function of time, while operators such as A^S are time-independent. On the other hand, in the Heisenberg representation, the density matrix $\varrho^{\mathrm{H}} = \varrho^S(0)$ is time-independent, while operators such as $A^{\mathrm{H}}(t)$ are functions of time. The expectation values of operators in the two repesentations are related by

$$\left\langle A^S \right\rangle = \mathrm{Tr}\left[\varrho^S(t) A^S\right] = \left\langle A^{\mathrm{H}}(t) \right\rangle = \mathrm{Tr}\left[\varrho^S(0) A^{\mathrm{H}}(t)\right]. \tag{19.1}$$

19.1 Single Atom

19.1.1 Single-Mode Field

We consider first a single-mode field interacting with a single, stationary, two-level atom located at the origin. In the RWA, the Hamiltonian is given by equation (15.34),

$$H = \frac{\hbar\omega_0}{2}\sigma_z + \hbar\omega a^\dagger a + \hbar\left(g\sigma_+ a + g^* a^\dagger \sigma_-\right), \tag{19.2}$$

with

$$g = -i\mu_{21}\left(\frac{\omega}{2\hbar\epsilon_0\mathcal{V}}\right)^{1/2}. \tag{19.3}$$

We have dropped the boldface notation here for Heisenberg matrices since all operators in this chapter are matrices. The operators appearing in equation (19.2) can be taken to be either Schrödinger or Heisenberg operators. In this chapter, unless indicated otherwise, all operators are assumed to be time-dependent, Heisenberg operators, except for the density matrix, $\varrho^S(t)$. At time $t = 0$,

$$\sigma_+(0) = |2\rangle\langle 1|, \tag{19.4a}$$

$$\sigma_-(0) = |1\rangle\langle 2|, \tag{19.4b}$$

$$\sigma_1(0) = |1\rangle\langle 1|, \quad \sigma_2(0) = |2\rangle\langle 2|, \tag{19.4c}$$

$$\sigma_{ij}(0) = |i\rangle\langle j|, \tag{19.4d}$$

that is, the operators coincide with the Schrödinger operators. It is important to remember that these equations hold only at $t = 0$; in general, $\sigma_{ij}(t) \neq |i\rangle\langle j|$. On the other hand, *equal time* products of any Heisenberg operators are the same as their Schrödinger counterparts since

$$A(t)B(t) = e^{iHt}Ae^{-iHt}e^{iHt}Be^{-iHt} = e^{iHt}ABe^{-iHt}. \tag{19.5}$$

Thus, $\sigma_{ij}(t)\sigma_{kl}(t) = \sigma_{il}(t)\delta_{j,k}$.

The *equal time* commutators for the atomic and field operators are

$$[\sigma_1, \sigma_z] = [\sigma_2, \sigma_z] = 0, \tag{19.6a}$$

$$[\sigma_1, \sigma_+] = -\sigma_+, \quad [\sigma_1, \sigma_-] = \sigma_-, \tag{19.6b}$$

$$[\sigma_2, \sigma_+] = \sigma_+, \quad [\sigma_2, \sigma_-] = -\sigma_-, \tag{19.6c}$$

$$[\sigma_+, \sigma_z] = -2\sigma_+, \quad [\sigma_-, \sigma_z] = 2\sigma_-, \tag{19.6d}$$

$$[\sigma_+, \sigma_-] = \sigma_z, \tag{19.6e}$$

$$[a, a] = \left[a^\dagger, a^\dagger\right] = 0, \quad \left[a, a^\dagger\right] = 1, \tag{19.6f}$$

$$\left[\sigma_{ij}, a\right] = \left[\sigma_{ij}, a^\dagger\right] = 0. \tag{19.6g}$$

The time evolution of the operators, obtained from

$$\dot{\sigma} = \frac{1}{i\hbar}[\sigma, H], \tag{19.7}$$

is given by

$$\dot{\sigma}_1 = ig\sigma_+ a - ig^* a^\dagger \sigma_-, \tag{19.8a}$$

$$\dot{\sigma}_2 = -ig\sigma_+ a + ig^* a^\dagger \sigma_-, \tag{19.8b}$$

$$\dot{\sigma}_z = -2ig\sigma_+ a + 2ig^* a^\dagger \sigma_-, \tag{19.8c}$$

$$\dot{\sigma}_+ = i\omega_0 \sigma_+ a - ig^* a^\dagger \sigma_z, \tag{19.8d}$$

$$\dot{\sigma}_- = -i\omega_0 \sigma_- + ig\sigma_z a, \tag{19.8e}$$

$$\dot{a} = -i\omega a - ig^* \sigma_-. \tag{19.8f}$$

These are operator equations. Moreover, they are *nonlinear* equations, since they contain *products* of atomic and field operators. Thus, it would seem that the prospect of obtaining solutions of these equations is not good. However, the Hamiltonian (19.2) is just the Jaynes-Cummings Hamiltonian for which we have already obtained exact solutions in the n basis. Thus, while it appears that equations (19.8) have no obvious solutions, it must be true that it is possible to solve exactly for the expectation values of these operators using the n basis. Recall, however, that while $\sigma_+(0)a(0)\,|1,n\rangle = \sqrt{n}\,|2,n-1\rangle$, since $\sigma_+(0)$ and $a(0)$ coincide with the Schrödinger operators, $\sigma_+(t)a(t)\,|1,n\rangle \neq \sqrt{n}\,|2,n-1\rangle$ for $t \neq 0$. As a result, the solution in the n basis for the operators is not as transparent as it was in the Schrödinger representation for the state amplitudes. So far, the Heisenberg representation does not appear to have simplified our lives.

On the other hand, if we start in the state

$$\boldsymbol{\varrho}^S(0) = |\alpha\rangle\langle\alpha|\boldsymbol{\varrho}^S_{atom}(0), \tag{19.9}$$

where $|\alpha\rangle$ is a coherent state for the field, and $\boldsymbol{\varrho}^S_{atom}(0)$ is the initial state atomic density matrix operator, we can examine under what conditions the atoms evolve as if they were subject to a classical monochromatic field. To do so, we need to take expectation values of equations (19.8a) to (19.8e). It is simple to evaluate

$$\langle\sigma_1\rangle = \varrho_{11}(t), \quad \langle\sigma_2\rangle = \varrho_{22}(t), \quad \langle\sigma_+\rangle = \varrho_{12}(t), \quad \langle\sigma_-\rangle = \varrho_{21}(t). \tag{19.10}$$

In addition, however, we must evaluate terms such as

$$\langle\sigma_+ a\rangle = \mathrm{Tr}\left[\boldsymbol{\varrho}^S(0)\sigma_+ a\right] = \mathrm{Tr}\left\{\sigma_+ a(t)\left[|\alpha\rangle\langle\alpha|\boldsymbol{\varrho}^S_{atom}(0)\right]\right\}, \tag{19.11}$$

where, as a reminder, the explicit dependence of the Heisenberg operator a on the time has been indicated.

In order to be able use

$$a(0)|\alpha\rangle = \alpha|\alpha\rangle, \tag{19.12}$$

we integrate equation (19.8f) formally to obtain

$$a(t) = a(0)e^{-i\omega t} - ig^* \int_0^t e^{-i\omega(t-t')}\sigma_-(t')dt' \tag{19.13}$$

and substitute this result in equation (19.11) to arrive at

$$\langle\sigma_+ a\rangle = \mathrm{Tr}\left[\sigma_+(t)a(0)|\alpha\rangle\langle\alpha|\boldsymbol{\varrho}^S_{atom}(0)\right]e^{-i\omega t} \\ -ig^*\mathrm{Tr}\left[\int_0^t e^{-i\omega(t-t')}\sigma_-(t')dt'|\alpha\rangle\langle\alpha|\boldsymbol{\varrho}^S_{atom}(0)\right]. \tag{19.14}$$

With the aid of equation (19.1), the first term can be evaluated as

$$\mathrm{Tr}\left[\sigma_+(t)a(0)|\alpha\rangle\langle\alpha|\boldsymbol{\varrho}^S_{atom}(0)\right] = \alpha\mathrm{Tr}\left[\sigma_+(t)|\alpha\rangle\langle\alpha|\boldsymbol{\varrho}^S_{atom}(0)\right] = \alpha\mathrm{Tr}\left[\sigma_+(t)\boldsymbol{\varrho}^S_{atom}(0)\right] \\ = \alpha\mathrm{Tr}\left[\sigma_+(0)\boldsymbol{\varrho}^S_{atom}(t)\right] = \mathrm{Tr}\left[|2\rangle\langle 1|\boldsymbol{\varrho}^S_{atom}(t)\right] = \varrho_{12}(t), \tag{19.15}$$

and we find that equation (19.14) can be written in the form

$$\langle\sigma_+ a\rangle = \alpha e^{-i\omega t}\varrho_{12}(t) - ig^*\mathrm{Tr}\left[\int_0^t e^{-i\omega(t-t')}\sigma_-(t')dt'|\alpha\rangle\langle\alpha|\boldsymbol{\varrho}^S_{atom}(0)\right]. \tag{19.16}$$

If we can neglect changes in the field (assume that it remains in the initial coherent state $|\alpha\rangle$), then we can neglect the second term in equations (19.13) and (19.16). In that case, taking expectation values of the operators in equations (19.8), we recover the Bloch equations

$$\dot{\varrho}_{22}(t) = -ig\alpha e^{-i\omega t}\varrho_{12}(t) + ig^*\alpha^* e^{i\omega t}\varrho_{21}(t), \tag{19.17a}$$

$$\dot{\varrho}_{12}(t) = -i\omega\varrho_{12}(t) - ig^*\alpha^* e^{i\omega t}\left[\varrho_{22}(t) - \varrho_{21}(t)\right], \tag{19.17b}$$

$$\dot{\varrho}_{21}(t) = \varrho^*_{12}(t), \quad \varrho_{11}(t) + \varrho_{22}(t) = 1, \tag{19.17c}$$

if we make the association $g\alpha \Rightarrow \chi$. Equations (19.17) are valid only for sufficiently large α and sufficiently small times to ensure that fluctuations in the field do not play a role. In other words, they cannot reproduce the quantum collapse and revival we found in chapter 15. Formally, the equations are valid in the limit that $\alpha \to \infty$ and $g \to 0$, but the product $g\alpha$ goes to a constant.

19.1.2 General Problem—n Field Modes

Now we extend the calculation to an arbitrary number of field modes. In free space, the vacuum field always introduces an infinite number of field modes with which we need to contend. The Hamiltonian in this case is

$$H = \frac{\hbar\omega_0}{2}\sigma_z + \sum_j \hbar\omega_j a_j^\dagger a_j + \sum_j \hbar\left(g_j\sigma_+ a_j + g_j^* a_j^\dagger\sigma_-\right), \tag{19.18}$$

where

$$g_j = -i\boldsymbol{\mu}_{21}\cdot\boldsymbol{\epsilon}_j\left(\frac{\omega_j}{2\hbar\epsilon_0\mathcal{V}}\right)^{1/2}. \tag{19.19}$$

The equations for the operators are

$$\dot{\sigma}_1(t) = i\sum_j g_j\sigma_+(t)a_j - i\sum_j g_j^* a_j^\dagger \sigma_-(t), \tag{19.20a}$$

$$\dot{\sigma}_+(t) = -i\sum_j g_j^* a_j^\dagger (\sigma_2(t) - \sigma_1(t)) + i\omega_0\sigma_+(t), \tag{19.20b}$$

$$\dot{a}_j(t) = -i\omega_j a_j(t) - ig_j^*\sigma_-(t), \tag{19.20c}$$

with $\sigma_- = \sigma_+^\dagger$, $\sigma_2 = 1 - \sigma_1$.

19.1.2.1 *Atomic operators*

The formal solution of equation (19.20c) is

$$a_j(t) = a_j(0)e^{-i\omega_j t} - ig_j^* \int_0^t e^{-i\omega_j(t-t')}\sigma_-(t')dt'. \tag{19.21}$$

Substituting this equation and its adjoint into equation (19.20a), we find

$$\dot{\sigma}_1(t) = i\sum_j g_j\sigma_+(t)a_j(0)e^{-i\omega_j t} + \sum_j |g_j|^2\sigma_+(t)\int_0^t e^{-i\omega_j(t-t')}\sigma_-(t')dt' + \text{adj}. \tag{19.22}$$

In the Weisskopf-Wigner approximation, we can calculate the sum in the second term just as we did for spontaneous decay. Explicitly, we find

$$\sum_j |g_j|^2 e^{-i\omega_j(t-t')} = \gamma_2\delta(t-t'), \tag{19.23}$$

implying that the second term in equation (19.22) is

$$\frac{\gamma_2}{2}\sigma_+(t)\sigma_-(t) = \frac{\gamma_2}{2}\sigma_2(t), \tag{19.24}$$

[having used $\int_0^t \delta(t-t')dt' = 1/2$] and equation (19.22) reduces to

$$\dot{\sigma}_1(t) = i\sum_j g_j\sigma_+(t)a_j(0)e^{-i\omega_j t} - i\sum_j g_j^* a_j^\dagger(0)\sigma_-(t)e^{i\omega_j t} + \gamma_2\sigma_2(t). \tag{19.25}$$

Recall that $[\sigma_\pm(t), a(0)] \neq 0$ for $t \neq 0$.

In a similar fashion, we find

$$\dot{\sigma}_1(t) = \gamma_2\sigma_2(t) + i\sum_j g_j\sigma_+(t)a_j(0)e^{-i\omega_j t}$$
$$-i\sum_j g_j^* a_j^\dagger(0)\sigma_-(t)e^{i\omega_j t}, \quad (19.26a)$$

$$\dot{\sigma}_2(t) = -\gamma_2\sigma_2(t) - i\sum_j g_j\sigma_+(t)a_j(0)e^{-i\omega_j t}$$
$$+i\sum_j g_j^* a_j^\dagger(0)\sigma_-(t)e^{i\omega_j t}, \quad (19.26b)$$

$$\dot{\sigma}_+(t) = -(\gamma - i\omega_0)\,\sigma_+(t) - i\sum_j g_j^* a_j^\dagger(0)\sigma_z(t)e^{i\omega_j t}, \quad (19.26c)$$

$$\dot{\sigma}_-(t) = -(\gamma + i\omega_0)\,\sigma_-(t) + i\sum_j g_j\sigma_z(t)a_j(0)e^{-i\omega_j t}, \quad (19.26d)$$

$$\dot{a}_j(t) = -i\omega_j a_j(t) - ig_j^*\sigma_-(t). \quad (19.26e)$$

The equations for the field operators contain relaxation terms and fluctuation terms involving $a_j(0)$ or $a_j^\dagger(0)$. If the fluctuation terms were not present, the commutation relations for the operators, as well as the expressions for operator products such as

$$\sigma_{ij}(t)\sigma_{kl}(t) = \sigma_{il}(t)\delta_{j,k}, \quad (19.27)$$

would no longer be preserved. The presence of relaxation terms and fluctuation terms is a particular case of the *fluctuation dissipation theorem* [1].

Now, we *have* gotten somewhere using the Heisenberg representation, even if we cannot solve these equations. We see how spontaneous emission enters in a simple fashion within the context of the Weisskopf-Wigner approximation. Moreover, if need be, we can calculate the effect of quantum field fluctuations on the atoms, something that was impossible with the optical Bloch equations.

Something a bit surprising happens if we average equations (19.26) with an initial coherent state for the ith mode of the field. The quantum-mechanical average of the atomic state operator $\sigma_{\mu\nu}(t)$ appearing in equations (19.26) is given by

$$\langle\sigma_{\mu\nu}(t)\rangle = \mathrm{Tr}\left[\varrho^S(0)\sigma_{\mu\nu}(t)\right] = \mathrm{Tr}\left[\varrho^S(t)\sigma_{\mu\nu}(0)\right] = \mathrm{Tr}\left[\varrho^S(t)\,|\mu\rangle\,\langle\nu|\right] = \varrho_{\nu\mu}(t), \quad (19.28)$$

where $\varrho_{\nu\mu}(t)$ is now a *reduced* density matrix element for the atom. Similarly, one finds $\langle\dot{\sigma}_{\mu\nu}(t)\rangle = \dot{\varrho}_{\nu\mu}(t)$. In carrying out the average of equations (19.26), we also encounter terms such as

$$\langle\sigma_{\mu\nu}(t)a_j(0)\rangle = \mathrm{Tr}\left[\sigma_{\mu\nu}(t)a_j(0)\varrho^S(0)\right] = \mathrm{Tr}\left[\sigma_{\mu\nu}(t)a_j(0)|\alpha_i\rangle\langle\alpha_i|\varrho^S_{atom}(0)\right]$$
$$= \alpha_i\delta_{i,j}\mathrm{Tr}\left[\sigma_{\mu\nu}(t)\varrho^S_{atom}(0)\right] = \alpha_i\delta_{i,j}\varrho_{\nu\mu}(t). \quad (19.29)$$

As a consequence of these results, when equations (19.26) are averaged assuming that the initial state of the field is a single-mode coherent state, one recovers the optical Bloch equations for atomic density matrix elements $\varrho_{\mu\nu}(t)$ with $g_i\alpha_i \Rightarrow \chi = \Omega_0/2$. Amazingly, the atomic density matrix elements evolve as if the field is *not being depleted*. Moreover, there are no collapses or revivals. The origin of these phenomena is linked to the Weisskopf-Wigner approximation. In free space, $g_i \sim 0$, and one must take $\alpha_i \sim \infty$ to have a finite Rabi frequency, since $g_i\alpha_i \sim \Omega_0/2$. Given that $g_i \sim 0$, the revival time is infinite, and since $\alpha_i \to \infty$, the average number of photons in the field is infinite and cannot be depleted.

On the other hand, if we start in a Fock state for the ith mode of the field, then

$$\begin{aligned}\langle\sigma_{\mu\nu}(t)a_j(0)\rangle &= \mathrm{Tr}\left[\sigma_{\mu\nu}(t)a_j(0)\varrho(0)\right] = \mathrm{Tr}\left[\sigma_{\mu\nu}(t)a_j(0)|n_i\rangle\langle n_i|\varrho_{atom}(0)\right] \\ &= \sqrt{n_i}\delta_{i,j}\mathrm{Tr}\left[\sigma_{\mu\nu}(t)|n_i-1\rangle\langle n_i|\varrho_{atom}(0)\right]. \end{aligned} \tag{19.30}$$

It is no longer possible to express the average $\langle\sigma_{\mu\nu}(t)a_j(0)\rangle$ in terms of the reduced density matrix of the atoms. As a consequence, one is led to a sequence of coupled equations for atomic density matrix elements in which spontaneous emission results in a cascade down the number states of the fields. For finite n_i, this approach is strictly valid only if we quantize in a finite volume, since this is the only way in which $g_i\sqrt{n_i}$ is nonvanishing. As a consequence, the Weisskopf-Wigner approximation is valid only for times much less than the round-trip time in the cavity. In this limit, collapse and revival cannot occur.

19.1.2.2 *Field operators*

Equations (19.26) can also be used to see how the field develops in the presence of the atoms. The number operator for the field is

$$\begin{aligned}n_j(t) = a_j^\dagger(t)a_j(t) &= \left[a_j(0)e^{-i\omega_j t} - ig_j^*\int_0^t e^{-i\omega_j(t-t')}\sigma_-(t')dt'\right]^\dagger \\ &\times\left[a_j(0)e^{-i\omega_j t} - ig_j^*\int_0^t e^{-i\omega_j(t-t')}\sigma_-(t')dt'\right]. \end{aligned}\tag{19.31}$$

With

$$\sigma_-(t) = \sigma_-^I(t)e^{-i\omega_0 t}, \tag{19.32}$$

this equation becomes

$$\begin{aligned}n_j(t) &= n_j(0) + ig_j\int_0^t \sigma_+^I(t')e^{i\delta_j t'}a_j(0)dt' - ig_j^*\int_0^t a_j^\dagger(0)e^{i\delta_j t''}\sigma_-^I(t'')dt'' \\ &+ g_jg_j^*\int_0^t dt'\int_0^t dt'' e^{i\delta_j(t'-t'')}\sigma_+^I(t')\sigma_-^I(t''),\end{aligned}\tag{19.33}$$

where $\delta_j = \omega_0 - \omega_j$. Using the identity $\int_0^t \to \int_0^{t'} + \int_{t'}^t$ in the last term and changing the order of integration in the second integral, we obtain

$$n_j(t) = n_j(0) + B_j + B_j^\dagger, \tag{19.34}$$

where

$$B_j = ig_j\int_0^t e^{i\delta_j t'}\sigma_+^I(t')a_j(0)dt' + |g_j|^2\int_0^t dt'\int_0^{t'} dt'' e^{i\delta_j(t'-t'')}\sigma_+^I(t')\sigma_-^I(t''). \tag{19.35}$$

In a scattering problem, one often starts with one field mode excited and all others in the vacuum state. If field mode j is initially in the vacuum state, then $\langle n_j(0)\rangle = 0$,

$\langle\sigma_+(t)a_j(0)\rangle = 0$, and

$$\langle n_j(t)\rangle = 2\text{Re}|g_j|^2 \int_0^t dt' \int_0^{t'} dt'' e^{i\delta_j(t'-t'')} \langle\sigma_+^I(t')\sigma_-^I(t'')\rangle. \tag{19.36}$$

Thus, the rate at which radiation is scattered into mode j is proportional to

$$\begin{aligned}\langle\dot{n}_j(t)\rangle &= 2\text{Re}|g_j|^2 \int_0^t dt' e^{i\delta_j(t-t')} \langle\sigma_+^I(t)\sigma_-^I(t')\rangle \\ &= 2\text{Re}|g_j|^2 \int_0^t d\tau e^{i\delta_j\tau} \langle\sigma_+^I(t)\sigma_-^I(t-\tau)\rangle.\end{aligned} \tag{19.37}$$

In essence, equation (19.37) gives the scattered spectrum as the Fourier transform of the correlation function of atomic state operators.

One can show that for $t > t'$, $\langle\sigma_+(t)\sigma_-(t')\rangle$, $\langle\sigma_2(t)\sigma_-(t')\rangle$, $\langle\sigma_1(t)\sigma_-(t')\rangle$, $\langle\sigma_-(t)\sigma_-(t')\rangle$ obey equations having the same basic structure as equation (19.26). This is known as the *quantum regression theorem* [2] and can be used to calculate the spectrum of resonance fluorescence [3]. We use an alternative method for obtaining this spectrum in the next chapter but return to equation (19.37) to arrive at an expression for the frequency-integrated spectrum.

19.2 *N*-atom Systems

The proceeding treatment can be extended to a system of N stationary atoms. If the atoms are separated by less than an optical wavelength λ, there can be collective decay modes [4]. Even if the separation is greater than an optical wavelength, collective effects can be important for specific geometries (e.g., in a pencil geometry having length L, collective effects become important if the density $\mathcal{N}\lambda^2 L \gtrsim 1$ ($\mathcal{N}$ is the atomic density) [5]—we indicate how these effects arise, but then neglect them.

The Hamiltonian is

$$\text{H} = \sum_{j=1}^{N} \frac{\hbar\omega_0}{2}\sigma_z^j + \sum_{\mathbf{k}} \hbar\omega_k a_{\mathbf{k}}^\dagger a_{\mathbf{k}} + \sum_{j=1}^{N}\sum_{\mathbf{k}} \hbar\left(g_{\mathbf{k}}^j \sigma_+^j a_{\mathbf{k}} + g_{\mathbf{k}}^{j\,*} a_{\mathbf{k}}^\dagger \sigma_-^j\right), \tag{19.38}$$

where σ^j is an operator for atom j located at position $\mathbf{R}_j$,

$$g_{\mathbf{k}}^j = f_{\mathbf{k}} e^{i\mathbf{k}\cdot\mathbf{R}_j}, \quad f_{\mathbf{k}} = -i\left(\boldsymbol{\mu}_{21}\cdot\boldsymbol{\epsilon}_{\mathbf{k}}\right)\left(\frac{\omega_k}{2\hbar\epsilon_0\mathcal{V}}\right)^{1/2}, \tag{19.39}$$

and we now specify the field modes by $\mathbf{k}$ instead of j. To get the equations of motion, we proceed as before. That is, the equation for $\dot{a}_{\mathbf{k}}$ is

$$\dot{a}_{\mathbf{k}}(t) = -i\omega_k a_{\mathbf{k}}(t) - i f_{\mathbf{k}}^* \sum_j \sigma_-^j(t) e^{-i\mathbf{k}\cdot\mathbf{R}_j}, \tag{19.40}$$

or

$$a_{\mathbf{k}}(t) = a_{\mathbf{k}}(0)e^{-i\omega_k t} - i f_{\mathbf{k}}^* \sum_j \int_0^t \sigma_-^j(t') e^{-i\omega_k(t-t')} e^{-i\mathbf{k}\cdot\mathbf{R}_j} dt'. \tag{19.41}$$

The equations for the $\dot{\sigma}^j(t)$ are essentially the same as before. For example,

$$\dot{\sigma}_2^j(t) = -i\sum_{\mathbf{k}} f_{\mathbf{k}} e^{i\mathbf{k}\cdot\mathbf{R}_j}\sigma_+^j(t)a_{\mathbf{k}}(t) + i\sum_{\mathbf{k}} f_{\mathbf{k}}^* e^{-i\mathbf{k}\cdot\mathbf{R}_j}a_{\mathbf{k}}^\dagger(t)\sigma_-^j(t). \tag{19.42}$$

However, when equation (19.41) and its adjoint are substituted back into equation (19.42), we get some terms that we did not have before. The $a_{\mathbf{k}}(0)$ and $a_{\mathbf{k}}^\dagger(0)$ terms are unchanged in equation (19.26b), but the terms leading to the $-\gamma_2\sigma_2(t)$ term in equation (19.26b) are replaced by

$$-\sum_{\mathbf{k}} |f_{\mathbf{k}}|^2 e^{i\mathbf{k}\cdot\mathbf{R}_j}\sigma_+^j(t)\sum_{j'}\int_0^t \sigma_-^{j'}(t')e^{-i\omega_k(t-t')}e^{-i\mathbf{k}\cdot\mathbf{R}_{j'}}dt' + \text{adj}. \tag{19.43}$$

The $j = j'$ terms reproduce the $-\gamma_2\sigma_2(t)$ term in equation (19.26b), but there are *additional* terms for $j \neq j'$ that represent a type of collective decay mode produced by a coupling of the atoms by the field. For $|\mathbf{R}_j - \mathbf{R}_{j'}| \ll \lambda$, this directly modifies the decay rate, but even for $|\mathbf{R}_j - \mathbf{R}_{j'}| > \lambda$, this term does not necessarily vanish in selected directions of emission. For sufficiently low densities, $\mathcal{N}\lambda^2 L \ll 1$, these collective effects can be neglected. We assume that the density is sufficiently low to neglect all collective effects. Then

$$\begin{aligned}\dot{\sigma}_2^j(t) &= -\gamma_2\sigma_2^j(t) - i\sum_{\mathbf{k}} f_{\mathbf{k}} e^{i\mathbf{k}\cdot\mathbf{R}_j}\sigma_+^j(t)a_{\mathbf{k}}(0)\\ &\quad + i\sum_{\mathbf{k}} f_{\mathbf{k}}^* e^{-i\mathbf{k}\cdot\mathbf{R}_j}a_{\mathbf{k}}^\dagger(0)\sigma_-^j(t),\end{aligned} \tag{19.44a}$$

$$\begin{aligned}\dot{\sigma}_-^j(t) &= -(\gamma + i\omega_0)\sigma_-^j(t)\\ &\quad + i\sum_{\mathbf{k}} f_{\mathbf{k}} e^{i\mathbf{k}\cdot\mathbf{R}_j}\left[\sigma_2^j(t) - \sigma_1^j(t)\right]a_{\mathbf{k}}(0),\end{aligned} \tag{19.44b}$$

$$a_{\mathbf{k}}(t) = a_{\mathbf{k}}(0)e^{-i\omega_k t} - i f_{\mathbf{k}}^*\sum_j e^{-i\mathbf{k}\cdot\mathbf{R}_j}\int_0^t \sigma_-^j(t')e^{-i\omega_k(t-t')}dt', \tag{19.44c}$$

$$\sigma_1^j(t) + \sigma_2^j(t) = 1, \tag{19.44d}$$

$$\sigma_+^j(t) = \left[\sigma_-^j(t)\right]^\dagger, \tag{19.44e}$$

where $\gamma = \gamma_2/2$. These equations are the starting point for calculations of absorption, four-wave mixing, etc.

19.3 Source-Field Equation

Actually, one of the most important consequences of the operator equations (19.26) and (19.44) is that, in dipole approximation, they lead to a *source-field* expression in which the field operator is expressed in terms of atomic operators [6]. To see this, we start with the expression for the field operator,

$$\mathbf{E}(\mathbf{R}, t) = i\sum_{\mathbf{k},\lambda}\left(\frac{\hbar\omega_k}{2\epsilon_0\mathcal{V}}\right)^{1/2}\epsilon_{\mathbf{k}}^{(\lambda)}a_{\mathbf{k}_\lambda}(t)e^{i\mathbf{k}\cdot\mathbf{R}} + \text{adj.}, \tag{19.45}$$

where the field polarization index λ is now shown explicitly. The calculation is done for a single atom located at the origin, since this result is generalized easily to the N-atom case. However, we would like to generalize the results to allow for atomic states having arbitrary angular momentum; moreover, we do not want to make the RWA from the outset.

To do so, we replace the atom–field interaction appearing in equation (19.2) by

$$V_{AF} = -\boldsymbol{\mu}(t)\cdot\mathbf{E}(\mathbf{R}, t) = \sum_{\mathbf{k},\lambda} \hbar f_{\mathbf{k}_\lambda}(t) a_{\mathbf{k}_\lambda} + \text{adj.}, \tag{19.46}$$

where

$$f_{\mathbf{k}_\lambda}(t) = -i\left(\frac{\omega_k}{2\hbar\epsilon_0 \mathcal{V}}\right)^{1/2} \boldsymbol{\mu}(t)\cdot\boldsymbol{\epsilon}_{\mathbf{k}}^{(\lambda)}, \tag{19.47}$$

and $\boldsymbol{\mu}(t)$ is the (Hermitian) dipole moment operator. For this interaction, equation (19.41) is replaced by

$$a_{\mathbf{k}_\lambda}(t) = a_{\mathbf{k}_\lambda}(0)e^{-i\omega_k t} - i\int_0^t f^{\dagger}_{\mathbf{k}_\lambda}(t')e^{-i\omega_k(t-t')}dt'. \tag{19.48}$$

When equation (19.48) is substituted into equation (19.45), one finds that the field operator can be written as

$$\mathbf{E}(\mathbf{R}, t) = \mathbf{E}_0(\mathbf{R}, t) + \mathbf{E}_s(\mathbf{R}, t), \tag{19.49}$$

where

$$\mathbf{E}_0(\mathbf{R}, t) = i\sum_{\mathbf{k},\lambda}\left(\frac{\hbar\omega_k}{2\epsilon_0\mathcal{V}}\right)^{1/2} \boldsymbol{\epsilon}_{\mathbf{k}}^{(\lambda)} a_{\mathbf{k}_\lambda}(0)e^{-i\omega_k t}e^{i\mathbf{k}\cdot\mathbf{R}} + \text{adj.} \tag{19.50}$$

corresponds to the propagation of the free field in the absence of the source atom at the origin, and

$$\mathbf{E}_s(\mathbf{R}, t) = \sum_{\mathbf{k},\lambda}\left(\frac{\hbar\omega_k}{2\epsilon_0\mathcal{V}}\right)^{1/2} \boldsymbol{\epsilon}_{\mathbf{k}}^{(\lambda)} \int_0^t f^{\dagger}_{\mathbf{k}_\lambda}(t')e^{-i\omega_k(t-t')}dt' + \text{adj.} \tag{19.51}$$

represents the contribution to the field from the source atom. In what follows, we concentrate on this source field.

Using equations (19.51) and (19.47), and converting from a sum to an integral, we obtain

$$\mathbf{E}_s(\mathbf{R}, t) = i\left(\frac{1}{2\epsilon_0}\right)\frac{1}{(2\pi c)^3}\sum_{\alpha,\beta=1}^{3}\int_0^\infty d\omega_k \int d\Omega_k \omega_k^3 \left[\boldsymbol{\epsilon}_{\mathbf{k}}^{(\lambda)}\right]_\alpha \hat{\mathbf{u}}_\alpha \left[\boldsymbol{\epsilon}_{\mathbf{k}}^{(\lambda)}\right]_\beta e^{i\mathbf{k}\cdot\mathbf{R}}$$

$$\times\int_0^t \mu_\beta(t')e^{-i\omega_k(t-t')}dt' + \text{adj.}, \tag{19.52}$$

where

$$\hat{\mathbf{u}}_1 = \hat{\mathbf{x}},\;\; \hat{\mathbf{u}}_2 = \hat{\mathbf{y}},\;\; \hat{\mathbf{u}}_3 = \hat{\mathbf{z}}, \tag{19.53}$$

and all the components of $\boldsymbol{\epsilon}_{\mathbf{k}}^{(\lambda)}$ and $\boldsymbol{\mu}(t)$ are Cartesian components. If we expand

$$e^{i\mathbf{k}\cdot\mathbf{R}} = \sum_{\ell=0}^{\infty}\sum_{m=-\ell}^{\ell} 4\pi\, i^{\ell} j_{\ell}(kR) Y_{\ell m}(\theta,\phi) Y_{\ell m}^{*}(\theta_k,\phi_k), \tag{19.54}$$

where the $Y_{\ell m}$'s are spherical harmonics and the j_{ℓ}'s are spherical Bessel functions, we can use the polarization vectors (16.18) to evaluate the angular integrals in

$$M_{\alpha\beta}(\mathbf{R}) = \sum_{\ell,m} i^{\ell} j_{\ell}(kR) Y_{\ell m}(\theta,\phi) \int d\Omega_k \left[\boldsymbol{\epsilon}_{\mathbf{k}}^{(\lambda)}\right]_{\alpha} \left[\boldsymbol{\epsilon}_{\mathbf{k}}^{(\lambda)}\right]_{\beta} Y_{\ell m}^{*}(\theta_{\mathbf{k}},\phi_{\mathbf{k}}), \tag{19.55}$$

to obtain

$$M_{\alpha\beta}(\mathbf{R}) = f_{\alpha\beta}\, j_0(kR) + g_{\alpha\beta}(\theta,\phi)\, j_2(kR), \tag{19.56}$$

where

$$f_{\alpha\beta} = (2/3)\,\delta_{\alpha,\beta}, \tag{19.57a}$$

$$g_{11}(\theta,\phi) = -\frac{3\cos^2\theta - 1}{6} + \frac{\sin^2\theta\cos(2\phi)}{2}, \tag{19.57b}$$

$$g_{22}(\theta,\phi) = -\frac{3\cos^2\theta - 1}{6} - \frac{\sin^2\theta\cos(2\phi)}{2}, \tag{19.57c}$$

$$g_{33}(\theta,\phi) = \frac{3\cos^2\theta - 1}{3}, \tag{19.57d}$$

$$g_{12}(\theta,\phi) = g_{21}(\theta,\phi) = \sin^2\theta\sin(2\phi)/2, \tag{19.57e}$$

$$g_{13}(\theta,\phi) = g_{31}(\theta,\phi) = \sin\theta\cos\theta\cos\phi, \tag{19.57f}$$

$$g_{23}(\theta,\phi) = g_{32}(\theta,\phi) = \sin\theta\cos\theta\sin\phi. \tag{19.57g}$$

Thus,

$$\begin{aligned} \mathbf{E}_s(\mathbf{R},t) = i\left(\frac{1}{4\pi^2 c^3\epsilon_0}\right) \sum_{\alpha,\beta=1}^{3} \int_0^t dt' \int_0^{\infty} d\omega_k \omega_k^3 \\ \times \left[f_{\alpha\beta}\, j_0(kR) + g_{\alpha\beta}(\theta,\phi)\, j_2(kR)\right] \hat{\mathbf{u}}_{\alpha}\mu_{\beta}(t') e^{-i\omega_k(t-t')} + \text{adj}. \end{aligned} \tag{19.58}$$

If we were to make the rotating-wave approximation at this point by writing

$$\mu_{\beta}(t') = \tilde{\mu}_{\beta}(t')e^{-i\omega t'} + \tilde{\mu}_{\beta}^{*}(t)e^{i\omega t'} \tag{19.59}$$

and keep only the slowly varying parts in each of the integrands, we would arrive at a result that is not fully retarded, since the ω_k integral is from 0 to ∞ rather than $-\infty$ to ∞. However, it is *not* necessary to make the rotating-wave approximation to evaluate $\mathbf{E}_s(\mathbf{R},t)$. The adjoint term in equation (19.58) is proportional to

$$\begin{aligned} &-i\int_0^{\infty} d\omega_k \omega_k^3 \left[f_{\alpha\beta}\, j_0(kR) + g_{\alpha\beta}(\theta,\phi)\, j_2(kR)\right] e^{i\omega_k(t-t')} \\ &= i\int_{-\infty}^{0} d\omega_k \omega_k^3 \left[f_{\alpha\beta}\, j_0(kR) + g_{\alpha\beta}(\theta,\phi)\, j_2(kR)\right] e^{-i\omega_k(t-t')}, \end{aligned}$$

which implies that

$$\mathbf{E}_s(\mathbf{R},t) = i\left(\frac{1}{4\pi^2c^3\epsilon_0}\right)\sum_{\alpha,\beta=1}^{3}\int_0^t dt'\int_{-\infty}^{\infty} d\omega_k\omega_k^3 \times\left[f_{\alpha\beta}\,j_0(kR)+g_{\alpha\beta}(\theta,\phi)\,j_2(kR)\right]\hat{\mathbf{u}}_\alpha\mu_\beta(t')e^{-i\omega_k(t-t')}. \quad (19.60)$$

To proceed, we write

$$j_\ell(x) = \frac{h_\ell^{(1)}(x)+h_\ell^{(2)}(x)}{2} = \frac{c_\ell(x)e^{ix}+d_\ell(x)e^{-ix}}{2}, \quad (19.61)$$

where the h_ℓ's are spherical Hankel functions, and c_0 and c_2 are given by [7]

$$c_0(x) = -\frac{i}{x}, \quad c_2(x) = \frac{i}{x}-\frac{3}{x^2}-\frac{3i}{x^3}, \quad (19.62)$$

substitute this result into equation (19.60), and use the fact that $\omega_k^n e^{-i\omega_k(t-t')} = i^n d^n\left[e^{-i\omega_k(t-t')}\right]/dt^n$. The integration over ω_k leads to delta functions having argument $(t'-t+R/c)$ for the c_ℓ terms and $(t'-t-R/c)$ for the d_ℓ terms. Since $t'\le t$, only the c_ℓ terms are nonvanishing. In this manner, we can obtain

$$\mathbf{E}_s(\mathbf{R},t) = \left(\frac{1}{4\pi\epsilon_0}\right)\sum_{\alpha,\beta=1}^{3}\hat{\mathbf{u}}_\alpha\left\{-\frac{f_{\alpha\beta}-g_{\alpha\beta}(\theta,\phi)}{c^2R}\ddot{\mu}_\beta(t-R/c) + g_{\alpha\beta}(\theta,\phi)\left[\frac{3}{cR^2}\dot{\mu}_\beta(t-R/c)+\frac{3}{R^3}\mu_\beta(t-R/c)\right]\right\}. \quad (19.63)$$

In dipole approximation, the quantum properties of the field are determined totally by the quantum properties of the atom, evaluated at the retarded time. The result is purely retarded and has the same form as the field radiated by a classical dipole [8]. The Weisskopf-Wigner approximation was *not* needed for its derivation. This is a fairly remarkable result, and one that is not used all that often. The advantage of using Heisenberg operators is evident since the quantum field mirrors that of the corresponding classical field. The extension to a system of N atoms is

$$\mathbf{E}_s(\mathbf{R},t) = \left(\frac{1}{4\pi\epsilon_0}\right)\sum_{\alpha,\beta=1}^{3}\sum_{j=1}^{N}\hat{\mathbf{u}}_\alpha\left\{-\frac{f_{\alpha\beta}-g_{\alpha\beta}(\theta_j,\phi_j)}{c^2R}\ddot{\mu}_\beta^{(j)}(t-\left|\mathbf{R}-\mathbf{R}_j\right|/c) + g_{\alpha\beta}(\theta_j,\phi_j)\begin{bmatrix}\frac{3}{cR^2}\dot{\mu}_\beta^{(j)}(t-\left|\mathbf{R}-\mathbf{R}_j\right|/c)\\ +\frac{3}{R^3}\mu_\beta^{(j)}(t-\left|\mathbf{R}-\mathbf{R}_j\right|/c)\end{bmatrix}\right\}, \quad (19.64)$$

where (θ_j,ϕ_j) are angles measured from atom j to the observation point.

We want to emphasize that equation (19.64) is *exact*, in dipole approximation. Neither the Weisskopf-Wigner nor the RWA has been used in its derivation. For optical fields driving atomic transitions, some simplifications are possible. If we assume that the source atoms have a G–H transition that provides the major contribution to the field, then

$$\mu_\beta^{(j)}(t) = \sum_{m_G,m_H}\langle Gm_G|\,\mu_\beta\,|Hm_H\rangle\,\sigma_-^{(j)}(Gm_G,Hm_H;t)+\text{adj.}, \quad (19.65)$$

where G and H are the angular momenta of the lower- and upper-state manifolds, respectively, and $\sigma_-(Gm_G, Hm_H; 0) = |Gm_G\rangle\langle Hm_H|$ is a lowering operator. In the RWA, $\sigma_-^{(j)}(Gm_G, Hm_H; t)$ oscillates as $e^{-i\omega_s t}$, where ω_s is some characteristic frequency (natural frequency, laser field frequency, etc.) in the problem that is much larger than any decay rates, detunings, or Rabi frequencies. With these assumptions, in the RWA, the positive frequency component of the field (19.64) can be written as

$$\begin{aligned}\mathbf{E}_s^+(\mathbf{R}, t) = {} & \left(\frac{1}{4\pi\epsilon_0}\right) \sum_{m_G, m_H} \sum_{\alpha,\beta=1}^{3} \sum_{j=1}^{N} \langle Gm_G|\, \mu_\beta \,|Hm_H\rangle\, \hat{\mathbf{u}}_\alpha \\ & \times \left\{ \omega_s^2 \frac{\left[f_{\alpha\beta} - g_{\alpha\beta}(\theta_j, \phi_j)\right]}{c^2 R} + g_{\alpha\beta}(\theta_j, \phi_j) \left(-\frac{3i\omega_s}{cR^2} + \frac{3}{R^3}\right)\right\} \\ & \times \sigma_-^{(j)}\left(Gm_G, Hm_H; t - \left|\mathbf{R} - \mathbf{R}_j\right|/c\right). \end{aligned} \tag{19.66}$$

In what follows, we will limit our discussion to the radiation zone, where

$$(\theta_j, \phi_j) \approx (\theta, \phi) \tag{19.67}$$

for a sample of finite size. Physically, this implies that we make observations so far from the sample that the angle to the observation point from any point in the sample is essentially the same. In the radiation zone, equation (19.66) reduces to

$$\begin{aligned}\mathbf{E}_s^+(\mathbf{R}, t) = {} & \left(\frac{\omega_s^2}{4\pi\epsilon_0}\right) \sum_{m_G, m_H} \sum_{\alpha,\beta=1}^{3} \sum_{j=1}^{N} \langle Gm_G|\, \mu_\beta \,|Hm_H\rangle \\ & \times \frac{f_{\alpha\beta} - g_{\alpha\beta}(\theta, \phi)}{c^2 R} \hat{\mathbf{u}}_\alpha \sigma_-^{(j)}\left(Gm_G, Hm_H; t - \left|\mathbf{R} - \mathbf{R}_j\right|/c\right). \end{aligned} \tag{19.68}$$

To see that this equation reduces to something familiar, imagine that $G = 0$, $H = 1$, and the field polarizations are such that only $\sigma_-^{(j)}(00, 10; t) \equiv \sigma_-^{(j)}(t)$ is of importance for the problem. Making use of equations (19.57) and (16.18b), we find that equation (19.68) can be written as

$$\mathbf{E}_s^+(\mathbf{R}, t) = -\left(\frac{\omega_s^2}{4\pi\epsilon_0 c^2 R}\right) \mu_{12} \sin\theta \sum_{j=1}^{N} \sigma_-^{(j)}(t - \left|\mathbf{R} - \mathbf{R}_j\right|/c)\hat{\boldsymbol{\theta}}, \tag{19.69}$$

with

$$\mu_{12} \equiv \langle 00|\, \mu_z \,|10\rangle. \tag{19.70}$$

This equation is recognized as the radiation field associated with dipoles oscillating in the z direction.

19.4 Source-Field Approach: Examples

The source-field approach is versatile and simple, as is illustrated in the following examples.

19.4.1 Average Field and Field Intensity in Spontaneous Emission

We calculated the spectrum and angular distribution of radiation emitted in spontaneous emission in chapter 16. Here, we are interested primarily in the *temporal* dependence of the emitted signal from a single atom located at the origin. We assume that the atom is prepared at $t = 0$ into some arbitrary superposition of Zeeman sublevels m_H in excited level H and undergoes spontaneous emission to the sublevels m_G in level G.

From equation (19.68), we see that the average field amplitude $\left\langle \mathbf{E}_s^+(\mathbf{R}, t)\right\rangle$ depends on $\langle \sigma_-(Gm_G, Hm_H; t - R/c)\rangle$. Using equation (19.26d), generalized to include magnetic state sublevels, we find that

$$\begin{aligned}\langle \sigma_-(Gm_G, Hm_H; t)\rangle &= \langle \sigma_-(Gm_G, Hm_H; 0)\rangle\, e^{-\gamma_H t/2}\Theta(t)\\ &= \operatorname{Tr}\left[\varrho^S(0)\,|Gm_G\rangle\,\langle Hm_H|\right] e^{-\gamma_H t/2}\Theta(t) = 0,\end{aligned} \tag{19.71}$$

where $\Theta(t)$ is the Heaviside step function. The trace vanishes, since

$$\varrho^S(0) = \varrho_{Hm_H,Hm'_H}(0)\,|Hm_H\rangle\left\langle Hm'_H\right|. \tag{19.72}$$

If the atom is prepared without any ground-state–excited-state coherence, the average field vanishes, since the phase of the field is random.

On the other hand, the average field intensity, given by $\left\langle \mathbf{E}^-(\mathbf{R}, t)\cdot \mathbf{E}^+(\mathbf{R}, t)\right\rangle$, is nonvanishing. In the RWA, the positive frequency component of the field in the radiation zone is

$$\begin{aligned}\mathbf{E}_s^+(\mathbf{R}, t) \approx \left(\frac{\omega_0^2}{4\pi\epsilon_0}\right)\sum_{\alpha,\beta=1}^{3}\frac{f_{\alpha\beta} - g_{\alpha\beta}(\theta,\phi)}{c^2 R}\\ \times \sum_{m_G,m_H}\langle Gm_G|\,\mu_\beta\,|Hm_H\rangle\\ \times \sigma_-(Gm_G, Hm_H; t - R/c)\,\hat{\mathbf{u}}_\alpha.\end{aligned} \tag{19.73}$$

Using equation (19.68), with $\omega_s = \omega_0 = \omega_{HG}$, the fact that $\hat{\mathbf{u}}_\alpha \cdot \hat{\mathbf{u}}_{\alpha'} = \delta_{\alpha,\alpha'}$ and

$$\begin{aligned}&\left[\sigma_-\left(Gm'_G, Hm'_H; t\right)\right]^\dagger \sigma_-(Gm_G, Hm_H; t)\\ &= \left|Hm'_H\right\rangle\langle Hm_H|\,\delta_{m_G,m'_G}e^{-\gamma_H t}\Theta(t),\end{aligned} \tag{19.74}$$

we find

$$\begin{aligned}&\left\langle \mathbf{E}^-(\mathbf{R}, t)\cdot \mathbf{E}^+(\mathbf{R}, t)\right\rangle\\ &= \left(\frac{\omega_0^2}{4\pi\epsilon_0 c^2 R}\right)^2 e^{-\gamma_H(t-R/c)}\Theta(t - R/c)\sum_{\alpha,\beta,\beta'}\left(f_{\alpha\beta} - g_{\alpha\beta}\right)\left(f_{\alpha\beta'} - g_{\alpha\beta'}\right)\\ &\times \sum_{m_G,m_H,m_{H'}}\langle Gm_G|\,\mu_\beta\,|Hm_H\rangle^*\,\langle Gm_G|\,\mu_{\beta'}\left|Hm'_H\right\rangle \varrho_{Hm_H;Hm'_H}(0).\end{aligned} \tag{19.75}$$

This is the intensity associated with the one-photon wave packet emitted by the atom. The wave front is located a distance ct from the atom, and the pulse duration is of order γ_H^{-1}. The angular distribution and polarization of the emitted radiation depends on the state preparation through the factor $\varrho_{Hm_H;Hm'_H}(0)$. Specific examples are left to the problems.

Equation (19.75) contains an interesting and even perplexing result. Nowhere in its derivation did we make any assumption regarding the ground-state–excited-state coherence produced by the excitation field. In other words, imagine that the atom is excited with a $\pi/2$ pulse. As a result of the excitation, both dipole (ground–excited state) and excited-state Zeeman coherence is produced. In this case, the average field emitted by the atom does not vanish; however, the radiated *intensity* depends in no *direct* way on the dipole coherence. The average intensity depends only on the excited-state density matrix elements created by the field. You might think that the *spectrum* of the radiation would provide a signature of the dipole coherence, but alas, this is not the case. Recall that the spectrum is proportional to $\sum_{m_G} |b_{Gm_G;\mathbf{k}_{\epsilon_\mathbf{k}}}(\infty)|^2$, the probability to find the atom in its ground state with a photon emitted in direction $\mathbf{k}$ having polarization $\boldsymbol{\epsilon}_\mathbf{k}$. You have seen in chapter 16 that this quantity depends only on excited-state density matrix elements. Thus, we are led to a somewhat paradoxical conclusion that the field intensity radiated by a single atom that has been prepared by an excitation pulse depends in no direct way on its dipole coherence. To prove that the dipole coherence is nonvanishing, one can interfere the radiated signal with a reference field. A nonvanishing dipole coherence becomes much more important when we consider the field emitted by *more* than one atom. A simple example to illustrate this is given in the problems, and an example involving coherent transients is given in section 19.4.3.

19.4.2 Frequency Beats in Emission: Quantum Beats

As a second example involving emission from a single atom, we consider frequency beats in emission [9]. Qualitatively, it is usually possible to distinguish two types of situations in which beats occur. The first corresponds to a beat frequency in the intensity radiated by two classical dipoles having different natural frequencies. This could be referred to as "classical" beats. However, as you saw earlier, dipole coherence can play no role in the intensity radiated by a single atom. Thus, even if a single atom is excited into a superposition of states giving rise to a dipole coherence that oscillates at two distinct frequencies, there can be no classical beats in the emitted intensity. One could detect the presence of the two field frequencies by heterodyning (beating) the radiated field with a reference field.

Thus, any beat in the field radiated by a single atom that has been prepared by an excitation pulse must originate from *excited-state coherence* between two excited-state manifolds differing in frequency (or between Zeeman sublevels in a single manifold that have been split in energy by an external magnetic field). The resultant *quantum beats* are a purely quantum phenomena.

The calculation given earlier for the field intensity in spontaneous emission is generalized easily to the case of two excited-state manifolds H and H', separated in frequency by $\omega_{HH'}$. One simply replaces

$$\sum_{m_G,m_H,m_{H'}} \langle Gm_G|\,\mu_\beta\,|Hm_H\rangle^* \langle Gm_G|\,\mu_{\beta'}\,\left|Hm'_H\right\rangle \varrho_{Hm_H;Hm'_H}(0)$$

appearing in equation (19.75) by

$$\sum_{F=H,H'} e^{-i\omega_{FF'}t} \sum_{m_G,m_F,m_{F'}} \langle Gm_G|\, \mu_\beta \,|Fm_F\rangle^* \,\langle Gm_G|\, \mu_{\beta'} \,\left|F'm'_F\right\rangle \varrho_{Fm_F;F'm'_F}(0). \quad (19.76)$$

Clearly, if coherence is created between the H and H' manifolds by the excitation pulse, there will be frequency beats at frequency $\omega_{HH'}$ in the field intensity. Generally speaking, there are two conditions needed to produce and detect quantum beats. The excitation pulse must be sufficiently broadband to excite the $H - H'$ coherence (it must contain frequency components at $\omega_{HH'}$), and the detector must be sufficiently broadband so as *not* to detect both emitted frequencies. If the detector could distinguish the frequencies, we would know from which level the radiation was emitted; such "which path" information destroys the interference pattern.

19.4.3 Four-Wave Mixing

As a third example of the source-field approach, we look at a coherent transient problem in which three cw fields having propagation vectors $\mathbf{k}_1, \mathbf{k}_2, \mathbf{k}_3$ and frequencies $\omega_1, \omega_2, \omega_3$ are incident on an ensemble of N two-level atoms. You may recall that we solved this problem in perturbation theory in chapter 10, using the Maxwell-Bloch equations to calculate the emitted field from the atomic polarization. In that case, the three incident classical fields led to polarization components with different propagation vectors, each corresponding to a possible direction for the radiated field. Phase matching was needed to produce a signal that varied as the density squared.

The analogous quantized field calculation begins with equations (19.44) for an ensemble of two-level atoms interacting with the fields. To calculate the radiated field, we need to find $\sigma_-^j(t)$ for each atom. One term in the "steady-state" (after all transients have died away) perturbation solution of equations (19.44) to lowest order in the product of the three incident field creation or annihilation operators is of the form

$$\sigma_-^j(t) \sim e^{i\mathbf{k}_s\cdot\mathbf{R}_j} e^{-i\omega_s t} a_1(0)a_2^\dagger(0)a_3(0)\,, \quad (19.77)$$

where

$$\mathbf{k}_s = \mathbf{k}_1 - \mathbf{k}_2 + \mathbf{k}_3\,, \quad (19.78a)$$

$$\omega_s = \omega_1 - \omega_2 + \omega_3. \quad (19.78b)$$

Of course, there are additional contributions as well, but we concentrate on this one, which will lead to a signal in the $\mathbf{k}_s = \mathbf{k}_1 - \mathbf{k}_2 + \mathbf{k}_3$ direction. The multiplicative factor involving decay constants and detunings that is given in equation (7.14) is omitted here, since we are concerned mainly with the dependence of the signal on atom density.

If we assume that the incident fields are all z-polarized, that $G = 0$ and $H = 1$, and that $\ddot{\mu}_\beta^j \approx -\omega_s^2\mu_\beta^j$, then we can combine equations (19.69) and equation (19.77) to obtain

$$\mathbf{E}_s^+(\mathbf{R}, t) \sim a_1(0)a_2^\dagger(0)a_3(0)\sin\theta \sum_{j=1}^{N} e^{i\mathbf{k}_s\cdot\mathbf{R}} e^{-i\omega_s(t-|\mathbf{R}-\mathbf{R}_j|/c)}\,\hat{\boldsymbol{\theta}}\,. \quad (19.79)$$

In the radiation zone, $\left(k_s R_j^2/R\right) \ll 1$, implying that

$$\omega_s|\mathbf{R}-\mathbf{R}_j|/c \approx \omega_s\mathbf{R}/c - \hat{\mathbf{R}}\cdot\mathbf{R}_j\omega_s/c,$$

leading to

$$\mathbf{E}_s^+(\mathbf{R},t) \sim e^{i\mathbf{k}\cdot\mathbf{R}}a_1(0)a_2^\dagger(0)a_3(0)\sin\theta\sum_{j=1}^{N}e^{i(\mathbf{k}_s-\frac{\omega_s}{c}\hat{\mathbf{R}})\cdot\mathbf{R}_j}e^{-i\omega_s t}\,\hat{\boldsymbol{\theta}}. \tag{19.80}$$

If $k_s R_j \gg 1$, there is destructive interference unless $\hat{\mathbf{R}} = \hat{\mathbf{k}}_s$ *and* $k_s = \omega_s/c$—these are the *phase-matching* conditions that have been obtained previously.

The average field intensity is

$$\left\langle \mathbf{E}_s^-(\mathbf{R},t)\mathbf{E}_s^+(\mathbf{R},t)\right\rangle \sim n_1n_2n_3\sin^2\theta\sum_{j,j'=1}^{N}e^{i(\mathbf{k}_s-\frac{\omega_s}{c}\hat{\mathbf{R}})\cdot(\mathbf{R}_j-\mathbf{R}_{j'})}, \tag{19.81}$$

where $n_j = \langle a_j^\dagger(0)a_j(0)\rangle$. In the phase-matched direction, the fields add coherently, and the intensity goes like N^2; for other directions, the fields add incoherently (only $j = j'$ contributes in the sum), and the signal varies as N. Note that the average signal intensity depends only on the average number of photons in each field.

19.4.4 Linear Absorption

As a final example of the source-field approach, we consider linear absorption for an optically thin sample. A monochromatic field having frequency ω propagates in a medium. After all transients have died away, there is a scattered field in all directions except the incident field direction $\hat{\mathbf{k}}_0$. In this direction, the initial field amplitude is reduced. To see this, we must keep the initial field term in the expression for the total field. In other words, for a $G = 0$ and $H = 1$ transition and an initial field that is z-polarized, we write

$$\begin{aligned}\langle\mathbf{E}^+(\mathbf{R},t)\rangle &= e^{i(\mathbf{k}_0\cdot\mathbf{R}-\omega t)}\langle E^+(0)\rangle\hat{\mathbf{z}} - \sin\theta\left(\frac{k_0^2}{4\pi\epsilon_0 R}\right)\mu_{12}\\ &\times\sum_{j=1}^{N}\langle\sigma_-^j(t-\left|\mathbf{R}-\mathbf{R}_j\right|/c)\rangle\hat{\boldsymbol{\theta}},\end{aligned} \tag{19.82}$$

where

$$E^+(0) = i\left(\frac{\hbar\omega}{2\epsilon_0\mathcal{V}}\right)^{1/2}a, \tag{19.83}$$

$k_0 = \omega/c$, and μ_{12} is given by equation (19.70). In perturbation theory and assuming that the field amplitude is approximately constant in the optically thin medium, it follows from equation (19.44b) that, in steady state,

$$\langle\sigma_-^j(t)\rangle = -\frac{i}{\hbar}\frac{\mu_{21}e^{-i\omega t}e^{i\mathbf{k}_0\cdot\mathbf{R}_j}}{\gamma+i\delta}\langle E^+(0)\rangle \tag{19.84}$$

or

$$\langle \sigma_-^j(t - |\mathbf{R} - \mathbf{R}_j| / c) \rangle = -\frac{i}{\hbar} \frac{\mu_{21} e^{i(k_0 R - \omega t)}}{\gamma + i\delta} e^{i(\mathbf{k}_0 - \mathbf{k}) \cdot \mathbf{R}_j} \langle E^+(0) \rangle , \tag{19.85}$$

where

$$\mathbf{k} = k_0 \hat{\mathbf{R}} . \tag{19.86}$$

By combining equations (19.82) and (19.85), we obtain

$$\langle \mathbf{E}^+(\mathbf{R},t) \rangle = e^{i(\mathbf{k}_0 \cdot \mathbf{R} - \omega t)} \langle E^+(0) \rangle \hat{\mathbf{z}} + i \sin\theta \left(\frac{k_0^2}{4\pi\epsilon_0 \hbar R} \right) \times \sum_{j=1}^{N} \frac{|\mu_{21}|^2 e^{i(k_0 R - \omega t)}}{\gamma + i\delta} e^{i(\mathbf{k}_0 - \mathbf{k}) \cdot \mathbf{R}_j} \langle E^+(0) \rangle \hat{\boldsymbol{\theta}} . \tag{19.87}$$

In a small cone centered about the forward direction $\mathbf{k} = k_0 \hat{\mathbf{R}} = \mathbf{k}_0$, the scattered field interferes with the incident field and leads to a decrease in the field intensity. In directions other than $\mathbf{k}_0$, the contributions from different atoms add incoherently, and the scattered intensity is proportional to N. Of course, the energy lost from the incident field is simply scattered into previously unoccupied modes of the radiation field. This constitutes the *optical theorem* [10].

19.5 Summary

The Heisenberg representation offers advantages over the Schrödinger representation, especially when one is concerned with atomic operators only or field operators only. We derived nonlinear coupled equations for the atomic and field operators and looked at limits in which the equations reduced to the conventional Bloch equations. Although the equations are nonlinear, they can often be solved easily in perturbation theory. Thus, in four-wave mixing, it is not difficult to solve for the atomic operators to lowest nonvanishing order in the field amplitudes. What one loses in such an approach is detailed information about the entanglement of atomic states with the generation of new field modes. We have seen some of the power of the Heisenberg representation in applications involving the source-field expression that was derived. In the next chapter, we further exploit this equation to analyze scattering of cw fields by atoms.

Problems

1. Calculate $\langle a \rangle$ in the Jaynes-Cummings model for an arbitrary initial state using the state vector given in equation (15.36). This helps to explain why the solution for the operators is not overly transparent in the Heisenberg picture.
2. Evaluate equation (19.75) when $\varrho_{Hm_H;Hm'_H}(0) = \delta_{m_H,0}\delta_{m'_H,0}$, $H = 1$, and $G = 0$. Use the expression for the field operator to show that the field is linearly polarized in the $\hat{\boldsymbol{\theta}}$ direction and has the characteristic $\sin^2\theta$ dependence of a z-polarized dipole.

3. Evaluate equation (19.75) when $\varrho_{Hm_H;Hm'_H}(0) = \delta_{m_H,1}\delta_{m'_H,1}$, $H = 1$, and $G = 0$. Use the expression for the field operator to show that the field is circularly polarized if the emission is along Z.
4. Consider two atoms located at $\mathbf{R}_1 = -R_{21}\hat{\mathbf{x}}/2$ and $\mathbf{R}_2 = R_{21}\hat{\mathbf{x}}/2$, with $k_0 R_{21} \gg 1$ ($k_0 = \omega_{HG}/c$). The two atoms are prepared in superpositions of their $G = 0$ and $H = 1$, $m_H = 0$ states at $t = 0$. Calculate the average radiated field intensity in the radiation zone and for $R \gg R_{21}$, assuming that the radiated field is the sum of the fields radiated by each atom. (In other words, neglect any scattering of the field emitted by one atom at the site of the other atom, based on the assumption that $k_0 R_{21} \gg 1$.) Show that the intensity consists of the sum of the intensities of the two atoms separately, plus an interference term that depends on the dipole coherence created by the excitation field.
5. Continue problem 4 and show that the total energy emitted by the atoms is equal to

$$W = \hbar\omega_0 \left\{ \varrho_{22}^{(1)}(0) + \varrho_{22}^{(2)}(0) + 3\frac{\sin x}{x}\mathrm{Re}\left[\varrho_{12}^{(1)}(0)\varrho_{21}^{(2)}(0)\right] \right\},$$

where state 2 is the $H = 1$, $m_H = 0$ state, and $x = k_0 R_{21}$. Thus, it would seem that the energy can be *greater* or *less* than the total energy $\hbar\omega_0[\varrho_{22}^{(1)}(0) + \varrho_{22}^{(2)}(0)]$ stored in the atoms. Actually, even though cooperative decay effects are small, they add a correction that cancels the last term in the expression for W.
6. In the linear absorption problem, assume that the beam has cross-sectional area A and that the medium has length L. Starting with equation (19.87), prove that the rate at which energy is being lost by the field is equal to the power radiated into previously unoccupied vacuum field modes. Assume that the average spacing between the atoms is much greater than a wavelength, and neglect any cooperative decay effects.

7–8. Use the source-field approach to calculate the signal emitted by atoms contained in a cylinder of cross-sectional area A and length L that have been excited by a pulse propagating along the axis of the cylinder. Assume that cooperative effects can be ignored ($\mathcal{N}\lambda^2 L \ll 1$, where $\mathcal{N}$ is the density) and that $\lambda^2 \ll A$. The emitted field consists of an incoherent part emitted into 4π solid angle and the free polarization decay signal emitted primarily in the forward direction. Calculate both contributions to the signal. (To calculate the free polarization decay signal, it will help to replace the sums over particles to integrals over the volume.) Show that the total energy in the field for times much longer than the excited-state decay time associated with the *incoherent* part of the signal is equal to the original energy stored in the medium. As in problem 4, cooperative decay effects play a small role, but provide just the right correction to allow for conservation of energy. Assume that the atoms are stationary and that the incident field is z-polarized and drives a transition between a $G = 0$ ground state and $H = 1$ excited state. Show that the free polarization decay signal is of order $\mathcal{N}\lambda^2 L$ that of the incoherent signal.

9. Starting with equation (19.75), show that the total energy radiated by the field is $\hbar\omega_0$. You should note that, with the definition of $\mathbf{E}^+$, the time-averaged Poynting vector is $2\epsilon_0 c \left\langle \mathbf{E}^-(\mathbf{R}, t) \cdot \mathbf{E}^+(\mathbf{R}, t) \right\rangle$.

References

[1] H.B. Callen and T.A. Welton, *Irreversibility and generalized noise*, Physical Review **83**, 34–40 (1951).

[2] M. Lax, *Quantum noise. XI. Multitime correspondence between quantum and classical stochastic processes*, Physical Review **172**, 350–361 (1968).

[3] C. Cohen-Tannoudji, *Atoms in strong resonant fields,* in *Frontiers in Laser Spectroscopy*, Les Houches, Session XXVII, vol. 1, edited by R. Balian, S. Haroche, and S. Liberman, (North Holland, Amsterdam, 1975) pp. 88–98; C. Cohen-Tannoudji, J. Dupont-Roc, and G. Grynberg, *Atom-Photon Interactions* (Wiley-Interscience, New York, 1992), complement A_V.

[4] R.H. Dicke, *Coherence in spontaneous radiation processes*, Physical Review **93**, 99–110 (1954).

[5] D. Polder, M.F.H. Schuurmans, and Q.H.F. Vrehen, *Superfluorescence: quantum-mechanical derivation of Maxwell-Bloch description with fluctuating field source*, Physical Review A **19**, 1192–1203 (1979).

[6] R. Loudon, *The Quantum Theory of Light*, 3rd ed., (Oxford University Press, Oxford, UK, 2003), sec. 7.8, and references therein.

[7] M. Abramowitz and I.A. Stegun, *Handbook of Mathematical Functions with Formulas, Graphs, and Mathematical Tables* (U.S. Department of Commerce, Washington, DC, 1965), chap. 10.

[8] J.D. Jackson, *Classical Electrodynamics*, 3rd ed., (Wiley, New York, 1999) sec. 9.2.

[9] A. Corney and G.W. Series, *Theory of resonance fluorescence excited by modulated or pulsed light*, Proceedings of the Physical Society **83**, 207–216 (1964); S. Haroche, J.A. Paisner, and A.L. Schawlow, *Hyperfine quantum beats observed in Cs vapor under pulsed dye laser excitation*, Physical Review Letters **30**, 948–951 (1973).

[10] See, for example, R. Newton, *Scattering Theory of Waves and Particles*, 2nd ed. (Springer-Verlag, New York, 1982; Dover, New York, 2002), sec. 1.3.9.

Bibliography

For an excellent review of the Heisenberg approach to atom-field dynamics, see

P. Milonni, *Semiclassical and quantum electrodynamical approaches in nonrelativistic radiation theory*, Physics Reports **25**, 1–81 (1976), and references therein. Milonni is a strong proponent of the use of the Heisenberg equations of motion.

Some early papers on superradiance are listed here:

C.M. Bowden and D.W. Howgate, Eds., *Cooperative Effects in Matter and Radiation* (Plenum Press, New York, 1977).

R. Friedberg and S.R. Hartmann, *Frequency shifts in emission and absorption by resonant systems of two-level atoms*, Physics Reports **7**, 101–179 (1973).

I.P. Herman, J.C. MacGillivray, N. Skribanowitz, and M.S. Feld, *Self-induced emission in optically pumped HF gas: the rise and fall of the superradiant state*, in *Laser Spectrocopy*, edited by R. G. Brewer and A. Mooradian (Plenum Press, New York, 1974), 379–412.

N.E. Rehler and J.H. Eberly, *Superradiance*, Physical Review A **3**, 1735–1751 (1979).

In addition to the quantum optics texts listed in chapter 1, some additional references for noise properties of the field are:

H. Haken, *Laser Theory* (Springer-Verlag, Berlin, 1984).

M. Lax, *Fluctuation and coherence phenomena* in *Classical and Quantum Physics, Brandeis University Summer Institute* in *Theoretical Physics, 1966, Statistical Physics, Phase Transitions and Superfluidity*, edited by M. Chrétien, E. P. Gross, and S. Deser, vol. II (Gordon and Breach, New York, 1968), pp. 270–478.

W.H. Louisell, *Quantum Statistical Properties of Radiation* (Wiley, New York, 1973).

L. Mandel and E. Wolf, *Optical Coherence and Quantum Optics* (Cambridge University Press, Cambridge, UK, 1995), chap. 17.

20

Light Scattering

Light scattering is responsible for many pleasures of life. The color of the sky, beautiful sunsets, and colors in insect wings are attributable to light scattering. Moreover, light scattering represents a fundamental physical process that illustrates many interesting features of the interaction of radiation with matter. It would not be difficult to write an entire book on this subject, so our discussion will, of necessity, be limited in scope.

With the development of laser sources, new classes of light scattering experiments became possible, owing to the relatively large power of the laser fields. In the late 1960s and 1970s, the problem of light scattering received considerable attention. The question was simple. If an intense, monochromatic radiation field interacts with a two-level atom, what is the spectrum of radiation scattered by the atom? Although the initial interest in this problem has subsided somewhat, it is still rewarding to review the underlying physics associated with light scattering. Often, this problem is discussed in terms of quantized dressed states of the atoms and the field. Such a treatment allows one to identify the basic structure of the light scattering spectrum and to easily evaluate the *integrated* intensity of the various components of the spectrum. To get the line strengths themselves involves a more complicated calculation using dressed states. A standard method for treating this problem is the *quantum regression theorem* [1,2]. We present what we believe to be a somewhat simpler calculation of the scattering spectrum using semiclassical dressed states, although we show how to use quantized dressed states to calculate the integrated intensity of the spectral components of the scattered field.

For the most part, we consider scattering by a single atom on a $G = 0$ to $H = 1$ transition that is driven by a cw laser field of arbitrary strength polarized in the $\hat{\mathbf{z}}$ direction. As such, the problem reduces to that of scattering of light by a two-level atom. In somewhat qualitative terms, we then discuss the correlation function of the scattered field, as well as scattering on transitions having arbitrary angular momentum.

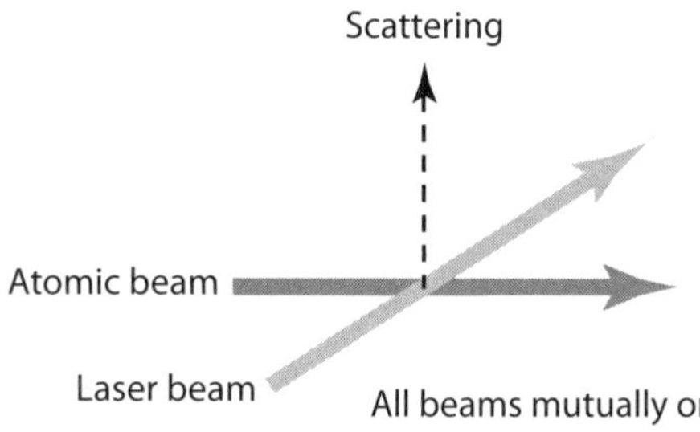

Figure 20.1. Schematic diagram illustrating a geometry in which light scattering can be observed without Doppler broadening.

20.1 General Considerations: Perturbation Theory

As noted earlier, we consider first scattering by a single atom on a $G = 0$ to $H = 1$ transition that is driven by a cw laser field of arbitrary strength, polarized in the $\hat{\mathbf{z}}$ direction. We label the $G = 0$ state as $|1\rangle$ and the $H = 1$, $m_H = 0$ state as $|2\rangle$, with $\omega_{21} \equiv \omega_0$.

The scattering geometry is represented schematically in figure 20.1. A single-mode laser beam is incident on an atomic beam, and the spectrum of radiation scattered perpendicular to the beam is observed. The incident field and scattered field are both perpendicular to the direction of motion of the atom. If we neglect any recoil the atom undergoes as a result of the scattering, the problem is equivalent to that of scattering by a stationary atom, provided that the atom stays in the beam for a sufficiently long time to establish a constant scattering rate. We make this assumption and limit this discussion to scattering by a stationary atom.

In a strong field, an amplitude approach is all but useless, since many fluorescence photons are emitted, and each one repopulates the ground state. However, some insight into the problem can be obtained by looking at a perturbative solution using an amplitude approach.

The Hamiltonian in the RWA for the atom–field system is

$$\begin{aligned}\mathrm{H} = {} & \frac{\hbar\omega_0}{2}\boldsymbol{\sigma}_z + \sum_k \hbar\omega_k a_{\mathbf{k}}^\dagger a_{\mathbf{k}} + \hbar\chi\left(\boldsymbol{\sigma}_+ e^{-i\omega t} + \boldsymbol{\sigma}_- e^{i\omega t}\right) \\ & + \sum_{\mathbf{k}} \hbar\left(g_{\mathbf{k}}\boldsymbol{\sigma}_+ a_{\mathbf{k}} + g_{\mathbf{k}}^* a_{\mathbf{k}}^\dagger \boldsymbol{\sigma}_-\right), \qquad (20.1)\end{aligned}$$

where

$$g_{\mathbf{k}} = -i\mu_{21}\sin\theta_{\mathbf{k}}\left(\frac{\omega_k}{2\hbar\epsilon_0\mathcal{V}}\right)^{1/2}, \qquad (20.2)$$

all other symbols have their conventional meanings, and $\chi = -\mu_{21}E_0/2\hbar$ (assumed real) is one-half the Rabi frequency associated with a field having amplitude E_0 and frequency ω driving the 1–2 transition. This "two-state" atom can interact with only one polarization component (the $\boldsymbol{\epsilon}_{\mathbf{k}}^{(1)} = \hat{\boldsymbol{\theta}}_{\mathbf{k}}$ component) of the vacuum field owing to the selection rules for the atomic transition. The angular dependence of the scattered radiation is contained in the $\sin\theta_{\mathbf{k}}$ factor.

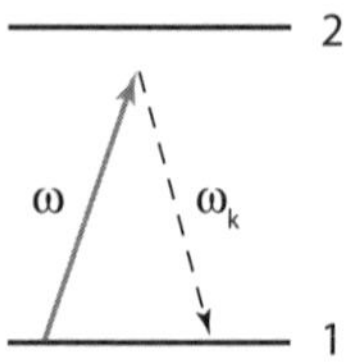

Figure 20.2. Scattering to first order in the applied field.

To lowest order in the applied fields, the only type of diagram that contributes is shown in figure 20.2. Starting from the ground state, the incident radiation is scattered into a previously unoccupied mode of the vacuum field. From conservation of energy, one must have $\omega_k = \omega$ *exactly*, independent of the excited state width, given the fact that we neglect any recoil that the atom undergoes as a result of the scattering. This is nothing more than Rayleigh scattering, corresponding to a two-photon transition in which the initial and final state widths are equal to zero.

The Rayleigh scattered spectrum can be calculated without much difficulty. To this order, the only nonvanishing state amplitudes for the atom–field system are $c_{1;0}$, $c_{2;0}$, $c_{1;\mathbf{k}}$, where the first subscript is the atomic state and the second the field state, all in the standard interaction representation. The equations for these state amplitudes, obtained using the Hamiltonian (20.1), are

$$\dot{c}_{2;0} = -\frac{\gamma_2}{2}c_{2;0} - i\chi e^{i\delta t}c_{1,0}, \tag{20.3a}$$

$$\dot{c}_{1,\mathbf{k}} \approx -ig_{\mathbf{k}}e^{-i\delta_k t}c_{2,0}, \tag{20.3b}$$

where $\delta = \omega_0 - \omega$, and

$$\delta_k = \omega_0 - \omega_k. \tag{20.4}$$

There should be an additional term in equation (20.3b), $-i\chi e^{-i\delta t}c_{2,\mathbf{k}}$, but it is is neglected in this order. The $-\gamma_2/2$ term was obtained using the Weisskopf-Wigner approximation.

In perturbation theory, we set $c_{1,0} = 1$ and calculate $|c_{1,\mathbf{k}}|^2$ for $t \gg 1/\gamma_2$. Only terms in $|c_{1,\mathbf{k}}|^2$ proportional to t are of interest here—the rest are transients. The solution of equation (20.3a) is

$$c_{2,0}(t) \sim -i\chi \int_0^t e^{-\frac{\gamma_2}{2}(t-t')}e^{i\delta t'}dt' = \frac{-i\chi}{\frac{\gamma_2}{2} + i\delta}\left(e^{i\delta t} - e^{-\gamma_2 t/2}\right). \tag{20.5}$$

The $e^{-\gamma_2 t/2}$ term is a transient that does not contribute a term linear in t to $|c_{1\mathbf{k}}|^2$ and can be dropped. As a result, we find

$$\begin{aligned} c_{1,\mathbf{k}} &= -ig_{\mathbf{k}}\int_0^t e^{-i\delta_k t'}c_{2,0}(t')dt' \approx -\frac{g_{\mathbf{k}}\chi}{\frac{\gamma_2}{2} + i\delta}\int_0^t e^{i\Delta_k t'}dt' \\ &= -\frac{g_{\mathbf{k}}\chi}{\frac{\gamma_2}{2} + i\delta}\frac{2\sin(\Delta_k t/2)}{\Delta_k}e^{i\Delta_k t/2} \end{aligned} \tag{20.6}$$

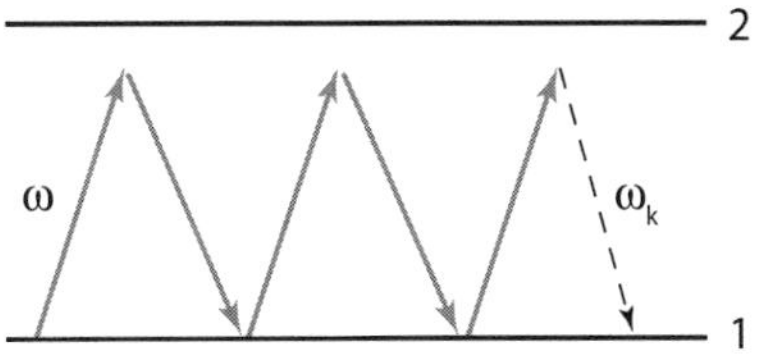

Figure 20.3. A higher order diagram that gives rise to elastic scattering.

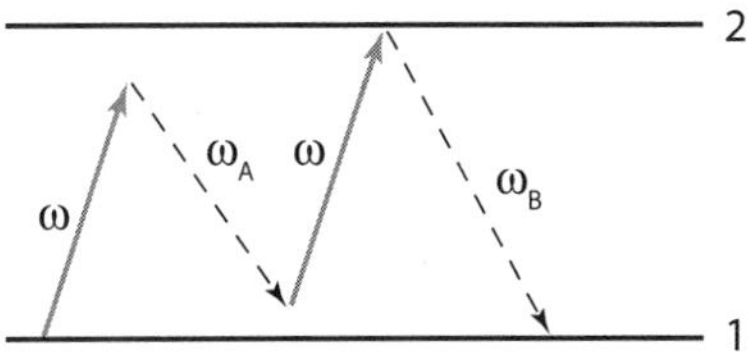

Figure 20.4. Scattering to second order in the applied field.

and

$$|c_{1,\mathrm{k}}|^2 = \frac{|g_\mathrm{k}\chi|^2}{\gamma^2+\delta^2}\frac{\sin^2[\Delta_k t/2]}{(\Delta_k/2)^2}, \tag{20.7}$$

where $\gamma = \gamma_2/2$, and the detuning

$$\Delta_k = \omega_k - \omega \tag{20.8}$$

is measured relative to the incident field frequency. In the limit that $t \to \infty$,

$$\sin^2[\Delta_k t/2]/(\Delta_k/2)^2 \sim 2\pi t \delta_D(\Delta_k), \tag{20.9}$$

$$|c_{1\mathrm{k}}|^2 = \frac{|g_\mathrm{k}\chi|^2}{\gamma^2+\delta^2} 2\pi t \delta_D(\Delta_k), \tag{20.10}$$

and the scattered spectrum $\mathcal{I}_\mathrm{k}$ is

$$\mathcal{I}_\mathrm{k} \equiv \lim_{t\to\infty}\frac{d}{dt}|c_{1\mathrm{k}}|^2 \sim \frac{2\pi|g_\mathrm{k}\chi|^2}{\gamma^2+\delta^2}\delta_D(\Delta_k), \tag{20.11}$$

where δ_D is the Dirac delta function.

As predicted, the scattered radiation is at the frequency of the incident radiation, independent of the excited-state width. In higher order, terms (diagrams) like those in figure 20.3 involving no intermediate-state resonance, still contribute to the δ_D function; however, they are saturation-type terms that reduce the overall amplitude of this contribution. Scattered radiation at the laser frequency having zero width is referred to as *elastic scattering*. In effect, the solution (20.11) corresponds to scattering by a classical dipole that has been excited by the incident field.

The lowest order diagram involving *two* scattered photons is shown in figure 20.4 (reference [3]), with an overall resonance condition

$$2\omega - \omega_A - \omega_B = 0, \tag{20.12}$$

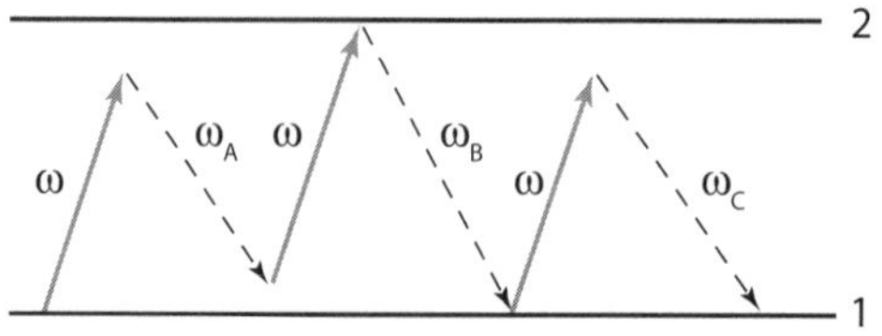

Figure 20.5. Scattering to third order in the applied field.

where ω_A and ω_B are the frequencies associated with the scattered radiation. Now we can have an *intermediate-state resonance* if

$$|2\omega - \omega_A - \omega_0| \lesssim \gamma, \tag{20.13}$$

which implies that

$$|\omega_B - \omega_0| = |2\omega - \omega_A - \omega_0| \lesssim \gamma. \tag{20.14}$$

Thus, there are resonances at $\omega_k = \omega_B$ and $\omega_k = \omega_A = 2\omega - \omega_0$. In terms of δ, these resonances occur at $\omega_k = \omega + \delta$ and $\omega_k = \omega - \delta$. The two components of the spectrum are symmetric about the *laser* frequency and not the atomic frequency; this feature is consistent with the scattering picture of "absorption" we have adopted throughout this book. The width (FWHM) of these *inelastic* spectral components is 2γ, arising from the intermediate state resonance in the scattering interaction. Since one peak is centered at the atomic frequency in this low-intensity approximation, its position could be used as a measure of the atomic transition frequency.

The lowest contribution involving three scattered photons is shown in figure 20.5. The overall resonance condition is

$$3\omega - (\omega_A + \omega_B + \omega_C) = 0, \tag{20.15}$$

with intermediate resonances at

$$|2\omega - \omega_A - \omega_0| \lesssim \gamma, \tag{20.16a}$$

$$|3\omega - \omega_A - \omega_B - \omega_0| \lesssim \gamma. \tag{20.16b}$$

Equations (20.15) and (20.16b) together result in $|\omega_C - \omega_0| \lesssim \gamma$ or

$$|\omega_C - (\omega + \delta)| \lesssim \gamma. \tag{20.17}$$

Equation (20.16a) gives $|\omega_A - (2\omega - \omega_0)| \lesssim \gamma$ or

$$|\omega_A - (\omega - \delta)| \lesssim \gamma. \tag{20.18}$$

Last, equations (20.16a) and (20.16b), *taken together*, give rise to the resonance condition

$$|\omega_B - \omega| \lesssim 2\gamma. \tag{20.19}$$

As a consequence, there are three components, symmetric about the laser frequency. The central component has twice the width of the sidebands in this low-intensity approximation.

Higher order terms do not change the number of resonances, but simply alter the positions of the sidebands, the widths of the inelastic components, and the relative amplitudes of the inelastic and elastic components. The three components are often referred to as the *Mollow triplet* [4]. To calculate the weights and widths of the resonances, a calculation to all orders in χ is needed. Here, we use a density matrix approach and semiclassical dressed states [5].

20.2 Mollow Triplet

Let $\varrho_{1,0;1,0}$ be the probability for the atom to be in state 1 with no photons of type **k** present in the field. This is not the probability to be in state 1 with no photons present—only no photons of the type **k** we are looking for. This is the "trick" we can use. The probability for scattering into a *given* mode **k** is always negligibly small and can be treated by perturbation theory, even if the *sum* over all these modes repopulates the ground state. Thus, we can define our spectrum as the trace over atomic states of the time rate of change of the probability to have a photon of type **k** present—namely,

$$\mathcal{I}_{\mathbf{k}} = \frac{d}{dt}\left(\varrho_{1,\mathbf{k};1,\mathbf{k}} + \varrho_{2,\mathbf{k};2,\mathbf{k}}\right), \tag{20.20}$$

which involves a trace over atomic states. We can equally well use semiclassical dressed states and take

$$\mathcal{I}_{\mathbf{k}} = \frac{d}{dt}\left(\varrho_{I,\mathbf{k};I,\mathbf{k}} + \varrho_{II,\mathbf{k};II,\mathbf{k}}\right), \tag{20.21}$$

where the dressed states are defined via equation (2.126). We are interested in these expressions for times $t \gg \gamma_2^{-1}$ when all transients have died away. The Hamiltonian is given by equation (20.1). Thus, the goal of the calculation is to evaluate equation (20.21).

For the moment, we neglect relaxation. The only states that we need consider are $|1,0\rangle$, $|2,0\rangle$, $|1,\mathbf{k}\rangle$, $|2,\mathbf{k}\rangle$. In a field interaction representation, the state vector for the system can be written as

$$\begin{aligned} |\psi(t)\rangle &= \tilde{c}_{1,0}(t)e^{i\omega t/2}|1,0\rangle + \tilde{c}_{2,0}(t)e^{-i\omega t/2}|2,0\rangle \\ &\quad + \tilde{c}_{1,\mathbf{k}}(t)e^{-i\omega t/2}|1,\mathbf{k}\rangle + \tilde{c}_{2,\mathbf{k}}(t)e^{-3i\omega t/2}|2,\mathbf{k}\rangle. \end{aligned} \tag{20.22}$$

In this basis, the Hamiltonian (20.1) can be represented as the 4×4 matrix

$$\tilde{\mathbf{H}} = \hbar \begin{pmatrix} -\delta/2 & \Omega_0/2 & 0 & 0 \\ \Omega_0/2 & \delta/2 & g_{\mathbf{k}} & 0 \\ 0 & g_{\mathbf{k}}^* & -\delta/2 + \Delta_k & \Omega_0/2 \\ 0 & 0 & \Omega_0/2 & \delta/2 + \Delta_k \end{pmatrix}, \tag{20.23}$$

where $\Omega_0 = 2\chi$.

The transformation to the dressed-state basis $|I,0\rangle$, $|II,0\rangle$, $|I,\mathbf{k}\rangle$, $|II,\mathbf{k}\rangle$ is given by

$$\mathbf{H}_d = \begin{pmatrix} \mathbf{T} & 0 \\ 0 & \mathbf{T} \end{pmatrix} \tilde{\mathbf{H}} \begin{pmatrix} \mathbf{T}^\dagger & 0 \\ 0 & \mathbf{T}^\dagger \end{pmatrix}, \tag{20.24}$$

where

$$\mathbf{T} = \begin{pmatrix} c & -s \\ s & c \end{pmatrix}, \qquad (20.25)$$

$$c = \cos\theta, \quad s = \sin\theta, \qquad (20.26)$$

and θ is the dressed-state angle defined by equation (2.135), $\tan(2\theta) = \Omega_0/\delta$. Carrying out the transformation on the Hamiltonian (20.1), we find

$$\mathbf{H}_d = \hbar \begin{pmatrix} -\Omega\sigma_z/2 & \mathbf{V} \\ \mathbf{V}^\dagger & -\Omega\sigma_z/2 + \Delta_k \mathbf{1} \end{pmatrix}, \qquad (20.27)$$

where $\Omega = (\delta^2 + \Omega_0^2)^{1/2}$ is the generalized Rabi frequency of the classical field, $\mathbf{1}$ is the 2×2 unit matrix, and

$$\mathbf{V} = g_{\mathbf{k}} \begin{pmatrix} -sc & -s^2 \\ c^2 & sc \end{pmatrix}. \qquad (20.28)$$

The energy levels of the dressed Hamiltonian consist of two doublets of levels that are coupled by the vacuum field; these are analogous to the types of coupling we describe in section 20.2.1 in terms of quantized-field dressed states.

Using equations (20.27), (20.28), and $\dot{\varrho}_d = (i\hbar)^{-1}[\mathbf{H}_d, \varrho_d]$, it follows that the spectrum (20.21) is given by

$$\begin{aligned} \mathcal{I}_{\mathbf{k}} &= \lim_{t \gg \gamma_2^{-1}} \frac{d}{dt}\left(\varrho_{I,\mathbf{k};I,\mathbf{k}} + \varrho_{II,\mathbf{k};II,\mathbf{k}}\right) \\ &= -ig_{\mathbf{k}}^* \lim_{t \gg \gamma_2^{-1}} \left(c^2 \varrho_{II,0;I,\mathbf{k}} - s^2 \varrho_{I,0;II,\mathbf{k}} + sc\mathcal{D}_{\mathbf{k}}\right) + \text{c.c.}, \end{aligned} \qquad (20.29)$$

where

$$\mathcal{D}_{\mathbf{k}} = \varrho_{II,0;II,\mathbf{k}} - \varrho_{I,0;I,\mathbf{k}}. \qquad (20.30)$$

The inclusion of relaxation does not modify equation (20.29), since the relaxation terms cancel in the summation. The beauty of this expression is that each term can be identified with one component of the scattering triplet, *even when these components are not resolved*. The problem reduces to finding the density matrix elements $\varrho_{II,0;I,\mathbf{k}}$, $\varrho_{I,0;II,\mathbf{k}}$, $\varrho_{II,0;II,\mathbf{k}}$, and $\varrho_{I,0;I,\mathbf{k}}$.

The density matrix elements are obtained by solving

$$\dot{\varrho}_d = (i\hbar)^{-1}[\mathbf{H}_d, \varrho_d] + \text{relaxation terms}, \qquad (20.31)$$

where the relaxation terms are given in equations (3.103). We adopt the simplest possible relaxation model, in which there are no collisions ($\gamma = \gamma_2/2$). With this relaxation model, we can use equations (20.31), (20.27), (20.28), and (3.103) to obtain

$$\dot{S}_k = -(\epsilon - i\Delta_k)S_k + ig_k sc(\varrho_{II,II} - \varrho_{I,I}) + ig_k\left(c^2\varrho_{I,II} - s^2\varrho_{II,I}\right), \quad (20.32a)$$

$$\dot{D}_k = -\left[(c^4+s^4)\gamma_2 - i\Delta_k\right]D_k - (c^2-s^2)\gamma_2 S_k + ig_k sc - ig_k\left(c^2\varrho_{I,II} + s^2\varrho_{II,I}\right) + sc\left(c^2-s^2\right)\gamma_2\left(\varrho_{I,0;II,k} + \varrho_{II,0;I,k}\right), \quad (20.32b)$$

$$\dot{\varrho}_{II,0;I,k} = -\left[\left(\frac{1}{2}+s^2c^2\right)\gamma_2 - i(\Delta_k - \Omega)\right]\varrho_{II,0;I,k} + ig_k\left(c^2\varrho_{II,II} - sc\varrho_{II,I}\right) + \frac{1}{2}sc\left(c^2-s^2\right)\gamma_2 D_k + sc\gamma_2 S_k - s^2c^2\gamma_2\varrho_{I,0;II,k}, \quad (20.32c)$$

$$\dot{\varrho}_{I,0;II,k} = -\left[\left(\frac{1}{2}+s^2c^2\right)\gamma_2 - i(\Delta_k + \Omega)\right]\varrho_{II,0;I,k} - ig_k\left(s^2\varrho_{I,I} - sc\varrho_{I,II}\right) + \frac{1}{2}sc\left(c^2-s^2\right)\gamma_2 D_k + sc\gamma_2 S_k - s^2c^2\gamma_2\varrho_{II,0;I,k}, \quad (20.32d)$$

where

$$S_k = \varrho_{I,0;I,k} + \varrho_{II,0;II,k}, \quad (20.33a)$$

$$D_k = \varrho_{II,0;II,k} - \varrho_{I,0;I,k}, \quad (20.33b)$$

$\varrho_{\alpha,\beta} \equiv \varrho_{\alpha,0;\beta,0}$, and we used the fact that $(\varrho_{I,I} + \varrho_{II,II}) = 1$. In equation (20.32a), the factor $-i\Delta_k$ was replaced by $(\epsilon - i\Delta_k)$ to avoid divergences. Eventually, we will take the limit where $\epsilon \to 0$ to recover the elastic contribution to the spectrum.

The $\varrho_{\alpha,\beta}$ are obtained by solving equations (3.103), which are reproduced here as

$$\dot{\varrho}_{I,I} = -\gamma_2 s^4\varrho_{I,I} + \frac{sc}{2}\gamma_2\left(1-2c^2\right)\left(\varrho_{I,II} + \varrho_{II,I}\right) + \gamma_2 c^4\varrho_{II,II}, \quad (20.34a)$$

$$\dot{\varrho}_{II,II} = -\gamma_2 c^4\varrho_{II,II} + \frac{sc}{2}\gamma_2\left(1-2s^2\right)\left(\varrho_{I,II} + \varrho_{II,I}\right) + \gamma_2 s^4\varrho_{II,II}, \quad (20.34b)$$

$$\dot{\varrho}_{I,II} = -\left[\left(\frac{1}{2}+s^2c^2\right)\gamma_2 - i\Omega\right]\varrho_{I,II} + (sc/2)\gamma_2\left(1+2s^2\right)\varrho_{I,I} + (sc/2)\gamma_2\left(1+2c^2\right)\varrho_{II,II} - s^2c^2\gamma_2\varrho_{II,I}, \quad (20.34c)$$

$$\dot{\varrho}_{II,I} = -\left[\left(\frac{1}{2}+s^2c^2\right)\gamma_2 + i\Omega\right]\varrho_{II,I} + (sc/2)\gamma_2\left(1+2s^2\right)\varrho_{I,I} + (sc/2)\gamma_2\left(1+2c^2\right)\varrho_{II,II} - s^2c^2\gamma_2\varrho_{I,II}. \quad (20.34d)$$

We have not yet made the secular approximation so that these equations are exact. It can be deduced from equations (20.29) and (20.32) that there are three features in the spectrum, centered near $\delta_k = 0, \pm\Omega$. Equations (20.32) and (20.34) can be solved in steady state, with the results substituted into equation (20.29), to obtain the spectrum.

If $\Omega \gg \gamma_2$ (secular approximation), equations (20.32) simplify considerably. In steady state, they become

$$(\epsilon - i\Delta_k)S_{\mathbf{k}} = ig_{\mathbf{k}}sc\left(\varrho_{II,II} - \varrho_{I,I}\right), \tag{20.35a}$$

$$\left[\left(c^4 + s^4\right)\gamma_2 - i\Delta_k\right]D_{\mathbf{k}} + \left(c^2 - s^2\right)\gamma_2 S_{\mathbf{k}} = ig_{\mathbf{k}}sc, \tag{20.35b}$$

$$\left[\left(\frac{1}{2} + s^2c^2\right)\gamma_2 - i(\Delta_k - \Omega)\right]\varrho_{II,0;I,\mathbf{k}} = ig_{\mathbf{k}}c^2\varrho_{II,II}, \tag{20.35c}$$

$$\left[\left(\frac{1}{2} + s^2c^2\right)\gamma_2 - i(\Delta_k + \Omega)\right]\varrho_{I,0;II,\mathbf{k}} = -ig_{\mathbf{k}}s^2\varrho_{I,I}, \tag{20.35d}$$

where we have set $\varrho_{II,I} \approx 0$, $\varrho_{II,I} \approx 0$, and

$$\varrho_{I,I} = \frac{c^4}{c^4 + s^4}, \qquad \varrho_{II,II} = \frac{s^4}{c^4 + s^4}, \tag{20.36}$$

consistent with equations (3.63). It is clear that $\varrho_{I,I}$ is the source of the sideband at $\Delta_k = \Omega$, $\varrho_{II,II}$ is the source of the sideband at $\Delta_k = -\Omega$, and $(\varrho_{II,II} - \varrho_{I,I})$ is the source of the central inelastic component of the spectrum. In writing equation (20.35b), we were able to neglect the last term in equation (20.32b) since it is small near $\Delta_k = 0$.

Solving equations (20.35), substituting the results into equation (20.29), and taking the limit $\epsilon \to 0$, we obtain the scattered spectrum or Mollow triplet,

$$\mathcal{I}_{\mathbf{k}} \simeq \frac{|g_{\mathbf{k}}|^2 s^2 c^2}{c^4 + s^4}\left[\begin{array}{c}\frac{2s^2c^2\gamma_s}{\gamma_s^2+(\Delta_k+\Omega)^2} + \frac{2s^2c^2\gamma_s}{\gamma_s^2+(\Delta_k-\Omega)^2}\\ + \frac{8s^4c^4}{c^4+s^4}\frac{\gamma_0}{\gamma_0^2+\Delta_k^2} + \frac{2\pi\left(c^2-s^2\right)^2}{c^4+s^4}\delta_D(\Delta_k)\end{array}\right], \tag{20.37}$$

where

$$\gamma_s = \left(\frac{1}{2} + s^2c^2\right)\gamma_2, \tag{20.38a}$$

$$\gamma_0 = \left(c^4 + s^4\right)\gamma_2. \tag{20.38b}$$

For a resonant driving field, $\delta = 0$, $c = s = 1/\sqrt{2}$, the ratios of each sideband to the central component are the following: height ratio, 1:3; width ratio, 3:2; and integrated signal ratio, 1:2. For $\delta \neq 0$, the relative contribution of the elastic component decreases with increasing field strength. If $\delta = 0$, the elastic component vanishes identically in the secular approximation, since this corresponds to a limit in which the dipole coherence goes to zero. In weak fields ($s \ll 1$), the widths agree with those predicted from perturbation theory in section 20.1. A three-peaked scattering spectrum of this type was first observed by Wu et al. [6].

For a $G = 0$ to $H = 1$ transition and an incident field polarized in the $\hat{\mathbf{z}}$ direction, the scattered spectrum varies as $\sin^2\theta_{\mathbf{k}}$, since $g_{\mathbf{k}}$ is proportional to $\sin\theta_{\mathbf{k}}$ and the scattered field is polarized in the $\hat{\boldsymbol{\theta}}_{\mathbf{k}}$ direction. For transitions having different values of G and H, the *elastic* component of the scattered spectrum is always proportional to $\sin^2\theta_{\mathbf{k}}$ and is polarized in the $\hat{\boldsymbol{\theta}}_{\mathbf{k}}$ direction, but the inelastic component of the spectrum has an isotropic, unpolarized component, as was discussed in chapter 16.

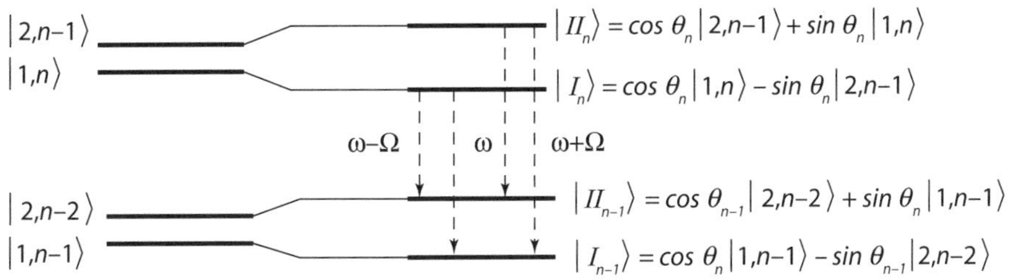

Figure 20.6. Quantized atom–field states including the atom–field interaction. The dashed arrows indicate possible spontaneous emission frequencies.

20.2.1 Dressed-State Approach to Mollow Triplet

A simple picture of the Mollow triplet can be given using quantized dressed states (see figure 20.6). Recall that the definition of the quantized dressed states involves an angle θ_n that is a function of n. In the following, we assume that the driving field is a coherent state of the radiation field with $\langle n\rangle \gg 1$ so that we can neglect differences between Ω_n and Ω_{n-1}; we evaluate them at $\Omega_{\langle n\rangle} \equiv \Omega$ and $\theta_{\langle n\rangle} \equiv \theta$. The incident radiation is scattered into modes of the vacuum field that are originally unoccupied. The vacuum field couples states $|2, n-1\rangle$ to $|1, n-1\rangle$, since spontaneous emission does not change the number of photons in the coherent field state. (In *subsequent* interactions with the coherent-state field, $|1, n-1\rangle$ is driven to state $|2, n-2\rangle$, which involves a reduction in photon number of the single-mode field.) This is a key point to remember—it is only in excitation of the atom that the number of photons in the applied field is reduced, not as a direct result of spontaneous emission. Spontaneous emission couples states with the same n. Since $|I_n\rangle$ and $|II_n\rangle$ each contain admixtures of states $|1, n\rangle$ and $|2, n-1\rangle$, and since $|I_{n-1}\rangle$ and $|II_{n-1}\rangle$ contain admixtures of $|1, n-1\rangle$ and $|2, n-2\rangle$, spontaneous emission couples each state in a doublet to each of the states in a doublet immediately below it.

As a consequence, one sees immediately from figure 20.6 that the frequency of the scattered spectrum can occur at frequencies ω, $\omega \pm \Omega$, the same result we found using semiclassical dressed states. The dressed picture lets us determine easily the positions of possible resonances.

Spontaneous emission does not significantly modify the dressed-state evolution only if $\Omega \gg \gamma_2$ (secular approximation). If this inequality does not hold, states $|I_n\rangle$ and $|II_n\rangle$ are mixed by spontaneous emission. For example, as a result of spontaneous emission,

$$\dot{\varrho}_{2,n-1;2,n-1} = -\gamma_2 \varrho_{2,n-1;2,n-1}. \tag{20.39}$$

In the dressed basis, this leads to coupling of ϱ_{II_n,II_n} to ϱ_{I_n,I_n}, ϱ_{I_n,II_n}, and ϱ_{II_n,I_n}. There is an advantage to using the dressed-state picture only if $\Omega \gg \gamma_2$. In that case, $\varrho_{I_n,II_n} \simeq 0$, and we can calculate the effect of spontaneous emission using simple rate equations.

Suppose that $\Omega \gg \gamma_2$. Then the steady-state values of the dressed-state populations in each doublet is the same (neglecting variations in n). Steady state is reached in time of order γ_2^{-1}. The relative populations of each states is the same in all the doublets, neglecting variations in n. Thus, the steady-state populations,

$$\varrho_{I,I} = \frac{c^4}{c^4 + s^4}, \tag{20.40a}$$

$$\varrho_{II,II} = \frac{s^4}{c^4 + s^4}, \tag{20.40b}$$

are exactly what we found using semiclassical dressed states, where

$$\begin{aligned} \varrho_{I,I} &= \sum_n \varrho_{I_n,I_n}, \\ \varrho_{II,II} &= \sum_n \varrho_{II_n,II_n}. \end{aligned} \tag{20.41}$$

It is now an easy matter to calculate the *integrated* rates for spontaneous emission of the three components using simple rate approximations. The total emission rate is simply equal to the population of a given state times the square of the matrix element taking it to the next lower doublet. For example, consider the $(\omega + \Omega)$ component that results from a transition from state $|II_n\rangle$ to $|I_{n-1}\rangle$. The vacuum field couples the $c|2, n-1\rangle$ component of $|II_n\rangle$ to the $c|1, n-2\rangle$ component of $|I_{n-1}\rangle$, with a matrix element that is proportional to c^2. As a consequence,

$$\mathcal{I}_{\omega+\Omega} = K\left(c^2\right)^2 \varrho_{II,II} = K\frac{c^4 s^4}{c^4 + s^4}, \tag{20.42}$$

where K is some constant. Similarly, one finds for the $(\omega - \Omega)$ component,

$$\mathcal{I}_{\omega-\Omega} = K\left(-s^2\right)^2 \varrho_{I,I} = K\frac{c^4 s^4}{c^4 + s^4}, \tag{20.43}$$

and for the ω component (which contains two contributions),

$$\mathcal{I}_{\omega} = K\left[(cs)^2 \varrho_{II,II} + (-cs)^2 \varrho_{I,I}\right] = Kc^2 s^2. \tag{20.44}$$

In effect, these results can be read directly from figure 20.6.

The relative strengths of the integrated lines are

$$\mathcal{I}_{\omega} : \mathcal{I}_{\omega+\Omega} : \mathcal{I}_{\omega-\Omega} = \frac{c^4 + s^4}{c^2 s^2} : 1 : 1, \tag{20.45}$$

consistent with equation (20.37). If $\delta \sim 0$, $\theta \sim \pi/4$, and this ratio is 2:1:1. In weak fields, the integrated sidebands are proportional to s^4, which varies as the field intensity squared, while the integrated central component varies as s^2, owing to the elastic component. The integrated *inelastic* central component varies as s^6 [as can be calculated from equation (20.37)], consistent with the perturbation theory result that this amplitude involves the scattering of three driving-field photons.

To get the widths and weights of various components requires more work [2], since cascades in the dressed-state ladder need to be considered, in which coherence is maintained in the cascade process. The semiclassical dressed-state calculation given earlier maintains the positive aspects of a dressed atom approach, while providing a somewhat more direct method for obtaining the spectrum.

20.2.2 Source-Field Approach to Mollow Triplet

We can also use the results of the source-field theory to obtain expressions for the integrated spectral components. Equation (19.37) for the scattered spectrum is

$$\begin{aligned}\mathcal{I}_{\mathrm{k}} = \langle \dot{n}_{\mathrm{k}}(t)\rangle &= 2\mathrm{Re}|g_{\mathrm{k}}|^2 \int_0^t dt' e^{i\delta_k(t-t')}\langle \sigma_+^I(t)\sigma_-^I(t')\rangle \\ &= 2\mathrm{Re}|g_{\mathrm{k}}|^2 \int_0^t d\tau e^{i\delta_k\tau}\langle \sigma_+^I(t)\sigma_-^I(t-\tau)\rangle \end{aligned} \tag{20.46}$$

in the limit that $\gamma_2 t \sim \infty$. Let

$$S_\pm(t) = \tilde{\sigma}_\pm(t) - \langle \tilde{\sigma}_\pm(t)\rangle, \tag{20.47}$$

where

$$\tilde{\sigma}_\pm(t) = \sigma_\pm(t)e^{\pm i\omega t} = \sigma_\pm^I(t)e^{\pm i\delta t}, \tag{20.48}$$

and $\delta = \omega_0 - \omega$. Then

$$\langle S_\pm(t)\rangle = 0, \tag{20.49}$$

and

$$\mathcal{I}_{\mathrm{k}} = \lim_{t\gg\gamma_2^{-1}} \mathrm{Re}\left\{2|g_{\mathrm{k}}|^2 \int_0^t d\tau e^{-i\Delta_k\tau}\left[\langle S_+(t)S_-(t-\tau)\rangle + \langle\tilde{\sigma}_+(t)\rangle\langle\tilde{\sigma}_-(t-\tau)\rangle\right]\right\}, \tag{20.50}$$

where $\langle S_+(t)S_-(t-\tau)\rangle$ gives rise to the inelastic part (fluctuations) and $\langle\tilde{\sigma}_+(t)\rangle \times \langle\tilde{\sigma}_-(t-\tau)\rangle$ to the elastic part of the scattered spectrum.

By integrating equation (20.50) over Δ_k (and evaluating ω_k in g_{k} at ω), we find

$$\begin{aligned}\mathcal{I}_{\mathrm{inelastic}} &= \lim_{t\gg\gamma_2^{-1}} K'\langle S_+(t)S_-(t)\rangle = K'\left[\langle\sigma_2\rangle - |\langle\tilde{\sigma}_+\rangle|^2\right] \\ &= K'\left(\varrho_{22} - |\tilde{\varrho}_{12}|^2\right), \end{aligned}\tag{20.51a}$$

$$\mathcal{I}_{\mathrm{elastic}} = \lim_{t\gg\gamma_2^{-1}} K'|\langle\tilde{\sigma}_+(t)\rangle|^2 = K'|\langle\tilde{\sigma}_+\rangle|^2 = K'|\tilde{\varrho}_{12}|^2, \tag{20.51b}$$

where K' is a constant, and $\tilde{\sigma}_+$, σ_2, $\tilde{\varrho}_{12}$, and ϱ_{22} are the steady-state values of these quantities. Thus, we find the ratio

$$\frac{\mathcal{I}_{\mathrm{elastic}}}{\mathcal{I}_{\mathrm{inelastic}}} = \frac{|\tilde{\varrho}_{12}|^2}{\varrho_{22} - |\tilde{\varrho}_{12}|^2} = \frac{2(\gamma^2+\delta^2)}{|\Omega_0|^2}, \tag{20.52}$$

which is consistent with equation (20.37) in the secular limit $\gamma \ll \Omega$. To arrive at the last inequality in equation (20.52), we made use of equations (4.11) and (4.14). Equation (20.52) shows that for $\delta = 0$, $\mathcal{I}_{\mathrm{elastic}}/\mathcal{I}_{\mathrm{inelastic}} = 2\gamma^2/|\Omega_0|^2$; there is always a nonsecular contribution to elastic scattering, even for $\delta = 0$.

20.3 Second-Order Correlation Function for the Radiated Field

The quantum regression theorem can be used to obtain expressions for the first- and second-order correlation functions of the scattered field [1, 7, 8]. However, we can understand the qualitative behavior of these functions with the results that we have derived. The first-order correlation function is just the Fourier transform of the spectrum. For weak incident fields, the spectrum is proportional to $\delta_D(\Delta_k)$, implying that $|g^{(1)}(\tau)|=1$; the scattered field is first-order coherent. This result is not surprising; for weak fields, the scattering is elastic and resembles scattering by a classical oscillator. With increasing field strength, the three components of the Mollow triplet get resolved, and the spectrum consists of a delta function component plus three resolved Lorentzians. As a consequence, the first-order correlation function has four components. Neglecting the overall phase factor of $e^{-i\omega\tau}$, these components consist of a constant term corresponding to the delta function in the spectrum, plus three decaying exponentials corresponding to the three Lorentzians. For $\gamma\tau \gg 1$, the exponentials decay away, and we are left with [see equations (20.51)]

$$
\begin{aligned}
g^{(1)}(\tau) &\sim \frac{\mathcal{I}_{\text{elastic}}}{\mathcal{I}_{\text{elastic}}+\mathcal{I}_{\text{inelastic}}} = \frac{|\tilde{\varrho}_{12}|^2}{|\tilde{\varrho}_{12}|^2+\left(\varrho_{22}-|\tilde{\varrho}_{12}|^2\right)} \\
&= \frac{\gamma^2+\delta^2}{\gamma^2+\delta^2+|\Omega_0|^2/2} = \frac{1}{1+\frac{|\Omega_0|^2/2}{\gamma^2+\delta^2}}.
\end{aligned}
\tag{20.53}
$$

The second-order correlation function is somewhat more interesting. Recall that

$$
g^{(2)}(\tau) = \frac{\left\langle E^-(t)E^-(t+\tau)E^+(t+\tau)E^+(t)\right\rangle}{\langle E^-(t)E^+(t)\rangle\,\langle E^-(t+\tau)E^+(t+\tau)\rangle}. \tag{20.54}
$$

Using our source-field expression (19.69) and neglecting retardation, this can be written in terms of atomic operators as

$$
g^{(2)}(\tau) = \frac{\langle \sigma_+(t)\sigma_+(t+\tau)\sigma_-(t+\tau)\sigma_-(t)\rangle}{\langle \sigma_+(t)\sigma_-(t)\rangle\,\langle \sigma_+(t+\tau)\sigma_-(t+\tau)\rangle}. \tag{20.55}
$$

If $\gamma t \gg 1$, one achieves a stationary result, and this expression reduces to

$$
g^{(2)}(\tau) = \frac{\langle \sigma_+(t)\sigma_+(t+\tau)\sigma_-(t+\tau)\sigma_-(t)\rangle}{\varrho_{22}^2}, \tag{20.56}
$$

where ϱ_{22} is the steady-state, excited-state population.

We see immediately that

$$
g^{(2)}(0) = \frac{\langle \sigma_+(t)\sigma_+(t)\sigma_-(t)\sigma_-(t)\rangle}{\varrho_{22}^2} = 0, \tag{20.57}
$$

since the *equal time* product of $\sigma_+(t)\sigma_+(t)$ is proportional to $(|2\rangle\langle 1|)(|2\rangle\langle 1|) = 0$, where 2 refers to the $H=1, m_H=0$ state and 1 to the $G=0$ state. Since $g^{(2)}(0)<1$, the scattered field cannot correspond to a classical field. The scattered field exhibits *antibunching*. For weak incident fields, $g^{(2)}(\tau)$ starts at 0 when $\tau=0$ and rises to a value of unity for $(\chi^2/\Omega)\tau \gg 1$. In the secular limit corresponding to strong fields,

$g^{(2)}(\tau)$ rises more rapidly (in a time of order Ω^{-1}) and oscillates about unity as it approaches this value asymptotically. Antibunching of this nature was first observed experimentally by Kimble, Dagenais, and Mandel [9] (see also reference [10]).

The weak field result is somewhat surprising since, in weak fields, the scattering is elastic. One might have thought that the field scattered by the atom should resemble that of the field scattered by a classical oscillator. It is sometimes stated that quantum fluctuations in the scattered signal resulting from spontaneous emission are responsible for the quantum nature of the scattered field. Alternatively, it is argued that once a photon is detected at the photodetector, we know that the atom is in its ground state and it takes some time for the atom to be re-excited before we can get the next count at the photodetector. We prefer the latter explanation, since it reflects the fact that $g^{(2)}(\tau)$ is defined as the joint probability to destroy "photons" at the detector at times t and $t + \tau$; however, this argument must be supplemented with the fact that it is not possible for a single atom to absorb two photons of the field, in the two-level approximation that has been assumed.

It is interesting to compare the result for $g^{(2)}(\tau)$ for an atom with that of an oscillator. The source-field approach works equally well for classical or quantized sources. For a classical oscillator, the operators $\sigma_{\pm}(t)$ are replaced by c numbers corresponding to that part of the oscillator's dipole moment varying at $e^{\mp i\omega t}$. As such, $g^{(2)}(\tau) = 1$ in steady state; the scattered field is coherent since the classical dipole oscillates at the driving-field frequency. On the other hand, what would you expect for a quantum oscillator? Since the oscillator is linear, you might think that $g^{(2)}(\tau) = 1$, as for the classical oscillator. Conversely, since the oscillator starts from its ground state and scatters "photons," you might think that $g^{(2)}(0) = 0$ for weak incident fields, since one could use the same argument we used for the atom. You are asked to resolve this question in the problems.

20.4 Scattering by a Single Atom in Weak Fields: $G \neq 0$

In the weak field limit, scattering from a $G = 0$ ground state is always elastic, since it corresponds to a two-photon transition that begins and ends on the same zero-width state. When one considers ground states with $G > 0$, qualitatively new features appear in the scattered spectrum as a result of optical pumping and light shifts. Some of the mathematical details of the calculation of the scattered spectrum are given in a review of light scattering for weak fields [11] and in an article by Gao [12] for fields of arbitrary intensity. We avoid such mathematical details and are content to explain the physical origin of the new features. Moreover, we assume that the driving field is polarized in the $\hat{\mathbf{z}}$ direction and that $|\delta| \gg \gamma, \chi$, so that a weak field limit is applicable.

Recall that the redistribution of ground-state sublevels occurs to *zeroth* order in the applied field intensity, even if it takes a time of order $(\Gamma')^{-1} \sim (|\chi|^2\gamma/\delta^2)^{-1}$ to reach equilibrium. As a consequence, there are a *number* of decay rates associated with the relaxation of the ground-state sublevels to their final values. How many such rates are there for a ground state having angular momentum G? Starting from an unpolarized ground state, the only ground- and excited-state density matrix elements that are modified with a z-polarized field are the diagonal elements $\varrho_{m_G m_G}$ and $\varrho_{m_H m_H}$. Spontaneous decay does not add any new ground-state elements, since $\varrho_{m_H m_H}$ does not couple to ground-state coherence via spontaneous emission.

Although there are $(2G + 1)$ ground-state populations, owing to the fact that $\varrho_{m_G m_G} = \varrho_{-m_G,-m_G}$ for excitation using z-polarized light, only $G+1$ $(G+1/2)$ of these elements are independent for integral (half-integral) values of G. As a consequence, there are $G + 1$ $(G + 1/2)$ independent decay parameters associated with optical pumping for integral (half-integral) values of G. One of these decay rates is zero, since the total ground-state population does not decay at all. This zero-width decay rate is responsible for the elastic component of the scattered spectrum. There remain G $(G-1/2)$ nonzero independent decay parameters associated with optical pumping for integral (half-integral) values of G.

Qualitatively, the scattered spectrum consists of two distinct types of contributions. The first corresponds to transitions that begin and end on the same sublevel and can be considered a "Rayleigh-like" contribution centered at $\Delta_k = 0$. This contribution itself consists of two components, an elastic component and an inelastic component. The structure of the inelastic component is determined by the G $(G - 1/2)$ independent decay parameters associated with optical pumping for integral (half-integral) values of G. Thus, for $G = 1/2$, there is no inelastic Rayleigh component, while for $G = 1$, the inelastic Rayleigh component is a simple Lorentzian, since only one nonzero decay parameter enters.

The second contribution consists of "Raman-like" processes, involving two-photon scattering processes that begin and end on magnetic state sublevels differing by one unit of angular momentum. The widths of the Raman components is given by one-half the sum of the rate at which the incident field depletes each of the levels. To determine the central frequency of the Raman components, one notes that the light shifts of the ground-state sublevels depend only on $|m_G|$; as a consequence, there are G pairs of Raman components symmetrically displaced about $\Delta_k = 0$ for G integral. For half-integral G, there is one "Raman" component centered at $\Delta_k = 0$ and $(G - 1/2)$ pairs of Raman components symmetrically displaced about $\Delta_k = 0$. Since the light shifts vary as $|\chi|^2/|\delta|$ and the widths as $\gamma|\chi|^2/|\delta|^2$, all but the Raman component centered at $\Delta_k = 0$ are resolved from the Rayleigh component.

To summarize, let us first consider G integral. For $H = G$ or $(G - 1)$, there is no steady-state scattering, since the atoms are pumped into a dark state by the incident field. For $H = (G + 1)$, there are an elastic and G inelastic components centered at $\Delta_k = 0$, and G pairs of Raman components symmetrically displaced about $\Delta_k = 0$. If $G = 1$, the structure is similar to that for the Mollow spectrum but has a totally different origin. For example, the sideband intensity in this case varies as the field intensity, whereas it varies as the field intensity squared in the Mollow triplet when $|\chi|^2/|\delta|^2 \ll 1$. For half-integral G, there are an elastic component, $(G-1/2)$ inelastic Rayleigh components, a Raman component centered at $\Delta_k = 0$, and $(G-1/2)$ pairs of Raman components symmetrically displaced about $\Delta_k = 0$.

The presence of inelastic components in the scattered spectrum implies that, even in weak fields, $|g^{(1)}(\tau)| \neq 1$, with the structure of $g^{(1)}(\tau)$ more closely resembling that found for scattering of an atom in intense fields.

20.5 Summary

We have examined the scattering of a monochromatic field by a single, stationary atom. The ensuing atom–field dynamics can be quite complicated. In general, the scattered field contains both elastic and inelastic components. The elastic component can be associated with the dipole moment created in the atom by the field. The inelastic components result from scattering processes involving emission into field modes having a frequency different from that of the incident field. Taken together, the spectrum and correlation functions provide a complete picture of the scattering process.

Problems

1. Consider scattering in the geometry of figure 20.1 in which the atomic beam velocity $\mathbf{v}_0$, the field propagation vector $\mathbf{k}_0$, and the propagation vector of the scattered radiation $\mathbf{k}$ are mutually perpendicular. Including the recoil that the atom undergoes as a result of the scattering, show that the frequency of the scattered light is shifted only by the recoil shift.

   In general, for an incident wave vector $\mathbf{k}_0$, a scattered wave vector $\mathbf{k}$, and an atom moving with velocity $\mathbf{v}$, show that the shift of the frequency of the scattered radiation from that of the incident field is $(\mathbf{k}-\mathbf{k}_0)\cdot\mathbf{v}$, neglecting recoil effects. Give a simple interpretation to this result.
2. Derive equations (20.32).
3. Derive equation (20.37).
4. Starting from equation (20.37), integrate over Δ_k to show that the ratio $\mathcal{I}_\omega/(\mathcal{I}_{\omega+\Omega}) = (4s^2c^2)/(c^4+s^4)$ and that $\mathcal{I}_{\text{elastic}}/\mathcal{I}_{\text{inelastic}} = (c^2-s^2)^2/2s^2c^2$.
5. For a $G=2$ ground state, an $H=3$ excited state, and a z-polarized driving field, calculate the central frequencies and widths of the Raman components in the scattered spectrum for a weak driving field in the limit $|\delta| \gg \gamma, \chi$.

6–7. Calculate $g^{(2)}(0)$ for the field scattered from a (damped) quantum oscillator by a monochromatic field in terms of the populations of the levels, and show that it is equal to

$$g^{(2)}(0) = \frac{\sum_{n=1}^{\infty} n(n+1)\varrho_{n+1,n+1}}{\left(\sum_{n=1}^{\infty} n\varrho_{n,n}\right)^2} = \frac{\langle n^2\rangle - \langle n\rangle}{\langle n\rangle^2}.$$

   To do this, find an expression for the dipole moment operator of the oscillator in the RWA, allowing you to find the positive and negative frequency components of the dipole operator, and then use equation (20.55) with the dipole operator you found and $\tau = 0$. Note that if the oscillator is in a coherent state, then $g^{(2)}(0) = 1$. Evaluate this expression for a weak field and show that it is equal to unity. What happened to the argument that measuring one photon requires additional time to measure a second photon?

   To solve this problem, take the Hamiltonian for the oscillator–classical field interaction Hamiltonian to be (including only the first three states)

$$\tilde{\mathrm{H}} = \hbar\begin{pmatrix} 0 & \chi & 0 \\ \chi & \delta & \sqrt{2}\chi \\ 0 & \sqrt{2}\chi & 2\delta \end{pmatrix}.$$

   You will also need the relaxation rates: $\dot{\varrho}_{00} = 2\gamma\varrho_{11}$; $\dot{\varrho}_{01} = -\gamma\varrho_{01} + 2\sqrt{2}\gamma\varrho_{12}$; $\dot{\varrho}_{11} = -2\gamma\varrho_{11} + 4\gamma\varrho_{22}$; $\dot{\varrho}_{02} = -2\gamma\varrho_{02}$; $\dot{\varrho}_{12} = -3\gamma\varrho_{02}$; and $\dot{\varrho}_{22} = -4\gamma\varrho_{22}$. Consider only the stationary regime–steady-state result.

References

[1] B. R. Mollow, *Pure-state analysis of resonant light scattering: radiative damping, saturation, and multiphoton effects*, Physical Review A **12**, 1919–1943 (1975).

[2] C. Cohen-Tannoudji, *Atoms in Strong Resonant Fields*, in *Frontiers in Laser Spectroscopy*, Les Houches Session XXVII, vol. 1, edited by R. Balian, S. Haroche, and S. Liberman (North Holland, Amsterdam, 1975), pp. 88–98; C. Cohen-Tannoudji, J. Dupont-Roc, and G. Grynberg, *Atom-Photon Interactions* (Wiley-Interscience, New York, 1992), complement A_V.

[3] C. Cohen-Tannoudji, *Atoms in Strong Resonant Fields: Spectral Distribution of the Fluorescence Light*, in *Laser Spectroscopy*, edited by S. Haroche, J. C. Pebay-Peyroula, T. W. Hänsch, and S. E. Harris (Springer-Verlag, Berlin, 1975), pp. 324–339; S. Reynaud, *La fluorescence de résonance: étude par la méthode de l'atome habillé* (Resonance fluorescence: study by the dressed atom method), Annales de Physique **8**, 315–370 (1983).

[4] B. R. Mollow, *Power spectrum of light scattered by two-level systems*, Physical Review **188**, 1969–1975 (1969).

[5] P. R. Berman, *Theory of fluorescence and probe absorption in the presence of a driving field using semiclassical dressed states*, Physical Review A **53**, 2627–2632 (1996).

[6] F. Y. Wu, R. E. Grove, and S. Ezekiel, *Investigation of the spectrum of resonance fluorescence induced by a monochromatic field*, Physical Review Letters **35**, 1426–1429 (1975); R. E. Grove, F. Y. Wu, and S. Ezekiel, *Measurement of the spectrum of resonance fluorescence from a two-level atom in an intense monochromatic field*, Physical Review A **15**, 227–233 (1977).

[7] H. J. Kimball and L. Mandel, *Theory of resonance fluorescence*, Physical Review A **13**, 2123–2144 (1976).

[8] H. J. Carmichael and D. F. Walls, *Proposal for the measurement of the resonant Stark effect by photon correlation techniques*, Journal of Physics B **9**, L43–L46 (1976); H. J. Carmichael and D. F. Walls, *A quantum-mechanical master equation treatment of the dynamical Stark effect*, Journal of Physics B **9**, 1199–1219 (1976).

[9] H. J. Kimball, M. Dagenais, and L. Mandel, *Photon antibunching in resonance fluorescence*, Physical Review Letters **39**, 691–695 (1977).

[10] M. Dagenais and L. Mandel, *Investigation of two-time correlations in photon emissions from a single atom*, Physical Review A **18**, 2217–2228 (1978).

[11] P. R. Berman, *Light scattering*, Contemporary Physics **49**, 313–330 (2008).

[12] B. Gao, *Effects of Zeeman degeneracy on the steady-state properties of an atom interacting with a near-resonant laser field: resonance fluorescence*, Physical Review A **50**, 4139–4156 (1994).

Bibliography

W. Heitler, *The Quantum Theory of Radiation*, 3rd ed. (Oxford University Press, Oxford, UK, 1954), sec. 20.

R. Loudon, *The Quantum Theory of Light*, 3rd ed. (Oxford University Press, Oxford, UK, 2003), chap. 8.

L. Mandel and E. Wolf, *Optical Coherence and Quantum Optics* (Cambridge University Press, Cambridge, UK, 1995), sec. 15.6.

B. R. Mollow, *Theory of intensity dependence resonance light scattering and resonance fluorescence,* in *Progress in Optics*, edited by E. Wolf, vol. XIX (North-Holland, Amsterdam, 1981), pp. 1–43.

R. G. Newton, *Scattering Theory of Waves and Particles*, 2nd ed. (Dover, New York, 2002).

Lord Rayleigh, *On the transmission of light through an atmosphere containing small particles in suspension, and on the origin of the blue of the sky*, Philosophical Magazine **XLVII**, 375–384 (1899).

21

Entanglement and Spin Squeezing

We have just about come to the end of the road. Our trip has taken us to many destinations, but there are many others that were given just a glance or not seen at all. Topics such as optical bistability, applied nonlinear optics, superradiance, pulse propagation, and the theory of laser operation were excluded. Moreover, many-atom systems in which correlations play a dominant role, such as Bose-Einstein condensates, were not discussed. Many of these topics are included in the bibliography entries that have been given throughout this book.

We would like to think that we have given the reader an introduction to the theory of atom–field interactions that will prove helpful. The techniques that were developed are quite general and can be applied to a wide range of problems. By this time, it is hoped that you are comfortable with the various representations that were used and that the choice of which representation to use has become "natural" for you.

One topic that was not discussed is *entanglement*. Entangled or correlated states of a multiple-atom system are states that cannot be factorized into a product of state vectors for the individual atoms. For example, the state function for a two-atom system

$$|\psi\rangle = \frac{1}{2}(|1\rangle_1 + |2\rangle_1)(|1\rangle_2 + |2\rangle_2) = \frac{1}{2}(|11\rangle + |12\rangle + |21\rangle + |22\rangle),$$

in which $|\alpha\rangle_j$ corresponds to the state vector for atom j to be in state α and $|\alpha\alpha'\rangle$ corresponds to the state vector for atom one to be in state α and atom two to be in state α', is not an entangled state since it can be factorized. On the other hand,

$$|\psi\rangle = \frac{1}{\sqrt{2}}(|12\rangle + |21\rangle)$$

is an entangled state since it cannot be factorized. Entangled states have become increasingly important with the development of quantum information and the quest for a quantum computer.

Of course, entanglement and correlated states are at the heart of quantum mechanics. In the spirit of this book, we finish this chapter by considering three cases in which a quantized field state becomes entangled with a many-atom state. *Direct* interactions between the atoms, such as those produced by electrostatic forces, are neglected; any entanglement of the atoms is produced by the field(s).

21.1 Entanglement by Absorption

The first case we look at involves the complete absorption of a single-photon pulse in an optically dense medium [1,2]. As the pulse is absorbed in the medium, an entangled state of the atoms and the field is produced; once the pulse is totally absorbed, the atoms are left in an entangled state involving a single excitation [3].

To simplify the problem, we restrict the discussion to an effective one-dimensional problem. That is, we neglect all modes of the radiation field having propagation vectors in other than the $\hat{\mathbf{x}}$ direction; we do not consider scattering of the incident field into transverse field modes. The neglect of scattering into transverse modes is valid provided that the calculation is limited to times for which $\gamma_2 t \ll 1$, where γ_2 is the excited-state decay rate. The process we consider is truly absorption by the atoms, and not the scattering process with which we have been dealing throughout most of this book. The energy lost by the field at any time is converted to excitation energy of the atoms.

It is assumed that the atomic medium is inhomogeneously broadened. It turns out that inhomogeneous broadening of the atomic transition frequencies provides an important simplification in this problem, provided that the inhomogeneous width is much larger than the single-photon pulse bandwidth. In that limit, all frequency components of the incident pulse are absorbed in an identical manner, resulting in a pulse that propagates without distortion. In the case of homogeneous broadening, there could be significant reshaping of the radiation pulse as it propagates in the medium.

The two-level atoms are uniformly distributed in the half-space $X > 0$, with the position of atom m denoted by $\mathbf{R}_m$. Owing to inhomogeneous broadening, these atoms have a distribution of atomic transition frequencies centered about some central frequency ω_0. That is, atom m has frequency ω_m, which is detuned from the central transition frequency by an amount

$$\Delta_m = \omega_m - \omega_0. \tag{21.1}$$

The inhomogeneous frequency distribution is denoted by $W(\Delta_m)$.

The positive frequency component of the electric field is

$$\mathbf{E}^+(X) = i\hat{\mathbf{z}} \sum_{j=-\infty}^{\infty} \left(\frac{\hbar\omega_j}{2\epsilon_0 AL} \right)^{1/2} e^{ik_j X} a_j, \tag{21.2}$$

where

$$\omega_j = \left| k_j \right| / c, \tag{21.3}$$

A is the cross-sectional area of the field pulse, and a_j is a destruction operator for field mode j. As noted earlier, we consider an effective one-dimensional problem in which the field propagates in the $\hat{\mathbf{x}}$ direction with polarization $\hat{\mathbf{z}}$. The quantization scheme is one in which the field is periodic over a distance L, implying that

$$k_j = 2\pi j/L, \tag{21.4}$$

where j is an integer (positive, negative, or zero). With this definition, the quantization volume is equal to AL, and the field modes are defined such that

$$2A\epsilon_0 \int_{-L/2}^{L/2} \mathbf{E}_j^-(X) \cdot \mathbf{E}_{j'}^+(X) dX = \hbar\omega_j a_j^\dagger a_j \delta_{j,j'}, \tag{21.5}$$

that is, the expectation value of the energy in mode j of the field is equal to $n_j \hbar\omega_j$, where n_j is the average number of photons in that mode. For future reference, we note that the prescription for transforming from discrete to continuum modes of the field is

$$\sum_j \to \frac{L}{2\pi} \int_{-\infty}^{\infty} dk. \tag{21.6}$$

In dipole and rotating-wave approximation, the Hamiltonian appropriate to the atom–field system is

$$\begin{aligned} H = & \sum_m \frac{\hbar\omega_m}{2} \sigma_z^m + \sum_j \hbar\omega_j a_j^\dagger a_j \\ & + \sum_{j,m} \hbar g_j \left(e^{ik_j X_m} \sigma_+^m a_j - a_j^\dagger e^{-ik_j X_m} \sigma_-^m \right), \end{aligned} \tag{21.7}$$

where

$$g_j = -i \left(\frac{\omega_j}{2\hbar\epsilon_0 AL} \right)^{1/2} \mu_{21} \tag{21.8}$$

is a coupling constant; $\sigma_\pm^m$ are raising (+) and lowering (−) operators for atom m; σ_z^m is the population difference operator, $(|2\rangle\langle 2| - |1\rangle\langle 1|)$, for atom m; and μ_{21} is the z component of the atomic dipole moment matrix element (assumed to be real).

Since there can be at most one atomic excitation starting from a state in which all atoms are in their ground states and the field is in a one-photon state, the state vector for the atom–field system can be written in the interaction representation as

$$|\psi(t)\rangle = \sum_j b_j(t) e^{-i\omega_j t} |G; 1_j\rangle + \sum_m c_m(t) e^{-i\omega_m t} |m; 0\rangle, \tag{21.9}$$

where $b_j(t)$ is the probability amplitude for the field to be in mode j and all the atoms to be in their ground states (corresponding to state $|G; 1_j\rangle$), while $c_m(t)$ is the probability amplitude for the field to be in the vacuum mode and the atoms to be in a state where atom m is in its excited state and all the other atoms are in their ground state (corresponding to state $|m; 0\rangle$). The initial state for the system is taken as

$$|\psi(0)\rangle = \sum_j b_j(0) |G; 1_j\rangle, \tag{21.10}$$

a single-photon state of the field with all atoms in their ground states, normalized such that $\sum_j |b_j(0)|^2 = 1$. In practice, the $|b_j(0)|$ are chosen to correspond to a pulsed field propagating in the $\hat{\mathbf{x}}$ direction with $k_j \approx k_0 = \omega_0/c$.

It is now a simple matter to write equations for the state amplitudes using the Schrödinger equation. We find that the equations of motion for the relevant state amplitudes are given by

$$\dot{b}_j = ig_j \sum_m e^{-ik_j X_m} e^{i(\delta_j - \Delta_m)t} c_m, \tag{21.11a}$$

$$\dot{c}_m = -i \sum_{j'} g_{j'} e^{ik_{j'} X_m} e^{-i(\delta_{j'} - \Delta_m)t} b_{j'}, \tag{21.11b}$$

where

$$\delta_j = \omega_j - \omega_0, \tag{21.12}$$

and Δ_m is given by equation (21.1).[1] By formally solving equation (21.11b) and substituting the result in equation (21.11a), we find

$$\dot{b}_j = g_j \sum_{m,j'} g_{j'} e^{-i(k_j - k_{j'})X_m} e^{i(\delta_j - \Delta_m)t} \int_0^t dt' \, e^{-i(\delta_{j'} - \Delta_m)t'} b_{j'}(t'). \tag{21.13}$$

Equations (21.11) and (21.13) are exact, giving rise to a rather complicated entangled state of the atoms and the field. Note that, in general, field mode j is coupled to all other field modes at *earlier* times. Moreover, the evolution of each spectral component depends on absorption at each atomic site. All is not lost, however, if one neglects fluctuations in both particle position and frequency. Normally, the average over particle position and frequency is carried out *after* one obtains expressions for expectation values of quantum-mechanical operators. However, if fluctuations are neglected, one can carry out this average directly in equation (21.13). This type of average is implicit in the Maxwell-Bloch equations derived in chapter 6. In general, for our one-dimensional geometry, fluctuations can be neglected if the number of two-level atoms in a slice of length $\lambda_0 = 2\pi/k_0$ is much greater than unity—that is, if

$$\mathcal{N} A \lambda_0 \gg 1, \tag{21.14}$$

where $\mathcal{N}$ is the atomic density. In this limit, the phase factor $e^{ik_j X_m}$ is approximately constant within a given slice.

If fluctuations in particle position and frequency are neglected, we can replace the summation over m in equation (21.13) by

$$\mathcal{N} A \int_0^{L/2} dX_m \int_{-\infty}^{\infty} d\Delta_m W(\Delta_m). \tag{21.15}$$

The integral over X_m would lead to a decoupling of the modes in the limit that $L \sim \infty$ *if* the integral were from $-L/2$ to $L/2$; however, this is *not* the case, since the integral goes from 0 to $L/2$. Instead, let us first integrate over Δ_m.

The exact form of the distribution $W(\Delta_m)$ is not of critical importance; for the sake of definiteness, we choose the Gaussian distribution,

$$W(\Delta_m) = \left(\frac{1}{\pi \delta_0^2} \right)^{1/2} e^{-\Delta_m^2/\delta_0^2}. \tag{21.16}$$

[1] The definition of δ_j differs in sign from that used in previous chapters.

Performing the integral over Δ_m in equation (21.13), we find

$$\dot{b}_j = \mathcal{N} A g_j \int_0^{L/2} dX \sum_{j'} e^{-i(k_j - k_{j'})X} g_{j'} \int_0^t dt' \, e^{-\delta_0^2 (t-t')^2/4} e^{i\delta_j t} e^{-i\delta_{j'} t'} b_{j'}(t'). \quad (21.17)$$

The temporal dependence of the field amplitude is not local, a consequence of the fact that different frequency components of the field are absorbed differently by the atoms. If the inhomogeneous width δ_0 is much larger than the pulse bandwidth τ^{-1}, however, then all frequency components in the field are affected in an identical manner by the atoms. We will assume this to be the case, $\delta_0 \tau \gg 1$, allowing us to approximate

$$e^{-\delta_0^2 \tau^2/4} \sim \frac{2\sqrt{\pi}}{\delta_0} \delta_D(\tau), \quad (21.18)$$

where δ_D is a Dirac delta function. In this limit,

$$\dot{b}_j = \frac{\sqrt{\pi}}{\delta_0} g_j \mathcal{N} A \int_0^{L/2} dX \sum_{j'} g_{j'} e^{-i(k_j - k_{j'})X} e^{i(\delta_j - \delta_{j'})t} b_{j'}. \quad (21.19)$$

To arrive at this expression, we set $\int_0^t dt' \delta_D(t - t') = 1/2$.

To make further progress, we go over to continuum states using equation (21.6) and replace $b_j(0)$ with

$$b_j(t) \to \sqrt{\frac{2\pi}{L}} b(k, t), \quad (21.20)$$

k_j by k, ω_j by $\omega_k = kc$, and δ_j by $\delta_k = (\omega_k - \omega_0)$. In this manner, equation (21.19) is transformed into

$$\dot{b}(k, t) = \frac{\sqrt{\pi} L g^2}{2\pi \delta_0} \mathcal{N} A \int_0^{L/2} dX \int_{-\infty}^{\infty} dk' e^{-i(k-k')X} e^{i(\delta_k - \delta_{k'})t} b(k', t). \quad (21.21)$$

In writing this equation, we have set $g = g_k(\omega_k = \omega_0)$ and extended the k' integral to $-\infty$, since the major contributions from both k and k' are in the vicinity of k_0.

These equations are integral equations with a kernel that is a function of $(k - k')$. As such, they can be solved using Fourier transform techniques. Defining

$$b(k, t) = \frac{1}{\sqrt{2\pi}} \int_{-\infty}^{\infty} d\rho e^{-ik\rho} B(\rho, t), \quad (21.22)$$

we can convert equation (21.21) to the form

$$\dot{B}(\rho, t) = \frac{\sqrt{\pi} L g^2}{\delta_0} \mathcal{N} A \int_0^{L/2} dX \delta_D (X - ct - \rho) B(\rho, t). \quad (21.23)$$

It is now possible to carry out the integral over X. In the limit that $L \sim \infty$, we find

$$\dot{B}(\rho, t) = -\frac{\alpha c}{2} B(\rho, t) \Theta(\rho + ct), \quad (21.24)$$

where

$$\alpha = \frac{\sqrt{\pi} \mathcal{N} \omega_0 \mu^2}{\hbar \epsilon_0 c \delta_0}, \quad (21.25)$$

and Θ is a Heaviside function. As a consequence, the final solution for the Fourier transform of the field amplitude is

$$B(\rho, t) = \begin{cases} B(\rho, 0) & \rho < -ct \\ e^{-\alpha c(t+\rho/c)/2} B(\rho, 0) & -ct \leq \rho < 0 \\ e^{-\alpha ct/2} B(\rho, 0) & \rho \geq 0 \end{cases}, \tag{21.26}$$

assuming $t \geq 0$.

By taking the inverse transform, one obtains the field probability amplitude

$$b(k, t) = \frac{1}{\sqrt{2\pi}} \int_{-\infty}^{-ct} d\rho e^{-ik\rho} B(\rho, 0) + \frac{1}{\sqrt{2\pi}} \int_{-ct}^{0} d\rho e^{-ik\rho} e^{-\frac{\alpha c}{2}\left(t+\frac{\rho}{c}\right)} B(\rho, 0)$$
$$+ \frac{1}{\sqrt{2\pi}} e^{-\frac{\alpha c}{2}t} \int_{-ct}^{\infty} d\rho e^{-ik\rho} B(\rho, 0). \tag{21.27}$$

We assume that, at $t = 0$, the single photon pulse envelope is centered at $X = -X_0 < 0$ and has a spatial extent $(\Delta X) \ll X_0$. Moreover, we assume that $\alpha (\Delta X) \ll 1$, so the pulse travels several pulse widths in the medium before it is absorbed. Since $|B(\rho, 0)|^2$ corresponds to the pulse envelope at $t = 0$, peaked at $\rho = -X_0$, it follows from equations (21.26) and (21.27) that, for $ct < X_0 - (\Delta X)/2$ (before the pulse enters the medium),

$$b(k, t) \sim \frac{1}{\sqrt{2\pi}} \int_{-\infty}^{-ct} d\rho e^{-ik\rho} B(\rho, 0) \approx \frac{1}{\sqrt{2\pi}} \int_{-\infty}^{\infty} d\rho e^{-ik\rho} B(\rho, 0) = b(k, 0) \tag{21.28}$$

and, for $ct > X_0 - (\Delta X)/2$ (once the pulse is entirely in the medium),

$$b(k, t) \sim \frac{1}{\sqrt{2\pi}} \int_{-ct}^{0} d\rho e^{-ik\rho} e^{-\frac{\alpha c}{2}\left(t+\frac{\rho}{c}\right)} B(\rho, 0)$$
$$\approx \frac{e^{-\frac{\alpha c}{2}(t-X_0/c)}}{\sqrt{2\pi}} \int_{-\infty}^{\infty} d\rho e^{-ik\rho} B(\rho, 0) = e^{-\frac{\alpha c}{2}(t-X_0/c)} b(k, 0), \tag{21.29}$$

where the assumption that $\alpha (\Delta X) \ll 1$ was used to set $e^{-\alpha\rho/2}$ equal $e^{\alpha X_0/2}$ in the integrand. The result is remarkably simple. The pulse propagates without distortion and without loss until it enters the medium, and then decays away exponentially in time as it propagates in the medium.

Combining equations (21.11b) and (21.27), and going over to continuum states, we find that the atomic state amplitude evolves according to

$$\dot{c}_m = -ig \frac{L}{2\pi} \sqrt{\frac{2\pi}{L}} \int_{-\infty}^{\infty} dk e^{ikX_m} e^{-i(\delta_k - \Delta_m)t} b(k, t)$$
$$= -ig \frac{1}{2\pi} \sqrt{L} \int_{-\infty}^{\infty} d\rho \int_{-\infty}^{\infty} dk e^{-ik\rho} e^{ikX_m} e^{-i(\delta_k - \Delta_m)t} B(\rho, t)$$
$$= -ig\sqrt{L} e^{ik_0 X_m} e^{i\Delta_m t} \int_{-\infty}^{\infty} d\rho e^{-ik_0\rho} \delta_D (X_m - ct - \rho) B(\rho, t)$$
$$= -ig\sqrt{L} e^{i\Delta_m t} e^{ik_0 ct} B(X_m - ct, t)$$
$$\approx -ig\sqrt{L} e^{-\alpha X_m/2} e^{i\Delta_m t} e^{ik_0 ct} B(X_m - ct, 0), \tag{21.30}$$

where the fact that $X_m > 0$ has been used. The last line follows from the properties of $B(\rho, t)$ given in equation (21.26) and the assumption that $|B(\rho, 0)|$ is sharply peaked at $X = -X_0 < 0$, implying that only the second line in equation (21.26) enters. An atom in the medium is excited by the pulse as it passes by, but the amplitude of the pulse has decreased exponentially owing to absorption in the medium.

If we choose

$$b(k, 0) = \left(\frac{1}{\pi (\Delta k)^2}\right)^{1/4} e^{-(k-k_0)^2/2(\Delta k)^2} e^{ikX_0}, \tag{21.31}$$

with $(\Delta k) = 1/(\Delta X)$, then

$$B(\rho, 0) = \pi^{1/4}\sqrt{\Delta k/\pi}\, e^{-(\rho+X_0)^2(\Delta k)^2/2} e^{ik_0(\rho+X_0)} \tag{21.32}$$

and

$$\dot{c}_m = -\pi^{1/4}\sqrt{\frac{\Delta k \omega_0 \mu^2}{2\pi \epsilon_0 A\hbar}}\, e^{-\alpha X_m/2} e^{i\Delta_m t} e^{ik_0(X_m+X_0)} e^{-(X_m-ct+X_0)^2(\Delta k)^2/2}. \tag{21.33}$$

Using the time integral of equation (21.33) and equation (21.18), it is easy to verify that

$$\mathcal{N} A \int_0^\infty dX_m \int d\Delta_m W(\Delta_m)\, |c_m(t \gg \alpha/c)|^2 = 1; \tag{21.34}$$

all the field energy transmitted into the medium is transferred to the atoms when the pulse is fully absorbed.

Explicitly, one finds the atomic state amplitude at time t is given by

$$c_m(X_m, t) = -\pi^{1/4}\sqrt{\frac{\omega_0 \mu^2}{2c\epsilon_0 A\hbar}}\sqrt{\frac{\tau}{2}}\, e^{-\alpha X_m/2} e^{i(X_m+X_0)(k_0+\Delta_m/c)} e^{-(\Delta_m \tau)^2/2}$$
$$\times \left[\Phi\left(\frac{\frac{t}{\tau} - \frac{X_m+X_0}{\Delta X} - i\Delta_m \tau}{\sqrt{2}}\right) + \Phi\left(\frac{\frac{X_m+X_0}{\Delta X} + i\Delta_m \tau}{\sqrt{2}}\right)\right], \tag{21.35}$$

where $\tau = 1/[c(\Delta k)]$ and Φ is the error function. For sufficiently long times, $\alpha c(t - X_0/c) \gg 1$, all the energy initially in the single-photon pulse has been transferred to the atoms and the atoms are in an *entangled state* $|\psi(t)\rangle = \sum_m c_m(t)e^{-i\omega_m t}\,|m; 0\rangle$ with a spatial phase factor $e^{ik_0 X_m}$ that has been imprinted by the field, along with an exponentially decaying factor $e^{-\alpha X_m/2}$, resulting from absorption in the medium.

The entanglement results from the atoms sharing a single excitation. This entanglement depends critically on the quantum properties of the incident pulse. Had we taken a *classical* field, the excitation in the sample would decrease exponentially, but there would be no entanglement in the atomic state vector. We would run into problems, however, if we took a classical pulsed field having energy $\hbar\omega_0$. Such a field has some probability for exciting two atoms, as well as a relatively large probability, $1/e$, of leaving all the atoms in their ground states, despite the fact that the pulse is totally absorbed. Such contradictions are avoided if we replace the classical field pulse by a quantized, multimode coherent state for the field having *average* energy $\hbar\omega_0$. It does not appear that this multimode coherent state produces any entanglement of the atoms, despite the fact that there are number state fluctuations in the field.

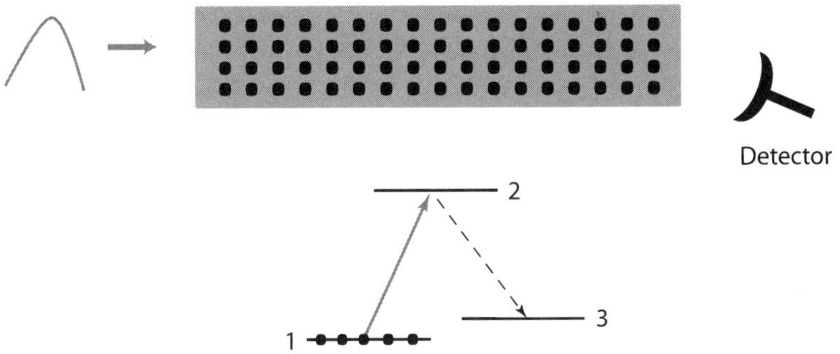

Figure 21.1. A schematic representation of the Duan, Lukin, Cirac, Zoller protocol. Atoms are prepared in state 1. If scattered radiation on the 2–3 transition is detected, the system is projected into an entangled state.

21.2 Entanglement by Post-Selection—DLCZ Protocol

As a second example of entanglement, we consider atomic state entanglement produced by *post-selection* in the so-called Duan, Lukin, Zoller, Cirac (DLCZ) protocol [4]. The mechanism responsible for entanglement by post-selection is the projection of a quantum system into an entangled state that results from a measurement on the system [5–9].

The DLCZ protocol is indicated schematically in figure 21.1. An ensemble of three-level atoms having a Λ configuration is prepared in level 1. A *weak* pulsed field is incident on the atomic sample. A detector is positioned to collect any radiation scattered on the 2–3 transition in the nearly forward direction. In most cases, the pulsed field travels through the sample without scattering, and no signal is observed at the detector. However, occasionally, the detector is triggered. When this happens, it is certain that one photon in the pulsed field has been scattered by the atoms, and the ensemble is left in a state where at least one atom is in level 3. Since it is impossible to know *which* atom is in level 3, the ensemble is left in an entangled state involving a single excitation into level 3.

If a single-photon pulse is sent into the medium, then there is at most one excitation into level 3. On the other hand, if a weak coherent state of the field is incident on the medium, there is always the possibility that there are two excitations of the system. The incident field intensity must be weak to avoid this outcome. As a consequence, the probability that the detector is triggered is very small and the experiment must be carried out thousands of times, on average, before detecting a "click" at the detector. The triggering of the detector acts as the post-selection that projects the ensemble into an entangled state.

Not only can the information in the incident pulse be stored in the atomic medium, but it can also be read out at later times with high fidelity using a control pulse, similar to the situation we discussed in chapter 9 for slow light and electromagnetic-induced transparency (EIT). Several proof-of-principle experiments of this nature have been carried out [5–9]. To enhance the scattering, optically dense media and resonant fields are used. To avoid problems associated with absorption of the field as it propagates in the medium, auxiliary levels and fields are employed, allowing the signal field to propagate without loss owing to an EIT window that is created for the fields [10–12].

21.3 Spin Squeezing

As a third example of entanglement, we consider the transfer of entanglement from a quantized cavity field state to the quantum state of a pair of atoms [13].

21.3.1 General Considerations

In a Ramsey fringe experiment, one measures the upper-state population as a function of atom–field detuning. This is the basic type of measurement needed for optical clocks to determine the frequency of a transition. The precision of such a measurement, as well as its accuracy, is of critical importance in metrology. Quantum fluctuations evidently play a role in the precision that can be achieved.

For a single atom, a quantum measurement of the excited-state population yields either a 1 (if the atom is measured in its excited state $|2\rangle$) or a 0 (if the atom is *not* measured in its excited state $|2\rangle$). In an ensemble of N atoms, excited in the same manner by the fields giving rise to the Ramsey fringe signal, one measures an excited-state population equal to $N\varrho_{22}$, *on average*. However, owing to quantum fluctuations in the measurement process, there is usually an uncertainty of order $(N\varrho_{22})^{1/2}$ associated with the measurement referred to as the *standard quantum limit* (SQL). Since the precision of clock measurements is limited by this factor, there has been considerable interest is determining whether these fluctuations can be reduced by *spin squeezing* [14–18], the matter analogue of optical squeezing. An excellent introduction to this topic is given in an article by Wineland et al. [19]. Here, we give a very brief discussion of spin squeezing.

We have seen already that the state of a two-level quantum system can be represented in terms of the Pauli matrices. That is, the operator for a two-level atom, in the field interaction representation, can be written as

$$\mathbf{s} = s_x\mathbf{i} + s_y\mathbf{j} + s_z\mathbf{k}, \tag{21.36}$$

where the operators

$$s_z = (|2\rangle\langle 2| - |1\rangle\langle 1|)/2, \tag{21.37a}$$

$$s_+ = |2\rangle\langle 1|, \quad s_- = |1\rangle\langle 2|, \tag{21.37b}$$

$$s_x = (s_+ + s_-)/2, \quad s_y = (s_+ - s_-)/2i, \tag{21.37c}$$

are closely related to the Pauli spin matrices. Suppose that we choose our coordinate system with $\langle \mathbf{s}\rangle$ along the z axis, such that $\langle s_x\rangle = \langle s_y\rangle = 0$. In that case, the quantum fluctuations in the measurement of $\langle \mathbf{s}\rangle$ occur in the xy plane, with

$$\left\langle s_x^2\right\rangle = \left\langle s_y^2\right\rangle = 1/4. \tag{21.38}$$

There can be no spin squeezing for a single atom, since both the $\langle s_x^2\rangle$ and $\langle s_y^2\rangle$ variances are constant.

If we now consider N atoms interacting uniformly with the external fields, we can define collective operators

$$S_\alpha = \sum_{j=1}^{N} s_\alpha^j, \quad \alpha = x, y, z, +, -. \tag{21.39}$$

It is an easy matter to show that these collective operators obey the same commutation relations as the single-atom operators,

$$[S_x, S_y] = iS_z, \quad [S_y, S_z] = iS_x, \quad [S_z, S_x] = iS_y, \tag{21.40}$$

but that the *product* of collective operators does *not* obey the same type of relationship that holds for the single-atom operators. That is, while $s_x^2 = 1/4$,

$$\begin{aligned} S_x^2 &= \sum_{i,j=1}^{N} s_x^i s_x^j = \sum_{i=1}^{N} \left(s_x^i\right)^2 + \sum_{i,j\neq i}^{N} s_x^i s_x^j \\ &= N/4 + \sum_{i,j\neq i}^{N} s_x^i s_x^j. \end{aligned} \tag{21.41}$$

Imagine that the atoms have been prepared with all the spins "down." This represents a product state for all the atoms with $\langle S_z \rangle = -N/2$, $\langle S_x \rangle = \langle S_y \rangle = 0$, and

$$|\langle \mathbf{S} \rangle| = \sqrt{\langle S_x \rangle^2 + \langle S_y \rangle^2 + \langle S_z \rangle^2} = N/2. \tag{21.42}$$

Such a state is referred to as a *coherent spin state* [20]. It is convenient to use angular momentum notation to label the N-atom states. The states can be labeled by the total angular momentum S and the z component of angular momentum S_z quantum numbers. In coupling N spins, one finds that S can vary from 0 (or 1/2 if N is odd) to $N/2$. Clearly, the coherent state has maximal $S = N/2$, since all the spins are aligned.

In a coherent state, $\langle S_x \rangle = \langle S_y \rangle = 0$, and the variance of S_x is

$$\begin{aligned} \left\langle S_x^2 \right\rangle &= \frac{1}{4} \left\langle (S_+ + S_-)^2 \right\rangle \\ &= \frac{1}{4} \left\langle \frac{N}{2}, -\frac{N}{2} \right| (S_+^2) + (S_-^2) + S_+S_- + S_-S_+ \left| \frac{N}{2}, -\frac{N}{2} \right\rangle \\ &= \frac{1}{4} \left\langle \frac{N}{2}, -\frac{N}{2} \right| S_-S_+ \left| \frac{N}{2}, -\frac{N}{2} \right\rangle = N/4, \end{aligned} \tag{21.43}$$

where we have used the relation

$$S_\pm |S, m\rangle = \sqrt{(S \mp m)(S \pm m + 1)}\, |S, m \pm 1\rangle. \tag{21.44}$$

Similarly, $\langle S_y^2 \rangle = N/4$. We can define a parameter

$$\xi_x = \frac{\sqrt{N}\sqrt{\langle S_x^2 \rangle}}{|\langle \mathbf{S} \rangle|}, \tag{21.45}$$

which is a measure of the precision. A value of $\xi_x = 1$ corresponds to the SQL. If we can find a spin component perpendicular to the direction of $\langle \mathbf{S} \rangle$ that is less than this value, the state is said to be *spin-squeezed*.

To determine the conditions needed to generate spin squeezing, we start from an average spin aligned along the z axis, and use the fact that $\Delta S_x \Delta S_y \geq |\langle S_z \rangle| /2$ to write equation (21.45) as

$$\xi_x = \sqrt{N}\Delta S_x/|\langle S_z \rangle| \geq \sqrt{N}/(2\Delta S_y) = \left[1 + \frac{4}{N}\sum_{j,j' \neq j}^{N} \left\langle S_y^{(j)} S_y^{(j')} \right\rangle \right]^{-1/2}. \tag{21.46}$$

For correlated states, the sum can be positive and one cannot rule out the possibility that $\xi_x < 1$. On the other hand, for uncorrelated states, using the fact that $\langle S_y \rangle^2 = 0$, it follows that

$$1 + \frac{4}{N}\sum_{j,j' \neq j}^{N} \left\langle S_y^{(j)} S_y^{(j')} \right\rangle = 1 - \frac{4}{N}\sum_{j}^{N} \left\langle S_y^{(j)} \right\rangle^2. \tag{21.47}$$

As a consequence, $\xi_x \geq 1$, and there is no spin squeezing for unentangled states. We are led to the important conclusion that entanglement is necessary for spin squeezing. An entangled state does not necessarily lead to spin squeezing, but spin squeezing cannot occur unless there is entanglement.

You might think that, just as in the optical case, it is possible to squeeze one component of the spin to an arbitrarily small value. However, this is not the case. Since $\Delta S_y \leq N/2$, it follows from equation (21.46) that the minimum value of ξ_x that can be reached is $1/\sqrt{N}$, a value referred to as the *Heisenberg limit*, since it is a limit imposed by the Heisenberg uncertainty relation.

For $N = 3$ $(S = 3/2)$, an example of a squeezed state is

$$|\psi\rangle = 0.935\,|3/2, 3/2\rangle + 0.354|3/2, -1/2\rangle, \tag{21.48}$$

for which $\xi_x = 0.76$, not quite at the Heisenberg limit of $\xi_x = 1/\sqrt{3} = 0.577$. For $N = 2$ $(S = 1)$, the state

$$|\psi\rangle = \frac{1}{\sqrt{2\cosh(2\theta)}}(e^{-\theta}\,|1, 1\rangle + e^{\theta}\,|1, -1\rangle), \tag{21.49}$$

approaches the Heisenberg limit as $\theta \sim 0$ [21].

To construct spin-squeezed states, one has essentially two options to produce the needed entanglement. One can use interactions between atoms to provide the entanglement, or couple the atoms to a quantized state of the radiation field and transfer the quantum properties of the field to the atoms. We look at the latter of these possibilities [13].

21.3.2 Spin Squeezing Using a Coherent Cavity Field

In dipole approximation and RWA, the Hamiltonian for an ensemble of two-level atoms (lower state $|1\rangle$, upper state $|2\rangle$, transition frequency ω_0) interacting with a resonant cavity field having frequency $\omega = \omega_0$ is

$$H = \hbar\omega S_z + \hbar\omega a^{\dagger}a + \hbar g(S_+ a + S_- a^{\dagger}), \tag{21.50}$$

where a and $a^\dagger$ are annihilation and creation operators for the field, and g is a coupling constant (assumed to be real). Constants of the motion are $S^2 = S_x^2 + S_y^2 + S_z^2$ and $\left(S_z + a^\dagger a\right)$. The Hamiltonian (21.50) is referred to as the *Tavis-Cummings Hamiltonian* [22]. If, initially, all spins are in their lower energy state, then $\langle S_z + a^\dagger a\rangle = S(S+1) + n_0$, where $S = N/2$, and n_0 is the average photon number in the initial field. Since $S = N/2$, states of the atom plus field can be labeled by $|S_z, n\rangle$, with $-S \leq S_z \leq S$.

We note two general conclusions that are valid for arbitrary N. First, if we were to replace the cavity field by a classical field, the Hamiltonian would be transformed into

$$H_{class} = \sum_j \left\{\hbar\omega S_z^{(j)} + \hbar g'[S_+^{(j)}e^{-i\omega t} + S_-^{(j)}e^{i\omega t}]\right\}, \tag{21.51}$$

where g' is a constant. Since the Hamiltonian is now a sum of Hamiltonians for the individual atoms, the wave function is a direct product of the wave functions of the individual atoms. As a consequence, there is no entanglement and no spin squeezing for a classical field. Second, if the initial state of the field is a Fock state, although there is entanglement between the atoms and the field, there is no spin squeezing. There is no spin squeezing unless the initial state of the field has coherence between at least two states differing in n by 2.

It is convenient to carry out the calculations in an interaction representation with the wave function expressed as

$$|\psi(t)\rangle = \sum_{m=-N/2}^{N/2}\sum_{k=0}^{\infty} c_{mk}(t)\, e^{-i\omega(m+k)t}|m, k\rangle, \tag{21.52}$$

where m labels the value of S_z and k labels the number of photons in the cavity field. In this representation, the Hamiltonian governing the time evolution of the $c_{mk}(t)$ is given by

$$H = \hbar g(S_+ a + S_- a^\dagger). \tag{21.53}$$

We consider only the two-atom case, $N = 2$, $S = 1$. For other N, see reference [13]. If the spins are all in their lower energy state at $t = 0$, the initial wave function is

$$|\psi(0)\rangle = \sum_{k=0}^{\infty} c_k|-1, k\rangle, \tag{21.54}$$

where the c_k are the initial state amplitudes for the field. Solving the time-dependent Schrödinger equation with initial condition (21.54), we find

$$c_{-1,k}(t) = \frac{1}{(2k-1)}\left[k - 1 + k\cos(\sqrt{4k-2}gt)\right] c_k, \tag{21.55a}$$

$$c_{0,k}(t) = -i\sqrt{\frac{k+1}{2k+1}}\sin(\sqrt{4k+2}gt)c_{k+1}, \tag{21.55b}$$

$$c_{1,k}(t) = \frac{\sqrt{(k+1)(k+1)}}{2k+3}\left[-1 + \cos(\sqrt{4k+6}gt)\right] c_{k+2}. \tag{21.55c}$$

These state amplitudes can be used to calculate all expectation values of the spin operators.

If the initial state of the cavity field is a coherent state, then

$$c_k = \alpha^k e^{-|\alpha|^2/2}/\sqrt{k!}, \tag{21.56}$$

and the average number, n_0, of photons in the field is given by $n_0 = |\alpha|^2$. For simplicity, we take α to be real. We limit our discussion to a weak coherent state of the field, $|\alpha|^2 \ll 1$.

Keeping terms to order α^2, we find from equations (21.55) and (21.56) that the only state amplitudes of importance are

$$c_{-1,0}(t) = (1 - \alpha^2/2), \tag{21.57a}$$

$$c_{-1,1}(t) = \alpha \cos(\sqrt{2}gt), \tag{21.57b}$$

$$c_{-1,2}(t) = \frac{\alpha^2}{3\sqrt{2}}\left[1 + 2\cos(\sqrt{6}gt)\right], \tag{21.57c}$$

$$c_{0,0}(t) = -i\alpha \sin(\sqrt{2}gt), \tag{21.57d}$$

$$c_{0,1}(t) = -\frac{i\alpha^2}{\sqrt{3}} \sin(\sqrt{6}gt), \tag{21.57e}$$

$$c_{1,0}(t) = -\frac{\alpha^2}{3}\left[1 - \cos(\sqrt{6}gt)\right]. \tag{21.57f}$$

Using equations (21.52) and (21.44), with $S = 1$ ($N = 2$), and the fact that $S_z\,|S, m\rangle = m\,|S, m\rangle$, we find that the spin components' expectation values are

$$\langle S_x \rangle = 0, \qquad \langle S_y \rangle = \sqrt{2}\alpha \sin(\sqrt{2}gt), \tag{21.58a}$$

$$\langle S_z \rangle = -\left[1 - \alpha^2 \sin^2(\sqrt{2}gt)\right]. \tag{21.58b}$$

We need to calculate the spin variances in a plane perpendicular to $\langle \mathbf{S} \rangle$. The motion of the average value for the spin vector operator is in the yz plane, with the length of the vector always equal to unity, to order α^2. Since $\langle S_x \rangle = 0$, the plane in which we look for spin squeezing is the one defined by the x axis and an axis orthogonal to both $\hat{\mathbf{x}}$ and the instantaneous direction of the spin. Making an appropriate rotation in the yz plane to define a y' axis perpendicular to $\langle \mathbf{S} \rangle$ and $\hat{\mathbf{x}}$, and afterward choosing an arbitrary direction defined by an angle ϕ in this plane (see figure 21.2), we find that $\xi_\phi \geqslant \min\{\xi_x, \xi_{y'}\}$, which implies that the best squeezing is to be found in either the x or y' directions. The analytic expressions for ξ_x, $\xi_{y'}$, obtained using equations (21.45) and (21.57) are

$$\xi_x = \sqrt{2}\frac{\Delta S_x}{|\langle \mathbf{S} \rangle|} \simeq 1 + \alpha^2 \left[\frac{1}{2}\sin^2(\sqrt{2}gt) - \frac{2}{3}\sin^2(\sqrt{6}gt/2)\right], \tag{21.59a}$$

$$\xi_{y'} = \sqrt{2}\frac{\Delta S_{y'}}{|\langle \mathbf{S} \rangle|} \simeq 1 + \alpha^2 \left[-\frac{1}{2}\sin^2(\sqrt{2}gt) + \frac{2}{3}\sin^2(\sqrt{6}gt/2)\right]. \tag{21.59b}$$

The lowest possible value for the squeezing occurs in the x direction and is equal to

$$\xi_{\min} = 1 - \frac{2}{3}\alpha^2 \tag{21.60}$$

at a time when $\sin(\sqrt{2}gt) = 0$ and $\cos(\sqrt{6}gt) = -1$.

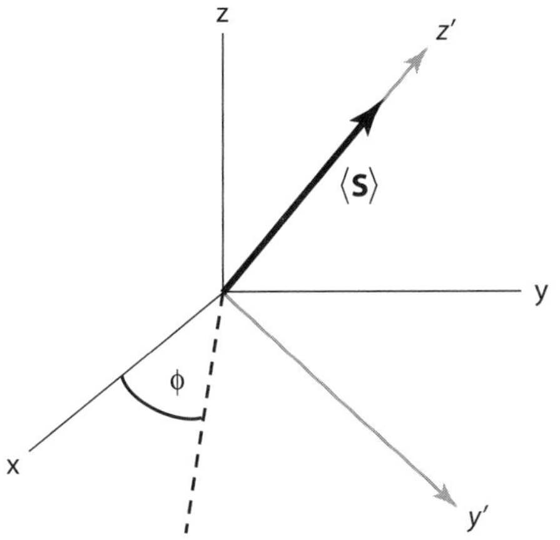

Figure 21.2. The average spin vector is in the y-z plane. We rotate the axes about the x axis so that the average spin is along the z' axis and then measure fluctuations in the x-y' plane at some angle ϕ relative to the x axis.

For larger values of N and α, there can still be spin squeezing provided that the average number of photons in the field is less than the number of atoms. In this limit, the atoms can "deplete" the field, implying that field fluctuations are important.

21.4 Summary

In this final chapter, we have looked at two mechanisms that lead to entangled atom states. The entanglement of the atoms is produced in an intermediate step in which the atoms become entangled with a quantized state of the field.

Problems

1. Derive equation (21.23) from equation (21.21).
2. Prove that a necessary condition for squeezing of a single-mode radiation field is that there is a coherence between photon occupation states differing by 2, provided that $\langle a_1 \rangle = 0$. If $\langle a_1 \rangle \neq 0$, you can translate that state back to a state with $\langle a_1 \rangle = 0$, and the condition for squeezing will be that the translated state has coherence between photon occupation states differing by 2.
3. Now consider an N atom system labeled by states $|S = N/2, S_z\rangle$. Prove that to have spin squeezing in this system, there must be a coherence between S_z states differing by two for states in which $\langle S_x \rangle = \langle S_y \rangle = 0$.
4. Prove that $[S_x, S_y] = iS_z$, but that $S_z S_+ \neq S_+/2$, even though $s_z s_+ = s_+/2$.

5–6. Derive equations (21.59).

7. In the perturbation theory limit $z \ll 1$, the squeezing operator acting on the vacuum state produces a state of the field

$$|\psi\rangle \approx |0\rangle + \beta|2\rangle,$$

where $\beta = -z/\sqrt{2}$. Assuming this to be the cavity field with z real, calculate the spin squeezing that this cavity field produces in a two-atom system. Although spin-squeezed fields transfer squeezing to the atoms, they cannot produce states at the Heisenberg limit [13].

8. Consider two, *noninteracting*, two-level atoms driven by the same atom–field interaction. The atoms start in state $|11\rangle$, and the field only couples states $|11\rangle$, $(|12\rangle + |21\rangle)/\sqrt{2}$, and $|22\rangle$, since

$$\langle 11|\, V(|12\rangle - |21\rangle)/\sqrt{2} = (\langle 11|V|12\rangle - \langle 11|V|21\rangle)/\sqrt{2} = 0,$$

because the interaction is the same for both atoms. Thus, it would appear that a weak incident field pulse that is resonant with the 1–2 transition in each atom excites only the entangled state $(|12\rangle + |21\rangle)/\sqrt{2}$. Prove that this is the wrong conclusion—there is no entanglement following the pulse.

9. Now consider an additional atom–atom interaction term in problem 8 that shifts level $|22\rangle$ by frequency Δ. Show that in this case, it is possible to create entanglement, since the atom can be *blockaded* in state $(|12\rangle+|21\rangle)/\sqrt{2}$. What are the conditions on the pulse width to produce this entanglement?

References

[1] A. E. Kozhekin, K. Mølmer, and E. S. Polzik, *Quantum memory for light,* Physical Review A **62**, 033809, 1–5 (2000).

[2] S. A. Moiseev and S. Kröll, *Complete reconstruction of the quantum state of a single-photon wave packet absorbed by a Doppler-broadenened transition,* Physical Review Letters **87**, 173601, 1–4 (2001).

[3] P. R. Berman and J.-L. Legouët, *Quantum information storage: a Schrödinger equation approach,* Physical Review A **79**, 042314, 1–13 (2009).

[4] L.-M. Duan, M. D. Lukin, J. I. Cirac, and P. Zoller, *Long-distance quantum communication with atomic ensembles and linear optics*, Nature **414**, 413–418 (2001).

[5] A. Kuzmich, W. P. Bowen, A. D. Boozer, A. Boca, C. W. Chou, L.-M. Duan, and H. J. Kimble, *Generation of nonclassical photon pairs for scalable quantum communication with atomic ensembles*, Nature **423**, 731–734 (2003).

[6] B. Julsgaard, J. Sherson, J. I. Cirac, J. Fiurášek, and E. S. Polzik, *Experimental demonstration of quantum memory for light,* Nature **432**, 482–486 (2004).

[7] C. W. Chou, S. Polyakov, A. Kuzmich, and H. J. Kimble, *Single photon generation from stored excitation in an atomic ensemble*, Physical Review Letters **92**, 213601, 1–4 (2004).

[8] C. W. Chou, H. de Riedmatten, D. Felinto, S. V. Polyakov, S. J. van Enk, and H. J. Kimble, *Measurement-induced entanglement for excitation stored in remote atomic ensembles*, Nature **438**, 828–832 (2005).

[9] R. Zhao, Y. O. Dudin, C. J. Campbell, D. N. Matsukevich, T.A.B. Kennedy, A. Kuzmich, and S. D. Jenkins, *Long-lived quantum memory*, Nature Physics **5**, 100–104 (2008).

[10] D. A. Braje, V. Balić, S. Goda, G. Y. Yin, and S. E. Harris, *Frequency mixing using electromagnetically induced transparency in cold atoms*, Physical Review Letters **93**, 183601, 1–4 (2004).

[11] T. Chanelière, D. N. Matsukevich, S. D. Jenkins, S.-Y. Lan, T.A.B. Kennedy, and A. Kuzmich, *Storage and retrieval of single photons transmitted between remote quantum memories,* Nature **438**, 833–836 (2005).

[12] D. N. Matsukevich, T. Chanelière, M. Bhattacharya, S.-Y. Lan, S. D. Jenkins, T.A.B. Kennedy, and A. Kuzmich, *Entanglement of a photon and a collective excitation,* Physical Review Letters **95**, 040405, 1–4 (2005).

[13] C. Genes, P. R. Berman, and A. Rojo, *Spin squeezing via atom-cavity field coupling,* Physical Review A **68**, 043809, 1–10 (2003).

[14] J. M. Radcliffe, *Some properties of coherent spin states*, Journal of Physics A **4**, 313–323 (1971).

[15] B. Yurke, S. L. McCall, and J. R. Klauder, *SU(2) and SU(3) interferometers,* Physical Review A **33**, 4033–4054 (1986).

[16] D. J. Wineland, J. J. Bollinger, W. M. Itano, F. L. Moore, and D. J. Heinzen, *Spin squeezing and reduced quantum noise in spectroscopy,* Physical Review A **46**, R6797–R6800 (1992).

[17] M. Kitagawa and M. Ueda, *Squeezed spin states*, Physical Review A **47**, 5138–5143 (1993).

[18] G. S. Agarwal and R. R. Puri, *Atomic states with spectroscopic squeezing,* Physical Review A **49**, 4968–4971 (1994).

[19] D. J. Wineland, J. J. Bollinger, W. M. Itano, and D. J. Heinzen, *Squeezed atomic states and projection noise in spectroscopy,* Physical Review A **50**, 67–88 (1994).

[20] F. T. Arecchi, E. Courtens, R. Gilmore, and H. Thomas, *Atomic coherent states in quantum optics*, Physical Review A **6**, 2211–2237 (1994).

[21] M. A. Rashid, *The intelligent states. I. Group-theoretic study and the computation of matrix elements; The intelligent states. II. The computation of the Clebsch–Gordan coefficients,* Journal of Mathematical Physics **19**, 1391–1396; 1397–1402 (1978).

[22] M. Tavis and F. W. Cummings, *Exact solution for an N-molecule–radiation-field Hamiltonian*, Physical Review **170**, 379–384 (1968).

Bibliography

Some general references on quantum information, including discussions of entanglement, are listed here:

I. Bengtsson and K. Zyczkowski, *Geometry of Quantum States: An Introduction to Quantum Entanglement* (Cambridge University Press, Cambridge, UK, 2006).

D. Bouwmeester, A. K. Ekert, and A. Zeilinger, Eds., *The Physics of Quantum Information: Quantum Cryptography, Quantum Teleportation, Quantum Computation* (Springer-Verlag, Berlin, 2000).

D. Esteve, J.-M. Raimond, and J. Dalibard, Eds., *Quantum Entanglement and Information Processing*, vol. LXXIX Les Houches Summer School 2003 (Elsevier, Amsterdam, 2004).

Z. Ficek and R. Tanas, *Entangled states and collective nonclassical effects in two-atom systems*, Physics Reports **372**, 369–443 (2002).

R. Horodecki, P. Horodecki, M. Horodecki, and K. Horodecki, *Quantum entanglement,* Reviews of Modern Physics **81**, 865–942 (2009).

M. A. Nielsen and L. Chuang, *Quantum Computation and Quantum Information* (Cambridge University Press, Cambridge, UK, 2000).

Some early experiments on spin squeezing are listed here:

J. Hald, J. L. Sørensen, C. Schori, and E. S. Polzik, *Spin squeezed atoms: a macroscopic entangled ensemble created by light,* Physical Review Letters **83**, 1319–1322 (1999).

A. Kuzmich, N. P. Bigelow, and L. Mandel, *Atomic quantum nondemolition measurements and squeezing,* Europhysics Letters **42**, 481–486 (1998).

A. Kuzmich and L. Mandel, *Quantum nondemolition measurements of collective atomic spin,* Physical Review A **60**, 2346–2350 (1999).

A. Kuzmich, L. Mandel, and N. P. Bigelow, *Generation of spin squeezing via continuous quantum nondemolition measurement,* Physical Review Letters **85**, 1594–1597 (1999).

A. Kuzmich, K. Mølmer, and E. S. Polzik, *Spin squeezing in an ensemble of atoms illuminated with squeezed light,* Phyical Review Letters **79**, 4782–4785 (1997).

V. Meyer, M. A. Rowe, D. Kielpinski, C. A. Sackett, W. M. Itano, C. Monroe, and D. J. Wineland, *Experimental demonstration of entanglement-enhanced rotation angle estimation using trapped ions,* Physical Review Letters **86**, 5870–5873 (2001).

J. L. Sørensen, J. Hald, and E. S. Polzik, *Quantum noise of an atomic spin polarization measurement,* Physical Review Letters **80**, 3487–3490 (1997).

Index